GRANT BYERS
42-A-2
  201-231
  85-101

Read - p 116 → 135
Problems - 3-12
           3-13
           3-14
           3-16
           3-19

WATTS COSTUMES
E 6TH AVE

GARDAL COSTUMES

INTRODUCTORY SOIL MECHANICS AND FOUNDATIONS

George B. Sowers
Consulting Engineer, Lt. Colonel, Corps of Engineers, AUS (Res)

George F. Sowers
Regents Professor of Civil Engineering, Georgia Institute of Technology; Senior Vice President and Consultant, Law Engineering Testing Company

INTRODUCTORY

SOIL MECHANICS AND FOUNDATIONS

THIRD EDITION

THE MACMILLAN COMPANY
COLLIER-MACMILLAN LIMITED
LONDON

© Copyright, The Macmillan Company, 1970

All rights reserved. No part of this book may be reproduced or transmitted in any form or by any means, electronic or mechanical, including photocopying, recording or by any information storage and retrieval system, without permission in writing from the Publisher.

PRINTING   8910   YEAR   456789

Earlier editions copyright 1951 and © 1961 by The Macmillan Company.

Library of Congress catalog card number: 76-88843

THE MACMILLAN COMPANY
8o6 Third Avenue, New York, New York 10022

COLLIER-MACMILLAN CANADA, LTD., TORONTO, ONTARIO

Printed in the United States of America

Preface to the Third Edition

Nearly twenty years have passed since the first edition appeared in 1951. During that period there have been many changes—changes that became educational challenges that generated new textbooks. Among these was our second edition in 1961, and now we introduce our third.

One aspect of change has been the rapid growth in engineering technology: (1) more realistic theories, (2) better techniques and tools for experimental research, (3) more and better-correlated data on the real behavior of full scale soil and rock masses, and (4) more elegant devices for making rapidly and economically what had been tedious, complex computations.

A second aspect of change is the increasing demands of society upon engineers. As man runs out of space on this planet and commences exploring others, he finds that structures and machines become heavier, the tolerances to undue movement become more critical, and the construction sites that are left become increasingly poorer. At the same time the economic need of building more at a lower cost, the social responsibility of conserving and improving the total environment, the political demand to build before the technology is available, and the legal liabilities of possible errors add new dimensions to the challenges that have always faced the civil engineer.

A third aspect of change is the student, with his better command of mathematics and science and his growing sense of responsibility toward society as a whole. However, his increasing insulation from the earth in cities of concrete and steel or in machines of steel, operating on ribbons of concrete or asphalt and screened by billboards and fences, has reduced his feel for reality. To some students engineering is a textbook exercise, to be solved like a move in chess, rather than a phase in man's struggle to cope with the physical demands of his planet.

This third edition is our response to the challenge of these changes. Advanced theories as well as the impact of computer technology on engineering analysis are included. The requirements of the allied disciplines, such as structural engineering, are emphasized. Most important, engineering geology and real problems in engineering and construction are emphasized so that the student (and the practicing engineer) will maintain a realistic outlook on the impact of their endeavors—technically, economically, and socially. Engineering, construction, and society are parts of a system; they must work together for the good of all.

Finally, we wish to convey to our readers a sense of destiny. Engineering is the most exciting endeavor that confronts a man—the challenge of the new, the unknown, and the risk of failure for the good of mankind. We, your authors and counselors, bring you our eighty-five years of professional experience in helping our fellow man cope with the physical problems of his environment. We have taken the oldest material, the earth, and have helped fashion it into the most esoteric devices. It is our hope that you, the reader, will add your genius and sweat to this common goal of engineering—to apply the knowledge of science and the understanding of experience to the problems that will face us tomorrow.

The authors are indebted to the many users of the text in the academic, design, and construction professions who have offered helpful criticisms and suggestions. Particularly, Professor R. D. Barksdale and Professor B. B. Mazanti of the Georgia Institute of Technology and Mr. C. M. Kennedy and the staff of the Law Engineering Testing Co. have made many concrete suggestions after reviewing portions of the text. Professor Sowers wishes to express his thanks to Dr. Arthur Hansen, President, and Dr. W. M. Sangster, Director of the School of Civil Engineering, Georgia Institute of Technology, for the reduction in academic responsibilities that made this revision possible. Finally, the writing would never have been completed without the continued help of Frances L. Sowers, who reviewed the entire text, and the patience of the remainder of the Sowers families who suffered the rebirth pains.

G. B. S.
G. F. S.

Contents

Nomenclature and Symbols   xi
Useful Equivalents   xv

1 THE NATURE OF SOILS AND ROCKS   1

1.1 Definition of Soil and Rock / 1.2 Development of Soil and Rock Engineering / 1.3 Phases of Soil and Rock Composition / 1.4 Solids: Rock Minerals and Weathering / 1.5 Clay Minerals / 1.6 Organic, Precipitation, Volcanic, and Man-Made Solids / 1.7 Grain Size / 1.8 Particle Shape / 1.9 Interaction of Water and Solid Phase / 1.10 Plasticity and the Atterberg Limits / 1.11 Micro Structure or Fabric

2 ROCKS, SOILS AND GROUND WATER   40

2.1 The Cycle of Soil and Rock Formation / 2.2 Igneous Rocks / 2.3 Transported and Deposited Soils / 2.4 Clastic Sedimentary Rocks / 2.5 Calcareous Sedimentary Rocks / 2.6 Metamorphic Rocks / 2.7 Profile Development—Pedogenesis / 2.8 Ground Water / 2.9 Rock Classification and Description / 2.10 Soil Classification / 2.11 Soil Identification and Description / 2.12 Soil and Rock Names

3 THE ENGINEERING PROPERTIES OF SOIL AND ROCK   85

3.1 Surface Tension / 3.2 Capillary Tension in Soils / 3.3 Permeability / 3.4 Permeability of Soil and Rock / 3.5 Stress and Effective Stress / 3.6 Compressibility and

Settlement / 3.7 Compressibility of Soils and Rocks /
3.8 Time Rate of Compression / 3.9 Shrinking, Swelling, and
Slaking / 3.10 Combined Stresses / 3.11 Strain and
Failure / 3.12 Methods of Making Shear Tests / 3.13 Strain
and Strength of Dry Cohesionless Soils / 3.14 Shear in Wet
Cohesionless Soils / 3.15 Strain and Strength of Saturated
Cohesive Soils / 3.16 Strength of Partially Saturated Cohesive
Soils / 3.17 Strength of Cemented Soil and Rock /
3.18 Creep / 3.19 Engineering Properties of the Mass

4 SEEPAGE, DRAINAGE AND FROST ACTION 161

4.1 Saturated Flow: The Flow Net / 4.2 Seepage Effects /
4.3 Seepage Control / 4.4 Capillary Moisture and Flow /
4.5 Drainage / 4.6 Drainage Systems / 4.7 Frost Action

5 EARTH CONSTRUCTION: COMPACTION AND STABILIZATION 201

5.1 Soils and Rocks as Construction Materials / 5.2 Theory of
Compaction / 5.3 Evaluation of Materials / 5.4 Excavation,
Placement, and Compaction / 5.5 Hydraulic Filling /
5.6 Soil Stabilization / 5.7 Grouting—Injection Stabilization /
5.8 Subgrades and Pavements

6 UNDERGROUND INVESTIGATION 255

6.1 Planning an Investigation / 6.2 Reconnaissance / 6.3 Air
Photo Interpretation / 6.4 Exploratory Investigation /
6.5 Boring and Sampling / 6.6 Penetration Tests / 6.7 Ground
Water / 6.8 Geophysical Exploration / 6.9 Analyzing the
Results of an Exploratory Investigation / 6.10 Intensive
Investigation / 6.11 Undisturbed Sampling / 6.12 Field
Tests / 6.13 Laboratory Testing and Evaluation /
6.14 Feedback During Construction

7 MASS RESPONSE TO LOAD 315

7.1 The Mass Load—Body Forces / 7.2 Elastic Equilibrium /
7.3 Plastic Equilibrium / 7.4 Rheology / 7.5 Solution of
Problems in Mass Behavior

CONTENTS   ix

8 PROBLEMS IN EARTH PRESSURE 336

8.1 Theory of Earth Pressure / 8.2 Deformation and Boundary Conditions / 8.3 Computing Earth Pressure / 8.4 Retaining Walls / 8.5 Excavation Bracing / 8.6 Anchored Bulkheads / 8.7 Underground Structures

9 FOUNDATIONS 388

9.1 Essentials of a Good Foundation / 9.2 Stability—Bearing Capacity / 9.3 Stress and Settlement / 9.4 Settlement Observations / 9.5 Allowable Pressure on Soil / 9.6 Rational Procedure for Allowable Pressure and Design / 9.7 Footings and Mats / 9.8 Special Problems in Shallow Foundation Design / 9.9 Foundations on Rock / 9.10 Foundations Subject to Vibrations

10 DEEP FOUNDATIONS 445

10.1 Development and Use of Piles / 10.2 Pile Driving / 10.3 Pile Capacity / 10.4 Dynamic Analysis of Pile Capacity / 10.5 Pile Groups / 10.6 Lateral Loads / 10.7 Types of Piles and Their Construction / 10.8 Design of Pile Foundations / 10.9 Pier Foundations / 10.10 Anchors / 10.11 Underpinning

11 STABILITY OF EARTH MASSES 504

11.1 Analysis of Stability / 11.2 Open Cuts / 11.3 Embankments / 11.4 Embankment Foundations / 11.5 Earth and Rockfill Dams / 11.6 Earth Movements in Nature

APPENDIX 541

Unit Costs 541
Geologic Age 543

INDEX 545

Nomenclature and Symbols

The following symbols are used throughout this text. They conform generally to ASTM D653-67, "Terms and Symbols Relating to Soil Mechanics," and *Technical Terms, Symbols and Definitions*, 3rd ed., International Society of Soil Mechanics and Foundation Engineering, Zurich (published by the Swiss National Society, SM and FE, 1967). If the two disagree or more than one symbol is commonly used in literature, the first one shown is used in this text.

a	acceleration (ft per sec^2 or cm per sec^2)
A	area (sq ft or sq cm)
a_v	coefficient of compressibility (sq ft per lb or sq cm per kg)
B	width (ft or cm)
C_c	compression index (dimensionless)
C_u	uniformity coefficient (dimensionless)
c	shear strength of saturated clays in quick shear or apparent cohesion (lb per sq ft or kg per sq cm)
c_a	adhesion of saturated clay to a surface (lb per sq ft or kg per sq cm)
c_v	coefficient of consolidation (sq cm per sec, sq ft per min, sq ft per day)
d	diameter of capillary tube (ft or cm)
D	diameter of soil particle (mm); diameter of pile or anchor (in. or cm)
D_R	relative density (also I_D, R_D)
D_{10}	diameter of 10 per cent size (mm)
E	modulus of elasticity or modulus of deformation (lb per sq ft or kg per sq cm)
e	void ratio (dimensionless)
e_0	void ratio at time compression or shear begins
F	force (lb, kips, or kg)
G_s	specific gravity of solids (dimensionless)
g	acceleration of gravity (ft per sec^2 or cm per sec^2)
H	height of retaining wall, thickness of soil stratum (ft or cm)

h	head (ft or cm)
i	hydraulic gradient (dimensionless)
K_0	coefficient of earth pressure at rest (dimensionless)
K_A	coefficient of active earth pressure (dimensionless)
K_P	coefficient of passive earth pressure (dimensionless)
k	coefficient of permeability (ft per min or cm per sec)
k_s	coefficient of subgrade reaction (lb per in.^{3} or kg per cm^3)
L	length or distance (ft or cm)
LL	liquid limit (dimensionless) (also L_W, W_L)
m	stability number (dimensionless)
n	porosity (dimensionless)
P	resultant of earth pressure over an area (lb, kips, or kg)
P_A	resultant of active earth pressure (lb, kips, or kg)
P_P	resultant of passive earth pressure (lb, kips, or kg)
PL	plastic limit (dimensionless) (also W_p, P_W)
PI	plasticity index (dimensionless) (also I_p)
p	pressure (lb per sq ft, kips per sq ft, or kg per sq cm), also normal stress on failure plane
p', $\bar{p}$	effective stress at failure, effective pressure (lb per sq ft or kg per sq cm)
Q	total load (lb, kips, or kg)
q	pressure on soil (lb per sq ft, kips per sq ft, or kg per sq cm); discharge (cu ft per sec or cu cm per sec)
q_a	allowable soil pressure (lb per sq ft, kips per sq ft or kg per sq cm)
q_0	ultimate bearing capacity of soil (lb per sq ft or kg per sq cm)
q_r	compressive strength from triaxial shear test (lb per sq ft or kg per sq cm)
q_s	safe bearing capacity (lb per sq ft, kips per sq ft, or kg per sq cm)
q_u	unconfined compressive strength (lb per sq ft or kg per sq cm)
r	radius
S	degree of saturation, shear force (lb)
SF	safety factor (also F_s)
s	shear strength (lb per sq ft or kg per sq cm)
T	time factor
T_0	surface tension (lb per ft or gm per cm)
t	time (sec, min, or days)
U	per cent consolidation, also uplift force (lb or kg)
u	neutral stress (lb per sq ft or kg per sq cm)
V	volume (cu ft or cu cm)
W	weight (lb, gm, or kg)
w	water content
z	depth (ft or cm, measured positive downward)
α	angle of plane with major principal plane, also angle of failure plane with major principal plane

NOMENCLATURE AND SYMBOLS

β	angle of face of earth dam, embankment or backfill behind a retaining wall with respect to a horizontal plane
γ	unit weight (lb per cu ft or gm per cu cm) also shear strain (dimensionless)
γ_w	unit weight of water (lb per cu ft or gm per cu cm)
γ'	unit weight of soil submerged in water (lb per cu ft or gm per cu cm)
γ_d	weight of soil solids in cubic foot of soil (lb per cu ft or gm per cu cm)
Δ	change or increment (used as prefix)
δ	angle of wall friction
ε	strain (dimensionless)
ν	Poisson's ratio
ρ	contact settlement (ft or cm)
σ	normal stress (lb per sq ft, kips per sq ft, or kg per sq cm)
$\sigma', \bar{\sigma}$	effective normal stress
$\sigma_1, \sigma_2, \sigma_3$	principal stresses
σ'_c	preconsolidation load
σ_h	horizontal stress (also σ_x, σ_y)
σ_0	initial stress
σ_z, σ_v	vertical stress
σ_α	normal stress on plane making angle of α with major principal plane
τ	shear stress
τ_α	shear stress on plane making angle of α with major principal plane
ϕ	angle of internal friction

Useful Equivalents

Three systems for expressing the fundamental units of measurement are in use today among engineers: the English, based on the foot and pound of force, the Metric, based on the meter and the kilogram of force, and the SI, based on the meter and the kilogram of mass. This text employs both the English and Metric engineering systems that use the foot and meter as fundamental units of length and the pound and kilogram as units of force. Because of the growing body of international literature using all three systems, it is necessary for the engineer to convert between them.

Dimension	Equivalent	Useful Approximation
Length	1 ft = 30.5 cm	3 m = 10 ft
	1 in. = 2.54 cm	
	1 A = 1 × 10^{-7} mm = 1 × 10^{-10} m	
	1 m = 3.28 ft	
Volume	1 cu yd = 0.765 cu m	1 cu m = 4/3 cu yd
	1 cu m = 35.3 cu ft	
Density	62.4 lb per cu ft = 1 gm per cu cm	
Force	1 lb = 454 gm	
	1 lb = 4.48 Newtons	
	1 Newton = 10^5 dynes	
	1 kip per sq ft = 0.488 kg per sq cm	1 tsf = 1 kg per sq cm
	1 kip per sq ft = 0.0488 t_m per sq m	1 ksf = ½ kg per sq cm
	1 lb per sq in. = 0.0704 kg per sq cm	
	1 kip per sq ft = 4.7 × 10^4 Newton's per sq m	
	1 lb per sq in. = 0.689 Newton's per sq cm	
	1 Bar = 10 Newton's per sq cm	

Velocity	1 ft per min = 0.509 cm per sec	1 cm per sec = 2 ft per min
	1 ft per year = 0.966×10^{-6} cm per sec	1 cm per sec = 1×10^6 ft per year
Acceleration	$g = 32.2$ ft per sec^2 = 9.81 m per sec^2	
Flow	1 cu ft per sec = 0.0283 cu m per sec	
	1 cu ft per min = 0.472 liters per sec	

CHAPTER 1
The Nature of Soils and Rocks

THE SITE FOR A SUPERMARKET WAS STRATEGICALLY LOCATED AT THE INTERsection of two main streets, although most of the property was a hillside. In order to provide level space for the building and for the parking of automobiles, a wide cut was made at the toe of the slope. Of course this resulted in a steeper hillside, but since the soil appeared to be very stiff, the builder assumed that it would be safe. A few months later the owner of the property noticed that the rear corner of his new building, 20 ft from the toe of the cut, was rising. At the same time the driveway between the building and the hill was narrowing. The builder diagnosed the cause as sliding of the earth at the toe of the slope, and he erected a concrete retaining wall to restrain the movement. Instead of stopping the slide, the wall moved with the hill toward the building. Frantically the contractor drove steel sheet piling between the building and the hillside to support the concrete wall and the hill. The movement continued at the same rate. Finally, in desperation, he constructed a horizontal reinforced concrete beam against the sheet piling and supported the beam with steel H-piles driven at an angle and bearing on rock. The hillside, the wall, the sheeting, and the beam continued to advance on the building.

An investigation of soil conditions disclosed that the stiff clay in the hillside absorbed water and expanded when the weight on the soil was reduced by the cutting at the toe. The expansion took place slowly, and therefore the freshly excavated slope was stable. The expanded soil was much weaker than in its original state and was incapable of supporting itself on the new-cut slope. The retaining wall and the sheeting were designed on the basis of the ordinary formulas and were incapable of resisting the unsupported mass in motion on the hillside.

The project was not large and the cost of the excavation and the building was just over $100,000. The cost of the retaining wall, the sheet piling, and the concrete beam was $80,000, or nearly as much as the original project, but they were of no value in correcting the difficulty. At that point the entire project was a financial disaster because correction of the unstable hillside would be more expensive than the value of the building.

The owner decided that the only alternative to bankruptcy was to try to operate the store. He rented a power shovel and a truck, and each week removed the sliding soil from the toe of the slope. He hauled it to the top of the slide area and dumped it on the moving mass. His objective was to fill the depression formed at the top of the slide and to protect a city street and some houses still higher on the hill. The refilling only aggravated the movement and failed to support the land above the slide. First the street, then a gas main and finally a house beyond were destroyed. The store owner ultimately paid for them. The business at his location was so good and his increasing damages so high that he could not afford to stop. Finally, after three years he secured professional advice and installed a drainage system that gradually reduced the water in the hillside. The movement virtually ceased. The cost of the engineered solution to the problem was a fraction of the amount he had spent.

Such failures are not uncommon and they illustrate the need for careful, scientific soils engineering on even small projects. While the soil conditions encountered in the preceding example were unusual, they could have been detected by an investigation costing less than $1,000, and a safe design for the hillside and the structure could have been prepared within the economic limits of the project.

Traditionally, soil problems have not received the attention they deserve from engineers and constructors. Too often designs have been based on handbooks written years ago, experiences with other sites that are not representative, or even on guesses about the soil properties. Only extremely generous safety factors (as high as 20 in some cases) have prevented serious failures. Too often construction operations involving soil have been based on blind trial and error, but costly failures and dead workmen are a high price to pay for experience.

1:1 Definition of Soil and Rock

The reason for the lack of a rational approach to problems of soil design and construction has been a lack of understanding of the complex nature and behavior of soils. Until the twentieth century engineers considered soil to be a sort of witch's brew—a mysterious mixture that was incapable of scientific study. Great advances in the techniques of studying soils have led to a better understanding of their nature and to rational methods in design and construction.

DEFINITION OF SOIL / To a farmer, soil is the substance that supports plant life, whereas to the geologist it is an ambiguous term meaning the material that supports life plus the loose material or mantle from which it was derived. To the engineer the term *soil* has a broader meaning.

Earth, or *soil*, in the engineering sense is defined as *any unconsolidated material composed of discrete solid particles with gases or liquids between.* The maximum particle size that qualifies as soil is not fixed but is defined by the function involved. For footing excavations and trenches where hand excavation is employed, and for the construction of fills in layers, the limiting size is 12-in. diameter (about 85 lb or 40 kg), the maximum size a man can lift. Where power shovels are used for excavation, the limit is sometimes given as $\frac{1}{2}$ cu yd (about 1 ton).

Soils include a wide variety of materials such as the gravel, sand, and clay mixtures deposited by glaciers, the alluvial sands and silts and clays of the flood plains of rivers, the soft marine clays and beach sands of the coast, the badly weathered rocks of the tropics, and even the cinders, bed springs, tin cans, and ashes of a city dump. Soils can be well-defined mixtures of a few specific minerals or chaotic mixtures of almost anything.

DEFINITION OF ROCK / Rock is defined by the engineer as any indurated material that requires drilling, wedging, blasting, or other methods of brute force for excavation. The minimum degree of induration that qualifies as rock has sometimes been defined by a compressive strength of 200 psi. In all cases the dividing line between soil and rock is not definite; there is a continuous series of materials from the loosest soil to the hardest rock, and any division into two categories must be arbitrary. In preparing engineering documents, such as specifications, the engineer must define the limit so that all who are affected will be in agreement.

The definition of rock from an engineering or functional viewpoint is complicated by structure and defects. A rock that is hard but fractured may be easier to excavate than a softer but more coherent material. Although the fractured rock is easier to excavate, it may require bracing to support it in a deep excavation, whereas the soft rock may stand without support.

Problems in soil and rock engineering, therefore, can seldom be solved by blind reliance on either emperical data gathered on past projects or on the most sophisticated computerized analyses. Each situation is unique; it requires careful investigation and thorough scientific analysis tempered with engineering judgement based on varied experience. Most important, soil and rock engineering requires imagination, intuition, initiative and courage: Imagination to visualize the three-dimensional interplay of forces and reactions in the complex materials; intuition to sense what cannot be deduced from scientific knowledge or past experience; initiative to devise new solutions for old and new problems; and courage to carry the work to completion in spite of skeptics and in the face of the ever-present risk of the unknown. It is this continuing challenge that makes real engineering intriguing and rewarding.

1:2 Development of Soil and Rock Engineering

Construction involving soil began before the dawn of history. Some of the first construction operations were the digging of holes to bury the dead or to dispose of excrement and the building of earth mounds for worship and burial. Earth, in the form of sun-dried brick or of mud daubed on interlaced sticks or reeds, was used in building houses.

The builders of the ancient civilizations, such as those of India and Babylon, have left numerous examples of their ability to handle soil problems. Some earth dams in India have been storing water for more than 2000 years. The cities of Babylon were placed on fills to raise them above the flood plains, and the buildings were erected on stone mats that spread the loads to the weak soils below. These builders were skilled artisans who learned from their own bitter experiences or from the success and failures of others. During the Middle Ages craftsmen improved the art of construction involving soils by the process of trial and error. They received little help from the early scientists, who felt that problems involving the earth were beneath the dignity of a gentleman.

Since the eighteenth century, however, the need for better structures led scientists and engineers to study soil problems and to try to analyze them like other problems in structural design. Such eminent investigators as Coulomb and Rankine, who are well known in the fields of physics and applied mechanics, turned their attention to the mechanics of soil masses. They started from mathematical expressions of soil strength or from crude experiments on piles of sand and from them developed expressions for earth pressure on walls and the bearing capacity of foundations. This procedure was logical and when applied to other problems in mechanics led to theories that are still in good repute. The theories they developed for soils, however, often proved to be dismal failures. Retaining walls failed, buildings settled, and excavations caved in when designed in accordance with theory.

The tremendous increase in the size of the structures in the early twentieth century and the need for greater economy in their construction forced many brilliant engineers to re-evaluate the work of the earlier investigators and to develop new and more realistic methods of analyzing soil masses. The work of Fellenius in Sweden, Kogler in Germany, Hogentogler in the United States, and above all the contributions of Karl Terzaghi in both Europe and the United States brought about the birth of a new phase of civil engineering—*soil mechanics*, and its application to practical problems, *soils engineering*. Since the middle 1930's soil mechanics has become an indispensible tool to the planner and designer and an aid to the builder who must work with the ground.

Rock mechanics as an applied science has had a parallel development. Shafts and tunnels were excavated through rock during the early days of Greece and Rome. Quarry techniques were evolved from the demand for accurately shaped rock to be used in constructing the architectural masterpieces and the marble sculpture of that era. Knowledge was gained by experience and not by scientific reasoning until the late nineteenth century. It is interesting that many of the early mechanical engineering advances, including pumps and the steam engine, were generated by the needs of the mines; however, the rock itself was given little scientific study. Modern scientific rock mechanics has developed from the needs of both the mining and the construction industries. Although soil mechanics and rock mechanics outwardly appear different, the differences are in emphasis and in terminology. The art of rock mechanics is mine oriented. The scientific analysis has emphasized the behavior of relatively rigid masses laced with cracks. Soil mechanics is construction oriented, and has emphasized the behavior of weak, compressible materials. In the all over view there is little difference scientifically and practically. The civil engineer, the builder, and the miner are concerned with the full spectrum of materials, from the softest soil to the hardest rock; any separation between soil and rock mechanics is in the point of view of the engineers and artisans involved.

TYPES OF PROBLEMS IN SOIL AND ROCK ENGINEERING / Two distinct types of problems are involved in soils engineering. The first type deals with soils and rocks as they actually occur in nature. Buildings ordinarily are founded on undisturbed materials, excavations and highway cuts are made through natural deposits, and drainage networks are constructed to remove water from existing soil masses. The second type of problem involves soils or rocks as raw materials for construction. Fills for highways and railroads, earth dams and levees, and airport and highway subgrades use the earth as a source of construction materials. The characteristics of the soil and rock are altered to form new materials similar to that in which sand, cement, and stone are made into concrete.

In handling either type of problem, the engineer must keep one thought uppermost in his mind—he is dealing with highly complex materials and with variable ingredients that at times will appear to defy all laws of nature. With careful study based on scientific analysis and sound judgment, even the most difficult problems can be analyzed. The numerical results often are not accurate in more than one or two significant figures, but in most cases the accuracy is as good as that obtained in calculating the stresses in a structure caused by assumed live loads.

1:3 Phases of Soil and Rock Composition

THREE-PHASE COMPOSITION / Since soils by definition include all unconsolidated materials, we may expect them to be composed of many different ingredients in all three states or phases of matter—solid, liquid, and gaseous. The same applies to many rocks. Although they are consolidated and indurated, most contain some liquid and gaseous matter.

The interrelations of the weights and volumes of the different phases are important because they help define the condition or the physical makeup of a soil. The definitions and terms attached to these relations must be clearly understood before the engineer can gain an understanding of the properties of soils and rocks.[1:1]

The volumes and weights of the different phases of matter in a soil can be represented by a block diagram or graph such as Fig. 1.1.[1:2] The total volume or weight is denoted by the entire block; the solids, by the lower section; the liquids, by the middle section; and the gases, by the upper section. For all practical purposes the gases may be considered to be air (although methane is occasionally found in some soils containing decaying organic matter), and the liquids are ordinarily water (although in some instances the water may contain small quantities of dissolved salts). The composition of the solids, however, varies considerably and will be discussed in detail in Section 1:4.

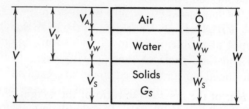

Figure 1.1 Block diagram showing relationship of weights and volumes of solids, air and water in a mass of soil or rock.

VOLUME RELATIONSHIPS / The volume of the solids in a mass of soil or rock is denoted as V_s, the volume of the water V_w, and the volume of the air V_a. The volume of the mass of soil, including air, water, and solids is V. The space between the solid particles that is occupied by air and water is called the *voids* and its volume is denoted V_v. The ratio of the volume of voids to the volume of solids is expressed by the *void ratio*, e, as

$$e = \frac{V_v}{V_s}. \tag{1:1}$$

The void ratio is always expressed as a decimal. A second way of expressing the relation between voids and solids is the *porosity*, n, which is defined by

$$n = \frac{V_v}{V} \times 100 \text{ per cent} \tag{1:2}$$

and is always expressed as a percentage.

The voids in rock are in two forms: discrete voids like tiny bubbles completely surrounded by solids, and interconnected or open voids, similar to those in soils. It is the changes that occur in the interconnected voids that play the dominant role in physical behavior. The *apparent void ratio* or *apparent porosity* of the connected open voids is less than the *true* or *total porosity* that includes all voids. The total porosity is difficult to measure and of less significance from the view point of engineering behavior. In this text, void ratio or porosity of rock and soil refers to the interconnected voids only.

SEC. 1:3] PHASES OF SOIL AND ROCK COMPOSITION

(In many published works in rock mechanics, it is not always clear which porosity is utilized; the reader often must establish this by implication.)

The *degree of saturation*, S, expresses the relative volume of water in the voids and is always expressed as a percentage,

$$S = \frac{V_w}{V_v} \times 100 \text{ per cent.} \qquad (1:3)$$

A soil is said to be *saturated* if $S = 100$ per cent.

WEIGHT RELATIONSHIPS / The weight of solids in a mass of soil or rock is denoted by W_s; the weight of water, by W_w; and the weight of the entire mass, including water and solids, by W. (The air has negligible weight, so it is neglected.) The ratio of the weight of water to the weight of solids is termed the *water content*, w, and is expressed by the formula

$$w = \frac{W_w}{W_s} \times 100 \text{ per cent.} \qquad (1:4)$$

The *unit weight* of the mass is the ratio of the weight of the mass to the volume of the mass. It is denoted by the Greek letter γ (gamma) and is expressed in pounds per cubic foot or in grams per cubic centimeter:

$$\gamma = \frac{W}{V}. \qquad (1:5)$$

The *unit weight of water*, γ_w, is 62.4 lb per cu ft or 1 gm per cc.

The *specific gravity* of a substance is the ratio of its weight to the weight of an equal volume of water. The specific gravity of a mass of soil or rock (including air, water, and solids) is termed *mass specific gravity* or apparent specific gravity. It is denoted by G_m and may be expressed by the formula

$$G_m = \frac{\gamma}{\gamma_w} = \frac{W}{V\gamma_w}. \qquad (1:6)$$

The *specific gravity of the solids*, G_s, (excluding air and water) is expressed by

$$G_s = \frac{W_s}{V_s \gamma_w}. \qquad (1:7)$$

This is the weighted average of the soil minerals. The specific gravity of rock can be expressed two ways: including the discrete voids and any gas or liquid that fills them, or the specific gravity of the solid matter alone. It is usually expedient to include the effect of the discrete voids in the specific gravity of the rock. The specific gravity of a rock, therefore, may be somewhat less than that of the weighted average of its minerals. Specific gravity is a dimensionless ratio and therefore has no units. It is the same in both the metric and English systems of measurement. Numerically the specific gravity and unit weight are equal in the metric system but not in the English system.

COMPUTATIONS INVOLVING RELATIONSHIPS / The relations between volumes and masses are very important in many types of calculations such as those to determine the stability of a soil mass, to estimate building settlements, or to specify the amount of compaction necessary to construct an earth fill. These calculations are the arithmetic of soil mechanics and must be mastered before proceeding further. In making each calculation, a block diagram showing the relation of the different phases should be drawn and the different data entered on it as shown in the following examples.

Example 1 : 1

Calculate the unit weight, void ratio, water content, porosity, and degree of saturation of a chunk of moist soil weighing 45 lb and having a volume of 0.43 cu ft (Fig. 1.2). When dried out in an oven, the soil weighed 40 lb. The specific gravity of the solids was found to be 2.67.

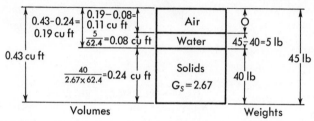

Figure 1.2 Block diagram showing computations for weights and volumes in Example 1 : 1.

1. $V = 0.43$ cu ft, $W = 45$ lb, $G_s = 2.67$, $W_s = 40$ lb.
2. $W_w = 45 - 40 = 5$ lb.
3. $\gamma = 45$ lb$/0.43$ cu ft $= 105$ lb per cu ft.
4. $G_m = 105/62.4 = 1.68$.
5. $V_w = \dfrac{5 \text{ lb}}{62.4 \text{ lb/cu ft}} = 0.08$ cu ft.
6. $V_s = \dfrac{40 \text{ lb}}{62.4 \text{ lb/cu ft} \times 2.67} = 0.24$ cu ft.
7. $V_v = 0.43 - 0.24 = 0.19$ cu ft.
8. $V_a = 0.19 - 0.08 = 0.11$ cu ft.
9. $e = 0.19/0.24 = 0.79$.
10. $n = (0.19/0.43) \times 100\% = 44\%$.
11. $S = (0.08/0.19) \times 100\% = 42\%$.
12. $w = (5/40) \times 100\% = 12\%$.

Example 1:2

Calculate the void ratio and specific gravity of solids of a saturated soil whose unit weight is 117 lb per cu ft and whose water content is 41 per cent (Fig. 1.3).

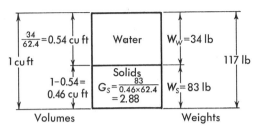

Figure 1.3 Block diagram showing computations for weights and volumes in Example 1:2.

1. $\gamma = 117$ lb/cu ft, $w = 41\%$, $S = 100\%$.
2. Assume a mass of soil of 1 cu ft; then $V = 1$ cu ft, $W = 117$ lb.
3. $W_w/W_s = 0.41$; $W_w + W_s = 117$ lb,
   $W_s + 0.41 W_s = 117$ lb,
   $W_s = 83$ lb; $W_w = 34$ lb.
4. $V_v = V_w = 34$ lb/62.4 lb per cu ft $= 0.54$ cu ft.
5. $V_s = 1.00 - 0.54 = 0.46$ cu ft.
6. $G_s = \dfrac{83 \text{ lb}}{0.46 \text{ cu ft} \times 62.4 \text{ lb/cu ft}} = 2.88$
7. $e = 0.54/0.46 = 1.17$.

1:4 Solids: Rock Minerals and Weathering

The solid phase plays a major part in determining the engineering behavior of soil and the dominant part in rock. According to the engineering definition, almost anything on the earth's crust is included in the definition of soil and rock. The most important solids fall into three classes: (1) minerals, (2) the products of organic synthesis and decay and, (3) man made materials. By far the most important constituents of soil and rock are minerals: naturally occurring chemical compounds of definite composition and crystal structure.

PREDOMINATE MINERALS / Although hundreds of different minerals are listed in mineralogy manuals, a relatively few make up the greater part of rock and soils. These minerals and some of their pertinent properties related to civil engineering are listed in Table 1:1.

TABLE 1:1 / MAJOR MINERALS IN ROCK AND SOILS

Mineral Group	Variety	Hardness*	Color	Cleavage	Specific Gravity
Silica	Quartz	7	Colorless-white	None	2.66
	Chert	7	Light	None	2.66
Feldspar	Orthoclase, Microcline	6	White-pink	Right-angle	2.56
	Plagioclase	6	White-gray	Right-angle, Striated surface	2.6–2.75
Mica	Muscovite	2–2.5	Silvery	Thin platy	2.75–3.0
	Biotite	2.5–3	Dark	Thin platy	
Ferro Magnesian	Pyroxene: Augite	5–6	Black	Right-angle	3.1–3.6
	Amphibole: Hornblende	5–6	Black	Oblique angle	2.9–3.8
	Olivene	6–5.7	Greenish		3.3
Iron Oxides	Limonite, Magnetite	5, 6	Red, yellow, black		5.4
Calcite†	Crystalline to earthy	3	White-gray	3 parallelogram faces	2.7
Dolomite‡	Crystalline to earthy	4	White-gray	3 parallelogram faces	2.8
Clay Minerals	Kaolinite, Illite, Montmorillonite	1	White	Earthy	2.2–2.6
Cellulose				Fibrous	1.5–2

NOTE: * Hardness: Finger nail = 2, Copper coin = 3, Pocket knife = 5, glass = 5.5.
† Will effervesce with cold hydrochloric acid.
‡ Weak effervescence with cold acid.

ROCK WEATHERING / Rock weathering is the breakdown of intact masses of rock into smaller pieces by mechanical, chemical, or solution processes. *Mechanical weathering*, or *disintegration*, is a combination of grinding, shattering, and breaking that reduces the rock to smaller and smaller fragments that have the same mineral composition as the original rock. It is caused by the freezing of water in cracks and pores, the impact of water, the abrasion of gravel and boulders carried by mountain streams and rivers, the pounding of water waves on beaches or cliffs, the sand blast of sand-laden desert winds, the expansion and contraction of rock by violent temperature changes, and by the plowing action of glaciers. *Chemical weathering*, or *decomposition*, is a chemical alteration of the rock minerals to form new minerals that usually have chemical and physical properties completely different from their parent materials. It is caused by the reaction of the minerals with water, dissolved carbon dioxide and oxygen from the air, organic acids from plant decay, and dissolved salts present in the water. *Solution* is the dissolving of soluble minerals from the rock, leaving the insoluble minerals behind as a residue. All three processes occur simultaneously but at different rates, depending on the climate, topography, and the composition of the original rock. In general, decomposition predominates in warm, humid regions and in areas with flat topography, and disintegration predominates in dry regions and areas with rugged topography. Solution obviously is predominant in humid regions underlain by soluble rocks.

MINERALS AND WEATHERING / *Silica* (silicon dioxide) is one of the most important constituents of many rocks and most soils. It occurs in two forms: crystalline (quartz) and noncrystalline (chert, flint, and chalcedony.) It is inert in chemical weathering and insoluble in water although it is slightly soluble in a basic environment. In the crystalline form and in most cases of the noncrystalline it is hard and tough with no cleavage, and resists mechanical weathering better than the other important rock minerals. It eventually breaks into tough, angular, irregular fragments that resist abrasion.

Feldspars constitute a second important group of rock-forming minerals that consist of potassium, sodium, calcium, or similar aluminium silicates. They are brittle, with pronounced planes of cleavage, and they break easily to form small prism-like particles. The feldspars are very susceptible to chemical breakdown, and the mechanical disintegration accelerates the chemical processes to such an extent that feldspar fragments are rarely found in soils in humid regions. The decomposition products of the feldspars are exceedingly variable, depending on the type of feldspar and on the weathering conditions, but they can be described in three groups; complex hydrous aluminium silicates, soluble or semi-soluble carbonates of sodium and similar metals, and silica (usually in a colloidal suspension). The hydrous aluminium silicates comprise a family called the *clay minerals*, that are physically very different from the feldspars from which they came.

Micas comprise a second family of silicate minerals that often contain iron and magnesium in addition to their potassium. The mica flakes are soft and resilient, with a pronounced cleavage. They split easily and break to form still smaller, thinner flakes. Their chemical breakdown is similar to that of the feldspars, producing the clay minerals, carbonates, and silica, but in addition various oxides of iron are formed from those containing iron. The chemical weathering of the micas is not so rapid as that of feldspar; and so mica is often present in soils in humid regions.

The *ferro-magnesian* family of minerals (including hornblende, olivine, pyroxene) are complex aluminum silicates that contain both iron and magnesium. They are moderately hard and tough, with no pronounced cleavage, and break mechanically into irregular dark-colored fragments. They alter chemically to form iron oxides, clay minerals, and the other products of silicate decomposition.

Iron oxides and hydroxides occur in various crystalline and non-crystalline forms and in both the ferrous and ferric state. They may be present in the original rock, but are also produced by the weathering of iron-bearing minerals such as biotite or the ferro-magnesian group. Iron is responsible for much of the coloration, from the greenish hues of ferrous iron in deeply submerged formations, through the yellows of surface soils in temperate zones to the bright reds and purples of the highly oxidized ferric materials of the tropics.

The carbonate minerals, calcite and dolomite, break down mechanically into both irregular and prism-like fragments, depending on the degree of crystallization of the rock. Carbonate fragments, particularly the smaller sizes, are found most frequently in arid and glaciated regions. In humid regions, chemical weathering takes the form of solution. Weak acids from organic decay and plant roots, but principally from carbon dioxide dissolved in water, produce the following reactions:

$$H_2O + CO_2 = H_2CO_3$$
$$2H_2CO_3 + CaCO_3 = Ca(HCO_3)_2 + H_2O.$$

The soluble bicarbonate is leached away with the ground water leaving behind any insoluble portions of the original rock, such as chert, quartz, clay minerals and iron oxides.

REPRECIPITATION / The soluble products of weathering, carbonates and bicarbonates, as well as silica, iron oxides and iron hydroxides in the colloidal state are carried away from their point of origin by percolating moisture. In a new remote environment, further chemical and physical changes can take place and cause precipitation of the materials, either in the colloidal or crystalline form. Frequently this occurs in soil voids or rock fissures, filling them and sometimes cementing the mass into a new material.

The weathering process is dynamic: rock minerals are broken physically, changed chemically, and dissolved in water. The end products, soils, consist

1:5 Clay Minerals

of a relatively small group: predominantly quartz and clay minerals, with varying amounts of mica, ferro-magnesium minerals, iron oxides, and carbonates. The alteration continues, with the changes in environment that occur naturally and that are induced by drainage, excavation, flooding, filling, and by the weight of structures.

1:5 Clay Minerals

The decomposition of the feldspars, micas, and ferro-magnesium minerals, all of which are complex aluminum silicates, occurs in many ways. The major factors are moisture, temperature, oxidation or reducing conditions, the ions present in solution (including those released by weathering), pressure, and time. The reactions are varied, but an over simplified form of decomposition of feldspar illustrates how they might occur.

$$2KAlSi_3O_8 + 2CO_2 + 3H_2O = 4SiO_2 + 2KHCO_3 + Al_2O_3 \cdot 2SiO_2 \cdot 2H_2O.$$

The first product, silica is in the form of a colloidal gel or suspension; the second product, potassium bicarbonate is in solution, and the third product is a hydrous aluminum silicate—a simplified clay mineral.

There are many forms of clay minerals, with some similarities and wide differences in composition, structure and behavior. All are extremely fine grained, with large surface areas per unit of mass. Probably all have definite crystal structures that include large numbers of atoms arranged in complex three dimensional patterns. All are electrically active.

SHEET STRUCTURE / Most clay crystals consist of atomic sheets, principally of two types: silica and alumina. The silica sheet is a repeating two dimensional linkage of silicons with valences of 4 and oxygens with valences of 2. Each silicon is surrounded by four oxygens, each of which contributes one valence link to the central silicon. Some of the remaining oxygen valence bonds are linked with adjoining silicons, as shown in Fig. 1.4a, but the oxygens on one side of the sheet are unsatisfied. The geometric shape is that of tetrahedrons with oxygens at the points, a silicon in each center,

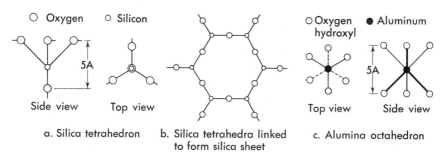

Figure 1.4 Atomic patterns in clay mineral sheets.

and the tetrahedrons arranged in a plane to form repeating hexagons, Fig. 1.4b. The height of the tetrahedron and therefore the sheet thickness is 5 Angstrom units or 5×10^{-7} mm.

The alumina sheet, Fig. 1.4c, is more complex. It consists of units of one aluminum surrounded by oxygens and hydroxyls (OH) that form octahedrons. Adjacent aluminums share oxygens and also OH groups, by alternating one to another. The thickness of this unit is likewise 5Å. The sheet formed by linking the unit octahedra does not balance the valences as does that of silica, so an occasional unit octahedra will contain no aluminum. This makes the sheet nonsymmetrical, and nonuniform.

The complexity of the sheet is enhanced by *isomorphous substitution*, the replacement of one or more aluminums by magnesium. The substitution of magnesium with a valence of 2 for aluminum with a valence of 3 creates an unbalance. This aggravates any local unbalance caused by the absence of aluminum in the octahedra. Similarly iron and other atoms can substitute for aluminum provided they physically fit the space; even aluminum can substitute for silicon.

Most clay minerals consist of silica and alumina sheets stacked together, Fig. 1.5, to form plates. The unsatisfied oxygens of the silica sheet are shared with the alumina sheet, to form a more or less balanced whole. Any remaining unbalance may be satisfied by cations supplied by salts in the surrounding water. In some instances cations are shared by adjacent plates. Similarly hydrogen atoms may switch back and forth between plates. The shared attraction, known as the *hydrogen bond*, can provide linkage to hold the plates together in stacks.

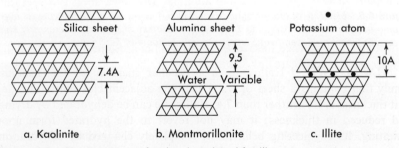

Figure 1.5 *Sheet structure of major clay mineral families.*

KAOLINITE / Kaolinites consist of alternating alumina and silica sheets that form a clay mineral whose nominal unit plate thickness is 7.4Å. There are several members of the kaolinite family, depending on variations in the alumina sheet. In general they are relatively well balanced electrically and exhibit only limited isomorphous substitution. The sheets are tightly bonded in the plates; moreover they stack like leaves in a book to form stacks as thick as 0.01 mm. Such a stack is shown in Fig. 1.6.

SEC. 1:5] CLAY MINERALS

Figure 1.6 Electron photomicrograph of kaolinite showing individual plates overlapping to form a stack, magnified 7400 diam. (*Courtesy of Electron Microscopy Laboratory, Engineering Experiment Station, Georgia Institute of Technology.*)

HALLOYSITE / Halloysite is a peculiar member of the kaolinite family that contains a sheet of water between adjacent clay units. The clay unit thickness is 10Å rather than 7.4. Halloysite can be dehydrated by drying, and reduced in thickness; it may not revert to the hydrated form upon rewetting. Its engineering behavior is completely changed by dehydration. Serious technical problems in embankment construction have been traced to halloysite that was tested in the laboratory after drying but used moist as a fill material.

MONTMORILLONITE / Montmorillonites or smectites are clay minerals consisting of one alumina sheet sandwiched between two silica sheets. The unit plate is 10Å thick, but can be as wide as 1×10^4Å. The units do not stack together readily, and when stacked, break apart easily. The montmorillonites are characterized by extensive isomorphous substitution, with each substitution theoretically producing a different mineral. The variety of cations that compensate for each substitution further multiplies the variety of montmorillonites. Many different minerals, such as nontronite, sauconite and saponite are included in this group but from the engineering point of view differentiation has little value. Generally montmorillonites form in regions rich in ferro-magnesian rocks (the source of magnesium) such as volcanics, and particularly in areas of high temperature and intense rainfall.

ILLITE / The illites, like smectites, consist of an alumina sheet between two silicas, 10Å thick. However, adjacent illite units are linked by their sharing of potassium atoms and thus stack together rather tightly. There is only limited isomorphous substitution in the alumina sheet; but there may be some substitution of aluminum for silicon in the silica sheet. The illites are often present in shales and other deposits that have been subject to a changing environment; they appear to be the alteration products of other clay minerals.

Other clay mineral groups include *chlorite* and *vermiculite* with mica-like structures. Vermiculite is similar to montmorillonite and is sometimes considered to be a part of that family. However, like halloysite it contains sheets of water between the vermiculite unit sheets. Both minerals exhibit isomorphous substitution but not to the same degree as montmorillonites.

The *attapulgite* minerals are different in that the silica tetrahedrons form a double layer chain with magnesium and aluminum atoms providing the link between the layers. The crystal is in the form of a long slat or curled ribbon that can contain water molecules. *Sepiolite* and *polygorkite* are similar. These clays appear to form from others when they are subjected to a highly saline environment.

IDENTIFICATION / Because of the extremely small size, identification of the clay minerals is difficult. Most soils contain several, so fine grained and so similar in their size and weight that separation is virtually impossible. One method of identification of mixed clay minerals is *differential thermal analysis*. The clay is heated slowly. At different temperatures, depending on the clay mineral, water is released. The heat required to release the water produces a characteristic change in the rate of heating. The clay is identified emperically and semi-quantitatively by comparing the rate of heating with standard curves for the pure minerals. The *x-ray diffraction* method subjects the clay to an x-ray beam at varying angles. The shadows produced by the atoms in the structure create a characteristic pattern from which the atomic arrangement can be deduced. This is probably the most reliable method for identification and can give both the clay mineral and the approximate amount present.

The electron microscope makes a shadow photograph of the mineral. The well-stacked kaolin particles can be easily identified as in Fig. 1.4, but the montmorillonites, whose unit plates separate easily, sometimes produce no identifiable pattern.

Identification of the clay minerals and appreciation of their structure are essential in gaining an understanding of the behavior of the soils which contain them. The character of the clay greatly influences a soil's drainage, strength, compressibility and particularly its reaction to changes in moisture. Any rational approach to improving the behavior of a soil containing clay minerals must be based on a knowledge of the physical and chemical make-up of these complex solids.

1:6 Organic, Precipitation, Volcanic, and Man-Made Solids

Other forms of solids, although scattered in occurrence and often only a small percentage of the total weight, can have an influence on the ultimate behavior of soil and rock mass that is far greater than their relative proportion would suggest.

Organic materials are present in many surface soils, particularly when the environment is not conducive to rapid decomposition. Fibrous mats of roots and partially decayed vegetation accumulate in swampy regions where the water is stagnant, or where the materials are buried in soils that limit the circulation of water and oxygen. The partial decay produces hydrogen sulfide gas that is a serious hazard to life in excavations and which induces corrosion of construction materials. The soluble organic products leach away as brown swamp water.

As decay continues the individual pieces lose their identity and a nearly structureless, fibrous peat is produced. This is largely cellulose, but is often mixed with mineral matter that is deposited simultaneously. The peat continues to decompose if it has some access to circulating ground water or air, producing methane gas. This presents an explosion hazard in excavations and tunnels. Soils containing large amounts of root matter or peat are so compressible that they are avoided or discarded if possible. However, it is essential to understand them and know how to use them, because there may be no alternatives.

Organic decay produces humic acids that reduce ferric iron to ferrous, and which aid in the decomposition of rock minerals and the development of kaolin-type clays. Concentrations of organic cations also play a part in clay mineral structure; organic clays behave differently than those that are inorganic. These humic acids often percolate deeply into soil and rock, causing local alteration at great depth and peculiar discontinuities in engineering behavior.

Organic growth produces soil and rock. The shells, shell fragments and coral formations of the sea are well-known examples of biologic calcareous deposits. The beach sands of some islands and even of mainlands, such as Florida, consist largely of shell fragments and ground up coral. Diatoms, minute sea and fresh water organisms, have skeletons of silica, like tiny snowflakes; accumulations of these are occasionally found in sedimentary soils. Iron-fixing organisms similarly are thought responsible for some accumulations of iron ore as well as iron cementations of other sediments.

Precipitation of carbonates and silica in the sea form great accumulations of limesands, and nodular silica. Some of the Bahama Islands appear to be forming in this manner. Replacement of calcite by less soluble silica entirely alters the character of the lime deposit, forming nodules of chert in the softer calcite mass.

Volcanic action generates soil directly when the lava explodes into *volcanic ash*. This is a porous, cinderlike material. It rapidly decomposes into soil, cements itself into a volcanic sandstone or *tuff*, or remains in its initial form, depending on the environment.

MAN-MADE MATERIALS / Man-made materials are becoming increasingly important soil and rock solids. The major source is waste of all kinds, ranging from uniform products of an industrial process to the heterogeneous accumulations of garbage, building debris, and metallic spoil from domestic and varied industrial sources. Each presents its special problems. Some industrial byproducts, such as slag, are excellent soil-like raw materials for embankment construction. Some even harden into rock-like masses that obstruct future work. Others, such as old mine tailings are hazards, as was the old pile of coal mine waste that demolished a Welsh school in the 1960's. Some can be utilized as foundations, but some wastes, like most sanitary land fills, must be kept covered but vented for years to minimize health and explosion hazards. It is particularly important to realize that most man-made materials are likely to be far more active chemically than natural minerals and react more quickly and more dramatically to changes in environment.

1:7 Grain Size

The range in the sizes of soil particles or grains is almost limitless; the largest grains are by definition those that can be moved by hand, while the finest grains are so small they cannot be identified by an ordinary microscope. The particles produced by mechanical weathering are rarely smaller than 0.001 mm in diameter and are usually much larger, for nature's grinding processes are not very efficient and the small grains often escape further punishment by slipping through the voids between the larger grains. The products of chemical weathering, including iron oxides and the clay minerals, are tiny crystals which occasionally are larger than 0.005 mm in diameter but which are usually very much finer.

GRAIN SIZE TESTS / Two methods are commonly used to determine the grain sizes present in a soil. Calibrated sieves or screens having openings as large as 4 in. and as small as 0.074 mm (U.S. Standard No. 200) are used for separating the coarser grains. Sieves with smaller openings are available but are impractical for soil work. The portions finer than 0.1 mm can be measured by sedimentation. This is based on the principle that the smaller the size of a particle, the more slowly it settles through water. This method is unsatisfactory for grains smaller than 0.0005 mm because such particles are kept in suspension indefinitely through molecular agitation. For particles that have a near-spherical shape, both the sieve and the sedimentation tests give the same results in the size range in which they overlap. For flat particles, however, the sieve measures the intermediate dimension, or width as 1.4 times the sieve size, while the sedimentation indicates the diameter of a sphere that settles at the same rate through water as the soil particle. This equivalent diameter is approximately the grain thickness. Particles smaller than 0.0005 mm can be measured by an electron microscope, but the data are of little use in soil engineering. Particles larger than 4 in. are measured by calipers or special sieves.

TABLE 1:2 / STANDARD SIEVE SIZES

U.S. Standard		British Standard		Metric Standard	
No.	Diameter (mm)	No.	Diameter (mm)	No.	Diameter (mm)
4	4.76	5	3.36	5000	5.00
6	3.36	8	2.06	3000	3.00
10	2.00	12	1.41	2000	2.00
20	0.84	18	0.85	1500	1.50
40	0.42	25	0.60	1000	1.00
60	0.25	36	0.42	500	0.50
100	0.149	60	0.25	300	0.30
200	0.074	100	0.15	150	0.15
		200	0.076	75	0.075

GRAIN SIZE SCALES / Because of the extreme range in grain sizes, scientists and engineers have attempted to break the entire scale into smaller divisions. Many methods have been proposed, but all are arbitrary and no one method is better than another. A convenient scale, adopted by the ASTM, is shown on Fig. 1.7. The coarsest division is *gravel*, which includes all soil grains larger than a No. 4 sieve. *Sand* includes all particles smaller than the No. 4 sieve and coarser than a No. 200. The grains smaller than the No. 200 sieve are the *fines*. This fraction is sometimes subdivided into *silt sizes*, the particles coarser than 0.002 mm, and *clay sizes*, the particles finer than 0.002 mm. Unfortunately any attempt to designate clay by particle size is misleading, for some soils finer than 0.002 mm contain no clay minerals and some clay mineral grains are larger than 0.002 mm.

GRAIN SIZE CHART / A better method of representing the different grain sizes in a soil is the logarithmic chart shown in Fig. 1.7. The entire range in grain sizes, plotted on a logarithmic scale, forms the horizontal divisions, while the percentages by weight of the soil grains that are finer

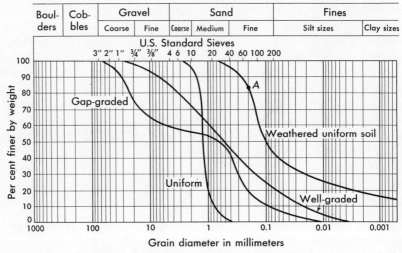

Figure 1.7 Grain size chart and ASTM-ASCE grain size scale.[1:1]

than a given size form the vertical divisions. For example, point A on Fig. 1.7 means that 83 per cent by weight of that soil is finer than 0.2 mm. A curve drawn through all the points that represent a single soil is known as the *grain size curve* of that soil. The interrelation of the different grain sizes in a soil can be seen from the shape of grain size curve. A steep curve indicates that the grains are nearly the same size, and such a soil is termed *uniform*. A flat curve shows a wide range in grain sizes; such a soil is termed *well-graded*. Humps in the curve indicate that a soil is composed of a mixture of two or more uniform soils. Such a soil is *gap-graded*. A steep curve in the sand range that gradually becomes a long, flat curve in the fines range may indicate a soil that was formed by mechanical weathering and later altered by chemical weathering.

EFFECTIVE SIZE AND UNIFORMITY / The *effective size* of the grains is defined as the size corresponding to 10 per cent on the grain size curve. It is given the symbol D_{10}.

Other sizes statistically defined that are useful include the *median*, D_{50}, the *finest quartile*, D_{25}, and D_{15}. The effective size, as well as D_{15}, has been found to be a major factor in the effective pore diameters and is related empirically to drainage and seepage of soil moisture.

The uniformity of the soil can be statistically defined in a number of ways. An old but useful index is the *uniformity coefficient* C_u. It is defined by the relation

$$C_u = \frac{D_{60}}{D_{10}}. \tag{1:8}$$

Soils with C_u less than 4 are said to be uniform; soils with C_u greater than 6 are well graded, provided the grain size curve is smooth and reasonably symmetrical. Another measure of uniformity, frequently encountered in geologic work, is the sorting coefficient S_0. It is defined by the relation

$$S_0 = \sqrt{\frac{D_{75}}{D_{25}}}. \qquad (1:9)$$

1:8 Particle Shape

The shape of the particles is fully as important as the size in determining the engineering behavior of soil and clastic rock. However, because shape is more difficult to measure and describe quantitatively, it is often neglected. Three classes of grain shapes have been defined: bulky grains, flakey or scale-like grains and needle-like grains. The first two are far more important, but all three are significant in their difference in physical behavior.

When the length, width, and thickness of the particles are of the same order of magnitude, the shape is *bulky*. Bulky grains are formed by the mechanical disintegration of rocks and minerals, by precipitation, and by volcanic action. They are rarely finer than 0.001 mm in diameter and most can be readily examined with a good magnifying glass or microscope, Fig. 1.8. A binocular-steroscopic microscope is better, because of the three dimensional representation.

SPHERICITY / Two aspects of the bulky shape are significant: The *sphericity* and the *angularity* or *roundness*.[1.7] The sphericity describes the differences between length L, width B, and thickness H. The equivalent diameter of the particle D_e is the diameter of the sphere of equal volume.

$$D_e = \sqrt[3]{\frac{6V}{\pi}}. \qquad (1:10a)$$

The sphericity, X, is defined as follows:

$$X = D_e/L. \qquad (1:10b)$$

A sphere has sphericity of 1, while flat or elongated particles have lower values. A second index is the *flatness* defined by:

$$F = B/H. \qquad (1:10c)$$

The elongation, E, is defined by:

$$E = L/B. \qquad (1:10d)$$

The ease of handling soil or broken rock, the ability to remain stable when subject to shock, and the resistance to breakdown under load are related to sphericity. The higher the sphericity or the less the flatness or elongation, the less the tendency of the particles to fracture or degrade into smaller particles under loading. Such materials are easier to manipulate in construction. Flat

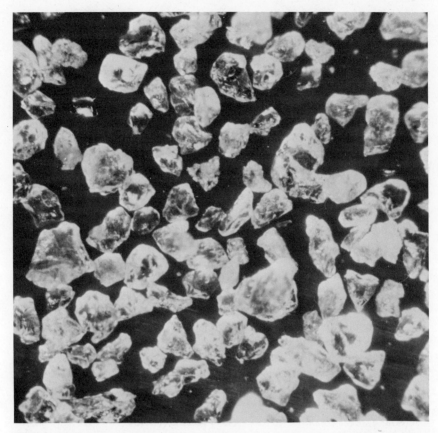

Figure 1.8 Photomicrograph of a bulky grained beach sand; magnified 54 diam. (Photo by G. B. Sowers)

and elongated particles tend to orient themselves so they are parallel when used in embankment or subgrade construction, forming planes of weakness. On the other hand, once interlocked by compaction they form a relatively stable mass.

ANGULARITY / Angularity or roundness, R, is a measure of the sharpness of the corners. It is quantitatively defined by:

$$R = \frac{\text{Average radius of corners and edges}}{\text{Radius of maximum inscribed sphere}}. \qquad (1:11)$$

Because it is difficult to measure, it is usually described qualitatively as shown in Fig. 1.9. When bulky grains are first formed by crushing or grinding of rocks, they are *angular*. After the sharpest edges become smoothed, they are *subangular*. When areas between the edges are somewhat smoothed and the corners begin to be worn away, the particle is *subrounded*. It is *rounded* when all the irregularities are nearly smooth, but the original shape can still be seen. When all traces of the original shape disappear, it is *well rounded*. Small sand particles close to their point of origin tend to be very angular, while the gravel and boulder sizes in the same environment are subrounded to rounded. Beach sands tossed by wind and wave are subangular to rounded, depending on the minerals and the distance to their origin. Wind-blown sands that roll and tumble in dunes become well rounded while the same sands in water are more angular. The microscopic examination of soils is a fascinating experience in which a common river sand is found to be shining quartz fragments and a dirty beach sand is found to be a collection of tiny shells, jewel-like garnets and zircons, on a background of crystal quartz.

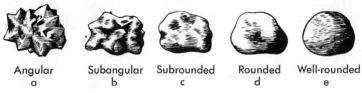

Angular　　Subangular　Subrounded　Rounded　Well-rounded
   a   b   c   d   e

Figure 1.9 Roundness of bulky particles.

The angularity has a profound influence on the engineering behavior. Under load the angular corners break and crush, but the particles tend to resist displacement. The smoother, more rounded particles are less resistant to displacement, but they are less likely to crush.

FLAKEY GRAINS / Flakey grains have very low sphericity (typically less than 0.01); they are thin, but not necessarily elongated. Generally they resemble a sheet of paper in their relative dimensions. They form from the mechanical weathering or disintegration of the micas, but the predominate flakey grains are the clay minerals. Compared to the bulky grains they are limber and resilient, like dry leaves. If the particles are randomly oriented, they may resist displacement; if stacked parallel, they resist displacement and shock perpendicular to their planes, but are easily displaced parallel to their surfaces. Small amounts of flakey mica can change the behavior of a predominately bulky grain soil. The flakes act like springs, separating the bulky grains and making the soil resilient and fluffy.

NEEDLE-LIKE GRAINS / Extremely elongated particles (E greater than 100) occur in some coral deposits and in the attapulgite clays. The particles are resilient and break easily under load.

EFFECT OF GRAIN SHAPE / Soils composed of bulky grains behave like loose brick or broken stone. They are capable of supporting heavy, static loads with little deformation, especially if the grains are angular. Vibration and shock, however, cause them to be displaced easily. Soils composed of flakey grains tend to be compressible and deform easily under static loads, like dried leaves or loose paper in a basket. They are relatively stable when subjected to shock and vibration. The presence of only a small percentage of flakey particles is required to change the character of the soil and to produce the typical behavior of a flakey material.

1:9 Interreaction of Water and Solid Phase 1:4

If equal volumes of dry sand and water are mixed together in a container, some of the water will fill the voids in the sand and the excess will merely rise above the sand surface. If more or less water is used, the effect on the sand will be the same (provided there is enough to fill the soil voids), and the only difference will be the amount of the excess water that covers the sand. The sand feels gritty, either dry or wet, and does not appear to be affected by the water. If equal volumes of dry montmorillonite clay (such as commercial bentonite) and water are mixed, the water will disappear and a sticky, greasy-feeling mass will be formed. If two volumes of water to one of montmorillonite are used, the result will be similar: The water will disappear and a sticky mass will be formed. The only difference is that the second mass will be somewhat softer than the first. In the case of the clay mineral, there is a reaction between the solid and the water, which results in a change in the characteristics of both. This phenomenon is termed *adsorption*, the binding of the water to the solid surface, and it has a profound effect on the physical properties of any soils containing minerals which exhibit it.

SURFACE CHARGE / The causes of adsorption are not fully understood, but they are definitely related to the electrical charges in the surface of the material. A number of mechanisms are responsible (1) Unsatisfied valence bonds at the edges of clay minerals and broken edges of other particles; (2) Unbalance caused by isomorphous substitution of low-valence atoms ($Mg = 2^+$) for one of higher valence ($Al = 3^+$) in the sheets of the clay minerals; (3) Nonuniform distribution of atoms and nonuniformity of the electrical charge in the surface; (4) Disassociation of ions such as hydrogen, hydroxyl, and cations from the clay surface in water. Both the electrical sign and the charge intensity vary with position. Clay minerals faces are generally negative, due to isomorphous substitution, and the edges positive; and there are localized areas of high and low charge. The total charge per unit of mass varies with the charge per unit of area and the ratio of area to mass. In most mineral particles with high sphericities and small intensities, the total charges are small. In the flakey clay minerals with relative intense surface charges, the total charges are high. The resulting electrostatic field is most intense close to the clay surface, but decreases rapidly with distance, as shown in Fig. 1.10a.

ADSORPTION OF WATER / Water is a peculiar molecule because of the nonsymmetrical distribution of the positive hydrogen atoms about the negative oxygen. The molecule, although neutral, is polar, with a positive charge on one side and a negative on the other. This polar molecule or *dipole*, is attracted to the surface of solids and particularly clay minerals with

their large electrostatic surface charges. It is held by a number of mechanisms:
1. The dipole—electrostatic attraction.
2. Hydrogen bonding—the sharing of hydrogen atoms with the clay.
3. Hydration of the cations that are attracted to the clay surface to compensate for isomorphous substitution.

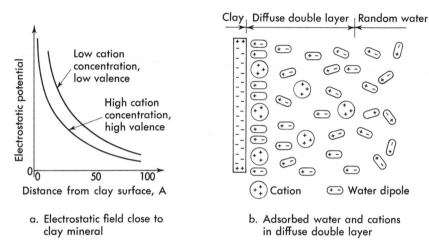

a. Electrostatic field close to clay mineral

b. Adsorbed water and cations in diffuse double layer

Figure 1.10 Force field of a clay mineral and the resulting adsorbed water.

The water collects close to the clay surface almost as iron filings are attracted to a magnet, Fig. 1.10b. The water closest to the surface is tightly held, and the molecules are oriented in the electrostatic field. The water closest to the clay surface appears denser than ordinary water. The attractive forces hold it to the clay and movement of the water is largely restricted to slow migration parallel to the surface. In this way the adsorbed water appears to have a higher viscosity than ordinary water. Further from the surface the attraction is less and the degree of orientation and apparent viscosity increase are smaller. The thickness of the innermost layer of water is probably 10Å (1×10^{-6} mm). The total thickness of water that is attracted to the clay may approach 400Å. This oriented water zone is termed the *diffuse double layer*.

EXCHANGEABLE CATIONS / Cations from soil moisture are attracted to the surface of clay minerals to balance the negative electrical charge produced by isomorphous substititution and possibly to balance the unsatisfied valence bonds at the particle edges. These are not fully integrated in the clay minerals; instead they can be replaced by other cations so long as the total valence balance is maintained. The cations present are those that comprise the soluble products of the weathering plus any that may be brought by percolating water. Sodium and potassium predominate, but calcium, aluminum, hydrogen, and even organic cations may be present depending on the environment.

The cations play a part in determining the clay behavior. Their positive screen reduces the effect of the negative clay charge and the behavior of the diffuse double layer. The higher the valence of the cation, the less the total attraction of the clay for water. The higher valence ions form a thinner layer with higher density positive charge than do the lower valence ions, and thus a more effective screen, Fig. 1.10a. The order of increasing cation effect is as follows:

$$\text{Li}(1^+) < \text{Na}(1^+) < \text{H}(1^+) \ll \text{Mg}(2^+) < \text{Ca}(2^+) \ll \text{Al}(3^+).$$

The cations are somewhat mobile, changing places with one another within the diffuse double layer. If the concentration of one cation, such as calcium, is increased by adding lime to a wet clay containing sodium, the calcium ions will replace the sodium ions in proportion to the relative concentrations. This is termed *base exchange* and the cations that take part are *exchangeable ions*. The exchange capability can be determined by leaching an ionic solution of a different cation through the clay and measuring the change. The amount that is exchangeable depends on the clay mineral. The kaolinites have low exchange capacity, the montmorillonites high exchange capacity.

CLAY PARTICLE INTERACTION / Clay particles in the presence of moisture exhibit greatly different behavior than do other minerals because of the interaction of the electrostatic fields and the diffuse double layers. Arranged face-to-face, they are held apart by their like electrical charges, with the diffuse double layers occupying the space between. At the same time they are locally attracted by unlike charges (such as at the edges), and the sharing of some hydrogens (*hydrogen bonding*) and possibly sharing of cations. A powerful attractive mechanism is the Van der Waals forces. This is essentially the dipole effect in which neutral molecules attract others because of their nonsymmetrical charges. These forces decrease rapidly with increasing spacing. Finally, there may be cementing by other minerals such as calcium or iron oxide between the grains.

The force system around the clay particles, including any external force, is in equilibrium. A large external force can move the particles closer, squeezing some of the water of the diffuse double layer away. A reduction in water content can reduce the diffuse double layer thickness and move the particles closer. The closer spacing increases the particle attraction and reduces potential movement between the particles. This gives rise to the phenomenon of plasticity in soils containing clay. A clay soil can deform plastically without cracking at varying water contents: the greater the moisture, the greater the particle spacing, the less the attraction and the greater the particle mobility. At low water contents the same clay particles are closer, exhibit more attraction, and form a more rigid mass.

COHESIVE AND COHESIONLESS SOILS / Soils in which the adsorbed water and particle attraction work together to produce a mass which holds together and deforms plastically at varying water contents are known as *cohesive soils* or *clays* (largely because this cohesive quality results from some proportion of clay minerals). Those soils that do not exhibit this cohesion are termed *cohesionless*. Soils composed of bulky grains are cohesionless regardless of the fineness of their particles. Many soils are mixtures of bulky grains and clay minerals and exhibit some degree of varying consistency with changes in moisture. These, too, are termed cohesive soils if the effect is significant. Obviously there is no sharp dividing line between cohesionless and cohesive, but it is often convenient to divide soils into these two groups for the purpose of study.

1:10 Plasticity and the Atterberg Limits

The Swedish soil scientist Atterberg developed a method of describing quantitatively the effect of varying water content on the consistency of fine-grained soils. He established stages of soil consistency and defined definite but arbitrary limits for each.

Each boundary or limit is defined by the water content that produces a specified consistency; the difference between the limits represents the range in water content for which the soil is in a certain stage or state.

TABLE 1:3 / ATTERBERG LIMITS

Stage	Description	Boundary or Limit
Liquid	A slurry; pea soup to soft butter; a viscous liquid	
		Liquid limit (LL)
Plastic	Soft butter to stiff putty; deforms but will not crack	
		Plastic limit (PL)
Semisolid	Cheese; deforms permanently but cracks	
		Shrinkage limit (SL)
Solid	Hard candy; fails completely upon deformation	

The *liquid limit* (LL) is defined as the water content at which a trapezoidal groove of specified shape, cut in moist soil held in a special cup, is closed after 25 taps on a hard rubber plate. The *plastic limit* (PL) is the water content at which the soil begins to break apart and crumble when rolled by hand into threads $\frac{1}{8}$ in. in diameter. The *shrinkage limit* (SL) is the water content at which the soil reaches its theoretical minimum volume as it dries out from a saturated condition. This is described in Section 3:9.

In themselves the Atterberg limits mean little, but as indexes to the significant properties of a soil they are very useful. The liquid limit has been found to be directly proportional to the compressibility of a soil. The difference between the liquid and plastic limits, termed the *plasticity index*

(PI), represents the range in water contents through which the soil is in the plastic state.[1:8].

The most important use of the Atterberg limits is in classifying fine-grained soils. In addition a number of relationships involving the Atterberg limits are useful in correlating soil behavior with simple test data. The activity, A, is the ratio of plasticity index to percentage of clay sizes, (finer than 0.002 mm)

$$A = \frac{\text{PI}}{(\% - 0.002 \text{ mm})}. \tag{1:12a}$$

The water-plasticity ratio or liquidity index, I_L, relates the water content of the soil to the liquid and plastic limits.

$$R_w = I_L = \frac{w - \text{PL}}{\text{PI}}. \tag{1:12b}$$

The Atterberg limits and their associated relationships are simple emperical expressions of the water adsorbing and absorbing ability of soils containing clay. They thus express both the clay-water behavior and also how that is diluted by the nonclay particles. The tests are usually standardized on the portion of the soil finer than 0.42 mm (passing No. 40 sieve: fine sand sizes and smaller). However, if little coarse sand is present, the tests are often made on the total sample, but with some differences in the results. The liquid limit is related to the total moisture potentially held in the diffuse double layer plus any water held by absorption; the plastic limit is related to the innermost moisture plus absorption. The plasticity index, the difference, is thus related to the potential water content changes of the diffuse double layer. The activity expresses the plasticity of the finest fraction, which is largely clay minerals. This is a measure of the water-holding ability of the clay minerals and also suggests whether the clay is a kaolinite (low activity, <1), a montmorillonite (high activity, >4), or illite (intermediate activity, 1–2).

1:11 Micro-Structure or Fabric

The mineral particles, water, and air are arranged in many different ways to form the materials we know as soils and rocks. In soil mechanics the term *structure* (more properly *micro-structure*) is used to describe the geometry of the particle-void formation. In rock mechanics the petrographic term *fabric* is used to denote the arrangement of the mineral grains, and structure is reserved for the larger features of the entire formation. The latter in soil and foundation engineering is sometimes termed *mass structure*. In agricultural engineering the structure means the pattern of the layering, cracking and agglomeration exhibited by soils close to the ground surface. In soil engineering this is sometimes termed *macro-structure*.

SEC. 1:11] MICRO-STRUCTURE OR FABRIC

The confusion of terminology merely adds to the problem of describing the endless variety of fabric or micro-structure possible, depending on the grain shapes, the interparticle forces, and the manner in which the soil or rock formed. For the purposes of study, most can be placed in four groups: cohesionless, cohesive, composite, and crystalline. The first three apply to soils and sedimentary rocks; the last applies to certain sediments and to igneous and metamorphic rocks. Many materials are encountered, however, which do not fit these simple basic patterns. Therefore, each soil or rock must be evaluated individually and unwarranted conclusions regarding engineering behavior should not be based on the description of the micro-structure.

COHESIONLESS STRUCTURES / The cohesionless soils are composed largely of bulky grains that can be represented by spheres or similar regular, equidimensional bodies. The simplest arrangement of such particles is similar to oranges stacked on a grocer's counter: each grain is in contact with those surrounding it. Such a structure is termed *single grain* and is typical of sands and gravels.

Depending on the relative positions of the grains, it is possible to have a wide range in void ratios. If we pack uniform, rounded grains in a box, with each directly above the one below as shown in Fig. 1.11a, a structure

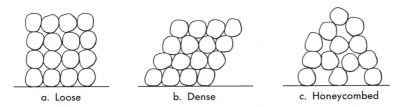

a. Loose   b. Dense   c. Honeycombed

Figure 1.11 Cohesionless soil structures.

with a void ratio of about 0.90 is formed. If we place them so that each succeeding layer falls into the depression between the spheres in the layer below, as in Fig. 1.11b, a structure with a void ratio of about 0.35 is formed. The arrangement corresponding to the higher void ratio is described as *loose*, and that corresponding to the lower is described as *dense*. Various arrangements of the same grains could be made to produce any void ratio between these two limits.

Similar variations in void ratio are possible in cohesionless soils having irregular grain shapes and mixed sizes. The highest void ratio possible for a given soil (and still have each particle touching its neighbors) is the *maximum void ratio*, e_{max}. The smallest void ratio is the *minimum void ratio*, e_{min}. The approximate minimum void ratio of a soil is determined by compacting it by combined tamping and vibration until no further densification is possible. The tamping must not be so vigorous that the soil grains are fractured, however. The approximate maximum void ratio is found by pouring the dry soil through a funnel into a graduated cylinder.

For uniform spheres, $e_{max} = 0.90$ and $e_{min} = 0.35$, and the range in e between the limits is 0.55. Soils with angular grains tend to have both higher maximum and minimum void ratios than spheres, but the range is usually somewhat smaller. Soils with mixed grain sizes, on the other hand, usually have lower values for e_{max} and e_{min}, and the range in e is also smaller than for uniform spheres.

Particles having low sphericity, particularly the slab-like fragments from layered rock do not form simple cohesionless structures. The slabs may bridge over large voids, yet be wedged tightly in a stable mass. When manipulated the slabs tend to orient themselves parallel to the direction of their movement. Such an *oriented structure* is *anisotropic* in its properties with entirely different behavior perpendicular to the orientation than parallel to it. Maximum and minimum void ratios probably have little significance in such materials.

Flakey particles, such as mica, with extremely low sphericity, similarly may form oriented structures. In soils formed entirely by decomposition and not transported from their point of origin, the orientation is a relic of the original rock fabric. Orientation in micaceous soils can also develop during sedimentation and from movements generated by shear or high pressure. Generally, oriented flakey structures have low void ratios. Nonoriented flakey grains may cause high void ratios when the flakes bridge over large voids. Maximum and minimum void ratios may be of little significance in micaceous soils. The heterogeneous arrangement of mica flakes wedged between bulky grains causes high void ratios but a rather stable arrangement. The more dense, oriented mica is highly anisotropic with a rather low resistance to displacement parallel to the planes of the particles. The maximum void ratio for soils containing even small amounts of mica is much higher than for the same soil without mica. The minimum void ratio depends on orientation; it is often impossible to determine with consistency.

TABLE 1:4 / TYPICAL VOID RATIOS AND WEIGHTS OF SINGLE GRAINED STRUCTURES

Soil Description	Void Ratio		Unit Weight (lb/cu ft)	
	Maximum	Minimum	Minimum	Maximum
Uniform subangular sand	0.85	0.50	118 (sat)	131 (sat)
			89 (dry)	110 (dry)
Well-graded subangular sand	0.70	0.35	123 (sat)	139 (sat)
			97 (dry)	122 (dry)
Very well-graded silty sandy gravel	0.65	0.25	125 (sat)	145 (sat)
			100 (dry)	132 (dry)
Micaceous sand and silt	1.25	0.80	110 (sat)	122 (sat)
			75 (dry)	94 (dry)

SEC. 1:11] MICRO-STRUCTURE OR FABRIC

The relation between the actual void ratio of a soil and its limiting values, e_{max} and e_{min}, is expressed by the *relative density* or relative void ratio D_r:

$$D_r = \frac{e_{max} - e}{e_{max} - e_{min}} \times 100 \text{ per cent.} \tag{1:13}$$

A natural soil is said to be loose if its relative density is less than about 50 per cent; dense, if it is higher.

The properties of loose, single-grained structures are greatly different from those of dense soil. The loose soil, with its grains perched directly on top of one another, is inherently unstable. Shock and vibration cause the grains to move and shift into a more dense, stable arrangment. The rounded particles are particularly unstable when loose, but even angular grains exhibit instability if their void ratios are sufficiently high. Dense, single-grain structures are inherently stable and are only slightly affected by shock and vibration. Both loose and dense structures are capable of supporting static loads with little distortion.

HONEYCOMBED STRUCTURE / Under some conditions it is possible to arrange cohesionless bulky grains in crude arches so that the void ratio exceeds the maximum for the single-grain arrangement. Such a structure has a negative relative density and is termed *honeycombed*, Fig. 1.11c. Honeycombed structures can develop when extremely fine sand or cohesionless silt particles settle out of still water. Because of their small size they settle slowly and wedge between each other without rolling into more stable positions, as do the larger particles. The structure also may develop when damp, fine sand is dumped into a fill or a pile without densification, a condition often termed *bulked*.

The honeycombed structure is usually able to support static loads with little distortion, similar to the way in which a stone arch carries its load without deflection. Under vibration and shock, however, the structure may collapse. In some cases this merely results in rapid settlement of the soil mass. In others the collapse sets off a chain reaction of soil failure that converts the entire mass momentarily into a heavy liquid capable of filling an excavation or swallowing a bulldozer. Fortunately such structures are not common and usually occur in lenses and pockets of limited extent. Because of the hazards involved, however, the engineer should view all water-deposited silts and very fine sands with suspicion until void-ratio tests prove them to be stable.

COHESIVE STRUCTURE—DISPERSION AND FLOCCULATION / In cohesive soils the structure is determined largely by the clay minerals and the forces acting between them. The clay particles in water are acted upon by a complex series of forces, some of which, including the universal attractive forces and mutual attraction to individual cations, tend to pull the particles together; others, such as the electric charge on each grain and the electric charges on the adsorbed cations, cause the particles to repel one another. The forces of both attraction and repulsion increase, but at different rates, as the distance between particles decreases. In a dilute suspension with wide particle spacings, the total repulsion usually exceeds the attraction. The particles remain apart and stay in suspension or settle very slowly while bounding about from the agitation of the water molecules, a motion termed *Brownian movement*. Such a system is termed *dispersed*. Dispersion can be increased by adding materials which increase the repulsion forces without increasing the attraction. Dispersing agents like sodium silicate and sodium tetraphosphate are used in the sedimentation test for soil grain size to ensure that the individual particles do not stick together and give a false indication of their equivalent diameter.

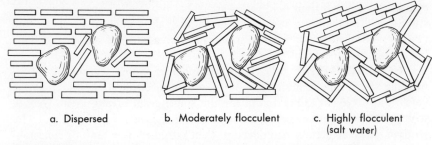

a. Dispersed   b. Moderately flocculent   c. Highly flocculent (salt water)

Figure 1.12 Cohesive soil structures. (*Adapted from T. W. Lambe.*[1:5] [1:6])

When the particle spacing is extremely small, as in a soil of a low water content, the attraction exceeds the repulsion and the particles hang together in a cohesive solid or semisolid, separated by their adsorbed layers. This effect can also be produced in a dilute suspension by reducing the repulsive forces. The addition of an electrolyte supplies ions to the soil particles, which partially neutralize the particle charges and thereby reduce their repulsion. The particles then attract each other even though widely spaced, and they move together and stick in a heterogeneous loose arrangement termed a *floc*. Such flocs often contain hundreds of individual particles and are sometimes visible to the naked eye.

DISPERSED STRUCTURES / The structural arrangement formed from a dispersed soil is shown diagrammatically in Fig. 1.12a. The repulsion between the particles as they come close causes each one to position itself for the maximum grain-to-grain distance in a given volume soil. The resulting structure is very much like flat stones laid on top of one another to form a wall. The bulky grains are distributed throughout the mass and cause localized departures from the pattern. This arrangement is termed an *oriented*, or a *dispersed*, structure. It is typical of soils that are mixed or remolded, such as by glacial action (glacial till), or of soils compacted under wet conditions in a man-made fill, or developed by sedimentation in the presence of a dispersing agent. Soils having a dispersed structure are likely to be dense and watertight. Typical void ratios are often as low as 0.5 but can be as high as 1 or 2, depending on the type of clay and the water content.

FLOCCULENT STRUCTURES / The arrangement in a flocculent structure is shown in Fig. 1.12b and c. It forms from a soil-water suspension which initially is dispersed, such as the suspended solids carried by a river. A sudden introduction of an electrolyte like salt water brings about flocculation. The particles, suddenly with less repulsion, fall together in a haphazard arrangement. There may be considerable interparticle contact between the positively charged clay mineral edges and the negative faces, producing strong bonds that resist displacement. Considerable free water is trapped in the large voids between the particles, in addition to the adsorbed water already immobilized by the clay. Flocculent structures are typical of water-deposited clays. The degree of flocculation depends on the type and concentration of clay particles and on the electrolyte. Deposits formed in the sea, which is a strong electrolyte, are frequently highly flocculent, with void ratios as large as 2 to 4. Fresh-water deposits, acted on by the weak electrolytes brought by rivers from different regions, are likely to be only partially flocculent or even dispersed. By way of contrast, organic acids from plant decay in shallow ponds and marshes may produce a high degree of flocculation.

Flocculent soils are light in weight and very compressible but are relatively strong and insensitive to vibration because the particles are tightly bound by their edge to face attraction. A peculiar characteristic is their sensitivity to remolding. If the undisturbed soil is thoroughly mixed without the addition of water, it becomes soft and sticky as though water had been added to it. In fact water has been added, for the bond between the particles has been destroyed so that the free water trapped between them has been released to add to the adsorbed layers at the former points of contact. This softening upon remolding is termed *sensitivity* and will be discussed at more length in Chapter three. Construction operations in flocculent clays are difficult because the soils become softer as equipment works on top of them and may develop into a sea of mud even in dry weather.

COMPOSITE STRUCTURES / Composite or cemented structures (Fig. 1.13) consist of a framework of bulky grains arranged like the cohesionless bulky grain arrangements and held together by a binding agent. A wide variety of such structures can develop, depending on the relative amounts of the binder and the bulky grains, the type of binder, and the method of deposition.

A number of different types of binder are found. Clay that has been highly compressed or dried, so that it is stiff or hard, and calcium carbonate are the most widespread. They are usually strong but may be weakened by water. Various iron oxides and colloidal silica from rock weathering also are encountered as binders and are relatively insensitive to softening by water.

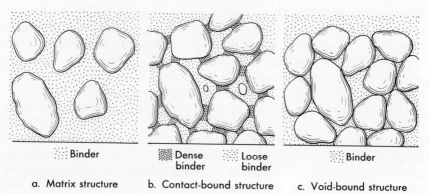

a. Matrix structure   b. Contact-bound structure   c. Void-bound structure

Figure 1.13 *Composite soil structures.*

In the *matrix* structure the volume of the bulky grains is less than about twice that of the binder so that the bulky grains float in a binder matrix, as shown in Fig. 1.13a. If the binder is clay, this is merely another form of a cohesive structure, and the physical properties are essentially those of a cohesive soil. With other binders the matrix structure is a form of rock whose physical properties depend on either the binder or the bulky grain, whichever is the weaker.

When the volume of the bulky grains is more than about twice that of the binder, skeletal structures develop. These take two forms: *contact bound* and *void bound*, depending on the position of the binder between the grains. In the contact-bound structure, the binder is concentrated between the points of contact of the bulky grains, holding them apart as stones are set in mortar. It can form in a number of ways. When bulky grains and clay settle simultaneously out of water, some of the clay is caught between the bulky grains and is compressed by the increasing weight of the sediment into a relatively rigid solid. A soft clay–water mixture occupies the voids between the grains but probably contributes little to the binding. The reweathering of a soil composed largely of quartz and with some feldspar, mica, or partially weathered clay can form a contact-bound structure in which the material in the voids is altered or leached away, leaving the material caught between the contact points of the grains largely unaltered. Contact-bound structures also form when large amounts of bulky grains and small amounts of clay are mixed and then consolidated or compacted. This occurs naturally by glacial action where the ice mass plows up and mixes the materials and the weight of the ice compacts the resulting *till* to a rocklike solid. It occurs artificially where clay–sand or clay–gravel mixtures are used in constructing highway or airfield subgrades.

Contact-bound structures are relatively rigid, incompressible, and resistant to shock and vibration as long as the binder remains strong. When the voids are large and open so that water can seep through them, the calcium carbonate and clay binders may soften. If the bulky framework is loose or honeycombed, the weakened soil will collapse like a loose cohesionless soil. If the bulky framework is dense, the softened binder will extrude into the voids, resulting in some settlement and weakening.

In the void-bound structure, the bulky grains touch each other and the binder occupies part or all the voids between them. This structure develops when the bulky grains are deposited first and the binder subsequently is deposited between them. Water seeping through a bulky grained soil can precipitate calcium carbonate, iron oxides, or silica to form a cemented sand or gravel which is rigid, strong, and dense. This structure also forms by the weathering of a rock such as granite, which consists of a framework of interconnected quartz grains with feldspar and micas between. The decomposition of silicate minerals leaves a quartz framework supported by clay minerals. Clay and fine bulky grains washed into a coarse sand or gravel deposit also can act as binders but not to the same degree as the binders that are precipitated. The void ratio of void-bound skeletal structures may be as low as 0.2, but typical values are 0.3 to 0.5. The soil is rigid and incompressible and not likely to be softened by water.

CRYSTALLINE STRUCTURES / Crystalline structures or fabrics form by crystal growth, from cooling of plastic or molten rock, through recrystallization produced by heat or pressure or by precipitation out of water. The mass consists of crystals meshed together into a more or less continuous material. The crystal shapes may be badly distorted, because they are not free to grow equally in all directions. The particles are bonded together by interlocking, sometimes by intergrowth, and by molecular bonds at points of contact. Surprisingly, the areas of true grain contact may be small, with minute voids separating many of the particle faces. These voids may be interconnected but most are independent.

Several details of crystalline fabric are of engineering significance. Oriented fabrics develop under heat, pressure and shear, producing anisotropic behavior. Shear oriented structures, with distorted broken crystals are termed *mylonitic*. A *porphoritic* fabric consists of large crystals or *phenocrysts*, in a matrix of finer grains. Such materials may exhibit great variations in behavior because the phenocrysts are points of structural discontinuity. A *vesicular* fabric contains bubbles, from gases in the molten rock. It is often light with both independent and some interconnected voids. *Vugs* are small cavities produced by solution or large gas voids.

MACRO-STRUCTURE / The continuous structure in fabric of natural soils and rock is frequently altered by local conditions to develop *macro-structure* or secondary structure. The principal cause is the continuing advanced weathering of the materials near the ground surface. The effects are particularly noticeable when soil materials are deposited in a greatly different environment from that in which they were formed or when there has been a significant change in environment.

The most prominent feature is cracking, caused by the shrinkage and expansion produced by moisture and chemical changes. These divide the soil into blocks whose dimensions range from a fraction of an inch to a foot or two. These macro-structures are termed *prismatic, blocky,* or *columnar* depending on their size and shape. A second form is aggregation in which clumps of grains accumulate because of reflocculation from organic acids. The resulting structure is termed *crumby*. A third form is the localized cementing of portions of the soil and porous or fractured rock by concentrations of humic acids from plants, calcium carbonate, or iron oxides to form *nodular* or *concretionary* structures or *hardpan* layers. A fourth form is *slickesided* cracking from local shear. A fifth form includes the filling of cracks and fissures with soil particles washed down from the surface or with materials deposited from solution. Soils with pronounced macro-structures are characterized by zones of hardness and weakness, and by patterns of color and texture which reflect the discontinuities.

REFERENCES

1:1 "Glossary of Terms and Definitions in Soil Mechanics," *Journal of the Soil Mechanics and Foundations Division, Proceedings, ASCE,* **84**, SM4, October 1958.

1:2 A. Casagrande, *Notes on Soil Mechanics, First Semester*, Harvard University, Cambridge, 1939.

1:3 R. E. Grim, *Clay Minerology*, McGraw-Hill Book Co., Inc., New York, 2nd ed., 1968

1:4 "Physico-Chemical Properties of Soil, A Symposium," *Journal of the Soil Mechanics and Foundations Division, Proceedings, ASCE,* **85**, SM2, April 1959.

1:5 T. W. Lambe, "The Structure of Inorganic Soil," *Proceedings, ASCE*, Separate 315, **79**, October 1953.

1:6 T. W. Lambe, "The Structure of Compacted Clay," *Journal of the Soil Mechanics and Foundations Division, Proceedings, ASCE,* **84**, SM2, May 1958.

1:7 W. C. Krumbein and L. L. Sloss, *Stratigraphy and Sedimentation*, W. H. Freeman, San Francisco, 2nd ed., 1963.

1:8 H. B. Seed, and R. J. Woodward, "Fundamental Aspects of the Atterberg Limits," *Journal of the Soil Mechanics and Foundations Division, Proceedings, ASCE,* **90**, SM6, November 1964 p. 75.

SUGGESTIONS FOR FURTHER STUDY

1. R. Y. Yong, and B. P. Warkenten, *Introduction to Soil Behavior*, The Macmillan Company, New York, 1966.
2. J. Feld, "Early History and Bibliography of Soil Mechanics,"

Proceedings, Second International Conference on Soil Mechanics and Foundation Engineering, **I**, Rotterdam, 1949.
3. References 1:3 and 1:4.
4. T. W. Lambe, and R. V. Whitman, *Soil Mechanics*, John Wiley & Sons, Inc., New York, 1969.
5. *Procedures for Testing Soils*, American Society for Testing Materials, Philadelphia, 5th ed., 1970.

PROBLEMS

1:1 The total weight of a chunk of moist soil is 330 lb. Its volume is 3 cu ft. The water content was found to be 27 per cent and the specific gravity of solids to be 2.72. Find e, n, S, and the weight per cubic foot.

1:2 A 50-cc sample of moist soil weighs 95 gm. It is dried out and found to weigh 75 gm. The specific gravity of solids is 2.67. Find e, n, w, S, and the weight per cubic foot of moist soil.

1:3 A 558-cc volume of moist soil weighs 1010 gm. Its dry weight is 918 gm and its specific gravity of solids is 2.67. Find e, n, w, S, and the weight per cubic foot of moist soil.

1:4 A 75-cc sample of moist soil weighs 120 gm. It is dried out and found to weigh 73 gm. The sample is assumed to be saturated, since it occurred below the ground water table. Compute its unit weight, w, e, n, and G_s.

1:5 A 120-gm sample of soil is 50 per cent saturated. The specific gravity of solids is 2.71 and the water content is 18 per cent. Compute the unit weight, e, and n.

1:6 A saturated soil has a water content of 38 per cent and a specific gravity of solids of 2.73. Find e, n, and the weight per cubic foot.

1:7 A saturated soil has a water content of 40 per cent and a unit weight of 114 lb per cu ft. Find e, n, and G_s.

1:8 A saturated soil has a water content of 47 per cent and a void ratio of 1.31. Find the weight per cubic foot and G_s.

1:9 A sand has a porosity of 37 per cent and a specific gravity of solids of 2.66.
   a. Compute e.
   b. Compute unit weight of sand if dry.
   c. Compute unit weight if sand is 30 per cent saturated.
   d. Compute unit weight if sand is completely saturated.

1:10 A soil has a unit weight of 109 lb per cu ft and a water content of 6 per cent. How much water in gallons should be added to each cubic yard of soil to raise the water content to 13 per cent? Assume that the void ratio remains constant.

1:11 A soil has a unit weight of 128 lb per cu ft and a water content of 12 per cent. What will be the water content if the soil dries out to a unit weight of 123 lb per cu ft and the void ratio remains unchanged?

1:12 A highly organic soil (peat) weighs 70 lb per cu ft saturated. The specific gravity of the solids is 2.35.
   a. Find e.
   b. Find the unit weight if the soil dries out without a change in void ratio.
   c. What would happen if the dried soil were subjected to a rising water table that reaches the ground surface?

1:13 How much difference in the unit weights and specific gravities of solids is there between a soil composed of pure quartz and a soil composed of 70 per cent quartz, 20 per cent mica, and 10 per cent iron oxide? Assume both soils are saturated and have void ratios of 0.63.

1:14 Plot on five-cycle semilog paper the grain size distribution curves from the following data. Compute the effective sizes and the uniformity coefficients of each. Record the percentage of sand, silt, and clay sizes according to the ASTM grain size scale.

PER CENT FINER BY WEIGHT

Sieve No.	Lagoon Clay, Beaufort, S.C.	Glacial Till Columbus, O.	Beach Sand, Daytona Beach, Fla.	River Sand-Gravel, Columbus, Ga.	Weathered Sandstone, Jasper, Ala.
½ in.	...	94	...	98	...
No. 4	...	68	...	86	100
10	...	50	...	60	82
20	...	35	100	39	76
40	...	22	98	26	70
60	100	18	90	4	60
100	95	15	10	...	43
200	80	11	2	...	27
0.045 mm*	61	10	...	...	23
0.010 mm*	42	7	...	...	13
0.005 mm*	37	5	...	...	8
0.001 mm*	27	2	...	...	3

* From sedimentation test.

1:15 A soil has a liquid limit of 56 and a plastic limit of 25. The water content of the soil as it is excavated for use in a fill is 31 per cent.
   a. Compute the *PI* of the soil.
   b. Is the soil likely to be stiff or soft when compacted at its existing moisture content in a fill?
   c. What would a light rain do to the consistency of this soil?

PROBLEMS

1:16 Compute the typical unit weight in pounds per cubic foot of:
   a. Dense, well-graded, subangular, dry sand.
   b. Dense, well-graded, subangular, saturated sand.
   c. Loose, uniform, rounded, dry sand.
   d. Loose, uniform, rounded, saturated sand.
   e. Honeycombed, saturated silt.
   (Assume typical values for e and G_s.)

1:17 A sand with a minimum void ratio of 0.45 and a maximum of 0.97 has a relative density of 40 per cent. The specific gravity of solids is 2.68.
   a. Find the unit weight dry and saturated in the present state.
   b. How much will a 10-ft thick stratum of this sand settle if the sand is densified to a relative density of 65 per cent?
   c. What will be the new unit weights, dry and saturated?

1:18 A sample of micaceous silt 10 cm in diameter and 2.5 cm thick is compressed to a 2-cm thickness without a change in diameter. Its initial void ratio is 1.35 and its specific gravity of solids is 2.70. Find the initial unit weight saturated, its void ratio after compression and its unit weight, after saturation, and the change in water content caused by compression. Assume that all of the compression is produced by a reduction in void ratio accompanied by a loss of water.

1:19 A soil has a specific gravity of solids of 2.72, a void ratio of 0.78 and a water content of 20 per cent.
   a. Compute its unit weight and degree of saturation.
   b. What will its new unit weight and void ratio be if it is compacted (a reduction in void ratio) without a loss of water until it becomes saturated?

1:20 A sample of volcanic ash weighs 40 lb per cu ft dry and 57 lb per cu ft saturated. When crushed and the specific gravity of solids is 2.75, find the total void ratio and the percentage of the voids that are not interconnected with the surface.

1:21 Two soils, clayey silty sands have identical particle sizes from a sieve test; both have 20 per cent finer than 0.074 mm. One dries out easily when exposed to air, the other does not. What explanation can you offer and how would you investigate this?

1:22 A soil has a void ratio of 0.95, a degree of saturation of 37 per cent and a specific gravity of solids of 2.72.
   a. Compute its water content, porosity, and unit weight.
   b. How much water in pounds must be added to a cubic foot of soil to increase its saturation to 100 per cent.

1:23 A cubic foot of saturated soil weighs 130 lb per cu ft. When oven dried it weighs 109 lb per cu ft. Compute the void ratio, porosity, water content and specific gravity of solids.

CHAPTER 2
Rocks, Soils and Ground Water

THE SITE SELECTED FOR A LARGE MANUFACTURING PLANT WAS A BROAD valley underlain by a limestone of Paleozoic age between ridges of sandstone and shale. The hard clay of the undulating valley floor was excavated from the high points to fill the many dimple-like swales that dotted the landscape. The structure was supported by spread footings. In spite of the hard clay above the rock, the site was well drained by numerous irregular pockets of chert within the clay that channeled water downward, apparently into cracks and fissures in the limestone. Rain water from roof drainage and even waste process water were diverted into the ground which provided such a simple means for disposal.

After the plant had been in operation a few years a large part of one wing suddenly subsided, virtually destroying the structure and damaging the machinery. The water had eroded soil from large fissures in the limestone leaving the plant supported by a blanket of clay that bridged across gaps in the rock. (This condition is illustrated in Fig. 2.8.) Eventually the clay bridge, weakened by continuing erosion and spalling of portions of the mass, collapsed, and the structure above tumbled with it.

Construction of a bridge foundation required excavation of clay underlain by a sand–gravel stratum. The data furnished to the contractor showed that the water table was close to ground surface within the clay stratum. Upon excavation, however, he found no water in the clay. He then proceeded to excavate downward as fast as his equipment would permit, congratulating himself in his good luck in not having to provide drainage. Suddenly, the bottom of the excavation erupted; clay chunks, sand and gravel boiled upward, engulfing the equipment in the excavation and seriously disrupting the entire construction operation. The contractor claimed he was beset by a mysterious changed condition that entitled him to extra compensation. Instead, he finally had to pay for a deeper pile foundation because his lack of ground water control had damaged a dense sand and gravel formation.

These case histories illustrate the problems that are generated by a lack

SEC. 2:1]   THE CYCLE OF SOIL AND ROCK FORMATION   41

of understanding of rock, soil and water as they occur in nature. Soil and rock formations comprise almost a limitless variety of structures that challenge the imagination. A rational solution to the engineering and construction problems involving these materials and their arrangement requires a thorough understanding of the formations and an accurate concise means of communicating that information to others.

2:1 The Cycle of Soil and Rock Formation

The earth is dynamic. The billion years or so of documented geologic history have seen changes brought on by both evolution and revolution that leave their imprints in the soils and rocks that comprise the earth's crust, Fig. 2.1.

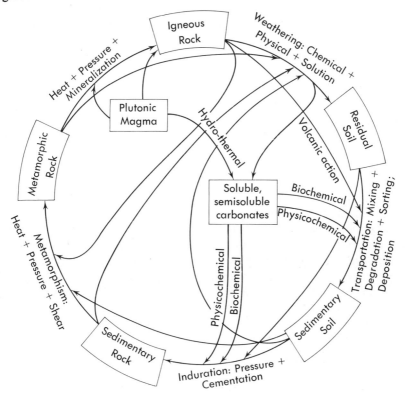

Figure 2.1 Cycles of change in soil and rock formation and alteration.

The entire earth's crust was probably at one stage a viscous liquid skin that slowly hardened into igneous rock. This crust is still being augmented by plutonic magma that occasionally flows or boils up from the depths.

The processes of weathering, aggravated by the wrinkling and cracking of the crust, attack the rock, creating *residual soils*: the in-place products of decomposition, solution and local disintegration. Some of these materials are transported by gravity through sliding and creep to make deposits nearby. Other materials are transported by wind, water and ice still greater distances. They are mixed together, sometimes sorted out by size, and then laid down in a new environment as *deposited soils*.

The deposited soils continue to weather, and some are re-transported and re-deposited in new formations. Others become indurated by consolidation and cementing into *sedimentary rocks*. Calcium and magnesium carbonates as well as calcium sulfate are precipitated from solution. Calcium carbonate and iron hydroxides biologically formed by marine organisms become sediments to add to the soils produced by weathering. They, too, can be hardened by cementing, consolidation and recrystallization into sedimentary rocks. The sedimentary rocks are subject to the same distortion and fracturing produced by tectonic movements as the igneous rocks. Environmental changes can similarly subject them to in-place weathering that produces new residual soils, followed by the processes of erosion and transportation that eventually create new deposited soils.

Instead of being exposed to weathering and erosion the sedimentary rocks can be buried beneath accumulating sediments and subjected to increasing heat, pressure and shear. The minerals are altered chemically and distorted or realigned physically to produce *metamorphic rocks*. The new rocks may resemble their ancestors but are usually more crystalline, denser, and harder. The metamorphic rocks are subject to weathering if exposed, forming residual soils that eventually are transported and mixed into new sedimentary deposits. Igneous rocks may also be metamorphosed by heat, pressure and shear, but the changes are generally less drastic. Finally, the metamorphic rock can be transformed back to igneous rock by heat, pressure and the addition of new minerals from molten masses below. The cycle is replete with short cuts and reversals; but it is continuous, with no definite beginning or end point. Engineering is merely another process added to the cycle, an insignificant one in the total pattern but locally often drastic. Therefore, all engineering works involving the earth must be evaluated in terms of their total impact on these dynamic processes.

2:2 Igneous Rocks

The ultimate source of igneous rock is still a cosmic mystery: molten magma wells up from the deeper part of the mantle bringing the mineral ingredients of igneous rock: principally quartz, orthoclase feldspar, plagioclase feldspar, muscovite and biotite mica, pyroxene and its variety augite, amphibole and its variety hornblende and magnetite. These, in varying proportions, form the igneous rocks. Nearly all are silicates, from the simple quartz to the complex pyroxene, but they vary particularly in the other metallic elements. Not all are present in any one rock: certain of the ferro magnesian mineral combinations are not likely in the presence of quartz; similarly large amounts of quartz and muscovite are not likely when the rock is rich in plagioclase feldspar.

The grain size or texture depends on the rate of cooling: rapid cooling means fine grains and slow cooling, coarse grains. Classification of igneous rocks, Section 2:9, is based on grain size and mineral content.

There are two primary forms of igneous structure; massive or *intrusive*, and *extrusive*. The intrusives include such regional masses as the Canadian shield of granite and more localized *domes* and *batholiths* (large knobs) such as Stone Mountain, Georgia, and Devils Tower, Montana. These are coarse grained rocks because of the slow cooling of the large volume of plastic or molten magma. Most intrusives are granite-like, rich in quartz, but large masses of coarse grained dark rocks like *gabbro* also occur. *Dikes* are wall-like intrusive bodies that fill fissures in older rock; and *sills* are similar intrusives between bedding planes. They frequently occur together, dividing the older rock with walls and sheets of the intrusion. Such smaller intrusions, which cool rapidly are finer grained than the large bodies.

The extrusive rocks form at the earth's surface by flow or volcanic eruption. The flows may form uniform strata covering wide areas or irregular waves of porous glassy boulders.

The extrusive rocks are generally very fine grained because of their rapid cooling. Rocks rich in ferromagnesian minerals are more common extrusives, the *basalts* and *dolerites*, (often collectively called *trap* rock) than those rich in quartz, the rhyolites. The *ash* and *bombs* formed by volcanic eruptions and explosions are another form of extrusive rock that falls to become sedimentary deposits. Often ash deposits and flows are interbedded; they later may be intruded with dikes and sills, to form solid walls and floors that enclose the more porous ash.

Large intrusive bodies are rather homogeneous. The smaller intrusives and the extrusives and the boundaries of the large intrusives exhibit numerous discontinuities. Joint cracks from cooling and flexure divide the large mass into blocks or prisms or varying sizes, weakening it. Blocks of older rock are frequently engulfed in the surface of intrusive bodies. Exposed at the ground surface, the rocks may exfoliate or spall in broad dome-like shells several inches thick with small gaps beneath. Because of this, foundations on "bed rock" occasionally settle, or move laterally. Joint cracks, shear cracks due to tectonic activity, and gas pockets in extrusives form paths for seepage and accelerate weathering from within the mass.

RESIDUAL SOILS FROM IGNEOUS ROCK / Residual soils are found wherever the rate of weathering exceeds the rate at which the products of weathering are removed by gravity, erosion and glacial action. In gently sloping terrain in the tropics they may be several hundred feet thick. In cool regions the weathering is slow and the residual blanket is thinner—only a few inches in Greenland. Moreover, in much of the Northern hemisphere, glacial action has plowed away the residual accumulations leaving ancient igneous rocks, such as the Canadian Shield, naked except for local pockets of soil cover.

Figure 2.2 Boulder like-blocks produced by weathering along joints.

The soils reflect the mineralogy of the parent rock. Granites produce tan and yellow sandy silts and silty sands, with varying amounts of mica and clays of the kaolinite family. The rocks rich in ferromagnesian minerals, such as basalt, yield highly plastic montmorillonite clays with iron oxide coloration ranging from deep red to the dark browns of the rich "black cotton soils" of India. The degree of weathering varies with depth. At the surface the feldspars, micas, and ferro magnesian minerals are largely converted into clay minerals, while with increasing depth they are only partially altered and still retain their inter-particle bonding. There is no well-defined boundary between soil and rock—only a transition. The weathering extends deepest and is most advanced along joints and in shear zones. In jointed rock, weathering from the joints inward creates plane sided "unweathered" blocks with rounded edges that resemble boulders, which float in a matrix of the more completely weathered soil, Fig. 2.2.

The deeper residual soils retain the fabric of the original rock in concentrations of minerals and orientation of grains. Residual soils exhibiting this relic micro-structure are termed *saprolites* (saprolites with strong relic structure are also derived from schists and gneisses).

Residual soils from granites vary in their foundation capabilities depending on weathering. They ordinarily are good construction materials.

Those derived from the ferromagnesian minerals are less desirable because of their montmorillonite clays.

The transition between the saprolite and the unweathered rock consists of irregular seams of partially and more completely weathered materials, in alternating seams or in the form of the boulder-like bodies previously mentioned. This zone is very porous, and is a cause of leakage of dam foundations.

2:3 Transported and Deposited Soils

GRAVITY TRANSPORTED SOILS / All soils are subjected, partially at least, to transportation by gravity alone. Residual soils in rolling areas tend to move slowly downhill—a process known as *creep*—but the general character of the soil deposit is usually not changed. Creep is important in that structures on shallow foundations may be moved out of position, and structures on deep, rigid foundations may be damaged by the pressure of the moving mass of soil.

A *talus* is an accumulation of fallen rock and rock debris at the bases of steep rock slopes and faces. It is composed of irregular, coarse particles and is very likely to be in an unstable condition. It is often a good source of broken rock and coarse-grained soil for construction.

Mud flows take place when loose, sandy, residual soils on relatively flat slopes become saturated. The soils flow like water and then come to rest in a more dense condition. The deposits are characterized by their heterogeneous composition and irregular surface topography.

RIVER DEPOSITS (ALLUVIUM) / Running water is one of the most active agents for soil transportation. As a transporting agent, water serves to mix soils from several different sources and then sorts and deposits them according to grain size. Small soil particles are lifted by the turbulence of the moving water and are carried downstream with little physical change, while the larger particles of sand, gravel, and even boulders are rolled along the stream bed to become ground down and rounded by abrasion.

The ability of running water to move solid particles is a function of the velocity and rate of flow. The total volume of particles that can be carried by a single cubic foot of water is proportional to the velocity squared. The volume of the largest particle that can be moved is proportional to the sixth power of the velocity. Therefore, during periods of high discharge, rivers carry tremendous volumes of coarse and fine particles; in periods of low flow, only small quantities of fine particles may be transported. If the stream velocity increases, such as where steeper portions of its channel are reached or when rainfall swells the flow, the river erodes its channel until its ability to transport materials is satisfied. If the stream velocity decreases because of flatter slopes or decreased flow, some of the transported particles are deposited, with the largest particles dropped first.

Streams in arid regions are characterized by flash floods and prolonged periods of little or no flow. Tremendous quantities of small boulders, gravel, and sand may be carried during periods of high water, but the volume of material transported in dry weather is negligible. The deposits formed in the steep portions of such a river fill the channel to great depths and also form narrow *terraces* of gravel and sand parallel to the low-water channel. Both shift and change during every flood season. At the point the river enters flat country, its velocity is sharply checked, and some of its load is deposited in the form of a flat, triangular mass termed an *alluvial fan*. As the fan builds up, the river shifts its course to build a succession of these masses.

Figure 2.3 Alluvial fan in a desert region, Death Valley, Calif. (from 35,000 ft).

The fans (Fig. 2.3) join to form a sloping *piedmont* at the foot of the mountains, a thick tilted undulating surface underlain by thick erratic deposits of silty sand and gravel.

The valley below the piedmont becomes filled with irregular lenticular masses of silt, sand and gravel brought by the rivers during flooding. The stream exhibits *braiding*, because it chokes its channel with solids after every period of high discharge that forces it to break through into a new course. These valley fills are usually very loose. They may develop some cementing because of continuing weathering, and precipitation of the soluble salts in hot desert climates. These cemented soils often collapse upon flooding, causing canals to slough and foundations to settle.

SEC. 2:3] TRANSPORTED AND DEPOSITED SOILS

Both the alluvial fans and the braided stream deposits in the valley fills are good sources of sand and gravel for construction.

Streams in humid regions are characterized by floods and sustained dry-weather flow. The particles carried by such streams are likely to be finer than those carried by streams in arid regions because the flood velocities tend to be smaller and because the greater degree of weathering in humid regions produces a much larger proportion of fines. The deposits in the steeper portions of streams in humid regions are similar to those formed by steep streams in arid regions but are smaller and less likely to shift during every period of high water. Where the rivers enter flat valleys they tend to form alluvial fans that are ordinarily broad and flat and composed largely of sands and fine gravels.

River deposits in flat valleys in humid regions are very important because valleys are often the sites of highways, railroads, airfields, industrial plants, and large cities. During periods of low flow, the stream is confined in its channels and deposition is balanced by erosion. During flood periods, however, it overflows its banks and floods the valley to form immense lakes and broad, flat sheets of slowly moving water. The velocity in the overflow areas is so much smaller than in the channel that deposition takes place along the banks of the channel, forming natural levees. The broad overflow areas act as settling basins in which the fine particles are deposited out of the slowly moving water. As the flood subsides, still finer particles are deposited until evaporation reduces the remaining puddles to dust. Flood-plain deposits (Fig. 2.4) consist of broad, flat, thin strata of very fine sands and

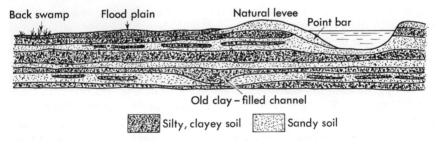

Figure 2.4 Cross-section of a flood plain deposit of an old river in a humid region.

clays with occasional elongated lenses of sand that formed in temporary channels or sloughs.

The lowest part of the flood plain is often farthest from the river and the last area to become dry after flooding. It is termed the *backswamp* because of the wet, soft soil and swampy organic matter that accumulates.

In old flat valleys the river meanders back and forth in sweeping S-curves. It erodes the outsides of the bends where the water velocity is highest and fills the insides with crescent shaped beaches of sand known as *point bar* deposits. The river loops migrate downstream, cutting and building simultaneously. Sometimes the river cuts across a large bend to leave the old channel behind as an *oxbow lake*. This eventually fills with flood plain silts and organic accumulations. Finally it becomes buried in the flood plain—an alien sinuous deposit of soft soils and organic matter trapped in the otherwise horizontal strata. These are hazards to heavily loaded foundations because they are weak discontinuities in otherwise regular formations.

The foundation capacity of flood-plain deposits is often limited, depending on the relative thickness and compressibility of the clay strata, and construction is often complicated by high ground water. The old river bottoms and the insides of bends are good sources of sand and gravel, and the plains are sources of sand, silt, and clay for construction.

LAKE (LACUSTRINE) DEPOSITS / Geologically, lakes are temporary basins of water supplied by rivers, springs, and the outflow from glaciers. They act as giant desilting basins in which the greater portion of the suspended matter carried by the streams that feed them is deposited.

Streams in arid regions carry great quantities of suspended, coarse sands during periods of high discharge. These are deposited at the point the stream enters the lake and form a *delta*. Deltas are characterized by uniform grain size and by bedding at angles of about $30°$. The finer suspended particles are carried out into deeper water where they settle out to form horizontal, thin strata of alternately coarse, then fine-grained particles. Lakes in arid regions soon fill with soil and become nothing more than shallow ponds that dry out in the hot summer sun. If they have no outlets, they are salty or alkaline, depending on the dissolved matter in the inflowing streams. The resulting deposits consist of thin strata of fine sands, silts, and sometimes clays that may be partially cemented with borax, gypsum, or calcium carbonate. At the edges of the deposits are thick, uniform beds of sand that represent the former deltas.

Lakes in humid regions also accumulate deltas at the mouths of the inflowing streams, but the deposits are likely to be finer-grained than those of arid regions. The finer particles are carried out into deeper water where the silt particles and the coarser clays are slowly deposited. During periods of low flow when there is little turbulence and when slight changes in the water may produce flocculation, the colloidal clays are deposited. The result is alternate thin strata of silt and clay. As the lake fills up and becomes shallow, plant life around the edges increases. Rotting vegetable matter produces organic colloids that are deposited with the silts and clays to form organic soils. Microscopic organisms called *diatoms* contribute their silica skeletons, and other organisms add their calcium carbonate shells to the deposit. Finally the lake chokes up with vegetation so thick and matted that only incomplete decomposition can take place. The result is a covering of fibrous organic matter known as *peat*, and at that stage the lake has become a swamp or bog. Lake deposits (Fig. 2.5) consist of alternate thin strata of

SEC. 2:3] TRANSPORTED AND DEPOSITED SOILS   49

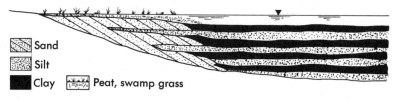

Figure 2.5 *Cross-section of a lake deposit in a humid region.*

silt and clay overlaid by organic silts and clays and finally topped with a stratum of peat. Thick beds of sand, the former deltas, are found at the edges of the deposit.

Lake deposits ordinarily make poor foundations because the soils are likely to be weak and compressible. The deltas sometimes provide good structural support, and they are sources of uniform sand for construction.

MARINE DEPOSITS / Marine soils include two groups: offshore deposits and shore deposits. The offshore conditions are similar to those in lakes in that deposition takes place in relatively still water below the zone of wave action. The degree of flocculation may be considerably greater because of the salt water, and calcium carbonate in the form of shells or microscopic particles may accumulate. Offshore deposits consist of horizontal strata of silt and clay that frequently have a highly flocculent structure. Occasional strata of shells or calcareous sands, silts, and clays termed *marls* may be formed that are partially cemented.

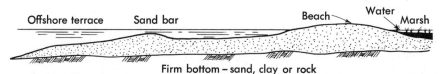

Figure 2.6 *Cross-section of a shore deposit.*

The shore deposits are highly complex, owing to the mixing and transporting activities of the many different shore currents and wave action. Materials brought to the sea by rivers and washed from the sea by wave action are swept along the shore by the shore currents, to be deposited in the form of *spits* or *bars* in areas where deep water or wide bays reduce the current velocity. These materials are reworked by the waves to form the offshore bar at the line of breakers and the beach itself (Fig. 2.6). The deposits continually move along the shore as *littoral drift*. The drift accumulates behind structures that extend out from the shore, leaving the shore beyond starved for sand and subject to accelerated erosion. Spits, bars, and the beach are composed of coarser soil particles—sands, fine gravels, and shell fragments—which are uniform at any one point but which may vary considerably in size throughout the deposit. The coarser particles of sand

and fine gravel may be subrounded to rounded, but the finer sand grains are usually subangular. Irregular beds of broken shells are often parts of beach deposits.

On many coastlines the spits or bars form barriers that eventually close off the beach from the sea and create shore lagoons. In some cases the lagoons are permanent lakes that rise and fall with the tide but in others they may be flat tidal marshes. Lagoon deposits are similar to the deposits in shallow lakes. The clay deposits are likely to be thick and have a highly developed flocculent structure. They often contain thick lenses of sand or shells that are washed into them during large storms. Marine sands and gravels and the cemented strata provide excellent foundation support and are good sources of cohesionless materials for construction. The clays are ordinarily weak and highly compressible and capable of supporting only light loads. They are too wet for use in construction.

WIND DEPOSITS / Wind is a highly selective agent of particle transportation. Particles coarser than 0.05 mm, such as sand, are rolled along by the wind or may be lifted a few feet from the ground during violent wind storms only to be deposited a short distance away. Sand deposits formed by wind action are known as *dunes*. They form in desert regions where mechanical weathering produces an abundance of coarse particles and along lake or sea shores where the sands have been concentrated in beaches or bars by wave action. The most important characteristic of sand dunes is their continual migration in the direction of the prevailing wind—a migration that man is often powerless to halt. The moving sands cover highways, railroads, farmlands, and even towns, and efforts to stop them with sand "fences" or by attempting to cover them with protective coverings of vegetation have met with only sporadic success. Dunes take the form of irregular hills or ridges with flat slopes on their windward sides and slopes equal to the angle of repose to the leeward. They are usually composed of relatively uniform rounded to subrounded particles of sand sizes and are a good source of such materials.

Wind has the ability to lift and transport particles that are smaller than fine sand. Wind erosion is largely limited to dry silts of the arid regions, however, since cohesive or moist soils resist wind erosion.

Wind-blown silt may be carried for many miles before being deposited. Thick beds of wind-blown silt ordinarily accumulate in the semiarid grasslands that border the arid regions. The deposits build up slowly; therefore the grass growth keeps pace with the deposition. The result is high vertical porosity and vertical cleavage combined with an extremely loose structure. Such soils are termed *loess*. Most loess soils are hard because of deposits of calcium carbonate and iron oxide that line the former rootholes, but they become soft and mushy when saturated. Loess deposits are characterized by their uniform grain size, their yellow-brown color, and their pronounced vertical cleavage. Stream banks, gullies, and cuts in loess assume nearly vertical slopes because of the cleavage and because the high vertical permeability permits rapid saturation from rainfall and consequent sloughing of the soil on vertical planes. Loess may be altered by weathering in humid regions, particularly if the soil grains consist of feldspars that were broken up by mechanical weathering alone. Such a soil is termed *loess loam* and it lacks the characteristic uniformity, high void ratio, and cleavage of true loess.

Loess provides good foundation support if it does not become saturated. It can be a source of fine-grained soil for construction if its structure can be broken down before use.

Volcanic ash may be grouped with wind-transported soils. It consists of small fragments of igneous rock blown out by the superheated steam and gasses of the volcano. Fresh volcanic ash is a lightweight sand or sandy gravel. The deposits may be stratified or may be well-graded mixtures. Volcanic ash soaks up water readily and decomposes rapidly. When partially decomposed and then dried, it cements to form a soft rock known as *tuff*. Complete decomposition of the ash results in the formation of highly plastic clays with extremely high void ratios and high compressibilities. Although thick deposits of such clays are uncommon, the extreme settlement of structures built on them, such as in Mexico City, makes them worthy of attention.

Cemented volcanic ash makes a good foundation. It is sometimes used as a construction material, but it tends to break down chemically and physically.

GLACIAL DEPOSITS / Ice, in the form of the glaciers which plowed up great portions of North America and Europe, has been a very active agent of both weathering and transportation. The expanding ice sheets planed off hill tops, ground up rock, and mixed the materials together as they pushed their way southward. Some of the materials were directly deposited by the moving ice, while the remainder were transported by water flowing from the ice to be deposited in the lakes along the face of the ice sheets or transported in the rivers flowing away from the ice.

The direct deposits of the glacier are usually termed *moraines*. They are composed of *glacial till*, which is a term applied to the heterogeneous mixtures of particles, ranging from boulders to clay, that the ice accumulated in its travels. *Terminal moraines* are irregular, low hills or ridges pushed up by the bulldozing action of the ice sheet. These mark the outermost limit of the glacier's travel, for they were left behind as the ice retreated. A *ground moraine* or *till plain* is the irregular veneer of till left on the areas once covered by the glacier. The upper surface of the ground moraine is undulating but rather level over broad areas; its thickness varies considerably, however, depending on the preglacial topography of the area. *Drumlins* are elongated low hills of till that point in the direction of the ice travel. They occur in areas of ground moraines and possibly represent deposits of soil that accumulated in crevasses in the ice.

The water-laid deposits of glaciers resemble those derived from mountain streams except that both the volume of water and the load of solids were considerably greater. *Eskers* are the remains of rivers that flowed in tunnels beneath the ice. When the ice retreated, the river bed materials formed sinuous ridges of coarse sands and gravels that resemble a crooked railroad embankment. *Kames* are terraces of coarse sand and gravel deposited in valleys along the margins of the ice sheets. Rivers flowing out of the edge of the ice sheet broke through the terminal moraine to deposit great quantities of sand and gravel in irregular, flat beds termed *outwash plains*. In many areas the glacial streams flowed into large lakes that formed in depressions left by the retreating ice. The deposits in these lakes are similar to those formed in other lakes except that they are more extensive. Great deltas of sand formed at the mouths of the rivers, and thick beds of silt and clay formed in the still, deep waters beyond the shores. Occasional boulders and gravel found in the clays are believed to have been dropped by floating pieces of ice as they melted. The silts and clays often formed in thin alternate strata which represent seasonal variations in the rate of ice melting and the resulting stream flow. The coarser particles were deposited in summer, during periods of high discharge, and the clays in winter. Such deposits are known as *varved clays* when the individual strata are more than $\frac{1}{8}$ in. thick and as *laminated clays* when the strata are thinner.

Glacial sands, gravels, and till usually make good foundations. They are also good sources of construction materials. The glacial clays are only moderately strong and are often compressible. They often are problems in foundation design and usually are too wet to be used as construction materials.

2:4 Clastic Sedimentary Rocks

The clastic sedimentary rocks are formed from soil deposits by some process of hardening or induration. The calcareous sediments also harden into sedimentary rock, but because of their special qualities they will be discussed separately in Section 2:5.

A number of different processes are involved in induration. Increasing overburden weight from glaciers or continued deposition as well as lateral stresses induced by tectonic movements can consolidate the grains into a denser structure. The pressure may also produce added interparticle attraction in silts and clays. Cementing is the most important mechanism of induration. Silica, calcium carbonate and iron oxides precipitated in the voids bind the solids together. Even clay in a dry climate can be a cementing agent. The degree depends on the amount and type of the cementing agent as well as the way in which it was precipitated.

STRUCTURE OF SEDIMENTARY ROCKS / The rock generally retains the structure of the original sediment. The shape of the mass may be somewhat distorted by pressure, and the particles of low sphericity may be broken and re-oriented with their longest dimensions perpendicular to the direction of greatest stress.

The formations may be tilted and folded by crustal movements. Flexure of stratified deposits produces two sets of cracks, or joints, one parallel to the axis of folding and the other at right angles to it, and both perpendicular to the stratification. The former are *strike joints* and the latter *dip joints*, because they point in the direction of the dip of the rock. The bedding planes may slide across one another during folding, so that the entire deposit becomes a mass of tightly packed, more or less rectangular blocks of hard rock that retains the sedimentary structure but which has lost its continuity completely.

Sedimentary rocks are sometimes invaded by igneous intrusions or blanketed by lava flows. The resulting structure can be a sandwich of rock forms, sliced apart by faulting and rejoined by intrusions.

TYPES OF SEDIMENTARY ROCKS / The classification of sedimentary rocks, Section 2:9, parallels the texture of the soil deposits. *Siltstones* and *claystones* are hardened silts and clays. If the micas and clay minerals become re-oriented so their surfaces are parallel, the claystone is *shale*. Indurated sand is *sandstone* and if the cementing is stronger than the sand particles it is an ortho-quartzite. Indurated gravel is *conglomerate*. All can occur independently, but are frequently interbedded, as in the original sediment.

Angular broken rock produced by faulting is not properly a sediment. However, any angular fragmented masses, from faulting, volcanic action or accumulation that become indurated are *breccia*. Indurated volcanic ash is *tuff* or *tuffaceous* sandstone or *tuffaceous* siltstone, depending on its texture.

RESIDUAL SOILS FROM SEDIMENTARY ROCK / Sedimentary rocks weather back into soils if a changing environment reverses the induration processes. The minerals are often changed by the reweathering process. The new soil is somewhat different from the original sediment, although it may retain much of the original structure. Generally induration followed by weathering causes fracture of the harder particles so that residual soils from sedimentary rocks are finer grained than their soil parents.

The depth of weathering of sedimentary rocks is generally less than for igneous rocks in the same environment because the mineral components had been weathered before induration. For example, in Georgia, the residual soil from porous sandstone may be 30 ft thick, that from less pervious shale only 15 ft thick, but from granite 50 ft or more, although all occur in the same vicinity and have been exposed for about the same period of time.

SEDIMENTARY ROCKS AS ENGINEERING MATERIALS / Sedimentary rocks provide good foundations, depending on their induration. Some claystones and shales are likely to disintegrate rapidly upon stress relief or exposure to air; others expand because of chemical reactions between soluble salts in the clay and oxygen. Even biochemical disintegration has been observed in shales containing organic colloids. The loss of induration of sedimentary rocks is a serious problem in excavations and tunnels. It can be predicted by exposing small samples and noting their expansion and breakdown. Those that break down must be immediately protected from air and surface moisture by plastic or cement mortar spray coatings to preserve their integrity.

The use of sedimentary rock as construction materials depends on their cementation; generally the sandstones and conglomerates make good fill. Shales, because of their tendency to swell and disintegrate after excavation and recompaction, should be considered suspect until tests show otherwise.

The residual soils are as variable in their structural qualities and their use in construction as the soil deposits from which they were derived. Generally, if better strength or rigidity is needed, slightly increased depth can solve the problem.

2:5 Calcareous Sedimentary Rocks

Limestones and dolomites occur in a wide variety of forms and degrees of induration, depending on their mode of deposition and history. They are precipitated in warm, shallow water, often with tiny shells to form granular sediments. Sometimes these are soft and chalky but often indurated by slight reprecipitation into hard rock. A second form consists of shells, coral and other calcareous debris that cements into porous *shellrock* or *coquina*. The third form appears to be the result of reprecipitation of the first two into a more crystalline rock, often with fossils embedded in the calcite or dolomite mass. A less common form is *travertine*, a soft limestone precipitated by hydro-thermal action.

Limestones and dolomites frequently are deposited with other sediments, commonly clay. The combination may be mixed, as an argillaceous limestone, or interlayered as alternate seams of claystone (or shale) and limestone. Calcareous sandstones and interbedded sandstones and limestones also form, but are less common. Chemical alteration often takes place during or after induration with portions of the calcareous matter replaced by silica to form *chert* or *flint* nodules. In some old limestones the chert may comprise a large part of the total volume.

STRUCTURE / Two forms of structural defects are prominent in limestones. The poorly indurated chalky, shell, and coral rocks are porous and are laced with interconnected voids of varying size, similar to soils. These voids are frequently enlarged by solution into networks of small cavities. The second form consists of joints and bedding planes that divide the mass into more or less prism-like blocks. These cracks are also enlarged by solution.

WEATHERING / The limestones and dolomites weather by solution leaving behind residual soils which comprised the insoluble parts of the original rock. The rate of solution depends on the acidity of the water, the carbon dioxide dissolved, the temperature, the water flow, the amount of

SEC. 2:5] CALCAREOUS SEDIMENTARY ROCKS

previous saturation of the water by carbonate minerals, and the solubility of the rock. The surface area of the rock exposed to solution is a major factor. Solid, impervious limestone weathers from the surface down creating an irregular, pitted rock surface. If the rock is jointed, solution proceeds deeper, enlarging the cracks and fissures into irregular slots that usually become narrower with increasing depth, Figs. 2.7 and 2.8. The slots may clog with the residual soil eroded down from above, redirecting the flow and the

Figure 2.7 Joints enlarged to slots filled with residual soil in limestone.

solution through smaller joints or across bedding planes. The rock mass is gradually changed into an irregular assortment of rock pinnacles and blocks with deep slots between and occasional horizontal caves and vertical chimney-like holes. The porous limestones also dissolve, with the enlargement of the voids into tortuous channels of varying size, and shape. Solution activity appears to be greatest just above the water table; however, it occurs wherever there is ground water movement. Moreover, limestones both above and far below the water table exhibit the results of solution produced during earlier geologic periods when the ground water flow was different.

RESIDUAL SOIL / The residual soil consists of all the insoluble impurities of the rock: clay from kaolinite to montmorillonite, silica in the form of chert, from boulder sizes down, silica sand and silt, and iron oxides. The residual soil blanket is of varying thickness, depending on the age, the intensity of weathering and the percentage of impurities. Some very cherty or clayey limestones accumulate more than 100 ft of residual cover while young or pure limestones in dry regions are bare. There is generally a sharp, although irregular line, between soil and rock, in contrast to other forms of weathering. The soil immediately above the rock, and particularly in the slots and pits, is usually soft and pasty. That above is drier, stiffer and sometimes partially indurated by cementing and desiccation. The residual soil mass is generally

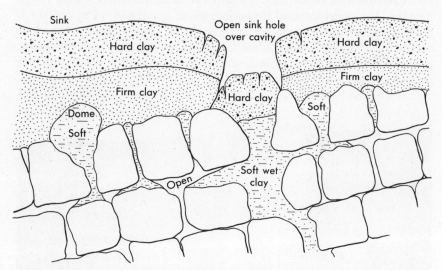

Figure 2.8 Cross-section of a residual soil with sink and cavity development in limestone.

structureless, except for the soft zone above the rock. Very sandy or shaley limestones sometimes reflect the original rock structure in distorted bands of gravel, sand, or plastic clay.

CAVITIES, RAVELING, AND SUBSIDENCE / The interplay between the solution slots and cavities in the rock, the sawtooth rock profile, and the blanket of residual soil above creates several serious engineering problems: (1) collapse of caves in the rock, (2) squeezing of the residual blanket at the points of support by pinnacles of rock, (3) raveling of the soil into the open cavities and slots below, and (4) crushing of the weakened rock. These can occur independently or simultaneously and are responsible for many structural catastrophies, such as the subsidence of the manufacturing plant described in the introduction to this chapter. The most common and insidious problem is raveling. Downward seepage erodes part of the soil into the rock cavities leaving a dome-shaped cavity in the lower surface of the residual blanket. This forces the seepage in that direction, aggravating the erosion. Pieces of the dome spall off when moisture weakens the dome surface, a process termed *raveling* or *roofing* in which the dome enlarges, Fig. 2.8. Depending on local weaknesses and on soil plasticity, the enlargement may be vertical to form a narrow chimney-like hole or lateral to form a broad dome. Eventually a dome becomes so large that the soil cannot span the opening and a truncated cone drops downward. This raveling is aggravated by changes in ground water, unusually dry weather, diversion of surface water into the ground, and sometimes through weakening the residual soil blanket by excavation.

ENGINEERING QUALITIES OF LIMESTONES / The harder, sound limestones and dolomites are among the best rocks for foundations, tunnels, and construction materials. The shell and coral rocks, shaley limestones, and the more earthy forms are variable; their qualities depend largely on the degree of cementation. Their crushing or disintegration must be evaluated from past experience or by tests before they can be used. The solution defects are major problems in foundations, reservoirs, dams and tunnels. It is best to assume that they are present until adequate investigation (Chapter 6) proves otherwise.

The sandy gravelly clay residual soils are usually good sources of clay for embankment construction provided montmorillonite is scarce. They ordinarily are good foundations where not undermined by raveling. Subsidence must be suspected, until investigation proves otherwise.

2:6 Metamorphic Rocks

Metamorphic rocks are produced by heat, pressure, and shear sufficient to alter the minerals in the original materials, to re-orient them, and to contort the rock mass as if it were a soft plastic. Generally clay minerals are recombined to produce new minerals such as chlorite. Other minerals are recrystallized and the crystals distorted by shear and pressure. Shales and mudstones with moderate metamorphism become *slates*, and *phyllites*. Higher levels of heat and pressure produce more profound changes. Extreme orientation and segregation of platey, micaceous minerals is characteristic of *schists*, with their laminated structure while *gneisses* exhibit segregation of minerals and a degree of orientation in their banding. Sandstones are altered into *quartzites*, which retain relics of the bedding, and which are among the toughest of rocks.

Limestones and dolomites are recrystallized into marble without significant changes in mineralogy. However, the impurities may be altered considerably.

STRUCTURE AND WEATHERING / Most metamorphic rocks exhibit joints and other structures similar to igneous rocks. In addition, their *schistosity* and *flow banding* are frequently surfaces of weakness that are twisted and contorted by the pressures which formed them, Fig. 2.9.

The marbles weather by solution similar to the non-porous limestones. The jointing is seldom as severe, and so the rate may be slower. The other metamorphic rocks weather in much the same way as the igneous rocks, with decreasing decomposition with increasing depth and no sharp boundary between residual soil and rock.

RESIDUAL SOILS / The soils range from silts to silty sands with varying amounts of mica in those derived from gneiss and schist. The deposits are extremely variable in composition and extent. The minerals are arranged in the same laminations or bands as in the original rock. These

Figure 2.9 Banding in a residual soil derived from the decomposition of gneiss.

soils which retain the relic structure and fabric of the rock are also termed *saprolites*. Within the mass of unweathered rock there are often seams of partially weathered material, from some less resistant band, or along old joints or fault zones. Within the soil there are frequently pinnacles or sawtooth projections of hard, slightly weathered rock. At one site in the Southeastern United States, the depth to sound rock varied from 5 to 70 ft in a distance of 200 ft. Seams of soft, highly kaolinized feldspar alternated with harder bands of quartz and thin seams of weak mica. Such a formation is often brightly banded as can be seen in Fig 2.9.

The residual soils from slates and phyllites are similar but more clayey, with slab like inclusions of less weathered rock alternating with clayey and silty strata.

Because of their variability and high mica and clay content, schists, gneisses and similar rocks as well as their residual soils require careful study. The residual soils derived from quartzites more nearly resemble those from granites, and residuals of marbles resemble those of limestones in their engineering behavior.

2:7 Profile Development—Pedogenesis [2:1, 2:2, 2:3, 2:4]

The continuing exposure of the uppermost soils develops a characteristic *profile* from the ground surface down. A number of different mechanisms are involved: the accumulation and decay of organic matter, leaching, precipitation, oxidation or reduction and additional weathering. The profile that ultimately develops depends on the *parent material* that comprises the initial deposit. Even more important are the environmental factors: temperature, amount and seasonal distribution of rainfall, ground surface slope, ground water level, and vegetation. These factors are interdependent to some extent and it is not always easy to identify the exact contribution of each. The degree of profile development depends on the environment and the period of time it has acted. In newly deposited materials the profile is shallow and poorly defined; in older deposits it may be as thick as 10 to 20 ft and clearly defined.

The science of profile analysis is termed *pedology* or *soil science*. It is one of the basic sciences of agronomy, where tilth and fertility are directly related to the soil profile. It is also vital to the soil engineer who utilizes the uppermost portions of a soil deposit as a foundation or as a source of construction materials.

In the tropics where the secondary weathering occurs rapidly and in regions that have been geologically stable for long periods, the profile development is so deep that it comprises the greater part of the soil mass, obscuring the nature of the original deposit. These are sometimes termed *pedo-genetic* soils or *geosols*.

PEDOLOGIC CLASSIFICATION[2:1] / The soil profile, in general and in detail is the basis for classifying soils pedologically. Those with similar profiles are members of the same *great soil group*. Those profiles whose formation are dominated by climate and vegetation are members of *zonal* groups. Within any of these groups the character of the weathering is more important than the composition of the parent material, and similar profiles can develop from a variety of different soil deposits. Those profiles in which local topography and drainage are the dominant factors are included in the *intra-zonal* groups that cross climatic boundaries. The peat deposits of low, wet areas are an example: the root mat in a Central American swamp has a striking resemblance to the *muskeg* of Canada. Some soils exhibit little profile development either because the profiled materials have been eroded, or because of insufficient time for development. These are *azonal* soils.

The great soil groups are divided into *series* in each of which the parent materials are similar. They are generally named for the locality in which they were first identified, similar to the naming of geologic formations. The series are then divided into types, based on the texture of the upper materials.

Pedalogic mapping has been conducted in many parts of the world, primarily for evaluating the agricultural potential. In some cases it has been general, only identifying the great soil groups; in others, series and even types have been mapped in considerable detail.

Empirical correlations have been developed between pedologic map units (types) and engineering behavior of the surface soils. Those include drainage, plasticity, texture and potential use as construction materials. The

U.S. Department of Agriculture county soil bulletins since 1960 have included engineering as well as agricultural data.

COOL AND TEMPERATE HUMID REGION PROFILE / In cool and temperate regions with humid climates there is an abundant growth of vegetation and an accumulation of dead leaves, plants, and other organic debris. The slow decomposition of these materials and the secretion of plant roots make weak acids which accelerate the weathering. The prevailing soil-moisture movement is downward to the water table, and this creates a profile with three distinct layers or *horizons*: A, B, and C, Fig. 2.10.

Horizon		Description	
O		Organic litter, humus	
A	A_1	Organic colloids + mineral	Zone of leaching or eluviation
	A_2	Light-colored leached	
	A_3	Leached, but transitional	
B	B_1	Accumulation, transitional	Zone of accumulation or illuviation
	B_2	Accumulation, clay formation, deep color	
	B_3	Transition, more color than C, Carbonate accum.	
C		Silica, carbonate, sulfate accumulation: Slight weathering	
D or R		No alteration	

Figure 2.10 *Soil profile in a region of mild temperature and moderate rainfall.*

The A-horizon is characterized by the chemical alteration of the soil materials in an acid, reducing environment. Clays of the kaolinite family, soluble carbonates, and semisoluble reduced iron minerals are likely to be produced. These are leached downward by the soil moisture, leaving the A-horizon deficient in them but rich in silica. As a result the lower part, designated the A_2-*horizon*, is usually very sandy and light-colored. The upper part, designated A_1, is dark-colored from its content of organic matter and has a spongy texture. The A_2-horizon is frequently a source of sandy soils in regions underlain by clays.

The leached materials accumulate in the B-horizon below. It is thicker than the A and contains a greater concentration of clay minerals, iron, and carbonates than does the original soil. The top part, B_1, is frequently partially cemented, and is deeply colored. The B_2 is rich in clay and soluble carbonates. The B_3 again suffers from downward leaching and is lighter colored. The B-horizon is the best source of clays in regions where they are scarce.

Below is the C-horizon, which is the slightly weathered parent material. When an unrelated stratum of different soil underlies the other materials, it is sometimes called the *R-horizon*. The D horizon is the unaltered parent.

HOT, HUMID REGION PROFILES / In hot, humid regions the upper parts of the deposit are also subject to alternate wetting and drying and to downward leaching. However, the climate is favorable to the rapid decay of organic matter and of its consumption by such insects as termites. There is little or no organic acid produced, and because of the formation of soluble carbonates, the silicate weathering proceeds in a basic environment. The soluble colloidal silica is leached downward. The aluminum and iron become highly oxidized, and are insoluble in the more basic environment. They remain to cement the quartz into a stiff rock-like solid. Advanced weathering causes the iron and aluminum to accumulate in nodules or concretions, giving the soil the texture of a loose, but cemented gravel. The color ranges from tan to bright red because of the highly oxidized iron, with mottling reflecting the local iron accumulations. The process is termed *laterization* and the well-indurated material is *laterite* or *ferricrete*. Well-developed laterites are strong and relatively incompressible although often light weight and porous. Some forms are sufficiently cemented to be used as a gravel base for road construction while less well-developed laterites may soften upon wetting. They are usually identified by their low ratio of silica to the aluminum and iron oxides.

DRY REGION PROFILE / In dry regions there is little or no organic matter, Any moisture movement is predominantly upward because of surface evaporation. This results in the accumulation of soluble materials such as carbonates near the surface and in the partial cementing of the soil.

Sometimes the carbonates are well distributed through the mass while in other cases they are concentrated in lenses or concretions at the level of evaporation of capillary moisture in the deposit.

These soils are usually strong and incompressible when dry. Upon saturation they weaken, and in some cases collapse with sudden loss of strength and rapid subsidence. They may be good construction materials depending on the parent material.

In extremely arid regions soluble salts are brought upward by capillarity following the brief periods of rainfall and are precipitated near and on the surface forming a saline or alkali topsoil and sometimes a white crust. Over-irrigation in very dry regions can cause the same salt accumulation and eventually make the ground infertile.

HUMID—POORLY DRAINED / The wet, poorly drained soils form intrazonal groups depending on the moisture. In a very wet environment, organic growth is rapid. Decay, however, is slow if the area is inundated and the water stagnant. The organic decay depletes the oxygen and the decay products inhibit further decay. The organic matter accumulates rapidly under such conditions. Higher rates of decay associated with fluctuating water levels produce nearly fiberless *mucks* while slower decay, associated with continuing stagnant inundation, produces *fibrous peat*. Because the decay tends to be slowest in cool regions, the thickest peat deposits are frequently found in the subarctic. Thick peat deposits also form in the tropics in slowly submerging deltas or in old oxbow lakes.

2:8 Ground Water

Water is one of the most important factors of soil strength, compressibility, and volume change. Although water is present in all soils, the term *ground water* is reserved for the continuous body of underground water in the soil voids that is free to move under the influence of gravity. The *water table* is the upper surface of a body of ground water. It is defined by the level of water in an open hole in the ground and is the level at which the pressure in the water is zero. Ground water is not a static body with a level surface as the name *ground water table* implies. Instead it is a moving stream with a sloping surface that takes many shapes, depending on the structure of the soils and rocks through which it flows.

The elevation of the water table at any one point is not constant. Water is supplied to the moving stream by percolation from the ground surface and it leaves the ground by evaporation and by seepage into rivers, lakes, and the ocean. When the rate of intake exceeds the rate of loss, as it does during wet weather, the water table rises; and when the intake decreases, as in dry weather, or when the loss increases because of pumping for water supply or because of drainage, the water table falls.

AQUIFERS / Aquifers are relatively pervious soil and rock strata that contain ground water. They are similar to the lake basins and river channels that contain surface water. The most familiar aquifer is a stratum of relatively pervious soil in which the ground water level rises and falls with the weather and with pumpage. The water table slopes in the same direction as the ground surface, but the slopes are more gentle and uniform. In soils that consist of alternate strata of pervious and impervious soils the ground water pattern becomes more complex. A sagging, impervious stratum creates a basin that may hold a small quantity of ground water perched above the general water table. *Perched water tables* (Fig. 2.11) occur rather frequently but are

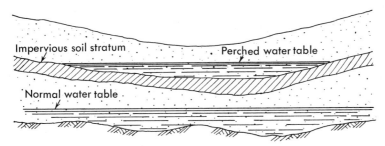

Figure 2.11 Perched water table.

ordinarily of limited extent. They may be drained by drilling a hole through the impervious basin, allowing the water to seep downward. When an aquifer is confined between two impervious strata, it is capable of carrying water under pressure. When it does, the elevation of zero pressure is above the upper surface of the water, and the ground water is said to be under *artesian* pressure. Artesian pressures are usually developed by sloping aquifers (Fig. 2.12) where the point at which the water enters the confined pervious stratum is higher than the point at which the pressure is measured. When a hole is drilled into an artesian aquifer, the water rises to the elevation of zero pressure. If this level is above the ground surface, a flowing artesian well results. Artesian aquifers may be local structures existing over an area of a few acres, or they may be continuous over large areas like the vast artesian sandstones in North and South Dakota. They are often troublesome to engineers because of the reduction of soil strength by neutral stress. Excavations that extend close to strata that are under artesian pressure may be damaged from *bottom blowouts*. The water pressure, which formerly was balanced by the weight of the overlying soil, causes the remaining soil to burst upward into the excavation, or if the soil is sand, it will create a "quicksand" condition (Chapters 3, 4).

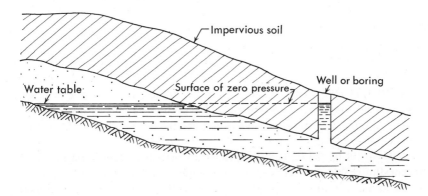

Figure 2.12 Artesian water table.

SPRINGS AND SWAMPS / When the ground water table intersects the ground surface on a hillside, a *spring* is formed, water trickles down the ground surface, and the soil may be softened by the added water and also by the seepage pressures. This may be corrected by intercepting the water with drains before it reaches the surface. The intersection of the water table and a level ground surface produces a *swamp*. During periods of wet weather and high water table the swamp may be partially covered with water, while during dry weather it may be relatively dry and firm. The upward seepage of water in some swamps produces a semiquick condition that is most pronounced during periods of a rising water table. Swamps are difficult to correct because they ordinarily occur in low areas in which there is no place to drain the excess water.

2:9 Rock Classification and Description

The study of rocks, including their classification, is termed *petrography*. Because of the complexities of rock formation and their many different mineral constituents, it is a complex science. However, from the standpoint of engineering behavior, detailed classification is seldom necessary. Table 2:1 is adequate for most engineering uses.

INDURATION-STRENGTH / The strength of the rock is of more importance in engineering than the texture or geologic classification. Tests of the intact rock, Chapter 3, are necessary for a quantitative determination. A standard for describing induration in terms of unconfined compressive strength, and simple field tests for estimating are given in Table 2:2.

TABLE 2:2 / DESCRIBING ROCK INDURATION
(Adapted from Duncan and Jennings)[2:6]

Description	Unconfined Compressive Strength	Field Test
Very Hard	20,000 lb/in.^{2} (1400 Kg/cm^2) or more	Difficult to break 4-in. piece with pick.
Hard	8–20,000 lb/in.^{2} (560–1400 Kg/cm^2)	4-in. piece broken with one hammer blow
Soft	2.5–8,000 lb/in.^{2} (175–560 Kg/cm^2)	Can be scraped, or dented slightly with pick point
Very Soft	1–2,500 lb/in.^{2} (70–175 Kg/cm^2)	Crumbles with pick, easily scraped with knife

STRUCTURE / The behavior of rock in engineering work is largely controlled by its mechanical structure. An infinite variety of arrangements occur; these can be grouped from the engineering point of view as described below.

Layering, Fig. 2.13a, is the segregation of like materials into more or less parallel sheets. It occurs in most sediments as deposit bedding and in metamorphic rocks due to pressure and flow. Many patterns are observed, thin layers of similar materials, hard layers over soft, soft over hard, alternating hard and soft, lying horizontally or dipping, straight or contorted, and even nonconformal. The character of the interface between layers is significant in establishing the behavior under load. Smooth surfaces slide more easily than rough ones in which the irregularities match or conform. Nonconforming irregularities contain points of stress concentration that can crush or tear under load. The orientation of the bedding with respect to the applied loads determines the stresses and the tendency for movement to develop. The scale, of the layering expressed as the ratio of the engineering structure width, B, to the layer thickness, H, establishes the all-over significance of the stratification and particularly of any weak interfaces.

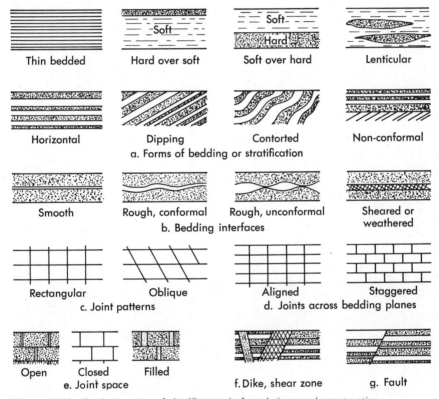

Figure 2.13 *Rock structure of significance in foundations and construction.*

The joints, Fig. 2.13c, are cracks more or less perpendicular to the bedding surfaces. They occur in groups or sets, with the joints of any one set approximately parallel and equally spaced. Typical joints lie parallel and at right angles to the axes of folds; these are *strike* and *dip* joints respectively. They divide the rock into rectangular prisms. Other sets may occur at oblique angles that divide the rock into parallepipeds or wedges. Joints in one layer aligned with those in the next, are favorable to joint movement. If they are staggered it is a more stable arrangement. The character of the joint itself is a major factor in its contribution to rock behavior. If closed,

TABLE 2:1 / SIMPLE ROCK CLASSIFICATION

Class	Type	Family	General Features
Igneous	Intrusive (Coarse Grained)	Granite	Light Colored 10–40% quartz, 40–60% O & M* feldspar, 2% Mica, 0–10 FM†
		Syenite	5–10% quartz, 25–50% O & M* feldspar, 0–20% O feldspar, 5–20 FM
		Diorite	0–5% quartz, 0–25% O & M* feldspar, 20–50% O feldspar, 20–30% FM
		Gabbro, ultra basics	Dark colored 0 quartz, 0 O & M* feldspar, 40–50% P feldspar, 30–50% FM
		Obsidian	Volcanic glass, often dark colored
Igneous	Extrusive (Fine Grained)	Rhyolite	Light-colored 10–40% quartz, 40–50% O & M* feldspar, 2% Mica, 0–10% FM, M
		Trachite	5–10% quartz, 25–50% O & M* feldspar, 0–20% P† feldspar, 5–20 FM‡
		Andesite	0–5% quartz, 0–25% O & M* feldspar, P† feldspar, 20–30% FM‡
		Basalt, Dolerite	Dark Colored 0 quartz, 0 O & M* feldspar, 40–50% P† feldspar, 30–50% FM‡
Igneous	Ejecta	Tuff (Volcanic Ash)	Cinder-like sand and silt-sized fragments, with occasional gravel-sized angular fragments
		Pumice (Porous)	Frothy or porous lava—usually light colored

				Range in texture: coarsely crystalline-
Sedimentary	Calcareous	Limestone	Calcium Carbonate	fossiliferous-granular-earthy.
		Dolomite	Calcium-magnesium carbonate	
Sedimentary	Silaceous	Siltstone, claystone, mudstone		Nonoriented, indurated silt and clay
		Shale		Oriented-laminated, indurated silt and clay
		Sandstone		Cemented sand, arkosic if appreciable feldspar present
		Conglomerate		Cemented sand-gravel or gravel
		Breccia		Cemented angular rock fragments
Metamorphic	Foliated	Slate		Finely foliated or oriented; fine grained; thin smooth cleavage
	(Oriented Grains)	Schist		Pencil-line to paper-thin foliation
		Gneiss		Bands of minerals 1/16 in. or under; rough cleavage
Metamorphic	Nonfoliated	Quartzite		Dense sand gravel structure that fractures through the grains
		Marble		Re-crystallized limestone or dolomite

* Orthoclase and Microcline feldspar.
† Plagioclase feldspar.
‡ Ferro-magnesian minerals.

the mass behaves like a continuous solid. If it is open, considerable movement will take place before there is rock-to-rock transfer of load across the joint. If the joint is filled with a soft material, the behavior of the rock mass is controlled by the resistance of the filling. The roughness of the joint surface, similarly, influences potential movement through or across the crack. Orientation of the joints with respect to the loading and the ratio of structure width B to joint spacing, S, is of similar significance as in the layers.

Defects in the rock mass, Fig. 2.13e–g, include shear zones, slickensides, dikes, cavities, solution channels, and porous zones. Some discontinuities are zones of weakness; the more rigid ones concentrate stresses. Their spacing and orientation are major factors in their engineering effects, similar to the layers and joints.

In each situation all pertinent details of the structure must be established. In most cases the structure is the dominate factor, and the nature of the rock between joints or bedding planes is of minor importance.

Systems for describing the structure quantitatively are described in textbooks of geologic structure and treatises on rock mechanics. Useful terms for describing spacing are given in Table 2:3.

TABLE 2:3 / SPACING OF STRUCTURE (After Deere)[2:7]

Descriptive Term Bedding	Descriptive Term Joints	Beds or Joints per Meter or Yd	Spacing	
			Cm	In.
Very Thin	Very Close	>20	Less than 5	Less than 2
Thin	Close	3–20	5–30	2–12
Medium	Medium	3–1	30–90	12–36
Thick	Wide	1–1/3	90–300	36–120
Very Thick	Very Wide	<1/3	Over 300	Over 120

2:10 Soil Classification

A soil classification system is an arrangement of different soils into groups having similar properties. The purpose is to make it possible to estimate soil properties or capabilities by association with soils of the same class whose properties are known, and to provide the engineer with an accurate method of soil description. However, there are so many different soil properties of interest to engineers and so many different combinations of these properties in any natural soil deposit that any universal system of classification seems impractical. Instead the groups or classes are based on those properties which are most important in that particular phase of engineering for which the classification was developed. For example, the Public Roads Classification System groups soils according to their suitability for road construction. The same properties may be of little use in classifying soils for earth dams. The soils engineer should be familiar with the purposes and particularly the limitations of the important soil classification systems. He should be able to develop new systems to fit new problems rather than try to adapt the old systems to situations where they do not apply. But the engineer must not lose himself in this, for as A. Casagrande has said, "Those who really understand soils can, and often do apply soil mechanics without a formally accepted classification."[2:8]

TEXTURAL CLASSIFICATIONS / Textural classifications group soils by their grain-size characteristics. The gravel and larger sizes are disregarded and the particles finer than 2 mm in diameter are divided into three groups: sand sizes, silt sizes, and clay sizes. The soils are then grouped by the percentage of each of these three components.[2:8]

Textural classification was developed by agricultural engineers who found that grain size was an indication of the workability of topsoils. A number of different textural classification schemes have been employed in engineering work, but they have been superseded by the more complete engineering classification systems described below.

BUREAU OF PUBLIC ROADS CLASSIFICATION SYSTEM / The Public Roads Classification is one of the oldest systems of grouping soils for an engineering purpose. Since its introduction in 1929 it has undergone many revisions and modifications and is widely used for evaluating soils for highway subgrade and embankment construction. The modification proposed in 1945 is termed the *Revised Bureau of Public Roads, Highway Research Board*, or *AASHO* (American Association of State Highway Officials) System.[2:9] This system divides all soils into three categories: granular, with 35 per cent or less by weight passing a No. 200 sieve (finer than 0.074 mm); silt-clay, with more than 35 per cent passing the No. 200 sieve; and organic soils. The first two categories are subdivided further, depending on their gradation and plasticity characteristics, as shown in Table 2:4. The symbols A-1 through A-8 are given to the classes which loosely indicate a decreasing quality for highway construction with increasing number. Some of the classes are subdivided, to indicate differences in plasticity, but the subdivisions are not an essential part of the system. The classification is supplemented by the *Group Index*, or GI:

$$\text{GI} = 0.2a + 0.005ac + 0.01bd, \qquad (2:1)$$

where a = percentage passing the No. 200 sieve greater than 35 and not exceeding 75; expressed as a whole number (0 to 40)
b = percentage passing No. 200 sieve greater than 15 and not exceeding 55; expressed as a whole number (0 to 40)
c = that portion of the liquid limit greater than 40 and not exceeding 60; expressed as a whole number (0 to 20)
d = that portion of the plasticity index greater than 10 and not exceeding 30; expressed as a whole number (0 to 20)

The value of the GI ranges from 0 to 20, with the low numbers indicating higher quality than the high numbers. The number is placed in parentheses following the class, such as A–2(0) or A–5(9).

Since the same basic symbols have been used for all the versions of the Public Roads System, the engineer should always note which he is using. The presence of the GI number, however, denotes the 1945 revision.

UNIFIED SOIL CLASSIFICATION SYSTEM / The Unified Soil Classification[2;10] is an outgrowth of the Airfield Classification (AC) system developed by A. Casagrande[2;8] as a rapid method for identifying and grouping soils for military construction. The soils are first divided into coarse-grained and fine-grained classes. The coarse-grained soils have over 50 per cent by weight coarser than 0.074 mm (No. 200 sieve). They are given the symbol G if more than half of the coarse particles by weight are coarser than 4.76 mm (No. 4 sieve) and S if more than half are finer. The G or S is followed by a second letter that describes the gradation: W, well graded with little or no fines; P, poorly graded, uniform, or gap-graded with little or no fines; M, containing silt or silt and sand; and C, containing clay or sand and clay. The fine-grained soils (over half finer than 0.074 mm) are divided into three groups: C, clays; M, silts and silty clays; and O, organic silts and clays. These symbols are followed by a second letter denoting the liquid limit or relative compressibility: L, a liquid limit less than 50; and H, a liquid limit exceeding 50.

The Casagrande plasticity chart (Fig. 2.14) is the basis for dividing the fine-grained soils. It also aids in comparing different soils. For example, clays having a similar geologic origin will usually plot in a narrow band parallel to the dividing line (often called the A-line) between the C and M-O soils. The different symbols, the soils they represent, and the classification criteria are given on Table 2:5. For border-line soils a dual classification is sometimes given, such as GW-GC. To the symbols should be added a description giving information on the grain shape, composition, color, macrostructure, and the soil strength or density in the ground.

HIGHWAY DEPARTMENT SYSTEMS / Several state highway departments have developed classification systems that are adapted to the soils which they regularly encounter. Most of these systems are based on empirical correlations between soil performance in highway construction and some simple laboratory tests such as grain size, plasticity, shrinkage, expansion, and density after compaction. Within the state for which they were developed, they can be a valuable guide to the selection of materials for highway work.

CIVIL AERONAUTICS (CAA) SYSTEM / The CAA system classifies soils according to their suitability for airfield subgrades. The soils are divided into groups E–1 to E–14 on the basis of grain size and plasticity, similar to the Public Roads system. Other tests are used to aid in the classification but are not essential.

TABLE 2:4 / REVISED BUREAU OF PUBLIC ROADS OR AASHO CLASSIFICATION[2:9]

Group	Sub-group	Per Cent Passing U.S. Sieve			Character of Fraction Passing No. 40 Sieve		Group Index No.	Soil Description	Subgrade Rating
		10	40	200	Liquid Limit	Plasticity Index			
A–1		50 max							
	A–1–a		50 max	25 max		6 max	0	Well-graded gravel or sand; may include fines	
	A–1–b		30 max	15 max		6 max	0	Largely gravel but can include sand and fines	
			50 max	25 max		6 max	0	Gravelly sand or graded sand; may include fines	Excellent to Good
A–2*				35 max			0 to 4	Sands and gravels with excessive fines	
	A–2–4			35 max	40 max	10 max	0	Sands, gravels with low-plasticity silt fines	
	A–2–5			35 max	41 min	10 max	0	Sands, gravels with elastic silt fines	
	A–2–6			35 max	40 max	11 min	4 max	Sands, gravels with clay fines	
	A–2–7			35 max	41 min	11 min	4 max	Sands, gravels with highly plasticclay fines	
A–3			51 min	10 max		Nonplastic	0	Fine sands	
A–4				36 min	40 max	10 max	8 max	Low-compressibility silts	Fair to Poor
A–5				36 min	41 min	10 max	12 max	High-compressibility silts, micaceous silts	
A–6				36 min	40 max	11 min	16 max	Low-to-medium-compressibility clays	
A–7	A–7–5			36 min	41 min	11 min	20 max	High-compressibility clays	
	A–7–6			36 min	41 min	11 min†	20 max	High-compressibility silty clays	
				36 min	41 min	11 min†	20 max	High-compressibility, high-volume-change clays	
A–8								Peat, highly organic soils	Unsatisfactory

* Group A–2 includes all soils having 35 per cent or less passing a No. 200 sieve that cannot be classed as A–1 or A–3.
† Plasticity index of A–7–5 subgroup is equal to or less than LL–30. Plasticity index of A–7–6 subgroup is greater than LL–30.

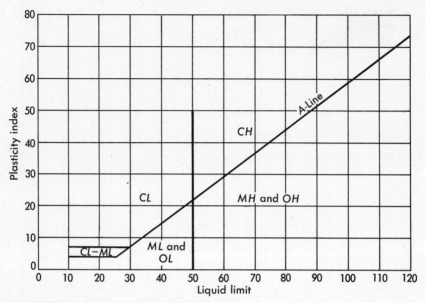

Figure 2.14 Plasticity chart for the classification of fine-grained soils. Tests made on fraction finer than No. 40 sieve. (After A. Casagrande[2:8] and the U.S. Waterways Experiment Station.[2:10])

The original AC system differs from the Unified in the grouping of the coarse-grained soils: the division between coarse and fine is 0.1 mm; the symbol M is replaced by F; and all the coarse-grained subdivisions, W, C, P, and F, have slightly different meanings. For most practical purposes, however, the systems are the same.

The Unified System has proved very useful in classifying soils for many different purposes such as highway and airfield construction, earth dams, embankments, and even for foundations. It is frequently supplemented by tables showing the typical properties of each group, such as the drainage characteristics, and as such is a valuable guide for design and construction. The system is simple. Many soils can be grouped visually and only tests for grain size and plasticity are necessary for accurate classification. It must be kept in mind, however, that no classification is a substitute for tests of the soil's physical properties and engineering analysis of the results.

2:11 Soil Identification and Description

The different soil classification schemes, while useful for grouping soils for a particular purpose, may be useless or misleading in other applications. In many fields, such as foundation engineering, there are so many significant soil properties that any scheme of soil classification would be very awkward. Instead an accurate description of the significant soil properties can convey

TABLE 2:5 / UNIFIED SOIL CLASSIFICATION
(After U.S. Waterways Experiment Station and ASTM D 2487-66T)

Major Division		Group Symbol	Laboratory Classification Criteria		Soil Description
			Finer than 200 Sieve %	Supplementary Requirements	
Coarse-grained (over 50% by weight coarser than No. 200 sieve)	Gravelly soils (over half of coarse fraction larger than No. 4)	GW	0-5*	D_{60}/D_{10} greater than 4, $D_{30}^2/(D_{60} \times D_{10})$ between 1 & 3	Well-graded gravels, sandy gravels
		GP	0-5*	Not meeting above gradation for GW	Gap-graded or uniform gravels, sandy gravels
		GM	12 or more*	PI less than 4 or below A-line	Silty gravels, silty sandy gravels
		GC	12 or more*	PI over 7 and above A-line	Clayey gravels, clayey sandy gravels
	Sandy soils (over half of coarse fraction finer than No. 4)	SW	0-5*	D_{60}/D_{10} greater than 4, $D_{30}^2/(D_{60} \times D_{10})$ between 1 & 3	Well-graded sands, gravelly sands
		SP	0-5*	Not meeting above gradation requirements	Gap-graded or uniform sands, gravelly sands
		SM	12 or more*	PI less than 4 or below A-line	Silty sands, silty gravelly sands
		SC	12 or more*	PI over 7 and above A-line	Clayey sands, clayey gravelly sands
Fine-grained (over 50% by weight finer than No. 200 sieve)	Low compressibility (liquid limit less than 50)	ML	Plasticity chart		Silts, very fine sands, silty or clayey fine sands, micaceous silts
		CL	Plasticity chart		Low plasticity clays, sandy or silty clays
		OL	Plasticity chart, organic odor or color		Organic silts and clays of low plasticity
	High compressibility (liquid limit more than 50)	MH	Plasticity chart		Micaceous silts, diatomaceous silts, volcanic ash
		CH	Plasticity chart		Highly plastic clays and sandy clays
		OH	Plasticity chart, organic odor or color		Organic silts and clays of high plasticity
Soils with fibrous organic matter		Pt	Fibrous organic matter; will char, burn, or glow		Peat, sandy peats, and clayey peat

* For soils having 5 to 12 per cent passing the No. 200 sieve, use a dual symbol such as GW-GC.

the necessary information without the restrictions of a definite classification scheme. The following soil properties are of significance in most soil problems and therefore form the basis of a complete soil description. They are also a required supplement to the Unified Classification.

1. Shear strength (cohesive soils).
2. Density (cohesionless soils).
3. Compressibility.
4. Permeability.
5. Color.
6. Composition (grain size, shape, plasticity, mineralogy).
7. Structure of soil.

For a precise description many of these properties must be determined by laboratory tests. An experienced soils engineer, however, can estimate most of these by careful field observation and examination of small samples of the soil.[2:13]

SOIL STRENGTH / Shear strength is a fundamental property of undisturbed cohesive soils, a knowledge of which is necessary in solving many problems. It is ordinarily defined in terms of unconfined compressive strength (Section 3:15) but may be estimated from the pressure required to squeeze an undisturbed sample between the fingers. If the soil is *brittle* (fails suddenly with little strain), *elastic* (rubbery), *friable* (crumbles easily), or *sensitive* (loses strength on remolding), these terms should be included in the description.

TABLE 2:6 / SOIL STRENGTH*

Term	Unconfined Compressive Strength (After Terzaghi and Peck)[2:12]	Field Test (After Cooling, Skempton, and Glossop)[2:14]
Very soft	0–0.5 kips per sq ft	Squeezes between fingers when fist is closed
Soft	0.5–1.0	Easily molded by fingers
Firm	1.0–2.0	Molded by strong pressure of fingers
Stiff	2.0–3.0	Dented by strong pressure of fingers
Very stiff	3.0–4.0	Dented only slightly by finger pressure
Hard	4.0 or more	Dented only slightly by pencil point

* (A method of estimating strength from sampling operations is given in Table 6:5)

DENSITY / Density is as important for cohesionless soils as strength is for cohesive. It can be found by comparing the soil's actual void ratio with the range in void ratio from loose to dense for that soil. It may be estimated from the ease with which a reinforcing rod penetrates the soil, or from Table 6:4.

TABLE 2:7 / SOIL DENSITY

Term	Relative Density	Field Test
Loose	0–50%	Easily penetrated with ½-in. reinforcing rod pushed by hand
Firm	50–70	Easily penetrated with ½-in. reinforcing rod driven with 5-lb hammer
Dense	70–90	Penetrated a foot with ½-in. reinforcing rod driven with 5-lb hammer
Very dense	90–100	Penetrated only a few inches with ½-in. reinforcing rod driven with 5-lb hammer

COMPRESSIBILITY / Compressibility is determined by direct laboratory tests (Section 3:6) or is estimated from the liquid limit and void ratio.

TABLE 2:8 / COMPRESSIBILITY

Term	Compression Index	Liquid Limit (approx.)
Slight or low compressibility	0–0.19	0–30
Moderate or intermediate	0.20–0.39	31–50
High compressibility	0.40 and over	51 and over

PERMEABILITY / Permeability is determined by direct laboratory and field tests or may be estimated from Table 3:1.

COLOR / Color, while not an important physical property in itself, is an indication of more important properties. For example, yellow and red hues indicate that a soil has undergone severe weathering, for the colors are iron oxides. A dark greenish brown is often an indication of organic matter. A change in color encountered during excavation often means a different soil stratum with different properties has been uncovered. Color is usually the easiest property of a soil for persons untrained in soil mechanics to identify; therefore a practical method of describing a certain soil to workers is by color. Soil color is described visually with the aid of the Munsel color charts.[2:15]

COMPOSITION / Composition includes the grain size, gradation, grain shape, mineralogy (of the coarser grains), and plasticity. Two groups of soils are recognized: predominantly coarse-grained (over 0.074 mm) and predominantly fine-grained (less than 0.074 mm), as in the Unified Classification. The coarse-grained soils are described primarily on the basis of the grain size, the fine-grained primarily on the basis of their plasticity. The amount of coarse or fine component required to predominate is not fixed, for it depends on the soil structure: If the coarse-grained particles can make contact with one another, the soil behaves essentially as a coarse-grained material; if they cannot touch but are separated by the fines, the fines predominate. The Unified Classification arbitrarily defines predominate grain size at over 50 per cent by weight. However, in soils containing clay minerals, the fines may predominate even though they comprise considerably less than 50 per cent by weight of the soil. Therefore no fixed point can be established, and the engineer must exercise his judgment.

The sizes of particles as defined by ASTM–ASCE and their visual indentification are given in Table 2:9.

TABLE 2:9 / GRAIN SIZE IDENTIFICATION

Name	Size Limits	Familiar Example
Boulder	12 in. (305 mm) or more	Larger than basketball
Cobbles	3 in. (76 mm)–12 in. (305 mm)	Grapefruit
Coarse gravel	¾ in. (19 mm)–3 in. (76 mm)	Orange or lemon
Fine gravel	4.76 mm (No. 4 Sieve)–¾ in. (19 mm)	Grape or pea
Coarse sand	2 mm (No. 10 Sieve)–4.76 mm (No. 4 Sieve)	Rocksalt
Medium sand	0.42 mm (No. 40 Sieve)–2 mm (No. 10 Sieve)	Sugar, table salt
Fine sand*	0.074 mm (No. 200 Sieve)–0.42 mm (No. 40 Sieve)	Powdered sugar
Fines	Less than 0.074 mm (No. 200 Sieve)	

* Particles finer than fine sand cannot be discerned with the naked eye at a distance of 8 in. (20 cm).

Gradation is estimated by the same criterion as for the Unified Classification (Table 2:5). A smooth grain-size curve and a uniformity coefficient of more than 6 for sands or 4 for gravels denotes a *well-graded* soil. An irregular gradation denotes *gap-graded*. A uniformity coefficient less than the above limits indicates a *uniform* soil. The term *poorly graded* is sometimes applied to either uniform or gap-graded soils.

Grain shapes are identified as angular to well-rounded, as shown in Fig. 1.9. In addition, elongated or platey particles can be identified.[2:13]

The mineral composition of the grains can often be determined by a microscopic examination. The carbonates are easily identified by a strong acid, which causes them to effervesce.

The fines are described on the basis of the Casagrande plasticity chart (Fig. 2.14). Soils above the A-line are clays and those below, silts. Soils that plot near the A-line are given a double designation: If the PI is less than 10 per cent above the A-line, the soil is described as a silty clay; if it is less than 33 per cent below the A-line, it is a clayey silt. A soil should not be termed a silty clay if its liquid limit is above 60, however.

Silts and clays can be identified in the field by the shaking test. A pat of wet soil (consistency of soft putty) is shaken in the hand. If it becomes soft and glossy with shaking or tapping the hand and then becomes hard, dull, and forms cracks when the pat is squeezed between the fingers, it has a *reaction to shaking*, or *dilatancy*. A rapid reaction indicates a nonplastic silt; a slow reaction means an organic silt, slightly clayey silt, or possibly a nonplastic silt with a very high liquid limit (over 100). No reaction indicates a clay or silty clay. (To be decisive the test should be made at different water contents.)

The toughness of the thread that forms when the soil is rolled at the plastic limit also helps identify the fines. Inability to form a thread or a very weak thread indicates an inorganic silt of very low plasticity (ML). A weak spongy thread indicates an organic silt or an inorganic silt having a high liquid limit but low plasticity (MH). A firm thread indicates a low-plasticity clay (CL), while a tough, rigid thread indicates a highly plastic clay (CH).

Plasticity is determined by the plasticity index or can be estimated from the strength of an air-dried sample. The sample is prepared by first removing

TABLE 2:10 / PLASTICITY

Term	PI	Dry Strength	Field Test
Nonplastic	0–3	Very low	Falls apart easily
Slightly plastic	4–15	Slight	Easily crushed with fingers
Medium plastic	15–30	Medium	Difficult to crush
Highly plastic	31 or more	High	Impossible to crush with fingers

all particles coarser than a No. 40 sieve and then molding a cube at the consistency of stiff putty, adding water if necessary. The cube is dried in air or sunlight and then crushed between the fingers.

Organic soils can be identified by their odor, which is intensified by heating, and their color, which is usually black, brown, dark green, or blue-black. However, some inorganic soils are black from certain iron, titanium, and ferromagnesian minerals. Organic matter can also be identified by oxidizing the soil with hydrogen peroxide and noting the loss in dry weight. It also can be identified by the loss of weight on ignition, provided no carbonate minerals and no adsorbed water or water of crystallization are present.

Soils containing fibrous organic matter are identified visually or by their loss of weight on ignition. A weight loss (organic content) of 80 per cent or more denotes *peat*. If less fibrous organic matter is present, the soils are described as peaty sand or peaty clay, as the case may be.

In forming the description, a predominately coarse-grained soil is termed either a *gravel* or a *sand*, depending on which component appears to be the more abundant. The less abundant component and the fines (either silt or clay) are used as modifiers, with the least important component first. For example, a soil with 30 per cent fines (silt), 45 per cent gravel, and 25 per cent sand would be described as sandy, silty gravel. The grain shapes and sizes precede the component to which they apply. A predominately fine-grained soil would be considered either silt or clay and the coarse components would be used as modifiers, with the least important first. For example, a soil with 70 per cent fines (clay), 20 per cent sand, and 10 per cent gravel would be described as gravelly, sandy clay.

DESCRIBING SOIL STRUCTURES / The structure of the soil must be determined by careful observation. The following descriptive terms may be used:

Homogeneous (uniform properties)
Stratified (alternate layers of different soils)
Laminated (repeating alternate layers less than $\frac{1}{8}$ in. thick)
Banded (alternate layers in residual soils)

It is important to recognize defects in a soil structure. The following are often observed:

Slickensides (former failure planes)
Rootholes
Fissures (cracks, from shrinkage, frost)
Weathering (irregular discoloration)

TABLE 2:11 / SOIL AND ROCK SHORTHAND FOR DESCRIPTION

Consistency		Color		Soil Texture, Composition	
Ls	Loose	Gy	Gray	Bldr	Boulder
Fm	Firm	Bn	Brown	Gvl	Gravel
Ds	Dense	Tn	Tan	Sa	Sand
VDs	Very Dense	Y	Yellow	Si	Silt
VSo	Very Soft	R	Red	Cl	Clay
So	Soft	Bl	Black	Org	Organic
Stf	Stiff	Gn	Green	Pt	Peat
Hd	Hard	W	White	Cal	Calcareous
		Ora	Orange	Shl	Shells
				Lat	Lateritic
				Cse	Coarse
				Fn	Fine
				Ang	Angular
				Rd	Rounded

Rock		Details	
SS	Sandstone	Sl	Slickensides
Shl	Shale	Sm	Seam
LS	Limestone	Lns	Lens
Gn	Gneiss	Por	Porous
Sch	Schist	WT	Water Table
Q	Quartzite	Wth	Weathered
Gr	Granite	Jt	Jointed
Bas	Basalt	Dec	Decomposed
Sl	Slate	Cav	Cavity
Dio	Diorite	Rt	Roots
Tf	Tuff	Con	Concretions
Cg	Conglomerate	Nod	Nodular
Lv	Lava	Mot	Mottled
		Fr	Fractured
		V	Varved
		Ban	Banded

SEC. 2:12] SOIL AND ROCK NAMES   79

WRITING A SOIL DESCRIPTION / The soil should be described in essentially the order of the significant properties given previously. As many of the properties as are of interest should be included. Examples might be:
1. Hard, moderately compressible, blue-gray, medium-plastic clay.
2. Dense, well-graded, clayey, sandy, well-rounded gravel.
3. Loose, brown, uniform, angular, fine sand.
4. Firm, black, slightly compressible silt and clay; laminated.
5. Loose, compressible, brown, micaceous, sandy silt; banded.

SHORTHAND FOR DESCRIPTION / Too often engineers resort to soil classification alone or to inadequate descriptions because of the limitations of time required for their writing or the space necessary on small scale drawings. A shorthand system, Table 2:11, gives the most useful abbreviations whose meanings are self evident to engineers accustomed to soil and rock terminology.

2:12 Soil and Rock Names

In addition to standardized terms used to describe soils, there are many names that have been given to soils, rocks and deposits. Many of these names, such as *gumbo* and *buckshot*, have originated through association with other more familiar objects and have essentially a local meaning. Others, such as *silt* and *clay*, have been applied to soils having such wide ranges in characteristics that the names often have no definite meaning. In some cases a soil name that is associated with soils of certain characteristics in one region refers to soils of completely different characteristics in other regions. The soil names defined below are often encountered in engineering literature.[2:16]

ADOBE refers to sandy clays of medium plasticity found in the semiarid regions of the southwestern United States. These soils have been used for centuries for making sun-dried brick. The name is also applied to some highly plastic clays of the West.

BENTONITE is a highly plastic clay resulting from the decomposition of volcanic ash. It may be hard when dry but swells considerably when wet.

BUCKSHOT is applied to clays of the southern and southwestern United States that crack into small, hard, relatively uniform sized lumps on drying.

BULL'S LIVER is inorganic silt of very low plasticity. In a saturated condition it quakes like jelly from shock or vibration and often becomes quick and flows like a fluid.

CALICHE is a silt or sand of the semiarid areas of the southwestern United States that is cemented with calcium carbonate. The calcium carbonate is deposited by the evaporation of ground water brought to the ground surface by capillary action.

CLAY is applied to any soil capable of remaining in the plastic state through a relatively wide range in water contents. As a soil name, the term *clay* has been badly abused. In some parts of the United States clay means a nonplastic silt; in other parts it is used to designate micaceous silts of low plasticity. Engineers should use the word clay in the restricted meaning given by Casagrande's plasticity chart (Fig. 2.14).

COQUINA is a soft, porous limestone made up largely of shells, coral and fossils.

DIATOMACEOUS EARTHS are silts containing large amounts of diatoms—the silaceous skeletons of minute marine or fresh water organisms.

FILL is any man-made soil deposit. Fills may consist of soils that are free of organic matter and that are carefully compacted to form an extremely dense, incompressible mass, or they may be heterogeneous accumulations of rubbish and debris.

FULLER'S EARTHS are soils having the ability to absorb fats or dyes. They are usually highly plastic, sedimentary clays.

GRAVEL means a soil composed largely of particles from 4.76 mm to 3 in. in diameter. In some parts of the United States the term *gravel* is restricted to rounded gravel.

GUMBO is a fine-grained, highly plastic clay of the Mississippi Valley. It has a sticky, greasy feel and forms large shrinkage cracks on drying.

HARDPAN is a term that should be avoided by the engineer. Originally it was applied only to a soil horizon which had become rocklike because of the accumulation of cementing minerals. True hardpan is relatively impervious and does not soften upon exposure to air or water. Unfortunately the term is also applied to any hard or highly consolidated soil stratum that is excessively difficult to excavate. Many lawsuits have centered about the meaning of hardpan because of its ambiguity. The name implies a condition rather than a type of soil.

KAOLIN is a white or pink clay of low plasticity. It is composed largely of minerals of the kaolinite family.

LATERITES are residual soils formed in tropical regions. The cementing action of iron oxides and hydrated aluminum oxides makes dry laterites extremely hard.

LOAM is a surface soil that may be described as a sandy silt of low plasticity or a silty sand that is well suited to tilling. It applies to soils within the uppermost horizons and should not be used to describe deep deposits of parent materials.

LOESS is a deposit of relatively uniform, wind-blown silt. It has a loose structure, with numerous rootholes that produce vertical cleavage and high vertical permeability. It consists of angular to subrounded quartz and feldspar particles cemented with calcium carbonate or iron oxide. Upon saturation it becomes soft and compressible because of the loss of cementing. Loess altered by weathering in a humid climate often becomes more dense and somewhat plastic. It is known as *loess loam*. *Swamp loess* is water-deposited loess. It does not have the loose structure or vertical cleavage of loess.

MARL is a water-deposited sand, silt or clay containing calcium carbonate. Marls are often light to dark gray or greenish in color and sometimes contain colloidal organic matter. They are often indurated into soft rock.

MUCK OR MUD is extremely soft, slimy silt or organic silt found on river and lake bottoms. The terms indicate an extremely soft consistency rather than any particular type of soil. Muck implies organic matter.

MUSKEG is peat found in northwest Canada. The bogs in which the peat forms are often termed *muskegs*.

PEAT is fibrous, partially decomposed organic matter or a soil containing large amounts of fibrous organic matter. Peats are dark brown or black, loose (void ratio may be 5 to 10), and extremely compressible. When dried they will float. Peat bogs often emit quantities of inflammable methane gas.

QUICKSAND is a condition and not a soil. Gravels, sands, and silts become "quick" when an upward flow of ground water takes place to such extent that the particles are lifted.

ROCK FLOUR is extremely fine-grained silt formed by the grinding action of glaciers.

SAND is a soil composed largely of particles from 0.074 mm to 4.76 mm in diameter. DIRTY SAND means a slightly silty or slightly clayey sand.

SILT is any fine-grained soil of low plasticity. Often it is applied to fine sands. In some parts of the United States the term *silt* is applied only to organic silts. Casagrande's plasticity chart may be used to differentiate between silt and clay.

STONE is sometimes used to designate angular gravel. It is more properly applied to gravel manufactured by crushing rock.

TILL is a mixture of sand, gravel, silt, and clay produced by the plowing action of glaciers. The name *boulder clay* is often given such soils, particularly in Canada and England.

TOPSOILS are surface soils that support plant life. They usually contain considerable organic matter.

TRAP includes all dark colored fine-grained intrusive rocks, usually in the form of dikes or sills. The most common trap rock is basalt.

TUFF is the name applied to deposits of volcanic ash. In humid climates or in areas in which the ash falls into bodies of water, the tuff becomes cemented into a soft, porous rock.

TUNDRA is the mat of moss peat and shrubby vegetation that covers a gray, clayey subsoil in arctic regions. The deeper soil is permanently frozen, while the surface soil freezes and thaws seasonally. (See Section 4:6.)

VARVED CLAYS are sedimentary deposits consisting of alternate thin layers of silt and clay. Ordinarily each pair of silt and clay layers is from $\frac{1}{8}$ in. to $\frac{1}{2}$ in. thick. They are the result of deposition in lakes during periods of alternately high and low water in the inflowing streams and are often formed in glacial lakes.

REFERENCES

2:1 *Soil Classification, a Comprehensive System, the Seventh Approximation*, U.S. Dept. of Agriculture, Washington, 1960.

2:2 *Soils and Man*, Yearbook of American Agriculture, U.S. Dept. of Agriculture, 1938.

2:3 *Soil Survey Manual*, Handbook 18, U.S. Dept. of Agriculture, Washington, 1951 (and 1962 Supplement).

2:4 "Glossary of Pedologic and Landform Terminology," *Special Report 25*, Highway Research Board, Washington, 1957.

2:5 "Engineering Use of Agricultural Soil Maps," *Bulletin 22*, Highway Research Board, Washington, 1949.

2:6 N. Duncan, "Rock Mechanics," a lecture delivered to the Southern Assoc., Institution of Civil Engineers (Great Britain) 1968.

2:7 D. U. Deere, "Technical Description of Rock Cores for Engineering Purposes," *Rock Mechanics and Engineering Geology*, **1**, 1, 1963.

2:8 A. Casagrande, "Classification and Identification of Soils," *Transactions*, ASCE, 1948, p. 901.

2:9 "Classification of Highway Subgrade Materials," *Proceedings, Highway Research Board*, Washington, 1945.

2:10 *Unified Soil Classification System*, Technical Memorandum 3-357, U.S. Waterways Experiment Station, Vicksburg, 1953.

2:11 ASTM Standard, D 2487-66T, "Classification of Soils for Engineering Purposes," *ASTM Standards*, Part 11, ASTM, Philadelphia, 1969.

2:12 K. Terzaghi and R. B. Peck, *Soil Mechanics in Engineering Practice*, John Wiley & Sons, Inc., New York, 1948, p. 31.

2:13 ASTM Standard, D 2488-66T, "Description of Soils," *ASTM Standards*, Part 11, ASTM, Philadelphia, 1969.

2:14 L. F. Cooling, A. W. Skempton, and R. Glossop, *Discussion*, of Reference 2:8.

2:15 *Munsell Soil Color Charts*, Munsell Color Company, Inc., Baltimore, 1954.

2:16 W. L. Stokes and D. J. Varnes, *Glossary of Selected Geologic Terms*, Colorado Scientific Society Proceedings, **16**, Denver, 1955.

2:17 *Glossary of Geology and Related Sciences*, American Geologic Institute, National Academy of Sciences, Washington, 1957.

SUGGESTIONS FOR ADDITIONAL STUDY

1. References 2:1, 2:2, 2:7.
2. *PCA Soil Primer*, Portland Cement Association, Chicago.
3. D. P. Krynine and W. R. Judd, *Principles of Engineering Geology and Geotechnics*, McGraw-Hill Book Co., Inc., New York, 1957.

SUGGESTIONS FOR ADDITIONAL STUDY

4. R. F. Leggett, *Geology and Engineering*, McGraw-Hill Book Co., Inc., New York, 1962, 2nd ed.
5. D. S. Jenkins, D. J. Belcher, L. E. Gregg, and K. B. Woods, "The Origin, Distribution and Airphoto Identification of U.S. Soils," U.S. Department of Commerce, CAA Technical Development Report 52, 1946.
6. D. M. Burmister, "Identification and Classification of Soils," *Symposium on Identification and Classification of Soils*, ASTM Special Publication 113, Philadelphia, 1951.
7. *The Identification of Rock Types*, Bureau of Public Roads, U.S. Government Printing Office, Washington, 1950.
8. *Quarternary Soils*, Proceedings, **9**, Seventh Congress, Center for Water Resources Development, University of Nevada, Reno, 1967.
9. K. W. John, "An Approach to Rock Mechanics," *Journal of the Soil Mechanics and Foundations Division, Proceedings, ASCE* **88**, SM 4, August 1962.
10. L. Obert and W. I. Duvall, *Rock Mechanics and the Design of Structures in Rock*, John Wiley and Sons, Inc., New York, 1967.
11. *ASTM Standards, Part 11*, "Bituminous Materials, Soils, Skid Resistance," ASTM, Philadelphia, 1969.

PROBLEMS

2:1 Prepare a log of the soils you might expect when making an excavation in the following deposits:
   1. Flood plain of a flat river in Mississippi.
   2. Coastal swamp in North Carolina.
   3. Edge of swampy lake in the glaciated portion of Michigan.
   4. A dried-up lake bed in Nevada.
   5. Sand dune in Indiana.
   6. Prairie in North Dakota.
   7. River valley in Connecticut.
   8. Coastal plain of Texas.

2:2 Classify the following soils according to the Revised Public Roads System and find the GI number:

| Soil No. | Per Cent Passing | | Characteristics of —40 Fraction | |
	No. 40	No. 200	LL	PI
1	95	57	37	18
2	72	48	31	4
3	100	97	73	45
4	18	0		
5	63	8		
6	97	65	50	6
7	45	18	14	3
8	70	30	17	5

2:3 The following data were obtained by mechanical analysis and plasticity tests of soil samples. (Percentages finer than given size noted.)

Size	Sample 1	Sample 2	Sample 3	Sample 4	Sample 5	Sample 6
No. 10	—	—	—	—	—	100
20	86%	98%	93%	99%	98%	
40	72	85	79	94	95	86
60	60	72	68	89	92	
100	45	56	56	82	86	
200	35	42	42	76	83	9
0.05 mm	33	41	41	74	82	
0.01 mm	21	20	11	38	57	0
0.002 mm	10	8	4	23	36	
LL	19	44	30	40	67	NP
PI	0	0	0	12	27	NP

Plot the grain size curve.
a. Classify each using the Unified System.
b. Classify each using Revised Public Roads System.

2:4 a. Describe visually a sample of soil you have secured from an excavation or highway cut.
b. Estimate the soil classification according to the Revised Public Roads System and the Unified System.

CHAPTER 3
The Engineering Properties of Soil and Rock

AN EXCAVATION FOR A PUMPING STATION HAD REACHED THE FINAL LEVEL, 30 ft below the ground surface and 10 ft below the ground water table in a micaceous silt residual soil, Fig. 3.1a. Stakes were set late in the afternoon at the proper elevation for pouring the concrete mat foundation. The next morning, during the placement of the reinforcing steel, the stakes were found to be nearly a foot high. The surveyor was thoroughly reprimanded for his error, the excavation was again brought to grade and new stakes were set. The next morning these, too, were high. An investigation of the problem showed that the surveyor had been right all along. The soil in the bottom of the excavation, relieved of the weight of 30 ft of overburden and subject to an unbalanced internal water pressure, slowly expanded. The bottom of the excavation rose, like a compressed sponge rubber pillow rebounds, lifting the stakes with it. The problem was eventually solved by lowering the ground water table well below the excavation bottom in order to partially reload the soil by means of capillary tension.

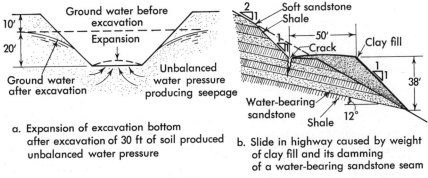

a. Expansion of excavation bottom after excavation of 30 ft of soil produced unbalanced water pressure

b. Slide in highway caused by weight of clay fill and its damming of a water-bearing sandstone seam

Figure 3.1 *The effects of stress changes and pore water pressure on the engineering behavior of soil and rock.*

A highway was constructed across a steep hillside underlain by gently sloping alternate seams of weak sandstone and shale by cutting into the hill on one side and filling with clay on the other, Fig. 3.1b. Suddenly the fill began to move downward. More fill was placed at the toe of the hill to help hold the soil in place but this only aggravated the movement. Water trickled slowly out of the toe of the fill and out of the cut face and into the ditch on the high side of the road. The rate of movement was irregular, increasing slightly after each rain. A study of the rock showed that the strength of the sandstone decreased sharply with increased water pressure. The clay fill acted as a dam, preventing the free drainage from the sandstone. Water pressure built up because of downward percolation through joint cracks in the shale and sandstone above the cut. The problem was solved by draining the sandstone seams and removing the excess fill that had been placed to bolster the original embankment and which had only aggravated the problem.

These case histories illustrate two significant aspects of soil and rock behavior; first these materials change properties drastically with changes in the state of stress, far more than the simpler steel or concrete involved in man-made structures; and second, the importance of the water phase. The design of every engineering structure requires a knowledge of the properties of the materials subjected to stress. These stresses occur in both the water and solid phases; therefore a study of the physical properties must begin with the water phase and the movement of water through the soil. Changes in soil and rock volume under load are involved in problems of structural settlement. Finally, the strain and failure resulting from shear stresses are vital in the bearing capacity of foundations, the design of retaining walls and the stability of slopes.

3:1 Surface Tension

Moisture in soil and porous rock can be present in two forms: the adsorbed films surrounding the grains, and free water occupying part of or all the voids between the grains. If the voids are completely filled with water, the material is saturated and the moisture is said to be continuous; if the voids are only partly filled, the moisture is discontinuous forming wedges of water between adjacent grains and moisture films around them (Fig. 3.2). The boundary between air and water in the voids is particularly important. The unbalanced molecular attraction of the water at this boundary gives rise to *surface tension*, a force acting parallel to the surface of the water in all directions, similar to the tension in a tightly stretched rubber membrane.

SURFACE TENSION PHENOMENA / Surface tension has many manifestations in soils. If a hole is dug in the ground, the soils encountered will be found to be saturated long before the ground water table (the level of water in a large pit or well) is reached. This results from capillary rise of water

SEC. 3:1] SURFACE TENSION

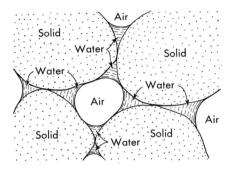

Figure 3.2 *Moisture wedges between soil grains.*

in the soil voids which form irregular tubes or pores. If a sample of a saturated clay is dried, it decreases in volume in the process. Surface tension acting in the soil voids acts to compress the soil structure and decrease the volume of the sample. A dried soil will absorb water rapidly and often will disintegrate in the process. The force of surface tension is partly responsible. Dry sand will run out between the fingers if one tries to mold it into a ball, but moist sand can be packed and formed easily. The tension in the moisture films between the grains is responsible for this moist strength. If the moist sand is immersed in water, the tiny moisture films will no longer exist, and the sand will again run between the fingers.

CAPILLARY TENSION / Surface tension can be visualized as a force in the boundary between air and water acting parallel to the water surface. It has a magnitude of about

$$T_o = -0.005 \text{ lb/ft},$$
$$T_o = -0.075 \text{ gm/cm}$$

at 70 F. It decreases with increasing temperature and increases with decreasing temperature. (In soil mechanics *tension* is denoted by a negative sign.) The attraction of water for soil particles or clean glass is of the same or greater magnitude as surface tension, so there is a tendency for the boundary to extend itself along any such solid material. If the extension of the air–water boundary surface is prevented by some force such as gravity, the surface is stretched, forming a curved surface called a *meniscus* and developing a tensile stress in the water. This stress is known as *capillary tension*, and it may be computed for a cylindrical tube by considering the force developed by the stretched meniscus (Fig. 3.3). If α is the angle of contact between the meniscus and the solid material and d is the diameter of the tube, the total unbalanced force, F, developed along the perimeter of the meniscus (assuming air pressure is zero) is

$$F = \pi d T_o \cos \alpha. \tag{3:1a}$$

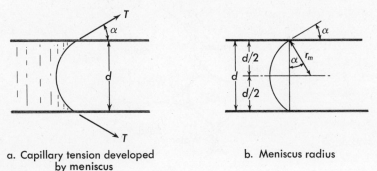

a. Capillary tension developed by meniscus

b. Meniscus radius

Figure 3.3 *Capillary tension and meniscus radius in a uniform tube of diameter, d*

Since the area of the tube A is

$$A = \frac{\pi d^2}{4}, \tag{3:1b}$$

the capillary stress, u, is found to be

$$u = \frac{F}{A} = \frac{4T_o \cos \alpha}{d}. \tag{3:2}$$

(Note that u has a negative sign when the value of T_o is substituted in the equation.)

MENISCUS RADIUS / The relation of capillary tension to the radius of the meniscus, r_m, can be found by considering the geometry of the meniscus (Fig. 3.3):

$$\frac{d}{2} = r_m \cos \alpha. \tag{3:3}$$

If the expression for meniscus radius is substituted in the formula for capillary tension, then

$$u = \frac{2T_o}{r_m}. \tag{3:4}$$

It can be seen that for water in contact with air, the capillary tension is dependent only on the meniscus radius, and it varies inversely with it.

MAXIMUM TENSION / The maximum tension will occur with a minimum meniscus radius. This will occur when the surface of the meniscus is tangent to the soil particles and $\alpha = 0$. In a cylindrical tube, therefore, the minimum r_m equals half the tube diameter. The maximum tension, therefore, will be

$$u_{\max} = \frac{4T_o}{d}. \tag{3:5}$$

SEC. 3:2] CAPILLARY TENSION IN SOILS

LIMITING TENSION / In large tubes water cannot sustain true tensile stresses. As soon as the pressure falls to the vapor pressure of water, (at 68 F or 20 C equal to -14.4 psig or $+0.3$ psi abs; -1.012 kg per sq cm gage or $+0.0235$ kg per sq cm abs) a bubble of vapor forms which expands upon any tendency of the pressure to drop lower. In a small tube, however, the radius of the largest bubble is equal to the tube radius. In the small tube, therefore, water can sustain stress limited only by the vapor pressure in the largest bubble that can occur. If the vapor pressure is u_v and the limiting stress is u_L, then

$$u_L = u_v + u_{\max}. \qquad (3:6)$$

If $u_{\max}$ (a negative stress) is greater than the vapor pressure in the bubble, the value of u_L will be negative, denoting a true tensile stress in the water. *Note:* u_L will be expressed in either gage or absolute pressure the same as u_v. For gage pressure u_L will always be negative.

3:2 Capillary Tension in Soils

The interconnected pores or voids in the soil form irregular but definite capillary tubes, Fig. 3.4. The maximum tension that can develop will vary

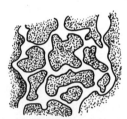 

a. Labyrinth path of a capillary tube formed by voids

b. Cross section of a capillary tube formed by void

Figure 3.4 *Capillary tubes formed by soil pores.*

from point to point, depending on the pore diameter and the degree of saturation. In a saturated pore in a cohesionless soil, experiments show that the effective pore diameter for capillary tension is approximately one-fifth D_{10}.

If a soil mass is completely saturated and inundated, the air–water boundaries disappear and capillary tension becomes zero. When a saturated soil is exposed to open air, capillary tension develops as soon as evaporation creates meniscuses at the surface. Since the soil moisture in a saturated soil is continuous, the water-tension stress developed at the air–water boundary is felt throughout the mass in a way similar to the way pressure applied at one point in a continuous body of fluid is transmitted throughout the body. The water, however, also obeys the law of hydrostatics:

$$\Delta u = \gamma_w \Delta z, \tag{3:7}$$

where z is measured positive downward. Therefore, below the elevation of the meniscus, the tension decreases; or, in other words, the pressure increases.

The capillary rise of water in a soil above the ground water table illustrates the combined effect of capillary tension and hydrostatic pressure. At the ground water elevation, a free surface, the water pressure u is zero. Below the water table the pressure increases in accordance with Equation 3:7. Above the water table it decreases in the same way as shown in Fig. 3.5.

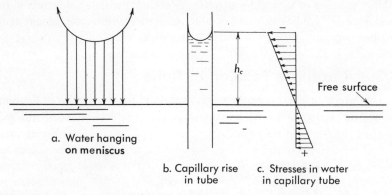

Figure 3.5 Capillary rise of water in a tube.

The greatest negative stress in the water is the maximum capillary tension, $u_{\max}$. The height, h_c, to which the water can rise above the surface of zero pressure is found by equating the expression for hydrostatic pressure with that for maximum capillary tension:

$$h_c = \Delta z, \quad \Delta u = u_{\max} = \frac{4T_o}{d},$$

$$h_c \gamma_w = \frac{4T_o}{d},$$

$$h_c = \frac{4T_o}{\gamma_w d}. \tag{3:8}$$

In a partially saturated soil the moisture may be either *continuous* or *discontinuous*, depending on whether the moisture wedges (Fig. 3.2) are interconnected or discrete. If it is continuous, the variation of water stress with elevation follows Equation 3:7, and the meniscus radius in each wedge adjusts itself to conform to the water stress. If the moisture is discontinuous, the stress in the water at any point is independent of the elevation and is determined only by the meniscus radius.

The height of capillary saturation in soils varies from a few inches in sands to over 100 ft in some clays, and a zone of partial saturation extends even higher, as will be discussed in Chapter 4. If the soil is homogeneous, the approximate height of rise can be computed from the effective grain size by using the relationship between pore diameter and grain size. The actual capillary rise and the maximum tension are seldom as great as the computations indicate, since in natural soils, cracks and seams are present which are so large that they rather than the soil pores control the capillary tension.

3:3 Permeability

The voids in a soil are not isolated cavities that hold water like storage reservoirs but are interconnected, small, irregular passageways (Fig. 3.4) through which water can flow in the same way as it flows through other conduits.

LAMINAR AND TURBULENT FLOW / Two completely different types of flow can exist. *Turbulent flow* is characterized by chaotic, irregular movements of the fluid particles and by energy losses that are roughly proportional to the square of the velocity of flow. This type takes place at relatively high velocities in large diameter conduits such as in pipes carrying air or water. In *laminar flow* the water particles move in a smooth, orderly procession in the direction of flow, and the energy losses are directly proportional to the velocity. Laminar flow takes place at low velocities in small conduits and is characteristic of all soils except the coarsest gravels.

DARCY'S LAW / A French physicist, Darcy, studied flow of water in soils by using apparatus similar to Fig. 3.6. He placed a sample of length L and cross-sectional area A in a tight-fitting tube with open ends. A reservoir of water was connected to each end of the tube. The level of the water in one reservoir was a distance of Δh above that in the other. (The term *head loss* is often applied to this difference in level, Δh.) He found by experiments

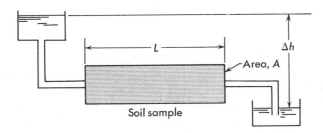

Figure 3.6 *Simple permeability test.*

that the flow of water, q, in cubic centimeters per second was directly proportional to the area A and to the ratio $\Delta h/L$ (which is termed the *hydraulic gradient* and given the symbol i). This relation is expressed by

$$q = kiA, \qquad (3:9)$$

in which k is the constant of proportionality and is given the name *coefficient of permeability*, or simply *permeability*. This formula is true as long as laminar flow exists; it has been found to apply to all soils finer than coarse gravel, as long as the hydraulic gradient is less than 5.

COEFFICIENT OF PERMEABILITY / The coefficient of permeability is a constant (having the dimensions of a velocity) that expresses the ease with which water passes through a soil. Ordinarily it is reported with the dimensions of centimeters per second or feet per minute but occasionally for very impervious soils, feet per day may be used.

The magnitude of the permeability coefficient depends on the viscosity of the water and on the size, shape, and area of the conduits through which the water flows. Viscosity is a function of the temperature: the higher the temperature, the lower the viscosity and the higher the permeability. Ordinarily the permeability is reported at 68 F (20 C). At 32 F (0 C) it is 56 per cent, and at 104 F (40 C) it is 150 per cent of the value at 68 F. The influence of the factors that determine the size and shape of the conduits is less specific and no valid mathematical expression for their effect has been derived. For clean cohesionless soils the permeability varies approximately as $(D_{10})^2$. Hazen's formula[2.1] for permeability of clean sands is

$$k = C(D_{10})^2, \qquad (3:10)$$

in which k is given in centimeters per second, D_{10} in millimeters, and C is a constant whose value ranges between 1 and 1.5. At the best, this formula gives only an indication of the order of magnitude of the permeability of clean sands. In soils having cohesion, the effect of grain size is even more pronounced, for part of the soil moisture around the fine clay particles is immobilized in the adsorbed layers. Void ratio is a factor in most soils, with the permeability approximately proportional to e^2. Grain shape and gradation are also important, particularly in the coarser soils, but it is difficult to express their effects quantitatively. The degree of saturation is a major factor because air in the voids reduces the cross-sectional area and may block some voids completely.

Clay minerals greatly influence permeability because some of the adsorbed water is so tightly bound to the clay surfaces that it may not move from one particle to another, without very high gradients. The permeability of a homogeneous clay, therefore, will usually be far lower than the grain size or void ratio alone suggests. Furthermore, the permeability is virtually zero at low gradients, but increases with increasing gradients.

SEC. 3:4] PERMEABILITY OF SOIL AND ROCK

PERMEABILITY TESTS / Because of the numerous, complex factors that influence the permeability coefficient, only crude estimates of its magnitude can be made from a knowledge of the character of the soil. Therefore tests must be performed to obtain the coefficient with any certainty. The simplest test is the *constant head*, shown diagrammatically in Fig. 3.6. It is used primarily on sands and gravels. For fine sands and silts *the falling head* test is used. The upper reservoir of Fig. 3.6 is replaced with a vertical standpipe. During the test, the level of water in the standpipe falls, and the volume of water that flows is equal to the volume difference in the standpipe. Extreme care is essential in testing fine-grained cohesionless soils to avoid migration of the soil particles caused by excessive hydraulic gradients. For clay soils either the constant or falling head test is employed. The quantity of seepage is so small that great care is necessary to avoid leaks and evaporation which could be many times greater than the flow through the soil.

3:4 Permeability of Soil and Rock

The range in permeability of natural soils is even greater than the range in grain size. The following table can be used as a standard for describing a soil's permeability and as a guide for rough estimates.

TABLE 3:1 / RELATIVE VALUES OF PERMEABILITY*
(After Terzaghi and Peck)[3:2]

Relative Permeability	Values of k (cm/sec)	Typical Soil
Very permeable	Over 1×10^{-1}	Coarse gravel
Medium permeability	1×10^{-1}–1×10^{-3}	Sand, fine sand
Low permeability	1×10^{-3}–1×10^{-5}	Silty sand, dirty sand
Very low permeability	1×10^{-5}–1×10^{-7}	Silt, fine sandstone
Impervious	Less than 1×10^{-7}	Clay

* (To convert to feet per minute, multiply above values by 2; to convert to feet per day, multiply above by 3×10^3)

VARIATION OF k IN A REAL MASS / In most soils the value of k depends on the direction in which the water is traveling. The k in the direction parallel to the bedding planes or planes of stratification is usually from 2 to 30 times that in the direction perpendicular to the bedding or stratification, because of the layers of soils with relatively low permeabilities. In soil deposits with erratic lenses of either coarse, pervious materials or fine, impervious materials, the permeability varies greatly from point to point and is extremely difficult to determine.

Soils and rock in which there is an orientation of flakey or slab-like particles exhibit higher permeabilities parallel to the aligned faces than perpendicular to them. Similar anisotropic permeability is typical of some compacted soils.

In soils of low permeability and in most rocks, the permeability of the mass is governed by the cracks and fissures present. The effective permeability will be far greater than that of the intact material between the cracks. On the other hand, cemented seams within a generally pervious formation will make the effective permeability across the seams very low. Because of anisotropy and the effects of nonhomogeneous defects, a large number of tests, with flow in several directions is necessary for realistic values for the permeability coefficient.

3:5 Stress and Effective Stress

The reaction of soil or rock to stress is the major factor in the design of foundations, embankments, slopes and earth retaining structures. Because of their three-phase composition soils and rocks do not always behave in the same way as simple one-phase materials such as steel. The solids are relatively incompressible; they will support static shear stresses (although they may distort, and if the shear is sufficiently large, they will fail). The water is likewise relatively incompressible, but will offer only viscous (time-dependent) resistance to shear. The gaseous phase is compressible and has little viscous resistance to shear. Because each phase reacts differently to load, the distribution of stress between the phases must be determined to establish the effect of that stress on the mass.

EFFECTIVE STRESS / A load Q or Q' uniformly distributed over a layer of equal spheres with open voids between, Fig. 3.7a is not distributed uniformly throughout the mass. Instead the stress will vary from point to point within each sphere; it will be highest at the points of contact which have small areas and will be lowest at the center of each sphere. In a soil or rock consisting of irregular solids with voids between, the problem is complicated by the varying shapes and sizes of the particles as well as their geometric arrangement, Fig. 3.7b. In any case the stress is extremely high at the points of contact between the grain, and less between. The exact stress

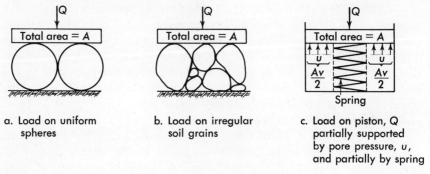

a. Load on uniform spheres

b. Load on irregular soil grains

c. Load on piston, Q partially supported by pore pressure, u, and partially by spring

Figure 3.7 Distribution of a load, Q or Q', to soil grains.

SEC. 3:5] STRESS AND EFFECTIVE STRESS

at any point within the solid phase is indeterminate except for the simplest case, Fig. 3.7a. Therefore it is convenient to express the stress in the solid phase in terms of the gross area A, and total load supported by the solid phase Q',

$$\sigma' = \frac{Q'}{A}. \qquad (3:11a)$$

The stress σ', sometimes written $\bar{\sigma}$, is the *effective stress*. It is less than the average stress in the solids at any level, and far less than the actual stress at most points within the particles. The name implies effectiveness in producing strain and failure in the solid phase.

NEUTRAL STRESS / If a total load Q is applied to a soil or rock consisting of solids and voids, with a pressure of u within the voids, the load distribution is more complex. If the area of voids in contact with the load is A_v then a force of uA_v will support part of the load. This is analogous to a piston supported by a spring in a closed cylinder, Fig. 3.7c. If the total load is Q, the area of voids (or cylinder) is A_v, and the pressure within the void (or cylinder) is u, then the load distribution is defined by,

$$Q = Q' + uA_v. \qquad (3:11b)$$

In this expression, Q is the total load and Q' that part of the load supported by the spring or solid phase. Dividing both sides of the expression by A, the gross area, gives

$$\frac{Q}{A} = \frac{Q'}{A} + u\left(\frac{A_v}{A}\right)$$

$$\sigma = \sigma' + u\left(\frac{A_v}{A}\right). \qquad (3:12a)$$

This is the *effective stress equation* and is fundamental to an understanding of the stress behavior of soil. The pore pressure, u, is frequently termed *neutral stress* because it is incapable of supporting shear.

VOID AREA / The void area over which the neutral stress or pore pressure acts is defined by the neutral stress ratio or coefficient, N

$$N = A_v/A. \qquad (3:12b)$$

For soils, the area of contact between grains is small, and the ratio, N, is approximately 1. The equation (3.12a) for soils simplifies to

$$\sigma = \sigma' + u. \qquad (3:12c)$$

It was this equation, first recognized by Terzaghi, that made it possible to analyze the deformation and strength of soil scientifically.

For concrete and rocks in which the solids are interconnected by crystal growth or cementing, the ratio is less than 1. Very little data is available for such materials, but the following values are suggested,

TABLE 3:2 / NEUTRAL STRESS RATIO

Material	$N^* = A_v/A$
Concrete	0.5 to 0.75
Fractured Rock	0.75 to 1
Porous Sandstone	0.75 to 1
Marble, Granite	0.1 to 0.5

* Skempton has used the same concept in terms of the fraction, a, of total area that constitutes grain contact in any surface through the soil. In this case $a = (1-N)$.

SIGNIFICANCE OF EFFECTIVE STRESS / Effective stress, as will be discussed in the remainder of the text, is a key factor in solving most engineering problems in moist or saturated soils and rocks. A simple example will illustrate its importance.

Example 3:1

An excavation is made in a clay soil overlying a sand stratum close to the river, Fig. 3.8. The clay weighs 120 lb per cu ft and is virtually impervious

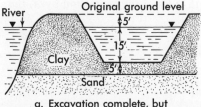
a. Excavation complete, but full of water

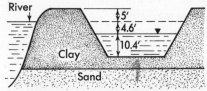
b. Excavation water level pumped down 4.6 ft, the point at which heave begins in the clay on the excavation bottom

Figure 3.8 Example 3:1 *Heave and a bottom blow out produced by pumping water out of an excavation below the ground water level.*

compared to the sand. The soil is removed with a clamshell and the hole remains nearly full of water during digging. The water is then pumped out. Suddenly the clay in the excavation bottom heaves upward, and then bursts. Why does this occur and at what level is the water in the excavation when it takes place?

1. At the beginning of the excavation the vertical stress at the boundary between clay and sand is as follows,

$$\sigma = 25 \times 120 = 3000 \text{ psf}$$
$$u = 20 \times 62.4 = \underline{1248} \text{ psf}$$
$$\sigma' = 1752 \text{ psf.}$$

SEC. 3:6] COMPRESSIBILITY AND SETTLEMENT

2. After the excavation and before pumping the effective vertical stress at the top of the sand, Fig. 3.8a, is

$$\sigma = 5 \times 120 + 15 \times 62.4 = 600 + 936 = 1536 \text{ psf}$$
$$u = 20 \times 62.4 = \underline{1248 \text{ psf}}$$
$$\sigma' = 288 \text{ psf}$$

The effective stress, the part of the load carried by the sand grains is reduced drastically at this point; however the sand is still effectively confined.

3. After pumping the water level down 4.6 ft the effective vertical stress at the top of the sand is, Fig. 3.8b:

$$\sigma = 5 \times 120 + 10.4 \times 62.4 = 600 + 649 = 1249 \text{ psf}$$
$$u = 20 \times 62.4 = \underline{1248 \text{ psf}}$$
$$\sigma' = 1 \text{ psf}$$

At this point the effective stress at the sand surface is practically zero. Further pumping reduces the total load still further. The pore pressure then exceeds the total stress and the clay heaves and breaks, allowing water to flow through the pervious sand upward into the excavation.

3:6 Compressibility and Settlement

The problem of building settlement has plagued builders for centuries. Many of the architectural masterpieces of the Middle Ages disappeared long ago because excessive settlement broke them apart. Others, such as the Leaning Tower of Pisa, became famous through their defects. But extraordinary settlements have not been limited to the Dark Ages; nearly every city of the world not founded on solid rock has its examples of buildings cracked and distorted through excessive settlement.

Until the twentieth century engineers vaguely attributed the cause of settlement to a squeezing of softer soils from beneath the structure. However, this did not explain the continuing settlement of such broad areas as the Mexico City Basin which drops at a rate of about 10 in. or 25 cm per year. Furthermore, samples of soil strata made in the vicinity of bad settlements did not show the bulges that might have been expected if squeezing had taken place. Samples made directly beneath the zones of greatest settlement indicated the following:

1. Some soil strata, particularly soft clays, had decreased in thickness an amount equal to the settlement.
2. Those soil strata that had decreased in thickness had smaller void ratios than the same strata outside the zones of settlement.

The evidence showed that a soil stratum, confined by the surrounding mass and subjected to a relatively uniform stress increase, decreased in void ratio significantly. Although this should have been suspected for a material that consisted of discrete, irregular particles, with small areas of contact and large voids between, it remained for Karl Terzaghi to clarify the mechanics of this process in *Erdbaumechanik* published in 1925.

COMPRESSIBILITY TEST / Terzaghi reasoned that squeezing or lateral movement of soft soil beneath a building was prevented by the strata of nonyielding soils interbedded with them. He developed a test to reproduce this for laboratory study by using a soil sample encased in a ring (to prevent lateral movement) and sandwiched between two porous plates, as shown in Fig. 3.9. In making the test, a vertical pressure or stress, σ' (sigma), is applied

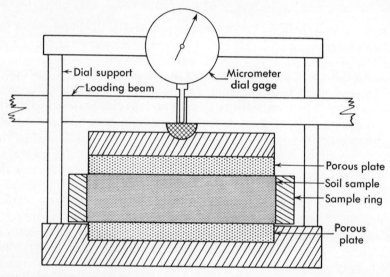

Figure 3.9 Consolidometer.

to the plates in order to produce a compression of the sample. It is allowed to remain until the compression virtually ceases, and then a much larger stress is added. This is repeated for the range in stresses to which the soil is likely to be subjected under the structure. The compression of the soil is measured with a micrometer dial in order to compute the void ratio corresponding to each stress.

The results are presented in several forms. The simplest is vertical strain as a function of effective stress, Fig. 3.10a. The deformation, ΔH of a stratum of thickness H is expressed in terms of strain, ε

$$\varepsilon = \frac{\Delta H}{H} \text{ or } \varepsilon_z = \frac{\Delta H}{H}. \tag{3:13}$$

SEC. 3:6] COMPRESSIBILITY AND SETTLEMENT 99

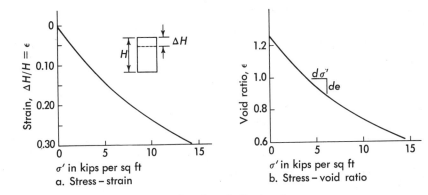

Figure 3.10 *Consolidation as a function of effective stress.*

Another way of expressing the results is the void ratio as a function of the effective stress, Fig. 3.10b. The curves are similar in shape for this simple condition of one-dimensional strain. The stress-strain form is more convenient for engineering computations of settlement while the void ratio form gives a better insight into the mechanics of the process.

COMPRESSION OF SOIL LAYER / The compression of a soil layer with an initial void ratio of e_0 can be found by proportion from a sample (Fig. 3.11) with a volume of solids of 1 and a cross-sectional area of 1. The height of the solids is 1, and the height of voids e_0. If the change in void ratio is Δe and the initial thickness of the stratum of the soil is H, then the decrease in stratum thickness can be found by the proportion:

$$\frac{\Delta H}{H} = \frac{\Delta e}{1 + e_0},$$

$$\Delta H = \frac{H \Delta e}{1 + e_0}. \tag{3:14}$$

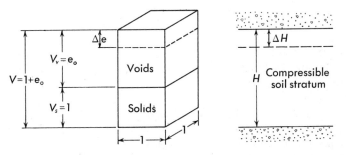

Figure 3.11 *Comparison of settlement of a soil stratum or sample of thickness of H with change in height of a sample whose initial height is $1 + e_0$.*

STRESS VOID-RATIO RELATION / The shape of the curve is concave upward, indicating a decreasing rate of compression with increasing stress. If the stress is increased to a certain point, σ_c', and then released, the soil will not swell to its original void ratio. Instead it will increase in volume gradually along a flat, concave upward curve (Fig. 3.12a) called the *decompression* curve. If stress is again applied, the recompression of the soil will follow a flat curve that is concave downward until the stress is nearly equal to σ_c'. At this point a more rapid decrease in void ratio takes place until the recompression curve practically joins the original curve. (The original curve is often termed the *virgin* curve.) Soil compression is not an elastic, reversible process; once compressed, a soil tends to remain so, even though the stress-producing compression may be removed.

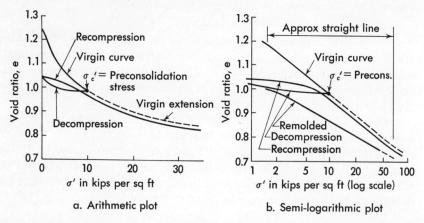

Figure 3.12 *Stress-void ratio: Compression, decompression, recompression and preconsolidation stress.*

The slope of the curve is a measure of the soil compressibility or relative strain. In the stress-strain form it is termed the *coefficient of volume compressibility*, m_v, because if no lateral movement occurs, the vertical strain equals the volume strain.

$$m_v = \frac{d\varepsilon}{d\sigma'}. \qquad (3:15a)$$

This is the reciprocal of the modulus of deformation or modulus of elasticity commonly used to describe stress-strain relationships in solids.

The second form is the *coefficient of compressibility*, a_v,

$$a_v = \frac{de}{d\sigma'} = m_v(1+e). \qquad (3:15b)$$

Neither m_v nor a_v are constants: both decrease with increasing stress. Both are influenced by the past stress or stress history; the unloading and reloading curves are both different from the virgin curve, as shown in Fig. 3.12a.

LOGARITHMIC REPRESENTATION / If the stress-void ratio curve is plotted with the logarithm of stress as the abscissa and the void ratio as ordinate (Fig. 3.12b) it will be seen that the virgin compression curve forms a straight line through the greater part of its range. The logarithmic representation has become the standard form for presenting soil compression data because it affords a convenient means for expressing the compressibility of soils by comparing the slopes of their virgin curves—the steeper the slope, the more compressible the soil.

The equation of the straight-line portion of the curve is

$$\Delta e = -C_c \log_{10} \frac{\overline{\sigma}_0 + \Delta \sigma}{\overline{\sigma}_0}. \tag{3:16}$$

The term C_c is the compression index, which expresses the slope of the virgin curve in the semi-log plot. It is a convenient index for comparing the compressibilities of soils in the virgin compression range. It can also be used to compute void ratio changes for stresses exceeding the preconsolidation load.

THE MECHANISM OF COMPRESSION / The mechanism of soil compression includes a number of phenomena.[3:3,3:4] A major component in rock, cohesionless, organic, and micaceous soils (and probably to some extent in clays) is the bending and distortion of the solid phase. This is largely elastic and the compression that results from it is reversible. Fracture of the solids, particularly at their points of contact, is probably a factor in the compression of all soils and porous rocks. This is not reversible, and partially accounts for the decompression being less than the original compression. The electrical repulsion between particles having like charges, or surrounded by cations with like charges which holds them apart, is a major factor in clays and is probably reversible. Reorientation of the grains occurs in all soils to some degree. In order for the grains to move, the bond or attraction between points of contact of clay particles must be overcome, which is a factor in all compression but particularly important at high stresses. Such structural changes or breakdowns are not reversible and also are partially responsible for the decompression being less than the compression. The adsorbed water affects both the electrical repulsion and the attraction. Whether it has an independent effect is not known. All these phenomena are related to the physicochemical properties and the structure of the soil or rock, which are difficult to evaluate and describe quantitatively. Most of them are related to the previous stress conditions, which usually are unknown. Therefore it is necessary to perform a consolidation test to determine the volume change of a soil or rock under load.

3:7 Compressibility of Soils and Rocks

Although the compressibility of a soil cannot be predicted from its other physical properties, similarities between certain groups of soils and weathered rocks and some empirical relationships involving compressibilities have been observed that are useful in interpreting test results and making estimates when test data are not available.

LOW TO MODERATE PLASTICITY, NORMALLY CONSOLIDATED / Soils of low to moderate plasticity include clays, silts, organic and micaceous silts with plasticity indexes up to 30, as well as sands, gravels, and porous rocks. When they have never been subjected to stresses greater than their present overburden load, they are termed *normally loaded* and exhibit the characteristic semilogarithmic stress-void ratio curve of Fig. 3.12b. The greater part of the curve is approximately straight, and the slope is expressed by the compression index, as was previously described. Terzaghi and Peck[3:5] have derived an expression for the compressibility of such clays from the work of Skempton.[3:6]

$$C_c = 0.009(\text{LL} - 10). \qquad (3:17)$$

For soils of very low plasticity and porous rock, Sowers[3:7] has found the compression index to be related to the undisturbed void ratio

$$C_c = 0.75(e - a), \qquad (3:18)$$

where a is a constant whose value is from 0.2 for porous rock to 0.8 for highly micaceous soils. Both relationships are approximate at the best, and considerable variations should be expected from the computed value of the compression index. Normally loaded clays can often be identified from their water content. Usually it is near the liquid limit.

The same soil remolded suffers a structural breakdown and a reduction in void ratio. It becomes somewhat less compressible, but exhibits a straight-line stress-void ratio curve on semilogarithmic coordinates, as shown in Fig. 3.12b.

A normally consolidated soil undergoes decompression during sampling. The stress can reduce to zero, or may be partially retained by capillary tension. In either case, the laboratory stress void ratio curve represents recompression until the initial stress in the ground, σ_0', is exceeded.

Most natural soils are somewhat disturbed by the sampling process. The stress void ratio curve found in the laboratory, therefore, lies somewhere between that for the remolded soil and the true pressure void ratio curve in nature.

The increments in stress employed in the laboratory test ($\Delta\sigma_2' = 2\Delta\sigma_1$ typically) are usually much larger than those produced by structural loads. In some soils these large increments cause disturbance and increased settlement.

Because of these differences between compression of the undisturbed soil in place and the laboratory test, the settlement estimates based on the test results tend to be excessive. The errors can be minimized by careful workmanship and by a test procedure that simulates as far as is practical the stress changes that will be produced in the field. A simple method for minimizing the disturbance error is to load the laboratory specimen to the

SEC. 3:7] COMPRESSIBILITY OF SOILS AND ROCKS

overburden stress σ'_0, and allow it to recompress. The soil is then unloaded to the minimum stress to be expected in the future. Finally it is reloaded in increments to stresses that exceed the future stresses and that define the shape of the virgin compression curve. Although it can be argued that this introduces a cycle of unloading and loading that does not occur in the actual soil deposit, laboratory tests show that a second cycle of unloading and loading is not greatly different in slope from the first cycle. Therefore, changes in void ratio or settlement computed from the second cycle are close to those that would occur in the first.

LOW TO MODERATE PLASTICITY—PRECONSOLIDATED / A *preconsolidated* soil is one which has been subjected to a stress, σ'_c or $\bar{\sigma}_c$, which exceeds the present overburden pressure. Most undisturbed soils are preconsolidated to some degree and exhibit the characteristic stress-void ratio curve of Fig. 3.13a. Preconsolidation is produced in a number of ways.

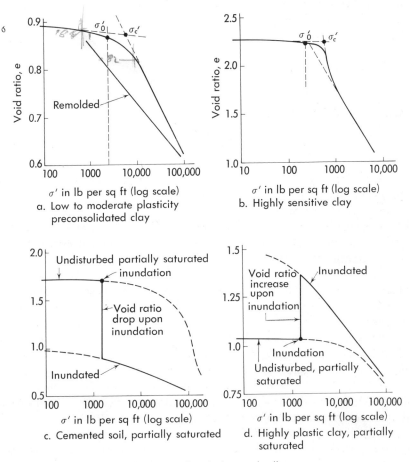

Figure 3.13 *Stress-void ratio curves of typical natural soils.*

Removal of overburden by erosion or excavation leaves the soil preconsolidated. This load can be estimated if the previous overburden thickness is known. The most important and widespread cause is capillary tension arising from desiccation or drying of the soil. Stresses greater than 10,000 psf due to desiccation have been observed in arid and warm regions. The preconsolidation load caused by desiccation often decreases with increasing depth below the ground surface. The amount produced in this way cannot be predicted. A water table rising above a compressible stratum causes preconsolidation by reducing the load carried by the soil through buoyancy. The amount of the reduction is seldom very great, but it can be computed as described in Chapter 9.

Chemical alteration produces the effect of preconsolidation by changing the physicochemical bonds between the clay particles or by introducing stresses by the expansion or contraction of the grains during the alteration process. Most weathered rocks and some partially indurated rocks exhibit preconsolidation from this source. Leaching that removes salt or high concentrations of cations may have the same effect in some clays, especially those deposited in salt water.

The preconsolidation load σ_c' can be estimated from the stress-void ratio curve shown in Fig. 3.13a. Tangents are drawn to the initial, flat section and to the steep straight-line portion of the curve. Their intersection is approximately the preconsolidation load.

Preconsolidated clays can be recognized from their water contents, which are usually much less than their liquid limits. In heavily preconsolidated soils (particularly if desiccation is the cause) the water contents are sometimes less than the plastic limits. Their compression indexes (above the preconsolidation load) can be found from the same relationships used for normally consolidated clays.

Preconsolidation is extremely important in foundation engineering. A soil that is inherently compressible usually will not settle appreciably until the stress imposed by the structure exceeds the preconsolidation load. If the natural preconsolidation load is not sufficient, it is sometimes possible to preconsolidate the soil by piling earth on the site until the soil compresses. This is often time consuming, but the process can be accelerated as will be described in Chapter 4.

SENSITIVE CLAYS / Sensitive clays and other soils with a flocculent or highly developed skeletal structure exhibit a third characteristic curve, Fig. 3.13b. Ordinarily such soils are preconsolidated to some degree. The curve is flat up to the preconsolidation load; then it drops sharply and gradually flattens to form a straight line on semilogarithmic coordinates. It is suspected that the sharp drop in void ratio reflects a structural breakdown in which the bonds between the particles are broken and the grains rearrange themselves into a more dense orientation. The same soil badly disturbed or remolded has a straight or slightly curved stress-void ratio curve that is slightly flatter than the virgin curve at high stresses. The water content of most highly sensitive clays exceeds their liquid limit, which aids in their recognition.

CEMENTED SOILS / Cemented soils and similar soft rocks have stress-void ratio curves similar to preconsolidated soils. The appearance of preconsolidation probably reflects the bonding (and the interparticle stresses) between the grains. In some cases the bonding breaks abruptly with increasing load, producing a stress-void ratio curve similar to that for a sensitive clay, Fig. 3.13b. For such materials two pressure void ratio curves can be developed: undisturbed (usually moist) and inundated as shown in Fig. 3.13c. The drop from the higher, undisturbed value will occur upon inundation under very small stresses, such as produced by the weight of the soil itself. Serious sudden settlements have occurred in dry regions where cemented soils have been saturated by constructing dams or by introducing irrigation. Inundation produces near-saturation; however, complete saturation probably does not occur.

EXPANSIVE SOILS / Partially saturated highly plastic clays and some highly micaceous soils may be heavily preconsolidated by desiccation. If they have access to water at low stresses, they will adsorb water and expand, Fig. 3.13d. Two stress-void ratio curves can be developed for such a soil: undisturbed but partially saturated, and inundated. The inundated soil approaches saturation; however there is little evidence that 100 per cent saturation occurs.

PEAT / Peats and highly organic soils are extremely compressible, depending on the void ratio. The pressure void ratio curves resemble those for clays, with slight preconsolidation. The virgin curve on a semi-log plot is seldom very straight. Compression indexes as high as 15 have been found for some peats of very high void ratio.

SANDS / Sands consolidate largely by grain re-orientation and fracture, accompanied by some elastic distortion of the grains. The compression indexes are usually less than 0.1, and the curves resemble those for clays.

COMPRESSIBILITY OF ROCK / Well indurated rocks with low void ratios (less than 0.2) are virtually incompressible. More porous rocks, including shale, tuff and porous limestones consolidate similar to soils. The mechanism appears to be a combination of elastic distortion of the solid framework between the voids and a crushing of the rock where it is locally highly stressed. The pressure void ratio curves for porous rocks resemble those for preconsolidated soils.

Continued leaching and weathering will weaken the rock and permit further consolidation. This has been observed in soft limestones and is suspected in tuff. The leaching appears to be aggravated under stress, and thus the settlement increases with increasing load as well as with the passage of time. This can only be evaluated by simulating the leaching under stress in the laboratory.

106 THE ENGINEERING PROPERTIES OF SOIL AND ROCK [CH. 3

Broken rock, such as used in rock fills, consolidates by re-orientation of the grains and by local crushing at the points of contact. The pressure void ratio curves resemble those for virgin compression, Figs. 3.10 and 3.12.

3:8 Time Rate of Compression

The compression of a soil stratum does not occur suddenly. In fact it often takes place so slowly that it is difficult to believe that any settlement is taking place. Buildings in Chicago continue to settle for 50 years, and the Leaning Tower of Pisa, commenced in 1174, is still moving.

The settlement begins rapidly and becomes slower with increasing time. As shown in Fig. 3.14, it can be divided into three stages: *initial*, *primary* or *hydrodynamic*, and *secondary*. The sum of the initial and primary stages is the settlement computed from the laboratory stress-void ratio curve. The secondary is of importance principally in highly organic, highly micaceous, and highly sensitive soils.

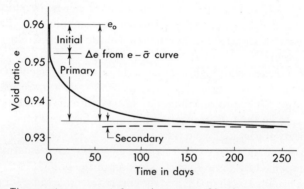

Figure 3.14 *Time-settlement curve for an increment of load.*

INITIAL CONSOLIDATION / The initial stage occurs as soon as the load is applied. It occurs largely by compression and solution of the air in the soil voids. It also includes small amounts of compression of the solid phase and the soil water.

The percentage of initial consolidation of soil is largely controlled by the degree of saturation. It is virtually 0 in a saturated soil, but may be as great as 50 per cent when the degree of saturation is 90 per cent. It decreases with increasing load because of the smaller void ratio (and thus higher saturation). Ordinarily, the initial compression is determined from the time-settlement data from the laboratory test and furnished as a part of the test results as in Fig. 3.19a.

SEC. 3:8] TIME RATE OF COMPRESSION

NEUTRAL STRESS / The time-rate of primary or hydrodynamic compression is controlled by the escape of water from the soil voids. The water is squeezed out by an external force applied to the soil mass. This force increases the water pressure beyond the hydrostatic, developing a hydraulic gradient that produces flow. The mechanics of this process can be demonstrated by the analogy of a spring, piston, and cylinder. A spring (Fig. 3.15a) has attached to its top a piston whose cross-sectional area is 1 sq ft and whose weight is 100 lb. Under this weight the spring has a length of 1.0 ft. When an added load of 50 lb is placed on the spring (Fig. 3.15b), it compresses to a length of 0.8 ft. The compression is instantaneous as soon as the load is applied. Suppose that instead of in the open air, the piston and spring is placed in a tight-fitting cylinder (Fig. 3.15c) with the space below the piston filled with water. The spring is compressed by the weight of the piston, 100 lb, and the water is under no pressure at all.

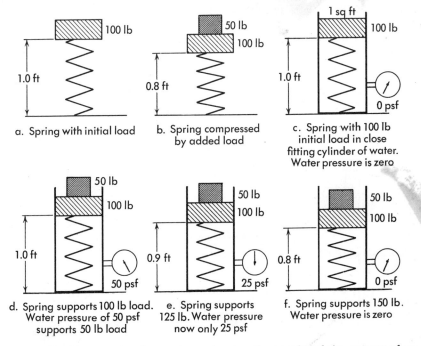

Figure 3.15 *Piston and spring analogy, showing the transfer of the support of an added 50-lb load from water pressure (neutral stress) to the spring.*

If the 50-lb weight is now added on top of the piston (Fig. 3.15d), the spring cannot compress because the water below the piston cannot escape. The spring still supports the 100-lb piston but offers no support to the added weight. The 50 lb are supported by water pressure on the piston of 50 psf. If the total load of 150 lb is denoted by σ, and the actual spring load by σ', then the following equation describes the way the total load is supported:

$$\sigma = \sigma' + u, \qquad (3:12c)$$
$$150 = 100 + 50.$$

If 0.1 cu ft of water leaks out because of the pressure from beneath the piston, the spring will be compressed to a length of 0.9 ft. The spring will now support 125 lb, and the neutral stress will be reduced to 25 psf, as shown in Fig. 3.15e:

$$150 = 125 + 25.$$

When an additional 0.1 cu ft of water leaks away, the spring will carry 150 lb and the neutral stress will be zero, as in Fig. 3.15f. The spring compression is a process of transferring to the spring the 50 lb of added load that was supported initially by water pressure.

Consolidation of soil is similar to the above analogy. The resilient grain structure is represented by the spring, and the voids filled with water are represented by the cylinder. When a load is placed on the soil, the grain

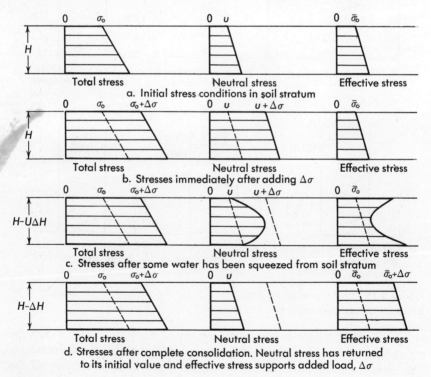

Figure 3.16 *Stresses during consolidation. The soil stratum is drained on both its top and bottom faces.*

SEC. 3:8] TIME RATE OF COMPRESSION

structure cannot immediately support it because compression cannot occur. Neutral stress therefore supports the load. As the water seeps out and the soil compresses, the grain structure assumes the load and the neutral stress becomes zero. The adjacent sequence (Fig. 3.16) illustrates the way in which the stress transfer occurs in a soil stratum that is bounded by pervious strata above and below. The initial stress in the soil is denoted by σ_o; the added stress due to the weight of a structure by $\Delta\sigma$; neutral stress by u, and the stress in the grain structure by $\bar{\sigma}$ or σ'.

PERCENTAGE OF CONSOLIDATION / The *percentage of consolidation*, U, is defined as the average percentage of the added stress $\Delta\sigma$ that is supported by increased effective stress. It represents the percentage of the total or ultimate compression that has already occurred in the stratum. The percentage of consolidation depends on several factors:

1. The soil's permeability, which governs the rate of flow of the water.
2. The thickness of the stratum, which influences both the volume of water that must seep out and the distance it must travel and the hydraulic gradient.
3. The number of pervious boundaries of the stratum from which the water can leave, which influences the distance the water must travel and the gradient.
4. The void ratio and the rate of change of void ratio with pressure, which influence both the volume of water and the way the neutral stress will decrease with a loss of water.

A mathematical analysis of the percentage of consolidation leads to the following equation:

$$U = f*\left(\frac{t(1+e)k}{(H/N)^2 \gamma_w a_v}\right), \quad (3:19a)$$

$$U = f(T), \quad (3:19b)$$

where t is time; e is the void ratio; k is the permeability; H is the stratum thickness; N is the number of horizontal pervious boundaries of the stratum (either 1 or 2); γ_w is the unit weight of water; and a_v is the rate of change of void ratio with changes in pressure, $\Delta e/\Delta\sigma$. The term T is a dimensionless ratio known as the *time factor* and is defined by the expression:

$$T = \frac{t(1+e)k}{(H/N)^2 a_v \gamma_w}. \quad (3:20)$$

The relation between U and T is expressed by a mathematical function that is independent of the soil characteristics or the amount of compression. Instead it depends on the variation of $\Delta\sigma$ with depth throughout the stratum and on whether the stratum is bounded by a pervious layer both top and bottom or just on one surface.

* Function of.

110   THE ENGINEERING PROPERTIES OF SOIL AND ROCK   [CH. 3

The theoretical analysis of consolidation is based on several assumptions, including a homogeneous, saturated soil, a constant value of the term

$$\frac{k(1+e)}{a_v}$$

during compression, and vertical movement of the water. Because of these assumptions great accuracy cannot be expected in estimates of the time rate of compression. The chart (Fig. 3.17) shows the relation of U to T for most

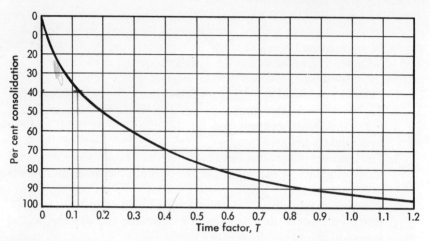

Figure 3.17 *Time-rate consolidation for a stratum drained on both faces and any distribution of stress increase or for a stratum drained on one surface and a uniform stress increase.*

practical problems. It is inaccurate, however, where great variations in $\Delta\sigma$ with depth occur in a stratum that is bounded by an impervious surface, either top or bottom.

The characteristics of a soil that govern its rate of consolidation may be expressed by the *coefficient of consolidation*, c_v, which is defined by

$$c_v = \frac{(1+e)k}{\gamma_w a_v}. \tag{3:21}$$

This coefficient varies somewhat, depending on the soil stress σ.

At stresses below the preconsolidation load it is high, indicating rapid consolidation. It falls to a minimum value at the preconsolidation load, and then increases slowly with increasing stress, as shown in Fig. 3.19b.

Example 3:2

Find the time required for 50 per cent consolidation to take place in the following soil stratum.

    $k = 0.0000001$ cm/sec.
    $e = 1.5$.

$a_v = 0.0003$ cm²/gm.
$\gamma_w = 1$ gm/cm³.
Thickness is 900 cm, with pervious strata top and bottom.
1. For $U = 50\%$, $T = 0.2$.
2. For 900 cm stratum with two pervious surfaces
$H/2 = 900/2 = 450$ cm.

3. $0.2 = \dfrac{t(1 + 1.5)0.000\,000\,1}{450^2 \times 0.0003 \times 1}$.

4. $t = \dfrac{0.2 \times 203000 \times 3 \times 1}{0.001 \times 2.5} = 48.7 \times 10^6$ sec.

5. $t = 565$ days.

SECONDARY COMPRESSION / After the excess hydrostatic pressure has been dissipated, the compression does not cease. Instead it continues very slowly at an ever-decreasing rate indefinitely. This is *secondary compression*. This appears to be the result of a plastic readjustment of the soil grains to the new stress, of progressive fracture of the interparticle bonds, and progressive fracture of the particles themselves.

The secondary compression can be identified on a plot of settlement as a function of logarithm of time (Fig. 3.18). The secondary appears as a straight

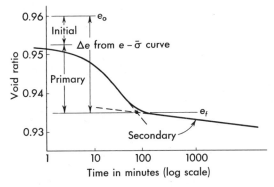

Figure 3.18 Secondary compression; semilogarithmic time-settlement curve for an increment in stress.

line sloping downward or, in some cases, as a straight line followed by a second straight line with a flatter slope. The void ratio e_f corresponding to the effective end of primary consolidation can be found from the intersection of the backward projection of the secondary line with a tangent drawn to the primary curve, as indicated on Fig. 3.18. The rate of secondary compression depends on the increment of stress increase, $\Delta\sigma$, and on the characteristics of the soil. For inorganic soils of low to moderate compressibility, secondary compression is seldom important. It can be a major part of the compression of highly compressible clays, highly micaceous soils, fills of broken rock and organic materials.

The equation for the rate of secondary compression can be approximated from the straight line on the log time plot

$$\Delta e = -\alpha (\log_{10} t_2 - \log_{10} t_1)$$

$$\Delta e = -\alpha \log_{10}\left(\frac{t_2}{t_1}\right). \tag{3:22a}$$

In this expression t_1 is the time required for hydrodynamic compression to be virtually complete and t_2 any later time.[3:8] This ignores the secondary compression that occurs during the hydrodynamic phase but the error is probably not serious. The value α is a coefficient expressing the rate of secondary compression. The coefficient of secondary compression, C_α is another way of expressing the same thing in terms of percentage of settlement:

$$C_\alpha = \alpha/(1 + e) \tag{3:22b}$$

$$\frac{\Delta H}{H} = -C_\alpha \log_{10}\left(\frac{t_2}{t_1}\right). \tag{3:22c}$$

Generally α and C_α increase with increasing stress, Fig. 3.19c.

The secondary compression is often irregular and only roughly can be approximated by the straight line on the logarithm of time graph. Therefore estimates of secondary compression are seldom accurate.

REPRESENTING TIME-RATE DATA / All three phases of the time rate are dependent on the stress level. Therefore it is convenient to express all three as functions of the final stress, as in Fig. 3.19. The functions are not unique, because they also depend on the stress increment, $\Delta\sigma'$. However if the increment is similar to that in the prototype, the values will be sufficiently realistic for engineering estimates.

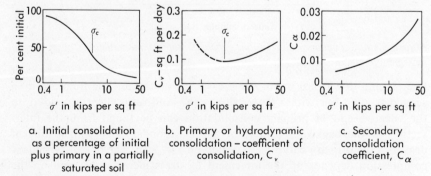

Figure 3.19 Plots of factors and coefficients describing time-rate of consolidation as functions of stress. (Different soils.)

3:9 Shrinking, Swelling, and Slaking

Soils undergo volume changes that are not produced by external loads. Instead they are caused by changes in water content and in the internal stresses affected by the water.

SHRINKAGE / Shrinkage is caused by capillary tension. When a saturated soil dries, a meniscus develops in each void at the soil surface. This produces tension in the soil water and a corresponding compression in the soil structure. This can be expressed quantitatively from equation 3.12c, where the external stress σ equals zero:

$$0 = \sigma' + u,$$
$$\sigma' = -u.$$

(Since u from capillary tension is negative, then $-(-u)$ is positive and the effective stress σ' is positive.) This compressive stress is just as effective in producing soil compression as an external load, and pressures of several thousand pounds per square foot can be produced in fine-grained soils.

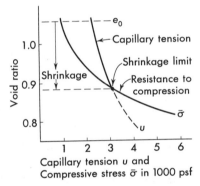

Figure 3.20 Shrinkage limit as a function of capillary tension and resistance to compression.

SHRINKAGE LIMIT / During the shrinkage process, the voids become smaller and the potential maximum capillary tension increases. This is shown graphically in Fig. 3.20. The resistance to compression, the stress-void ratio curve, is shown also. During the drying and shrinkage process, the void ratio decreases and with it both the maximum capillary tension and the resistance to compression increase, but at different rates. The soil remains saturated, for the water loss causes an equal reduction in void ratio. A void ratio is reached, however, where the maximum tension equals the resistance. Further drying cannot produce a reduction in void ratio because the resistance exceeds the tension. At this point, known as the *shrinkage limit*, the reduction in the void ratio largely ceases. The meniscus in each void begins to retreat from the soil surface. The soil surface no longer has a damp appearance but now looks dry, and the soil mass ceases to be saturated.

The shrinkage limit is defined as the water content at the point that shrinkage ceases and the soil is no longer saturated. It can be found by drying the soil slowly, visually observing the color change, and determining the moisture at this point. It can also be found by drying a saturated soil completely. If the weight and total volume at the beginning of shrinkage are W_1 and V_1 and at the end of shrinkage (oven dried) are W_2 and V_2, then the following can be derived by assuming that there is no volume change after the shrinkage limit is reached and that the loss of weight by evaporation up to the shrinkage limit is accompanied by a corresponding loss of volume:

$$W_2 = W_s,$$

$$(V_1 - V_2)\gamma_w = W_w \text{ (lost up to shrinkage limit)},$$

$$\text{SL} = \frac{W_1 - W_2 - (V_1 - V_2)\gamma_w}{W_2} \times 100. \qquad (3:23)$$

Beyond the shrinkage limit the capillary tension can increase in the smaller voids. On the other hand the tension is released in some of the larger ones. Some soils on drying, particularly those containing fibrous organic matter and mica, may expand beyond the shrinkage limit; others may shrink further. These changes are ordinarily insignificant.

Soil shrinkage results in settlement of compressible soils. Since the capillary tension is exerted in all directions, shrinkage occurs horizontally as well as vertically, causing shrinkage cracks to form. Cracks 1 ft wide and 15 ft deep have been observed in highly compressible clays. Reoccurring shrinkage brought on by desiccation during dry weather will produce networks of shrinkage cracks in all directions and a blocky macrostructure in the soil.

SWELLING / Some soils not only shrink on drying but also swell when the moisture is allowed to increase. The mechanism is more complex than shrinkage and is caused by a number of different phenomena: the elastic rebound of the soil grains, the attraction of the clay minerals for water, the electrical repulsion of the clay particles and their adsorbed cations from one another, and the expansion of air trapped in the soil voids. In soils that have been precompressed by load or shrinkage, all these factors probably contribute. In soils that have never been precompressed, probably the attraction of the clay minerals for water and the electrical repulsion of the clay particles surrounded by water are major factors.

High pressures can be developed if the soil has access to water but is prevented from swelling by confinement. If precompression is the cause, the swell pressure can be nearly as great as the preconsolidation load. Where adsorption and repulsion predominate, as in clays of the montmorillonite family, pressures of several thousand pounds per square foot can develop.

Swell is produced in some soils and rocks by chemical changes. Strong bases, leaking from chemical processes, can cause expansion of layered minerals such as micas and clays. Oxidation of iron pyrite in shale, freshly exposed to air, will cause it to swell and damage structures.

PREDICTING SWELLING AND SHRINKAGE / It is difficult to predict shrinkage and swelling quantitatively, for they depend on the character of the soil and on the moisture changes. Shrinkage can be found by merely drying the soil and computing the relation between saturated water content and volume. In general the lower the shrinkage limit, the greater the potential shrinkage of the soil. Swelling can be estimated by tests resembling consolidation. The expansion (free swell) is found by flooding the soil when it is acted upon by a constant nominal pressure (such as 100 psf). The swell pressure is found by inundating the soil and measuring the pressure required to prevent its expansion. It has been found that the shrinkage limit and the plasticity index are some indication of the potential volume change, as given in Table 3:3.

TABLE 3:3 / VOLUME CHANGE POTENTIAL
(Adapted from Holtz and Gibbs[3:10])

Volume Change	Shrinkage Limit	Plasticity Index
Probably low	12 or more	0–15
Probably moderate	10–12	15–30
Probably high	0–10	30 or more

The amount of swelling and shrinking, or the swell stress if the soil is confined, depends on the initial moisture in the soil.[3:11] If the soil is drier than the shrinkage limit, further drying will not produce significant additional shrinkage. If it is wetter, then the maximum possible shrinkage will be equivalent to the difference between the actual water content and the shrinkage limit.

Similarly, limited data indicate that little swell will occur after the soil moisture reaches the plastic limit, or slightly more, equivalent to a water-plasticity ratio of about 0.25. Smaller water contents produce higher swell and swell pressures as shown in Fig. 3.21.[3:11]

The major problem in predicting the amount of swell and shrink in soil deposits is that few data are available to define the potential changes in moisture content. Continuing field observations for several years may be necessary to determine the moisture fluctuations that occur in any region, and these must include the effects of any structure or any changes in environment that are produced by man.

SLAKING / If a soil that has dried well beyond the shrinkage limit is suddenly inundated or immersed in water, it may disintegrate into a soft wet mass, a process known as *slaking*.

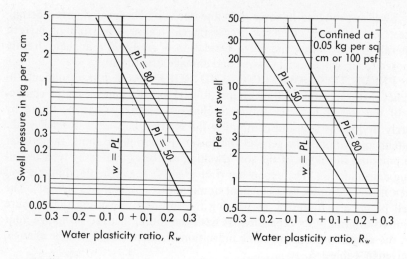

a. Swell pressure, completely restrained b. Swell, with nominal confinement

Figure 3.21 *Swell and swell pressure as a function of the water-plasticity ratio.*

Two factors are involved in slaking. First, the unequal expansion of the soil as the water penetrates from the surface causes pieces of soil to flake off the mass. Second, when the soil dries beyond the shrinkage limit, some of the voids fill with air. When the dried soil is immersed, water enters these air-filled voids on all sides. The air is trapped between the meniscuses of the entering water, and its pressure builds up as water fills the void. The result is an explosion of the void and disintegration of the soil. Both the flaking and the air bubbles can be seen by placing a lump of dried clay in a glass of water.

3:10 Combined Stresses

While the simple case of one dimensional stress and compression previously discussed (Sections 3:6–3:8) adequately describes many problems of settlement, many other problems of soil deformation and failure involve three dimensional stresses. The simple cases of one direction of compression and tension, so useful in structural design of steel and concrete, are of little importance in the soil mass whose weight is a substantial part of the total load and where the structural loads are introduced in several directions and at different levels within the mass. Therefore analysis of the effects of stress must begin with the entire stress field in three dimensions.

A stress is defined as a force per unit of area. A stress applied to a plane surface of a solid can be resolved into two components: one perpendicular (normal) to the plane known as the *normal* stress, σ (sigma), and one acting in the surface of the plane known as the *shear* stress, τ (tau), as shown in

SEC. 3:10] COMBINED STRESSES

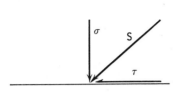 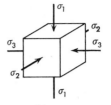

a. Shear and normal stresses   b. Principal stresses on a cube

Figure 3.22 *Shear, normal and principal stresses.*

Fig. 3.22a. When the stress acting on a plane consists only of a *normal* component and $\tau = 0$, that normal stress is termed a *principal* stress, (Fig. 3.22b).

When a cube of rock or mortar is to be tested to determine its strength, it is placed in a testing machine and gradually increasing compressive forces are applied to its top and bottom faces. The compressive forces result in compressive stresses in the faces to which they are applied. These stresses are principal stresses, and the horizontal planes in which they occur are called *principal planes*. Although it is rarely done, it would be possible to introduce compressive forces on the other two pairs of cube faces. These also would result in principal stresses in the faces to which they were applied, and these faces would also be considered principal planes. It can be shown that there are three independent, perpendicular, principal stresses acting on three perpendicular, principal planes. The largest of the three principal stresses is known as the *major principal* stress and is denoted σ_1. The smallest is known as the *minor principal* stress σ_3, and the third is called the *intermediate principal* stress σ_2.

In soil mechanics, tensile stresses are comparatively rare; therefore, to avoid many negative signs, compressive stresses are considered positive. In the case of the cube of mortar in an ordinary compression test, the compressive stress applied to the top and bottom faces is σ_1, and the other two principal stresses, σ_2 and σ_3, are zero.

If an inclined plane cuts through the cube, it is possible to compute the shear and normal stresses on that plane from the three principal stresses and the laws of statics. The general case is quite complicated, for it involves the direction cosines of the plane from the principal planes. In many problems in soil mechanics, however, we are interested in stresses on planes perpendicular to the intermediate principal plane which reduces the problem to two dimensions.

The direction of an inclined plane that is perpendicular to the intermediate principal plane is defined by α (alpha), the angle the plane makes with the plane of the major principal stress, as shown in Fig. 3.23.

Shear and normal stresses on the plane can be computed by the laws of statics from σ_1 and σ_3. If the cube is assumed to have dimensions

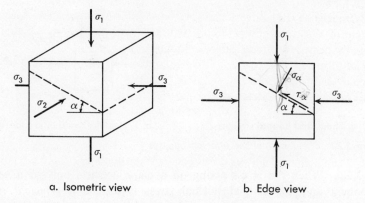

a. Isometric view b. Edge view

Figure 3.23 *Stresses on a cube that is cut by a plane which is perpendicular to the plane of σ_2 and which makes an angle of α with the plane of σ_1.*

$1 \times 1 \times 1$, then the forces acting on the plane in the directions of σ_1 and σ_3 are respectively

$$F_1 = \sigma_1 \times \text{area},$$
$$F_1 = \sigma_1 \times 1 \times 1,$$
$$F_3 = \sigma_3 \times 1 \times 1 \tan \alpha.$$

The sum of the components of these forces normal to the plane is

$$F_n = F_1 \cos \alpha + F_3 \sin \alpha,$$
$$F_n = \sigma_1 \cos \alpha + \sigma_3 \tan \alpha \sin \alpha.$$

The sum of the components parallel to the plane is

$$F_s = \sigma_1 \sin \alpha - \sigma_3 \tan \alpha \cos \alpha.$$

The area of the plane is $1/\cos \alpha$; therefore the normal stress on the plane σ_α, is

$$\sigma_\alpha = \frac{\sigma_1 \cos \alpha + \sigma_3 \tan \alpha \sin \alpha}{1/\cos \alpha},$$

$$\sigma_\alpha = \sigma_1 \cos^2 \alpha + \sigma_3 \sin^2 \alpha,$$

$$\sigma_\alpha = \frac{\sigma_1 + \sigma_3}{2} + \frac{\sigma_1 - \sigma_3}{2} \cos 2\alpha. \qquad (3\!:\!24\text{a})$$

In the same way the shear stress in the plane, τ_α, is

$$\tau_\alpha = \frac{\sigma_1 - \sigma_3}{2} \sin 2\alpha. \qquad (3\!:\!24\text{b})$$

SEC. 3:10] COMBINED STRESSES

By means of the above formulas, the stresses on any inclined plane at an angle of α can be computed, or if the stresses on any two planes are known, the principal stresses can be computed.

The formulas lead to the following conclusions, which should be kept in mind in analyzing stresses:

1. The maximum shear stress occurs when $\sin 2\alpha = 1$ or $\alpha = 45°$ or $135°$ and is equal to $\dfrac{\sigma_1 - \sigma_3}{2}$.
2. The maximum normal stress occurs when $\cos 2\alpha = 1$ and $\alpha = 0$.
3. The minimum normal stress occurs when $\cos 2\alpha = -1$ and $\alpha = 90°$ and the plane is parallel to the minor principal plane.
4. Shear stresses are equal in magnitude on any two planes perpendicular to each other.

MOHR'S CIRCLE / A German physicist, Otto Mohr, devised a graphical procedure for solving the equations for shear and normal stress on a plane perpendicular to one principal plane and making an angle α with the larger of the two other principal planes. A system of coordinate axes is established (Fig. 3.24a) where the x-distances represent normal stresses and the y-distances represent shear stresses. Compressive (positive) normal stresses are plotted to the right; tensile, to the left. Shear stresses may be

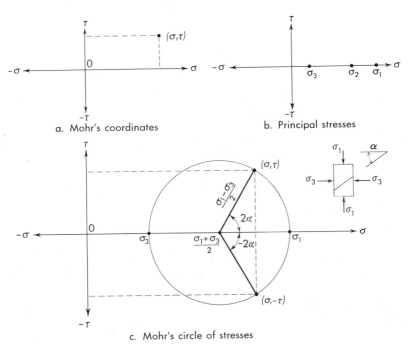

a. Mohr's coordinates
b. Principal stresses
c. Mohr's circle of stresses

Figure 3.24 *Mohr's coordinates and Mohr's circle of stresses.*

plotted either upward or downward, as their sign has no meaning. The coordinates of a point (σ, τ) represent the combination of shear and normal stress on a plane regardless of the plane's orientation.

On this diagram are plotted the coordinates of σ_1 and σ_3 (Fig. 3.24b). Both lie on the σ axis, since the shear stresses on the principal planes are zero. Through these points a circle is drawn whose center is also on the σ axis (Fig. 3.24c). The center of this circle is at the point $\left(\dfrac{\sigma_1 + \sigma_3}{2}, 0\right)$ and its radius is equal to $(\sigma_1 - \sigma_3)/2$. A radius is drawn at an angle of 2α measured counterclockwise from the σ axis. The x-coordinate of a point on the circle at the end of the radius is

$$\frac{\sigma_1 + \sigma_3}{2} + \frac{\sigma_1 - \sigma_3}{2} \cos 2\alpha,$$

which is σ on a plane that is inclined at an angle of α with the major principal plane. The y-coordinate of the point is

$$\frac{\sigma_1 - \sigma_3}{2} \sin 2\alpha,$$

which is τ on the same plane. Therefore the circle represents the possible stress conditions on any plane perpendicular to the intermediate principal plane. The stresses on a particular plane at angle α can be found from the construction. From this construction it can be shown that maximum τ occurs on a plane with an angle of $2\alpha = 90°$ and is equal to $(\sigma_1 - \sigma_3)/2$, or half the difference between the major and minor principal stresses. Also the shear stresses on two planes perpendicular to each other are equal.

The same construction can be applied to stresses on a plane that is perpendicular to the major principal plane, by using σ_2 and σ_3, or to a plane perpendicular to the minor principal plane, by using σ_1 and σ_2. Since the circle is symmetrical about the x-axis, ordinarily only the top half of the circle is plotted. The bottom half can be plotted, using -2α (measured clockwise from the axis) and negative values for the shear stress.

The orientation of the planes is not directly shown on the Mohr diagram, but it should be pictured separately, Fig. 3.24c.

In three dimensions, Fig. 3.25, there are three circles, each representing the stresses on a plane perpendicular to one of the principal planes. The area between the circles represents the combined stresses on planes that are oblique to all three principal planes. These can be computed analytically using the same reasoning as employed for Equations (3:24a) and (b), as described in texts on the theory of elasticity and advanced strength of materials. The highest shear stress is defined by the $\sigma_1 - \sigma_3$ stress circle that describes the stress combinations in any plane that is oblique to the major

SEC. 3:10] COMBINED STRESSES

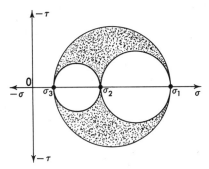

Figure 3.25 *Mohr's circles for 3-dimensional stress field defining all possible combinations of normal and shear stress.*

and minor principal planes and perpendicular to the intermediate. Therefore the $\sigma_1 - \sigma_3$ circle, and the planes it represents are of greatest significance in most real problems of soil strength and failure.

Mohr's circle of stresses was derived by the laws of statics and applies regardless of the material involved. Although the discussion was limited to stresses acting on the faces of cubes, it applies equally well to the infinitesimal cube we call a point.

Example 3:3

Given: $\sigma_1 = 10$ kg/cm² and $\sigma_3 = 2$ kg/cm²
Find: σ and τ on a plane making an angle of 30° with the major principal plane and perpendicular to the intermediate principal plane.
  1. Plot σ_1 and σ_3 on the σ-axis (Fig. 3.26a).
  2. Draw circle through points with center on axis.
  3. Construct radius with $2\alpha = 60°$.
  4. Scale σ and τ from diagram.
      $\sigma = 8$ kg/cm²; $\tau = 3.5$ kg/cm².

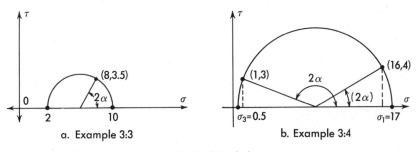

Figure 3.26 *Computing stresses with Mohr's circle.*

Example 3:4

Given: The normal and shear stresses on a plane are 16 psf and 4 psf, respectively; normal and shear stresses on a second plane are 1 psf and 3 psf, respectively. Compute the major and minor principal stresses and find the angle between the two planes.

1. Plot the stress coordinates of both planes (Fig. 3.26b).
2. Draw circle with center on axis through two points. (Center at intersection of perpendicular bisector of a line joining two points, with the σ-axis.)
3. σ_1 and σ_3 may be scaled directly from diagram.
   $\sigma_1 = 17$ psf; $\sigma_3 = 0.5$ psf.
4. The 2α of each plane is found from the diagram. The angle between the planes is $\frac{1}{2}(2\alpha_1 - 2\alpha_2) = 65°$.

3:11 Strain and Failure

When stresses are applied to any material, including soil, it first undergoes deformation. The nature of the deformation depends on the resistance of the material and the combination of stresses. If the stresses are increased further, a point is reached at which the material fails to resist the increase. At this point, termed *failure*, different materials, including soils, react differently—some disintegrate while others deform continuously with little or no stress increase. Soils with their three-phase composition exhibit a wider variety of deformation characteristics and more complex failure behavior than do the simpler engineering materials such as steel or concrete. Deformation is not always proportional to stress; it changes with time and environment. The resistance to failure depends on the stress field, the environment and time; further it is often difficult to define a point of failure. In spite of these differences between soil and rock and the simpler materials, the idealized concepts of applied mechanics such as modulus of elasticity, Poissons ratio, and Mohr's rupture theory can be utilized in soil and rock engineering. These concepts must be viewed as approximations, but if their limitations are understood, they are useful in solving real problems.

ELASTICITY / If a normal stress increment, $\Delta\sigma_z$ is applied to a prism of soil, Fig. 3.27a, it will deform an amount ΔH in the direction of the stress increase. The direct strain increment $\Delta\varepsilon_z$, is found from the deformation

$$\Delta\varepsilon_z = \frac{\Delta H}{H}. \tag{3:25a}$$

This is usually represented in a stress-strain curve, Fig. 3.27b. The ratio of the stress increment to the strain it produces is the *modulus* of *elasticity*, E. The general expression is

$$E = \frac{\Delta\sigma}{\Delta\varepsilon} \text{ or } \frac{d\sigma}{d\varepsilon} \quad \left(\text{or } E_z = \frac{\Delta\sigma_z}{\Delta\varepsilon_z} \right). \tag{3:25b}$$

SEC. 3:11] STRAIN AND FAILURE

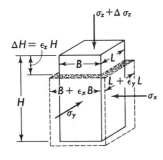

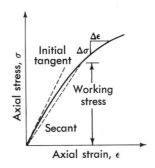

a. Strain under axial load increase, $\Delta\sigma_z$; σ_x and σ_y constant

b. Stress–strain curve, with initial tangent and secant modulus of elasticity as slopes of tangent and secant respectively

Figure 3.27 Strain and stress-strain relationships.

Geometrically it is the slope of the stress-strain curve. The modulus of elasticity of most soils and rocks is not a constant throughout the possible stress range and not quite the same for unloading as for loading. For this reason, the term modulus of deformation, M, defined in the same way as E, is used by some writers to call attention to the inelastic behavior of soils and rock. In this text, E will be used with the understanding that it is neither a constant nor a unique function of the load applied. The range in E for soils and rock is almost limitless, from virtually 0 for peats to higher than concrete for sound rock.

The value of E at very small strains, the *initial tangent*, E_T, is defined by a tangent to the stress-strain curve at its beginning, Fig. 3.27b. The average value of E for a specified range of stress is the *secant modulus*, E_s. The stress range is that for the particular problem (the working stress). Because E is a variable it is always necessary to define the stress field and stress range for which a particular value applies.

POISSON'S RATIO / The stress increment $\Delta\sigma_z$, also produces a bulging in the lateral dimensions, ΔB and ΔL, and corresponding lateral strains ε_x and ε_y. The ratio of the lateral to the direct strain is Poisson's ratio, ν

$$\nu = \frac{-\varepsilon_x}{\varepsilon_z} = \frac{-\varepsilon_y}{\varepsilon_z}. \qquad (3:26)$$

The range in Poisson's ratio for ideal elastic materials is between 0 and 0.5. The value 0.5 implies a material whose volume does not change under load, like jelly. The value 0 implies no bulging under load, like a prism of cork or sponge rubber. Rocks and soils generally lie between these limits, but ν exceeds 0.5 in certain cases.

FAILURE / Otto Mohr also contributed to engineering science a theory of the failure of materials that represents more nearly the true stresses involved than do the theories involving simple stresses alone. The theory has been found to apply particularly well to soils and to materials such as concrete and stone.

Mohr reasoned that yield or failure within a material was not caused by normal stresses alone reaching a certain maximum or yield point, or by shear stresses alone reaching a maximum, but by critical combinations of both shear and normal stresses. The failure is essentially by shear, but the critical shear stress is governed by the normal stress acting on the potential surface of failure.

The critical combinations of shear and normal stress, when plotted on the σ, τ coordinates form a line known as *Mohr's envelope of failure* (Fig. 3.28a). Failure will occur if for a given value of σ the shear stress exceeds that shown by the envelope.

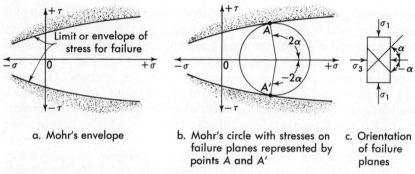

a. Mohr's envelope

b. Mohr's circle with stresses on failure planes represented by points A and A′

c. Orientation of failure planes

Figure 3.28 *Mohr's envelope of failure.*

If the stresses on any two planes through a point are known, the stresses can be found on any other planes by means of Mohr's circle. Since the circle represents *all* possible combinations of shear and normal stress at that point, failure will occur on the plane represented by the point of intersection with the envelope (Fig. 3.28b).

Example 3:5

A cylinder of soil cement has applied to it a minor principal stress of zero and a slowly increasing major principal stress. If the rupture envelope passes through (0,4000) at a slope of 20° upward to the right, compute: (a) the maximum value of axial load when failure occurs, (b) shear and normal stress on failure plane, and (c) the angle of failure plane.

1. At $\sigma_1 = \sigma_3 = 0$, Mohr's circle is a point. As σ_1 increases slightly, the circle enlarges (Fig. 3.29b).

SEC. 3:11] STRAIN AND FAILURE

2. At the instant the circle through (0,0) (σ_1,0) touches the envelope, failure will occur on the plane through the cylinder corresponding to the point of tangency (Fig. 3.29b).
3. From Mohr's diagram, $\sigma_1 = 11,500$ at failure.
4. $\tau = 5400$ and $\sigma = 3900$ on the failure plane. It is found that $2\alpha = 110°$; so, the failure plane makes an angle of $55°$ with σ_3 or an angle of $35°$ with the axis of the cylinder (Fig. 3.29c).

Note: The normal and shear stresses on the *failure* plane are usually designated p and s, respectively. The difference between the principal stresses, $\sigma_1 - \sigma_3$, at failure is termed q_c, the compressive strength.

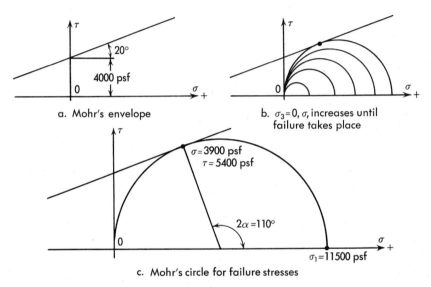

Figure 3.29 Stresses during compression test of soil cement.

ADVANCED FAILURE CRITERIA[3:12] / The Mohr theory of rupture implies that the intermediate principal stress has no influence on failure. High pressure testing has shown that this is not always true. Although the effect is small, research requires a method of representing failure that includes the intermediate. One form is the principal stress plot, Fig. 3.30 in which all three principal stresses are shown. The stress field can be represented by normal and shear stresses on the surface of an octahedron. The octahedral normal stress, σ_{oct}, is defined by,

$$\sigma_{oct} = \frac{\sigma_1 + \sigma_2 + \sigma_3}{3}. \tag{3:27a}$$

This is represented by a vector equidistant from the three perpendicular axes, the *hydrostatic axis*. The octahedral normal stress is equal to the length

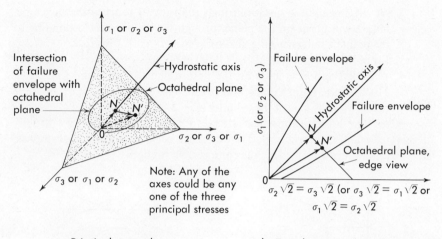

Figure 3.30 *Principal stress axes and stress representations.*

of the vector, ON, multiplied by $1/\sqrt{3}$. The second component of stress is the *octahedral shear*, τ_{oct}:

$$\tau_0 = 1/3\sqrt{(\sigma_1 - \sigma_2)^2 + (\sigma_2 - \sigma_3)^2 + (\sigma_1 - \sigma_3)^2}. \qquad (3:27b)$$

This can be represented by a vector perpendicular to the hydrostatic axis, NN'. The value $\tau_0 = NN'/\sqrt{3}$. The plane perpendicular to the hydrostatic axis is the *octahedral plane*. For any three dimensional system, eight such planes are possible. These define an octahedron, from which the name is derived.

For any given octahedral normal stress, failure occurs when the octahedral shear stress reaches a limiting value, determined by experiment. The limiting values define a three-dimensional envelope, centered about the hydrostatic axis, but not necessarily symmetrical to it.

Because of the problems of representing the three dimensional plot on paper, two-dimensional "cross sections" are frequently employed, Fig. 3.30b. These are limited to the cases where two of the principal stresses are equal.

The Mohr criteria for failure and the Mohr's envelope are sufficient for the solution of most problems in soil and rock engineering. The more sophisticated representations make it possible to evaluate the effects of the intermediate principal stress, and to determine the possible error involved in the Mohr approximation.

3:12 Methods of Making Shear Tests [3:13] [3:14]

Because of the complex nature of the shearing resistance of soils, many methods of testing have been tried with varying success. The principal shear tests in use today are *direct, ring or double direct*, and *triaxial*. Of these, triaxial testing gives the most consistent and reliable results with varying soils.

DIRECT SHEAR TEST[3:13] / One of the earliest methods for testing soil strength, used extensively today, is direct shear. A sample of soil is placed in a rectangular box (Fig. 3.31a) the top half of which can slide over

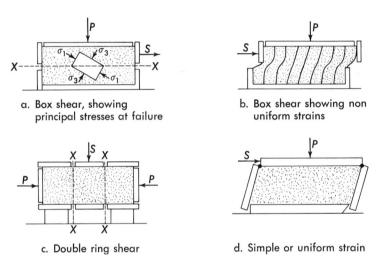

Figure 3.31 Direct shear tests.

the bottom half. The lid of the box is free to move vertically, and to it is applied the *normal load*, P. A *shearing force*, S, is applied to the top half of the box, shearing the sample along line x—x. In practice, the top and bottom of the box may be either porous plates to permit changes in the water content of the sample or projecting vanes to help develop a uniform distribution of stress on the failure surface. The test utilizes a relatively thin sample which consolidates rapidly under load (when such consolidation is required). The sample preparation and test operation are simple in most soils, which makes the test attractive for routine work.

Inherent shortcomings limit the reliability of the test results. First there is an unequal distribution of stresses over the shear surface; the stress is more at the edges and less at the center as shown in Fig. 3.31b. The result is progressive failure. In materials with highly developed structures, such as flocculent clays and cemented or very loose cohesionless soils, the strength indicated by the test will often be too low. Second the soil is forced to shear on a predetermined plane which is not necessarily the weakest one. The strength given by the test, therefore, may be too high. Finally it is difficult to control drainage or changes in water content during the test, which limits its usefulness in wet soils.

RING SHEAR / The ring shear is a double-direct shear test. A cylindrical sample is supported laterally by a close-fitting metal tube (Fig. 3.31c). Normal pressures are applied to the sample by pistons on the ends. A section of the tube is forced downward, shearing the soil on two surfaces, $x-x$. This equipment makes it possible to control sample water content changes more closely than in the single direct shear, and in addition small-diameter samples can be used. It suffers from the same limitations of non-uniform stress distribution and a forced failure plane as does the direct shear test.

SIMPLE SHEAR / Several devices have been designed to produce more uniform shear strains in direct shear. One form, Fig. 3.30d, employs tilting sides on the shear box. While it does develop uniform strain, it does not permit the localized increase in strain that accompanies failure. Further, it introduces stress concentrations within the mass. Another device employs a reinforced rubber tube for the walls of the shear box. This allows the irregular strains that develop in natural soils, but does not fully prevent lateral expansion. Both are research tools, and not well adapted to general shear testing.

PLANE STRAIN / The strains in the direct shear tests take place in two directions, vertically and parallel to the direction of shear. This condition is termed plane strain. It is similar to the strains developed in many real problems such as very long foundations or long walls which can be represented by a two-dimensional cross section.

TRIAXIAL SHEAR TEST[3:16] / The most reliable shear test is the triaxial direct stress (Fig. 3.32). A cylindrical sample is used with a diameter of 1.4 in., 2.8 in., or more and a length of at least twice the diameter. The sample is encased in a rubber membrane, with rigid caps or pistons on both ends. It is placed inside a closed chamber and subjected to a confining pressure σ_3 on all sides by air or water pressure. An axial stress σ_1 is applied to the end of the sample by a piston. Either the axial stress can be increased or the confining pressure decreased until the sample fails in shear along a diagonal plane or a number of planes. The Mohr circles of failure stresses for a series of such tests, using different values for σ_3, are plotted, and the Mohr envelope drawn tangent to them (Fig. 3.32c).

An alternate procedure is to hold the axial stress constant and increase the confining pressure until the sample bulges upward in the axial direction. In this form, the *triaxial extension* test, the confining pressure is $\sigma_1 = \sigma_2$ and the axial stress is σ_3. This is sometimes used to simulate the effect of a lateral thrust on a mass of soil. The Mohr envelope is similar to that for the compression test in a homogeneous isotropic soil; in stratified materials it is often different.

SEC. 3:12] METHODS OF MAKING SHEAR TESTS

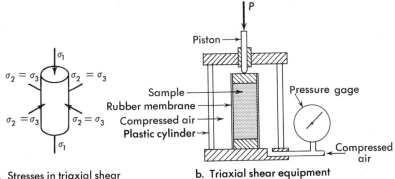

a. Stresses in triaxial shear

b. Triaxial shear equipment

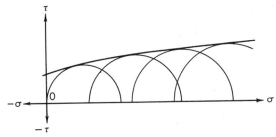

c. Mohr envelope drawn tangent to Mohr's circles of failure

Figure 3.32 *Triaxial shear test.*

A special case of the triaxial shear is the *unconfined compression test*, in which $\sigma_3 = 0$.

The important advantages of the method are the relatively uniform stress distribution on the failure plane and the freedom of the soil to fail on the weakest surface. Furthermore, water can be drained from the soil or forced through the soil during the test to simulate actual conditions in the ground. Sample preparation is simple, and small-diameter cylindrical samples can be used. The chief disadvantage is the elaborate equipment required, including sample membranes, compressed air or water pressure equipment, the triaxial cell itself, and auxiliary devices to measure the volume change of soil during testing. The conventional triaxial test utilizes rigid end caps. These restrain the shear and cause stress concentrations that change the conditions in failure. It is limited to values of $\sigma_2 = \sigma_3$ or $\sigma_2 = \sigma_1$ in compression and extension respectively.

Special triaxial tests have been developed that utilize rectangular specimens with independently controlled stresses on all three principal planes.[3:17] One variant of such testing prevents deformation in the direction of the intermediate principal stress, a state of plane strain useful in simulating the loading conditions in two-dimensional problems. A triaxial test in which the sample is in the form of a hollow cylinder can induce a variable intermediate principal stress by maintaining the inside pressure different from the outside. Such hollow cylinders have also been subjected to torsion in order to measure the shear modulus of elasticity directly.[3:18]

The rectangular, plane strain, and hollow cylinder triaxial tests are primarily research tools for studying the mechanisms of soil behavior, and minimizing some of the shortcomings of the conventional triaxial. In spite of the limitations, the triaxial tests utilizing $\sigma_2 = \sigma_3$ and axial compression continues to be the most useful method for simulating soil and rock behavior for a wide range of engineering problems.

3:13 Strain and Strength of Dry Cohesionless Soils

Cohesionless soils are composed of bulky grains, ranging in shape from angular to well rounded. A simplified representation of such a material subject to normal and shear forces, P and S, is shown in Fig. 3.33. The particles are in contact at only a few points at which the stresses are extremely high, far greater than the average stresses on the mass, $\sigma = P/A$ and $\tau = S/A$.

If the shear stress is increased the soil particle system responds in several ways simultaneously, Fig. 3.33. First, the particles deform more or less elastically. While it might seem that a solid grain of quartz is exceedingly

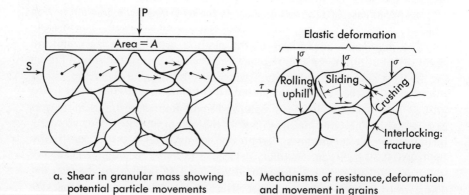

a. Shear in granular mass showing potential particle movements

b. Mechanisms of resistance, deformation and movement in grains

Figure 3.33 *Mechanisms of deformation and shear in a mass of bulky grains.*

rigid, even small stress changes in the mass induce high localized stresses and strains in each particle. Second, there is local crushing at the most highly stressed points of contact. Third, both the elastic distortion and the crushing cause slight translation and rotation of the grains, increasing the size of some of the voids and decreasing others. The vector sum of all the small movements of each particle is the deformation of the mass, ordinarily described in terms of strain.

Both the previous confining stress and the stress level at the beginning of the stress increment influence the strain. The higher the degree of confinement, the greater the previous crushing and local adjustments, and therefore the less the additional strain produced by additional increment of shear stress.

If the shear stress is increased still further, two additional responses are evident. First, the particles tend to roll across one another, Fig. 3.33b. The resistance depends on their angle of contact and is proportional to the confining stress σ. The total resistance to rolling is the statistical sum of the behavior of all the particles: some roll up, some roll down, but all do not move simultaneously. The second mechanism is sliding of one grain across the other. The resistance to sliding is essentially friction, which is proportional to the confining stress. A third mechanism involves the interference and interlocking of the corners of the more angular, irregular particles.

If the shear stress becomes sufficiently large, the statistical effect of the distortion, crushing, shifting, rolling and sliding of the grains will be continuous movement and distortion of the mass, or *shear failure*.

Moisture does not directly influence these mechanisms in most cohesionless soils because the intense stresses at the contact points between grains force the water molecules aside. In exceptional soils such as porous volcanic ash, or sands containing talc or chlorite, the particles may be weakened by moisture and thus their resistance to stress will be altered.

From the engineering point of view it is not necessary to identify nor evaluate quantitatively the contribution of each mechanism. Instead their combined effect can be described in terms of the stress-strain curve and the Mohr envelope of the soil as a whole.

STRESS-STRAIN / Typical stress-strain curves for a cohesionless soil subject to an increasing shear stress with a constant confining stress, σ_3, are shown in Fig. 3.34. Both exhibit strains that are approximately proportional to stress at low stress levels, suggesting a large component of elastic distortion. If the stress is reduced, the unloading stress-strain curve is nearly the same as the loading. Not all the strain is recovered upon unloading, indicating some particle re-orientation and point crushing. The *hysteresis loss*, the area of the stress-strain loop, represents the energy lost in crushing and re-positioning. The loose soil with larger voids and fewer points of contact exhibits greater strains and less recovery of strain upon unloading than the dense soil.

At higher shear stresses the strains are proportionally greater, indicating greater crushing and re-positioning. A plot of void ratio as a function of strain, Fig. 3.34 shows significant changes at higher stresses: a general expansion of a dense soil and contraction of a loose soil coinciding with the increasing strains of the stress-strain curves.

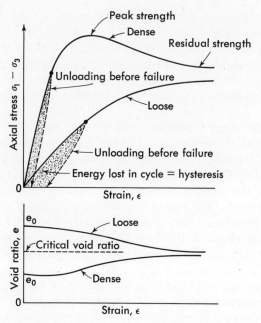

Figure 3.34 *Stress-strain-void ratio in a cohesionless soil at constant confining stress, σ_3.*

ELASTICITY / As defined earlier the modulus of elasticity is the slope of the stress-strain curve. From Fig. 3.34 it can be seen that for low shear stresses E is nearly constant, but with increasing shear it becomes smaller. The modulus of elasticity is also dependent on confining stress: the higher the stress level the greater the E for any given shear stress. A complete picture of E as a function of σ_3 and the shear stress (expressed in terms of σ_1/σ_3) for a typical sand is shown on Fig. 3.35. This can be approximated by the equation

$$E = C\sigma_3^n \qquad (3\!:\!28a)$$
$$C,n = f(\sigma_1/\sigma_3, D_r). \qquad (3\!:\!28b)$$

Typical values of n, the exponent, are between 0.8 and 0.5 for the range in stresses involved in engineering problems. Because of the wide variety of soils and stress conditions, it is not possible to estimate the value of E from a description of the soil or from simple tests such as grain size. It should be determined directly by tests that simulate the soil-structure stress field, and the level of strains that will be involved in any real situation.

POISSON'S RATIO / Like E, the Poisson's Ratio is not a constant. For the stress ranges involved in engineering (and well below failure) typical values lie between 0.25 and 0.40. Poisson's ratio must also be determined by tests utilizing the soil structure, stress level and strains involved in the real problem.

SEC. 3:13] STRAIN AND STRENGTH OF DRY COHESIONLESS SOILS

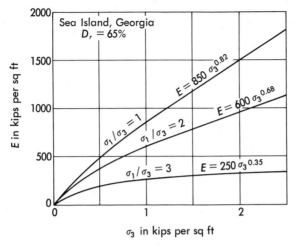

Figure 3.35 *Modulus of elasticity of a beach sand.*

STRENGTH / The stress conditions producing very large strains or failure can be defined by the Mohr envelope, Fig. 3.36. The results of innumerable tests on dry cohesionless soils show that the shear stress of failure, termed the *shear strength*, s, is nearly proportional to the normal effective stress on the failure surface, p'. (Note that the letters s and p' are used for the limiting values of τ and $\bar{\sigma}$.) The Mohr envelope for the test is approximately a straight line through the origin which makes an angle of φ with the σ-axis. The equation for soil strength is given by

$$s = p' \tan \varphi. \tag{3:29a}$$

The angle φ is termed the *angle of internal friction*. It is analogous to the angle of friction between two sliding bodies whose φ is the angle of sliding friction and $\tan \varphi$ the coefficient of friction. The angle of the failure plane

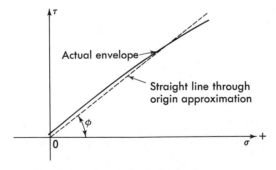

Figure 3.36 *Mohr's envelope for a cohesionless soil.*

can be found graphically from the Mohr's circle of failure or analytically from the geometry of the Mohr's circle:

$$\alpha = 45 + \frac{\varphi}{2}. \quad (3:29b)$$

As previously discussed, continuous strain or failure is the sum of grain distortion, crushing, shifting, rolling and sliding. The major factors in failure are rolling and sliding. The sliding resistance of the grains is determined by the effective stress, the coefficient of friction between the minerals, the surface roughness, and the angle of contact between the grains. These in turn depend on the grain shape and the soil structure as reflected in the relative density. The resistance to rolling depends on the particle shape, the gradation, and the relative density. As a result the angle of internal friction is greater than the angle of friction between the minerals, and it varies with grain shape, gradation, and relative density. The Mohr envelope is not always perfectly straight nor does it always pass through the origin because the resistance to rolling is present even with no confinement. At very high stresses the envelope may be curved concave downward owing to fracture of some of the grains. Typical values of the angle of internal friction are given in Table 3.4.

TABLE 3:4 / ANGLE OF INTERNAL FRICTION OF COHESIONLESS SOILS COMPOSED LARGELY OF QUARTZ

Description	Angle of Internal Friction	
	D_r less than 20	D_r over 70
Rounded, uniform	29	35
Rounded, well-graded	32	38
Angular, uniform	35	43
Angular, well-graded	37	45

At very high confining stresses, shear fracture of the individual grains becomes significant. The grain strength is more dependent on the shape of the particle and on the geometry of the load concentrations than on the confinement of the mass; therefore the fracture strength does not increase in proportion to the confining stress. The Mohr envelope flattens appreciably at high stresses and appears to approach a limiting shear strength. In quartz sands, broken hard rock and gravels, the flattening occurs when σ_3 lies between 500 and 1500 lb per sq in. or 35 to 100 kg per sq cm, approximately. If the grains are weak as in volcanic ash, cinders or weathered rock, the flattening of the envelope begins at lower stresses.

VOLUME CHANGES DURING SHEAR / As described by Figs. 3.33 and 3.34, grain movement accompanies strain and becomes greater as shear failure occurs. Although individual voids both enlarge and contract as the strain increases, the all-over effect depends on the initial density. As can be seen by the simplified representation of Fig. 3.37, the loose soil contracts and the dense soil expands. If shearing continues beyond the point of failure in the dense sand, and beyond the point of greatest change of strain, both the loose and dense sand approach the same void ratio, termed the *critical void ratio* or critical density, Fig. 3.34. A soil with an initial void ratio close to the critical does not change volume appreciably during shear. If the void ratio is less than the critical, the soil is properly described as *dense*; if the void ratio is more than the critical the soil is *loose*, and tends to be unstable.

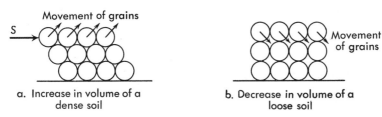

Figure 3.37 *Volume changes that accompany shear in cohesionless soils.*

The critical void ratio depends on the confining stress, but data relating the critical void ratio to σ_3 are conflicting. For stress levels involved in foundations, the critical void ratio corresponds to relative densities between 50 and 60 per cent.

PEAK AND RESIDUAL STRENGTH / The dense soil reaches a peak strength followed by a loss of strength (and increase in void ratio) with continuing strain. The strength remaining after large strain is termed the *residual* strength. The complete picture of the strength of a cohesionless soil that is initially dense requires two Mohr envelopes: one representing the peak strength, and a second, lower envelope representing the residual.

3:14 Shear in Wet Cohesionless Soils

As stated earlier, the total stress applied to a soil is sustained by grain structure stress which is σ' (*effective stress*) and by *neutral* or *water stress, u*. At any time the stresses can be represented by the relation

$$\sigma = \sigma' + u. \qquad (3:12c)$$

Since shearing resistance is a friction phenomenon, it depends on grain structure stress; therefore at failure the equation for shearing strength of moist sand must be written:

$$s = p' \tan \varphi, \qquad (3:29)$$

$$s = (p - u) \tan \varphi, \qquad (3:30)$$

where p' is the *effective normal stress* at failure. Failure to recognize this relationship has resulted in many misconceptions of the variation of internal friction with moisture. Experiments have shown φ to be almost unchanged by moisture—the real change occurs in the effective stresses that produce friction and shearing resistance.

HYDROSTATIC NEUTRAL STRESS / Many cases of hydrostatic neutral stress exist in nature, largely owing to the water table and its fluctuations. In a dry sand the vertical normal stress is caused by the unit weight of the sand itself; so, at a depth z the resistance to horizontal shear is given by:

$$\sigma' = \sigma = \gamma z, \tag{3:31a}$$

$$s = \gamma z \tan \varphi. \tag{3:31b}$$

If the water table rises through this soil, the total unit weight of the soil increases to γ_s, the saturated unit weight, but now

$$\sigma' = \gamma_s z - u, \tag{3:31c}$$

$$s = (\gamma_s z - u) \tan \varphi. \tag{3:31d}$$

If u is sufficiently large, the shear strength may be reduced to a negligible amount. Landslides are often caused by neutral stress that builds up until the soil shear strength is incapable of sustaining the applied loads. A shear failure can result and large masses of soil can be placed in motion under such circumstances.

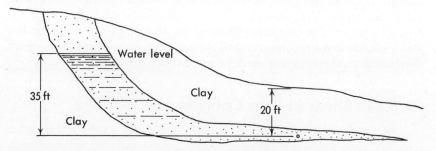

Figure 3.38 *Hydrostatic neutral stress in a hillside* (Example 3:6).

Example 3:6

Given: Clay $\gamma_s = 110$ lb/cu ft. Compute shear strength of sand at point x.

1. $\sigma = 20 \times 110 = 2200$ psf.
2. $u = 35 \times 62.4 = 2180$ psf.
3. $\sigma' = 20$ psf.
4. $s = 20 \tan \varphi$ (practically zero).

SEC. 3:14] SHEAR IN WET COHESIONLESS SOILS

NEUTRAL STRESSES ACCOMPANYING SEEPAGE / High neutral stress often accompanies seepage because of the head differences described by Darcys law, Equation 3:9. If the effective stress should become zero as a result, the soil will lose all its strength and become *quick*. In this condition it is like a heavy fluid. It will not resist stress, and if not confined, will flow. This is defined as *quicksand*. It is not a particular type of soil, but instead is a condition, often temporary but always dangerous.

A quick condition is most likely to develop in fine sands and cohesionless silts because only a small quantity of seepage is involved with a large head difference. However, it can form in coarse sands and gravels if the flow is sufficiently great.

Because of their peculiar behavior, quicksands are thought by many people to be endowed with magical powers. For example, horror movies sometimes show the villan slowly sinking out of sight in quicksand. Instead, a person will float easily in quicksand. The human body with its unit weight of about 63 lb per cu ft or 1.01 g per cu cm is buoyed up by the saturated sand that typically weighs 100 to 130 lb per cu ft or 1.6 to 2.1 g per cu cm.

Example 3:7

An excavation 20 ft deep in an alluvial deposit 30 ft thick is protected by steel sheet piling driven through silt underlain by sand to a pervious gravel and boulder stratum, Fig. 3.39. The boulders prevent driving the piling

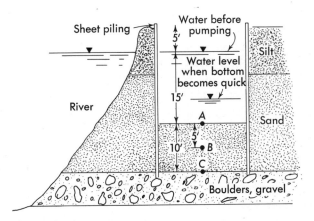

Figure 3.39 Upward seepage and quicksand.

deeper. A river near the excavation intersects the gravel-boulder stratum and water seeps upward through the sand into the excavation so that the water levels are equal. The excavation is made by a clamshell bucket working in water. After the final level is reached, the contractor attempts to pump the excavation dry. However, when the water level has been lowered 10 ft the sand in the excavation becomes quick. What is happening? The silt and sand weigh 125 lb per cu ft and the gravel is so pervious that head loss through it is negligible.

1. After excavation the vertical stresses in the sand are as follows:

Point	Total, psf	Neutral, psf	Effect., psf
A	$15 \times 62.4 = 936$	$15 \times 62.4 = 936$	0
B	$15 \times 62.4 + 5 \times 125 = 1621$	$20 \times 62.4 = 1248$	373
C	$15 \times 62.4 + 10 \times 125 = 2186$	$25 \times 62.4 = 1562$	624

2. After pumping the water down 10 ft the vertical stresses are:

Point	Total, psf	Neutral, psf	Effect., psf
A	$5 \times 62.4 = 312$	$5 \times 62.4 = 312$	0
B	$5 \times 62.4 + 5 \times 125 = 937$	$1562 - \frac{1}{2}(1562 - 312)^* = 937$	0
C	$5 \times 62.4 + 10 \times 125 = 1562$	$25 \times 62.4 = 1562$†	0

* Uniform gradient between the boulder gravel and excavation bottom, through the sand.
† From river (no head loss in the boulder-gravel stratum).

The entire sand stratum has zero effective vertical stress and therefore zero strength. This also is a *quick condition* and the sand is *quicksand*. Quicksand caused by construction without proper water pressure control is a relatively common occurrence. Such failures can be prevented by reducing the neutral stress by drainage, by cutting off the connection to the river by driving the sheet piling through the boulder-gravel or by sealing the boulder-gravel with a grout injection. Problems of neutral stress and accompanying seepage are discussed more fully in Chapter 4.

NEUTRAL STRESS THROUGH VOLUME CHANGE / As a cohesionless soil is sheared, its volume changes; if the soil is saturated, the volume change must be accompanied by a change in the distribution of water in the voids. If shear and the change in volume occur so slowly that the movement of water requires negligible head, there will be only insignificant changes in neutral stress. Quick shear, however, requires rapid changes in water content that develop tremendous neutral stresses. This is particularly important in the very fine-grained soils of low permeability.

In dense cohesionless soils void expansion accompanies shear as shown in Figs. 3.34 and 3.37. Negative neutral stresses are developed in the pore water in the same manner as negative pressures are produced by pulling a cork from a tightly stoppered bottle. The limiting negative stress in a saturated soil is the maximum capillary tension as determined by the soil void diameter, Eq (3:5). The effect of the negative neutral stress is a temporary increase in the strength of the soil.

In loose sands the opposite occurs. The voids decrease in volume, inducing a positive neutral stress. The limiting value of u is the confining pressure on the soil, p. The limiting effective stress (p-u) is zero, and the strength is zero. This is another case of quicksand, temporary, but just as serious as that produced by hydrostatic conditions.

In either loose or dense soils, the pore pressure change induced by shear is initially limited to the zone of shear. The temporary gain or loss of strength is thus, at first, localized. Depending on the soil permeability and the length and geometry of potential seepage paths, the neutral stress can persist and spread through the mass until eventually it is dissipated by seepage. This is a serious problem in a very fine loose sand or silt. A localized shear failure within the mass generates an increased pore pressure. The pore pressure increase is transmitted through the continuous body of water in the voids beyond the limits of the initial failure. This weakens the soil beyond the zone of initial failure, and creates additional failure and more pore pressure. Thus failure propagates throughout a mass of loose saturated soil although the initial stress conditions produced failure at only an isolated point. This is essentially a chain reaction or " snow balling " effect that can prove disastrous. Some of the most devastating landslides, such as the talus avalanch that demolished a large town in the Peruvian Andes, have commenced with local shearing in a mass of loose, saturated broken rock. On a smaller scale, a failure commenced as the result of an insignificant sewer trench excavation, but the chain reaction caused sliding of an entire hillside, and destruction of the construction site.

Increases in neutral stress leading to failure can also be generated by repeated small loads.[3:19] Each load repetition produces a small strain which, in turn, causes an increment of neutral stress. When the soil is fine-grained and the mass sufficiently large that the neutral stress cannot be immediately dissipated, the small, individually negligible increments of stress add up to significant changes. If the soil is initially very dense, the negative neutral stress increases its resistance. If the soil is loose, and (in some cases even moderately dense) the cumulative small increases in pore pressure weaken the soil, permitting larger strains and increasingly larger volume changes that eventually cause failure.

The cyclic loading produced by vibrating machinery, even when small, can induce a substantial pore pressure build up and eventual loss of soil strength. The repeated strains accompanying earthquakes have similarly produced increasing pore pressures in sand deposits and sufficient loss in strength that large masses of soil have become quick or *liquified*. In the large earthquake at Niigata, Japan in 1963, the liquefaction of a sand deposit caused a large apartment building to drop suddenly about one storey and to tilt more than 30 degrees. Similarly, other disastrous earth movements accompanying earthquakes have been traced to sand liquefaction.

The number of cycles of load required to produce liquefaction in a cohesionless mass depend on a number of factors: the soil's initial relative density, the stress increment compared to failure stress, the permeability, and the geometry of the seepage paths. In general the greater the stress increment and the looser the soil, the fewer the cycles of load required for liquefaction. Limited data suggest that there is little possibility of liquefaction regardless of the number of cycles or the load increment if the relative density exceeds 70 per cent.

Blasting and the shock waves generated by the impact of excavating machinery also can cause liquefaction in very loose sands and cohesionless silts where only a few cycles of repeated stress are sufficient to build up large pore pressures. Such deposits must be considered as potential hazards to any construction operation. They must be corrected for any permanent structure.

NEUTRAL STRESS FROM CAPILLARY TENSION / Capillary tension can be the cause of negative neutral stress that increases soil shear strength. Moist sand owes its ability to pack and maintain a shape to capillary tension in thin water films between the grains. The small meniscus radii develop high tensile stresses in the moisture wedges that hold the grains in rigid contact (Fig. 3.40).

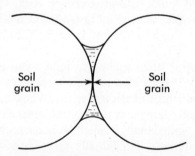

Figure 3.40 Neutral stress from capillary tension producing compression between soil grains.

Fine sand and silt above the ground water table within the zone of capillary rise owe their strength to the capillary tension and the resulting effective stresses in the soil structure. Frequently, deep excavations can be made in such soils with very steep side slopes because of this strength. If the soil should dry out completely or should become inundated, the capillary tension will be destroyed and the strength reduced. Many failures can be traced to such a loss of strength from a reduction in capillary tension.

A sample of saturated fine sand or silt will hold its shape when subjected to an unconfined compression test because capillary tension produces a positive effective σ_3. If loose, it will finally collapse when the load causes a reduction in the volume of the voids and a buildup of neutral stress. If dense, it will expand and develop even greater capillary tension.

3:15 Strain and Strength of Saturated Cohesive Soils [3:3, 3:16, 3:12]

Shear in a saturated cohesive soil (a clay) is more complex than in a sand or gravel. Like the cohesionless soil, the clay is made up of discrete particles which must slide or rotate for shear to take place. However, there are a number of significant differences. First the soil is relatively compressible; therefore, when a load is applied to the saturated clay, it is initially supported by neutral stress and is not transmitted to the soil structure. Second the permeability of the clay is so low that the neutral stresses produced by the load are dissipated very slowly. Therefore it may be months or even decades before the soil structure feels the full stress increase. Third there are significant forces developed between the particles of clay by their mutual attraction and repulsion.

RATE OF LOADING / Because of the slow changes in the neutral stress and the corresponding slow changes in effective stress, the strength of clays is defined in terms of neutral stress dissipation. Three basic conditions are defined.

DRAINED (ALSO TERMED CONSOLIDATED-DRAINED OR SLOW (S)) SHEAR: The confining and the shear stresses are applied so slowly that the neutral stress is not changed by the added loads; the applied stress produces an equal increase in effective stress; and the soil consolidates fully.

CONSOLIDATED-UNDRAINED (ALSO TERMED CONSOLIDATED-QUICK (R) SHEAR: The confining stress is applied so slowly that the neutral stress is not changed and the soil consolidates fully under the increased effective stress. The shear stress, however, is applied so quickly that neutral stress carries all this change and there is no further consolidation or increase in effective stress.

UNDRAINED (ALSO TERMED UNCONSOLIDATED-UNDRAINED, OR QUICK (Q)) SHEAR: Both the confining and shearing stresses are applied so rapidly that the neutral stress carries all the added load and there is no change in the water content.

DRAINED SHEAR—CONSOLIDATION AND STRESS-STRAIN / In drained shear there is no neutral stress change and any increase in total stress produces a corresponding increase in effective stress. The soil consolidates, reducing the void ratio and water content.

The consolidation of the soil takes place in two stages: first during the addition of the confining stress and second during the addition of the axial stress that produces shear. Although the mechanism is similar to the one-dimensional consolidation discussed in Section 3:6, the stress-void ratio curve is different because the stress fields are different. In the conventional consolidation test where $\varepsilon_2 = \varepsilon_3 = 0$, the lateral stress, $\sigma_2 = \sigma_3$ is a constant fraction, K, of the vertical stress σ_1. Successive Mohr circles for the increasing loads in such a consolidation test are given in Fig. 3.41a. The

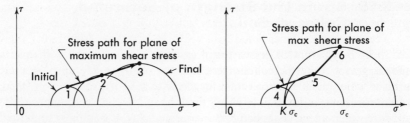

a. Successive stresses for increasing load with a constant ratio of axial to confining stress, as in simple consolidation test, 1-2-3

b. Increasing load with constant ratio of axial to confining stress, 4-5 followed by increasing axial stress at constant confining, 5-6

Figure 3.41 *Stress path for plane of maximum shear stress in triaxial compression.*

successive positions of a point on the circles that represent a particular plane in the soil are termed the *stress path* for that plane.[3:20] The stress path is the locus of the combinations of shear and normal stresses resulting from a loading sequence on the soil. The particular plane of interest may be the plane of maximum shear (Line 1–2–3) or the plane on which failure will eventually occur.

In the drained test the stress path can be different. The initial consolidation is either hydrostatic, with $\sigma_1 = \sigma_2 = \sigma_3$, or anisotropic with $\sigma_1 > \sigma_2 = \sigma_3$. After consolidation the lateral stress is held constant, and the vertical increased, Fig. 3.41b. The stress paths for the plane of maximum shear stress (Line 4–5–6) (or for the potential failure plane) are different from the path for simple one dimensional consolidation.

The results of the three-dimensional consolidation can be expressed in several ways, Fig. 3.42: a stress settlement curve similar to Fig. 3.10b, or a

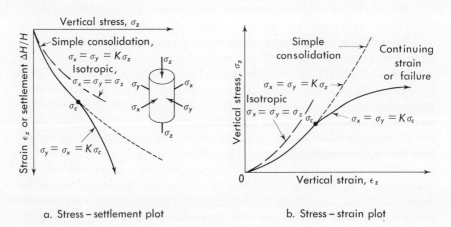

a. Stress – settlement plot

b. Stress – strain plot

Figure 3.42 *Stress-strain in drained or slow shear: Simple consolidation followed by increasing axial stress.*

stress-strain curve similar to that for a cohesionless soil. The strain includes both the elastic deformation of the soil mass and the void ratio change.

The simple one-dimensional consolidation test involves a special stress path, Fig. 3.41a, that describes the stress increase in relatively thin compressible strata confined between more rigid strata. The stress path for thick strata of compressible soils is more nearly like that of Fig. 3.41b, and the stress settlement curve, Fig. 3.42a, is somewhat different. Therefore, if the laboratory test is to give an accurate indication of real settlement, the stress path of the laboratory test should be the same as for the field load. Following the real stress path in the laboratory is not always practical. The conventional single dimension test is used with due allowance for any errors involved.

STRENGTH IN DRAINED SHEAR / As a result of consolidation the water content and interparticle spacings are reduced and interparticle bonds are increased in proportion to the confining stress that overcomes the resistance to compression. The strength, therefore, increases in proportion to the effective confining stress increase. The Mohr envelope is a straight line through the origin, Fig. 3.43.

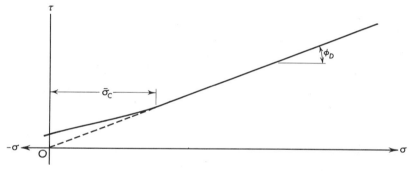

Figure 3.43 Mohr's envelope for saturated clay in drained (consolidated-drained or slow, S) shear.

The angle of the Mohr envelope is termed the *angle of shear resistance* or *apparent internal friction* and is denoted φ_D or φ_{CD}. Typical values lie between 15° and 30°. The higher angles are usually associated with clays having plasticity indexes of 5 to 10, while the lower values are for clays having plasticity indexes of from 50 to 100. This is a verification of the effect of particle repulsion and adsorbed water on the interparticle bonds, for the high PI indicates high adsorption and repulsion, large interparticle spacing, and correspondingly less interparticle attraction.

When a clay has been preconsolidated to a stress of σ'_c and then unloaded, the particles do not return to their original spacing and previously higher void ratio. As a result the interparticle attractive force is not reduced and the strength at stresses less than the preconsolidation load is no longer proportional to the effective confining pressure but is somewhat higher (Fig. 3.43).

The strength above the preconsolidation load is given by the expression

$$s = p' \tan \varphi_D. \tag{3:32a}$$

Below the preconsolidation load the strength must be obtained directly from the Mohr diagram. This curved portion of the envelope can be approximated by a straight line having the equation

$$s = c' + p' \tan \varphi'. \tag{3:32b}$$

In this expression c' is the intercept on the τ axis and φ' is the angle that the straight line makes with the σ axis.

The drained shear condition represents the strength of the soil developed by a long-term stress change. However, it can be used for any problem involving shear in saturated clays by determining the effective stress at failure from the total and neutral stress. It is particularly useful in analyses involving complex changes in loading and water pressure. The drained shear test is time consuming; however, the Mohr envelope for this condition can usually be approximated from the consolidated-undrained test results.

CONSOLIDATED-UNDRAINED SHEAR / In consolidated-undrained shear, the soil consolidates completely under the confining stress σ_3, with a corresponding reduction in void ratio and water content. The axial load is then increased rapidly by an amount $\Delta\sigma_1$ without further changes in void ratio or water content until failure occurs. The total major principal stress at failure is $\sigma_1 = \sigma_3 + \Delta\sigma_1$, and the total minor principal stress is σ_3. Since no drainage or consolidation occurs from the added load $\Delta\sigma_1$, it is supported entirely by neutral stress, or $\Delta u = \Delta\sigma_1$. Water pressure at any point is the same in all directions, according to the laws of hydrostatics; therefore the neutral stress produced by $\Delta\sigma_1$ is exerted in the direction of both σ_1 and σ_3. The effective stresses at failure are therefore

$$\sigma'_1 = \sigma_1 - \Delta u = \sigma_3 + \Delta\sigma_1 - \Delta\sigma_1 = \sigma_3,$$
$$\sigma'_3 = \sigma_3 - \Delta u = \sigma_3 - \Delta\sigma_1.$$

A plot of these effective stresses will give the Mohr envelope for drained shear. However, if the total stresses are plotted (Fig. 3.44), a different envelope will be produced because the circles are shifted horizontally to the right by $\Delta\sigma_1$. The apparent Mohr envelope of total stresses will also be a straight line through the origin above the preconsolidation load and will have an apparent angle of shear resistance, φ_{CU}, which is about half of φ_D. The equation for shear strength above the preconsolidation load is

$$s = p \tan \varphi_{CU}. \tag{3:33}$$

A consolidated-undrained test is frequently employed to obtain φ_D, using a plot of effective rather than total stresses. The test is less time consuming than the drained, and the computed effective envelope is approximately the same as that obtained in drained shear. Measurements of neutral stress can

SEC. 3:15] STRAIN AND STRENGTH OF SATURATED COHESIVE SOILS

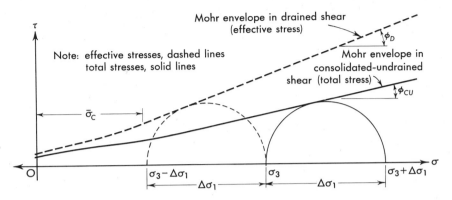

Figure 3.44 *Mohr's envelope for saturated clay in consolidated-undrained (consolidated-quick, R) shear.*

be made in the soil during the test to aid in plotting the effective envelope, but these require considerable skill and complex apparatus.

Consolidated-undrained shear represents a clay soil that is first fully consolidated by the weight of a structure which is later subjected to a sudden increase in stress by the construction of an addition or by an abnormal live load. It is frequently employed in the analysis of embankment foundations where construction lasts longer than the period required for the soil to consolidate significantly.

PORE PRESSURE / The previous analysis of effective stresses assumes that the axial stress increment $\Delta\sigma_1$ is entirely reflected in pore pressure change, Δu. This implies that the volume change of the water under load is negligible compared to that of the soil structure. However in some soils many rocks and certain conditions of loading this is not true. A more general relationship, including the effect of a changing σ_3 is

$$\Delta u = A\,(\Delta\sigma_1 - \Delta\sigma_3). \tag{3:34}$$

In this relationship A is the *pore pressure coefficient* that describes the effect of the changing difference between the principal stresses.[3:16] For many unconsolidated saturated clays A is approximately 1. For highly overconsolidated clays or dense clay-sand mixtures, an increasing shear stress, described by $\Delta\sigma_1 - \Delta\sigma_3$, produces an increase in volume similar to the volume increase of a dense cohesionless soil during shear. For such a soil, $A < 0$. For slightly overconsolidated clays and some compacted fills A lies between 0.25 and 0.75. In clays with a highly developed flocculent structure, shear can produce a structural breakdown similar to the shear of a loose cohesionless soil. In such sensitive clays, the value of A can be greater than 1. The correct value of A for a soil engineering problem can be determined only by tests that measure the neutral stress due to a loading that approximates the stress path of the soil mass.

UNDRAINED SHEAR / In undrained shear, both the confining and shear stresses are applied so rapidly that no consolidation takes place. The soil void ratio and water content remain unchanged, and neutral stress supports all the added loads. The soil initially supported an overburden pressure, σ'_0 (or a preconsolidation load σ'_c), under which it consolidated to establish its void ratio, water content, and interparticle spacing. The soil strength resulting from this initial effective stress can be obtained from σ'_0, using the Mohr envelope of drained shear shown in Fig. 3.45. An increased

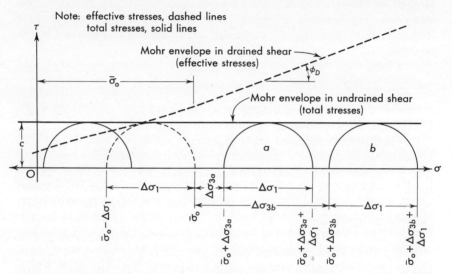

Figure 3.45 Mohr's envelope for saturated clay in undrained (unconsolidated-undrained, quick, Q) shear.

confining pressure, $\Delta\sigma_3$, is supported by neutral stress, and the void ratio, interparticle spacing, and resulting soil strength remain unchanged. An increased axial load, $\Delta\sigma_1$, also is supported by neutral stress, and likewise produces no change in void ratio or water content. The stress conditions during loading are tabulated below, assuming that the pore pressure coefficient, $A = 1$.

Loading	Total Stress	Neutral Stress	Effective Stress
Overburden	$\sigma_1 = \sigma'_0$	$u = 0$	$\sigma'_1 = \sigma'_0$
	$\sigma_3 = \sigma'_0$*	$u = 0$	$\sigma'_3 = \sigma'_0$*
Adding confinement, $\Delta\sigma_3$	$\sigma_1 = \sigma'_0 + \Delta\sigma_3$	$u = \Delta\sigma_3$	$\sigma'_1 = \sigma'_0$
	$\sigma_3 = \sigma'_0 + \Delta\sigma_3$	$u = \Delta\sigma_3$	$\sigma'_3 = \sigma'_0$
Added axial load, $\Delta\sigma_1$	$\sigma_1 = \sigma'_0 + \Delta\sigma_3 + \Delta\sigma_1$	$u = \Delta\sigma_1 + \Delta\sigma_3$	$\sigma'_1 = \sigma'_0$
	$\sigma_3 = \sigma'_0 + \Delta\sigma_3$	$u = \Delta\sigma_1 + \Delta\sigma_3$	$\sigma'_3 = \sigma'_0 - \Delta\sigma_1$

* In many cases the minor principal stress from the overburden load will be less than the major principal stress, but this does not alter the neutral stress effects described in the table.

The effective minor principal stress is independent of the added confining stress σ_3, and therefore the effective major principal stress at failure and the strength depend only on the original overburden stress σ'_o and the effective (drained shear) envelope. A plot of the total stresses, the solid lines on Fig. 3.45, shows a series of Mohr circles. All have the same diameter (since they are in reality the same circle), and the resulting envelope of total stresses is a horizontal straight line. As can be seen from the diagram, the intercept of the envelope on the τ-axis is approximately equal to the shear strength of the soil in its original condition, consolidated by the overburden stress, σ'_o. The intercept is denoted c and is called the *cohesion* of the soil. The strength of the soil under undrained conditions can be expressed by the equation

$$s = c. \tag{3:35a}$$

The apparent angle of friction, φ_U, is zero. However, the angle of the failure plane α is determined by Equation 3:29b, using φ_D, and is not 45° as might be assumed with $\varphi_U = 0$.

The undrained strength represents the existing strength of a natural soil. Since most construction proceeds rapidly compared with the rate of clay consolidation, undrained strength is used in most problems of design. Even where construction is so slow that some strength increase will develop, the undrained strength is frequently used because it is the minimum strength and therefore conservative. Caution must be exercised in using the undrained shear in the analysis of problems where the final stress is less than the original overburden load, such as the design of excavation bracing or in the study of landslides. For short-term conditions where the soil does not have sufficient time to expand, the undrained strength applies, but for long-term conditions, the soil becomes weaker and the use of undrained strength is unsafe.

The undrained strength depends on the original overburden stress, σ'_o or σ'_c and on the drained Mohr envelope. In a compressible soil such as a clay the overburden stress is related to the void ratio by the stress-void ratio curve. As a result the undrained strength of a saturated clay increases with decreasing void ratio and also decreasing water content. For normally consolidated clays, a graph of the logarithm of undrained strength plotted as a function of either water content or void ratio is approximately a straight line.

UNCONFINED COMPRESSION TEST OF SATURATED CLAY / Since the undrained strength is the same regardless of the confining pressure, the strength can be determined with zero confining pressure. The strength is given by

$$s = \frac{1}{2}q_u = c, \tag{3:35b}$$

where q_u, the unconfined compressive strength, is equal to σ_1 at failure when $\sigma_3 = 0$.

SENSITIVITY / If a sample of undisturbed saturated clay is completely remolded without changing its water content, and then tested, it will be found that the undrained strength has been reduced. This is caused by a breakdown in the soil structure and a loss of the interparticle attractive forces and bonds. In clays with a dispersed structure, the loss is small, but in clays with a highly flocculent structure or soils with a well-developed skeletal structure, the loss in strength can be large. The ratio of the undisturbed to the remolded strength is defined as the *sensitivity*, S_t:

$$S_t = \frac{c \text{ (undisturbed)}}{c \text{ (remolded)}} = \frac{q_u \text{ undisturbed}}{q_u \text{ remolded}}. \qquad (3:36)$$

TABLE 3:5 / TYPICAL VALUES OF SENSITIVITY

Clays of medium plasticity, normally consolidated	2–8
Highly flocculent, marine clays	10–80
Clays of low to medium plasticity, overconsolidated	1–4
Fissured clays, clays with sand seams	0.5–2

The sensitive clay reaches a *peak strength*, similar to that of a dense sand, and then becomes weaker with increasing strain, Fig. 3.46. The strength remaining after large strains is the *residual strength* and is approximately equal to the remolded.

STRESS-STRAIN IN UNDRAINED SHEAR / The relation of strain to stress in undrained shear involves little or no volume change, but only distortion of the mass (Poisson's ratio is nearly 0.5). The shape of the curve, Fig. 3.46, depends largely on the interparticle bonding imposed by preconsolidation and by the structure. For undisturbed clays the initial portion of the curve is straight, probably reflecting the distortion of the bonds. The curve flattens as increasing numbers of bonds are broken. The same soil remolded without a water content change has a flatter stress-strain curve.

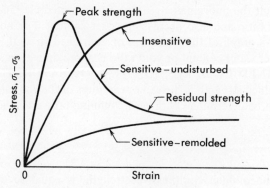

Figure 3.46 Stress-strain in clays in undrained shear.

The difference between the undisturbed and remolded soil is greatest in sensitive clays. Because of their flocculent structure and edge to face bonding, they are relatively rigid and have much higher values of E for a given water content than an insensitive clay with a more oriented structure. When the structure is broken by remolding, the E is only a fraction of the undisturbed value.

ANISOTROPIC STRENGTH / Many soils, because of stratification or orientation of the minerals, are stronger and more rigid in one direction than another. This anisotropy must be considered in evaluating the elasticity and strength. The soil is tested in different orientations; a minimum program includes shear parallel to and perpendicular to the stratification. The strength is represented by two Mohr envelopes, a minimum, usually for shear parallel to the orientation and a maximum, perpendicular to it. Others can be determined for intermediate conditions.

CLAYS WITH FISSURES / Some clays in nature develop cracks or fissures from desiccation, high overburden stresses which produce local fracture, or physicochemical alteration and weathering. Often the fissures appear to be closed, but they still remain as planes of weakness and paths of seepage. The strength of such clays is dependent on the orientation of the cracks and fissures and on the effect of changing stress and water percolation on the clay along the fissures. Tests of the intact clay between the fissures are misleading; large enough samples must be used so that the fissures are included. Drained tests are best, since the fissures permit more rapid dissipation of neutral stress than in ordinary clays. If the cracks have a particular orientation, minimum and maximum Mohr envelopes can be developed similar to those for anisotropic materials.

3:16 Strength of Partially Saturated Cohesive Soils

Shear of partially saturated cohesive soils involves the same forces as for saturated cohesive soils. However, the neutral stress in the soil pores is a complex combination of capillary tension and gas pressure which depends on the degree of saturation and the size of the voids. The effective stresses are difficult if not impossible to determine in partially saturated soils, and so the envelope of total stresses is usually employed to express test results (Fig.3.47). The Mohr envelope is ordinarily curved, with an intercept on the τ-axis and with a decreasing slope at increasing normal stresses. The intercept is probably the combined result of capillary tension in the voids and interparticle bonds from preconsolidation. The initially steep slope results from soil consolidation under the increasing confining pressure and is comparable to the drained shear of a saturated clay. As the soil consolidates under increasing pressure, however, the degree of saturation increases, capillary tension decreases, and positive pore pressures eventually develop. This is

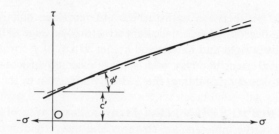

Figure 3.47 *Mohr's envelope of total stresses for partially saturated clay.*

comparable to undrained shear and results in the Mohr envelope approaching a horizontal asymptote. The strength for any confining pressure is read directly from the envelope. For convenience the curved envelope is often approximated by a straight line having the equation

$$s = c' + p \tan \varphi', \qquad (3:37)$$

where c' is the intercept on the τ-axis and φ (or sometimes φ') is the *angle of shear resistance*. More than one straight line can be used to approximate any given curved envelope, depending on which part of the envelope is of most importance in that particular case. Therefore c' and φ' should be considered to be empirical constants and not properties of the soil. The approximate angle of the failure plane, α, can be found graphically or by Equation 3:29b.

Partially saturated clays often become saturated from high rainfall or a rising ground water table. Therefore the strength of a partially saturated clay should not be used in analyzing practical problems unless the soil remains in that condition. Frequently partially saturated clays are first soaked in water and then tested as saturated clays to obtain data for design. Saturation is aided by *back pressure*, neutral stress added with increasing confinement.

3:17 Strength of Cemented Soil and Rock [3:18, 3:22]

Most rocks and cemented soils consist of relatively rigid mineral grains bonded directly by interparticle crystal bonds or by a cementing material between the grains. When load is applied the bonds distort. The grains also distort, but generally to a smaller degree.

In all cases there will be some reduction in volume. The reduction will be far greater in those materials having significant open voids than in the solid rocks such as marble in which the volume change occurs in the mineral crystals.

Increasing shear stress produces increasing strains with a high, nearly constant E for early stages in loading, reflecting the stretching of the crystal or cementing bonds. Failure generally occurs rather suddenly with breaking of the bonds.

SEC. 3:17] STRENGTH OF CEMENTED SOIL AND ROCK

STRENGTH OF ROCK / The Mohr envelope for most intact rocks and cemented soils, Fig. 3.48a is similar to that for partially saturated soils, but it is initially steeper and more sharply curved with some tensile strength and a substantial shear strength with no confining pressure. This envelope can be approximated by a straight line, with strength parameter c' and φ' as for the partially saturated clay. Sometimes the envelope is so curved that a reasonable approximation requires two straight lines and two sets of parameters, c_1 and φ_1 as well as c_2 and φ_2, for two ranges in normal stress.

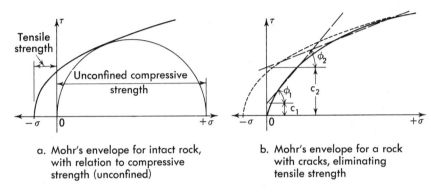

a. Mohr's envelope for intact rock, with relation to compressive strength (unconfined)

b. Mohr's envelope for a rock with cracks, eliminating tensile strength

Figure 3.48 Mohr's envelope for rock.

EFFECT OF DEFECTS / Many rocks are anisotropic. Moreover, the stress-strain behavior and failure of most rock formations is controlled by the joints, bedding planes and other defects or discontinuities. Therefore a realistic evaluation of both strength and elasticity must include these defects, and consider their orientation. Families of Mohr envelopes are defined as for fissured or anisotropic soils. Across the joints the tensile strength is zero, and the shear strength without confinement is small, Fig. 3.48b. At high levels of confinement the effect of the defects is small unless the cracks are filled with soil or other materials different from the rock.

Shear through unweathered joints and bedding planes resembles that for dense cohesionless soils. The peak strength reflects the irregularities of the surfaces and the residual strength reflects the rock to rock friction.

The stress-strain behavior may be largely the result of closing of the cracks, particularly at low confining stresses. In such materials the value of E is erratic and low at low confining pressures. It may suddenly increase when the mass strain is sufficient to close the cracks.

EFFECT OF PORE PRESSURE / There is little information on the effect of pore pressure on rocks. The more porous rocks such as sandstones with void ratios exceeding 0.2 behave as soils. The neutral stress coefficients, N, range between 0.75 and 1. The pore pressure coefficient, A, is usually substantially less than 1 because of the rigidity of the grain structure.

Intact solid rocks such as marble and granite with void ratios of less than 0.03 show little effect of pore pressure, suggesting that the neutral stress coefficient, N, is close to zero. For intermediate void ratios, there are probably intermediate values of N; however there are insufficient data to define the transition. The problem is largely academic, because of the cracks in most rocks. The rock as a whole, including cracks, behaves as if N is 1.

3:18 Creep

Most of the data available on strain and strength of soils and rock is based on laboratory tests conducted for a relatively short period of time, generally a few minutes to a few hours; rarely more than a week. In real situations, however, the stress is maintained for years or centuries. Such long-term loads have been observed to produce continuing small strains, a phenomenon termed *creep*.

Creep is observed in most materials under sustained high stress. For metals it is a critical factor in high temperature design. For soils so many environmental factors are significant in producing strain that creep, although suspected, is difficult to identify. However, limited research has shown that continued strain at constant stress levels does occur in soils and rocks and must be considered in design.

A typical plot of strain as a function of time is shown on Fig. 3.49.[3:23] For a low level of constant shear stress, the strain increases at a decreasing rate, and approaches a limit as shown by Curve A. The strain measured in a conventional laboratory test is only a portion of the ultimate strain. The modulus of elasticity for the conventional test, E, is therefore greater than the ultimate, E_u. Limited data suggest that for low stresses E_u/E is a constant.

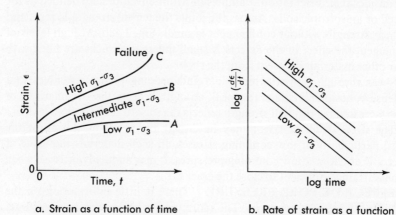

a. Strain as a function of time for three levels of constant shear stress

b. Rate of strain as a function of time on a log-log plot

Figure 3.49 *Creep in soils or rock.*

SEC. 3:19] ENGINEERING PROPERTIES OF THE MASS 153

At higher stresses the strain continues indefinitely at a decreasing rate as shown by Curve *B*. A plot of the logarithm of strain as a function of the logarithm of time is approximately a straight line for any given stress level, Fig. 3.49b.

At stresses close to failure as defined by a conventional test, the strain rate first decreases with increasing time and then increases sharply, ending in failure, as shown in Curve *C*, Fig. 3.49a.

The stress range subject to creep, τ_c, appears to be approximately the same for many clays.

$$0.3s < \tau_c < 0.9s. \tag{3:38}$$

In this expression *s* is the shear strength for a given confining stress as determined by conventional tests lasting a few hours at the most. The shear stress at which creep becomes significant, $\tau_c = 0.3s$ is sometimes termed the *threshold* stress. Generally, the stresses employed in foundations are below $0.3s$ to slightly above it. Therefore, creep is seldom significant in properly designed foundations. The stress levels in earth retaining structures, open excavations, and embankments, however, are usually within the creep range. Therefore, the effect of continuing strain must be considered in design.

3:19 Engineering Properties of the Mass

The solution of engineering problems involving soil and rock requires quantitative data on the engineering properties of the mass involved. Most testing, however, is limited to a portion of the mass. Generally that portion is infinitessimal compared to the total. Two questions are involved, therefore, in evaluating the mass properties from those of the test samples.

First, are the samples tested representative of the mass? This can only be answered by a sufficient number of tests to define the range of values for any property statistically. This is seldom possible, however. Instead, the testing should concentrate on the poorer materials, for they are likely to control any design. If a sufficiently large body of data is available, some statistical treatment is justified. Design for safety is frequently based on the poorest 10 to 25 percentile of the values obtained: one-tenth to one-fourth of the data are poorer and the remainder better. This, of course, is advisable only if the poor and good materials have a random distribution through the mass. Otherwise, if the poorest occur in a well-defined pattern, the poor zone is treated as a separate entity. If the probable performance of the mass is required and the variations occur at random, then the median of the data are utilized. Of course, this is meaningless unless the possible deviation from the median is also defined.

The second is what are the combined effects of the stratification, cracks, planes of weakness and other geometric and structural aspects of the mass?

These are seldom evaluated in the testing. Large-scale field tests can sometimes determine the effect of these structural details, or they can show how the defects distort the engineering properties deduced from the laboratory tests. In most cases, experience and intuition (judgment) are required to interpret the laboratory test data so they can be used in the solution of engineering problems.

REFERENCES

3:1 A. Hazen, "Water Supply," *American Civil Engineers Handbook*, John Wiley & Sons, Inc., New York, 1930.

3:2 K. Terzaghi and R. B. Peck, *Soil Mechanics in Engineering Practice*, John Wiley & Sons, Inc., New York, 2nd ed., 1967.

3:3 T. W. Lambe, "The Engineering Behavior of Compacted Clay," *Journal of the Soil Mechanics and Foundations Division, Proceedings, ASCE*, **84**, SM 2, May 1958.

3:4 T. K. Tan, "Discussion on Structure Mechanics of Clay," *Proceedings, Fourth International Conference on Soil Mechanics and Foundation Engineering*, **3**, London, 1957, p. 87.

3:5 Reference 3:2, p. 73.

3:6 A. W. Skempton, "Notes on the Compressibility of Clays," *Quarterly Journal of Geological Society*, **100**, London, 1944, p. 119.

3:7 G. F. Sowers, "Soil and Foundation Problems in the Southern Piedmont Region," *Proceedings, ASCE*, **80**, Separate 416, 1953.

3:8 G. F. Sowers, "Shallow Foundations," *Foundation Engineering*, McGraw-Hill Book Co., Inc., New York, 1961.

3:9 T. H. Wu, W. D. Resendiz, R. J. Neukischner, "Consolidation by a Rate Process Theory," *Journal of the Soil Mechanics and Foundations Division, Proceedings, ASCE*, **92**, SM 6, November 1966.

3:10 W. G. Holtz and H. J. Gibbs, "Engineering Properties of Expansive Clays," *Transactions, ASCE*, **120**, 1956.

3:11 G. F. Sowers and C. M. Kennedy, "High Volume Change Clays of the Southeastern Coastal Plain," *Proceedings, Third Pan American Conference on Soil Mechanics and Foundation Engineering*, **II**, Caracas, 1967, p. 99.

3:12 R. N. Yong and B. P. Warkentin, *Introduction to Soil Behavior*, The Macmillan Company, New York, 1966.

3:13 G. F. Sowers, "Strength Testing of Soils," *Laboratory Shear Testing of Soils*, Special Technical Publication 361, ASTM, 1963, p. 3.

3:14 *Laboratory Shear Testing of Soils*, Special Technical Publication 361, ASTM, Philadelphia, 1963.

3:15 *Symposium on Direct Shear Testing of Soils*, ASTM Special Technical Publication 131, American Society for Testing Materials, Philadelphia, 1953.

3:16 A. W. Bishop and D. J. Henkel, *The Measurement of Soil Properties in the Triaxial Shear Test*, 2nd ed., Edward Arnold, Ltd., London, 1962.

3:17 H. K. Ko and R. F. Scott, "Deformation of Sand at Failure," *Journal of the Soil Mechanics and Foundations Division, Proceedings, ASCE*, **94**, SM 4, July 1968.

3:18 B. B. Mazanti and G. F. Sowers, "Laboratory Testing of Rock Strength," *Testing Techniques for Rock Mechanics*, Special Technical Publication 402, ASTM, Philadelphia, 1966.

3:19 H. B. Seed and K. L. Lee, "Cyclic Stress Conditions Causing Liquefaction," *Journal of the Soil Mechanics and Foundations Division, Proceedings, ASCE*, **93**, SM 1, January 1967.

3:20 T. W. Lambe, "Stress Path Method," *Journal of the Soil Mechanics and Foundations Division, Proceedings, ASCE*, **93**, SM 6, November 1967, p. 304.

3:21 H. B. Seed and C. K. Chan, "Structure and Strength Characteristics of Compacted Clays," *Journal of the Soil Mechanics and Foundations Division, Proceedings, ASCE*, **85**, SM 5, October 1959.

3:22 *Testing Techniques for Rock Mechanics*, Special Technical Publication 402, ASTM, Philadelphia, 1966.

3:23 A. Singh and J. K. Mitchell, "General Stress-Strain-Time Function for Soils," *Journal of the Soil Mechanics and Foundations Division, Proceedings, ASCE*, **94**, SM 1, January 1968, p. 21.

SUGGESTIONS FOR ADDITIONAL STUDY

1. D. W. Taylor, *Research on the Consolidation of Clays*, Massachusetts Institute of Technology, Cambridge, 1942.
2. *Proceedings of Fifth International Conference on Soil Mechanics and Foundation Engineering*, **1**, Paris, 1961.
3. *Proceedings of Sixth International Conference on Soil Mechanics and Foundation Engineering*, **1**, Montreal, 1965.
4. *Proceedings of Seventh International Conference on Soil Mechanics and Foundation Engineering*, **1**, Mexico City, 1969.
5. *Proceedings of the Research Conference on Shear Strength of Cohesive Soils, ASCE*, 1960.
6. T. W. Lambe, *Soil Testing for Engineers*, John Wiley & Sons, Inc., New York, 1951.
7. *Proceedings of Specialty Conference: Design of Structures to Control Settlement, ASCE*, New York, 1964; See also *Journal of the Soil Mechanics and Foundations Division, Proceedings, ASCE*, **90**, SM 5, September 1964.
8. R. F. Scott, *Principles of Soil Mechanics*, Addison-Wesley Publishing Co., Reading, Massachusetts, 1963.

9. T. W. Lambe and R. V. Whitman, *Soil Mechanics*, John Wiley & Sons, Inc., New York, 1969.

PROBLEMS

3:1 a. Compute the maximum capillary tension in grams per square centimeter and pounds per square foot in a tube 0.001 mm in diameter.

b. Compute the height of capillary rise in the tube in feet.

3:2 Compute the capillary tension in pounds per square foot and the theoretical height of capillary rise in feet in a soil whose D_{10} is 0.002 mm if the effective pore diameter is about $1/5 D_{10}$.

3:3 Compute the height of capillary rise in feet in a sand whose D_{10} is 0.2 mm if the effective pore diameter is $1/5 D_{10}$.

3:4 A sample of soil in a permeability test is 5 cm in diameter and 12 cm long. The head difference is 25 cm and the flow is 1.5 cm^3 in 5 min. Compute coefficient of permeability in centimeters per second and feet per the minute.

3:5 Given a block of soil 12 cm long and 6 sq cm in cross-section. The water level at one end of the block is 20 cm above a fixed plane and at the other end is 3 cm above the same plane. The flow rate is 2 cm^3 in 1.5 min. Compute the soil permeability in feet per minute.

3:6 A canal and a river run parallel, an average of 150 ft apart. The elevation of water in the canal is El. 618 and in the river El. 595. A stratum of sand intersects both the river and the canal below their water levels. The sand is 5 ft thick and is sandwiched between strata of impervious clay. Compute the seepage loss from the canal in cubic feet per second per mile if the sand's permeability is 0.063 cm per sec.

3:7 A wood crib filled with earth serves as a temporary cofferdam across a river to lower the water level in a construction site. The water level upstream is 20 ft above the rock stream bed and downstream is 5 ft above the stream bed. The cofferdam is 200 ft long across the river and is 30 ft wide upstream to downstream. It is filled with well-graded, silty, sandy gravel having a coefficient of permeability of 0.0005 cm per sec. Estimate the seepage through the cofferdam in gallons per hour (the unit in which construction pumps are rated). *Hint:* Assume that the average cross-section of the water flowing through the cofferdam is the average of the intake area (20 × 200) and the outlet (5 × 200).

3:8 The stress void-ratio curve for a saturated clay is shown on Fig. 3.13a. Compute compression index C_c. Find the change in void ratio from the curve if the stress increases from 1000 to 10,000 psf. Find the change in void ratio from the curve if the stress changes from 10,000 to 100,000 psf. Recompute the change in void ratio in both cases, using Equation 3.16, and compare with the values found directly from the curve. Explain the differences.

3:9 A consolidation test had the following results:

σ	e	σ	e
250 psf	0.755	4000 psf	0.740
500	0.754	8000	0.724
1000	0.753	16,000	0.704
2000	0.750	32,000	0.684

a. Plot the stress void-ratio curves on semilog coordinates.
b. Compute the compression index.
c. If the initial soil stress is 1400 psf and the soil stratum is 8 ft thick, how high can the stress become before the ultimate settlement is $\frac{3}{4}$ in. ?

3:10 Consolidation tests on samples of soil yield the following void ratios for 100 per cent consolidation:

σ	e	σ	e
100 psf	1.85	16,000 psf	1.22
500	1.82	32,000	1.05
1000	1.77	10,000	1.10
2000	1.68	2000	1.20
4000	1.56	500	1.28
8000	1.39	100	1.38

a. Plot the stress void-ratio curves on both arithmetic and semi-log coordinates.
b. Compute the compression index, C_c.
c. Find the change in void ratio when the soil stress is raised from 1650 psf to 2700 psf.
d. If the soil stratum in (c) is initially 6.8 ft thick, compute its settlement.
e. If the soil has a coefficient of consolidation of 0.02 sq ft per day and the stratum in (d) is drained on both sides, compute the time required for 25, 50, and 75 per cent consolidation.

3:11 A soil has a compression index, C_c, of 0.31. Its void ratio at a stress of 2600 psf is 1.04, and its permeability is 3.5×10^{-8} cm per sec.
a. Compute the change in void ratio if the soil stress is increased to 3900 psf.
b. Compute the settlement in (a) if the soil stratum is 16 ft thick.

c. Find the time required for 25, 50, 75, and 90 per cent of settlement in (b) to occur.

3:12 Given a major principal stress of 8 kg per cm² and a minor principal stress of 1 kg per cm²; draw the Mohr's circle. Find the maximum shear stress and the normal and shear stresses on a plane that makes an angle of 60° with the major principal plane.

3:13 Given a major principal stress of 12,000 psf and a minor principal stress of 3000 psf; draw the Mohr's circle. Find the maximum shear stress and the normal and shear stresses on a plane that makes an angle of 60° with the minor principal plane.

3:14 The normal stresses on two perpendicular planes are 18 and 3 kg per cm² and the shear stresses are 6 kg per cm². Find the major and minor principal stresses graphically.

3:15 Given a major principal stress of 4 kip per sq ft and a minor principal stress of 1 kip per sq ft, find the maximum shear stress and the angle of the plane on which it acts.

3:16 The shear and normal stresses on one plane are respectively 2000 psf and 7000 psf, and on a second plane, 4000 psf and 3000 psf.
   a. Find the principal stresses.
   b. Find the shear and normal stresses on a plane making an angle of 30° with the major principal plane.

3:17 Given the normal stresses on two perpendicular planes as 3500 psf and 1200 psf, the shear stresses on each as 2300 psf, draw Mohr's circle.
   a. Can tension occur on any plane with this stress condition?
   b. Find the principal stresses.
   c. What are the shear and normal stresses on a plane making an angle of 74° with the direction of the major principal stress?

3:18 Given a major principal stress of 7500 psf, find the minimum value of the minor principal stress to limit shear stresses to 3200 psf.

3:19 A cylinder of concrete is tested in the ordinary manner and is found to have a "compressive" strength of 3450 psi. The failure plane makes an angle of 63° with the major principal plane.
   a. Draw Mohr's circle for the concrete at failure. (The minor principal stress is zero.)
   b. Draw the Mohr rupture envelope, assuming it to be a straight line.
   c. Find the "compressive" strength (difference between the principal stresses) if the minor principal stress is 1000 psi.

3:20 Given φ of a sand, derive the algebraic relation between φ and α.

3:21 Given φ of a sand, derive the algebraic expression for the ratio of the major principal stress to the minor principal stress when failure in the sand occurs.

PROBLEMS

3:22 A sample of sand subjected to a triaxial shear test failed when the minor principal stress was 3200 psf and the major principal stress was 11,500 psf. Draw Mohr's circle, and find φ and α.

3:23 A sample of sand in a direct shear test fails when the normal stress is 6 kip per sq ft and the shear stress is 4 kip per sq ft. Find the angle of internal friction and the principal stresses at failure.

3:24 Given the following stress conditions in a dense, angular, well-graded sand:

	Plane A	Plane B
Shear stress	1 kip/sq ft	1 kip/sq ft
Normal stress	3.5	2.2

Will failure occur?

3:25 A cylindrical sample of saturated rock flour composed of extremely fine-grained bulky particles is subjected to an unconfined compression test. The minor principal stress is developed by capillary tension in soil pores that have an effective diameter of 0.00075 mm. The angle the failure plane makes with the minor principal stress is 65°.
  a. Draw Mohr's circles for both total and effective stresses.
  b. Find φ and the compressive stress necessary to produce failure.

3:26 A soil stratum 30 ft thick overlies a bed of shale. The water table is 15 ft above the surface of the shale and the height of capillary rise is 10 ft. The soil has a void ratio of 0.35 and a specific gravity of solids of 2.65. Draw diagrams showing the total, neutral, and effective vertical stresses in the deposit. (Remember that above the water table the neutral stress is negative, which denotes tension.)

3:27 A thin seam of sand lies inclined at an angle of 30° and intersects the base of a cliff. The drainage of the sand is stopped by accumulated talus. The sand is overlaid by clay 50 ft thick and 3 ft of topsoil. The ground surface slopes at 30° and there is a deep vertical crack extending through the clay 60 ft from the face of the cliff. The clay and topsoil weigh 110 lb per cu ft and the angle of internal friction of the sand is 40°. How high must the water rise in the sand before the block of clay slides upon the sand layer?

3:28 The sand in Example 3:7 has a k of 75×10^{-3} cm per sec. Compute the pumping rate that will produce a quick condition, and that required to lower the water level to 5 ft below the surface. The excavation is 20 ft by 20 ft.

3:29 A saturated clay in consolidated-undrained shear had a φ_{CU} of 12°. Find the approximate value of φ_D graphically.

3:30 A saturated clay in drained shear was found to have a φ_D of 25°. Find the approximate value of φ_{CU} and the approximate unconfined compressive strength if the overburden pressure is 1200 psf.

3:31 Derive by means of Mohr's circle the relation between the major and minor principal stresses when given c from a quick shear test of a saturated clay.

3:32 Given the following data from an unconfined compression test of saturated clay:

Stress (psf)	Strain
0	0
2000	0.0035
4000	0.0080
6000	0.0170
7000	0.0270
8000	0.0650

a. Plot the stress-strain curve.
b. Find the shear strength, c.
c. Find the average modulus of elasticity for 40 per cent of the failure stress.

3:33 A sample of sandstone was tested in triaxial shear and found to have a cohesion, c, of 2000 psi and an angle of internal friction of 37 deg.

a. What will its compressive strength be in an unlined tunnel where the confining stress is 0 and the water pressure is 150 lb per sq in.?
b. If the water pressure should suddenly increase to 300 lb per sq in. because of the water hammer generated by a sudden shut down of the turbines, what would the compressive strength be?

3:34 A sample of closely laminated schist has a strength defined by $c = 600$ lb per sq in. and $\varphi = 42$ deg for shear perpendicular to the laminations and a lower strength defined by $c = 400$ lb per sq in. and $\varphi = 33$ deg for shear parallel to the laminations.

a. Plot the Mohr envelopes for this rock.
b. Find the range in the angle α between the major principal plane and the plane of the laminations for which failure when the minor principal stress is 1000 lb per sq in. will occur parallel to the laminations. (Hint: Draw failure circles for both conditions. The values of α for the larger circle that intersect the lower envelope define the range where failure will occur with equal ease by both mechanisms.

CHAPTER 4
Seepage, Drainage and Frost Action

THE DISTRICT REPRESENTATIVE OF A NATION-WIDE MANUFACTURING CONcern purchased what appeared to be an ideal site for a warehouse and distribution center. It was in a small valley close to a railroad and an arterial highway. After construction began, however, it was found that a high ground water table hampered work, and ground water flowing into some of the footing excavations made concreting impossible. The first contractor quit the job in despair, and another contractor, who tried to take over the work, finally lost his business. The manufacturing concern had no alternative but to drop the entire project, abandon the property, and purchase a new site. The cause of all this trouble was uncontrolled ground water—a difficulty that the businessman untrained in engineering could scarcely have recognized, and a difficulty that might have been corrected by proper seepage control.

A warehouse for frozen orange juice in a hot climate suddenly began to rise above its original level. The movement was not uniform, and it tore the walls and damaged the insulated floor which was supported by a 3-ft-thick fill of fine sand on the silty sand virgin soil. The damage to the floor insulation aggravated the rate of movement, and a heave of more than 12 in. developed within a year. The floor insulation had been incorrectly installed and the soil below became chilled. This increased the capillary tension, causing moisture to flow upward and freeze. Ice lenses accumulated in the soil, and eventually the ground became frozen to a depth of 12 ft. Reversal of the heat flow, produced by heating the ground with hot air ducts corrected the difficulty temporarily. The ultimate solution of the problem was to replace the insulation and to maintain some flow of heat into the soil to balance the loss into the warehouse.

Water is the ingredient of soils that fluctuates with time and the season; as it changes, the soil's strength or volume may change correspondingly. Control of the water content, control of the movement of water, and prevention of the damage caused by the movement of water in soils are vital aspects of soils engineering. They present problems in making excavations, constructing roads and airports, designing earth dams and levees, and building safe foundations.

The energy possessed by a particle of water is in three forms: *potential energy*, owing to its height; *pressure energy*, owing to the pressure; and *kinetic energy*, owing to its velocity. (In the flow of water through soils, the velocities are so low that the kinetic energy is practically zero.) Energy in water is usually expressed as *head*—a linear dimension such as feet that actually means foot-pound per pound. Since energy is only relative, head must always be expressed with relation to some fixed point, usually an arbitrary datum plane. The head possessed by water in soils is manifested by the height, h, to which water will rise in a small tube or standpipe above the fixed datum plane, as shown in Fig. 4.1a. This height to which water rises is often termed the *piezometric level, piezometric surface* or *phreatic surface* and it is a measure of the total energy of the water.

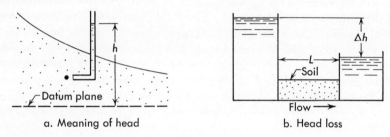

a. Meaning of head   b. Head loss

Figure 4.1 *Head and head loss.*

If at two different points within a continuous mass of water there are different amounts of energy, then there will be movement of the water particles toward the point of lesser energy, and the difference in head (energy) is used up in the work of moving the water. Darcy's law expresses the head loss, Δh, required to move water through soil at a rate q, a distance of L, by the formula

$$\Delta h = \frac{qL}{kA}, \qquad (4:1)$$

which is simply formula 3:9 rewritten (Fig. 4.1b). Of course this implies laminar flow, which is ordinarily the case in all soils except coarse gravels.

4:1 Saturated Flow: The Flow Net

The flow of water through saturated soil can be represented pictorially by *flow lines* (Fig. 4.2a), which are the paths taken by the moving particles of water. Water tends to follow the shortest path from point to point but at the same time makes only smooth curves when it changes direction. The flow lines, therefore, are curved, somewhat parallel lines, like loosely

SEC. 4:1] SATURATED FLOW: THE FLOW NET

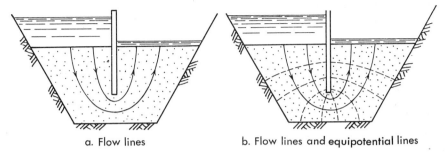

a. Flow lines   b. Flow lines and equipotential lines

Figure 4.2 *Flow net of seepage beneath sheet piling.*

stretched bundles of rubber bands, that extend from points of greater head to points of lesser head. In many cases the curves are segments of ellipses or parabolas.

The different amounts of energy or head can be represented on the same picture by *equipotential lines* (Fig. 4.2b), which are lines that indicate points having equal heads. The equipotential lines can be thought of as contours of equal energy; the flow lines cross them at right angles, since the water moves from higher energy levels to lower energy levels along paths of maximum energy gradient in the same way water flows down a hillside from higher levels to lower levels, following the steepest paths.

The pattern of flow and equipotential lines is termed the *flow net*, and it is a powerful tool for the solution of seepage problems.

DERIVATION OF THE FLOW NET / The mathematical expression for the flow net is derived on the basis that the soil is saturated, that the volume of water in the voids remains the same during seepage, and that the coefficient of permeability is the same at all points and in any direction at any point. The basic equation of seepage, Darcy's law (Eq. (2:8)), is resolved into *x*- and *y*-components:

$$q_x = k i_x A_x,$$
$$q_y = k i_y A_y,$$
$$i = \frac{\Delta h}{\Delta L} = \frac{dh}{dL}.$$

The seepage velocity v is the rate of seepage divided by the area of flow, and so the equations can be rewritten:

$$v = \frac{q}{A},$$
$$v_x = k \frac{\partial h}{\partial x},$$
$$v_y = k \frac{\partial h}{\partial y}.$$

The flow through a small element of soil having the dimensions of dx, dy, and 1 is shown in Fig. 4.3a and is expressed as follows:

In: $\qquad v_x\, dy + v_y\, dx.$

Out: $\qquad \left(v_x + \dfrac{\partial v_x}{\partial x}\, dx\right) dy + \left(v_y + \dfrac{\partial v_y}{\partial y}\, dy\right) dx.$

If the volume of water in the voids remains constant, then the flow *in* equals the flow *out*; so, equating the above expressions and collecting terms,

$$\frac{\partial v_x}{\partial x} + \frac{\partial v_y}{\partial y} = 0.$$

By substituting the equations for velocity, the relation becomes

$$\frac{\partial^2 h}{\partial x^2} + \frac{\partial^2 h}{\partial y^2} = 0. \qquad (4:2)$$

This is the Laplace equation of mathematical physics which describes the energy loss through a resistive medium. It represents two sets of lines, each set containing an infinite number of parallel curves and with each curve of one set intersecting each curve of the other at right angles, as shown in Fig. 4.3b. The *equipotential lines* comprise one set and the *flow lines* the other, and the entire pattern is the *flow net*.

FLOW NET CONSTRUCTION / The two-dimensional flow net derived above is a useful representation of the seepage patterns through earth dams, into large excavations, and below retaining walls and masonry structures. Unfortunately the Laplace equation can be integrated mathematically for only a few very simple conditions, and in practice the flow net must be obtained by other methods.

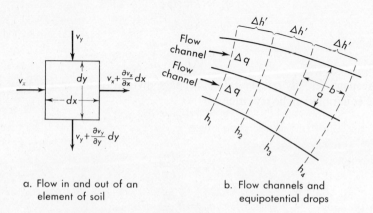

a. Flow in and out of an element of soil

b. Flow channels and equipotential drops

Figure. 4.3 *Physical meaning of the flow net.*

SEC. 4:1] SATURATED FLOW: THE FLOW NET

The graphical procedure of Forcheimer is simple and is applicable to any problem of steady flow in two dimensions. The space between any pair of flow lines is a *flow channel*. If a certain number of flow channels, N_f, is selected so that the flow through each, Δq, is the same, then

$$\Delta q = \frac{q}{N_f}.$$

The head loss between any pair of equipotential lines is the *equipotential drop* $\Delta h'$. If a certain number of equipotential drops are selected, N_D, so that all are equal,

$$\Delta h' = \frac{\Delta h}{N_D}.$$

The width of any one element of such a flow net is a and the distance between the equipotential lines is b, as shown in Fig. 4.3b. (The third dimension is 1.) The gradient and discharge are given by

$$i = \frac{\Delta h'}{b} = \frac{\Delta h / N_D}{b},$$

$$\Delta q = k\left(\frac{\Delta h / N_D}{b}\right) a.$$

The total discharge for the net, whose third dimension is 1, is expressed by

$$q = \Delta q N_f = k\,\Delta h \left(\frac{a}{b}\right) \frac{N_f}{N_D}. \tag{4:3a}$$

The ratio of (a/b) is fixed by the ratio of N_f/N_D and is the same throughout the net. If N_f and N_D are selected so that $a = b$, the equation for discharge (for a unit dimension perpendicular to the flow net) is

$$q = k\,\Delta h \frac{N_f}{N_D}. \tag{4:3b}$$

This is termed a *square net* because all the intersections between the sides are at right angles and the average length and width are equal. However, it should be noted that the term *square* is used in a descriptive sense because the opposite sides of the figures are not necessarily equal and they are seldom straight lines.

The first step in constructing a flow net is to make a scale drawing (Fig. 4.4a) showing the soil mass, the pervious boundaries through which water enters and leaves the soil, and the impervious boundaries that confine

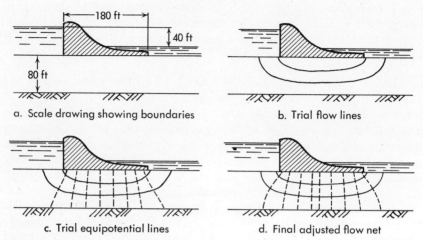

Figure 4.4 Steps in constructing a flow net.

the flow. Second, two to four flow lines are sketched, entering and leaving at right angles to the pervious boundaries and approximately parallel to the impervious boundaries (Fig. 4.4b). Third, equipotential lines are drawn at right angles to the flow lines (Fig. 4.4c) so that the length and width of each figure will be equal. Of course this will be impossible on the first attempt because the positions of the flow lines were only guessed, but the resulting net will guide the second attempt. Fourth, the flow lines and equipotential lines are readjusted so that all the intersections are at right angles and the length and width of each figure are equal (Fig. 4.4d). Between one pair of equipotential lines the figures may work out to be rectangles. However, each rectangle should have the same ratio of a/b. The resulting equipotential drop is a fraction of the others.

The quantity of seepage is computed by Equation (4:3b), using the values of N_f and N_D found by the graphical trial and revision. This is multiplied by the third dimension, perpendicular to the plane of the flow net, to get the total seepage.

Much practice is necessary to develop skill in drawing flow nets, and many cycles of trial and revision are required for an accurate solution.

Example 4.1

Compute the quantity of seepage under the dam in Fig. 4.4 if $k = 0.0003$ fpm and the level of water upstream is 60 ft above the base of the dam and downstream is 20 ft above the base of the dam. The length of the dam (perpendicular to the direction of seepage) is 850 ft.

1. From the flow net $N_f = 3$ and $N_D = 9.5$.

2. q per foot $= 0.0003 \times \dfrac{3}{9.5}(60 - 20)$,

   q per foot $= 0.0038$ cu ft per min.
3. $q = 0.0038 \times 850 = 3.2$ cfm.

FLOW NET WITH FREE SURFACE / In some cases, such as the flow of water through earth dams (Fig. 4.5a), one boundary flow line may be a free water surface that is not fixed by any solid, impervious mass. This is analogous to the free water surface in open channel flow and is a more difficult problem to solve with the flow net. The upper boundary flow line is called the *line of seepage*. It is also the piezometric surface. It must satisfy all the requirements of any flow line, and in addition its intersections with the equipotential lines must be vertically spaced a distance equal to $\Delta h'$ (Fig. 4.5b). Considerable juggling is necessary to construct such a net correctly, but in many practical problems even a rough net will be sufficiently accurate.

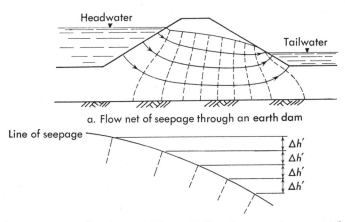

a. Flow net of seepage through an earth dam

b. Intersections of equipotential lines with line of seepage (uppermost flow line)

Figure 4.5 *Flow net with a free surface.*

The line of seepage intersects the downstream face of an embankment, Figs. 4.5 and 4.6, tangent to the surface. Below that point, however, the embankment face is not a line of seepage. The face cuts across the "squares" of the flow net above any tailwater level, and each equipotential line intersects the face at the elevation equivalent to its piezometric level.

If tailwater is present, the intersection of the line of seepage with the embankment face must be above the tailwater level as shown in Fig. 4.6a. If the line of seepage dropped to the tailwater level, Fig. 4.6b, the seepage conditions could not satisfy the requirements of the flow net. Below the free surface the embankment face is an equipotential line. If the line of seepage

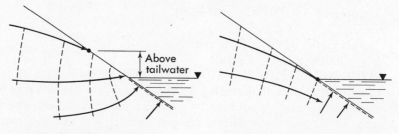

a. Correct intersection of line of seepage with embankment face, adjacent to tailwater

b. Impossible intersection of line of seepage with embankment face at tailwater level

Figure 4.6 *Intersection of line of seepage and flow net at embankment face, adjacent to tailwater.*

dropped to the tailwater level, the flow lines below tailwater would be at right angles to those above, an impossible condition as can be seen in 4.6b. The correct intersection, found by trial is shown in Fig. 4.6a.

A comprehensive paper describing methods of constructing flow nets with free surfaces and for constructing nets in soils whose permeability is not the same in all directions or at all points has been presented by A. Casagrande.[4:1]

OTHER METHODS OF ANALYSIS / Other methods are sometimes used to obtain the flow net. Seepage models can be constructed that are similar to hydraulic models. The soil or porous rock is modeled in sand that is coarse enough to minimize capillary rise but fine enough for laminar flow. There is no need to reproduce or model the permeability in the prototype, because the flow net shape is independent of the value of k. Layers or zones of different permeability can be represented by the correct ratio of k's in the model, but a ratio of 20 to 1 or greater is impractical and rarely of value. The flow line can be traced in a glass walled flume by injecting dye at points on the intake surface of the soil. Piezometric levels can be observed by miniature piezometer tubes forced into the soil.

Viscous fluids such as oil can be used to reduce velocities in models of unsteady flow. The most important uses of models are in studying complex three dimensional seepage, unsteady flow, changing free surface, or non-homogeneous permeability within the soil deposit. While accurate measurements are seldom possible, the results are ordinarily as good as the knowledge of the soil deposit and the boundary conditions justify.

Analogs are models employing phenomena that produce the same patterns of potential and flow as seepage. The flow of electricity through a semiconductor is also described by the Laplace equation. An electric potential applied to a graphite-coated paper produces the same potential distribution as water pressure applied to a soil cross section of the same

SEC. 4:1] SATURATED FLOW: THE FLOW NET

shape. The equipotential lines plotted on the graphite paper from voltage measurements are identical in shape to those in the corresponding soil.

A network of electrical resistors similarly can represent the resistances of soil voids to seepage. Varying permeabilities can be simulated by different resistances. The potential can be measured at points throughout the network as with the semiconductor. The same thing can be done analytically using a digital computer with sufficient memory to describe the entire network. Such techniques are useful in solving problems of complex nonhomogeneity where the graphical analysis would be tedious. Their precision is limited only by the number of elements in the network. Mathematical networks are also used to solve the Laplace equation, where the digital computer can undertake the repetitive processes of trial and correction for the numerous elements of the net necessary to describe a flow system.

WELL DRAWDOWN / Flow into a well and the drop in the piezometric surface that results (termed *drawdown*) is a complex problem in three-dimensional seepage as well as unsteady flow. The three dimensional flow can be analyzed by models or by mathematical approximations. Special cases of multiple wells closely spaced along a straight line can be approximated by a two-dimensional flow net drawn perpendicular to the line of wells. Although model studies and mathematical approximations are available for certain cases of unsteady flow, the results are not generally applicable to real problems. However, for many situations the simple approximation of steady flow is adequate.

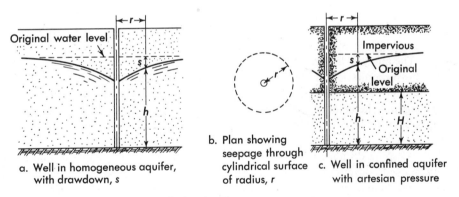

a. Well in homogeneous aquifer, with drawdown, s

b. Plan showing seepage through cylindrical surface of radius, r

c. Well in confined aquifer with artesian pressure

Figure 4.7 *Seepage into an isolated well.*

The flow into an isolated well is depicted in Fig. 4.7. A pervious aquifer lies above an impervious, level stratum, Fig. 4.7a, or is confined between two impervious strata (an artesian aquifer) Fig. 4.7b. The height of the piezometric surface above the impervious base at any distance from the well, r, is h. It is assumed that the average gradient, i, at any radius, r, can be approximated by dh/dr, the slope of the piezometric surface. So long as the slope of the piezometric surface is no steeper than about 25 deg. this is reasonably correct. At any radius, r, the seepage in the homogeneous aquifer takes place normal to a cylinder whose area is $2\pi rh$. The equation for steady state seepage can be derived as follows:

$$q = ki\,A,$$

$$q = k\,\frac{dh}{dr}\,2\pi rh,$$

$$\frac{q}{\pi k}\,\frac{dr}{r} = 2h\,dh,$$

$$\frac{q}{\pi k}\,\log_e r = h^2 + C. \tag{4:4a}$$

The constant of integration, C, is evaluated from the water levels, h_1 and h_2 measured at two points on the piezometric surface, r_1 and r_2.

$$\frac{q}{\pi k}\,\log_e\left(\frac{r_2}{r_1}\right) = h_2^2 - h_1^2. \tag{4:4b}$$

A similar analysis can be made for the confined aquifer whose thickness is H, provided the piezometric level does not fall below the top of the aquifer.

$$q = kH2\pi r\,\frac{dh}{dr},$$

$$\frac{q}{2\pi kH}\,\log_e\left(\frac{r_2}{r_1}\right) = h_2 - h_1. \tag{4:4c}$$

The analysis is frequently used with the well radius as r_1, and h_1 the level of the piezometric surface at the edge of the well. So long as the water surface slope is no steeper than 25 deg and the inflow into the well occurs throughout the height, h_1, the approximation is adequate. It must be kept in mind that the level of the water inside the well must be somewhat lower than h_1, however.

At some distance from a well, the ground water level remains virtually unchanged, despite the well pumping, due to recharge from a river or seepage from adjoining hills. In that case r_2 and h_2 are fixed by the site geology and the equations can be used to approximate the steady state water levels for any given rate of pumping, q.

4:2 Seepage Effects

Uncontrolled seepage results in two types of trouble: Too much seepage causes excessively wet excavations or a loss of water through dams; excessive water pressure causes heave or loss in strength of the soil and failure.

SEC. 4:2] SEEPAGE EFFECTS

Control of seepage is complicated by the fact that correction at one point can aggravate the conditions at another.

NEUTRAL STRESS / The water pressure change in still water can be found from the law of hydrostatics:

$$\Delta u = \gamma_w \, \Delta z. \qquad (3:7)$$

When the water is moving, no matter how slowly, this no longer applies, and the pressure must be computed from the flow net. The total head h at any point is given by the equipotential line. If the elevation of that point is z, then the pressure head is $h - z$. The water pressure is

$$u = \gamma_w \, (h - z). \qquad (4:5a)$$

The pressure is the same in all directions at any one point but not necessarily the same at different points all at the same level.

UPLIFT ON A STRUCTURE / When a structure rests on a soil, a part of the structure is in contact with the soil grains while the remainder bridges over the voids, as shown in Fig. 4.6a.

As was described in Section 3:5 the actual contact areas of the soil grains with another solid are small, Fig. 4.8a. The neutral stress acts over a portion of the total area A, described by the neutral stress ratio, N:

$$U = NAu = \gamma_w \, (h - z) \, NA. \qquad (4:5b)$$

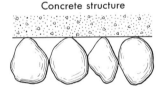
a. Soil against a concrete structure

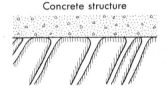
b. Rock against a concrete structure

Figure 4.8 Uplift and contact with a structure.

As given in Table 3:2, the value of N for soils is virtually 1; for jointed rock bonded to a concrete structure, the value of N can be as low as 0.75. Although lower values of N are theoretically possible for intact impervious rock such as granite and marble, it is unlikely that a perfect bond between structure and the rock can be realized. The minimum N used in design, therefore, should not be less than 0.5 for unjointed rock.

When the uplift on the base of a structure exceeds the downward force, owing to the weight of the structure and the loads it carries, the structure will rise or heave. In one case the empty concrete sedimentation basin for a sewage treatment plant under construction rose with the rising water table following a period of rain. Basement floors heave or even blow up as if they were blasted when subjected to excessive uplift.

HEAVE AND BOILING IN A SOIL / Uplift develops within a soil mass in a way similar to the way in which it occurs between soil and a structure.

Because the areas of contact between grains are very small the neutral stress ratio, N, is practically 1 and the uplift force, $U = uA$. If the upward force over an area A equals or exceeds the total load P of soil, water and structure, a zone of instability and potential failure is created. At the point of failure,

$$P = U \quad \text{and} \quad \frac{P}{A} = \frac{U}{A},$$

$$\sigma = u.$$

If the area is sufficiently great, any excess water pressure will force the overlying mass of soil and water to rise, a process called *heave*. Example 3:1 describes heave on a clay stratum overlying a sand layer. Similar heave can occur within a sand or silt stratum at any point where $\sigma = u$. The soil expands with a decrease in void ratio, and in some cases a blister of water forms within the soil mass (Fig. 4.9a). The roof of the blister falls to its bottom

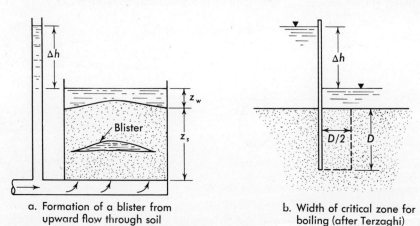

a. Formation of a blister from upward flow through soil

b. Width of critical zone for boiling (after Terzaghi)

Figure 4.9 *Development of heave and boiling.*

and by this process, termed *roofing*, the blister rises to the surface. The soil surface bulges upward and then appears to explode as the blister reaches the top. Finally the soil seethes and bubbles in a *boil* as if it were cooking. Heave can occur in any soil, but roofing and boiling are limited to cohesionless materials.

If the water pressure, u, is distributed so that a uniform upward gradient occurs, the heave may develop uniformly throughout the mass with many

SEC. 4:2] SEEPAGE EFFECTS

enlarged voids but without the formation of a big blister. When the soil is cohesionless, the heave is accompanied by a quick condition:

$$\sigma = u$$
$$\sigma - u = 0$$
$$s = (p - u)\tan \varphi = 0. \tag{3:30}$$

When the uplift and total stress are just in balance or when u is slightly less than σ, the mass may look deceptively stable. However, a machine or structure on the soil surface will sink slowly if its unit weight exceeds that of the saturated soil.

Boiling is merely aggravated quicksand, in which excess neutral stress causes concentrations of heave. Small sand volcanoes often rise to throttle the flow at one point and cause the boiling to shift to a weaker zone elsewhere.

CRITICAL GRADIENT / The hydraulic gradient associated with heave or boiling near an unrestrained soil surface is termed the *critical gradient*, i_c. For upward flow, Fig. 4.9a, the total and neutral stresses at the bottom of the sand are as follows:

$$u = \gamma_w (z_w + z_s + \Delta h)$$
$$\sigma = \gamma_w z_w + \gamma z_s.$$

At the instant heave and the quick condition develop, $\sigma = u$

$$\gamma_w z_w + \gamma_w z_s + \gamma_w \Delta h = \gamma_w z_w + \gamma z_s$$
$$\gamma_w \Delta h = \gamma z_s - \gamma_w z_s = z_s (\gamma - \gamma_w)$$
$$i_c = \Delta h / z_s = \frac{\gamma - \gamma_w}{\gamma_w}. \tag{4:6}$$

For a typical saturated sand, $\gamma = 125$ pcf and $\gamma_w = 62.4$ pcf so $i_c = 1$ approximately (in upward seepage). For seepage toward an unrestrained sloping surface the value of i_c is less. If the surface slopes at the angle of internal friction of the sand, the critical gradient for seepage toward that surface is essentially zero.

Heave without boiling results in an increase in void ratio, a decrease in strength, and a great increase in compressibility.

Terzaghi[4:2] has stated that heaving ordinarily will not take place unless the instability occurs over a width of $D/2$, where D is the depth of soil above the level of instability (Fig. 4.9b). The average neutral stress over different widths $D/2$ for different assumed depths D can be computed from a flow net. Where the average neutral stress equals or exceeds the stress because of the weight of the overlying soil and water, there is a possibility of heave or boiling. It must be remembered that stability computations are approximate at best, and that a large safety factor should be used to be certain that boiling will not occur.

PIPING AND SEEPAGE EROSION / If the soil within the zone of boiling is washed away by the flowing water, an open pit will be created. This causes a concentration of flow into the pit and an increase in the hydraulic gradient because the seepage path is shortened. Consequently the boiling is even more fierce and the pit becomes deeper, working its way upstream at an increasing speed toward the source of the water, as shown in Fig. 4.10. An opening or *pipe* is developed in the soil, and the process of continued backward erosion is called *piping*.

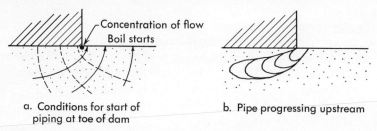

Figure 4.10 Piping beneath a masonry dam on a sand foundation.

Piping also begins from very localized boiling or concentrations of seepage, as shown in Fig. 4.10a. When the upward hydraulic gradient approaches 1 at the soil surface, a small surface boil can form, and if the soil is carried away, a pit will develop. This pit works its way upstream, becoming larger and moving faster as the seepage path is shortened.

If the seepage is horizontal, toward the sloping face of an excavation or downstream face of an earth dam or downward into an improperly protected drain or leaking sewer, piping sometimes will develop from very small gradients. Extensive cavities have been created where cohesive strata support the remainder of the soil mass across the opening. In one situation erosion of fine sand through a $\frac{1}{2}$-in. crack in a bulkhead created a cavity 5 ft deep beneath a concrete pavement which was not discovered until a loaded truck broke through into the crater below.

When the pipe approaches the source of water, there is a sudden breakthrough and a rush of water through the pipe, which enlarges it. One such pipe, a few inches in diameter through an earth dam, was enlarged to 10 ft in a few minutes after a breakthrough. Finally the enlarged hole collapses from lack of support, destroying part of the soil mass.

Cohesionless soils, particularly fine sands and silts, are most susceptible to piping failures. Clays resist piping because the interparticle bonds help prevent the particles from washing away; however, they are not immune. Soft rocks such as poorly cemented sandstones occasionally exhibit piping failures. Even shales, usually considered to be resistant, have developed piping under conditions of very high gradients.

DISRUPTIVE GRADIENTS / Heaving, boiling and most cases of piping involve insufficient restraint P, to withstand the uplift U. In cases where extremely high gradients are involved, $i \gg 1$, piping can develop even if $P > U$. The frictional drag of the water moves some of the finer soil particles even though most are restrained. Seepage then concentrates on the disrupted zone, aggravating the condition until true piping develops. Little information is available on disruptive gradients. Generally they are well above 1, and increase with σ_3 and interparticle attractive forces. For safety, gradients in restrained cohesionless soils are generally kept below 3 or 4. In clays the disruptive gradients may exceed 50, and are seldom critical in design.

4:3 Seepage Control

CONTROL OF SEEPAGE / Control of seepage involves reducing the flow, reducing the water pressure, or increasing the load that resists the water pressure. Excessive seepage is caused by high permeabilities or short seepage paths. If the soil mass through which the seepage occurs is man-made, like a dam, the permeability can be reduced by the proper selection of materials. For example, mixing a small amount of clay with the sand used for constructing a levee can reduce the permeability greatly. A natural deposit is difficult to change. Small amounts of a dispersing agent such as sodium tetraphosphate mixed in the surface of a flocculent structure clay, or injection of clay, chemicals, or cement into the voids of a coarse-grained soil, can reduce the permeability, but at considerable expense.

The seepage path can be lengthened, which will reduce the quantity of seepage and also reduce the water pressure at the downstream end of the flow. An impervious core in an earth dam (Fig. 4.11a) and an impervious cutoff trench in a pervious foundation for a dam can increase the path greatly. A complete cutoff (Fig. 4.11b) that extends to a deeper impervious stratum is more effective than the partial cutoff (Fig. 4.11a). Cutoffs are constructed of an impervious soil or steel sheet piling, depending on availability of materials and ease of construction. An impervious blanket of clay upstream (Fig. 4.11c) is also useful but must not be used downstream because it will increase the uplift.

A cutoff causes an increase in neutral stress upstream and a reduction downstream. A perfect, complete cutoff produces neutral stresses upstream that are equivalent to the headwater level. A cutoff located too far downstream will reduce the seepage quantity and eliminate piping only to create excessive uplift that destroys the structure by heave. An owner of a small dam attempted to correct piping that was slowly developing just downstream from the toe by driving sheet piling into the soil. He did this on the free advice of the piling salesman and against the warning of his engineer. The dam failed by shear in the downstream face that was weakened by the

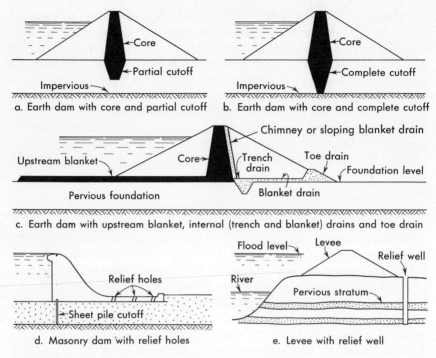

Figure 4.11 Seepage control measures.

increased neutral stress. The cutoff should be placed where the increased pore pressure is not harmful, at the center of the structure or upstream from the center and under the heaviest part of the structure, if possible.

Excessive water pressure can be controlled by drainage that short-circuits the flow and bleeds off the excess neutral stress at a point where it can do no harm. The *trench drain, blanket drain,* and *toe drain* (Fig. 4.11c) are used separately or in combination in earth dams to reduce neutral stresses in the downstream part of the embankment. Relief holes (Fig. 4.11d) reduce uplift on masonry dams. Relief wells (Fig. 4.11e) are used to reduce pressures in confined seams or pockets. Drainage has the disadvantage of shortening the seepage path and increasing the flow, but this can be corrected as previously described. It is essential that the drainage system be properly designed to avoid seepage erosion, as will be described in the section on filters.

FILTER DESIGN / A *filter* or *protective filter* is *any* porous material whose openings are small enough to prevent movement of the soil into the drain and which is sufficiently pervious to offer little resistance to seepage. Extensive experiments have shown that it is not necessary for a filter to screen out all the particles in the soil. Instead the filter openings need restrain only the coarsest 15 per cent, or the D_{85}, of the soil.[4:3] These coarser particles,

D_{85} and larger, will collect over the filter opening as shown in Fig. 4.12a. Their voids will create smaller openings to trap even smaller particles of soil. Therefore the diameter of the openings in the filter must be less than D_{85} of the soil. If the filter is a metal screen or holes in a perforated pipe, this limit fixes the finest soil that can be filtered by any given opening, or it establishes the largest opening that can be used with a given soil. Frequently a soil is employed as a filter. This means that the effective diameter of its voids must be less than D_{85} of the soil being filtered. Since the effective pore diameter is about $\frac{1}{5}D_{15}$, then

$$D_{15 \text{ (filter)}} \leqslant 5 D_{85 \text{ (soil)}}. \qquad (4\!:\!7a)$$

If the filter is to provide free drainage, it must be much more pervious than the soil. Since the permeability coefficient varies as the square of the grain size, then a ratio of permeabilities of over 20 to 1 can be secured by

$$D_{15 \text{ (filter)}} \geqslant 5 D_{15 \text{ (soil)}}. \qquad (4\!:\!7b)$$

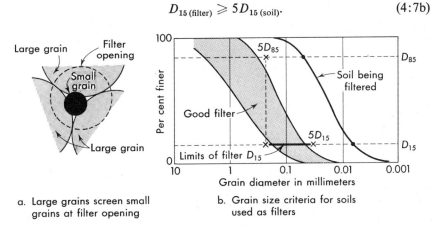

a. Large grains screen small grains at filter opening

b. Grain size criteria for soils used as filters

Figure 4.12 *Protective filter.*

These criteria (Fig. 4.12b) are the basis for filter design.[4:3] In general the filter soil should be well within these limits, and its grain-size curve should be smooth and parallel to or flatter than the soil. If the soil being filtered is very fine-grained, more than one filter layer will be required. The final filter layer is designed to fit between the openings in the conduit and the next finer filter. For many silty and clayey soils a well-graded concrete sand makes a satisfactory filter. A coarser pea-gravel second filter—usually described as No. 78 crushed stone—is then needed for the first.

If the soil being filtered is gap-graded, its grain size curve is redrawn considering only the portion of the soil finer than the gap to be the total soil being filtered, and disregarding the part of the soil coarser than the gap. The filter is designed to fit the redrawn curve.

There have been many attempts to devise a "universal filter" that is small enough to filter the finest soil and yet having a D_{85} large enough that it will not pass through the 5/16 in. (0.8 cm) perforations of commercial drain pipe. However such filter materials have such a wide range of sizes (high C_u) that the particles segregate during handling and construction. Therefore they should not be used unless care is taken to maintain their gradation.

The filter thickness is not critical; for small heads a few inches is sufficient, while in dams where the head differences are great, 2 to 10 ft or 0.6 m to 3 m are commonly used.

4:4 Capillary Moisture and Flow

Above the free water surface or water table the movement of moisture is more complex. The soil is saturated to the height of capillary rise but above that level the degree of saturation is less. Gravity and fluid friction still act on the soil moisture but the capillary forces are even more important. These include surface tension and the physicochemical bonds between water and soil. These forces are tensile and result in negative neutral stresses. The tension increases with decreasing temperature and with a decreasing degree of saturation. In the zone of partial saturation water also exists in the vapor phase. The vapor pressure decreases with decreasing temperature.

CAPILLARY EQUILIBRIUM / As described in Section 3.2, moisture rises above the free water surface as a result of capillary tension. When equilibrium is established, the soil moisture is distributed approximately as shown in Fig. 4.13. In the capillary zone the soil is saturated. The moisture is continuous and the neutral stress obeys the hydrostatic law. Above this zone is the *capillary fringe*. The degree of saturation falls off rapidly, but although the moisture does not fill the voids, it is still continuous in interconnected wedges between the grains. The effective stress is no longer equal

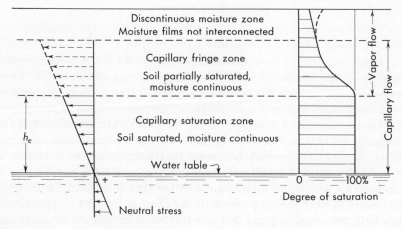

Figure 4.13 Capillary equilibrium and soil-moisture distribution.

to the total stress minus the neutral stress as given by Equation 3:12c because the neutral stress does not act over the entire void area. The degree of saturation becomes less with increasing height above the free surface until the moisture wedges are no longer interconnected. There is still neutral stress in the upper zone of *discontinuous moisture*, but it no longer follows the hydrostatic distribution. Each moisture wedge develops a different stress, depending on its radius, and although the stress can be very high, it acts over only a small fraction of the void.

VAPOR MOVEMENT / Moisture movement in the vapor phase occurs both in the fringe and discontinuous moisture zones. The difference in vapor pressures required to produce the flow can come about in a number of ways. Evaporation at the soil surface reduces the pressure and induces an upward movement. A sudden drop in temperature at the ground surface will also reduce the pressure and induce upward movement, while a sudden rise in temperature at the surface will produce a downward movement.

CAPILLARY FLOW / Capillary flow occurs in the zone of saturation and in the capillary fringe where the moisture is continuous. In the equilibrium condition, Fig. 4.13, the capillary tension, u, just balances the hydrostatic stress, $\gamma_w z$ and no movement occurs. If anything should change the capillary tension then flow will take place depending on whether the tension is increased or decreased compared to the hydrostatic gradient.

Evaporation of moisture in the fringe reduces the degree of saturation. As a result each meniscus radius at any given level in the fringe zone is reduced, increasing the tension. At the same time the level of capillary saturation is depressed to h_c'. At that level (the boundary between saturated and unsaturated voids), the capillary tension is unchanged. However, at the new lower level, the capillary tension now exceeds the hydrostatic stress $\gamma_w h_c'$ and upward flow is generated by the gradient.

$$i = \frac{\Delta u / \gamma_w}{h'_c}$$

as shown in Fig. 4.14a.

In arid regions the continuous evaporation in the fringe zone maintains the state of nonequilibrium and upward capillary flow shown in Fig. 4.14a. Dissolved salts are brought up from the zone of saturation and precipitated in the fringe when the water is evaporated. This concentration of precipitated salts has two significant effects. First, the fringe zone becomes cemented and eventually made impervious by the salts, including calcium carbonate and sulfate. A rock-like hardpan is created that presents serious problems in future excavation and which influences drainage and seepage if the area is inundated by ponding of a reservoir or if the area is irrigated for farming. The second, a more serious problem, is the effect of the salts on agriculture. They limit the crops to those that can survive in the salt environment and eventually make the land sterile. Vast areas of such lands of ancient irrigation

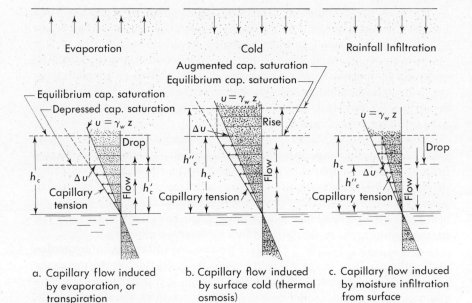

Figure 4.14 *Capillary flow in nonequilibrium conditions.*

as India lie fallow because of the destruction of fertility by irrigation. The process can be reversed by sufficient irrigation to flush the salts downward.

In humid regions, upward flow can be induced by loss of water from the capillary zone by the transpiration of crops and other vegetation. Accelerated evaporation, produced by heat can also increase capillary tension and upward moisture movement. If the soil is compressible, the increased capillary tension will produce shrinkage and settlement. Severe settlement from capillary desiccation is caused by some species of trees.

In dry regions, the continuous flow of moisture upward to be evaporated is sometimes stopped by construction of a building or a pavement that reduces the evaporation loss. Static equilibrium is produced with a rise in the level of the line of capillary saturation and an increase in the degree of saturation in the fringe zone. This can have two serious consequences. First, any cemented zone is weakened. Pavement rutting sometimes occurs due to a wet subgrade in a desert region. If the grain structure of the cemented soil is loose, there may even be subsidence of the ground surface and damage to structures. Second, the moisture increase will cause expansion of any highly plastic clays present. In arid regions this is severe, because the clays are initially highly desiccated.

Upward flow can be induced by a drop in temperature at the ground surface that increases the capillary tension, Fig. 4.14b. The increased tension, Δu, produces the gradient that maintains flow until a new equilibrium is reached at the higher level $h''_c = (u + \Delta u)/\gamma_w$. Temperature induced movement is sometimes termed *thermal-osmosis*. For example, construction work in the fall season when the air temperatures are dropping steadily is sometimes hampered by the increasing soil moisture in spite of no rainfall.

Downward flow is induced by an increase in the degree of saturation in the fringe accompanied by an increase in the radius of each meniscus in the fringe and on the boundary of capillary saturation. The tension is reduced, Fig. 4.12c, and the decrease, Δu, produces a gradient that causes downward seepage. In this way the ground water is replenished, although the water that reaches the water table may have come from the capillary zone. A similar downward movement can occur during periods of increasing ground surface temperature.

Capillary flow takes place horizontally as well as vertically if there are differences in capillary tension that induce a hydraulic gradient. The damage to the frozen orange juice warehouse described in the opening paragraphs of this chapter was caused by thermal osmotic flow upward and laterally to the cold floor with its inadequate insulation. Drying of soil in a deep excavation can induce capillary flow laterally to the exposed banks. Rain falling on these same banks can induce capillary flow into the soil mass.

The rate of capillary flow is proportional to three factors: the change in stress, Δu; the reciprocal of the distance through which flow occurs, $1/z$ and the permeability, k.

In sands, the stress changes, Δu, are small because the large voids limit the capillary tension. Therefore, although z is small and k large, the rate of flow is small. In clays with minute voids, the Δu is likely to be large, but z is also large and k extremely low. In clays, therefore, the rate of capillary flow is also very small. In silty soils the optimum combination of small voids producing moderately large values of Δu, and permeability coefficients that are not too small causes maximum rates of capillary flow.

SOIL MOISTURE ABOVE THE GROUND WATER TABLE / Because of the environmental changes that occur daily it is unlikely that capillary equilibrium exists for very long. Instead, the moisture in the capillary zone is constantly changing and with the changes there are profound variations in the engineering properties of the soils. These changes are most significant in the fringe zone, but the effect of the capillary stress changes are felt below the line of saturation. The engineering implications are greatest in the design of pavements and floor slabs that are supported directly on the ground surface, because not only are they directly affected by the changes in the engineering properties, but they also contribute to the changes. Shallow structural foundations, and the upper parts of structures deeply embedded in the ground, are also affected, but to a lesser extent.

Only limited data are available regarding the possible moisture content variations that occur beneath structures because so many factors are involved whose contribution cannot be evaluated quantitatively. The major factors are climate, potential capillary rise in the soil, and the water table position. When the water table is high in a relatively humid region (20 ft deep in clay, 10 ft in silts, 3 to 5 ft in sands) capillary saturation probably represents the limiting equilibrium moisture. With a deep water table in either a humid or dry region, equilibrium moisture eventually develops under pavements and wide structures. The moisture contents are somewhat less than saturation but relatively constant for any given soil in that region. Limited data suggest that for a small region of homogeneous topography and climate, there is a relatively constant water-plasticity ratio that develops at equilibrium. This can be established by testing the soil beneath existing structures and extrapolating to other structures in the area.

In extremely dry regions with a very deep water table the moisture content is not likely to be changed by a structure. The equilibrium soil moisture, therefore, will not be greatly different from that in the ground before the structure was built.

At any site the range in moistures must be established by field observations, correlated with the seasonal climatic conditions (chiefly rainfall and temperature) and with the soil properties.

4:5 Drainage

Drainage ordinarily means removal of water from the soil. It has two objectives: prevention of seepage out of the soil, such as into an excavation where it would be a nuisance or a hazard; and improvement of the soil properties, such as an increase in strength or a reduction in compressibility. Drainage is also employed to reduce water pressure in the soil. Usually this is accompanied by removal of water, but in fine-grained soils it can be effective even though little or no water is removed.

FORCES INVOLVED IN DRAINAGE / A number of forces are involved in the ease with which water drains from the soil. First is the resistance to seepage, as indicated by the permeability coefficient. Second is the effect of the drainage on the soil structure. If the soil is relatively incompressible, the water lost will be replaced by air in the voids. If the soil is compressible, the water loss can be accompanied by consolidation and the soil will remain virtually saturated. Third are the forces that restrain the water: capillarity and adsorption. Both the resistance to flow and the capillary retention become greater with decreasing grain size. Coarse-grained soils, such as gravel and coarse sand, drain rapidly and air replaces the water in the voids. Fine-grained soils that have low permeability and very high capillary retention drain very slowly and may lose only as much water as the consolidation will permit.

In order to remove water from the soil, the force producing drainage must be greater than the retentivity and the resistance to flow. *Gravity* is the force most often employed: Water moves from the soil into the drain under the influence of its own weight. This method is cheap and reliable but not

SEC. 4:5] DRAINAGE

strong enough in fine-grained soils. A *vacuum* can be used to add atmospheric pressure to the head produced by gravity. With its aid, finer soils such as silty sands can be drained. A direct electric current will induce a flow of water in the soil toward a negative electrode. This principle of *electro-osmosis* can be used to induce drainage of low-permeability soils such as silts.

Evaporation is ordinarily not considered a drainage method but it does cause a loss of water. It is a slow but powerful force that can drain even clays. *Consolidation* produced by a load on the soil mass is essentially a drainage process that is effective in compressible materials.

DRAINAGE AND SOIL TYPE / The ease of draining a soil and the forces that are effective in producing drainage can be estimated from laboratory tests for permeability, consolidation, and shrinkage. The grain-size distribution offers some indication of drainage properties, as shown in Fig. 4.15. Table 4:1 gives the drainage potential for the Unified Soil Classification.

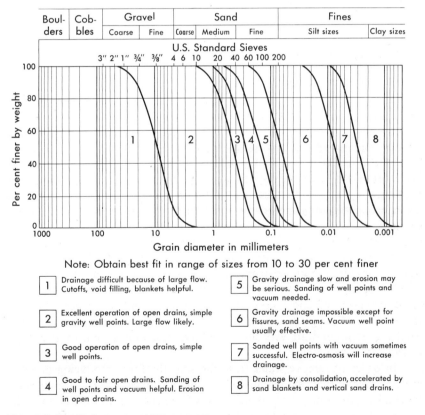

Figure 4.15 Drainage capabilities of soils.

TABLE 4:1 / DRAINAGE POTENTIALITIES, UNIFIED SOIL CLASSIFICATION[4:13]

Soil Class	Drainage Characteristics	Soil Class	Drainage Characteristics
GW	Excellent	ML	Fair to poor
GP	Excellent	CL	Impervious
GM	Fair to impervious	OL	Poor
GC	Poor to impervious	MH	Fair to poor
SW	Excellent	CH	Impervious
SP	Excellent	OH	Impervious
SM	Fair to impervious	Pt	Fair to poor
SC	Poor to impervious		

4:6 Drainage Systems

The design of a drainage system depends on the drainage characteristics of the soil, the length of time the system must operate, and the position of ground water. For temporary drains installed during construction, minimum interference with work and maximum effectiveness in a short time are essential. For permanent drains, long-term effectiveness and minimum maintenance are essential.

DRAINAGE LAYOUT / The position of the drainage system depends on the initial seepage pattern and the pattern that is to be established. If the drainage is installed in a dam, for example, the position of the water before and after drainage can be established by flow nets. If the drainage is for a building site, a highway, or an airfield, the initial ground water conditions must be established by exploration, as described in Chapter 10. A contour map of ground water elevation is prepared for the site and its surroundings. If the ground water level fluctuates appreciably, more than one such map will be necessary, each representing a different condition.

Three locations are possible, as shown by Fig. 4.16: *intercepting*, *site*, and *downstream*. The intercepting drain removes the water before it reaches

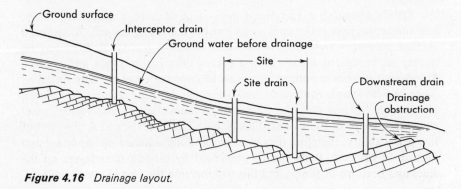

Figure 4.16 Drainage layout.

the site. It is particularly effective when the ground water surface slopes steeply or when confined pervious strata carry water under pressure. It does require use of land beyond the limits of the site, however. The site drain removes water directly from the area. In this way the quantity of water handled will be less and the drainage will be effective sooner than for the interceptor, but the drainage system may interfere with the work at the site. The downstream drain enables the water to leave the site more rapidly. It is most effective when an underground obstruction tends to dam up the ground water. In some cases one location is sufficient; in others, all three will be employed.

FILTERS, CONDUITS, DISPOSAL / A complete drain consists of three components: the filter, the conduit or collector, and the disposal system. The filter is essential for continued efficiency of the drain and to prevent seepage erosion when the hydraulic gradients are high.

The filter is pervious enough to permit the flow of water into the drain with little head loss and at the same time fine enough to prevent erosion of the soil into the drain. The proper filter is the key to a successful drainage system; an improper filter is the most important factor in drainage failures.

The drain conduit collects water from the filter and carries it away. The simplest is a ditch or pipe designed like any other hydraulic conduit. Ordinarily the conduit is several times larger than its hydraulics dictate to allow for silting. Typical collecting perforations in conduits are $\frac{5}{16}$ to $\frac{3}{8}$ in. in diameter and require a filter with a maximum size of $\frac{1}{2}$ in. A *French conduit* or *French drain* is made of coarse gravel or crushed rock; where the amount of water is small, it can be cheap and effective. It is not, however, a substitute for a filter, and if employed as one, it will soon clog.

The disposal system removes the water from the area. If possible, gravity is used because it is permanent and foolproof. However, the topography may make this impossible, particularly during wet weather when the drain is needed most. Pumping will remove the water faster, but the cost of power over a long period will be appreciable, and maintenance is often uncertain.

OPEN DRAINS / The oldest method of draining excavations, roads, and similar projects is the open drain—either a ditch or a sump. A *sump* is a shallow pit into which the ground water flows by gravity. A *ditch* may be merely an elongated sump. Both are very effective in sands and gravels. Sumps and ditches are cheap; they can be constructed easily with unskilled labor or with simple equipment, and ordinarily gasoline construction pumps are suitable for pumping the water out of them. Boiling and piping sometimes commence in sumps and ditches, particularly if the soil is fine sand of low permeability; therefore they must be carefully watched on important and hazardous projects. Boils can be prevented by placing filter layers on the sides and bottoms of sumps, but this will increase the cost of construction.

CLOSED DRAINS / When seepage erosion or piping is troublesome or where a permanent drain is desired, perforated pipe or open-jointed tile can be laid in the ditches and the ditch backfilled with a filter material. It is important that the pipe be surrounded by one or two filter layers, as required, to prevent soil from clogging the openings (Fig. 4.17a). The pipe should be laid in straight lines. Drains in silty soils should have an opening every 50 to 100 ft through which a fire hose can be inserted to flush out the pipe occasionally. Manholes should be provided at changes in direction and at intervals of 300 or 400 ft along straight sections.

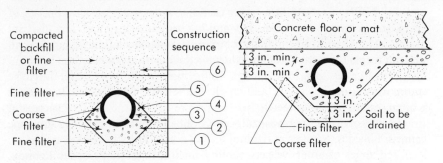

Figure 4.17 Simple closed drains.

BLANKET DRAINS / Continuous drainage blankets are sometimes provided beneath dams and basement floor slabs to reduce uplift pressures and beneath pavements to prevent capillary flow upward. The blanket consists of a filter layer in contact with the soil and a coarser collector layer which also serves as a second filter (Fig. 4.17b). The latter is placed in contact with the underside of a masonry dam or basement floor, or is sandwiched between two filter layers in the base of an earth dam. Water is removed from the collector by conduits.

DEEP WELLS / Deep wells, such as are used for water supply, are occasionally employed in temporary drainage. Diameters of 12 to 24 in. with spacings of 25 to 100 ft and depths of 100 ft or more have been used, depending on the size of the area to be unwatered and the amount the water table is to be lowered. They are also used in coarse soils and porous rock where the quantities of water drained are large.

The pump is in the bottom of the well, so the height the water is lifted is not limited to 25 to 28 ft, as with a suction well.

Like other drains, deep wells require filters. If the soil is coarse grained, or if a rock being drained resists erosion, a well screen may provide an adequate filter. The screen is placed in the drilled hole in direct contact with the pervious stratum. The screen is generally attached to a pipe casing that supports the hole through any impervious strata above. If the soil is too fine grained to be filtered by the screen alone, a *pack* may be used. This is a gravel-sand filter placed around the well screen that filters the soil and which is filtered by the screen. The well in this case is drilled 8 in. or 20 cm in diameter larger than the screen. The annular space between the hole and the screen is filled with the filter pack.

HORIZONTAL WELLS / Horizontal wells, about 3 in. or 8 cm in diameter and over 100 ft long, have been found useful in draining hillsides. The wells are installed by drilling into the hill at a slight upward angle to intercept water-bearing strata. The hole is then lined with a slotted or perforated pipe to keep it open and to carry the water out.[4:5]

Combinations of vertical wells and horizontal holes have been used to drain stratified soil and jointed or pervious rock formations. The vertical well intercepts the strata, draining them to the horizontal well at the bottom. The latter acts as a drain, collector, and disposal pipe. Because of the difficulties in making the two wells intersect, the vertical well (the easiest to drill) is often several feet in diameter.

Large horizontal drain tunnels have been employed to tap deep aquifers beneath hillsides. Smaller horizontal drains can be drilled from the tunnel to localized zones of excessive permeability such as large joints or fault zones. The tunnel is principally a collector and conduit; however it may directly drain the adjacent formations.

WELLPOINTS / Wellpoints are small diameter wells that are driven or jetted into the soil. Usually they are placed in straight lines along the sides of the area to be drained and are connected at their upper ends to a horizontal suction pipe called the *header*, as shown in Fig. 4.18a. Depending on the type of soil to be drained, one or two wellpoints are usually installed for each 8 ft of header. The header terminates in a self-priming pump specially designed for wellpoint work, and one pump is ordinarily used for each 50 to 100 points.[4:6]

Many different types of points have been devised (Fig. 4.18b). The drivepoint consists of a length of heavy-gauge, 2-in. pipe. To its lower end is attached a perforated section 2 to 4 ft long covered with a wire-gauze screen terminating in a conical steel tip. Points designed to be jetted into the soil are equipped with rubber ball valves at their lower ends. During jetting water is pumped into the well point and is directed out of the tip by the valve. This washes a hole in the soil allowing the wellpoint to sink into position. When the wellpoint is connected to the suction header, the ball valve closes and the wellpoint takes in water through the gauze screen at its lower end.

The wellpoint screen (typically 30 to 50 openings per in., 0.6 to 0.3 mm) is an adequate filter for medium sands. In finer soils, a sand filter is placed around the wellpoint to increase the effective area of the well, minimize seepage velocities, and provide a better filter. The filter is installed by drilling

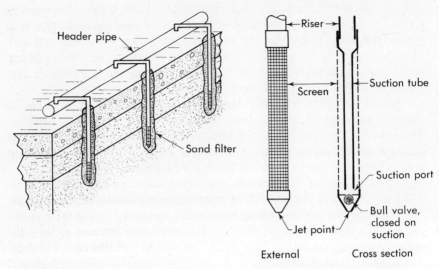

a. Cut-away drawing showing sanded wellpoints in ground attached to header

b. Construction of a wellpoint

Figure 4.18 Wellpoint installation and construction.

or jetting a hole about 1 ft in diameter. The wellpoint (2.5 in. O.D. or 7 cm in diameter) is centered in the hole and a graded clean sand such as concrete sand is placed around the screen. The *sanded* wellpoint can then be sealed with clay fill above the pervious strata if vacuum drainage is needed.

It has been found that the effectiveness of the points in fine-grained soils is increased by sealing the point into the uppermost soil strata with a plug of clay and maintaining a vacuum on the header at all times, even when little water is removed from the soil. With a vacuum applied within the soil mass, the atmospheric pressure tends to force the water out of coarser soils and to reduce the neutral stresses and decrease the void ratio of finer soil. Pumping units designed for wellpoint work usually include vacuum pumps.

Since most wellpoints are operated by suction, the maximum vertical distance from the pump intake to the water level at the points is from 20 to 25 ft or 6 to 8 m. If excavations extend more than 20 ft or 6 m below the water table, they must be unwatered in two or more stages as shown in Fig. 4.19. The first stage consists of a row of points that are set in the ground and placed in operation as soon as the ground water table is reached. Excavation is resumed as soon as the water table is lowered by the first points. A second row of points is then placed when the excavation again reaches the ground water table. Excavations as deep as 50 ft below the original water table have been made by using from three to four lift stages of points.

Special wellpoints are available that employ small water jet pumps at their bottoms. In this way the water is lifted out of the wellpoint rather than

SEC. 4:6] DRAINAGE SYSTEMS

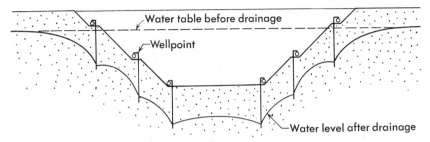

Figure 4.19 Multiple-stage wellpoint system.

sucked out. Lifts of 50 feet (15 m) or more are practical with such wellpoints. There are two disadvantages. First, two headers are required: one to supply water to the jet, and the second for discharge. Second, the system is less efficient mechanically and the flow capacity for each wellpoint is less than for the conventional points.

Wellpoints have proved very successful for draining soils of high and medium permeability, such as coarse sands and clean fine sands, and vacuum wellpoints have been used with some success in soils of low permeability, such as silty sands and sandy silts. Their success depends to a large extent on the experience and skill of the persons making the installation.[4:6, 4:7]

ADVERSE EFFECTS OF DRAINAGE / Drainage, in solving the problems of too much water or excessive water pressure, can create new problems. While reducing neutral stress increases the effective stress and the strength of a soil, the increased effective stress produces consolidation. If compressible strata are present, severe settlements can be caused in structures that are within the zone of drawdown.

Drainage of sands will temporarily enhance their strength by producing capillary tension and increased effective stress. As a result small excavations can be made with steep slopes without bracing. If the work is so prolonged that the sand dries, the capillary tension will be lost and the steep excavation slopes will collapse with loss of property or life.

Drainage without proper filters can produce piping, and even the formation of cavities in the soil. The cavities soon collapse, causing the destruction of anything above. Seepage erosion (piping) into drains and leaky sewers causes occasional drop-out of pavements and basement floors, particularly in sandy and silty soils.

RECHARGE / The adverse effect of drainage on adjacent structures can be controlled by *artificial recharge*. This is the pumping of water into the ground between the drainage system and the structure. Usually this is done with a system of well points, similar to that employed for drainage, but with a wider spacing. Water is pumped into the well points at a rate to maintain a constant ground water level under the endangered structure. This inflow rate is less than the drainage rate. Of course, the necessary drainage rate must be increased because of the high water table at the inflow wells. Continuing accurate control of inflow is necessary to avoid under or over recharging.

ELECTRO-OSMOSIS / If a direct current is passed through a soil of low permeability, the rate of drainage is increased greatly.[4:7] Wellpoints serve as the negative electrodes, and steel rods driven into the soil midway between the wells form the positive electrodes. From 20 to 30 amp of electric current are used per well, at voltages from 40 to 180. The amount of energy required varies from 0.5 kwhr to 10 kwhr per cu yd of soil drained.

Electro-osmosis requires expensive equipment and is relatively costly to operate; therefore it is used only when cheaper methods cannot produce sufficient drainage.

DESICCATION / Drainage of a soil by evaporation is an extremely slow process and is ineffective if the soil mass, by capillarity, can replace the moisture evaporated. Ventilation galleries have been used to dry out clay strata in hillsides when the clay is subject to swelling and loss of strength during wet seasons.

DRAINAGE BY CONSOLIDATION / Soft, wet, cohesive soils are impossible to drain by gravity methods or even by vacuum or electro-osmosis, yet they may require a reduction in their water contents before they have sufficient strength to support heavy, concentrated loads without undue settlement or failure. Consolidation—the removal of water by a reduction of the volume of the voids through compression—is an effective process in spite of its inherent lack of speed.

Consolidation is produced by loading the soil with earth, crushed stone, iron ore, or any heavy material that can conform to the settlement irregularity. If the loading material is impervious, a blanket of free-draining sand is placed below the fill to allow the water that is squeezed from the soil to escape.

The process can be accelerated two ways. One is to provide an overload or surcharge that exceeds the ultimate load on the soil. Partial consolidation by the overload, which can be effective in a limited time, is equivalent to a greater degree of consolidation for the ultimate load, which would require a greater length of time to take place. The consolidation rate can also be accelerated by vertical drains, known as *drainage wicks*, or vertical sand piles[4:8, 4:9] (Fig. 4.20). The sand pile is constructed by driving a pipe, 12 to 16 in. in diameter and with a removable bottom plate. The pipe is filled with sand and then withdrawn, leaving the sand in the hole. The drains are spaced 6 to 20 ft apart in both directions beneath the fill. Their upper ends connect with the sand-blanket drain at the top. Their purpose is to shorten the seepage path for the water being squeezed out of the soil and thereby increase the rate of consolidation. They are very effective in thick, homogeneous clay deposits where consolidation otherwise would be very slow. In deposits consisting of alternate strata of sand and clay, they are of little help because the seepage path is already short.

SEC. 4:7] FROST ACTION 191

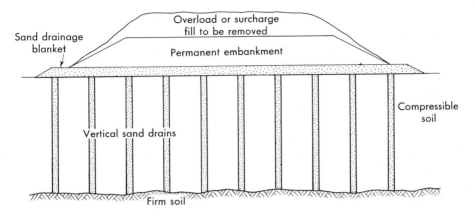

Figure 4.20 *Drainage by consolidation, accelerated by vertical sand drains and surcharge fill.*

In very soft soils it is impossible to add enough weight at one time to provide the necessary drainage because the soil is too weak to support the load. In such cases the load is applied in stages. The second stage is deferred until the soil has been consolidated and strengthened enough to support the entire load.

4:7 Frost Action

FROST ACTION AND ITS CONSEQUENCES / When the daily mean temperature remains below 32 F for a period longer than three or four days, the soil moisture at the ground surface freezes. The longer and the more intense the cold spell, the greater the depth to which the freezing extends. The result of the freezing is a rise of the ground surface, known as *frost heave*, that sometimes is as great as 12 in. or 30 cm in the northern parts of the United States. If an excavation is made into frozen ground that has heaved, it will be seen that the soil has changed considerably. Its average water content has been greatly increased, and much of the water is concentrated into ice layers or lenses that lie parallel to the ground surface. The amount of heave is rarely uniform, and the force exerted by the expanding soil may lift roads, walls, and buildings.

Frost heave is particularly damaging to highways and airfield pavements, as they are generally built directly on the surface of the ground. Unequal heave can crack concrete pavement slabs or tip the individual slabs at angles. Heave beneath flexible pavements causes bumps or waves in the surface. Small structures with shallow foundations, such as small bridges, culverts, walls, sewer inlets, and light buildings, often suffer if their foundations are above soils subject to frost heave. Cold storage warehouses are lifted and torn apart by unequal heave brought about by improperly insulated floors.

Frost action is not limited to the process of heave. When the weather becomes warm again, the frozen soil begins to thaw from the top down, where the warmer air and sunshine are in contact. The uppermost portions of the soil become wet and soft as the ice layers melt. They remain wet until the excess water can drain downward through the deeper strata when the frost disappears.

Thawing beneath a highway or airfield pavement converts the soil into a liquid that supports the pavement. The weight of a truck or airplane under such conditions causes the liquid soil to spurt up through the expansion joints in concrete pavements (often called *pumping*) or to form holes known as *mud boils* in flexible (asphalt) pavements. Thawing beneath structural foundations may result in failure since the soft, water-filled soil has little ability to support heavy loads.

Frost action, therefore, is a combination of two processes: first, freezing of the soil and the formation of layers of ice that cause frost heave; and second, the thawing of the *ice lenses*, which provides an excess of free water in the soil and results in a lowering of the strength of the soil.

MECHANICS OF FROST ACTION[4:10] / When water goes from the liquid to solid form it expands about 10 per cent. In a saturated soil with a void ratio of 0.7, this means there would be a 4 per cent increase in the volume of the soil, or about a 1-in. heave in a soil frozen to a depth of 25 in. However, the observed heave in such a situation might be as much as 4 or 5 in. Furthermore, since the average water content and void ratio of the soil increase during the freezing process, it must be concluded that expansion of water in the voids is not the primary cause of heave. Examination of frozen soils indicates that the heave is about equal to the total thickness of all the ice layers formed during the process. Therefore it is concluded that the formation of ice layers (and the increase in average soil water content) is the basic cause of heave.

The temperature deep in the ground remains nearly constant throughout the year, while the temperature just below the ground surface fluctuates with the air temperature. After a period of cold weather in which the air temperature is below freezing, a thermal gradient is established in which the 32 F point is below the ground surface, as shown in Fig. 4.21. This point defines the *frost line*. Neither the frost line nor the thermal gradient is fixed; both vary with the duration and intensity of the cold. The frost line is not found at a uniform depth, for it depends on the density, saturation, and composition of the soil.

Above the frost line the temperature is below the ordinary freezing point for water. However, in very small openings such as the voids of fine-grained soils, the freezing point may be depressed as low as 23 F. Thus, just above the frost line, water will freeze in the larger voids but remain liquid

SEC. 4:7] FROST ACTION

in the adjacent smaller ones. When water freezes in a larger void, the amount of liquid water at that point is decreased. The moisture deficiency and the lower temperature in the freezing zone increase the capillary tension and induce a flow toward the newly formed ice crystal. The adjacent small voids are still unfrozen and act as conduits to deliver the water to the ice. The ice crystal grows until an ice lens or layer forms. The capillary tension induced by the freezing and the low temperature sucks up water from the water table below or can even dehydrate and shrink adjacent compressible strata such as clays and micaceous silts when the water table is beyond reach. The result is a great increase in the amount of water in the frost zone, and *segregation* of the water into ice lenses.

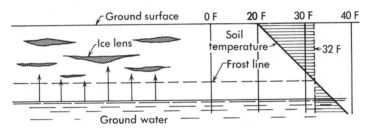

Figure 4.21 Formation of ice lenses in zone of freezing.

In order for the water to be drawn into the freezing zone by capillary forces, the soil must be saturated or approaching saturation. Partially saturated soils freeze, but the ice is scattered in tiny crystals and the heave is small. For continued heave, the freezing zone must be within the height of capillary rise above the water table so that water can be sucked up from below. If the freezing zone is saturated but above the height of capillary rise (for example, when the upper strata have been saturated by rain or leaking pipes), the segregation and heave will be limited by the amount of water that can be sucked from the adjoining soil.

Second the soil must be fine-grained. Segregation seldom occurs in coarse soils where the pores are so large that water freezes in them at the same temperature as it freezes in cracks or fissures. The water simply freezes in the soil pores, and the heave, if any, is limited to a 10 per cent expansion of the voids. On the other hand, very fine-grained soils are so impervious that the water migrates to the ice lenses very slowly. The most rapid segregation takes place in soils whose permeability is great enough to permit easy movement of the water.

Third the temperature gradient in the soil must be favorable. When the rate of change in temperature with depth (temperature gradient) is very rapid, the zone of soil in which the pore water is unfrozen but below 32 F is narrow. The ice layers formed under such conditions tend to be thin, and the amount of heave small. When the temperature gradient is small, the zone in which the pore water is unfrozen but below 32 F is wide, and the ice lenses tend to be thick and the amount of heave great.

A rapidly varying gradient with alternating freezing and thawing can aggravate frost heave in sands but will have little effect on silts and clays.

SOILS SUSCEPTIBLE TO FROST ACTION / The susceptibility of different soils to frost action has been studied by A. Casagrande and G. Beskow. They found that the coarse soils, sand and gravel containing no fines, are rarely subject to the formation of ice lenses and objectionable heave. On the other hand, fine sands and silts have the optimum combination of fine pores and relatively high permeability that results in maximum segregation and heave. Clays are usually considered to be frost susceptible because cracks and fissures may permit rapid movement of the water through them. According to Casagrande,[4:12] a uniform soil is susceptible to frost action if more than 10 per cent of its particles by weight are finer than 0.02 mm, and a well-graded soil is susceptible if more than 3 per cent of its grains are finer than 0.02 mm. Studies of frost action in Michigan, however, indicate that even sands may be susceptible to frost heave under some conditions. The potential frost action of the soil groups of the Unified Classification System are given in Table 4.2.

TABLE 4:2 / POTENTIAL FROST ACTION OF UNIFIED SOIL CLASSIFICATION[4:13]

Soil Class	Potential Frost Action	Soil Class	Potential Frost Action
GW	None to very slight	ML	Medium to very high
GP	None to very slight	CL	Medium to high
GM	Slight to medium	OL	Medium to high
GC	Slight to medium	MH	Medium to high
SW	None to very slight	CH	Medium
SP	None to very slight	OH	Medium
SM	Slight to high	Pt	Slight
SC	Slight to high		

DEPTH OF FROST PENETRATION / The depth below the ground surface to which a 32 F temperature extends is termed the *frost line*. Above the frost line, freezing occurs and ice lenses will form if the soil and water conditions are right. The depth of the frost line depends primarily on three factors: the air temperature, the length of time the air temperature is below 32 F, and the ability of the soil to conduct heat. The lower the air temperature and the longer it remains below 32 F, the greater the depth of the frost line; the higher the thermal conductivity of the soil, the greater the depth of frost penetration. The accompanying map (Fig. 4.22) shows the maximum depth of frost penetration in the United States. The map is only approximate: On

SEC. 4:7]   FROST ACTION

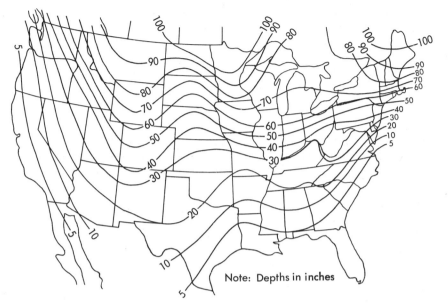

Figure 4.22   Maximum depth of frost penetration in the United States.

mountain tops the depths will be much greater; in highly organic soils or coarse gravels above the water table, the depth will be smaller.

PREVENTING FROST DAMAGE / Frost heave and frost damage may be prevented by correcting one or more of the factors responsible for the segregation of water and the formation of ice lenses: frost-susceptible soil, capillary saturation by rise of water from ground water table, and freezing temperatures in the soil.

One of the most effective methods of preventing frost heave is to remove the frost-susceptible soil throughout the depth of frost penetration and replace it with a soil that is not affected. In regions where large quantities of clean sands and gravels are readily available, soil replacement is an economical and permanent cure for frost heave.

Capillary saturation caused by the rise of water from the water table may be controlled by lowering the water table below the height of capillary rise or by obstructing the upward movement of the water. When the ground water table has a steep slope, or when it is perched on top of a saucer-shaped stratum of impervious soil, proper drainage may be very effective in preventing serious frost heave. In flat areas, and in areas in which excessive rainfall or snow melting are quickly followed by freezing, even extensive drains may not prevent a rise in the ground water table and the subsequent capillary saturation. An impervious blanket, such as asphalt, plastic, or commercial bentonite, may prevent movement of water upward into the zone of frost penetration. Such blankets are seldom used, for they are expensive and they tend to puncture from the weight of the soil above or to deteriorate. Blankets of coarse-grained soils such as clean coarse sand, gravel, or crushed rock (with fines removed) placed above the water are effective water stops, for they break the capillary tension. They should be thicker than the height of capillary rise through them and must be protected by filters so that the finer soils above or below do not penetrate their voids. Such blankets must be well drained because if they should fill with water, they would aggravate rather than prevent frost heave.

Insulating blankets between the ground surface and the frost-susceptible soil reduce the penetration of the freezing line. Ordinarily these are well-drained coarse sand and gravel. Often these are placed directly on the ground surface and are employed beneath pavements or the floors of cold-storage warehouses. Insulating blankets of foam glass, plastic foam and cork are used under very low-temperature storage facilities where their high efficiency justifies their high cost. Similar blankets are being used beneath pavements on a limited scale.

Chemical additives show promise in preventing frost heave. Dispersing agents, such as sodium polyphosphates, mixed with the soil cause higher densities and result in lower permeability and less heave. Waterproofing materials and chemicals that change the adsorbed cations of the clay minerals reduce heave by altering the attraction for water.

PERMAFROST / In North America, north of about the Arctic Circle, the soil remains frozen to great depths throughout the year. The condition of the permanently frozen ground is termed *permafrost*, and in some areas it is as deep as 1000 ft. The latitude of permafrost is not uniform and in parts of Canada and Alaska it is found as far south as latitude N. 62. Isolated islands of permafrost in high or sheltered areas are found south of the continuous zone and very localized pockets occur in high mountains throughout the world.

The upper few feet of permafrost may thaw in the summer time. This is the *active zone* and it frequently becomes a soft, soupy quagmire. Highways laid on permafrost fail and buildings settle in summer when the active zone softens. The heat of a building increases the depth of the active zone and aggravates the situation. Permafrost is often covered with an insulating blanket of moss and low, thickly matted vegetation called *tundra*. This minimizes the depth of the active zone, and removal of this natural insulation will greatly increase the active zone.

Foundations and subgrades must be placed below the active zone to avoid movement, or the soil in the active zone must be replaced with a non-frost-susceptible material like gravel or coarse crushed rock. Insulating blankets are sometimes used to minimize the thaw of the active zone. In extreme cases cold-air conduits through the soil and cooling systems are installed to offset the heat from buildings and boilers that would aggravate the thaw of the active zone.

REFERENCES

4:1 A. Casagrande, "Seepage Through Dams," *Journal New England Water Works Association*, July 1937.

4:2 K. Terzaghi, *Theoretical Soil Mechanics*, John Wiley & Sons, Inc., New York, 1943, p. 258.

4:3 T. A. Middlebrooks, "Seepage Control for Large Earth Dams," *Third Congress of Large Dams*, **2**, Stockholm, 1948.

4:4 N. F. Williams, personal communication with G. F. Sowers, 1950.

4:5 T. W. Smith and G. V. Stafford, "Horizontal Drains on California Highways," *Journal of the Soil Mechanics and Foundations Division, Proceedings, ASCE*, **83**, SM3, July, 1957.

4:6 C. I. Mansur and R. I. Kaufman, "Dewatering," Chapter 3, *Foundation Engineering*, McGraw-Hill Book Co., Inc., New York, 1962.

4:7 L. Casagrande, "Electro-Osmotic Stabilization of Soils," *Journal Boston Society of Civil Engineers*, **39**, January, 1952, p. 51.

4:8 F. E. Richart, "A Review of the Theories for Sand Drains," *Journal of the Soil Mechanics and Foundations Division, Proceedings, ASCE*, **83**, SM3, July, 1957.

4:9 *Vertical Sand Drains for Stabilization of Embankments*, Bulletin 115, Highway Research Board, Washington, 1955.

4:10 A. W. Johnson, "Frost Action in Roads and Airfields," *Special Report 1*, Highway Research Board, Washington, 1952.

4:11 G. Beskow, "Soil Freezing and Frost Heaving with Special Applications to Roads and Railroads," *Swedish Geotechnical Society 26th Yearbook*, Series C, No. 375, 1935 (translated by J. O. Osterberg, Northwestern University, 1957).

4:12 A. Casagrande, "Discussion on Frost Heave," *Proceedings Highway Research Board*, Washington, 1931, p. 168.

4:13 "The Unified Soil Classification System—Appendix B, Characteristics of Soil Groups Pertaining to Roads and Airfields," *Technical Memorandum 3-57*, Waterways Experiment Station, Vicksburg, March 1953.

4:14 T. W. Lambe, "Modification of Frost Heaving Soils with Additives," *Bulletin 135*, Highway Research Board, Washington, 1956.

SUGGESTIONS FOR FURTHER STUDY

1. References 4:1, 4:6, and 4:10 above.
2. D. W. Taylor, *Fundamentals of Soil Mechanics*, John Wiley & Sons, Inc., New York, 1949.

3. C. N. Zangar, "Theory and Problems of Water Percolation," *Engineering Monograph No. 8*, U.S. Bureau of Reclamation, Denver, 1953.
4. *The Moretrench Wellpoint System*, Moretrench Corporation, Rockaway, N.J., 1967.
5. C. I. Mansur and R. I. Kaufman, "Dewatering," Chapter 3, *Foundation Engineering*, McGraw-Hill Book Co., Inc., New York, 1962.
6. M. E. Harr, *Groundwater and Seepage*, McGraw-Hill Book Co., Inc., New York, 1962.
7. *Pore Pressure and Suction in Soils*, (Proceedings of a Conference by British National Society on Soil Mechanics and Foundation Engineering), Butterworths, London, 1961.
8. *Moisture Equilibria and Moisture Changes in Soils*, A Symposium, Butterworths, Inc., Sydney, Australia, 1965.
9. H. R. Cedergren, *Seepage Drainage and Flow Nets*, John Wiley & Sons, Inc., New York, 1967.
10. "Water and its Conduction in Soils," *Special Report 40*, Highway Research Board, Washington, 1959.

PROBLEMS

4:1   a. Draw a flow net for seepage under a vertical sheet pile wall penetrating 25 ft into a uniform stratum of sand 50 ft thick; $k = 0.03$ cm per sec.
     b. If the water level on one side of the wall is 35 ft above the sand and on the other side of the wall is 5 ft above the sand, compute the quantity of seepage per foot width of wall.

4:2   Draw the flow net for seepage under a concrete dam that is 120 ft long and rests on a 35-ft thick uniform stratum of silty sand. Bottom of the dam is 5 ft below upper surface of silty sand. Compute the quantity of seepage if the head on the dam is 65 ft and the permeability of the soil is 0.0003 cm per sec.

4:3   Draw the flow net for an earth dam 80 ft high, 15-ft crest width with a slope of 2.5 (horizontal) to 1 (vertical) upstream and 2 to 1 downstream. Dam rests on a 25-ft thick stratum of soil with the same permeability. Headwater level is 70 ft above base of dam; tailwater is 10 ft above base of dam.

4:4   Compute the hydraulic gradient required to produce a "quick" condition at the surface of a level mass of sand through which water flows vertically upward. The void ratio is 0.63 and the specific gravity of solids is 2.66.

4:5   Compute the safety against boiling in problem 4:1 if the soil void ratio is 0.42 and the specific gravity of solids is 2.67.

PROBLEMS

4:6 Specify the grain size distribution of soils that would serve as satisfactory filters for the soils given in problem 1:14.

4:7 Sketch a wellpoint system for draining an excavation that is 55 ft wide at the bottom and extends 57 ft below the water table. Show stages of construction by separate diagrams.

4:8 a. Which of the soils listed in problem 2:3 would you consider for drainage with wellpoints?
 b. Which of the soils listed in problem 1:14 could be drained with wellpoints?

4:9 A long excavation is 20 ft deep and 12 ft wide at the bottom. The ground water table is normally at a depth of 10 ft below the ground surface. A single line of wellpoints extends 10 ft below the bottom of the excavation and is 16 ft from the center line of the excavation. The soil is fine sand, with $k = 0.01$ cm per sec, and it is underlaid at a depth of 35 ft by rock. The ground water table remains at its original level 200 ft from the excavation and is lowered 15 ft below its normal elevation along the line of the points.
 a. Draw the flow net for seepage into the wells, assuming the line of wells to be one continuous slot in the ground.
 b. Compute the quantity of water pumped per well if the wells are 6 ft apart.
 c. At what level should the header be placed?

4:10 Which of the soils in problem 2:3 would you expect to be susceptible to frost action. Is this confirmed by Table 4:2?

4:11 a. Compute the theoretical critical gradient for a quick condition in Example 3:7.
 b. If 2 ft of pervious gravel were placed in the excavation bottom above the sand, how far down could the water level be pumped before the sand becomes unstable. The gravel weighs 133 lb per cu ft saturated.

4:12 a. What would be the maximum head on the under side of the sand in Example 3:1 that would permit pumping the excavation down to the sand surface, and which would maintain a ratio of total stress to uplift of 1.3 (a safety factor of 1.3).
 b. What would the seepage rate be under this condition?
 c. How could the seepage rate and uplift be controlled to maintain the safety factor described above?

4:13 A soil stratum overlies a bed of shale. The water table fluctuates greatly; the capillary rise is 8 ft. The soil has a void ratio of 0.48 and a specific gravity of solids of 2.67; it is 30 ft thick.
 a. Draw graphs of the vertical stress as a function of depth for total neutral and effective stress assuming the soil is dry.
 b. Draw similar graphs, assuming that the ground water table rises to the ground surface.

c. Draw similar graphs, assuming that the water table drops to 10 ft above the shale and that the soil is 30 per cent saturated above the height of capillary rise.

4:14 A sheet pile cofferdam 40 ft by 20 ft is driven into a river bottom to permit construction of a bridge pier. The river is 16 ft deep and the bottom is loose fine sand 8 ft thick underlain by coarse gravel. Unknown to the contractor, the fine sand merely fills a depression in the gravel; most of the river bottom at some distance from the cofferdam is gravel. The sheet piles are driven until they reach the gravel, where they can penetrate no further. (The contractor's foreman thought that he had reached bed rock.) The fine sand weighs 122 lb per cu ft saturated and has a permeability coefficient of 0.004 cm per sec. After driving the entire cofferdam box, pumping is commenced to remove the water and to expose the sand bottom.

a. When the water level reaches x ft above the fine sand the water turns muddy and commences to boil. What has happened?

b. Compute the level x and the rate of pumping required to maintain equilibrium at this instant. Neglect any head loss in the gravel.

CHAPTER 5
Earth Construction: Compaction and Stabilization

SOIL IS THE OLDEST OF CONSTRUCTION MATERIALS. MAN FOUND THAT HE could mold earth into blocks which he dried in the sun and laid up in walls for his house, and he learned how to mound earth up to bury his dead or to form an elevated platform for worship long before he discovered how to read or write. Similarly, rock debris from stream deposits or talus were heaped to form mounds, walls, and terraces. From this early mass-rubble, there evoked the more sophisticated coursed masonry construction. Although the techniques for executing masonry have improved through mechanization, the basic designs have changed little during the last millenium, a tribute to the ingenuity of the early builders and the integrity of this form of construction.

Soil is an enduring construction material, which is demonstrated by the prehistoric mud-walled cities and ancient mounds found in many parts of the world today. The scientific approach to the use of soil and rock as construction materials began in the 1930's and has revolutionized earthwork. Now it is possible to utilize them in the construction of dams, fills to support buildings and transportation avenues, and subgrades to resist concentrated wheel loads under conditions that were once thought to be unsuitable.

In spite of milleniums of successful experience, earth construction can lead to technical and financial fiascoes if care is not exercised in both design and construction. The following example demonstrates the problems that can arise when complete engineering control is relaxed.

In order to provide a level mall that would attract customers to a large shopping center in a hilly area, it was necessary to cut as deep as 40 ft or 12 m on the hill tops and fill 30 ft or 9 m in the low areas. The grading contract required placing the fill in 8 in. or 20 cm layers and compacting the soil to a specified density as measured by field tests. In order to advance the opening date and take advantage of the pre-Christmas rush, the owner, over the protests of his engineer, waived the compaction requirements. The work was done during the hot dry weather of late summer. The fill was placed in layers as thick as 18 in. or 45 cm and rolled with a heavy roller until it was hard. The owner was convinced of the quality of the fill when he found it impossible to drive a $\tfrac{3}{4}$ in. diameter rod more than a few inches with blows from a 10 lb sledge hammer.

December brought rain along with the crowds of people. Unpaved areas of fill soaked up water like a giant sponge. Settlement as great as 4 in. or 10 cm developed in floor slabs and in building columns whose foundations were supported on fill. Portions of the parking area pavement settled more than 6 in. or 15 cm, causing water to pond and patrons to complain.

Tests of the fill in the area of settlement found poor compaction. The hard appearance had been caused by lumps of dried clay that were rolled together into an open, but rigid mass. The lumps melted upon wetting, however, and the fill settled into a wet spongy mass. In the parking areas it was necessary to replace the fill. In the building areas, foundations required underpinning, and floor slabs were releveled by pumping mortar beneath them. The ultimate cost and delay far exceeded the time and money saved by neglecting the project specification.

Modern technology has made it possible to build larger structures of earth and rock quickly and economically, that will be even more durable than the monuments of the ancients if every step of the operation, from planning to construction, is properly engineered. However, if any phase of the work is allowed to get out of control, the result will only be larger problems as the previous example showed. The engineer must be intimately familiar with the materials, their ultimate behavior, and the construction operations that are necessary to obtain the required results if the product of construction is to perform as intended.

5:1 Soils and Rocks as Construction Materials

When soil or broken rock is used as a raw material for construction, it undergoes so many changes that it ultimately has little resemblance to its undisturbed state. Excavation is the first step in the process of change: The structure is broken down by blasting or the action of a shovel or scraper; the different strata become mixed; and the water content increases or decreases, depending on the weather. In some instances the material is purposely modified to improve its characteristics: It can be mixed with other soils; it can be supplied with chemical admixtures that change its chemical and physical properties; or it can be bound together by a cementing agent. The final step of placing the soil or rock in the structure involves still more changes: Mixing brings about a relatively uniform composition, and compaction produces a controlled void ratio that is often considerably less than that of the original soil and close to that of some rocks. Each step, from the undisturbed deposit to the finished structure, should be considered an engineering operation that is a part of a manufacturing process; each should be carefully planned and adequately controlled to ensure a satisfactory product.

USES OF SOIL AND ROCK IN CONSTRUCTION / The most important use of soil and broken rock is in the construction of fills. A *fill* is a man-made deposit used to raise the existing ground surface or sometimes used to dispose of waste such as industrial by-products, garbage or rubbish. The material from which it is constructed is termed *fill, fill material,* or *borrow*. Fills serve many purposes. Long narrow fills, termed *embankments*, carry railroads and highways across low areas or act as dams and levees to impound or confine water. Fills are used in building construction to provide level sites in hilly country. If properly constructed, they can support heavy structures with safety and only nominal settlement. Fills are placed behind retaining walls and bulkheads to secure the required ground surface contour and to bridge the gap between the wall and the original soil.

The foundation or supporting soil for a highway or airfield pavement is the *subgrade*. Subgrades can be the surface of the virgin soil or specially prepared, artificially compacted layers of soil or crushed rock.

Pavements themselves are sometimes constructed of soil or crushed rock. In such cases they often must be modified with admixtures and binders to give the pavements sufficient strength and resistance to deterioration.

In many countries earth is used for the walls of structures. Sun-dried bricks have been used for thousands of years in arid regions, while in regions of more moisture, walls of earth are built by ramming the soil between temporary forms. Earth buildings are practical in areas where cheap construction labor and abundant supplies of clays and sands are available. They are durable when well constructed and protected from rain wash and flooding.

Another ancient technique, followed today, is to use soil as a plaster or filling over a loose framework of wood or reeds. This prototype of reinforced concrete is termed *wattle and daub*. It is used where wood is plentiful and where the environment (rainfall and earthquake) are unfavorable to unreinforced earth.

CONSTRUCTION / Building structures of soil or rock requires a different approach from building structures of other materials. First, local materials must be used because it would cost too much to transport large quantities from a long distance away. For instance, in planning a highway, either a fill or a long bridge could be used to carry the roadway across a valley. Compared with the bridge, the fill would require enormously greater quantities of materials to do the same job, but if those materials were available locally at little cost, the fill would be cheaper than the bridge. If the fill had to be hauled a long distance, the cost would be so great that a bridge would be cheaper. The same reasoning often applies to the choice between an earth or concrete dam.

Second the design of the structure that is to be made of soil or rock must be very closely keyed to the construction. Ordinarily an engineer designs a structure with assumed values for strength and other material characteristics and then writes specifications that will ensure his getting what he assumed. When he contemplates using an earth structure, however, he first must investigate the materials available, determine their suitability for construction, and then design the structure to suit the actual characteristics of available soils or rock. Often the design must be modified after construction has started in order to compensate for some unforeseen changes in the materials. The design of the structure is controlled by the engineering problem of manufacturing a suitable, economical material for the structure. The engineering problems of selection and processing are discussed in this chapter; the structural design of earth structures, such as embankments and dams, is discussed in Chapter 11.

Finally the constructor must exercise engineering initiative in processing the materials, from excavation to compaction. He must guard against using methods that he found satisfactory on one job but which may not be suitable for handling the materials on a new project. For example, a contractor who has found that flooding helps to compact a damp sand too often wants to flood all fills with water. The result, if the soil is clay, is likely to be a soupy mass that will not support anything for years.

OBJECTIVES IN EARTH CONSTRUCTION / The product of earth construction, whether it be a fill for a highway, an embankment for a dam, the support for a building, or the subgrade for a pavement, must meet certain requirements:

1. It must have sufficient strength to support safely its own weight and that of the structure or wheel load on it.
2. It must not settle or deform under load so much that it damages the soil or the structure on it.
3. It must not swell or shrink excessively.
4. It must retain its strength and incompressibility permanently.
5. It must have the proper permeability or drainage characteristics for its function.

Strength is a major factor in the use of soil and rock in dams, high embankments, and subgrades. It depends on the nature of the material, the water content, and void ratio. In general, for any given earth material, the strength increases with a decrease in water content and with a decrease in void ratio (or increase in density). When the quality of the available material is poor, it is frequently possible to compensate for the deficiency by increased density.

Settlement from consolidation and elastic deformation is important in all applications of earth construction but particularly critical in subgrades and embankments that support pavements or structures. Settlement in the fill itself depends on the nature and density of the material of which it is composed. For most soil or broken rock, the elasticity increases and the compressibility decreases with increased density. By preconsolidation to a sufficiently high density through compaction, nearly every material can be made to support reasonable loads without undue settlement.

Shrinkage of soil can be a factor in the settlement of pavements and structures on fill and is sometimes a serious hazard in the leakage of earth dams. The amount depends on the soil character, the density, and the loss in water content after construction; the greater the density and the less the water content change, the less the shrinkage. Generally rock fills do not shrink.

Swelling is extremely hazardous because it disrupts the shape of the fill, damaging pavements and structures, and also because it is accompanied by a loss in strength. Swelling depends on the mineralogy of the rock or soil, the density, and the increase in moisture after construction. In general, the tendency to swell is increased with increased density and can be controlled best by the proper choice of soils and by preventing (where possible) increasing water contents.

Loss of strength or increase in compressibility are generally related to two mechanisms: the deterioration of the solid phase, or pore water pressure. Deterioration of the solids is a form of accelerated weathering induced by placing the material in a new environment. Moisture is a major factor in such physical and chemical deterioration: the clay minerals adsorb water, expand, and their bonding weakens; salts ionize to accelerate chemical reaction, and cemented bonds between particles soften.

Water pressure, as described in Chapter 3 is a direct factor in soil and rock strength. Seepage, induced by water pressure differences, can erode soil and some rocks and thus change them.

The losses in quality can be minimized by the selection of the material and by control of water as described in Chapter 4. Generally the higher the density the slower the deterioration. Furthermore, a dense material, even in the deteriorated state, is generally better than the same material loose. However, if the material swells in the presence of moisture, too much density can promote deterioration.

Permeability is a factor in fills subject to temporary inundation, subgrades which must drain, and in dams. It is a function of the character of the soil and must be controlled by the proper selection of the soil material.

OBTAINING THE REQUIRED CHARACTERISTICS / In order to obtain the required properties in the end product of earth construction the engineer must control the character of the material, the moisture, and the density. Control of the moisture is ordinarily possible during the construction period. Afterwards, however, the moisture is largely dependent on the environment and the use to which a structure is put, and often is not subject to control no matter how desirable it may be. The material in some uses, such as the upstream face of a dam, is certain to become saturated. In other applications, such as a subgrade, the material can ordinarily be protected by drainage, but there is still danger (however remote) that it could be saturated by abnormal rainfall. Proper water control by drainage (Chapter 4) is essential in the design of structures of earth in order to maintain the best properties under normal conditions. However, unless the engineer is certain that the drainage will always be effective, it is necessary to design earth structures on the basis that the moisture could increase to the point of saturation.

Control over the character of the soil or rock is ordinarily limited by the materials available and the cost of their excavation and hauling to the job site. Too often, only a limited range of materials is present in any locality, and the engineer must select the best of what there is. It is sometimes possible to alter the soil or rock by processing, such as mixing two soils to improve the gradation of both, or by adding a material that alters their physical or chemical nature. Such a change to improve the material is termed *stabilization* and will be described in Section 5:6.

The greatest control of the soil or rock properties is through densification. By densification it is usually possible to compensate for deficiencies in quality and for the deterioration in properties that results from increased moisture. The only property that is not improved by densification is the tendency to swell, and that must be controlled by proper selection.

5:2 Theory of Compaction

From prehistoric times builders have recognized the value of compacting soil to produce a strong, settlement-free, water-resistant mass. Earth has been tamped by heavy logs, trampled by cattle, or compacted by rolling for more than 2000 years, but the cost of such crude work was often more than the value of the compaction. On the other hand, earth that was merely dumped in place without compaction frequently failed under load and continued to settle for decades. It remained for R. R. Proctor to point the way to low-cost, effective densification.[5:1]

MECHANICS OF DENSIFICATION / Densification, or a reduction in the void ratio, occurs in a number of ways: reorientation of the particles; fracture of the grains or the bonds between them, followed by reorientation; and bending or distortion of the particles and their adsorbed layers. Energy consumed in this process is supplied by the *compactive effort* of the compaction device. The effectiveness of the energy depends on the type of particles of which the fill is composed and on the way in which the effort is applied. In a cohesive soil the densification is primarily accomplished by distortion and reorientation, both of which are resisted by the interparticle attractive forces of "cohesion." As the water content of the soil is increased, the cohesion is decreased, the resistance becomes less, and the effort becomes more effective. In a cohesionless soil or crushed rock the densification is primarily attained by reorientation of the grains, although fracture of the particles at their points of contact is an important secondary factor. The reorientation is resisted by the friction between the particles. Capillary

SEC. 5:2] THEORY OF COMPACTION

tension in moisture films between the grains increases the contact pressures and increases the friction. As the moisture content increases, the capillary tension decreases and the effort becomes more effective.

In some sands and in broken rock, the local crushing of the points of contact between particles is a major mechanism of densification. Moisture accelerates the crushing and thus aids compaction. The acceleration of crushing reduces future settlement after construction.

If the moisture content is very high, however, the densification and decrease in void ratio of both cohesionless and cohesive soils leads to saturation. The buildup of neutral stress prevents further reduction in void ratio so that additional effort is wasted. Saturation, therefore, is the theoretical limit for compaction at any given water content.

In coarse cohesionless soils and broken rock, the permeability is so great that saturation cannot occur during construction unless the mass is inundated. The limiting density is controlled by the particle geometry and the most favorable structural arrangement termed *packing*: the minimum void ratio.

MOISTURE–DENSITY RELATION / The importance of soil moisture in securing compaction is illustrated by the following experiments. A sample of soil is separated into six or eight portions. Each portion is mixed thoroughly with a different quantity of water so that each has a different water content, ranging from nearly zero to about midway between the liquid and plastic limits. Each portion is compacted in a container with *exactly the same compactive effort*; its water content and weight of solids per cubic foot of compacted soil, usually termed the *dry density* and denoted γ_d, are determined:

$$\gamma_d = \frac{W_s}{V}, \qquad (5:1a)$$

$$\gamma_d = \frac{\gamma}{1+w}. \qquad (5:1b)$$

If a graph is plotted with water content as the abscissa and the dry density as the ordinate, the resulting curve will be similar to Fig. 5:1. It will be seen that there is a particular water content, known as the *optimum moisture*, that results in the maximum dry density for the particular compaction method used. For a given soil, the greater the dry density, the smaller the void ratio, regardless of water content; so, maximum dry density is just another way of expressing the minimum void ratio or the minimum porosity.

For any given water content, perfect compaction would expel all air from the soil and produce saturation. If the dry densities corresponding to saturation at different water contents are plotted on the graph, the result will be a curve that lies completely above the first. This is known as the *zero air voids* curve and represents the theoretical densities obtained by

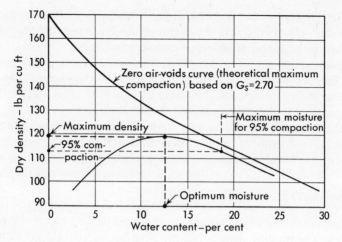

Figure 5.1 *Moisture-density curve of a cohesive soil for one method of compaction and maximum moisture for a specified degree of compaction.*

perfect compaction at different water contents. The theoretical maximum density of the zero air voids curve γ_z is computed from the specific gravity of the solids for each given moisture:

$$\gamma_z = \frac{\gamma_w}{w + (1/G_s)}. \tag{5:2}$$

At high water contents the theoretical dry density is low because much of the volume of the soil is occupied by water. At low water contents the theoretical density increases until at a water content of zero it becomes equal to $\gamma_w G_s$, the weight of the soil grains.

Because of the geometry of the grains there is some limiting density (or minimum void ratio) beyond which no further change is possible without a major breakdown of the grains. This point can be fairly well defined for sands, and is equivalent to the minimum void ratio, Section 1:11.

In materials having weak, porous grains, such as volcanic ash, coquina, and cinders, this limit cannot be defined and continued working will produce increased densities until the material approaches a solid with $\gamma_d \to \gamma_w G_s$.

The compacted density increases with increasing moisture, as would be expected from the mechanics of the process previously described. The increase is limited by saturation—the zero air voids curve—where neutral stress prevents a further reduction in void ratio without a reduction in moisture. If the moisture increases, the density must therefore decrease, the slight difference between the actual curve and the theoretical maximum being caused by air trapped in the voids. The optimum moisture is a compromise where there is enough water to permit the grains to distort and reposition themselves but not so much water that the voids are filled.

SEC. 5:2] THEORY OF COMPACTION

In clays the optimum moisture for compaction by rollers is often close to or slightly below the plastic limit. In sands the moisture density curve drier than optimum is poorly defined, Fig. 5.2. In some cases it rises toward the maximum density at very low moistures because there is little capillary tension to resist repositioning of the grains.

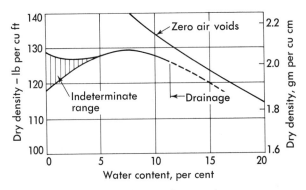

Figure 5.2 Moisture-density curve for a medium sand.

COMPACTIVE EFFORT / If a second set of soil samples is made up with different water contents, as previously described, and then compacted by a different effort, a similar moisture–dry-density curve will be produced but with a different optimum moisture and maximum density. The greater the effort, the higher the maximum density and the lower the optimum moisture.

The relation between effort and maximum density is shown in Fig. 5.3. It is not linear, and a large increase in effort is required to produce a small increase in density. The way in which the effort is applied has a significant

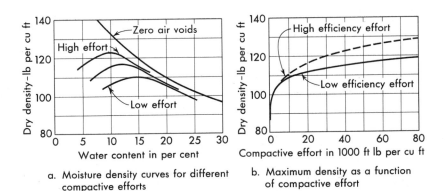

a. Moisture density curves for different compactive efforts

b. Maximum density as a function of compactive effort

Figure 5.3 Effect of compactive effort on moisture-density curves and maximum density.

effect on the density. In cohesionless soils as well as crushed or broken rock, vibration, which reduces the friction between the grains, is particularly effective. In cohesive soils, pressure that bends and forces the grains into new positions is better. A large number of applications of small pressure is not so effective as the same total effort applied in a single application, because a small force cannot overcome the cohesive resistance to grain movement, no matter how often it is applied. The duration of the effort is sometimes a factor in the density obtained. In coarse-grained soils the neutral stress that resists compaction at high moisture contents will not build up if the effort is applied so slowly that the water can drain away. In some clays a rapidly applied effort appears to mobilize viscous resistance in the water and is less effective than a slowly applied effort.

COMPACTION TESTS / A number of arbitrary standards for determining the optimum moistures and maximum densities have been established to simulate different amounts of effort as applied by the full-sized equipment used in soil construction. The simplest and the most widely used are the "Proctor tests" named for R. R. Proctor, who first developed the optimum-moisture–maximum-density concept.[5:1]

Standard Proctor (ASTM D 698, AASHO T 99. British Standard 1377: 1948):

Twenty-five blows of a 5.5 lb hammer falling 12 in. on each of 3 equal layers in a 4 in. diameter 1/30 cu ft cylinder.* The effort is 12,400 ft-lb per cu ft which is comparable to light rollers or thorough tamping.

Modified Proctor (ASTM D 1557, Modified AASHO):

Twenty-five blows of a 10 lb hammer falling 18 in. on each of 5 equal layers in a 4 in. diameter 1/30 cu ft cylinder.* The effort is 56,200 ft-lb per cu ft which is comparable to that obtained with the heaviest rollers under favorable working conditions.

In most soils the maximum density by the Modified method is from 3 to 6 lb per cu ft greater than that by the Standard. A number of other procedures for obtaining optimum moistures and maximum densities have been developed, such as static pressures and kneading pressures, more closely to simulate field conditions. Their use is not widespread, however, because the test equipment is more complex and the results not sufficiently different in many soils to justify the added expense.[5:3]

These laboratory tests are limited to particles finer than $\frac{3}{4}$ in. or 2 cm. Full scale field pilot tests are necessary for coarser materials such as broken rock and gravels.

PERCENTAGE OF COMPACTION—RELATIVE DENSITY / It is frequently convenient to express the dry density of a soil as a percentage of the maximum as defined by one of the two standards or by any other specified method. This is the *percentage of compaction*. It can exceed 100 per cent.

---

* If the soil contains many particles larger than a No. 4 sieve, a 6-in. diam cylinder of the same height is used and the blows increased to 55 per layer.

The maximum dry density from the laboratory compaction test is a function of the method of compaction. The minimum void ratio (and its equivalent dry density) as defined in the discussion of relative density of cohesionless soils, Section 1:11, is the limit for all methods of compaction that do not break the particles. Therefore, there is no fixed relation between the minimum void ratio and maximum density as defined by any given level of compaction. Generally the maximum densities for sands by the Modified Proctor standard are equivalent to relative densities of between 95 and 100 per cent.

5:3 Evaluation of Materials

The evaluation of the materials includes the determination of the quantity and quality of the materials available, the testing of the soils to find their physical properties when compacted, and the selection of the material and the degree of compaction to be used in construction.

FIELD SURVEY / The first step is a survey of all the soil deposits that can be used. The depth and extent of the different soil strata are determined by auger boring on a grid pattern, as described in Chapter 6. Samples weighing about $\frac{1}{2}$ lb or 200 g are secured of each different material in each boring to the depth that it appears likely the soil will be removed.

PRELIMINARY EVALUATION OF SOIL / A preliminary evaluation of the soil samples is made on the basis of past experience. For this purpose, classification by the Unified System[5:2] or the Public Roads System (Chapter 2) is helpful because both have been supplemented by performance ratings, such as Table 5:1.[5:2] A number of state highway departments have developed systems that are applicable to their own peculiar soil problems. Since these ratings have been based on field behavior, classification is a cheap, rapid method for estimating the properties of the compacted soil and determining its promise of suitability for construction.

COMPACTION STUDIES / Soils whose availability and suitability for fill purposes have been found satisfactory by the field survey and classification are sampled again to secure enough material (50 to 100 lb or 20 to 40 kg) for more extensive testing. In order to avoid two sampling operations, representative large samples are often made in the first place, but the handling and transportation of many samples weighing 50 lb to 100 lb is a problem.

Tests are made of each sample to determine its natural or *field moisture* and to obtain its moisture density curve by one of the standard procedures. Samples are then prepared at different percentages of the maximum density, such as 92, 95, and 97 per cent, usually at the highest moisture consistent with the degree of compaction.

Swelling and shrinking tests can be made by soaking compacted samples in water and determining their percentage volume increase and by drying samples to determine the percentage volume decrease. The sum of the volume increase and the volume decrease is called the *percentage volume change*. Fills made of soil having a volume change over 5 per cent may require special provisions to prevent their moisture content from changing enough to cause damage by swelling and shrinking.

Strength and consolidation tests can be made of the compacted samples, simulating the worst possible field conditions. Unless it is certain that the soil will never become saturated, the tests are made after soaking the compacted soil in water under the future confining load or saturating the soil under back pressure. The results of the tests can be presented on moisture–density coordinates in terms of "contours" of settlement under a given load or contours of strength under a given confining pressure, as shown in Fig. 5.4, or as functions of density, Fig. 5.5.

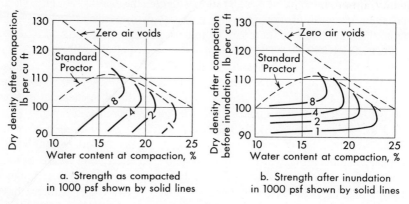

Figure 5.4 Relationship of unconfined strength of a compacted cohesive soil *(undrained shear)* to water content and dry density. *(After Seed and Chan.)*

STRENGTH AND COMPRESSIBILITY OF COMPACTED COHESIVE SOILS[5:4, 5:5, 5:6, 5:7] / The physical properties of a compacted soil depend largely on the soil material, moisture, and density. In addition the structure and the conditions of compaction that produced it are important in cohesive soils. When the cohesive soil is compacted at moisture contents less than optimum, an aggregated structure is formed; when the soil is compacted at high moisture contents, a dispersed structure is formed, with the flakey particles aligned in parallel.

The typical undrained strength of a cohesive soil, as compacted, at a constant confining pressure is shown in Fig. 5.4a. At constant density the strength decreases with increasing moisture; at constant moisture the strength increases with increasing density. An exception to the latter occurs as the zero air voids curve or saturation is approached. The increasing density brings about increased pore pressure and thereby decreased undrained strength. This sometimes occurs when a soil is compacted at a moisture content well above the optimum and can result in shear failure during construction. This is termed "overcompaction" and is prevented by proper moisture control.

The undrained strength of a compacted cohesive soil after inundation (under a constant confining pressure) is shown in Fig. 5.4b. Although the moisture content after inundation is nearly the same for equal densities, the strength decreases slightly with increasing initial (compaction) moisture, probably because the soil structure changes from the stronger aggregated to the weaker dispersed form. As before, close to the zero air voids curve the strength drops rapidly with increasing density because of the greater buildup of neutral stress in the dispersed structure.[5,6]

The strength as a function of density for various water contents is shown in Fig. 5.5a. Generally the undrained strength increases with increasing density, although it falls off as saturation is approached.

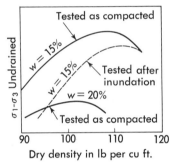
a. Undrained shear at constant σ_3 and two compaction moistures

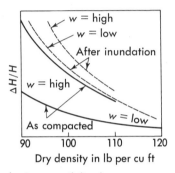
b. Compressibility for a constant $\bar{\sigma}_0$ and $\Delta\sigma$, compacted

Figure 5.5 Undrained compressive strength and compressibility as functions of compacted dry density.

The compressibility of a typical cohesive soil as a function of compaction is shown in Fig. 5.5b. In general compressibility decreases with increasing density (or decreasing void ratio) and decreasing compaction moisture. If the compacted soil is inundated before being subjected to consolidation, the compressibility is increased greatly. The soil that was compacted dry is still less compressible than that which was compacted wet of optimum because compaction of the soil in a dry state produces more strong edge to face bonding than wet compaction.

The potential swell of clays increases with density and decreases with the compaction moisture. Generally the swell is far greater for a soil compacted drier than the plastic limit than for one compacted wet. Swell can be minimized by limiting the degree of compaction and increasing the moisture content to the plastic limit or slightly above, (somewhat more than the optimum moisture by the Standard AASHO method).

TABLE 5:1 / CHARACTERISTICS AND RATINGS OF UNIFIED

Class (1)	Compaction Characteristics (2)	Maximum Dry Density Standard Proctor (pcf) (3)	Compressibility and Expansion (4)	Drainage and Permeability (5)
GW	Good: tractor, rubber-tired, steel wheel, or vibratory roller	125–135	Almost none	Good drainage, pervious
GP	Good: tractor, rubber-tired, steel wheel, or vibratory roller	115–125	Almost none	Good drainage, pervious
GM	Good: rubber-tired or light sheepsfoot roller	120–135	Slight	Poor drainage, semipervious
GC	Good to fair: rubber-tired or sheepsfoot roller	115–130	Slight	Poor drainage, impervious
SW	Good: tractor, rubber-tired or vibratory roller	110–130	Almost none	Good drainage, pervious
SP	Good: tractor, rubber-tired or vibratory roller	100–120	Almost none	Good drainage, pervious
SM	Good: rubber-tired or sheepsfoot roller	110–125	Slight	Poor drainage, impervious
SC	Good to fair: rubber-tired or sheepsfoot roller	105–125	Slight to medium	Poor drainage, impervious
ML	Good to poor: rubber-tired or sheepsfoot roller	95–120	Slight to medium	Poor drainage, impervious
CL	Good to fair: sheepsfoot or rubber-tired roller	95–120	Medium	No drainage, impervious
OL	Fair to poor: sheepsfoot or rubber-tired roller	80–100	Medium to high	Poor drainage, impervious
MH	Fair to poor: sheepsfoot or rubber-tired roller	70–95	High	Poor drainage, impervious
CH	Fair to poor: sheepsfoot roller	80–105	Very high	No drainage, impervious
OH	Fair to poor: sheepsfoot roller	65–100	High	No drainage, impervious
Pt	Not suitable		Very high	Fair to poor drainage

* Adapted from Reference 5:2.
† Not suitable if subject to frost.

SEC. 5:3] EVALUATION OF MATERIALS

SOIL SYSTEM CLASSES FOR SOIL CONSTRUCTION*

Value as an Embankment Material (6)	Value as Subgrade when Not Subject to Frost (7)	Value as Base Course (8)	Value as Temporary Pavement	
			With Dust Palliative (9)	With Bituminous Treatment (10)
Very stable	Excellent	Good	Fair to poor	Excellent
Reasonably stable	Excellent to good	Poor to fair	Poor	Fair
Reasonably stable	Excellent to good	Fair to poor	Poor	Poor to fair
Reasonably stable	Good	Good to fair†	Excellent	Excellent
Very stable	Good	Fair to poor	Fair to poor	Good
Reasonably stable when dense	Good to fair	Poor	Poor	Poor to fair
Reasonably stable when dense	Good to fair	Poor	Poor	Poor to fair
Reasonably stable	Good to fair	Fair to poor†	Excellent	Excellent
Poor stability, high density required	Fair to poor	Not suitable	Poor	Poor
Good stability	Fair to poor	Not suitable	Poor	Poor
Unstable, should not be used	Poor	Not suitable	Not suitable	Not suitable
Poor stability, should not be used	Poor	Not suitable	Very poor	Very poor
Fair stability, may soften on expansion	Poor to very poor	Not suitable	Very poor	Not suitable
Unstable, should not be used	Very poor	Not suitable	Not suitable	Not suitable
Should not be used	Not suitable	Not suitable	Not suitable	Not suitable

SELECTION OF SOIL AND DEGREE OF COMPACTION / The final selection of the soil depends on its availability and compacted characteristics, and the cost of excavation, hauling, and compaction. The density specified is the minimum percentage of the maximum dry density that will provide the necessary strength and incompressibility under the worst possible future moisture conditions. Table 5:2 will serve as a guide for preliminary estimates where soil test data are not yet available.

The amount of soil moisture to be used in construction must be consistent with the percentage of compaction specified. Some indication of the range can be seen in the moisture density curve: If a high percentage is required, the range in moistures is small; if a low percentage is required, the range is wider. The maximum possible moisture is the intersection of the horizontal line corresponding to the required compaction with the leg of the moisture density curve adjacent to the zero air voids curve. This point, for 95 per cent compaction, is shown on Fig. 5.1. This limit, or a moisture content 1 or 2 per cent below it, is sometimes specified to prevent useless attempts at compaction when the soil is too wet. Still lower limits for maximum moisture, such as the optimum, are sometimes specified to prevent the buildup of neutral stress by compaction.

TABLE 5:2 / TENTATIVE REQUIREMENTS FOR COMPACTION, UNIFIED SOIL SYSTEM CLASSES

Soil Class	Required Compaction—Percentage of Standard Proctor Maximum		
	Class 1	Class 2	Class 3
GW	97	94	90
GP	97	94	90
GM	98	94	90
GC	98	94	90
SW	97	95	91
SP	98	95	91
SM	98	95	91
SC	99	96	92
ML	100+	96	92
CL	100	96	92
OL	—	96	93
MH	—	97	93
CH	—	—	93
OH	—	97	93

Class 1 Upper 8 ft of fills supporting 1- or 2-story buildings
Upper 3 ft of subgrade under pavements
Upper 1 ft of subgrade under floors
Earth dams over 100 ft high

Class 2 Deeper parts of fills under buildings
Deeper parts (to 30 ft) of fills under pavements, floors
Earth dams less than 100 ft high

Class 3 All other fills requiring some degree of strength or incompressibility

SEC. 5:4] EXCAVATION, PLACEMENT, AND COMPACTION 217

EVALUATION OF ROCK FILL / The evaluation of rock and some industrial wastes for use in fills is more difficult because the experiences with them have been difficult to evaluate quantitatively. Two major factors are involved: (1) the physical changes, particularly the fragmentation and crushing during excavation, handling and compaction and (2) the deterioration or physical-chemical changes after placement. These same factors apply to soils; however they are more critical in rocks and industrial wastes. Most soils have already undergone major physical and chemical changes during weathering and are therefore at least partially adjusted to their environment. Freshly excavated rock and industrial wastes have not been subject to weathering; in their new environment of exposure to air, water, and physical punishment they may change drastically.

A preliminary evaluation of rock for construction purposes is given in Table 5:3. So many factors are involved, however, that such a classification can only indicate what problems are likely to be involved.

A more complete evaluation requires both field and laboratory tests. The field test is essentially a model of the proposed excavation and construction operation on a sufficiently large scale that it is realistic. The excavation, manipulation and compaction utilize a variety of techniques that simulate full scale construction. Tests of the behavior of the rock at all stages of the operation will show the degree of breakdown as well as the engineering properties.

Deterioration is more difficult to evaluate. The best method is to examine old fills, spoil areas, and waste heaps of the material, as well as exposures in open cuts that are several years old. Changes in texture, color and resistance to sliding with increasing depth below the surface are indications of deterioration.

Deterioration can be tested qualitatively by exposing identical samples of compacted rock to different numbers of cycles of wetting and drying or to different periods of exposure to the environment. After exposure the samples are tested for volume, gradation, and strength. A change in any of these properties with time is an indication of deterioration. An interpretation of such test data must be based on a correlation with the performance of fills made of the same material.

5:4 Excavation, Placement, and Compaction [5:8] [5:9]

Excavation of the raw materials, processing, hauling, placement, and compaction are important from the standpoint of cost and the time required for construction. All, and particularly compaction, are also vital in determining the quality of the completed structure.

EXCAVATION METHOD / The excavation procedure is usually selected by the constructor on the basis of the type of material and the layout of the borrow pit. However, since the excavation method affects the breaking of the material into small pieces and the way the material is mixed,

TABLE 5:3 / ROCKS AS CONSTRUCTION MATERIALS

Rock	Excavation Method	Fragmentation	Deterioration
Granite, Diorite	Requires blasting	Irregular fragments, depending on blasting pattern	Likely to be resistant
Basalt Tuff	Requires blasting Machines* to blasting	Irregular fragments depending on joints Irregular, often with excessive fines	Likely to be resistant Many forms deteriorate rapidly
Sandstone, Conglomerate	Machines* to blasting	Slabby, depending on bedding Excess fines, depending on cementing	Depends on character of cementing Some deteriorate into silty sand
Siltstone, Shale	Machines	Small blocks to thin slabs and chips	Many slake or break down rapidly into clay, accompanied by consolidation and loss of strength; should be considered suspect unless tests show otherwise
Limestone: Massive	Blasting	Irregular fragments, sometimes slabby	Shaley seams deteriorate, others resistant except to acids
Coquina Chalk	Machines	Porous fragments, excess fines common	Some porous forms soften on wetting, others become partially cemented with alternate wetting and drying
Quartzite	Blasting	Irregular, very angular	Likely to be resistant
Slate, Schist	Blasting	Irregular, slabby to flakey depending on laminations	Some may deteriorate on alternate wetting and drying
Gneiss	Blasting	Irregular fragments, sometimes elongated	Likely to be resistant
Mine Waste and Industrial Waste	Machinery	Depends on the material, in most cases irregular	Most forms (except igneous mine waste) should be considered susceptible until experience or tests show otherwise

* Blasting often helpful.

it is sometimes necessary for the engineer to specify the methods to be used, to ensure the desired result.

Hand excavation has been used throughout history and is still important in areas of cheap labor. Most soils can be excavated, but coarse gravel, boulders, and rock cannot be handled efficiently. Stratified materials can be either mixed or segregated by layers with equal ease, and small pockets of objectionable materials can be removed with little added expense. The soil is well broken up and usually requires little additional pulverization.

The *power shovel* is adapted to a wide variety of materials, from soft soil to boulders and layered or soft rock. It is also suitable for hard rock that has been broken by explosives. Stratified materials are easily mixed because the shovel makes a nearly vertical slice; but segregation by layers is difficult and expensive. At its most efficient output, the shovel is likely to excavate soil in large chunks, but it can be made to break the soil thoroughly, although there is some sacrifice in efficiency. The shovel is best adapted to deep borrow pits above the water table.

The *dragline* is adapted to most soils except tough clays and hard or cemented materials. Stratified materials can be either mixed or segregated if the layers are not thin. The soil is well broken up and mixed by the boiling action in the bucket. The dragline is most efficient when it excavates below its own base level, and for that reason it can excavate satisfactorily below water. Of course the materials would be waterlogged in that case.

The *scraper* or *pan* is adapted to most soils except very soft or sticky clays. It can excavate broken rock. Stratified materials are easily segregated because the scraper can excavate layers as thin as 6 in. Some mixing of strata is possible by making a deep cut or by making a sloping cut across several strata. The soil is broken up by the cutting blade and by the boiling action in the scraper bowl. The scraper is best in long borrow pits because it must travel in a straight line during excavation.

The *elevating grader* is similar to the scraper in the types of materials handled, but it tends to provide more pulverizing. A long level pit is required for best operation.

Continuous excavating machines originally developed for mining thin layers of coal have been adapted to large scale excavation of soil and soft rock. These consist of a wheel 10 to 15 ft in diameter, with teeth and shallow buckets on the perimeter. The wheel is operated in a vertical plane, at the end of a long boom that is pivoted so as to sweep across a pie-shaped sector of ground, at an ever increasing radius. The rotating wheel cuts the material into small fragments and drops them on to a belt conveyor mounted in the boom. The belt transfers the material to the operations center where it is transferred to the hauling equipment.

LOOSENING / Rock and the harder soils cannot be excavated without loosening. Thinly stratified rock and some hard soils can be loosened with *rippers* or *rooters*. Essentially such devices consist of hook-shaped blades that are towed by a tractor or mounted on a road grader. The blade lifts rock layers, breaking them into slab-shaped pieces, and tears hard soils into fragments that can be excavated by power shovels or scrapers. They are very effective in thinly bedded sandstones, shales, and shists, partially weathered rock, and hard, brittle soils.

Very hard soils and most rocks require blasting. This is a highly developed art that must be planned and executed by experienced personnel. However, the engineer must maintain control over this work because the engineering properties of the fragmented material are largely controlled by the blasting operations. The size and gradation of the particles depend on a number of factors: (1) the joint and bedding plane spacing; (2) the hardness of the rock; (3) the spacing of the blast holes; (4) the depth blasted at one time (termed the bench height); (5) the spacing of the explosive throughout the drill hole; and (6) the amount and type of explosive. To a great degree an experienced *powder man* can compensate for variations in the rock by his control over the blast hole spacing and the *powder factor* (weight of explosive per cubic yard of material). In most cases full-scale experimenting is necessary to establish the best combination of drill pattern, explosive and powder factor to produce the size and shape of particles needed for the subsequent construction operations.

The soil-rock boundary presents many problems in excavation, both technical and contractural. The rock surface is usually irregular and often hard strata are underlain by soft seams. The technical problem arises as soon as the earth moving equipment strikes the hard knobs or pinnacles in the rock surface: earth moving is either severely hampered or practically impossible. It may be necessary to resort to drilling and blasting of the knobs the same as in solid rock, although the total percentage of rock may be small. On the other hand, continuous rock consisting of alternate hard and soft seams sometimes can be loosened with rippers making it possible to excavate with equipment designed for soil. The contractural problem arises because of the change in procedure that is necessary and the greatly increased cost of work below the uppermost point of rock. It is essential that the contract documents describe the soil-rock interface as accurately as possible, and that they clearly define the limits of payment for each class of material. In some instances it is so difficult to do this that the contract lumps all excavation in one category, *unclassified*, that includes everything from muck to rock. Although this minimizes any controversies, it makes it necessary for the contractor to include a large contingency for the uncertain quantities of cheap soil and expensive rock work.

BORROW PIT CONTROL / Moisture control is often necessary in the borrow pit to obtain efficient operation of the excavation equipment and to condition the soil for future compaction. Predrainage is required in low, wet pits. Plowing and exposure to air and sunshine often help dry the soil. Moisture addition makes it easier to excavate hard, dry soils. If the soil is drier than optimum, water addition is necessary for compaction; and when the moisture is added at the borrow pit, the moisture is mixed by the subsequent handling and is likely to be uniformly distributed.

Some gradation control may be necessary in the borrow pit or rock quarry. A *rake* attached to a tractor can remove boulders or oversized rock fragments. A bulldozer or front end loader can push aside an occasional large lump of hard soil or boulder that will be objectionable in the fill. In extreme cases, a coarse screen is fabricated of structural shapes, and all the material dumped through it to *scalp* the unwanted sizes. Such a *grizzley* impedes the normal excavation process, and is used only when other methods of oversize control fail.

Supervision of the borrow pit is necessary to ensure the quality of the materials. This requires experienced technicians who can recognize the specified soils, and sometimes even a field laboratory for checking moisture, gradation, and plasticity.

HAULING / Transport of the soil and rock is primarily the concern of the contractor; while the operation is intimately involved in the cost of construction, it generally has no direct influence on the qualities of the soil or rock. However, if materials from several borrow pits are to be combined, or if different materials from a single borrow pit are to be segregated, such as sand for drainage layers and clay for impervious zones, then the hauling system must be compatible with these requirements. For example, a belt conveyor is not well adapted to transporting several different materials at the same time. Different materials can be easily segregated in single truck or scraper loads; if different materials from different sources are to be mixed, the proportioning can be controlled by the proportion of truck loads.

Some wetting or drying may occur during transportation; this is seldom significant, except for long belt conveyors. Addition of moisture to the trucks and scraper loads is sometimes practiced. Better mixing is usually possible when moisture is added before excavation or after spreading on the fill.

PLACEMENT AND PROCESSING / The placement of the fill materials depends on the method of hauling, the processing that is necessary before compaction, and the size of the area to be filled. The materials excavated by shovels, draglines, and elevating graders are hauled by trucks and wagons and *end-dumped* from the rear of the hauling unit in uniformly spaced piles, or *side-dumped*, or *bottom-dumped* from special units while moving to form long, narrow windrows. The scraper spreads the materials in ribbon-like layers the width of the blade.

The piles and windrows are spread out into uniform layers with a bulldozer or road grader. Some leveling of the layers spread by the scrapers is usually needed. At the same time objectionable materials such as roots, clumps of grass, and large stones are removed.

If the materials are too wet, they are cut and turned with a disk plow so that they will aerate and dry in the sun. Cutting is also necessary if the soil is in lumps that are too large to compact. If the soil is too dry, the correct amount of water (plus an allowance for evaporation) is added by sprinkling and mixed into the soil by plowing.

In some cases it is necessary to mix different soils, such as a wet and a dry soil, to obtain optimum moisture, or a sand and a clay to secure a sandy clay. If the materials are hauled by trucks, they are placed in alternate piles or windrows. They are then mixed by blading them across each other before spreading. Materials hauled by scrapers are placed in thin layers, one on top of the other, and are mixed vertically by plowing. Different types of traveling mixers are available to pick up the soils, mix and pulverize them, and place them back, ready for compaction.

COMPACTION CHARACTERISTICS OF SOILS / Compaction occurs by the reorientation of the particles or the distortion of the particles and their adsorbed layers. In a cohesionless soil, compaction is largely by reorientation of the grains into a more dense structure. Static pressure is not very effective in this process, for the grains wedge against each other and resist movement. If the grains can be momentarily freed, then even light pressure is effective in forcing them into a more dense arrangement. Both vibration and shock are helpful in reducing the wedging and aiding compaction. Flowing water will also reduce the particle friction and permit easier compaction. However, water in the voids also prevents the particles from assuming a more dense arrangement. For this reason flowing water can be used to aid compaction only when the soil is so coarse-grained that the water can leave the voids quickly.

In cohesive soils the compaction occurs by both reorientation and distortion of the grains and their adsorbed layers. This is achieved by a force that is great enough to overcome the cohesive resistance or interparticle forces. Vibration and shock are of little help, for although they provide a dynamic force in addition to the static, this is largely offset by the increased cohesive resistance that accompanies dynamic loading.

For greatest efficiency the compaction force must be high enough to distort the particles and shift the individual grains but not great enough to shear the mass. In a cohesionless soil the strength is dependent on confinement. This can be provided by a wide area of load application. In cohesive soils the strength is dependent on void ratio and moisture and largely independent of the confinement.

In cohesionless soils efficient compaction requires either a moderate force applied to a wide area, or shock and vibration. In cohesive soils efficient compaction requires higher pressure for dry soil than for wet, but the size of the loaded area is not critical. Increasing the pressure during compaction, as the density and strength increases, promotes efficiency.

COMPACTION METHODS / Many different compaction methods are in use, each with its adaptations and limitations that must be understood by the engineer. In many jobs the constructor claims that the specified degree of compaction is impossible, whereas it is only the wrong equipment or the wrong use of equipment that is responsible for the lack of density. Too often the available performance data on compaction equipment do not define the job conditions or the character of the soil accurately enough for the engineer to decide if the equipment will work in the new application. Therefore the engineer must study the mechanical behavior of the equipment and, if uncertain, conduct his own performance tests.

Tamping is the oldest method of compaction. It provides momentary pressure at the instant of impact and some vibration, and because of this dual action it is effective in both cohesive and cohesionless soils. The hand tamper, a block of iron or stone weighing 6 to 9 lb or 3 to 4 kg, is the simplest, but the compactive effort is so small that the soil must be tamped to 1- to 3-in. or 2.5 to 8 cm layers at a moisture 2 to 4 per cent above the Standard Proctor optimum. Furthermore, it is slow. Pneumatic tampers are much faster but only slightly more effective in producing high compaction.

Jumping tampers are actuated by gasoline-driven pistons that kick them into the air to drop back on the soil. The Barco Rammer (Fig. 5.6a) weighs 210 lb or 100 kg, jumps as high as 18 in. or 45 cm, and delivers a blow capable of compacting soils in layers 6 to 12 in. or 15 to 30 cm thick to the Standard Proctor maximum density at optimum moisture.

Rolling produces pressure that is applied for a relatively short time, depending on the roller speed. The *sheepsfoot roller* (Fig. 5.6b) consists of a steel drum with projecting lugs or feet.[5;9] It applies a high static pressure to a small area of 7 to 9 sq in. or an equivalent diameter of 3 to $3\frac{1}{2}$ in. or 7.5 to 8 cm. The pressure exerted depends on the number of feet in contact with the ground at any one time and on the roller weight (which can be varied by changing the water or wet sand ballast in the drum). Although pressures as low as 100 psi or 7 kg per sq cm and as high as 1200 psi or 84 kg per sq cm are available, most equipment now in use falls into two categories: light, with pressures of 150 to 300 psi or 10 to 20 kg per sq cm; and medium, with pressures of 350 to 700 psi or 25 to 50 kg per sq cm.

Many different foot shapes are used. There is little experimental evidence that favors any one of these.

Because of the small width of the loaded area, the sheepsfoot roller is adapted best to cohesive soils such as clays. The medium rollers are capable of producing densities greater than the Standard Proctor maximum in layers 6 to 12 in. or 15 to 30 cm thick (after compaction) and at moistures slightly below the optimum in six to eight passes over the surface. The light roller can produce 95 per cent of the maximum at optimum moisture with 4- to 6-in. layers.

A modified sheepsfoot roller with wider feet, up to 8 or 10 in. or 20 to 25 cm across, is far better for silty soils of low cohesion because the increased width of the loaded area produces greater confinement. This device, termed the *elephant's foot roller* by the authors, is made by removing some of

b. Sheepfoot rollers in tandem. (Courtesy Le Tourneau-Westinghouse, Inc.)

Figure 5.6 Soil compacting equipment.

a. Soil rammer. (Courtesy of Barco Manufacturing Co.)

c. Fifty-ton rubber-tired roller. (Courtesy of Bros, Inc.)

d. Fifteen-ton self-propelled vibrating drum roller.

the feet from the ordinary sheepsfoot and welding flat plates on those remaining.

Tamping rollers are often operated in tandem as shown in Fig. 5.6b. The trailing roller should have a different foot arrangement than the lead roller to prevent the trailing roller feet from following in the exact footsteps of the lead roller. The principal of increasing pressure can be realized by using larger feet and lower pressures on the lead roller.

Sometimes a sheepsfoot roller is employed to break up slabby fragments of soft rock, to pulverize a hard dry soil, or to aid in mixing or in the addition of water. Any compaction is incidental, and probably not of benefit in this stage of the operation.

The pneumatic tire has proved to be an excellent compactor for cohesionless and low cohesion soils, including gravels, sands, clayey sands, silty sands, and even sandy clays. It applies a moderate pressure to a relatively wide area so that enough bearing capacity is developed to support the pressure without failure. The *light rubber-tired roller*, also known as the *traffic roller*, employs from 7 to 13 wheels mounted in two rows and spaced so that the wheels of the rear row track in the spaces between those of the front row. The wheels are mounted in pairs on oscillating axles so that they can follow the ground irregularities. The tires are similar in size to those used on pickup trucks, and each exerts a load of up to 2000 lb or 1 ton on the soil, depending on the ballast in the box mounted above the wheels. With a tire contact pressure of 35 psi or 2.5 kg per sq cm the load is applied to an area with an equivalent diameter of nearly 9 in. or 20 cm.

The *heavy rubber-tired roller* (Fig. 5.6c) consists of four large tires mounted side by side on a suspension system that permits them to follow the ground irregularities. The load is provided by a ballast box, which is filled with earth, water, or even steel billets. A number of sizes are available, from a 35-ton maximum load to a 200-ton maximum load, with tire pressures of from 75 to 150 psi or 5 to 10 kg per sq cm. The widely used 50-ton roller applies its 25,000-lb wheel load at 100 psi or 7 kg per sq cm to an equivalent circle having an 18-in. or 45 cm diam.

The light rollers are capable of compacting soils in 4 in. or 10 cm thick layers to densities approaching the Standard Proctor maximum at optimum moisture in three or four passes. The heavy rollers can obtain densities far greater than the Standard Proctor maximum (as high as 105 per cent of the Modified Proctor maximum in one case) in layers up to 18 in. or 45 cm thick with four to six passes, with moistures slightly less than optimum.

The smooth-drum paving roller is sometimes used for compacting cohesionless soils. It is satisfactory if the layers are thin and well leveled, but it tends to bridge over low spots. A drum-type roller with a surface of circular segments with space between, called the *Kompactor*, provides low pressure over a wide area and is useful in compacting sand, gravel, and crushed rock. The grid roller employs a heavy steel grid in place of the smooth steel drum. It develops a relatively high pressure over small areas and has been found useful in compacting cohesive soils in thin layers.

Vibrators in a number of forms have been developed for compacting cohesionless soils. The surface plate form consists of a curved metal plate or shoe on which is mounted the vibrator. Within the vibrator are eccentric weights that rotate at speeds of from 1000 to 2500 rpm, depending on the particular design, and produce an up-and-down vibrating impulse. These are available in single units that propel themselves slowly across the surface and in multiple units, mounted on a self-propelled chassis. They are capable of compacting cohesionless soils to densities as high as the Standard Proctor maximum in layers 6 to 12 in. thick.

Vibrating rollers have been developed to provide both greater weight and greater intensity of vibration. One form consists of a steel drum 4 ft to 5 ft in diameter and weighing several tons, Fig. 5.6d. A vibrating unit driven by a gasoline engine is mounted on the roller and delivers the impulse to the drum. A second form consists of a two-wheel rubber-tired roller. A gasoline-engine-driven vibration unit is attached to the axle, and the ballast box is spring-suspended so that the axle is free to vibrate unhampered by the mass above. Both types are capable of compacting cohesionless soils in layers from 1 to 3 ft or 30 cm to 1 m thick to the Standard Proctor maximum in two or three passes. The largest drum vibrators, weighing up to 15 tons are very effective in compacting broken rock up to 2 feet or 60 cm in diameter in layers as thick as 3 ft or 90 cm. Significant density increases have been measured 10 ft or 3 m below the surface after 3 to 6 passes of these machines.

The *vibroflotation* process employs a giant cylindrical vibrator that is suspended from a crane. Water jetted from within the vibrator loosens the soil and permits the vibrator to penetrate as deep as 40 ft or 12 m. The vibration, which is in the horizontal direction, compacts the soil in a cylindrical column 8 to 10 ft or 2 to 3 m in diameter. The device is slowly withdrawn and at the same time sand is placed in the annular space between the vibrator and the soil, filling the hole left behind. The water opens the hole for the vibrator and in some cases helps the particles reorient themselves. The method is suitable only for free-draining soils where the water in the voids will not hamper compaction. It can be used below the water table.

Pile driving is a very effective way of compacting loose, cohesionless soils throughout great depths. The hammer vibration coupled with the displacement force of the pile is ideal, but the method is slow and expensive. The method ordinarily used is the same as for driving uncased concrete piles, but instead of concrete, sand can be used. The resulting piles are called *sand piles*.

The treads of a crawler tractor are efficient compactors of cohesionless soils placed in layers no thicker than 3 or 4 in. or 8 to 10 cm. The action produced is light, static pressure combined with vibration.

Explosives can be used to densify very loose cohesionless soils, utilizing both the vibration and the transient pressure produced by the detonation.[5:10] The typical spacing of charges is 10 to 25 ft or 3 to 8 m, on a grid pattern, with the explosive at a depth below the surface equal to or slightly less than the spacing. From 3 to 5 lb or 1.5 to 2 kg of dynamite are used per charge. The method is most effective in either dry or saturated soil, in damp sand capillary tension resists densification. Although dramatic increases of density are possible in some loose sands, the results are seldom uniform or as reliable as vibratory compaction.

The *vibratory tampers* consist of a curved shoe mounted with an eccentric weight vibrator and powered by a small, one-cylinder gasoline engine. These provide vibration and high-speed tamping and are useful in compacting cohesionless and slightly cohesive soils. Although theoretically they should produce large impulse forces, much of their energy is dissipated in elastic rebound, and they can produce up to 95 per cent of the Standard Proctor maximum in layers 2 to 3 in. or 5 to 8 cm thick.

Jetting and flooding have been used with some success in cohesionless soils of high permeability. Flooding destroys the capillary tension, which prevents the grain from moving into more compact arrangements. Jetting with water under pressure in some cases provides flooding and a little vibration that helps to compact medium sands, but the results are extremely erratic. Since modern, controlled compacting methods are available, jetting and flooding should not be used.

FILL OPERATION[5:9] / The selection of the proper compaction method, layer thickness, contact pressure, and moisture is the joint responsibility of the engineer and the constructor, for these affect the quality and uniformity of the soil structure and the speed and cost of construction.

The most important factor is the pressure in the layer being compacted. As shown in Fig. 5.7, the pressure beneath the compaction device decreases with depth. A high pressure applied over a small area, such as by the

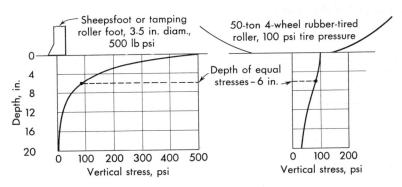

Figure 5.7 Comparison of vertical stresses in soil below a sheepsfoot roller and a heavy rubber-tired roller.

sheepsfoot roller, decreases rapidly, while a moderate pressure over a large area produces a more uniform pressure throughout the layer. The average pressure in a layer can be increased by decreasing the layer thickness or increasing the surface load. Ordinarily the best compaction efficiency is obtained with the maximum possible pressure that will not produce bearing-capacity failure. This is found by experiment. Bearing failure is indicated by rutting of the surface by rubber tires or failure of the sheepsfoot roller to rise out of the ground or "walk out" with each successive pass.

A loose, uncompacted soil has a low bearing capacity even when the compacted soil has high capacity. Heavy equipment operating on the loose soil is likely to create bearing failure and accompanying loss of compaction efficiency until the soil is densified enough to support the loading. If the soil is first partially compacted by a light roller, the bearing can be increased enough to support the heavy roller. Such *stage compaction* can be very effective where high densities are required.

The best moisture for compaction is the optimum for that particular method. The laboratory optimum has no exact counterpart in the field, but it serves as a guide. The Standard Proctor optimum is indicative of the needs of light rollers; the Modified Proctor optimum (a few per cent less than the Standard) is indicative of the needs of the heaviest equipment.

On large projects a test area is constructed to try different combinations of equipment, pressure, layer thickness, and moisture to determine the best for each different soil. Such testing ultimately saves much time and money.

ROCK FILLS / Both coarse gravel and broken rock make excellent fills because they are free draining and frost free. However, they require compaction to obtain their best strength and to minimize settlement. If the largest particles are smaller than about 12 in. or 30 cm, the fill is compacted by 50- to 100-ton rubber-tired rollers.

The 5-ton and heavier vibratory rollers are even more effective. As mentioned previously, layers as thick as 3 ft or 90 cm and particles as large as 2 ft or 60 cm can be compacted.[5:11]

Wetting promotes the compaction of most rock and minimizes future settlement. Two mechanisms are involved. First water jets wash the fines from between the contact points of the larger particles allowing them to wedge more tightly. Second, water helps break down the highly stressed projections of the rock, and allows them to crush and produce closer contact. The amount depends on the rock permeability; as little as $\frac{1}{5}$ volume of water to 1 of rock is required for fine rock while as much as 1 to 1 may be necessary for very coarse rock.

Very large rock, up to 20 ton pieces, can be compacted by dumping and sluicing. The rock is dumped from a hillside or from one end of a completed embankment so that it slides and rolls down the slope. The momentum wedges the pieces together. At the same time water jets (3 in. nozzles at 100 psi or more) wash the fines inward so that the coarser particles make intimate contact. Dumped rock fills are strong but tend to settle from $\frac{1}{2}$ to 1 per cent of their height over a period of five to ten years.

DENSITY CONTROL / Continuous testing of the moisture and density of the compacted fill is essential to ensure that the finished product meets the requirements.

The density test procedure depends on the character of the soil. If little gravel is present, a thin-walled tube 4 in. in diameter and 5 in. long with a sharp cutting edge (Fig. 5.8), is used to secure a sample of known volume

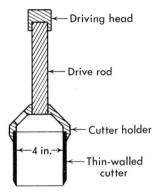

Figure 5.8 *Field density sampler.*

which is weighed and then tested for moisture to determine the dry density. For gravelly soils the *sand cone method* is often used. A hole 4 to 6 in. or 10 to 15 cm in diameter and 6 in. or 15 cm deep is dug, and the soil from it is weighed and tested for moisture. The volume of the hole is measured by filling it with loose, dry sand that falls from a fixed height through a cone-shaped stand. An alternate method of measuring the hole volume is to line it with a thin rubber membrane and fill it with water. A number of different devices, called *balloon density* meters, incorporate a retracting rubber membrane and self-indicating water supply to expedite measuring the hole volume.

The moisture content can be measured in the ordinary way by oven-drying at 105 C. A number of alternate methods are used to speed the drying, including burning out the moisture with alcohol, heating the soil at a temperature of about 200 C, and by direct-reading moisture meters.

These use either gas pressure produced by chemical reaction or electrical resistance to measure the water content. These quick alternates are most accurate in cohesionless soils. In clays and soil containing interparticle moisture they can be seriously in error.

Moisture and density determinations can be made by measuring the effects of nuclear radiation through the compacted soil. Both surface tests of the uppermost foot of fill and deep tests throughout the fill depth are possible.

and the results are available immediately. The results should be checked by direct density tests from time to time because in some soils there is a tendency toward erratic calibration.

QUALITY EVALUATION[5:12, 5:13, 5:14] / The requirements for compaction are ordinarily specified in terms of the percentage of compaction. For cohesionless soils, the requirements alternatively can be defined by relative density. These requirements are established either by experience or by tests in order to produce a mass that as a whole will have sufficient strength or incompressibility. Within the mass there will be variations in quality due to differences in the soil composition, moisture, and the compactive effort. Any evaluation must concern itself with the effect of these variations on the behavior of the total.

Sufficient tests must be made to define the range of quality. In large fills with uniform compaction, one test for each lift of fill and for every 10,000 to 20,000 sq ft of each different soil is adequate. In small fills where the uniform operation of compaction equipment is impossible, 2 to 3 tests per lift in each area, however small, are necessary.

The results are best depicted on a frequency distribution graph, Fig. 5.9. The average and *median* quality as well as the variations can be easily determined.

Ordinarily the required quality is assured by requiring that all test results exceed a specified minimum. The median quality, therefore, is greater than the specified minimum. Statistically, however, this approach can be misleading. Areas found to be less than the specified density are

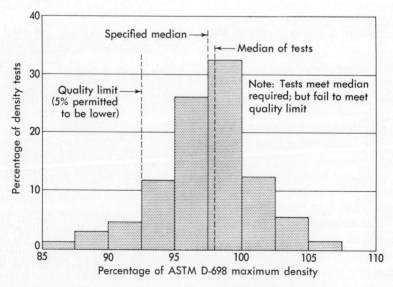

Figure 5.9 *Frequency distribution of percentage of compaction.*

recompacted until they meet the requirements. However, assuming a random distribution of tests and a random variability, there are probably other points, not tested, at which the compaction is less, and which will not be recompacted.

Requiring a minimum compaction level generally assures adequate fill performance because the median is well above the minimum. Furthermore, if a sufficient number of tests have been made, areas of unseen poorer compaction should be scattered, and have little detrimental effect on the whole.

For large fills a statistical requirement for compaction is more realistic. A median density level is specified to assure the all-over behavior of the fill. A limit of variation or *quality limit* is permitted, usually 3 to 5 per cent of the maximum less than the median, with no more than 5 per cent of the tests values below the limit if the low densities occur at random. If the low densities are concentrated in one layer or area, that area must be recompacted until the median meets the requirements.

5:5 Hydraulic Filling

When large volumes of soil must be excavated and transported, hydraulic methods may be economical. This is particularly true in water-front construction or dam construction where sufficient quantities of water are available.

HYDRAULIC EXCAVATION AND TRANSPORTATION / Hydraulic excavation can be employed in cohesionless or slightly cohesive soils. Jets of water, forced by pressures as high as 150 psi through 2-in. to 4-in. nozzles, wash the soil from the borrow pit into sluices. The mixture of soil and water can then be pumped and transported by pipe for miles.

Suction dredging excavates sands below the water surface. The sand is literally sucked up through a large flexible pipe attached to a powerful centrifugal pump. To the suction head can be attached a power-driven rotating cutter that loosens sands and even makes it possible to remove slightly cohesive soils. Clays can be excavated by using a rotating cutter that slices the soil into chunks about the size of a large grapefruit. These chunks can pass through the pump and hydraulic lines without clogging.

The larger cutter-head dredges are capable of excavating gravel and softer rocks such as coquina, shale and siltstone. The rock pieces are carried by turbulence through the suction intake and the pump and are discharged through the pipeline with the finer material. The maximum particle is typically $\frac{1}{3}$ to $\frac{1}{2}$ the suction diameter.

The soil-water mixture is typically 10 per cent solids, by weight. Too much solid matter increases the pumping head loss and the abrasion; too little means inefficient excavation. The suspension can be pumped a mile or two across level terrain, using only the dredge pump. Greater distances are possible, but require booster pumps.

FILL CONSTRUCTION / If a hydraulic pipe line discharges directly on the ground, a fan-shaped mound is formed, with the pipe at its focus. The coarsest particles settle out close to the outlet while the finer ones are carried away until the water velocity is low enough that they are deposited. Such a *run-out fill* has surface slopes between 5(H) to 1(V) and 10(H) to 1(V) in coarse sand, gravel or clay balls, near the discharge, and slopes between 20(H) and 1(V) and 40(H) to 1(V) in the fine sands in the broader parts of the fan. Wide, level fills are often constructed by direct discharge in this manner. The discharge pipe is moved periodically, to form a relatively flat surface. The fill edges feather out on flat slopes, as described.

To confine the fill with steeper slopes, diking is employed, Fig. 5.10a. An initial dike, typically 6 ft or 2 m high, is placed around the area to be filled. The hydraulic pipe rests on the dike or extends across the enclosed area with discharge valves or sluices at regular intervals. The suspension fills the

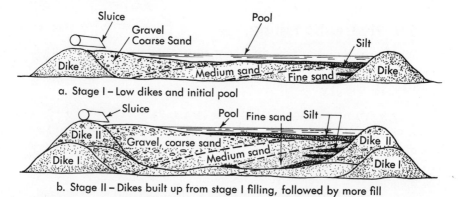

Figure 5.10 *Hydraulic filling with dikes.*

dike area; the solids settle out with the coarsest close to the dike, and the fines in the center. Unwanted silt and clay in suspension are discharged with the water over small spillways spaced around the perimeter dikes. The proportion of fines that settle can be controlled by the proportion of water to solids, and the spillway levels: The less the sedimentation time the greater the water velocity and the coarser will be the fill. When the pond is full, the dike is raised by excavating the coarser material with a dragline. The process can be repeated in 6 to 10 ft (2 to 3 m) lifts. Heights greater than 100 ft are practical with good control. In constructing dams by the hydraulic fill method, it is possible to make use of these different rates of sedimentation to create an impervious core of fines with shells of coarse, pervious soil. The outlets are placed at the outside of the fill where the coarse materials settle out, and the fines for the core settle in the pool in the middle.

COMPACTING HYDRAULIC FILLS / Hydraulic fills of sand and gravel are formed in a relatively loose state. Since it is impractical to compact them during filling, they are ordinarily loose throughout their entire depth. The methods of compaction are pile driving, explosives, and vibroflotation, and it is difficult to get uniformly high densities.

Surface vibration with the larger drum types can be effective in stabilizing loose sand fills to a depth of 10 ft or 3 m. The work must be done in increments of one or two passes, with a day or two for drainage in between. Too much vibration of the loose saturated sand at one time will create a quick condition and such a drastic loss of strength that the equipment will be swallowed up with little warning.

Silt placed by hydraulic methods is usually in an extremely loose condition. Pile driving is effective in compacting deep silt fills, but the results may be somewhat erratic. If the silt has cohesion, pile driving will be somewhat less successful, but compaction by consolidation may be possible. A thick layer of sand or gravel on top of the fill and the weight of the fill itself will provide the consolidation load. In order to increase the rate of consolidation, well-points or vertical sand drains may be inserted into the soil. They provide short paths for drainage of the water expelled from the soil during the consolidation process.

Hydraulic fills of impervious clays consist of clay balls in a matrix of very soft clay. The clay balls are concentrated near the sluice discharge points, creating irregular impervious dams with the other solids, sand, silt and clay slurry, respectively, at increasingly greater distances. Water is thus trapped in the sandier zones by the less pervious clay balls and the clay slurry, creating pockets of loose saturated cohesionless soil that can become quick upon slight vibration.

The first step in stabilizing the mass is to drain the sandier zones with ditches (or even well-points). This permits capillary action to stabilize the sand temporarily and to accelerate consolidation of the clays. Surcharge loading with pervious fill will further accelerate the stabilization of the fill; however it will never be structurally as sound as a densified hydraulic fill of sand and gravel.

Hydraulic methods of fill manufacture are very cheap and have been used for many years in marine and dam construction. Unfortunately the quality of the resulting fill is often poor, and expensive methods of compaction may be required to make the fill suitable for its purpose.

5:6 Soil Stabilization

Frequently the soils available for construction cannot meet the requirements, such as strength and incompressibility, imposed by their use in embankments or subgrades. The process of improving the soil so that it can meet the requirements is known as *stabilization*. In its broadest meaning, stabilization includes compaction, drainage, preconsolidation, and protection of the surface from erosion and moisture infiltration. However, the term *stabilization* is gradually being restricted to one aspect of soil improvement: the alteration of the soil material itself.

REQUIREMENTS OF STABILIZATION[5:17] / The mode of alteration and the degree of alteration necessary depend on the character of the soil and on its deficiencies. In most cases additional strength is required. If the soil is cohesionless, this can be provided by confinement or by adding cohesion with a cementing or binding agent. If it is cohesive, the strength can be increased by making the soil moisture-resistant, altering the clay-adsorbed water films, increasing cohesion with a cementing agent, and adding internal friction. Reduced compressibility can be obtained by filling the voids or cementing the grains with a rigid material or by altering the clay-mineral adsorbed water forces. Freedom from swelling and shrinking can be provided by cementing, altering the clay-mineral water-adsorbing ability, and by making the soil resistant to moisture changes. Permeability can be reduced by filling the voids with an impervious material or by altering the clay-mineral adsorbed-water structure to prevent flocculation. It can be increased by removing fines or creating an aggregated structure.

Many different methods for stabilization have been proposed. From the standpoint of their function or effect on the soil they can be classified as follows:

1. Moisture-holding: retain moisture in soil.
2. Moisture-resisting: prevent moisture from entering soil or from affecting clay materials.
3. Cementing: binding the particles together without their alteration.
4. Void-filling: plugging the voids.
5. Mechanical stabilization: improving the soil gradation.
6. Physicochemical alteration: changing the clay mineral or the clay-mineral adsorbed-water system.

A satisfactory stabilizing agent must provide the required qualities and in addition must satisfy the following criteria: (1) compatible with the soil material; (2) permanent; (3) easily handled and processed; and (4) low cost. Many materials have been employed but with varying degrees of success. No one material meets all the requirements and most are deficient in the last criterion—cost. The principal methods and materials and their typical applications are described in the subsequent paragraphs.

MOISTURE-HOLDING ADMIXTURES / Moisture in the soil provides some cohesion in sands and silts by capillary tension and prevents dust in all materials. It prevents shrinkage and cracking of cohesive soils and thereby reduces their disintegration in the first rain following a dry spell. Ordinary salt is an excellent moisture-holding material in hot but relatively humid regions; it is applied at a rate of about 25 lb per cu yd or 15 kg per cu m. Calcium chloride, at 15 lb per cu yd, is very effective, particularly

SEC. 5:6] SOIL STABILIZATION

in dry areas, because it is deliquescent (capable of taking moisture from the air).

MOISTURE-RESISTING ADMIXTURES / Moisture-resisting or waterproofing materials help keep water away from the soil particles and prevent softening or swelling. This is done in two ways: coating the grains or by preferential adsorption. Bituminous materials, such as RC-3 or 5 or MC-3 or 5 cutback asphalts, are the most widely used waterproofing agents. The amounts required vary from 2 to 7 per cent, increasing with the percentage of fines. They are most successful in soils of low plasticity but in many soils tend to lose their effectiveness with time. Resinous waterproofing agents have been used but are expensive and likely to deteriorate badly.

Hydrophobic agents, which are adsorbed on the particles in preference to water, show considerable promise. Silicones and stearates have been used for this purpose and appear to be relatively permanent in their effect. Processing is difficult, for each clay particle must be treated. The method is promising but requires development.

CEMENTING / A wide variety of cements or binding agents are employed in cementing, the most widely used and most successful method for stabilization. While the most pronounced benefit is an increase in strength by cohesion, there is also a reduction in permeability of most soils through filling of the voids with the cementing agent. When the cementing agent is relatively rigid, the modulus of elasticity of the soil can be increased and its compressibility decreased.

SOIL CEMENT / *Soil-cement stabilization* employs portland cement to form a mixed-in-place concrete in which the soil is the aggregate. It has proved very successful in making low-cost pavements for light traffic and rigid base courses for the heaviest traffic.

The proper mix is determined by a trial procedure to obtain the required durability and strength.[5:18, 5:19] Samples of soil are prepared with different amounts of cement and compacted at optimum moisture by the Standard Proctor method.

Higher densities are occasionally used; however there is little evidence to show there is a significant improvement in the soil-cement properties.

The soil cement is cured under moist but not saturated conditions, because field curing is usually limited to retaining the compaction moisture by an impervious membrane or asphaltic seal coat. The typical curing period is 7 days, for Type I cement, because the greatest strength gain is in that period.

Evaluation of the compacted soil-cement depends on its use. For base courses or other exposed applications, durability is the factor that controls the minimum cement content. Test samples are subjected to 12 cycles of freezing and thawing or 12 cycles of wetting and drying. The maximum volume change (swell plus shrink) permitted is 2 per cent. The maximum loss of weight permitted (after brushing) ranges from 7 per cent for A-6 and A-7 soils to 14 per cent for A-1, A-2, and A-3 soils. In some cases a minimum unconfined compressive strength is also specified. The required cement content is the least percentage which will satisfy these criteria. Typical cement requirements are 8 per cent for sandy soils to 15 per cent by weight for clayey soils.

Typical unconfined compressive strengths for soil-cement contents that provide the standard durability are from 200 psi or 14 kg per sq cm for clayey soils to 800 psi or 56 kg per sq cm for sandy soils. Higher strengths can be obtained by increasing the cement content, more or less in accordance with the water-cement ratio rules used in concrete mix design. Higher cement contents, however, increase the optimum moisture.

The key to successful soil cement stabilization (as well as other stabilization) is thorough mixing. Most problems with soil-cement arise from poor mixing of cohesive soils where clay lumps containing no cement are locked in a matrix of soil containing excess cement. Proper compaction is also essential. Ordinarily this is done in the same way as the soil alone would be compacted. The surface is finished by rubber-tired rolling in all cases. Soil cements are not highly abrasion resistant. For pavements and areas exposed to percolation, surface sealing with a thin bituminous coating is necessary. This is usually applied after compacting the soil cement so that it also serves as the curing membrane.

Occasionally, organic colloids in the soil inhibit hardening. All organic soils should be considered suspect until tests prove otherwise. Sodium hydroxide treatment is often effective in correcting the organic effect before the cement is added. High sulfate contents in the soil are also damaging and even sulfate resistant cements may not prevent trouble.

The uses of soil-cement are rapidly increasing. By far the widest application is for highway base courses. Tests indicate a 6-in. soil-cement base course is as effective in spreading wheel loads as 12 to 15 in. of crushed stone or bituminous concrete. Soil-cement linings for irrigation and drainage ditches are proving economical and desirable. Soil-cement erosion protection on dam faces has replaced rip-rap in areas where good stone is expensive.

Soil-cement modification uses about one-fifth the usual amount of cement to improve the strength and rigidity of soils that do not require complete stabilization. This is particularly useful when proper soil compaction is prevented by excessive soil moisture, and when sufficient time is not available to wait for dry weather.

Lime-fly ash stabilization resembles soil cement in that a pozzolanic cement is created by the reaction of lime on the silica of the fly ash. The proportions of lime to fly ash and the amount of cementing material are found by trial. Typical requirements are 10 to 15 per cent of a mix consisting of two parts of ash to one part of lime mixed with the soil and compacted in the same way as portland cement.

BITUMINOUS CEMENTING / *Bituminous binders*, usually asphaltic cutbacks such as RC-1, RC-3, MC-1, and MC-3, have been used for both subgrades and low-cost pavements. Emulsified asphalts are also used, but they require a long period of dry weather to permit the mix to cure properly. The amount of bitumen is determined by trial or from past experience, and usually is from 4 to 7 per cent by weight. Bituminous stabilization finds its widest use in sandy soils having little or no clay, such as SW, SP, and SM classes.

CHEMICAL CEMENTING[5:20, 5:22] / Chemical cementing consists of bonding the soil particles with a cementing agent that is produced by a chemical reaction within the soil. The reaction does not necessarily include the soil particles, although the bonding does involve inter-molecular forces of the soil. (Soil-cement properly is classified as chemical cementing but is generally given a separate category.)

The first chemical cements were the soluble silicates, usually sodium silicate solution. In the presence of a weak acid or metallic salts the silicate breaks down into sodium hydroxide and colloidal silica gel or an insoluble silicate. The hypothetical reactions are as follows:

$$Na_2SiO_3 \cdot 2H_2O + 2HCl = 2NaCl + SiO_2 \cdot 2H_2O$$
$$Na_2SiO_3 \cdot 2H_2O + CaCl_2 = 2NaCl + SiO_2 \cdot H_2O + Ca(OH)_2$$
$$Na_2SiO_3 \cdot 2H_2O + Ca(OH)_2 = CaSiO_3 + 2H_2O$$

The silica gel, $SiO_2 \cdot H_2O$, is a viscous, jelly-like mass that solidifies into silica, with the release of water. The calcium silicate is an insoluble precipitate, reached after rapidly passing through the silical gel phase. The silical gel-soil mixture can be manipulated and rolled to form a membrane that becomes hard and impervious.

The calcium silicate precipitate fills the soil voids with an impervious binder in a flash reaction that does not permit any manipulation. This form is utilized for injection into large voids to block the flow of water.

A delayed reaction is possible with the use of an organic reagent, *formamide*, that slowly breaks down to form the acid that produces the colloidal silical gel. The time of gelling can be controlled so that a period of from several minutes to several hours is available in which the soluble silicate and the reagent remain in their initial condition. This process is used in injection stabilization where the lower viscosity of the ungelled silicate permits greater penetration of low permeability soils.

The silicates have had their widest use in injection stabilization of sands and open jointed rock to provide both strength and reduction in water flow. Although they shrink on drying and become brittle, they appear to be relatively permanent in a moist state.

Organic monomers include a wide range of complex chemicals that initially are water soluble. A water solution of typically 10 per cent can be readily mixed with the soil or, in the most common application, injected into the voids. A second chemical, termed the activator or catalyst, causes the molecules of the monomer to link together, termed *polymerization*. The resulting polymer consists of a lattice of the linked organic molecules with water trapped between. A similar reaction involves separate soluble ingredients which react together in solution in the soil to form the monomer which subsequently polymerizes. In either case, the polymer is an elastic solid, by weight largely water, whose strength and rigidity is controlled by its chemistry and concentration.

The first of these to be extensively used in soil stabilization has the trade name *AM-9*. It is a mixture of acrylamide-methylene-bisacrylamide used in a 10 per cent solution, that has a viscosity only 1.5 times that of water. It polymerizes into a rubbery gel similar to very stiff gelatine. The rate of reaction can be controlled from a few minutes to 10 hours by the choice and proportion of activators.

A second such system, *Terranier*, is a water soluble, low molecular weight phenolic flavonoid monomer derived from pine bark. Two forms are available, *Terranier A* with a viscosity 20 to 30 times that of water and *Terranier C* with a viscosity of 2 times water. The catalyst solutions of iron sulfate and formaldehyde produce polymerization with the gel times controllable from a few minutes to several hours. The dark colored gel is rubbery to semi-rigid.

A third system with a trade name *Cynaloc* is a white viscous liquid. Diluted with an equal volume of water, the usual concentration, it is 10 to 20 times more viscous than water. The polymerization can be controlled between a few minutes and 1 hour. The resulting material is a relatively rigid solid resin.

A fourth system is *chrome-lignin*, which utilizes the waste lignin black liquor from a sulfite paper manufacturer. Potassium or sodium dichromate reacts with the lignin to form an organic monomer, chrome-lignin, which slowly polymerizes into a brown gel. The typical concentrations are from 10 to 20 per cent, by weight. The rate of gel formation is controlled by temperature and concentration; typical times are 15 minutes to 1 or 2 hours.

The solidified water-polymers generally shrink greatly and lose their continuity upon drying, but maintain their strength if continuously wet. The exception is Cynaloc which shrinks only slightly to a rigid strong solid resembling a white plastic. The long-term stability of the organic polymers is yet to be established. Case histories of chrome lignin of more than a decade and of AM-9 for nearly a decade indicate no deterioration in a wet environment.

Analine, a liquid coal-tar derivative, and *furfural*, an organic liquid from corn products refining, in the ratio of two parts to one, react to form a deep-red viscous resin that hardens slowly by polymerization to a solid. One liquid is mixed with the soil, followed by the other, after which the soil must be immediately compacted. About 5 per cent by weight of the resin has proved ample for rapidly stabilizing loose sands and similar soils so they can be used as roads within a few hours of processing.

Chemical reaction with the soil particles to form a cementing agent shows some promise. Phosphoric acid with a wetting agent acts on the clay minerals to form aluminum phosphates. Acids can react with carbonates and silicates to dissolve them and then reprecipitate them as binders between the grains.

FREEZING / *Freezing* the soil is, in effect, cementing with ice. Coaxial pipes are forced into the soil and a refrigerant at 0 F is forced down the center pipe and up the outside one to cool the soil. In this way the softest soil can be quickly converted into a strong, impervious, incompressible solid. Because of the high cost and its obvious impermanence, the method is employed only in emergencies or where nothing else can do the job. In one case, freezing was used to create an arch dam to halt a landslide, and in another to solidify the soil beneath a skyscraper that was in danger of failing because of excessive settlement.

VOID FILLING / The filling of the voids in cohesionless soils reduces the permeability and at the same time maintains the soil strength by reducing the penetration of water. A number of materials are used for this purpose, including portland cement, soluble silicate gels, the organic monomer polymers, and chrome-lignin, which are also used in other applications as cementing agents.

Fine-grained materials such as silt, fly ash, and clay can also be used in void filling. They may gradually wash out under high gradients. Swelling materials with fine particles that can lodge in the voids and later expand to fill the voids have been developed. An example is a highly plastic clay that is treated so as to delay its water adsorbency. Emulsified asphalt or latex can be treated so that the emulsion breaks in the soil voids and creates an impervious mass. Most of these materials are placed by injection to seal pervious, natural soil formations where excessive seepage could be hazardous or expensive.

MECHANICAL STABILIZATION[5:20, 5:21] / *Mechanical stabilization* is the improvement of the soil by changing the gradation. It usually consists of mixing two or more natural soils to secure a composite material that is superior to any of its components, but it also includes adding crushed rock or slag, or involves the screening of the soil to remove certain particle sizes.

The soil is considered to be made up of two components. The *aggregate* includes all particles coarser than some arbitrary limit, such as a No. 40 sieve (0.42 mm) or a No. 200 sieve (0.074 mm), and consists of predominately bulky grains. The *binder* is the finer fraction and includes the finer bulky grains and clay minerals. The aggregate provides internal friction and incompressibility and ideally consists of well-graded, strong, angular particles. The binder provides cohesion and imperviousness. It should have sufficient plasticity to develop high cohesion but not so much that it tends to swell. Based on experience, the best binders (finer than the No. 40 sieve) are CL soils with liquid limits less than 40 and plasticity indexes between 5 and 15.

The relative amounts of aggregate and binder determine the physical properties of the compacted stabilized soil. With no binder, the soil has high internal friction and is relatively incompressible because the loads are carried by grain-to-grain contact of bulky particles, but the cohesion is negligible. With small amounts of binder, some is trapped between the bulky grains and is highly compressed by compaction, while the remainder partially fills the voids. The result is a sharp increase in cohesion, a slight decrease in the angle of internal friction, slightly greater compressibility, and relatively high permeability (and potential softening from water circulating in the voids). The optimum amount of binder is reached when the compacted binder fills the voids without destroying all the grain-to-grain contact of bulky particles. Increasing the binder beyond this point results in a sharp drop in the internal friction, a small increase in cohesion, and greater compressibility.

The design of a mechanically stabilized mixture is the determination of the proportions that will provide the optimum binder. A number of standard gradation specifications have been developed for this purpose, based on past experience, but most do not consider the effect of the grain shape and the volume of water adsorbed in the clay. A rational procedure[8:19] separates each soil into aggregate and binder. The aggregates and the binders are compacted separately, to determine the volume of voids in the compacted aggregate and the density of the compacted binder. The mix is proportioned so that the total binder (from all the ingredients) is from 75 to 90 per cent of that required to fill the voids. Typical binder requirements for maximum strength are 20 to 27 per cent and are somewhat less than the amounts that result in the maximum compacted density. Mechanical stabilization is primarily used for pavement subgrades and for low-cost pavements where some improvement in the soil is needed but where great expense is not justified.

PHYSICOCHEMICAL ALTERATION / Physicochemical alteration, including *chemical stabilization*, consists of changing the properties of the soil grains, principally the clay minerals, and their adsorbed water. Ion exchange or *base exchange* is the changing of the cations in the adsorbed water films. The plasticity of the clay tends to decrease with increasing valence of the cations. By the addition in sufficient concentrations of chemicals with high valence cations, the soil is forced to exchange with a resulting lowering of its plasticity. Lime and calcium chloride provide calcium ions with a valence of 2 which bring about a marked improvement of high plasticity clays having sodium or potassium cations. Aluminum sulfate and certain organic chemicals have also been used for this purpose experimentally.

SEC. 5:6] SOIL STABILIZATION

The amount of the chemical required is small, as low as 0.1 per cent in some cases.

Electrochemical stabilization involves base exchange induced by an electric current. Aluminum cations leave a positive electrode of aluminum in the soil and migrate toward the negative, and in the course of their movement, bring about base exchange. At the same time electro-osmotic drainage toward the negative electrode in the form of a well helps harden the soil.

Dispersing agents, such as sodium silicate and sodium polyphosphate, increase the repulsion in the clay-adsorbed water layers and cause the soil to develop a dispersed or oriented structure. The liquid limit, plasticity index, and permeability are reduced, and the maximum compacted density is increased. The procedure is inexpensive, for it requires only from 0.1 to 0.2 per cent of cheap chemicals. The method has been employed successfully for sealing leaky ponds.

Coagulating or aggregating chemicals provide the opposite effect to dispersion. The particles link themselves together in chains with large voids between. The permeability and plasticity are increased and the maximum compacted density is decreased. They improve cultivation and drainage and are used to help establish grass on steep slopes.

Surface active agents and enzymes (also used in laundry detergents) increase the ability of water to wet a material. If the soil is too dry for compaction, a small amount of the agent, generally less than 0.1 per cent, mixed with the added water will aid in uniform diffusion of the moisture. The maximum density may be increased slightly by their use and the strength and incompressibility improved somewhat.

Thermal alteration is the application of intense heat to desiccate the soil and even produce limited fusion and some vitrification. The brick-like mass, 7 to 10 ft in diameter, that forms is permanently stabilized.[5:23]

A hole 4 to 8 in. in diameter is drilled vertically through the depth requiring stabilization. A burner utilizing fuel oil or gas with compressed air is introduced near the bottom to create a column of burning gases that heats the walls of the hole to temperatures exceeding 2000 F or 1100 C. In porous dry soils such as loess, it is claimed that the burning gases penetrate into the mass. The walls of the hole rapidly vitrify into a glassy cylinder, and thereafter heat is diffused into the soil only by conduction. Within a few feet of the hole the soil becomes partially vitrified into a brick-like mass that does not deteriorate. Beyond, the soil is stabilized by desiccation. The cost of thermal stabilization depends on the soil moisture, the heat conductivity, and the availability of fuel. In dry soils in areas of abundant cheap fuel, it is apparently economical.

STABILIZATION PROCESSING / Processing is the most critical part of stabilization because the effectiveness of any method depends on what proportion of the soil particles are treated. The ease of processing depends on the cohesion: Cohesionless soils break up readily and are easily processed, whereas cohesive soils tend to form impenetrable lumps that defy treatment. Dry materials such as cement are spread mechanically on each soil layer, while soluble materials and liquids such as asphalt are added by sprinkling.

Proportioning of dry materials is usually by *layering*. A layer of soil 6 to 8 in. or 15 to 20 cm thick is placed loosely but as uniformly as possible. The weight of soil per unit of area is established by testing selected samples. The dry stabilizer is spread across the surface at the weight per unit of area required by the soil. Two methods are used. The batch method applies a measured weight to each unit of area defined by a grid laid out on the soil layer. The spreader method utilizes a mechanical device that spreads a thin uniform layer over the soil surface. This may be calibrated for a particular stabilizer, such as cement, but must be readjusted by trial for variations in stabilizer texture. Liquid admixtures are applied to the soil layer by spray, utilizing metering pumps. In either case variations in the proportioning of ± 10 per cent are normal and ± 20 per cent are not unusual. If the soil is placed in windrows, the same approach is utilized. If the amount of stabilizer is large, it is placed in a parallel adjacent windrow. If the quantity is small, the stabilizer is placed on the windrow directly.

Mixing of the stabilizer into the soil requires either extensive manipulation of the layer or the use of a travelling mixer. Manipulation requires cutting the stabilizer into the layer with a disc plow or harrow that penetrates the entire thickness. Many passes are required in alternating and different directions so as to obtain mixing in three dimensions. Finally the layer is turned over back and forth with a road grader to insure that there are no zones that are under or over stabilized. Travelling mechanical mixers are more effective and can produce better blending of the materials. These consist of high speed rotating blades that accurately cut through and lift the entire layer, throw it back and forth between blades or paddles within the machines, and then deposit the soil-stabilizer mix in a uniform layer. Similar devices are adapted to mixing windrows. In some cases portable mixing plants similar to those used for asphalt paving are used.

After mixing, the soil layer is compacted as previously described. Compaction must follow stabilization before any permanent hardening develops; otherwise the compaction will break down any stabilization that has occurred. Cement and emulsified asphalt stabilization require curing. Ordinarily a bituminous seal coat is applied to soil cement immediately after compaction to hold the mixing moisture, which is adequate for curing. Curing, ranging from a week or two for soil cement to a few days for the RC bitumen, is necessary for stabilized base courses. Most of the other stabilizing mixtures develop their strength within a day, and special curing is not needed.

5:7 Grouting—Injection Stabilization [5:24, 5:25, 5:26, 5:27]

Injecting the stabilizing agent into the soil, termed *grouting*, makes it possible to improve the qualities of natural soil and rock formations as well as existing fills without excavation, processing, and recompaction. Grouting ordinarily has two objectives: to improve the structural properties or to reduce permeability. This is done by filling cracks, fissures and cavities in rock and the voids in soil with a stabilizer that initially is in a liquid state or in suspension, and which subsequently solidifies or precipitates.

GROUTING MATERIALS / Many of the stabilizing agents previously discussed, and particularly the cements are utilized in grouting. Grouting mixtures must meet a number of requirements:
1. Sufficiently liquid to be pumped.
2. Viscosity and particle size compatible with the size of opening to be filled.
3. Reaction or hardening time compatible with both the pumping requirement and the diffusion through the soil or dilution by ground water.

The properties of the grout must fit the soil or rock formation being injected. The dimensions of the pores or fissures determine the size of grout particle that can penetrate. Generally the D_{85} of the grout must be smaller than $\frac{1}{3}$ the crack width or diameter of the smallest pores. If a soil is being grouted, the effective pore diameter is about $1/5\ D_{15}$.

$$D_{85}\ (\text{Grout}) < 1/15\ D_{15}\ (\text{Soil}), \tag{5:3a}$$

$$D_{85}\ (\text{Grout}) < 1/3\ B\ (\text{Fissure}). \tag{5:3b}$$

The ratio of the D_{85} of the grout to the D_{15} of the soil is sometimes termed the groutability ratio, R_g.

$$D_{85G}/D_{15S} = R_g. \tag{5:4}$$

Although the ratio should exceed 15, penetration is sometimes successful with ratios of 5 and 10.

The viscosity determines the rate of grout penetration under a given gradient produced by the grout pressure. Viscosities as low as possible are necessary in fine grained soils or thin fissures. In large voids or cavities a high viscosity is desirable to restrict grout flow to the area where stabilization is required. The rate of hardening also controls the penetration. Rapid hardening restricts flow to large voids while slow hardening permits maximum penetration through small voids.

The grout must not be unduly diluted or washed away by ground water. Insoluble or rapid setting grouts are used in large water-bearing voids or cavities to restrict the grout loss by dilution and to stop the flow.

The typical range of use of grouts is shown in Fig. 5.11.

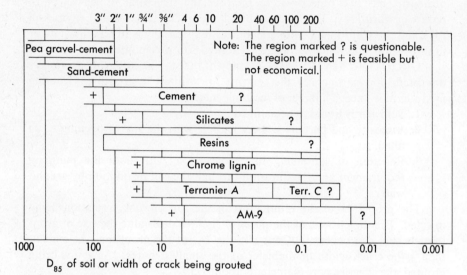

Figure 5.11 *Groutability of soil and rock.*

INJECTION METHOD / The grouting plant includes the material handling system, mixers, pumps, and delivery pipes or hoses. Slow setting grouts such as portland cement utilize a single mixing and pumping system. Rapid setting or controlled time-of-set grouts require two systems, one for each component. The mixing of components is done by a proportioning valve or pump at the point of injection.

The injection point for soil consists of a perforated pipe with a conical point that is driven to the level at which the grout is to be injected. In rock or hard soil a hole is drilled to the grouting level. A grout pipe is sealed into the hole by an expanding gasket, termed a *packer*, just above the level to be grouted. If the rock is too broken or soft to hold a packer, the grout pipe is sealed into the formation by a quick setting cement.

The injection pattern depends on the purpose. If an area is to be stabilized to increase its bearing capacity or reduce its compressibility (known as consolidation grouting) a grid pattern is used. Typical spacings are 20 to 50 feet or 6 to 15 m initially. If an impervious barrier is to be constructed, one or two parallel lines of injection are used. The line spacing may be 20 to 40 feet or 6 to 12 m and the hole spacings along the lines 10 to 20 feet or 3 to 6 m initially.

SEC. 5:7] GROUTING—INJECTION STABILIZATION

In either case each hole is grouted in increments of depth of from 10 to 50 feet or 3 to 15 m using sufficient pressure to force the grout into the voids and fissures, but not high enough to damage the formation. Generally the pressure measured at the ground surface is initially limited to 1 psi per foot of depth below the surface. Higher pressures may be used if no ground heave occurs. Generally each level is grouted until the volume of grout is sufficient to fill the voids in a hypothetical cylinder of soil, Fig. 5.12a, or until the resistance builds up to the pressure limit.

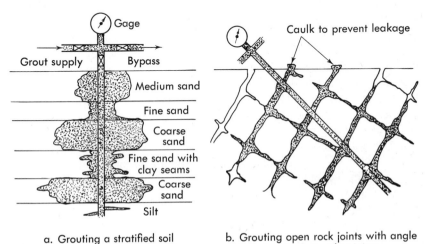

a. Grouting a stratified soil

b. Grouting open rock joints with angle hole to intersect widest joints

Figure 5.12 Grouting pervious formations.

After the first pattern is complete, its effectiveness is tested by drilling *split spacing* or intermediate holes halfway between the initial grid holes or halfway between the holes on each grout line. If the intermediate holes take no grout, that portion of the grout grid or line is considered complete. If grout can be pumped, each new hole is grouted as before. A second split spacing is employed between all holes taking grout on the first split. Sometimes four or five splits are required to effectively grout an area.

The hole orientation is designed to fit the porosity of the formation. In porous or horizontally stratified soils, vertical holes are used. In jointed rock, the holes are inclined to intersect as many fissures as possible, Fig. 5.12b. In cavernous limestone holes 24 to 30 in. in diameter may be drilled to permit direct access and the pumping of concrete through a hose into the larger cavities.

GROUTING SILT AND CLAY / Because of the low permeability, it is impossible to secure much penetration of the most fluid grouts into the voids of silts and clays. Instead, the grout forms irregular fingers and sheets that penetrate weaker seams and force them apart. This fingering may produce some consolidation of poorly consolidated partially saturated soils. Their strength will be increased and compressibility reduced. Grouting of poorly compacted clay fills has arrested building settlements and has reduced lateral soil movement. However, the results are usually unpredictable. Therefore, grouting silts and clays always should be considered experimental, until the results show otherwise.

PLANNING AND EXECUTING GROUTING[5:28] / Grouting is a highly specialized art requiring an intimate familiarity with the structure of the soil or rock formation and a high degree of experience with the materials, equipment, and procedures that might be used. Any grouting program must be considered tentative, and must be revised as the work progresses. Changes in materials, equipment and procedures are made as the conditions dictate, which means the work must be supervised by experienced engineers who can make the required decisions.

GROUTING CAVITIES / Enormous quantities of grout are required to fill cavities and old mine chambers both because of the size of the opening and because the grout flows easily far beyond the area where it is needed. Running water aggravates the grout loss. Hot asphalt grout solidifies in contact with the water, forming an elastic membrane that confines the liquid yet allows the mass of grout to expand and plug the opening. Cloth bags attached to a grout pipe and filled with cement grout also can plug openings with running water. Pea gravel washed into the cavity through a 6 in. or 15 cm diameter hole will partially plug the cavity, and can be later grouted with sand-cement to form a solid mass. Cavity fill grout is generally placed in batches of 100 to 1000 cu ft and then allowed to harden before a second batch is introduced. In this way the loss beyond the area needing grout can be limited.

5:8 Subgrades and Pavements[5:29]

A *pavement* is a structure whose primary function is to spread the concentrated load of a vehicle wheel sufficiently that the underlying soil can support it without failure or excessive deflection. In addition the pavement should provide a smooth, nonskid, running surface that resists weathering. Finally it protects the underlying soil from loss of its qualities due to exposure to sun, rain, and cold.

The earliest pavements were narrow treads of stone spaced as far apart as the chariot and cart wheels, which allowed ancient military vehicles to service remote outposts in the worst weather. In heavily forested areas, 4- to 6-in. diam poles were laid side to side across the roadway to form a stable but rough "corduroy" pavement. The ancient road building was climaxed by the Romans, who laid large slabs of stone in overlapping layers on beds of broken stone and natural cement mortar. While the

SEC. 5:8] SUBGRADES AND PAVEMENTS

Roman road was strong and durable, it was so expensive that such construction was abandoned in the Middle Ages. Modern pavement design began with the Scottish engineer MacAdam, who formulated rules for drainage of the pavement and for developing load-spreading ability through the use of interlocking bulky fragments.

COMPONENTS OF PAVEMENT SYSTEMS / Two different systems of load spreading are in use: *rigid* and *flexible*. The rigid pavement utilizes beam action of a relatively rigid member to distribute the load. The overlapping stone slabs and the wood poles of the primitive roads and the concrete slab of modern construction function in this way. The flexible pavement system distributes the load by particle-to-particle contact throughout its thickness.

The load is spread by the transfer of stress from particle to particle in the base course by interlocking, friction and cohesion. The sub-base (sometimes omitted) also spreads the load. In addition it serves as a transition filter to prevent the base course materials from punching down into a fine-grained subgrade, as an insulator to prevent frost action in the subgrade, and as a drain. The subgrade is the ultimate support of the loads and is either natural soil or compacted fill.

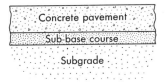
a. Rigid pavement system

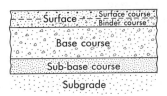
b. Flexible pavement system

Figure 5.13 Pavement systems.

As shown in Fig. 5.13a the rigid system includes three components. The pavement slab provides the riding surface and the load-spreading slab. The pavement is in effect a beam that spreads the load through its flexural rigidity.

The *sub-base* (sometimes omitted) serves a number of purposes: insulation against frost penetration, drainage, and a layer resistant to erosion and *pumping*. The *subgrade* is the underlying natural soil or the surface of a compacted fill that ultimately supports the load.

The flexible pavement (Fig. 5.13b) includes four components. The surface course provides the riding surface and an impervious membrane that sheds water. The *base course* is the main load-spreading layer. It is strong enough to withstand the shear stresses produced by the wheel, incompressible, and rigid enough to distribute the load over the underlying layers.

SUBGRADES / In general a subgrade has the same requirements as a fill: strength, incompressibility, and freedom from swelling and shrinking. In addition it must be rigid enough to prevent excessive deflection under live loads. Susceptibility to frost action is important in regions where the depth of frost penetration is as great as the pavement thickness.

The investigation of materials for proposed use in subgrades is similar to that for fills. Large numbers of small samples are secured in prospective borrow areas and of the upper layers of virgin soil where no fill is to be used. These are classified visually and by laboratory tests. A preliminary evaluation of their qualities is made on the basis of empirical correlations of performance with the soil classes. Such a rating, based on the Unified System, is shown in columns 4, 5, 6, and 7 of Table 5:1. Similar ratings have been developed for other classification systems; the choice of which to use depends on the experience of the engineer.

Special tests for subgrades have been developed by a number of highway departments. The Georgia Highway Department first obtains the maximum dry density by the Standard Proctor method. Two samples are then compacted to maximum density and one is allowed to shrink and the other to swell. The soil is rated on the basis of its compacted density and percentage of swell, plus the percentage of shrinking.

SUBGRADE EVALUATION FOR PAVEMENT DESIGN / The design of the pavement is essentially the fitting of the wheel load to the capabilities of the subgrade. While this can be done by empirical rules based on past experience, economical engineering practice requires an evaluation of the significant physical properties of the subgrade and the design of a pavement to fit. A brief discussion of the evaluation of the subgrade follows. The design procedures can be obtained from texts on highway engineering and pavement design.

The significant property of the subgrade for the design of a rigid pavement is the deflection under short-term loads. This can be approximated by a plate load test that simulates the loading of the rigid pavement. A circular plate 30 in. or 76 cm in diameter is seated on the subgrade and loaded to a pressure σ of 10 psi or 0.7 kg per sq cm. The deflection ρ of the plate is measured. The modulus of subgrade reaction, k_s, is defined by

$$k_s = \frac{\sigma}{\rho} \qquad (5:5)$$

and has the dimensions of pounds per cubic inch. In most cases the field test is not conducted under the worst possible condition of saturation. To simulate the effects of saturation, two samples of the subgrade are subjected to a short-term laboratory consolidation test of 10 psi, one in the original condition and one inundated. The ratio of the "as is" settlement to the inundated sample is multiplied by the field k_s factor to obtain a k_s value that is corrected for saturation.

A number of methods are used to evaluate subgrades for flexible pavement design. The California bearing ratio, usually abbreviated CBR, is a semiempirical index of the strength and deflection characteristics of a soil that has been correlated with pavement performance to establish design curves for pavement thickness.[5:29] The test is performed on a 6-in. or 15 cm diam, 5 in. or 7.5 cm thick disk of either compacted or undisturbed soil that is confined in a steel cyclinder. Before testing, the sample is inundated under a confining pressure equivalent to the weight of the future pavement in order to determine the potential swelling and to simulate the worst possible condition of moisture that could occur in the field. A piston approximately 2 in. or 5 cm in diameter is then forced into the soil at a standard rate to determine the resistance to penetration. The CBR is the ratio (expressed as a percentage) of the actual load required to produce a 0.1-in. or 2.5 mm (or 0.2-in.) deflection to that required to produce the same deflection in a certain standard crushed stone. A field version of the test is used to determine the existing CBR of subgrades by in-place tests without soaking. The CBR test and its accompanying design curves have been widely used for designing flexible pavements for highways and particularly for airfields in many parts of the world. It has been criticized as being overly conservative in requiring inundation of the soil before testing. While this is admittedly a severe requirement, so little is known about the actual maximum subgrade moisture conditions that this is probably justified.

Plate load tests, similar to the subgrade modulus test, have been used in a rational evaluation of the deflection characteristics of the soil. A number of different procedures have been developed, each to simulate certain field conditions such as load or load repetition. The results are analyzed on the basis of deformation settlement, using elastic theories derived for uniform, horizontal layers.

Rational analyses, based on the strength and deflection properties of the soil as determined by triaxial shear tests, are proving to be a useful approach to design because it is possible to reproduce different conditions of load and soil moisture at a low cost and in a short time. It is likely that such rational methods will eventually replace the semiempirical and empirical designs.

REFERENCES

5:1 R. R. Proctor, "Fundamental Principles of Soil Compaction," *Engineering News Record*, August 31, September 7, September 21, and September 28, 1933.

5:2 "The Unified Classification: Appendix A—Characteristics of Soil Groups Pertaining to Roads and Airfields, and Appendix B—Characteristics of Soil Groups Pertaining to Embankments and

Foundations," *Technical Memorandum 357*, U.S. Waterways Experiment Station, Vicksburg, 1953.
5:3 "Factors Influencing Compaction Test Results," *Bulletin 319*, Highway Research Board, Washington, 1962.
5:4 T. W. Lambe, "The Engineering Behavior of Compacted Clay," *Journal of Soil Mechanics and Foundation Division, Proceedings, ASCE*, **84**, SM2, May 1958.
5:5 H. B. Seed and C. K. Chan, "Structure and Strength Characteristics of Compacted Clays," *Journal of Soil Mechanics and Foundation Division, Proceedings, ASCE*, **85**, SM5, October 1959.
5:6 H. B. Seed and C. K. Chan, "Undrained Strength of Compacted Clays After Soaking," *Journal of Soil Mechanics and Foundation Division, Proceedings, ASCE*, **85**, SM6, December 1959.
5:7 G. A. Leonards, "Strength Characteristics of Compacted Clays," *Transactions, ASCE*, **120**, 1955.
5:8 H. K. Nichols, *Moving the Earth*, North Castle Books, Greenwich, Conn., 1955.
5:9 "Factors That Influence Field Compaction of Soils," *Bulletin 272*, Highway Research Board, Washington, 1960.
5:10 B. J. Prugh, "Densification of Soils by Explosive Vibrations," *Journal, Construction Division, Proceedings, ASCE*, **89**, CO1, March 1963.
5:11 F. A. Robeson and R. L. Crisp, "Rockfill Design, Carters Dam," *Journal, Construction Division, Proceedings, ASCE*, **93**, CO3, September 1966.
5:12 F. C. Walker and W. G. Holtz, "Control of Embankment Materials by Laboratory Testing," *Transactions, ASCE*, **118**, 1953, p. 4.
5:13 W. J. Turnbull, J. R. Compton and R. G. Ahlvin, "Quality Control of Compacted Earthwork," *Journal of Soil Mechanics and Foundation Division, Proceedings, ASCE*, **92**, SM1, January 1966.
5:14 J. L. Beaton, "Statistical Quality Control in Highway Construction," *Journal, Construction Division, Proceedings, ASCE*, **94**, CO1, January 1968.
5:15 H. F. Winterkorn, "Soil Stabilization," *Proceedings, Second International Conference on Soil Mechanics and Foundation Engineering*, **5**, Rotterdam, 1948, p. 209.
5:16 "Stabilization of Soils," *Bulletin 98*, Highway Research Board, Washington, 1955.
5:17 "Soil and Soil Aggregate Stabilization," *Bulletin 108*, Highway Research Board, Washington, 1955.
5:18 "Soil Stabilization With Portland Cement," *Bulletin 292*, Highway Research Board, Washington, 1961.

5:19 *Soil Cement Laboratory Handbook*, Portland Cement Association, Chicago.
5:20 "Chemical and Mechanical Stabilization," *Bulletin 129*, Highway Research Board, Washington, 1956.
5:21 E. A. Miller and G. F. Sowers, "Strength Characteristics of Soil Aggregate Mixtures," *Bulletin 183*, Highway Research Board, 1958.
5:22 T. W. Lambe, "Soil Stabilization," Chapter 4, *Foundation Engineering*, McGraw–Hill Book Co., Inc., New York, 1962.
5:23 I. M. Litvinov, "Discussion on Thermal Consolidation," *Proceedings, Fourth International Conference on Soil Mechanics and Foundation Engineering*, 3, London, 1957, p. 169.
5:24 T. B. Kennedy and W. F. Swiger, "Symposium on Grouting," *Transactions, ASCE*, **127**, 1962, p. 1337.
5:25 W. E. Perrott, "British Practice on Grouting Granular Soils," *Journal of Soil Mechanics and Foundation Division, Proceedings, ASCE*, **91**, SM6, November 1965.
5:26 H. B. Erickson, "Strengthening Rock by Injection of Chemical Grout," *Journal of Soil Mechanics and Foundation Division, Proceedings, ASCE*, **94**, SM1, January 1968, p. 159.
5:27 R. H. Karol, "Chemical Grouting Technology," *Journal of Soil Mechanics and Foundation Division, Proceedings, ASCE*, **94**, SM1, January 1968, p. 175.
5:28 "Guide Specifications for Chemical Grouts," *Journal of Soil Mechanics and Foundation Division, Proceedings, ASCE*, **94**, SM2, March 1968, p. 345.
5:29 E. J. Yoder, *Principles of Pavement Design*, John Wiley & Sons, Inc., New York, 1959.

SUGGESTIONS FOR ADDITIONAL STUDY

1. References 5:3, 5:9, 5:18., 5:29
2. *Journal of Soil Mechanics and Foundation Division, Proceedings, ASCE*, from 1960.

PROBLEMS

5:1 The following data were secured from a moisture-density test. The soil $G_s = 2.71$.

Water Content	Unit Weight	Water Content	Unit Weight
10	97	20	127
13	105	22	126
16	117	25	121
18	123		

    a. Plot the moisture-dry density curve. Find the maximum density and optimum moisture.
    b. Plot the zero air voids curve.
    c. If the contractor is required to secure 90 per cent compaction, what is the range in water contents that would be advisable?

5:2 Compaction tests were made on the same soil, using first the Standard Proctor Method and second the Modified Proctor Method. The following results were obtained:

Standard Proctor		Modified Proctor	
Water Content	Dry Density (lb/cu ft)	Water Content	Dry Density (lb/cu ft)
6	102	6	107
9	106	9	113
12	108	12	118
14	109	13	118
16	108	14	117
19	105	16	112
22	100	18	108

    a. Plot both curves on the same graph and determine maximum density and optimum moisture for each.
    b. Plot the zero air voids curve if the specific gravity of solids is 2.67.
    c. How much increase in maximum density results from the modified compaction? What decrease in optimum moisture occurs when using modified compaction?
    d. The soil is classed CL by the Unified System. What densities would be required for a highway fill 40 ft high? What would the range in permissible moisture contents be if the field methods were comparable to the effort of (1) the Standard Proctor Test, and (2) the Modified Proctor Test?

5:3 A Standard Proctor Test on a ML soil having a specific gravity of solids of 2.68 was

PROBLEMS 253

Water Content	Dry Density (lb/cu ft)	Water Content	Dry Density (lb/cu ft)
12	86	24	91
15	89	27	88
18	91	30	84
21	93		

    a. Plot the moisture density and zero air voids curve. What is the maximum degree of saturation of the soil?

    b. The soil is to be used in a fill 15 ft high that supports a one-story building. What densities should be specified, based on Table 5:2?

    c. The soil moisture is 25 per cent. The constructor is able to obtain a dry density of 86 lb per cu ft, using a sheepsfoot roller developing 700 psi. The roller fails to walk out. What should he do to obtain the required density (1) in the deeper part of the fill, and (2) in the upper part?

5:4 Make an estimate of the suitability of each of the soils listed in problem 3:3 for
    a. Fill for a highway.
    b. Subgrade for an airfield pavement in Illinois.
    c. Core of an earth dam.
    d. Shell (structural supporting part) of an earth dam.

5:5 List in the order of their importance the properties necessary for a soil to be used in the following ways:
    a. Highway fill.
    b. Railroad embankment.
    c. Earth dam.
    d. Subgrade for major airport.
    e. Surface for a secondary road.

5:6 Prepare a report on available compaction equipment, showing width compacted, pressure, coverage for
    a. Sheepsfoot rollers.
    b. Heavy rubber-tired rollers.

5:7 Prepare a report, based on an article appearing in an engineering or construction journal, that describes the construction of a large fill. Include the following points:
    a. Soil description.
    b. Method of excavation.
    c. Method of compaction.
    d. Control of compaction.

5:8 Prepare an outline of the procedure required for testing soil-cement mixtures. (Secure information from bulletins of the Portland Cement Association.)

5:9 Prepare a table similar to 5:1 showing the probable adaptability of each of the soil groups to stabilization by the different methods given in Section 5:6
   a. Mixed-in-place stabilization of a subgrade.
   b. Injection stabilization.
   c. Where any soil or material is not included, state why.

5:10 From Fig. 5.9 compute the median and average compacted densities.
   a. How do the median and average compare?
   b. What percentage of tests fell below the quality limit?
   c. If the soil moistures are approximately the optimum, what might be done to make the compaction meet the requirements?
   d. Would these steps change the median? Why?

5:11 A contractor is compacting a soil classified as MH using a sheepsfoot roller with a 750 lb per sq in. foot pressure, using layers 24 in. thick. The optimum moisture is 22 per cent but the soil moisture is 30 per cent. The density obtained is only 93 per cent of the maximum (ASTM D 698) although 98 per cent is specified.
   a. List all the things that are wrong with this situation.
   b. What should be done to obtain the required strength if the layer is to be a subgrade for a pavement.

CHAPTER 6
Underground Investigation

CONSTRUCTION OF AN ACCESS ROAD TO A MOUNTAIN-TOP RADAR STATION in the tropics was at a standstill. The pioneer road had been blocked in several locations by landslides and the permanent construction could not proceed because the soils were far too wet to be compacted in the embankments. Data on the soil and rock conditions along the route had not been obtained because of the high cost of moving equipment through the jungle growth on the rugged mountainside. Instead, the designer examined the nearby mountain roads, looked at a diorite outcrop near the entrance and studied the topography from aerial photographs. The resulting design required cuts and fills as deep as 75 feet. The slopes were steepest in the deepest cuts, reflecting the designer's assumption that the more rugged, steep terrain reflected reasonably sound rock while the more gentle slopes were underlain by soil. The fills were to be constructed of the cut material, and compacted to 95 per cent of the "Modified Proctor" maximum.

After two years of construction little had been accomplished. The cut slopes failed repeatedly. During one rainy week more than thirty separate landslides occurred, some small but one of more than 40,000 cu yd. Instead of sound diorite rock the deepest cuts encountered badly weathered volcanic ash, laced with water-bearing fissures. Excavation reduced the confining pressure on the ash, and the high pore water pressures in the fissures propelled large masses of the material onto the road.

The water content of the weathered ash averaged 50 per cent, more than twice the optimum. No compaction was possible, much less the specified 95 per cent.

A complete redesign of the road was necessary. The cuts were widened and fills eliminated. The drainage was revised to remove moisture from the slopes. The road was finally completed, over two years behind schedule and at triple the original estimated cost.

This was an expensive lesson in the need for an adequate investigation before the design is conceived for all concerned in the project. The owner suffered in the two-year delay and in the expensive lawsuit brought by the contractor. The engineer suffered from embarrassment and from the cost of the complete redesign of the road. The contractor paid in the delay of his other work on the project and in not being able to undertake new projects with the forces tied up on the road. Moreover, although he was awarded substantial damages in compensation, they did not fully cover his losses. The cost of an adequate investigation, while greater than for similar projects in more favorable terrain, would have been less than 5 per cent of the extra costs incurred.

A new water treatment plant for a small city was to be constructed on a gently sloping hillside. The designer requested that the city obtain borings to define the underground conditions. After obtaining proposals for the work from reputable firms, the city retained a local man of doubtful qualifications, but good political connections. He borrowed a drill rig team from the state highway department one weekend and made some auger borings. They were stopped by rock at the level of the bottom of the filter building and the sedimentation basins. On that basis the designer planned for all foundations to be supported on rock.

Excavation for the basin and the filter building foundations disclosed knobs of rock with soil between. The designer had considered that the rock might not be continuous and had provided enough reinforcing for the basin bottom that it could span over soil pockets of the size encountered. Therefore, there was no further study of the rock-soil pocket character and no change in the design.

Three significant events during construction were recorded by the resident engineer but ignored by the designer. First, a bulldozer sank several feet into one of the soil pockets. Second, a large water main from a 50,000-gallon elevated tank was broken by a power shovel, and all the water disappeared into the ground in a few minutes. Third, all surface drainage disappeared in the excavation.

One month after the plant went into operation the bottoms suddenly dropped out of the three sedimentation basins. Five minutes later the filter building cracked apart. A subsequent investigation showed that the rock found in the auger boring and the knobs exposed in the excavation consisted merely of limestone boulders in the throat of an old sink hole. The water loss from draining the 50,000-gallon tank probably reactivated the underground erosion that had initially caused the sink. The surface drainage into the excavation and possibly the inevitable leakage from the water pipes kept the erosion active. The failure was the climax of the process set in motion by the construction and aggravated by water.

This failure was the result of an inadequate investigation and a slipshod engineering evaluation of the site. First, the region and that hillside in particular had a history of local sinkhole subsidences that was well known to the city officials. They failed to inform the designer; however, he made no inquiries regarding past experiences with foundations in the area. Second, the underground investigation was entirely inadequate: the driller was selected on the basis of cost (and politics), and the results were interpreted in terms of the cheapest design. Third, there was no reevaluation of the site conditions during construction, although the loss of the bulldozer and the disappearance of water were symptoms of the underground cavity development. In this case the cost of the failure was shared equally between the city, the designer, and the contractor: the city for its negligence in not informing the designer of previous trouble at the site and for making an inadequate investigation; the designer, for failing to insist on proper data and for not reevaluating the design in the light of construction experience, and the contractor for breaking the water main and for allowing the surface water to drain into the ground.

The designer of a steel structure cannot proceed without knowing the physical properties of his steel; the designer of reinforced concrete must know the physical characteristics of both steel and concrete; yet both often complete their superstructure designs without quantitative data on the material that supports these structures—the earth. The soil and rock formations are just as much a part of the structure as the concrete and steel above. However, whereas the designer can control the character of the man-made materials and may specify them to suit his needs, he has little control over the character of the soil and rock masses. Therefore, either the design must be adapted to the site conditions or else the site conditions must be altered. In either case it is imperative that these conditions be evaluated accurately.

The designer is not the only one concerned with underground conditions as the two case histories show. Even before a site is purchased, the prospective owner should determine if the property is suitable for the purpose. For example, expensive pile foundations were required for a bargain building site in a marsh. A more expensive site nearby with a stable sand foundation proved cheaper when the total development cost was evaluated.

The builder must understand the site conditions in order to plan earth moving and foundation construction. Furthermore, superstructure construction is often controlled by the time and operation sequence required for foundations. It is not surprising therefore that the largest part of the builder's allowance for the unknown (the bid *contingency*) is often in the "below ground" work. When the soil and rock are significantly different from what was reasonably anticipated, either from the underground investigation or from past experience in the area, this may constitute a *changed condition*. Many contracts include provisions for added time and payment for changed conditions; in other cases these conditions have generated tedious, expensive lawsuits.

The need for data on the underground conditions at a site is generally recognized by the engineer and the contractor and to varying degrees by the prospective owner. The differences between evaluating site conditions and obtaining data on other engineering materials are not always understood, however.

When a structural material, such as steel is purchased, the specifications define the minimum quality. The manufacturer is responsible for meeting the specification; his quality control measures limit the variability of the material, and independent laboratory tests establish that at least the minimum quality is furnished.

The soil and rock, however, did not form under rigid quality control. Defects are frequently hidden from view with blankets of topsoil and thick vegetation, and the manufacturer cannot be called upon to meet a specification or to certify a minimum quality. Therefore, evaluating the quality of the underground conditions at a site is far more difficult and leaves a much greater margin for uncertainty than establishing the properties of the other materials of construction.

The responsibility for meeting this uncertainty lies with all three parties to construction: the owner, the engineer, and the builder. The builder must plan his operations to allow for the unknown, the engineer and architect must design the structures with enough margin for safety to cover the variable conditions, and they must be prepared to revise the design when unforeseen site conditions are encountered. The owner, whose peculiar site includes the unknowns, is responsible for the extra costs resulting from these conditions, as well as for the cost of investigating his site.

Because nature rarely provides assurances of site quality, the engineer-architect must plan a program of site investigation that will identify the significant underground conditions and define the variability as far as is practical. *There will always be some risk of unknown conditions*; it can be minimized by a more intensive investigation but never eliminated. This risk is inherent in all human endeavor: no project of any kind is undertaken with a guarantee of no trouble. The degree of success reflects the skill and imagination of those involved, but it also depends on circumstances beyond their control.

6:1 Planning Investigation[6:1]

INFORMATION REQUIRED / A complete investigation of underground conditions includes the following points:
1. Nature of the deposits (geology, recent history of filling, excavation, and flooding; possibilities of mineral exploitation).
2. Depth, thickness, and composition of each soil and rock stratum.
3. The location of ground water and variations of ground water.
4. The engineering properties of the soil and rock strata that affect the performance of the structure.

In many cases all this information is not necessary and in others estimates will suffice. The best investigation is the one that provides adequate data at the time it is needed and at a cost consistent with the value of the information.

COSTS / The value of an investigation can be measured by how much money might be spent for the structure if no investigation were made. When a designer is confronted with inadequate data, he compensates for the lack by overdesign; when a contractor is furnished with incomplete information, he increases his estimates to allow for possible trouble. In most cases the cost of inadequate data is considerably more than the cost of the investigation. When unforeseen soil conditions necessitate a change in design or a construction procedure, the cost of the structure increases rapidly. If a failure of the structure should result, the entire project may become a loss. In such cases, the cost of an adequate investigation would be but a small fraction of the money lost.

The cost of an adequate investigation (including laboratory testing) has been found to be from 0.05 per cent to 0.2 per cent of the total cost of the entire structure. For bridges and dams, the percentage is somewhat higher, from 0.5 to 1 per cent.[6:1]

PROCEDURE FOR INVESTIGATION / A complete investigation consists of three steps:
1. Reconnaissance, to determine the nature of the deposit and to estimate the soil conditions.
2. Exploratory investigation, to determine the depth, thickness, and composition of the soils, the depth to water and to rock, and to estimate the engineering properties of the soil.
3. Intensive detailed investigation, to secure accurate information about critical strata from which design computations can be made.

In some instances, such as for small buildings, only a minimum of reconnaissance and exploration will be necessary; in others, such as for large bridges and heavy power plants, extensive reconnaissance, exploration, and detailed investigation will be required to secure adequate data for economical, safe design.

Planning and conducting a soil investigation is among the most intricate of engineering problems. Careful coordination is necessary between the engineer, the laboratory, and the men in the field in order to secure the best information in the least time and at the lowest cost. If the field men send soil samples to the laboratory immediately, time-consuming tests can be started before the field work is finished. If the soils engineer is promptly furnished with test data, he can make changes in the field and laboratory procedures without expensive delays and without having to repeat some operations.

6:2 Reconnaissance

GEOLOGIC STUDY / A geologic study, no matter how brief it may be, is essential in planning and interpreting a complete soil investigation.[6:2] The primary purpose of such a study is to establish the nature of the deposits underlying the site. The types of soil and rock likely to be encountered can be determined, and the best methods of underground exploration can be selected before boring, sampling, or field testing is commenced. The geologic history may reveal changes such as faulting, flooding, or erosion that have taken place and which have changed the original character of the soil or rock. The possibility of defects in the rock, such as cracks, fissures, dikes, sills, sinkholes, and caves, may be indicated. This information will greatly aid in the interpretation of the results of sampling and field tests. Another important function of geologic studies is to establish the possible presence of minerals having economic value. If there is a likelihood of future mining or well drilling on the site, this must be considered both in the design of the structure and in planning for the use of the site. Legal and engineering problems have arisen where structures settle because of the collapse of mine workings beneath them or where valuable minerals are discovered below expensive buildings.

SEISMIC POTENTIAL / Potential seismic activity is a major factor in structural design in many regions of the world. Even in regions that have rarely experienced an earthquake, potential seismicity must be considered in the design of structures such as dams and nuclear reactors whose damage or failure could cause widespread injury or loss of life. Earthquakes are generally the result of deep seated accumulated strains in the crust, which are climaxed by their release in cracking or faulting. Earthquake damage to foundations occurs in two forms: the direct tearing of structures that lie on the fault and the acceleration of structures within the zone of more intense motion. Rough estimates of the potential seismicity of U.S. sites can be had from a map of earthquake zoning, Fig. 6.1, that reflects the experience in the area.[6:3] However, for the more active areas and for the more critical structures, a two pronged evaluation is necessary. First is a geologic study of the region to identify all faults that might extend within a few miles of the site. Previous published studies, supplemented by studies of outcrops and well records is ordinarily required. The second phase is an examination of past earthquake history of the region, within a radius of several hundred miles. On this basis geologists and seismologists as a team can estimate the earthquake intensity that is likely to occur as well as the likelihood of active faulting on the site.

SOURCES OF GEOLOGIC INFORMATION / Geologic studies have been made of many parts of the earth by state and national geologic surveys, oil companies, mining interests, and industrial concerns. Water-well or oil-well records are frequently available that will show the depth of soil and whether it is sand, gravel, or clay. In many cases, soil and rock profiles along highway or railroad cuts may be studied. Geologic maps often show ancient shore lines and river and lake locations, with their terraces, deltas, and fills that are now soil strata of gravel, sand, and clay. The "deltas" of Mississippi and Louisiana and the shore lines of glacial Lake Maumee in Ohio are examples of such deposits.

SEC. 6:2] RECONNAISSANCE

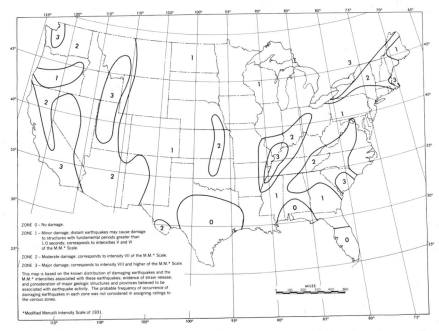

Figure 6.1 Seismic risk in the United States. (Courtesy of S.T. Algermissen, U.S. Coast and Geodetic Survey.[6:3])

The U.S. Geological Survey, state geological surveys and mining departments, the U.S. Department of Agriculture, and state highway departments have collected data dealing with soils. Bulletins and special reports may be secured by writing for them. Most of the state geological surveys have libraries that include the publications applicable to their regions. Out of print publications can be examined, and copies made for engineering studies.

SOIL SURVEY DATA[6:4] / The U.S. Department of Agriculture has been making surveys of the upper horizons of soil in the United States. These have been published by counties (and occasionally other land subdivisions) since 1899. The data include a map, delineating the surface soils identified, and descriptions of the horizons. Ordinarily the data are limited to the A, B, and upper C horizons and shallow ground water levels, although the regional geology is usually described. Since 1957 the data have been expanded to include representative engineering properties such as the liquid and plastic limits, grain size, compacted maximum density, estimates of drainability, and similar qualities of importance in potential land use as well as the agriculture-related properties discussed in earlier bulletins. The scale of the modern maps is 3 in. to 1 mile or 4 in. to 1 mile approximately. These maps are helpful in estimating the conditions on large sites where the uppermost 5 to 10 ft of materials are most significant, such as for highways and real estate developments involving limited cutting and light loads.

SITE INSPECTION / An examination of the site and the adjacent areas will reveal much valuable information. The topography, drainage pattern, erosion pattern, vegetation, and land use reflect the underground conditions, particularly the structure and texture of the soil and rock. Highway and railroad cuts and stream banks often disclose the cross-section of the formations and indicate the depth of rock. Outcrops of rock or areas of gravel and boulders may indicate the presence of dikes and more resistant strata. Ground water conditions are often reflected in the presence of seeps, springs, and the type of vegetation. For example, marsh grass on what appears to be a dry hillside shows that the area is wet during the growing season. The water levels in wells and ponds often indicate ground water, but these also can be influenced by intensive use or by nearby irrigation.

The shape of gullies and ravines reflects soil texture. Gullies in sand tend to be V-shaped, with uniform straight slopes. Those in silty soils often have a U-shaped cross-section. Small gullies in clay often are U-shaped, while deeper ones are broadly rounded at the tops of the slopes.

Special features such as sinkholes, sand dunes, old beach ridges, and tidal flats are often obvious to the layman. More obscure forms require intensive geological training for their recognition.

Valuable information about the presence of fills and knowledge of any difficulties encountered during the building of other nearby structures may be secured by talking to old residents of the area adjacent to the site. Settlement cracks in nearby buildings often indicate that poor foundation conditions will be encountered. However, it must be remembered that good soil conditions at one site do not necessarily mean good conditions at an adjacent site.

AERIAL RECONNAISSANCE / Examination of the site from the air can reveal the broad patterns of topography and land form, drainage, and erosion even more effectively than site inspection on the surface. Features that are obscured because they cover too large an area or because of poor access are easily observed from the air. Large areas can be inspected in a short time, especially if the site is in rugged country.

A study of aerial photographs permits reconnaissance in any weather and without exciting local residents. Mosaics covering large areas are good base maps for extensive surface reconnaissance. Low altitude photographs or enlarged high altitude photographs are better base maps for site inspection than the usual property maps. Surface features can be studied in detail at leisure, and if there is sufficient overlap between adjacent photographs, a three-dimensional examination is possible. More intensive photo study, termed *air photo interpretation* is discussed in Section 6:3.

A personal inspection from a small aircraft or helicopter often permits examination of outcrops as close as 100 to 200 ft, as well as observation of the site as a whole from altitudes as high as 8000 ft. A permanent record of the inspection made with a good camera, particularly in color, is extremely useful in later detailed studies of the soil and rock conditions.

THE VALUE OF RECONNAISSANCE / A reconnaissance investigation establishes the probable soil conditions at the site. If the site appears to be unsuitable for the structure, then it may be abandoned without further study.

6:3 Air Photo Interpretation[6:5, 6:6, 6:7]

Air photo interpretation is the estimation of underground conditions from their reflection in aerial photographs. It consists of three steps. First is the identification of the geologic and man-made features, both on a broad, regional scale and in detail. The second step is a grouping of these features according to the geologic formations with which they are usually associated. Finally, the probable geologic formation (soil and rock) is determined from the total pattern of associations. In many instances the sequence of soil and rock strata can be deduced and sometimes even certain engineering properties of the soils estimated from the details that are observed.

PHOTO MATERIALS / The photographs are the same as those commonly obtained for photogrammetry. They are made in sequence by flying in more or less straight lines across a site with a two-thirds overlap in the direction of flight and one-quarter overlap between successive flight lines. The scale depends on the purpose: for general mapping, 1:20,000 or 3 in.= 1 mile approximately on 9-in. contact prints is adequate; for more detailed work larger scales obtained by low altitude photography are necessary. Generally black and white photography is adequate but for special purposes infra-red and color are useful supplements.

PRINCIPALS OF ANALYSIS / Analysis consists of an identification of all of the natural and man-made features and their grouping by geologic association. These features include the following:

1. Topography.
2. Stream patterns.
3. Erosion and gully details.
4. Gray tones or color.
5. Vegetation.
6. Micro details in topography.
7. Culture—man-made features.
8. Natural and man-made boundaries.

The topographic study defines the shape of the ground surface, such as hills, valleys, terraces, and similar features both on a large scale involving several miles and a small scale of a few hundred feet. The shape, size, slope and the sequence or interrelationship of adjacent shapes are identified.

The drainage or stream pattern is a fundamental unit of topographic shape. The largest or *primary* streams frequently are clues to geologic age while the *secondary* streams or tributaries and the *tertiary* streams (creeks and smallest permanent water courses) often reflect structure and the sequence of geologic events. For example, parallel streams usually reflect a gradual tilting of the ground surface. Streams that flow parallel, with right-angle tributaries, and which join together at right-angles in a trellis pattern reflect long parallel folds. Sharp bends with straight reaches between them in primary and secondary streams sometimes indicate the major joint sets in the rocks below. Lazy meandering loops with swampy zones and sand deposits on the insides of the bends are typical of flood plains of an old river, Fig. 6.2; sharp changes in direction of adjacent streams may indicate a

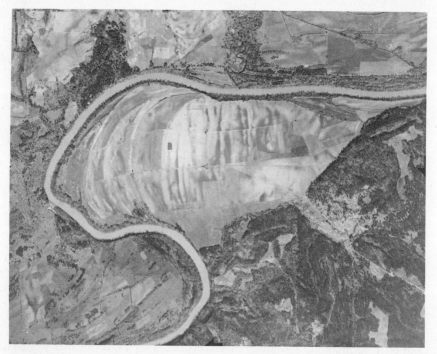

Figure 6.2   Air photo of a flood plain of an old river in a humid region.

fault or shear zone if all the changes are in the same direction and along the same straight or curved line, or *lineation*.

Gulley profiles and erosion details reflect the permeability and the strength of the surface materials. In clean sand the few gullies are short with uniformly sloping sides, and stream banks slope at approximately the angle of internal friction. In silts, weak sandstones and clayey sands of low cohesive strength, the gullies are long and deep. The gulley and stream banks drop vertically, from tension cracking, while the gulley bottoms are rounded from accumulating sloughed soil. The U-shaped cross-section is often interrupted by isolated pinnacles of the original bank that either became detached and slid partially downward or that resisted erosion. In clays and shales the gullies are long but shallow with the tops of the banks rounded by the progressive surface softening of the materials.

The color or gray tones reflect the color of the formations where exposed: the reds and blacks appear dark while the tans and yellows, light. Damp materials appear darker than dry materials. Differences in vegetation are reflected in color differences: pine trees appear darker green than deciduous trees and blacker in the black and white photographs. Growing crops are darker than dryer vegetation.

Vegetation differences frequently reflect both drainage and soil character. For example, pine trees require drainage while cypress trees flourish in swamps. Some vegetation prospers on particular soils: the eastern cedar is often associated with residual soils derived from limestone and particularly phosphatic limestone. Differences in vegetation are easily observed and sometimes denote geologic boundaries; the implications of particular types require the aid of agriculturalists.

Micro details include limited features such as sink holes, rock outcrops, and boulder accumulations. These usually require on-foot examination to define their significance.

Differences in land use and man-made boundaries may be the result of arbitrary decisions but they also can reflect differences in the character of the underlying strata. For example, an irregular shaped pasture land surrounded by rectangular cultivated fields may be the result of too rough a surface or too shallow soil for cultivation. An abrupt jog in a fence line could indicate a rock outcrop, a swampy area, or a sink hole. Natural boundaries between river patterns, vegetation, or color tone usually reflect major differences in geology.

LAND FORM / The land form is the basic unit of association of the various features, particularly topography and stream pattern. It represents the total effect of environment and geologic history on the underlying soil and rock formations. The study of the evolutionary processes that produce a given land form is termed *geomorphology*. Once the land form is identified, the geologic associations are defined. From this the geologic structure can be established and the probable soil and rock stratification estimated. Ground water levels and even soil texture can be deduced from tone, drainage density and erosion patterns.

Air photo interpretation requires a thorough grounding in geology and geomorphology as well as an understanding of related fields such as agriculture and hydrology. The results indicate which areas are favorable for development, the locations where trouble is likely to occur, and the best places to look for materials of construction. The technique is a valuable preview and supplement to site reconnaissance, particularly in the planning stages where large areas for potential development must be compared without the time for extensive field work.

6:4 Exploratory Investigations[6:1]

PLANNING EXPLORATORY WORK / The purpose of the exploratory investigation is to secure accurate information about the actual soil and rock conditions at the site. The depth, thickness, extent, and composition of each stratum; the depth of the rock; and the depth of ground water are the primary objectives of exploration. In addition approximate data on the strength and compressibility of the strata are secured in order to make preliminary estimates of the safety and settlement of the structure.

A carefully planned program of boring and sampling is the best method of obtaining specific information at a site and is the heart of an exploratory investigation. Many different methods have been developed for doing this work, and construction organizations, well drillers, and commercial laboratories offer such services. Too frequently exploratory work is so poorly planned, carelessly performed, incompletely reported, and incorrectly interpreted that the results are inadequate or misleading. Soil or rock boring and sampling to obtain information that will give an accurate picture of underground conditions is an engineering problem requiring resourceful, intelligent personnel trained in the principles of geology and soil mechanics.

SPACING OF BORINGS / It is impossible to determine the spacing of borings before an investigation begins because the spacing depends not only on the type of structure but also on the uniformity or regularity of the soil deposit. Ordinarily a preliminary estimate of the spacing is made; this is decreased if additional data are necessary or is increased if the thickness and depth of the different strata appear to be about the same in all the borings. Spacing should be smaller in areas that will be subjected to heavy loads; greater, in less critical areas. The following spacings are often used in planning boring work.

TABLE 6:1 / SPACING OF BORINGS

Structure or Project	Spacing of Borings	
	(ft)	(m)
Highway (subgrade survey)	1000–2000	300–600
Earth dam, dikes	100–200	30–60
Borrow pits	100–400	30–120
Multistory buildings	50–100	15–30
One-story manufacturing plants	100–300	30–90

For uniform, regular soil conditions the above spacings are often doubled; for irregular conditions they are halved.

DEPTH OF BORINGS / In order to furnish adequate information for settlement predictions, the borings should penetrate all strata that could consolidate materially under the load of the structure. For very important heavy structures, such as large bridges or tall buildings, this means that the borings should extend to rock. For smaller structures, however, the boring depth may be estimated from geologic evidence, the results of previous investigations in the same vicinity, and by considering the extent and weight of the structure.

Experience indicates that damaging settlement is unlikely when the added stress in the soil due to the weight of the structure, $\Delta\sigma$, is less than 10 per cent of the initial stress in the soil due to its own weight, $\bar{\sigma}_0$. A rule adopted by E. De Beer of the Geotechnical Institute of Belgium requires that borings penetrate to the depth where $\Delta\sigma = 0.1\bar{\sigma}_0$.[6:8] Typical depths for exploratory boring, based on the above stress relationship, are given in Table 6:2.

TABLE 6:2 / DEPTH FOR EXPLORATORY BORING

Building Width	Boring Depth (ft)				
	Number of Stories				
	1	2	4	8	16
100 ft	11	20	33	53	79
200 ft	12	22	41	68	108
400 ft	12	23	45	81	136

An old rule is that the boring depth should be twice the building width; this is ridiculously deep for wide, one-story structures such as modern industrial plants and far too shallow for narrow towers. A better simple rule for structures, such as hospitals and office buildings, relates the estimated boring depth, z_b, to the number of stories, S.

Condition	Feet	Meters	
Light steel or narrow concrete	$z_b = 10S^{0.7}$	$z_b = 3S^{0.7}$;	(6:1a)
Heavy steel or wide concrete	$z_b = 20S^{0.7}$	$z_b = 6S^{0.7}$.	(6:1b)

For dams and embankments the depth ranges between half the height to twice the height depending on the strength, compressibility and permeability of the foundation. The borings for deep excavations should extend 5 to 15 ft (1.5 to 5 M) below the excavation bottom and deeper if soft clay or loose sand and silt are encountered. Borrow pit borings should extend to the anticipated depth of excavation or even deeper if artesian water is suspected. If the ground water level is reached and no drainage is planned, the borrow pit boring will stop at that level.

Example 6:1

A manufacturing plant 200 by 800 ft will be a one-story building, with the floor constructed on a fill 4 ft thick. The floor load averages 450 psf. What would be a reasonable boring layout and depth for this structure:

1. The 4-ft fill is equivalent to 2 stories of building load; the floor load equivalent to 2 to 3 stories of building load. Therefore the depth should be that for a 5 to 6 story building.
2. The boring plan depends on the site geology:
    (*a*) For a flood plain 8 to 10 borings in two rows might be sufficient.
    (*b*) For irregular deposits, 20 to 25 borings in three rows would be more realistic.
3. The boring depth nominally should be 40 to 60 ft; the first three borings, spaced along the building length, might go 60 ft. The remainder would be shallower (or possibly deeper) depending on the first borings.

6:5 Boring and Sampling[6:8, 6:9]

Many different exploratory techniques have been developed, some suitable for a wide range of site conditions while others are limited to special cases. A summary of the uses of the principal methods is given in Table 6:3. These methods are discussed in more detail in the following pages.

AUGER BORING / The *soil auger* (Fig. 6.3) is the simplest equipment for making a shallow hole in the ground and securing samples of the soil material in a much loosened condition. Several different styles are available. The stump auger, which resembles a long wood-auger bit and was originally designed for drilling dynamite holes beneath stumps and boulders, is useful

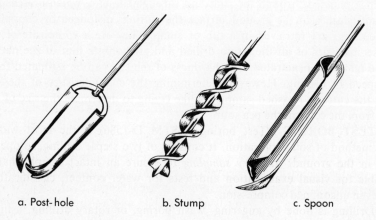

a. Post-hole   b. Stump   c. Spoon

Figure 6.3 *Soil hand augers.*

in drilling holes up to 10 ft or 3 m deep and about $1\frac{1}{2}$ in. or 4 cm in diameter. The most effective hand auger is the post-hole auger, consisting of two curved blades that retain the soil as it is cut. These are available in sizes from 2 to 6 in. in diameter. Small earth augers are generally fitted with handles, so that they can be turned by hand, and extensions can be added to the handle so that depths as great as 30 ft can be reached. Motor-driven augers are available which are capable of drilling holes in some soils as deep as 80 ft in a few minutes.

The power auger can drill continuously and the helical flutes will eventually bring the soil to the surface. However, the time lapse between cutting and the appearance of the pulverized soil at the surface is too great to estimate the depth at which that soil occurred. Therefore, it is essential to drill in increments of 3 to 5 ft or 1 to 1.5 m and then remove the auger. The strata depths can be determined by the amount of soil retained in the auger helix compared to the increment of depth drilled.

Soil augers have the advantage of obtaining a dry hole until the water table is reached and of providing easy visual recognition of changes in soil composition. On the other hand, they are difficult to use in soft clays and coarse gravel and impossible to use in most soils below the water table. Hand augers are seldom economical when boring deeper than 20 ft.

The auger sample is a well-disturbed mixture of all the materials penetrated. It is useful for determining the average water content, grain size, and plasticity characteristics and is sufficient for most borrow pit exploration. It gives little information on the character of the undisturbed soil.

WASH BORING / Wash boring was once widely used in soil exploration work and is still useful when only limited information, such as the depth to a hard stratum, is needed. The soil is drilled with a combination of jetting and chopping by the use of a chisel-shaped bit attached to hollow drill rods. Water is pumped through the rods and through the bits, which loosens the soil and washes to the ground surface the cuttings dislodged by chopping. The cuttings are retrieved in a tub or sump. They are a composite of the coarser particles of all the strata drilled and give only a hint of the nature of the materials penetrated. The hardness of the soil can be estimated from the speed of drilling. However, determining the characteristics of the soil from the cuttings (washed sample) is like trying to determine the size of the peas from the color of the pea soup.

TEST BORING / Test boring (ASTM D-1586) is the most widely used method of soil exploration. It consists of two steps: drilling, to open a hole in the ground; and *dry sampling* to secure an intact sample that is suitable for visual examination and tests for water content, classification, and even unconfined compression.

Drilling is done by augering, wash boring, or rotary drilling, using a high-speed revolving cutter and circulating water to remove the cuttings in

TABLE 6:3 / SUBSURFACE EXPLORATION—EXPLORATORY BORING METHODS

Method	Procedure	Use	Limitations
Auger Boring[6;9] ASTM D-1452	Hand or power auger with removal of material at regular short intervals.	Identify changes in soil texture above water table. Locate groundwater.	Grinds soft particles—stopped by rock, etc.
Test Boring ASTM D-1586	Drill hole, sample at intervals with 1.4 in. I.D., 2 in. O.D., split barrel sampler driven 18 in. in 3-6 in. intervals by 140 lb hammer falling 30 in. Below water, maintain hydrostatic balance with fluid.	Identify texture and structure; estimate density or consistency in soil or soft rock.	Gravel, hard seams
Continuous Core: Soil	Force and/or twist tube into soil until resistance prevents further movement. Remove cuttings with air or water.	Identify soil texture and structure continuously in cohesive soils.	Gravel, hard seams, sands. Misleading squeeze in some clays.
Bore hole Camera, TV	View inside of bore hole.	Examine stratification in place above water table.	Textural changes indistinct.
Continuous Core: Rock ASTM D-2113	Rotate tube with diamond-studded bit to cut annular hole. Cuttings removed by circulating water. Core retained in tube by cylindrical wedge. Best with stationary inner tube to protect core.	Identify rock strata and structural defects continuously.	No data on soft seams, etc.
Dynamic Sounding	Drive enlarged disposable point on end of rod with weight falling fixed distance, in increments of 6 in. to 1 ft.	Identify significant changes in density or consistency of materials.	Misleading in gravel.

Static	Force enlarged cone (Dutch Cone: 1.4 in. diameter 60° Angle) on end of rod into soil, measuring resistance of point at regular intervals.	Identify significant changes in density or consistency. Possibly identify soil by ratio of point load to skin friction.	Stopped by hard strata, misleading in gravel.
Pits, Trenches	Excavate pit or trench, by hand, large auger and by excavator.	Visual examination of structure and stratification above water table.	Caving of walls, ground water.

The following drilling methods are frequently used to advance the hole in test boring and core drilling; for quick but crude exploration, they are occasionally used independently.

Wash Boring; Rotary Wet Drilling	Chop with chisel bit or rotate toothed cutter. Cuttings washed to surface by circulating water or mud through bit.	Identify coarser fraction from cuttings, hardness from drilling rate.	Misleading if appreciable fines present.
Churn or Cable Drilling	Pound and churn soil boulders and rock to slurry by dropping heavy chisel bit in wet hole. Bail slurry at intervals.	Drill and identify broken rock, etc., from cuttings.	Strata difficult to define. Quick condition formed in sands.
Percussion Drilling	Impact—drill with jack hammer; remove cuttings with compressed air.	Identify rock from cuttings, hardness from rate.	Plugged by wet soil.

the same way as in wash boring. In firm soils the hole remains open by arching. In soft clays and in sands below the water table, it is kept open by inserting steel tubing (casing) or preferably by keeping the hole filled with a viscous fluid known as *drilling mud*. Drilling mud, usually a mixture of bentonite clay and water, has the advantage of supporting both the walls and the bottom of the hole. The mud also serves as the circulating liquid in wash and rotary boring and maintains a cleaner hole by washing out coarse sand and gravel which tend to accumulate in the bottom.

The *sampler* (Fig. 6.4), also called a split spoon, consists of a thick-walled steel tube split lengthwise. To the lower end is attached a cutting shoe; to the upper end, a check valve and connector to the drill rods. The standard size is 1.4 to 1.5 in. ID and 2 in. OD, but similar samplers with 2 in. ID × 2.5 in. OD and 2.5 in. ID × 3 in. OD are occasionally used.

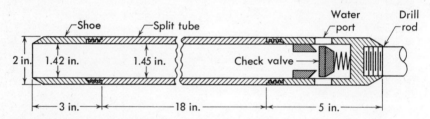

Figure 6.4 Standard split barrel sampler. (Courtesy of Law Engineering Testing Co.)

The hole is drilled as previously described until a change in the soil is detected. The drill tools are removed and the sampler lowered to the bottom of the hole by attaching it to drill rods. It is first driven 6 in. into the soil to ensure that the cutting edge is seated in virgin material.

It is then driven 12 in. (30 cm) in 6 in. (15 cm) increments with a 140-lb (63.5 Kg) hammer falling 30 in. (76 cm). The number of blows for each 6 in. or 15 cm are recorded. The *standard penetration resistance*, N, is the sum of the blows for the second and third increments.[6,9] Standard penetration sampling is shown in Fig. 6.5.

The sample is examined and classified by the field technician in charge of boring and then sealed in a glass or plastic container for shipment to the laboratory. The sample maintains the water content, composition, and stratification of the soil, although there may be appreciable distortion of the structure. Good samples can often be used for unconfined compression tests but are not of sufficient quality for triaxial testing.

The penetration resistance is an indication of the density of cohesionless soils and of the strength of cohesive soils. In effect, it is an in-place dynamic shear test. Tables 6:4 and 6:5 have been proposed to describe density and strength from the standard penetration test results.

a. Power auger drilling prior to sampling.

b. Driving the split barrel sampler with a 140 lb hammer.

c. Split barrel sampler disassembled after driving, showing soil in one half of the tube (in the foreground), soil in the cutting shoe on the right, and the check valve head and drill rod connector at the right rear.

Figure 6.5 Standard penetration sampling. (Courtesy of Law Engineering Testing Co.)

TABLE 6:4 / RELATIVE DENSITY OF SAND*
(After Terzaghi and Peck[6:11])

Blows	Relative Density
0–4	Very loose
5–10	Loose
11–20	Firm
21–30	Very firm
31–50	Dense
Over 50	Very dense

* Measured with 1.4 in. ID, 2 in. OD sampler driven 1 ft by 140-lb hammer falling 30 in.

TABLE 6:5 / CONSISTENCY OF COHESIVE SOILS*
(After Terzaghi and Peck[6:11])

Blows	Consistency
0–1	Very soft
2–4	Soft
5–8	Firm
9–15	Stiff
15–30	Very stiff
Over 30	Hard

* Measured with 1.4 in. ID, 2 in. OD sampler driven 1 ft by 140-lb hammer falling 30 in.

The resistances measured with a 2 in. ID, 2.5 in. OD sampler driven with a 300-lb hammer falling 18 in., as specified by some building codes, are roughly equivalent to those measured by the standard test.[6:1]

Test boring is the most widely used method for securing data on the depth, thickness, and composition of the soil strata and approximate information on the soil strengths. It is economical and rapid and adapted to most soils (except coarse gravel) and even to soft rock.

CORE DRILLING / When a soil boring encounters a material so hard that its penetration resistance exceeds 100 blows (measured with a 140-lb hammer falling 30 in. on a 1.4 in. ID spoon driven 1 ft), further progress with soil-boring equipment is difficult and often impossible. This resistance is termed *refusal*, and it may indicate a highly compacted soil, a boulder, or rock.

Core drilling is used to penetrate such hard materials in order to determine whether refusal indicates a hard lens or boulder underlaid by softer materials, or sound rock. Large-diameter holes (30 to 54 in.) drilled in rock permit an engineer or geologist to examine the strata in place, but the cost of drilling is great. Small-diameter cores that are brought to the surface

SEC. 6:5] BORING AND SAMPLING

make it possible to determine the composition, soundness, and defects of the rock for great depths at a moderate cost.

Diamond drilling is the most common method for obtaining small-diameter cores.

While the detailed procedures must be adapted to the type of rock and the fracture patterns, the ASTM standard D-2113 is suited to a wide range of conditions.

The sampler, or *core barrel*, is a piece of hardened steel tubing from 2 to 10 ft long with a *bit* attached to the lower end. The bit (Fig. 6.6) is ordinarily set with borts (black diamonds), although tungsten carbide or other very hard, tough materials can be used for drilling soft rocks. The six most popular sizes standard in the United States are given in Table 6:6.

Figure 6.6 Rock coring bits. (Courtesy of Law Engineering Testing Co.)

In drilling, the core barrel and bit rotate while water under high pressure is forced down the barrel and into the bit. The cuttings, ground to a powder, are carried up the hole with the wash water. The rock core extends upward into the barrel. The ratio of the length of core obtained to the distance drilled is known as the *core recovery* and is expressed as a percentage. The core recovery is an indication of the quality of the drilling and of the soundness of the rock; in homogeneous, sound rock a recovery of over 90 per cent may be expected; in rocks with seams a recovery of about 50 per cent is typical; however, in decomposed rock the recovery may be little or nothing.

TABLE 6:6 / SIZES OF DIAMOND BITS

Size	Outside Diameter (in.)	(mm)	Core Diameter (in.)	(mm)
EX	$1\frac{1}{2}$	38	$\frac{13}{16}$	21
AX	$1\frac{15}{16}$	49	$1\frac{3}{16}$	30
BX	$2\frac{3}{8}$	60	$1\frac{5}{8}$	41
NX	3	76	$2\frac{1}{8}$	54
$2\frac{3}{4} \times 3\frac{7}{8}$ in.	$3\frac{7}{8}$	98	$2\frac{11}{16}$	68
$4 \times 5\frac{1}{2}$ in.	$5\frac{1}{2}$	140	$3\frac{15}{16}$	100

To obtain good cores in soft or fractured rock, BX or larger is desirable.

Deere[6:13] proposed a modified recovery, RQD: the ratio of the length of intact rock in NX core sections longer than 4 in. to the distance drilled.* A ratio of 90 per cent or more denotes excellent rock, 75 to 90 per cent, good rock, 50 to 75 per cent, fair rock, and 25 to 50 per cent, poor rock.

In fractured or soft rock a double-tube core barrel is essential to obtain better core recovery. This employs a thin steel tube that fits snugly around the core and which remains stationary while the outer tube rotates. It protects the core from vibration and from erosion of the drilling wash water.

VISUAL EXPLORATION / It is difficult to determine the direction of dip of inclined strata as well as the orientation of defects from boring samples. Pits large enough for direct observation supplement the borings in providing a three-dimensional picture of the strata. Elongated pits or trenches are helpful to define erratic variations in the strata and to determine the continuity of thin seams or lenses. Color photographs provide a permanent record of the material exposed.

Sequence cameras and TV cameras have been developed to fit inside bore holes and record the strata exposed in the hole sides. These require special lenses to scan the entire perimeter or rotating lenses to picture a segment at a time. Although the records (including the TV data on magnetic tape) require considerable experience for interpretation, they make it possible to examine the strata below the practical limit for pit construction and even below ground water. Bore hole periscopes are also available for limited depths.

6:6 Penetration Tests[6:10, 6:12]

Changes in underground conditions can be identified by differences in the resistance of the strata to being pierced by a *penetrometer*. Ancient man who drove a pole into a soft marsh mud to locate a firm sand seam practiced this technique. Although the equipment is more sophisticated today, the principle is the same.

---

* Core sections that exhibit fresh breaks that were obviously produced during drilling are included in the intact lengths of core.

SEC. 6:6] PENETRATION TESTS

Most modern penetrometers consist of a conical point attached to a drive rod of smaller diameter. Penetration of the cone forces the soil aside, creating a complex shear failure, resembling the point penetration of a foundation pile. The test, therefore, is an indirect measure of the in-place shear strength of the soil.

Two forms of penetration are used: *static* and *dynamic*. In the *static* test the point is forced ahead of a controlled rate and the force required for movement is measured. In the *dynamic*, the penetrometer is driven a specified distance by hammer blows of equal energy. The number of blows or the total energy required for the specified distance is the measure of resistance. The static test is very sensitive to small differences in soil consistency. The test operation probably does not seriously change the structure of loose sands or sensitive clays. The dynamic test is adapted to a much wider range of consistencies and can penetrate gravels and soft rock that would stop a static device.

STATIC / The *Dutch Cone*, Fig. 6.7a is the most widely used static test. The cone has a 60 deg. point angle, a diameter of 1.4 in. (3.6 cm) and a projected area of 10 sq cm or approximately 0.01 sq ft. A number of variants are in use, differing in cone angle and diameter. In the form illustrated, Fig. 6.7a, an independent sleeve is attached behind the cone. The force

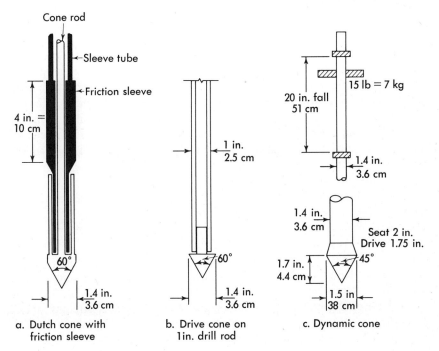

Figure 6.7 Penetrometers.

developed by friction between the sleeve and the soil can be measured independently of the cone resistance. The ratio of sleeve resistance to cone resistance is higher in cohesive soils than in cohesionless. This ratio can be used to estimate the type of soil. The mechanical systems for measuring the resistances vary with the manufacturer. They range from simple rack and pinion drives with spring balance weighing devices to automatic hydraulic driven machines with continuous load indicators. All are limited in the penetration force that can be developed: from half a ton in simple equipment to several tons for large machines that are anchored to the ground.

DYNAMIC / The dynamic test is utilized in many forms. The Standard Penetration Test has a dual function, penetration testing and sampling, that makes it possible to identify changes in the soil by two independent methods. It is for this reason it is such a useful tool in exploration. Cones and points of various size and shape for dynamic resistance measurement alone are also used because of their simplicity and adaptability to a wide range of conditions. In one form, Fig. 6.7b, an expendable cone point or spherical point, 1.4 in. (3.6 cm) in diameter is placed on a 1 in. OD drill rod and driven with a 140-lb hammer falling 30 in. The number of blows required to drive the cone a foot is comparable to the standard penetration resistance, N. The portable cone penetrometer, Fig. 6.7c, employs a 15-lb hammer falling 20 in. The number of hammer blows required to drive the sampler $1\frac{3}{4}$ in. is approximately the standard penetration resistance, N.[6:14] While the dynamic tests may disturb some soils by shock and vibration, they are simple and adapted to both very soft and very hard materials.

PROJECTILE TESTS / A projectile penetrometer is dropped onto the soil surface. The depth of penetration, related to the kinetic energy and geometry of the device, is an indication of the soil strength. One simple form is the pointed rod dropped through water to locate the boundary between soft recent silt accumulations and harder strata below. Similar sounding devices are employed by dredging contractors to estimate the character of the river and harbor bottoms they must excavate. The results of these simple tests are qualitative and the value depends largely on the experience of those who interpret them.

More advanced projectile devices have been designed to be dropped from aircraft. Their rate of deacceleration through the strata can be recorded automatically and even transmitted by radio to a remote recorder. The deacceleration rate and its changes can indicate the consistencies of the strata and their boundaries.

INTERPRETATION / A penetrometer is similar to a miniature pile foundation that forces the soil aside in a complex pattern of shear. While the force required to advance the point, Q_0, is related to shear strength, many other factors are involved, similar to those of importance in pile bearing capacity (Chapter 10). The most significant factors are:

1. c and ϕ of soil.

SEC. 6:6] PENETRATION TESTS

2. Overburden stress, γz.
3. Neutral stress, u.
4. Geometry of penetrometer.
5. Method of driving.
6. Effect of driving on ϕ, c, u.

For a clay in which the strength can be approximated by a single parameter s that is independent of changing confining stress and changing hydrostatic neutral stress, u, the resistance Q_0 can be approximated by

$$Q_0 = sN_p \cdot A \qquad (6:2a)$$

$$s = \frac{Q_0}{N_p \cdot A}. \qquad (6:2b)$$

In this expression the dimensionless penetrometer factor, N_p, embodies the shape of the device and the mode of driving, and A is the projected area in the direction driven. For the Dutch cone, values of N_p range between 5 for very sensitive soils to 9 for medium plasticity clays of low sensitivity.

For dynamic penetrometers, the energy of the hammer of weight W, falling a distance h with a total mechanical efficiency of m is mWh. If the distance penetrated by N hammer blows is S, then the static and dynamic resistance are related:

$$WhmN = Q_0 S$$

$$Q_0 = \frac{WhmN}{S}$$

$$Q_0 = sN_p A \qquad (6:2a)$$

$$s = \frac{WhmN}{SN_p A}. \qquad (6:3)$$

For the standard penetration test, limited data indicate that the value of m ranges between 0.1 and 0.5 with an average of about 0.25.

Numerous emperical expressions have been developed relating penetration resistance to engineering properties.[6:15,6:16,6:17] These are useful correlations but are only approximate. There is a considerable scatter of data, reflecting the many variable factors that are not included in each relation. Therefore, in any particular association of soils, it is best to collect sufficient data to verify the relationship that is used, or to adjust it to fit local conditions. For cohesionless soils, the angle of internal friction, Fig. 6.8a, is related to N. In cohesive soils, the in-place shear strength, s, at the test level is related to N, Fig. 6.8b.[6:1]

The modulus of compressibility in simple consolidation of sands, Fig. 6.8c, and the relative density above the ground water table, Fig. 6.8d, are also approximately related to the value of N. Below the ground water table both relations are reasonably valid for coarse sands if effective stresses

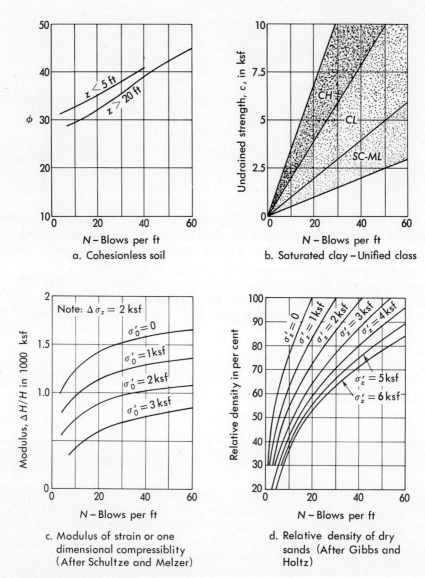

Figure 6.8 *Relation of Standard Penetration Resistance to soil properties for preliminary studies. Note: All the relationships are approximations.*

are used. In fine sands, the pore pressure effects in both loose and dense states make the results less reliable.

All penetration tests are only indirect indications of soil behavior. Therefore penetration testing should always be considered a supplement to direct methods of soil exploration such as boring and sampling. Once the

6:7 Ground Water

Determining the location of ground water is an essential part of every exploratory investigation. In most cases it is measured in the exploratory borings; however, it is frequently necessary to make borings expressly for ground water measurement when perched or artesian water is expected or if a drilling technique (such as the use of drilling mud) obscures the water.

The first hint of ground water may be wet samples or moisture trickling into the boring. Such observations are possible only with dry drilling methods. However, while these conditions must be recorded, they may represent only capillary saturation or a perched water table. Caving of uncased auger holes in sand is also indicative of ground water, but not conclusive. A more valid indication of the general ground water elevation is found by allowing the water in the boring to reach an equilibrium level. In sandy soils a few hours will be enough but in clays a week or more is required.

It is more difficult to identify perched water tables and artesian pressures, particularly if the aquifers have low permeabilities. A series of borings, each terminating in a different suspected pervious aquifer will generally find the perched water tables, although very localized perched water can be drained by the deeper borings themselves. In such cases sealed piezometers, discussed below, are necessary. Artesian water is probable when the water level in the boring suddenly increases, when drilling mud thins or increases in volume, or if deep borings find a higher water level after stabilization than nearby shallow borings. A sealed piezometer is essential in measuring the artesian pressure. These should be installed in all pervious strata in which abnormal pressures are suspected.

OBSERVATION WELLS / Observations for a year or more are required to show seasonal fluctuations of the ground water. In most cases it is necessary to provide casing to maintain an open hole and to assure that the water level in the hole does not lag behind the changing water levels in the ground. In sandy soils a simple well can be made from plastic pipe with a 1 in. (2.5 cm) or larger inside diameter. The lower end of the pipe is slotted with fine saw cuts throughout the level of the aquifer. The well is sealed with concrete at the top to keep out surface water, and fitted with a vented cap. In fine soils the construction is more elaborate. Perforated or slotted plastic pipe is wrapped with plastic screen. The annular space between the soil and the screen is filled with clean concrete sand throughout the aquifer thickness and to the maximum probable ground water level. The top of the well is sealed as previously described.

PIEZOMETER / In order to measure artesian pressure the well must be sealed into the impervious stratum immediately above the aquifer, to form a piezometer. It is extremely difficult to form a good seal, but without it the piezometer is of little value. Balls of clay or damp bentonite are dropped onto the annular sand fill in the aquifer. They are rammed into a continuous plug by a cylindrical drop weight that just fits around the plastic pipe and inside the bore hole. Several layers of rammed clay balls alternating with cement mortar are required. An observation well and a piezometer are shown in Fig. 6.9a and b.

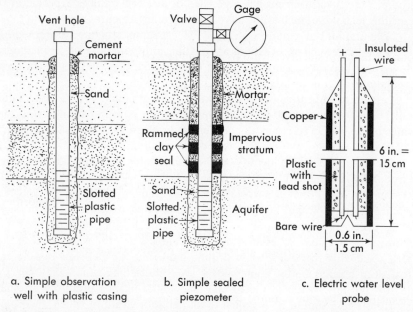

Figure 6.9 Ground water observations.

WATER LEVEL PROBE / The sensitivity of an observation well or piezometer to rapid changes in water level or pressure depends on the volume of water required to fill the well tube during a ground water rise or to drain from it during a fall. For aquifers of low permeability, a rapid response requires that the well tube be as small as practical, $\frac{1}{2}$ in. (1.3 cm) to 1 in. (2.5 cm). In order to detect the ground water level accurately a slender electric probe, Fig. 6.9c, is necessary. It consists of two insulated wires embedded in a weighted sleeve that will fit inside the piezometer tube. The wire ends, uninsulated, extend a fraction of an inch below the sleeve. When the wires touch water, there is sufficient conductivity that the current can be indicated by a milliameter. Recording probes of many forms are available for continuous readings.

6:8 Geophysical Exploration

In geophysical exploration the stratification is inferred from distortions of physical force patterns, either those inherent in the earth or imposed on the earth by the exploration work. In a theoretically homogeneous, isotropic mass, the shape of the pattern can be defined mathematically. Any deviation from the theoretical pattern, termed an *anomaly*, is the result of a nonhomogeneity, such as stratification. In many cases it is possible to interpret the anomalies in terms of the depth and thickness of the different strata and even to estimate some of the engineering properties of the materials.

Many different force systems have been investigated and most find some use in the study of geologic structure and the exploration for minerals. In most cases the anomalies are so large geographically that they are of little use in civil engineering. However, a number of techniques have been found valuable in site investigations, as summarized in Table 6:7.

REFRACTION SEISMIC / This method is based on the physical principle that an elastic shock wave in a homogeneous elastic material having a density γ and a modulus of elasticity E, travels at a velocity v, expressed by

$$v = C\sqrt{\frac{Eg}{\gamma}}, \qquad (6:4)$$

where C is a dimensionless constant involving v. While the densities of soil and rock vary only through narrow limits, the values of E vary greatly depending on the structural qualities of the material. Typical wave velocities are given in Table 6:8.

A small explosive charge is placed at or below the ground surface. Detectors, called *geophones*, are placed on line at increasing distances, d_1, d_2, ..., from the charge. The explosive is detonated, and the time required for the elastic wave to reach each detector is automatically recorded by a *seismograph*. The time required for the first shock to reach each geophone is plotted as a function of the distance from the charge as shown in Fig. 6.10. A simple interpretation is possible if each stratum is uniform in thickness, H_1, H_2, ..., and that each successively deeper stratum has a higher velocity of transmission: $V_2 > V_1$, ... The wave to the first few geophones travels directly through the upper stratum. Therefore the slope of the time distance graph is inversely equivalent to the velocity.

$$V_1 = \frac{d_2 - d_1}{t_2 - t_1}. \qquad (6:5a)$$

At the same time a shock wave is travelling down into stratum 2 where it is refracted to travel through stratum 2 and eventually return to the surface to be recorded by the geophones. Close to the explosive charge the time of wave travel is least by the more direct surface route. Eventually, if $V_2 > V_1$ a distance is reached where the time of travel by the longer route is less than

TABLE 6:7 / GEOPHYSICAL METHODS*

Method	Principle	Use	Limitations
Refraction Seismic	Shock wave by hammer impact or small explosive near ground surface. Time of travel to geophones at different distances measured. Shock may travel to distant phone faster through deeper hard strata than by shortest path.	Depth to ground water; depth to successively harder strata; possible estimate of rigidity and location of sink holes.	Interpretation questionable with irregular or poorly defined boundaries; will not identify soft strata under more rigid strata.
Electrical Resistivity	Electric current passed between electrodes at varied spacings. Potential drop between intermediate electrodes and current define apparent resistivity. Depth and resistivity of strata determined by resistivity-electrode spacing relations.	Depth to strata of different resistivities and groundwater. Location of masses of dry sands and gravels or hard rock.	Interpretation questionable with irregular or poorly defined boundaries.
Gravity	Measure earth's gravity by sensitive torsion balance.	Locate major structural anomalies: faults, domes, possibly large cavities.	Will not identify structures unless major differences in density involved.
Sonic	Time of travel of sound or super-sonic wave through water and soft silt and reflected upward by stratum changes.	Depth of water and soft silt above hard bottom.	Of little or no use in continuous soil or soil-rock.

* Reflection seismic, magnetic, self potential, and radiation sometimes useful in specialized applications such as well logging and mineral surveys.

TABLE 6:8 / SEISMIC WAVE VELOCITIES

Material	Velocity (ft per sec)	Velocity (m per sec)
Loose dry sand	500–1,500	150–450
Hard clay, partially saturated	2,000–4,000	600–1,200
Water, loose saturated soil	5,200	1,600
Saturated Soil Weathered Rock	4,000–10,000	1,200–3,000
Sound Rock	7,000–20,000	2,000–6,000

by the surface. The time-distance graph in this range is flatter than for the first, and V_2 can be computed similar to V_1 as shown in Fig. 6.10. The two lines intersect at a point equivalent to the distance, d' from the shot. The thickness of the stratum, H_1 is given by

$$H_1 = \frac{d'}{2}\sqrt{\frac{V_2 - V_1}{V_2 + V_1}}. \quad (6:5b)$$

The velocity and thickness of each successive stratum can be computed provided its velocity is higher than that of the one above. The method is

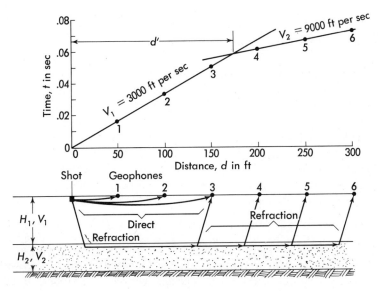

Figure 6.10 *Refraction seismic exploration.*

best adapted to horizontal or gently sloping strata with well defined contrasts in velocity, such as soil overlying rock, or loose dry sand overlying a sand saturated by the ground water table. Under ideal conditions it can define the depth of boundaries several hundred feet deep with an accuracy of 1 to 2 per cent.

ELECTRICAL RESISTIVITY / The electrical resistivity method is based on the fact that the conductivity of different strata varies with the ionized salts present. Dense rock with few voids, little moisture and little ionization will have high resistance while saturated clay will have low resistance. A number of procedures are used for measuring the resistivity of a soil mass. The Wenner method with four equally spaced electrodes is simple and widely used in site investigations. The four electrodes are placed in a straight line at equal distances, d, as shown in Fig. 6.11. An electrical

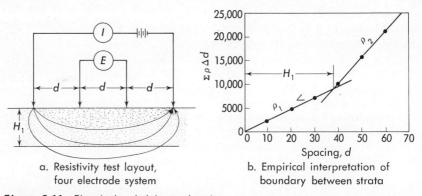

a. Resistivity test layout, four electrode system

b. Empirical interpretation of boundary between strata

Figure 6.11 Electrical resistivity exploration.

current, typically 50 to 100 milliamperes, is passed between the outer electrodes, and is precisely measured. The voltage drop in a segment of the mass is measured between the two inner electrodes by a null-point circuit that requires no current flow at the instant of measurement. Either alternating current with its less sensitive measuring systems or direct current with non-polarizing potential electrodes is required to avoid polarization (the accumulation of hydrogen ions at the negative electrode) and an error in potential.

In a semi-infinite homogeneous isotropic material the electrical resistivity, ρ, is given by the expression

$$\rho = \frac{2\pi d\, E}{I}, \qquad (6:6)$$

where I is the current in amperes, E, the potential difference between the center two electrodes and d, the spacing between electrodes. If the soil mass consists of strata of different resistivities, then the apparent resistivity as computed by this expression will be changed. The pattern of apparent resistivity as a function of electrode spacing or test location is the basis for interpretation.

The variable spacing technique is used to locate the depths of strata of different resistivity. A series of tests are made centered about one point, with increasing spacings, such as 10, 20, 30, ... ft. A plot of apparent resistivity as a function of spacing can be interpreted in terms of the depth of the boundary between strata, using theoretical standard curves. An emperical interpretation used in site exploration work plots the sum of the apparent resistivity values as a function of spacing as shown in Fig. 6.11b. The resulting curve consists of relative straight segments if the strata are horizontal and uniformly thick. Tangents are drawn to the curve. The spacings equivalent to the intersections of the tangents are the depths of strata boundaries. The slope of the curve is proportional to resistivity—a steep curve indicates dry soil or rock and a flat curve wet soils or other materials of low resistivity. Typical resistivity values are given in Table 6:9.

TABLE 6:9 / ELECTRICAL RESISTIVITIES OF SOILS AND ROCKS

Material	Resistivity in Ohm-centimeters
Saturated organic clay or silt	500–2000
Saturated inorganic clay or silt	1000–5000
Hard, partially saturated clays and silts, saturated sands and gravels	5000–15,000
Shales, dry clays and silts	10,000–50,000
Sandstones, dry sands and gravels	20,000–100,000
Crystalline rocks, sound	100,000–1,000,000

A different method is used in locating areas of shallow rock or high ground water level. A constant electrode spacing is used, approximately equal to the estimated depth of the material. Measurements are then made at widely spaced locations in a grid pattern. A contour map of apparent resistivity shows areas of highs and lows. Bedrock or dry sand and gravel are most likely to be found in the high areas and shallow ground water or clay in the low areas.

ADVANTAGES AND LIMITATIONS / The geophysical methods have two important advantages. First, they permit a rapid coverage of large areas at a relatively small cost. Second, they are not hampered by boulders or coarse gravel that obstruct borings. These qualities make them useful for both reconnaissance and exploration.

Their lack of unique interpretation is a distinct disadvantage. This is particularly serious when the strata are not uniform in thickness nor horizontal. Irregular or transitional contacts often are not identified and strata of similar geophysical properties sometimes have greatly different engineering properties. If the contact is very irregular, the boundary defined by resistivity is the average depth for a distance approximately equal to the depth. The same boundary defined by seismic refraction is the depth to the more or less continuous deeper stratum. For these reasons geophysical methods always must be used as a supplement to the direct methods, and the results verified by boring before definite conclusions can be reached.

6:9 Analyzing the Results of an Exploratory Investigation

LABORATORY TESTS / While a visual examination of the soil samples obtained from exploratory borings may provide the engineer with a preliminary picture of the soil conditions, a study of the results of laboratory tests clarifies the picture and makes it possible to analyze the soil conditions on the basis of factual data.

The samples are ordinarily described in the field by the engineer in charge of the boring and sampling work, but they should be re-examined in the laboratory and the field identifications should be verified. Tests can then be made on the samples to confirm their identification and to determine their physical properties. Table 6:10 summarizes the tests most useful in exploratory work.

Other tests, such as the loss of weight by ignition which identifies

TABLE 6:10 / LABORATORY TESTS FOR EXPLORATORY INVESTIGATIONS

Test	Types of Soils	Size of Sample (lb)	Type of Sample	Use of Data
Specific gravity of solids	All	$\frac{1}{10}$	Auger or split barrel	Determine composition, void ratio.
Grain size	Cohesionless (sands, gravels)	$\frac{1}{4}$	Auger or split barrel	Classification. Estimate permeability, shear strength, frost action, compaction.
Grain shape	Cohesionless (sands, gravels)	$\frac{1}{4}$	Auger or split barrel	Classification. Estimate shear strength.
Liquid and plastic limits	Cohesive (silts, clays)	$\frac{1}{4}$	Auger or split barrel	Classification. Estimate compressibility, compaction.
Water content	Cohesive	$\frac{1}{4}$	Auger or split barrel	Correlate with strength, compressibility, compaction.
Void ratio	Cohesive	$\frac{1}{4}$	Split barrel*	Estimate compressibility and strength.
Unconfined compression	Cohesive	$\frac{1}{4}$	Split barrel*	Estimate shear strength.

* Sample must be relatively undisturbed.

organic materials, or treatment with hydrochloric acid which indicates the presence of soluble carbonates, may be useful in identifying some soils. A microscopic examination of coarse soils and of the particles coarser than 0.074 mm in fine-grained soils is very useful in correlating similar strata in different borings.

PLOTTING BORING RECORDS / The first step in analyzing the data obtained by exploratory investigations is to plot the boring records graphically on a large *work sheet*, as shown in Fig. 6.12. Each boring is represented by a vertical bar graph, with the different soils indicated by appropriate symbols or abbreviations. All should be plotted to the same scale with elevation (above the site datum) as the vertical ordinate. If possible, borings that are adjacent on the site should be plotted adjacent to one another, but a space of 2 or 3 in. or 5 to 8 cm should be left between the plot of each boring to provide room for the laboratory data.

Although the soil descriptions can be shown by graphic symbols, a simple shorthand system is preferable as described in Chapter 2.

The soil penetration resistances are plotted as a broken-line graph next to the boring plot. This makes it possible to correlate the resistances of the different soils encountered. On the same graph can also be plotted the unconfined compressive strength data from the laboratory tests.

A second graph, also plotted adjacent to the boring record, shows water content and the liquid and plastic limits. The water content may be plotted as a broken-line graph and the limits as isolated points. The characteristics of cohesionless soils such as grain size and shape cannot be represented so conveniently on such a graph but may be indicated by notes or symbols.

PREPARING SOIL PROFILES / Soil profiles or geologic cross-sections (Fig. 6.13) for critical parts of the site are prepared by correlating the soils encountered in each of the borings. For example, a hard clay layer found in each of three adjacent borings at about the same elevation is probably the same continuous stratum, especially if the liquid limits and plasticity indexes are the same. By interpolating between borings, a reasonable soil profile may usually be established. In some very erratic soils, such as glacial moraines, interpolation may be dangerous, since apparently continuous strata may be discontinuous lenses. Silts and organic soils often occur in limited lens-shaped deposits and should be viewed with suspicion.

THREE-DIMENSIONAL REPRESENTATION / In most formations the soils vary not only with depth but also with location. A geologic cross-section or profile represents the changes between borings in one direction only. Therefore a number of cross-sections parallel and at right angles to one another may be necessary to depict the three-dimensional variations. Cross-sections are usually oriented with the structure, such as parallel to column lines, and thus may not always display the most critical orientation of the soils. For example, if the strata dip steeply, a cross-section parallel to the strike would depict horizontal boundaries, and might lead to a false

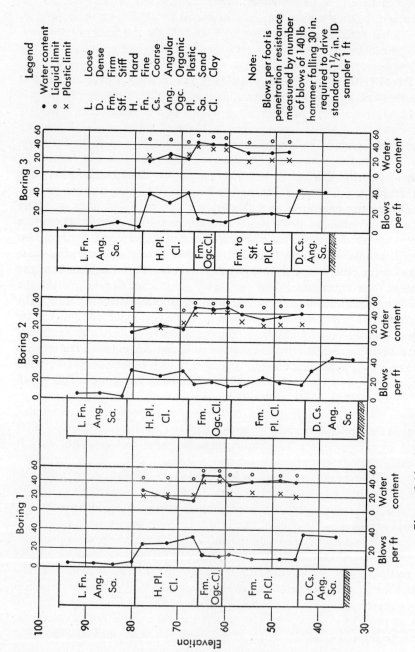

Figure 6.12 Plot of the results of the exploratory phases of a soil investigation.

SEC. 6:9] RESULTS OF AN EXPLORATORY INVESTIGATION   291

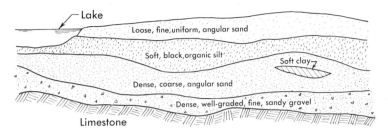

Figure 6.13   *Soil profile or geologic cross-section.*

sense of security. Cross-sections, therefore, should be oriented to show the most critical soil and rock variations.

Where the variations in soil and rock are irregular, such as where faulting and folding are significant, three-dimensional representations are helpful in visualizing the underground structure. Combined cross-sections plotted in a form of isometric projection termed *fence diagrams*, Fig. 6.14, convey a three-dimensional impression. Similarly, block diagrams can give a limited three-dimensional view.

Three-dimensional models permit a quantitative representation. The simplest is a *peg board*. This consists of a base board that represents the site plan at a reference level well below the zone of interest. Rods are set in this board at the boring locations, and the significant soil and rock features are marked at the appropriate levels on the rod above the base. The ground surface is shown by the upper end of the rod. The level of the building such

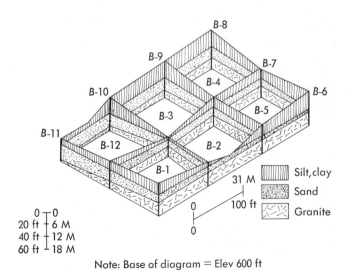

Note: Base of diagram = Elev 600 ft

Figure 6.14   *Fence diagram, showing site conditions in three dimensions.*

as a foundation mat can be indicated by a transparent plastic sheet pierced to fit over the rods at the proper depth. Boundaries between strata can be constructed with solid plastic foam or strings between the borings. If the horizontal and vertical scales are equal, complex problems in orientation can be solved quantitively with the model, and the results can be meaningfully presented to a layman. Unfortunately, models are bulky and not easily preserved in files or in reports.

In attempting to correlate the boring records and determine the soil profiles, the engineer often finds that additional borings would be very helpful. If the records are plotted at the same time the boring work is progressing, then the number or spacing of the borings can be changed to produce a clearer picture of the soil strata. In many cases low-cost auger borings can be used effectively to determine the extent of strata between the more expensive test borings, and in some cases geophysical methods or penetration tests prove very useful for the same purpose.

PRELIMINARY COMPUTATIONS / The unconfined compressive strength, void ratio, and compressibility of clays, and the unit weight and angle of internal friction of sands and gravels are necessary for most studies involving the safety and settlement of earth masses and the structures they support. The average values of these for each stratum in the soil profiles may be estimated from the laboratory data and the penetration resistance.

Preliminary computations for safety and settlement may be made by utilizing the soil profiles and the estimated soil properties. The results of these computations can be placed in four categories:

1. The structure is so safe from failure and excessive settlement that further study is unnecessary, and the estimated soil properties can be used as a basis of design without the sacrifice of economy.
2. The structure is safe and free from excessive settlement, but additional detailed soil studies may lead to a more economical design.
3. The structure appears to be unsafe or will probably settle too much; therefore additional detailed soil studies will be necessary before a satisfactory design can be developed.
4. The structure is so unsafe or will settle so much that further soil studies would be useless.

On the basis of these computations, therefore, the designer can decide whether to go ahead with his plans without further study, to secure additional, more accurate data, or to abandon the project as originally planned.

6:10 Intensive Investigation

The intensive investigation provides the engineering data on the soil and rock strata that are necessary for a quantitative design. Permeability, for projects involving seepage or drainage, strain and strength under changing loads, and volume changes produced by both stress and environment are as essential for the success of a project as the strength of the steel and concrete used in the superstructure. The more complex the soil conditions and the heavier the structure, the greater the cost savings possible with adequate data. The savings embrace both the construction operations and the ultimate structure. Moreover, when the soil conditions are marginal, estimates of soil behavior based only on exploratory data can lead to either ridiculously safe designs or to serious risks of future trouble.

The detailed investigation ordinarily focuses on the critical strata pinpointed by the exploratory work. For example, if a clay stratum is found that has a low penetration resistance, it is likely that its shear strength and compressibility will be significant factors in final design. Two approaches are possible in obtaining the needed data.

1. Secure representative samples of sufficient quality for laboratory testing.
2. Test the soil in place.

Testing the soil in the laboratory has the advantage that the environment, including stress, can be changed at will to simulate the changes produced by the construction and the future structure. Further, laboratory tests are available to measure nearly all the required soil qualities. The results are limited, however, by the soil samples: their alteration or disturbance produced by sampling, and their degree of representation of the total stratum. Testing the soil in the ground evaluates its behavior in its present environment. The test may integrate the effects of many variables that are difficult if not impossible to simulate in the laboratory, and the soil disturbance is limited to that produced by the test instrumentation. Unfortunately, the number of soil properties that can be evaluated and the number of changes in stress and other environmental factors that can be induced are severely limited. For most projects undisturbed sampling with laboratory testing is sufficient; for complex soil conditions a program of both is essential. The field test results can check the laboratory work and the laboratory tests expand the range of conditions possible in the field.

In both cases the intensive work can be done either by interrupting the exploratory boring and investigating each critical stratum at the time it is identified, or by a separate phase of work conducted after the exploration has been completed. Combining the exploratory and intensive phases of the investigation saves time and on small projects is more economical. In large projects a better integration of work is possible if the detailed program is planned from the correlated results of the exploratory phase.

DATA REQUIRED / Analysis of the safety of a structure and of limiting earth pressures requires the soil shear strength. The strength of cohesive soils can usually be measured adequately in the laboratory. Very soft or sensitive clays that cannot be sampled or that are easily disturbed are tested better in the field. The strength of cohesionless soils can be determined either by undisturbed samples or on disturbed samples reconstructed at the same relative density. Because of the difficulties in securing undisturbed cohesionless samples, and their tendency to change state during transit, it is often expedient to test the samples for density in a field laboratory. Otherwise direct field tests such as penetration tests will be more reliable, although they are limited to the present state of stress.

When settlement is critical, modulus of elasticity and compressibility data are essential. Laboratory consolidation and stress-strain tests are conducted on undisturbed cohesive samples and undisturbed or reconstructed cohesionless samples. Field tests for soil strain from changing stress can also be obtained from small or large scale field loading tests as well as from bore hole expansion. Permeability testing can be done on laboratory specimens; however very large numbers of tests are necessary to define the range in values. Permeability is so profoundly influenced by seemingly insignificant changes in soil stratification that field testing is advisable when possible. Field testing is limited by the present ground water conditions, and largely to horizontal seepage.

SAMPLES REQUIRED / In most investigations the critical strata prove to be composed of cohesive soils—clays, organic silts, and organic clays—which require undisturbed samples of sufficient size for laboratory tests. Table 6:11 lists the typical sizes of samples for testing,

TABLE 6:11 / SIZES OF SAMPLES FOR TESTING

Test	No. of Samples for One Test	Sizes of Samples Tested	
Unconfined compression	2	1.4 in. diam × 3 in. long	3.5 cm × 7.5 cm
	2	2.8 in. diam × 6 in. long	7 cm × 15 cm
Triaxial shear	4–6	1.4 in. diam × 3 in. long	3.5 cm × 7.5 cm
	4–6	2.8 in. diam × 6 in. long	7 cm × 15 cm
Direct shear	4–6	1 in. × 3 in. diam	2.5 cm × 7.5 cm
Consolidation	1	2.5 in. diam × 1 in. thick	6 cm × 2.5 cm
	1	4.25 in. diam × 1 in. thick	11 cm × 2.5 cm

The number of samples to be made depends on the uniformity of the stratum to be sampled. A perfectly homogeneous soil would require only one sample large enough for the necessary tests, but unfortunately, most actual soil deposits are far from uniform. The range in variation in soil properties can be determined from the results of the exploratory investigation. Typical points and extreme points within the stratum are selected from the boring logs and the plots of penetration resistance, water content, Atterberg limits, and unconfined compressive strength. The undisturbed samples are secured as close to these points as it is practical to do so. In many instances, however, it is necessary to secure an unbroken or continuous series of undisturbed samples throughout the depth of the critical stratum. Ordinarily, one series of undisturbed samples is made beneath each

TABLE 6:12 / METHODS FOR DEEP SAMPLING

Method	Equipment and Procedure	Type of Sample and Use	Limitations
Auger (ASTM D 1452)	Retain cuttings from short increments of auger boring.	Disturbed for soil identification, water content above water table.	Structure destroyed. Soil mixed with water below water table.
Split Barrel ASTM D-1586	Split barrel sampler, driven 18 in. into stratum, 1.4 in. I.D., by 2 in. O.D., 2 in. I.D., by 2.5 in. O.D., without or with liners for sample protection.	Intact but disturbed. Soil identification, structure, water content; density of very wide range of soils.	Sample distorted—Disturbance too great for strength, consolidation tests.
Thin Wall Tube ASTM D-1587	3 in. O.D., to 5 in. O.D. thin wall tube with sharp edge forced into soil, 10 to 20 diameters.	Relatively undisturbed sample for shear density, consolidation, etc., of most soils.	Sample lost in very soft clay or loose sand below water.
	Same—driven with hammer.	Stiff clays only	Slight disturbance.
Thin Wall Tube Fixed Piston	3 in. O.D., to 5 in. O.D. thin wall tube with sharp edge. Piston keeps cuttings out, remains stationary while tube advances and fixed after driving to help hold sample. Rod activated, internal hydraulic piston (Osterberg type), ratchet control (Hong).	Relatively undisturbed of very soft silts, clays, loose sands if hole filled with heavy drilling fluid.	Sample sometimes lost in soft clay, loose sand.
Swedish Foil	Thin strips of metal foil stored above cutting edge surround sample to prevent contact with sample tube and help hold sample.	Relatively undisturbed continuous (to 40 ft) sample of soft clay, for shear, consolidation, etc.	Gravel, appreciable sand or hard strata damage sampler.
Rotary Core: Soil	Outer tube with teeth rotates; stationary inner tube protects and holds soil. Cuttings removed by drill fluid. (Dennison type with fixed cutter; Pitcher type with automatic variation of cutter position.)	Relatively undisturbed of firm to stiff cohesive soils, soft rock, continuous.	Torsion failure in soft soil and sometimes in sands.
Rotary Core: Rock	Tube with diamond bit on end rotates. Core protected with stationary inner tube in double tube core barrels. Cuttings removed by drill fluid.	Continuous core in hard rock; nearly continuous in soft or fractured rock with M-type double tube.	Fractured or very soft rock not recovered.

important structure or beneath each different part of large structures, but more or less may be necessary, depending on the uniformity of the soil. Methods for deep sampling are summarized in Table 6 : 12.

6:11 Undisturbed Sampling

The most important step in the detailed investigation is securing a sample with as little disturbance as possible. Unfortunately it is impossible to secure a completely undisturbed sample. The removal of a portion of soil from the ground produces changes in the soil stresses which change the soil structure to some extent. The best "undisturbed samples" are those in which the soil water content and composition remain unchanged and the void ratio and structure are changed as little as possible.

CHUNK SAMPLING / A *chunk sample* excavated carefully by hand is usually the best undisturbed sample obtainable. Figure 6.15 illustrates the

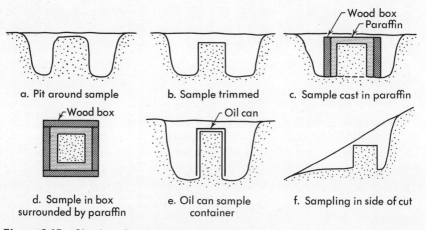

Figure 6.15 *Chunk undisturbed sampling.*

steps for securing a chunk sample. A pit or shaft is excavated in the proper location to the depth from which the sample is desired. The soil is carefully removed from around the sample, leaving it projecting in the side or bottom of the excavation like a small stump. If the sample is strong and rigid, it can be cut free with a flat shovel, then wrapped with plastic film to preserve its water content, placed in a substantial box for support, and transported to the laboratory by automobile. If the sample is weak, or if it must be transported by railroad or truck, additional protection is required. One very good method is to place a heavy wooden box, with lid and bottom removed, around the sample so as to leave a space of 1 in. on all sides. This space is filled with melted paraffin. The sample and the box are removed from the excavation, paraffin is poured on the top and bottom of the sample, and the top and bottom lids replaced. Another method is to slide a cylindrical container over the sample which has been carefully trimmed to make a snug fit. Four-quart oil cans with their tops removed, sections of stove pipe, and large water pipes have been used successfully. Wood or metal caps should be provided to protect the open ends of the samples from damage. A layer of paraffin should be poured over the open ends to prevent evaporation of the moisture. These containers should be placed in a substantial wooden box and surrounded with sawdust or shavings in order to protect them for shipment.

The test pit not only makes it possible to secure undisturbed samples, but also provides a "window" from which to observe the soil structure in place. A visual examination of the soil strata that are uncovered by excavation of the pit will disclose the arrangement, uniformity, and the inclination or dip of the strata—information that is not easily secured from borings. Photographs made of the sides of the pit will provide a permanent record of this information.

DEEP UNDISTURBED SAMPLING / Many types of equipment have been designed to secure undisturbed samples from deep bore holes, but not all types will be successful in every soil. The undisturbed quality of the sample has been found to be dependent on the following factors:

1. Displacement of the soil by the sampler.
2. Method of forcing sampler into ground.
3. Friction on inside of sample tube.
4. Squeezing of soil, owing to pressure of overburden.
5. Handling and storing of samples until tested.

The displacement of the soil by the walls of the sampler is probably the most important source of disturbance. The soil is forced aside and upward, which severely distorts it and changes its structure. This can be minimized by excavating around the sample, by using a long, thin cutting edge on the sampler, and by keeping the cross-sectional area of the walls of the sampler as small as possible. The relative displacement of the sampler can be expressed by the area ratio A_r:

$$A_r = \frac{D_0^2 - D_s^2}{D_s^2} \times 100 \text{ per cent.} \tag{6:7}$$

where D_0 is the outside diameter of the sampler and D_s the diameter of the sample. According to Hvorslev,[6:5] displacement disturbance is minimized by keeping the area ratio less than 10 to 15 per cent.

A thin sharp cutting edge minimizes displacement. This is easily damaged by gravel or hard seams, and therefore can be used only in continuously soft strata.

The method of driving the sampler into the ground is important in loose sands and very sensitive clays. In such soils continuous hammering and the accompanying shock and vibration are harmful, although a single blow of a heavy hammer does not appear detrimental. The best method is to force the sampler into the ground with a steady movement such as is provided by hydraulic-feed drilling machines or by blocks and tackle that pull against anchors set in the ground.

Friction between the sample and the walls of the sample tube produces disturbance in the edges of the sample that is readily visible in stratified or laminated soils but is present in all cases. This friction can be reduced by drawing in the cutting edge of the tube about $\frac{1}{50}$ in. The diameter of the sample is then about $\frac{1}{25}$ in. less than the inside diameter of the tube. If too much draw-in is used, the sample may expand, changing its structure; if too little is used, the sample will be distorted. Friction is also minimized by limiting the sample length and by excavating around the sample.

Removal of soil from the boring reduces the downward stress at the hole bottom without changing the lateral and upward stresses produced by the weight of the overburden. As a result the soil squeezes inward and upward, distorting itself in the process. This condition is very serious in soft clays but exists to some extent in all materials. In cohesionless soils any attempt to lower the water level in the hole below the ground water table by pumping or bailing will have an added effect because the unbalanced water pressure in the hole bottom creates a quick condition. The unbalanced stress can be overcome by keeping the hole filled with liquid at all times. Water is adequate in many soils, but in loose sands and soft clays, drilling mud (described in Section 6:5) is better. In extremely soft clays it is sometimes necessary to match the unit weight of the drilling fluid to that of the soil by adding baroids or iron filings to the mixture.

Suction due to removal of the sampler distorts the bottom of the sample or even pulls the soil out of the tube. This is difficult to prevent. A piston or check valve above the sample can help by developing comparable suction above if the sample tends to slide out. A small tube or duct fastened to the outside wall of the sampler and open at the cutting edge can allow the drilling fluid from above to fill the gap produced by removing the sampler. As an alternative air can be pumped through the tube from the ground surface. A simple suction release device is a $\frac{1}{4}$ in. diameter rod or bar welded to the side of the sampler. The sampler is rotated before extraction, creating a temporary duct in the soil. In any case the duct will increase the sampler displacement slightly.

Improper handling and storing cause shock, distortion, and drying. By sealing the sample immediately, packing it in a cushioned container, protecting it from temperature extremes, and exercising due care in transportation these causes of disturbance can be minimized.

Many types of equipment have been developed to minimize the disturbance; however, the relative importance of the factors differ with the soil, the depth and the ground water. Therefore, sampler design is a compromise.

SEC. 6:11] UNDISTURBED SAMPLING 299

The principal types are listed in Table 6:12 and the more versatile are described in the following paragraphs.

THIN WALLED SAMPLER / The simplest and most widely used deep undisturbed sampler is the *thin-wall* or *Shelby tube* (Fig. 6.16a). It is made of cold drawn steel tubing (sometimes known as *Shelby* tubing) from 2 in. to 5 in. in diameter and with walls of 18 gage ($\frac{1}{20}$ in.) or 1.2 mm for the 2-in. or 5 cm tube to 11 gage ($\frac{1}{8}$ in.) or 3 mm for the 5-in. or 13 cm. The lower end is beveled to form a tapered cutting edge, and it can be drawn in to reduce the wall friction. The upper end is fastened to a check valve that helps hold the sample in the tube when it is being withdrawn from the ground.

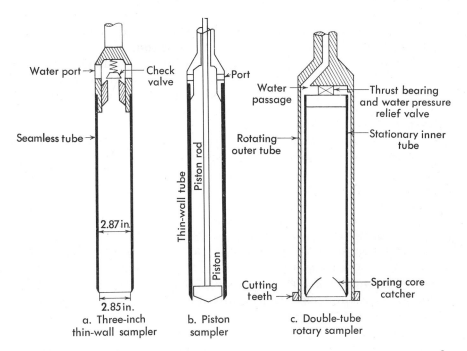

Figure 6.16 Undisturbed samplers.

The sampler is introduced in the bore hole and forced into the soil a distance of no more than 15 diameters so as to minimize friction between the sample and the walls of the tube. The sample is sealed in the tube with melted wax and shipped intact to the laboratory.

The thin-wall sampler minimizes the most serious sources of disturbance: displacement and friction. When used in a bore hole that is stabilized with drilling mud, excellent results can be had in a wide variety of soils.

PISTON SAMPLER / In extremely soft soils, even the small displacement of the thin wall tube tends to cause the soil to squeeze into the tube faster than the sampler is advanced, causing distortion. The distortion due to friction also can be serious, but if friction is reduced too far by draw-in and by limiting the sample length, the sample will slide out of the tube when it is being withdrawn from the ground.

These difficulties can be reduced by placing a piston in the thin-wall sampler, Fig. 6.16b. At the start of sampling, the piston is at the tube bottom and in contact with the soil surface. The piston is fixed in this position by its actuating rods, which extend to the ground surface and which are locked to a rigid support. The sample tube is driven ahead of the piston into the soil below. The fixed piston prevents the soil from squeezing upward. If the sample tends to slip out of the tube, a vacuum is created between the piston and the soil, which helps hold it. Good samples can be secured in the softest soils by this method but at a considerable increase in cost.

The Osterberg fixed piston sampler incorporates a hydraulic cylinder in the head of the sampling device to actuate the advance of the thin walled tube. Hydraulic pressure for operation is supplied through the hollow drill rods, eliminating the need for cumbersome double concentric drill rods, one to hold the piston and one to force the sampler.

A number of variations of the piston sampler are in use. The free piston is locked in the lower end of the tube while the sampler is being lowered into the hole. This keeps loose cuttings from choking the tube. When sampling begins, the piston is unlocked and floats up the tube on top of the soil. When the tube is withdrawn, the piston again locks and helps hold the sample in place. It does not, however, prevent squeezing of the soil during sampling. In another form the piston is torpedo-shaped, which allows the sampler to be advanced through very soft soil without a bore hole. None of these variations is so effective as the fixed piston.

FOIL SAMPLER / The Swedish foil sampler minimizes the friction between the sample tube walls and the soil by feeding thin strips of metal foil into the tube. These form a moving liner of thin metal that prevents the soil from touching the tube. The foil is pulled ahead by a piston at the same rate the sampler is advanced, so that there is no tendency to distort the soil by squeezing or wall friction. By this method continuous undisturbed samples as long as 50 ft or 15 m have been obtained in soft to firm clay or silts. It is of little use in sands and gravels, however.

The foil lining is made up of 16 strips, each nearly $\frac{1}{2}$ in. or 1 cm wide so as to nearly cover the surface of the 2.7-in. or 7 cm diam sample. The foil is supplied from rolls that fit in a retainer 12 in. or 30 cm above the cutting edge so that the displacement of the retainer will not cause undue soil disturbance. The strips feed to the sample 5 in. or 12 cm above the edge.

ROTARY SAMPLERS / The rotary samplers combine drilling and sampling, which minimizes disturbance due to sampler displacement. An example is the *Denison sampler* (Fig. 6.16c), which consists of two concentric tubes. The inner tube is actually the sampler and is a 3 to 5-in. or 7.5 to 13 cm diam, 24-in. or 60 cm long pipe with a thin steel liner and a heavy cutting shoe at the lower end. The outer tube rotates, cutting the soil, and the cuttings are washed up the bore hole by water that is pumped down the drill pipe and flows between the two tubes. The inner tube remains stationary and protects the sample from the wash water. At the same time, both tubes are forced downward into the soil. The samples secured by this device are excellent, particularly in hard clays and slightly cohesive sands that are difficult to sample in any other way. The equipment is comparatively expensive to buy and to operate.

A number of variants of the double tube rotary sampler are adapted to different conditions. The *M-series* of double tube diamond core drills was originally designed for sampling coal. The stationary inner tube extends almost to the cutting edge, and the cutting edge is altered. With it, good cores, 1.7 to 2.1 in. or 4 to 5 cm in diameter, can be made in very soft, easily eroded rock and hard soils. The *Pitcher sampler* employs a spring to control the advance of the inner tube. In hard formations, the tube retracts so that the rotating cutting edge leads. In soft formations, the inner tube, with a sharp edge, extends ahead of the rotating cutter. In this way the sampler resembles a thin-walled tube with a rotary cutter to minimize displacement slightly above the tube entrance.

DEEP SAMPLING IN SAND / It is difficult to secure undisturbed samples of cohesionless sand and gravel from bore holes, since the sampling operation may rearrange the grains and the samples often run out of the sampler. One method has been to freeze the soil and then drill through it with a rotary drill, but the disadvantage is that freezing increases the void ratio. Another method has been to sample with a thin-wall sample tube and then freeze the lower end of the sample to prevent the soil from escaping.

Undisturbed samples of sand can be secured below the ground water table if a very heavy drilling mud is employed. The mud forms a coating over the lower end of the sample, which prevents the sand from running out. The sample is drained of excess water before sealing so that capillary tension will help to maintain its structure during shipment. It is also helpful to measure the sample so that any change in density during shipment can be detected and the original void ratio can be computed from the weight as received in the laboratory.

The Bishop sand sampler utilizes a larger cylinder that is forced over the thin-walled sample tube after it penetrates the sand. Air is pumped into the larger cylinder so that the sampler with its loose sand is surrounded by air. If the replacement of water by air around the sample tube is sufficiently slow that the sand can drain partially, it will be held intact in the tube by capillary tension.

LINER TUBE SAMPLERS / An early type of undisturbed sampler that is still in use is similar to the split-barrel sampler except that it is equipped with a seamless liner tube that fits snugly in the barrel. The sample is shipped and stored in the liner to minimize handling disturbance. The large area ratio often causes severe displacement distortion, and for this reason such samplers are not widely used.

6:12 Field Tests

Testing the soil in place has the theoretical advantage of minimizing the disturbance caused by stress changes and similar sampling distortion as well as eliminating the shock and vibration of transport and subsequent handling. Moreover, the effect of associated features of the formation are included in the field test, so that it probably is the most realistic measure of the physical properties in the existing environment. The test itself, however, may provide some disturbance and it is usually impossible to evaluate the effects of drastic environmental changes. The methods are summarized in Table 6:13.

The various penetrometers described in Section 6:7 can be used as indirect field tests. The static cone is a direct measure of *end bearing* of the formation. Extrapolation of the results of full-sized foundations is discussed in Chapter 10; but as in all models the *scale effect* must be evaluated. The dynamic tests similarly can be related to bearing capacity. Because bearing capacity is a function of the soil's engineering properties (see Chapter 9) a relationship should exist between these properties and either static or dynamic penetration resistance. Typical empirical relationships involving the in-place strength of clays and the relative density and angle of internal friction of cohesionless soils are given in Section 6:7. However, because other factors are involved, these relationships are not well defined and should be used only as indications of the desired property. For any given site and stratum, the scatter of data is limited and well-defined relations can be derived for that one situation.

VANE SHEAR / The vane shear is adapted to testing soils below the bottom of a bore hole at great depths and with a minimum disturbance. Field vanes, Fig. 6.17a, employ two crossed blades attached to a vertical rod. The typical vane diameters are 2, 2.5, and 3 in. (5, 6.25, and 7.5 cm) with lengths of 3 to 5 diameters. The vane is forced into the soil so its top is 2 diameters below the bottom to the bore hole. Rotation of the vane shears the soil on a cylindrical surface. The torque required to initiate shear is measured, and often the increasing torsional strain is indicated as a function of torque. After failure, a second torque is made after several revolutions, to measure the soils remolded strength. The shear strength, s, for a long vane of diameter D, and length L, is found from the torque T, as follows

SEC. 6:12] FIELD TESTS 303

$$T = \frac{D}{2} \cdot \pi D L s,$$

$$s = 2T/(\pi D^2 L). \tag{6:8}$$

Corrections can be made for the end shear; if L/D exceeds 3, the correction is negligible.

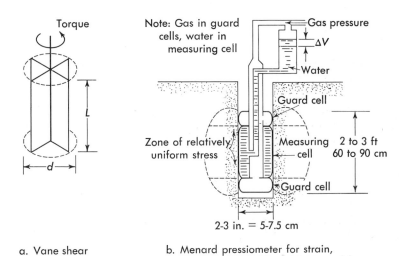

a. Vane shear

b. Menard pressiometer for strain, pseudo-strength (Note: Not to scale)

Figure 6.17 *In-place strength tests.*

INTERNAL OR BORE HOLE EXPANSION / The stress-strain character of soil or rock can be evaluated by the resistance of a bore hole to expansion from internal pressure. The simplest device is a rubber sleeve with rigid end caps that fit snugly in a portion of the hole. Pressure in the sleeve is transmitted to the soil or rock. The radial strain can be measured by internal probes or indirectly by the volume change of liquid in the sleeve. The Menard pressiometer, Fig. 6.17b, employs triple sleeves; all three with equal pressure. The stresses and strains around the two end or *guard cells* is three dimensional; they force two dimensional strain on the soil adjacent to the central cell. The radial strains corresponding to increasing internal pressures are measured by the volume increase in the central or measuring cell. From this a radial modulus of deformation is obtained corresponding to the initial and increased state of stress in the soil at that level. The modulus can be related empirically to the elastic modulus, E, and to soil compressibility, depending on the rate of test.

The initial state of stress in soil or rock can be investigated by one dimensional expansion. Reference points are placed in the surface of the mass. A slot is cut between them, and if there is sufficient internal stress,

TABLE 6:13 / FIELD TESTS
a. Direct Tests

Method	Procedure	Properties Measured	Limitations
Vane Shear	Rotate vane in undisturbed soil. Measure maximum and post maximum torque.	Undrained strength of soils difficult to sample (soft clay).	Progressive failure in sensitive soil—variable effect of gravel, questionable in sand.
Load Test ASTM D-1194 ASTM D-1143	Load large model to full-size footing or pile in increments; measure settlement.	Ultimate bearing; short-term deflection.	Interpretation in terms of prototype difficult.
Well Test (Field Permeability)	Pump water into or out of bore hole; measure drawdown in adjacent holes or rate of level change in well.	Effective horizontal permeability of mass.	Questionable above water table; not effective for vertical i.c.
Field Density D-2167 D-1556 (D-1587	Excavate hole; weigh soil; measure hole volume by volume of standard sand or volume of water-filled rubber membrane introduced into hole. Measure, weigh undisturbed Sample—D-1587)	Unit weight of soil, virgin or fill.	Limited by gravel, water.
Test Fill	Place fill equal in weight to structure on surface.	Settlement, pure pressure ultimate bearing.	Time, cost.
Test Pit	Excavate pit; measure hole volume, soil weight; obtain undisturbed samples and photographs.	Density of coarse gravel, broken rock; samples for detailed testing; in place dip, stratification, cracking.	Cost, ground water.

b. Indirect Tests

Dynamic Penetration	Drive standard sampler or cone point by hammer of known weight. Measure number of free-fall blows to drive a fixed distance. (ASTM D-1586 for "Standard Penetration Test").	Estimate strength and/or density. Estimate allowable bearing capacity.	Changes density of loose sands—sensitive to procedure changes.
Static Penetration	Force conical point into soil, measuring force to penetrate at fixed rate; rotate cone point with helical blade, measure force to advance cone ahead of auger.	Estimate strength and/or density. Estimate allowable bearing capacity.	Interpretation varies with soil; identification of soil questionable. Sensitive to procedure changes.
Borehole Expansion	Expand elastic tube in bore hole. Measure expansion volume.	Estimate soil strength and compressibility.	Interpretation controversial.
Electric Resistivity (see Geophysical)	Pass current between two electrodes; measure potential between intermediate electrodes; compute apparent resistivity.	Estimate soil identity (clay or sand). Estimate corrosion.	Interpretation of data controversial—rough estimates at best.
Elastic Wave (see Geophysical)	Measure time of travel of shock wave produced by impact or explosive. Compute velocity.	Estimate rigidity and identification.	
Vibration Response	Induce vibration by counter-rotating eccentric weights. Measure response by seismometers at varying distances.	Estimate dynamic response.	
Nuclear	Introduce nuclear radiation source into soil.	Density, water content.	Variable results.

the slot will close partially. A thin disk-shaped hydraulic jack is placed in the slot. (Very thin, wide jacks for this purpose are termed *flat jacks*.) The jack pressure is increased until the distance between the reference points returns to its initial value. If the response of the mass is elastic, the pressure required to force the mass back is equivalent to the initial stress in that direction.

FIELD PERMEABILITY / The permeability of the mass as a whole can be measured by test wells, in which water is introduced into or pumped out of the ground. The simple expressions for steady-state flow, derived in Section 4:2, can be used to evaluate either inflow or outflow wells. Better control of the operation is possible with pumping into the ground and a small cheap well is adequate. If the ground water level is initially high or if water to pump in is not available, pumping out is satisfactory. The test well, 3 to 6 in. in diameter penetrates (as far as practical) the aquifer thickness. Two and preferably three sets of observation wells are placed around the test well, Fig. 6.18. Typical distances of the sets from the test well are 25,

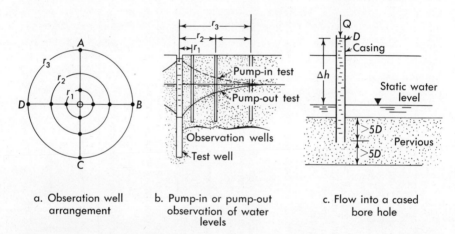

Figure 6.18 *Field permeability tests.*

50, and 100 ft. Pumping is continued in the test well at a constant rate until the levels in the observation wells stabilize. The average coefficient of permeability can be computed from the average of the draw-downs of any two sets. The differences in the permeabilities computed between different sets indicates variations in k with distance or insufficient time to establish equilibrium. If significant differences are noted, pumping should be continued until the results become more consistent.

The rate of flow into a cased hole can be used as an approximate measure of the permeability of a pervious stratum. The hole is cased into the pervious layer as shown in Fig. 6.18c, so that the end of the casing is no closer than 5 diameters from the stratum boundaries. The original ground

SEC. 6:12] FIELD TESTS

water level is measured. Water is then added to the boring to maintain it a constant level in the casing. When a constant rate of inflow, q, is established, the permeability is computed by the relation,

$$k = \frac{q}{2.75 \, D\Delta h}. \tag{6:9}$$

LARGE SCALE TESTS / Large scale tests of soil or rock can integrate the effects of stratification and structural defects and will provide a more realistic evaluation of the behavior of the total mass than can the smaller conventional tests. Because of time and expense, large tests are conducted only when the peculiarities of soil and rock structure prevent reliable conventional testing. The equipment arrangement and procedure is designed to fit the mass structure. Ingenuity and imagination must be exercised in devising each test program.

Shear tests of the mass can define the total significance of defects such as old shear zones, cracks, and seams of weakness. Direct shear is probably the simplest approach. Trenches are excavated in the mass to isolate a block which is representative of the whole. This may be several cubic yards or cubic meters in volume. The orientation of the trenches is such that thrust can be applied parallel to the weakness to be evaluated. A normal stress is applied by direct weighting if the potential shear surface is horizontal; otherwise the normal stress is developed by hydraulic jacks reacting against a beam anchored to the soil or rock beyond the test mass, Fig. 6.19a. Shear

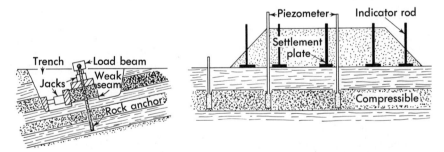

a. Direct shear of intact rock mass

b. Instrumented test fill loading

Figure 6.19 *Field tests.*

loads are applied by jacking against the trench walls, and strains measured by dial gages. The results are presented in the same form as for other direct shear tests. Such tests have been used in hard soils, jointed rock in dam foundations and tunnels, and to evaluate the sliding resistance of concrete (poured in a block) on a rock foundation.

Large scale bearing capacity and consolidation tests are conducted by test fills, or other weights, such as water-filled tanks. Settlement points consisting of steel plates 3 ft (1 mm) square with threaded indicator rods extending upward are placed on the ground. The plates are arranged in regular patterns both under and adjacent to the test area. Elevations are secured on the indicator rods and fill is added in increments. Settlement measurements are obtained on the rods at regular intervals as in a consolidation test. If the required load is great, extensions are added to the rods so that they are always above the fill surface. Piezometers are frequently installed in the critical strata to evaluate pore pressure increases, and samples of the critical strata are secured before and after loading to determine any differences in soil void ratio, water content or strength. Such tests are expensive and time consuming, but if properly executed they offer the most reliable test possible of the total soil or rock mass (short of the behavior of the structure itself).

6:13 Laboratory Testing and Evaluation

The objective of the detailed investigation, including the field testing, is to provide the quantitative data necessary for the engineering analyses and design of the project. The total structure of the soil and rock system is defined by the exploratory phase, Section 6:9; the final step is to evaluate the appropriate engineering properties of the critical strata that were previously identified, utilizing the results of laboratory tests of undisturbed samples and the field test data, correlated stratum by stratum with the identification tests, the geophysical parameters, and the penetration resistances.

SAMPLE PREPARATION / Undisturbed samples delivered to a laboratory are of two types: continuous and intermittent. The continuous samples are cut into sections representing 0.5 ft to 1.0 ft or 15 to 30 cm of depth. The samples are weighed and then small portions may be removed for visual classification and water content determinations. The average unit weight and void ratio for each section can be computed. The average unit weight and void ratio for each intermittent sample can also be determined. The test results are plotted on the same work sheet as the exploratory boring data.

A thin, vertical slice from each section or sample may be set aside to dry. This is examined at intervals to determine the presence of laminations or thin strata which ordinarily became more visible after partial drying. The disturbing effects of sampling can often be seen from the distortion of the strata. Finally the thickness of each distinctive soil seam is measured.

TESTS OF COHESIVE SOILS / A study of the classification test and void-ratio results will show that the soil characteristics vary within any one stratum. The points where the soil characteristics of a stratum appear to be both typical and extreme can be determined from the plots of water content, void ratio, plasticity, and penetration resistance. The undisturbed samples

SEC. 6:13] LABORATORY TESTING AND EVALUATION

to be tested for shear or consolidation are usually selected from the typical and extreme points in order to determine the variation of soil properties. Only a few samples are tested, for the cost of testing every one would be very great. The test results can be correlated with the other soil properties to determine the average strength or compressibility of the stratum.

TESTS OF COHESIONLESS SOILS / Strength tests of cohesionless soils are made at different void ratios with disturbed samples. The angle of internal friction, correlated with the relative density of the soil, can be used to estimate the strength of the actual soil deposit.

CORRELATING TEST RESULTS / Neither time nor money permit the extent of testing desired to establish the range of engineering properties (stress-strain, strength, volume change, permeability) in each critical stratum. By correlating the data with the total information available for each stratum, a more realistic appraisal can be made than from engineering tests alone. The factors suitable for correlation depend on the soil and rock and thus vary from project to project. Table 6:14 gives typical relationships that are frequently suitable.

TABLE 6:14 / CORRELATION OF TEST DATA

Simple Test	Possible Correlation
Water Content	Shear strength of clay.
	Compression index of clay.
Grain Size (D_{10}, D_{15}, C_u)	Permeability, strength and drainability of cohesionless soils.
Liquid Limit, LL	Compressibility.
Plastic Index	Swell-shrink, ϕ_D of clays.
Water plasticity ratio, R_w	Potential swell-shrink; preconsolidation load.
Void ratio, e, unit weight, γ	Compressibility, shear strength.
Relative Density, Dr	Strength, compressibility of cohesionless soil.
Seismic velocity, V	Modulus of elasticity; strength of rock.
Electrical resistivity, ρ	Water content, clay content, organic and salt content.
Penetration resistance, static and dynamic	Shear strength, relative density, modulus of compressibility.

PARAMETERS FOR ANALYSIS / In any stratum there is ordinarily a range of values for any given quality such as c, φ, c_c, C_v and k. Too often average values or median values are utilized. If the range in test results is narrow, this approach is adequate. In most strata the deviations from the average are so great that unsafe conclusions will be reached from averages. For design, therefore, the lower values are given emphasis. Although some designers argue that the design should be based on the poorest condition observed in each stratum, this is overly conservative because localized bad spots seldom control the behavior of the entire stratum. If the weak areas occur at random, a reasonable basis for design is the *lowest quartile*—the value for which 25 per cent of the data are poorer and 75 per cent better. If the range is very wide, the *lowest 10 or 20 per cent* is better. Moreover, the final design is analyzed using the poorest single value and the design possibly revised if the safety factor is less than one.

The results of all the testing, extended by correlations, are indexed on the underground cross-sections prepared from the exploratory phase of the investigation. Superimposed on the same sections are the outlines of the proposed structure, so that these sections become the starting point for the engineering analyses of stability, bearing capacity, and settlement.

6:14 Feedback During Construction

The underground investigation does not stop with the completion of design nor the commencement of construction. As was described in the introductory case histories, unforeseen conditions may be encountered or new evidence of underground conditions uncovered that could have a profound influence on both design and construction. Careful records of all soil, ground water and rock conditions should be kept, documented by photographs and by measurements. If these differ from what the earlier investigations indicated, the designer should be informed so that these deviations can be evaluated. If the conditions are the same, the records will do much to minimize unwarranted claims for changed conditions.

The very safety of the construction work, particularly of deep excavations and high embankments and dams, is often dependent on continuing observation of ground water elevations, piezometric levels in confined strata, and movements of the soil or rock mass, both lateral and vertical. Instrumentation for the regular observation of these features is routinely incorporated in earth dams and deep excavations. In even the smallest projects simple instrumentation, such as ground water observation wells and strategically placed bench marks are useful. The most important instrument, however, is an observant engineer who regularly inspects the site, notes all changes, and then interprets these in terms of the construction operations and their potential effect on the project. Such observations would have minimized the continuing trouble in the case of the mountain road in the tropics described in the introduction to this chapter and might have prevented trouble in the water treatment plant collapse.

REFERENCES

6:1 G. F. Sowers, "Modern Procedures for Underground Exploration," *Proceedings, ASCE,* **80**, Separate 435, May 1954.

6:2 R. F. Leggett, *Geology and Engineering*, 2nd ed., McGraw–Hill Book Co., Inc., New York, 1962.

REFERENCES

6:3 S.T. Algermissen "Seismic Risk Studies in the United States," *Proceedings, Fourth World Conference on Earthquake Engineering*, Santiago, Chile, 1969.

6:4 *List of Published Soil Surveys*, Soil Conservation Service, U.S. Dept. of Agriculture, April 1968.

6:5 D. R. Lueder, *Aerial Photographic Interpretation*, McGraw-Hill Book Co., Inc., New York, 1959.

6:6 *Manual of Air Photo Interpretation*, American Society for Photogrammetry, Washington, 1960.

6:7 T. E. Avery, *Interpretation of Aerial Photographs*, 2nd ed., Burgess Publishing Co., Minneapolis, 1968.

6:8 M. J. Hvorslev, *Subsurface Exploration and Sampling of Soils for Civil Engineering Purposes*, U.S. Waterways Experiment Station, Vicksburg, Miss., 1949. (Reprinted by Engineering Foundation, New York).

6:9 *ASTM Standards* for Exploration. (Issued annually—latest should be consulted.)
D-420, "Surveying and Sampling Soils for Highway Subgrades."
D-1452, "Soil Investigation and Sampling by Auger Borings,"
D-1586, "Penetration Test and Split Barrel Sampling of Soils."
D-1587, "Thin-Walled Sampling of Soils."
D-2113, "Diamond Core Drilling for Soil Investigation."
D-2573, "Field Vane Shear Test of Cohesive Soil."

6:10 G. Sanglerat, *Le Penetrometre et la Reconnaissance des Sols*, Dunod, Paris, 1965. (An excellent treatise on penetration tests, but entirely in French).

6:11 K. Terzaghi and R. B. Peck, *Soil Mechanics in Engineering Practice*, 2nd ed., John Wiley & Sons, Inc., New York, 1968, pp. 341 and 347.

6:12 G. Fletcher, "Standard Penetration Test, Its Use and Abuses," *Journal of Soil Mechanics and Foundation Division, Proceedings, ASCE*, **91**, SM4, July 1965.

6:13 D. U. Deere, "Technical Description of Rock Cores for Engineering Purposes," *Rock Mechanics and Engineering Geology*, **1**, 1, 1964, p. 17.

6:14 G. F. Sowers and C. S. Hedges, "Dynamic Cone for Shallow In Situ Penetration Testing," *Vane Shear and Cone Penetration Resistance Testing of Soil*, Special Technical Publication 399, ASTM, Philadelphia, 1966, p. 29.

6:15 G. G. Meyerhof, "Penetration Tests and Bearing Capacity of Cohesionless Soils," *Journal of Soil Mechanics and Foundation Division, Proceedings, ASCE*, **82**, SM1, January, 1956.

6:16 E. Schultze, "Determination of Density and Modulus of Compressibility of Non-Cohesive Soil by Soundings," *Proceedings of the Sixth International Conference on Soil Mechanics and Foundation Engineering*, **1**, Montreal, 1965, p. 354.

6:17 H. J. Gibbs and W. G. Holtz, "Research on Determining the Density of Sands by Spoon Penetration Testing," *Proceedings of the Fourth International Conference on Soil Mechanics and Foundation Engineering*, London, 1957.

SUGGESTIONS FOR ADDITIONAL STUDY

1. References 6:1, 6:2, 6:5, 6:8.
2. Reference 6:11, Art. 44, 45.
3. L. J. Goodman and R. H. Karol, *Theory and Practice of Foundation Engineering*, The Macmillan Company, New York, 1968, Chapter 2.
4. Reference 6:9 for latest modifications.

PROBLEMS

6:1 Prepare a typical soil profile for your locality. Secure data from contractors, city and county engineers, engineers specializing in underground exploration work, and from geologic and engineering reports.

6:2 A single test boring made for a water tank led to the following data:

Soil Data		Sampling Data			
Depth (ft)	Soil Stratum	Depth (ft)	Penetration Data	w	LL
0–4	Fill: cinders, brick	2	3	—	—
4–7	Hard, slightly plastic clay	5	32	21	44
7–25	Firm, uniform, coarse, subrounded quartz sand	8	15	—	—
		13	19	—	—
		18	27	—	—
		23	20	—	—
25–32	Firm, highly plastic clay	26	7	55	62
		31	6	57	64
32–51	Dense, uniform, fine, angular sand	33	35	—	—
		38	37	—	—
		43	40	—	—
		48	46	—	—
51–75	Dense to very dense graded, angular, coarse, sandy, subrounded, fine gravel. Coarser with depth	52	48	—	—
		57	55	—	—
		62	72	—	—
		67	68	—	—
		72	63	—	—
22	Ground water				

PROBLEMS 313

   a. Plot the boring log.
   b. Determine which strata, if any, require more detailed study. List tests necessary.
   c. Should more or deeper test borings be made? Tank weighs 450 tons and rests on four columns arranged in a square 30 ft apart.

6:3 Sketch and list the essential equipment for test boring, using both hand auger and wash drilling, and spoon sampling.

6:4 A 10-story office building with a basement floor 7 ft below the ground surface is 100 ft wide and 240 ft long. The dead load per floor is 120 psf and the live load 50 psf. The soil weighs 110 lb per cu ft.
   a. Show layout of borings for average conditions.
   b. Determine depth of boring, using the rule of 10 per cent of increase in effective stresses.

6:5 A one-story manufacturing plant is 120 by 600 ft. It rests on new fill 3.5 ft thick. The geology of the area indicates that it is underlain by horizontally stratified coastal lagoon deposits covered with an old man-made sand fill.
   a. Show the recommended boring layout.
   b. Estimate the required boring depth. Under what conditions encountered during boring should this depth be changed? When should undisturbed samples be made?
   c. Estimate the cost of making a soil investigation at this site, assuming that soil testing and engineering costs will be 40 per cent of the cost of boring and sampling.

6:6 A 20-story medical office building 100 × 140ft. in plan is centered in a 300 ft square 2-story shop and parking plaza. Both structures include a one-story basement. The site is located in an urban renewal area, once occupied by old houses and shops with basements, but now filled in for parking space. The city is located in a flood plain of a broad meandering river, with limestone cliffs overlooking the valley.
   a. Describe the steps advisable in a reconnaissance or preliminary evaluation of the site.
   b. Prepare a boring plan, and outline or guide to the procedures required for exploratory phase of the investigation.
   c. Estimate the cost of this investigation.

6:7 The exploratory borings for problem 6:6 found the following typical soil profile, relatively uniform across the site.
   0–20 ft.   Rubble.
   12–50 ft. Alternate seams of soft silty clay and clayey sand; N between 5 and 15 blows per ft.

50–70 ft. Limestone with clay seams: average core recovery 15 per cent.
70–90 ft. Hard limestone, average core recovery 95 per cent, dipping 60 deg with horizontal, joint spacing 3 ft.
20 ft. Ground water level—in September.

a. Draw a possible underground cross-section of this site.
b. Prepare an outline for the required final, detailed investigation including boring locations and undisturbed sampling methods.
c. List tests required for samples obtained in this phase.
d. Estimate the cost of this final phase of the investigation, excluding engineering work.

CHAPTER 7
Mass Response to Load

A DEEP OPEN EXCAVATION WAS MADE CLOSE TO AN OLD BUILDING supported by shallow spread footings. The engineer required that the excavation slopes be flatter than $2(H)$ to $1(V)$ to insure against shear failure of the banks and destruction of the building. The bank did not fail; however, the building cracked badly. The soil mass deformed elastically, with both outward and downward movement of the building wall closest to the excavation. This possibility was never considered by the engineer, yet it led to the condemnation of the building.

A new retaining wall 30 ft high formed one side of a large industrial plant under construction. The construction sequence required (1) building the wall, (2) placing the backfill behind it, and (3) constructing the remainder of the plant. The backfilling of the wall had just been completed when a crack was observed in the backfill surface about 20 feet from the wall. The engineer examined the wall and found that its top had tilted more than 2 inches away from the backfill. He blamed the trouble on faulty construction and ordered the wall and backfill reconstructed. The contractor refused and completed the building. In 1968, thirty-five years afterwards, the wall and the building stood intact; there had been no further movement of the wall nor additional cracking. The soil has sheared during backfilling and the wall had deflected, but both were normal, essential responses to the mass load that did not endanger the ultimate performance of the structure.

A deep excavation for a building was constructed without bracing for the banks, because a trial excavation demonstrated that the soil could stand on a vertical face without support. Five weeks later, a large part of one bank fell outward, killing two workmen and injuring five. The weather had been good, there had been no construction work in the vicinity of the bank that failed, and no apparent cause for the sudden soil shearing. Instead, it appeared to the layman that the soil had become tired of its continued load. In effect this was true; continued, but unnoticed, creep at high stress led to a delayed failure.

These three examples illustrate the extremes of soil mass response to stress changes. In the first case the soil deformed more or less elastically; although the soil was safe against failure, the structure was damaged beyond repair. In the second case the soil failed, but although the engineer was not aware of this, the soil failure was essential to the ultimate safe performance of the retaining wall. The earth pressure theory used in the wall design assumed shear failure in the soil; without that shear the pressure on the wall would have been twice as great. The third example illustrates the time-dependent nature of soil behavior, even when the environment remains unchanged. If the weather is unusually wet, or extremely hot and dry, or if freezing and thawing occur, the effect of time is more easily explained; the changes that occur with creep at high stresses can be just as serious, however.

Soil and rock mechanics deal with the response of the earth to the loads imposed. The principles of engineering mechanics apply to the solution of problems of soil and rock as they do to other engineering materials. The total problems, however, are usually more complex, because of three factors.

First, most problems include both the soil or rock mass and the engineering structure it supports. The two components of the system, the soil and the structure, obey the same physical laws. Although the structure is generally made of a stronger more rigid material than the earth, the two components respond or move together. Therefore, their interaction must be considered. The case of the soil slope that was safe although the structure failed illustrates the importance of interaction.

Second, the weight of the soil or rock mass is usually far more significant than the loads imposed by the engineering structure. For example, for a highway bridge across a river valley, the live load will be 20 to 40 per cent of the dead load of the structure. For a deep fill that supports the same highway across the same valley, the live load will be less than one per cent of the dead load of the fill. Because total weight of the soil mass is usually far greater than that of a structure it supports or of a live load, it must be given proportionally more consideration.

Finally, the response of the soil to load depends on a wide variety of factors, including the load itself, as described in Chapter 3. Although the properties of other engineering materials change, the changes in soil and rock are generally far greater, and have much greater effects on the response of the soil structure system.

The general approach to mass response to loads will be considered in this chapter. Specific cases will be covered in Chapters 8 through 11.

7:1 The Mass Load—Body Forces

The forces within a mass generated by the mass are termed *body forces*. Temperature stresses are an example of one form of important body force in steel or concrete; the most important body force in soil or rock mass is that of gravity. Certain simple cases will be considered below.

SEC. 7:1] THE MASS LOAD—BODY FORCES

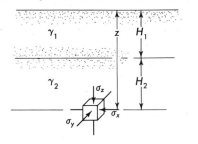
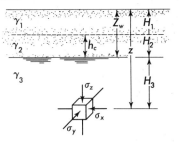

a. Vertical stress   b. Vertical stress with groundwater

Figure 7.1 *Vertical stresses due to soil or rock weight in a level stratified mass.*

VERTICAL STRESS IN A LEVEL MASS / The simplest case of a gravity body force is the vertical stress in an infinitely wide mass of level soil or rock. If the unit weight of a stratum of thickness H_1 is γ_1, Fig. 7.1a, then the total vertical stress σ_v (σ_z in a three dimensional coordinate system) at any depth z below the surface is

$$\sigma_v \text{ (or } \sigma_z) = (\gamma_1 H_1 + \gamma_2 H_2 \ldots). \tag{7:1a}$$

If the soil is saturated, with a ground water table, and a capillary rise zone of H_2, Fig. 7.2b, the vertical stress of any level can be separated into effective and neutral stress components. It should be kept in mind that the neutral stress or water pressure at any point is the same in all directions.

$$\sigma_v' = (\gamma_1 H_1 + \gamma_2 H_2 \ldots) - (z - z_w) \gamma_w. \tag{7:1b}$$

CHANGES IN STATIC VERTICAL STRESS / Changes in the static vertical stress occur in several ways: added surface load, such as a fill over a wide area, changes in the unit weight of the soil, and changes in neutral stress from water level fluctuations or capillary tension. These can cause significant decreases (or increases) in soil strength, breakdown of structure, and expansion or consolidation of compressible strata. The added stress produced by a uniform surface load is computed by Equation (7:1a) if the dimensions of the loaded area are much greater than the depth, z. If the area loaded is small the increase in stress is computed as shown in Chapter 9.

The effects of neutral stress changes are less obvious, but of great significance. If the ground water level falls by an amount Δz_w, the total stress at any point below the capillary fringe level will decrease slightly due to the reduced weight of water in the soil voids. However, the neutral stress will be reduced by $\Delta u = \gamma_w \Delta z_w$. The net effect will be an increase in effective stress (weight of water lost from the voids is seldom significant).

Example 7:1

Compute the change in vertical effective stress at the surface of the clay stratum in Fig. 7.2 if the water table falls 6 feet. The sand has a void ratio of 0.60 and a specific gravity of solids of 2.67. The height of capillary saturation is 2 ft. Above the capillary line the sand is 30 per cent saturated.

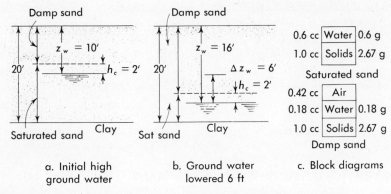

Figure 7.2 Example 7:1—Change in vertical stress on top of a clay stratum.

1. The sand saturated weighs $(2.67 + 0.60)/(1 + 0.60) = 2.045$ g per cu cm $= 127.5$ lb per cu ft.
2. The sand above the capillary line weighs $(2.67 + 0.3 \times 0.6)/1.6 = 1.78$ g per cu cm $= 111$ lb per cu ft.
3. Initially the vertical stresses at the clay surface are,

$$\text{Total} = 8 \times 111 + 12 \times 127.5 = 2418 \text{ psf}$$
$$\text{Neutral} = 10 \times 62.4 = 624$$
$$\text{Effective} = 1794.$$

4. After the water table drops, the vertical stresses at the clay surface are,

$$\text{Total} = 14 \times 111 + 6 \times 127.5 = 2320 \text{ psf (a loss of 98 psf)}$$
$$\text{Neutral} = 4 \times 62.4 = 250$$
$$\text{Effective} = 2070 \text{ psf (a gain of 276 psf)}.$$

Even if the sand above the fringe level is dry, the effective stress will increase below the water surface due to a ground water lowering.

The change in effective vertical stress is caused by water level changes of Δz_w can be expressed

$$\Delta \sigma'_v = -\Delta z_w [\gamma_w - (\gamma_{\text{sat}} - \gamma_{\text{drained}})]. \qquad (7:2)$$

The increase in effective stress of 276 psf for a 6 foot drop in water level is typical—approximately 46 psf per foot of water level change. A three foot drop in water level would then be the equivalent in effective stress of the weight of one foot of typical fill placed over the site or nearly the equivalent of the weight of a wide one-story building.

SEC. 7:1] THE MASS LOAD—BODY FORCES

EARTHQUAKE STRESS IN A LEVEL MASS / The dynamic effect of an earthquake can be approximated by the inertia stresses developed by the maximum acceleration. Consider a column of soil of depth z and length and width of L and B (Fig. 7.3). The weight of the column will be $LBz\gamma$, and the vertical stress at the bottom will be γz. In order to produce vertical acceleration of $\pm a_v$ it will be necessary to increase the vertical force by ΔF:

$$\Delta F = \pm M\, a_v$$

$$\Delta F = \frac{\pm LB\gamma z\, a_v}{g}.$$

The vertical stress change due to earthquake load, $\Delta\sigma_{ve}$, is

$$\Delta\sigma_{ve} = \pm \frac{LB\gamma z}{LB}\frac{a}{g} = \pm\gamma z\,\frac{a_v}{g}. \tag{7:3a}$$

The shear stress change, $\Delta\tau_e$, on the base of the column can be computed from a horizontal acceleration, a_h

$$\Delta F_h = \frac{LB\gamma z a_h}{g},$$

$$\Delta\tau_e = \frac{\gamma z a_h}{g}. \tag{7:3b}$$

The ratio a/g is often expressed as a percentage and in that form is termed the earthquake acceleration factor.

RESIDUAL STRESSES / Other sources of body stress include the effects of previous soil or rock loading that are retained because the mass is confined. Expansion and contraction due to long term temperature change or chemical alteration cause stresses in many rock masses and some soils. Tectonic movements and on a smaller scale, landslides, produce lateral stresses that can greatly exceed the vertical stress due to soil weight. Previous loads, now removed, can be maintained by mineral bonds or by capillary

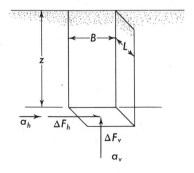

Figure 7.3 *Forces produced by acceleration of soil or rock at a depth of z.*

tension. The magnitude of these residual stresses seldom can be computed theoretically. Field measurements as described in Chapter 6 are necessary to determine their magnitude and large scale tests are required to evaluate their possible changes.

7:2 Elastic Equilibrium

As was discussed in Chapter 3, changes in stress acting on soil produce strains. When the stress is well below that required for failure there is a definite relation between a stress increment and the strain it produces. If the stresses acting on a mass of soil are in equilibrium, and if a definite relation between stress and strain is present throughout the mass, and if the deformations are compatible with the strains, the mass is in a state of *elastic equilibrium*. When the relationship between stress and strain can be defined mathematically, the effects of a system of stresses can be described by a *theory of elasticity*.[7:1,7:2] A number of such theories have been developed based on simple relationships between stress and strain. If the stress level in the soil is sufficiently low that one of the simple expressions for stress-strain is approximately correct, then that theory can be applied to the solution of soil engineering problems.

STRAINS / A portion of an element of soil is shown in Fig. 7.4. The angle of ABC, initially a right angle is displaced a small amount to a new position $A'B'C'$, as the result of a stress change. It can be seen that in the new position the sides of the angle have first moved by translation to $A'B'C'$ and then the ends have further rotated to $A''B'C''$. The x-displacement of B is u, and the y displacement is v (the z displacement is w). The x displacement of A to A'' because of its distance, dx, from point B is $u + (\partial u/\partial x)dx$; the y displacement of A to A'' is similarly $v + (\partial v/\partial x)dx$.

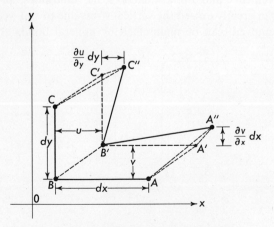

Figure 7.4 Linear and angular displacements in the x-y plane.

SEC. 7:2] ELASTIC EQUILIBRIUM

The strain in the x direction is equal to the elongation of line AB divided by the initial length, dx. Similar expressions can be derived for all three directions of linear strain.

$$\frac{A''B - AB}{dx} = \frac{dx + \frac{\partial u}{\partial x}dx - dx}{dx} = \varepsilon_x,$$

$$\varepsilon_x = \frac{\partial u}{\partial x} \quad ; \varepsilon_y = \frac{\partial u}{\partial y} \quad ; \varepsilon_z = \frac{\partial u}{\partial z}. \tag{7:4a}$$

The angular change of line AB, angle $A''B'A'$, can be found approximately from the y displacement of its end from A' to A'', $(\partial v/\partial x)dx$. The angular change in radians is

$$A'B'A'' = \frac{\frac{\partial v}{\partial x}dx}{dx} \qquad C'B'A'' = \frac{\frac{\partial u}{\partial y}dy}{dy}.$$

The angular strain, γ_{xy} is the sum of the two angles:

$$\gamma_{xy} = \frac{\partial v}{\partial x} + \frac{\partial u}{\partial y}; \quad \gamma_{xz} = \frac{\partial w}{\partial x} + \frac{\partial u}{\partial z}; \quad \gamma_{yz} = \frac{\partial v}{\partial z} + \frac{\partial w}{\partial y}. \tag{7:4b}$$

The complete picture of strains is described by the six equations (7:4a) and (b).

COMPATIBILITY / If the mass remains intact or a solid, then the angular strains must be compatible with the linear strains. This condition is illustrated by the distortion of the set of squares (7.5a) into (7.5b). If the strains are not compatible, the mass will be discontinuous, Fig. 7.5c. This is assured mathematically by taking the second derivatives of the linear strain

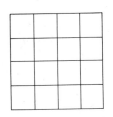
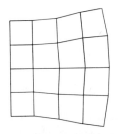
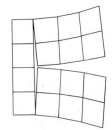

a. Solid before deforming   b. Compatible strains in deformed solid   c. Incompatible strains in deformed solid

Figure 7.5 *Significance of compatibility of strains.*

functions and equating them to the second derivative of the angular strain function:

$$\frac{\partial^2 \varepsilon_x}{\partial y^2} \quad \frac{\partial^2 \varepsilon_y}{\partial x^2} = \frac{\partial^2 \gamma_{xy}}{\partial x \partial y}, \quad (7:5a)$$

$$\frac{\partial^2 \varepsilon_y}{\partial z^2} + \frac{\partial^2 \varepsilon_z}{\partial x^2} = \frac{\partial^2 \gamma_{yz}}{\partial y \partial z}, \quad (7:5b)$$

$$\frac{\partial^2 \varepsilon_z}{\partial x^2} + \frac{\partial^2 \varepsilon_x}{\partial z^2} = \frac{\partial^2 \gamma_{xz}}{\partial z \partial x}. \quad (7:5c)$$

These are the *equations of compatibility* and must be satisfied in an intact mass.

ELASTICITY / An elastic theory assumes that the relationship between stress and strain can be defined mathematically. The only limitation imposed is that the relation be simple enough that it can be utilized in analysis and that it is valid for both increases and decreases in stress (hysteresis is negligible). The simplest form is when strain is proportional to stress, a linear function described by the modulus of elasticity, E, Fig. 7.6a

$$E = \frac{\Delta \sigma}{\Delta \varepsilon}. \quad (3:23b)$$

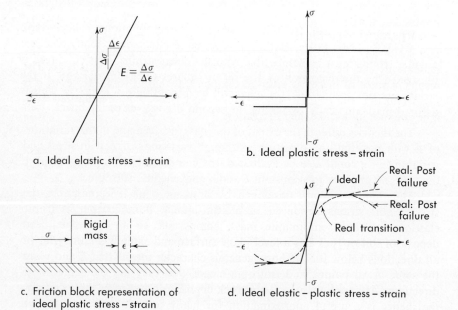

a. Ideal elastic stress – strain

b. Ideal plastic stress – strain

c. Friction block representation of ideal plastic stress – strain

d. Ideal elastic – plastic stress – strain

Figure 7.6 *Elastic and plastic stress-strain relationships.*

SEC. 7:2] ELASTIC EQUILIBRIUM

The modulus at one point may be the same in all directions, an *isotropic* material, or it may be different in which case E must be identified with a specific direction:

$$E_x = \frac{\Delta\sigma_x}{\Delta\varepsilon_x}; \quad E_y = \frac{\Delta\sigma_y}{\Delta\varepsilon_y}; \quad E_z = \frac{\Delta\sigma_z}{\varepsilon_z}. \tag{7:6}$$

This state is analogous to a spring that is infinitely long, and where deformation is proportional to stress, whether increasing or decreasing.

The lateral strains produced by an axial stress, $\Delta\sigma$, are defined by Poisson's ratio, ν

$$\nu = \frac{\varepsilon_L}{\varepsilon_D}. \tag{3:24}$$

In a three-dimensional stress field with x, y, z coordinates:

$$\Delta\varepsilon_z = \frac{\Delta\sigma_z}{E_2} - \frac{\nu\Delta\sigma_x}{E_x} - \frac{\nu\Delta\sigma_y}{E_y}. \tag{7:7}$$

The solution of elastic equilibrium problems requires combining the equations of stress-strain, (7:6) and (7:7) with the equations of compatibility. Equilibrium requires that the sums of the forces and moments acting on each portion of the mass be zero. The equations are integrated with the constants of integration determined by the boundary conditions. The results express the stresses, strains, and displacements in terms of the loads or stresses imposed on the mass.

ELASTIC THEORIES / Direct mathematical expressions of stress and strain have been developed only for certain special simple cases. The derivations can be found in textbooks on the theory of elasticity.[7:1] The results of some of these are given in Chapter 9 without proof. However, it is important that the engineer understands the basis for these equations even if he does not undertake their derivation.

The theories differ in the extent of the mass involved and in the variations of E with position and direction. Although variations of E with stress and variations in ν could also be evaluated, the complexities of integration have limited most analyses to a constant E and homogeneous, isotropic ν.

The simplest theory, developed by Boussinesq in 1888, describes the stresses and strains produced in a semi-infinite homogeneous isotropic elastic solid. The semi-infinite mass, comparable to an extensive, deep deposit of soil or rock has a level plane surface and extends infinitely far in all directions below it. The homogeneous character means that E and ν are the same at all points, and isotropic means that they are the same in all directions. Although many soil and rock deposits do not meet these criteria, the theory is a useful approximation for many real situations as will be discussed in Chapters 8, 9, and 10.

Other theories have been developed for layers having different E values, materials in which E varies with depth or with direction, and for a material in which lateral strain is prevented by thin layers of more rigid materials. The latter, developed by Westergaard, is an approximation of a soil mass consisting of elastic clay strata alternating with more rigid sand seams. It is useful in foundation analysis, Chapter 9. A more complete discussion of these theories and their equations for stress and strain can be found in the references.

LATERAL STRESS WITH NO STRAIN / A simple example of the application of elasticity to the computation of stresses in soil is the case of the uniform level mass described in Section 7:1. The vertical pressure at any depth, z, is equal to the overburden stress, $\Sigma \gamma H$. The lateral pressure produced by that vertical stress depends on the strain.

In the level mass of infinite extent, each little element of soil tries to bulge laterally under the vertical stress, σ_z, in accordance with the Poisson's ratio. One element cannot bulge without the neighboring element contracting. If the soil is homogeneous at that level and all elements are loaded equally, all must respond identically; therefore there can be no lateral expansion or contraction and all lateral stresses, σ_x and σ_y, are equal. Rewriting Equation (7:7) and then simplifying,

$$\Delta \varepsilon_x = \frac{+\Delta \sigma_x}{E_x} - \frac{\nu \Delta \sigma_y}{E_y} - \frac{\nu \Delta \sigma_z}{E_z} \text{ and if } E_x = E_y \text{ and}$$

$$0 = \frac{+\Delta \sigma_x}{E_x} - \frac{\nu \Delta \sigma_y}{E_x} - \frac{\nu \Delta \sigma_z}{E_z} \text{ and if } \sigma_x = \sigma_y$$

$$\frac{\Delta \sigma_x}{E_x} (1 - \nu) = \frac{\nu \Delta \sigma_z}{E_z}$$

$$\Delta \sigma_x = \left(\frac{\nu}{1 - \nu}\right)\left(\frac{E_x}{E_z}\right) \Delta \sigma_z. \tag{7:8a}$$

The relationship between the lateral stress $\sigma_x = \sigma_y$ and the vertical stress σ when there is no lateral strain depends on Poisson's ratio and the ratio of the lateral to vertical E. If the soil is isotropic, so $E_x = E_y = E_z$

$$\Delta \sigma_x = \left(\frac{\nu}{1 - \nu}\right) \Delta \sigma_z = K_0 \, \Delta \sigma_z \tag{7:8b}$$

$$K_0 = \frac{\nu}{1 - \nu}. \tag{7:8c}$$

The lateral stress, $\Delta \sigma_x$, with no strain is essentially the earth pressure developed against a very rigid wall, as will be described in Chapter 8. The ratio K_0 is the *coefficient of earth pressure at rest*.

SEC. 7:3] PLASTIC EQUILIBRIUM

APPLICATIONS AND LIMITATIONS OF ELASTIC THEORIES / The elastic theories have proved extremely useful in computing stresses and strains in a wide range of materials, including soil and rock masses. The validity of the theories is limited by how accurately they represent the actual material behavior. As was discussed in Chapter 3, the stress-strain relations for soils and rock are not linear nor the same in unloading as loading. Furthermore, the degree of linearity depends on the stress level and on the stress history. Small changes in stress compared with the total stress level, and repeated stresses are more nearly elastic than large single loadings that approach failure. In spite of their theoretical limitations, elastic analyses of stress and strain in soil and rock provide reasonable approximations when the stress levels are low and when the theory selected fits the geometry and the variations in E of the soil or rock mass.

7:3 Plastic Equilibrium

Plastic equilibrium is essentially a state of impending failure. The forces acting on each element of the mass are in equilibrium. The stresses within the zone of plasticity are those which produce failure. The strains, instead of being related to the stress, are indefinite.

The stress-strain curve for an ideally plastic material is given in Fig. 7.6b. The failure stress is reached without appreciable strain, after which the strain increases without a stress increase. This is analogous to a weight resting on a rough surface, and acted upon by an increasing load, σ as in Fig. 7.6b. There is no movement until σ overcomes friction, thereafter, the block moves without an increase in σ.

STRESSES IN A PLASTIC STATE / If the stresses required to produce failure can be defined mathematically, then the state of stress in plastic equilibrium can be described by the equations developed in Chapter 3. For many materials the Mohr envelope is a reasonably accurate criterion for the stresses on the surface of failure, and it can be approximated by the simple expression; originally proposed by the French engineer-physicist, Coulomb:

$$s = c + p \tan \varphi \text{ or} \quad (3:32b)$$
$$\tau_f = c + \sigma_f \tan \varphi$$
$$\alpha = 45 + \varphi/2. \quad (3:29)$$

The relationship between the principal stresses in the plastic state defined by the Coulomb equation can be found from the geometry of the Mohr circle, Fig. 7.7a, either graphically or by analysis:

$$\sigma_1 = \sigma_3 \tan^2 (45 + \varphi/2) + 2c \tan (45 + \varphi/2). \quad (7:9a)$$

Two sets of shear surfaces develop, at angles of $\pm\alpha$ between their normals and the direction of the major principal stress, as shown in Fig. 7.7b.

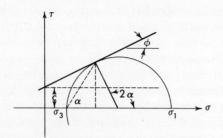 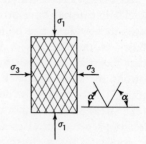

a. Mohr's circle solution of Coulombs equation

b. Shear surfaces in plastic equilibrium

Figure 7.7 *Plastic equilibrium and Mohr's circle.*

The earth pressure theory of Rankine, Section 8:1 is a direct application of this definition of plasticity, utilizing the simple case of plane rupture surfaces and the principal stresses at the same orientation throughout the mass.

Example 7:2

Compute major principal stress, which acts laterally in a mass of soil that is in a state of plastic equilibrium if the vertical (minor) principal stress is 250 psf. The soil is clay with an undrained strength of $c = 1000$ psf. From Mohr circle, Fig. 7.8,

$$\sigma_1 = \sigma_3 + 2c \qquad (7:9b)$$
$$\sigma_1 = 250 + 2000 = 2500 \text{ psf}.$$

The state of plasticity may involve a large part of the mass or only a limited zone. The boundary conditions, the load, and the trajectories of soil movement all must be compatible with the failure.

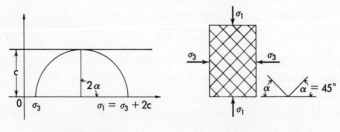

a. Mohr's Circle

b. Shear surfaces by plastic theory

Figure 7.8 *Plastic equilibrium in a saturated clay in undrained shear.*

PLASTICITY APPROXIMATIONS / In many practical problems it is possible to approximate the state of plastic failure by considering the failure stresses along the boundary of the zone of plasticity. The geometry of the boundary that satisfies the plastic state of stress as well as the equilibrium of the portion of the mass defined by the boundary is not necessarily known but can be found by trial. The circular arc analysis of slope stability, Chapter 11, is an example of such an approximation of plastic equilibrium.

APPLICATIONS AND LIMITATIONS OF PLASTIC THEORIES[7:3, 7:4] / The plastic theories define a state of failure or impending failure in a part of a mass. If the stress conditions for failure can be defined simply, as by the Coulomb equation 3:32b, the analysis may be a valid approximation of the conditions of failure in a mass. What occurs before the point of failure, and what the consequences of failure might be, such as amount or rate of movement, cannot be determined.

As with most approximations, the plastic theories neglect certain real conditions that influence the results of the analysis. First, the state of plasticity commences after the mass has been distorted by strains, Fig. 7.6c, rather than without strain as the definition of simple plasticity suggests, Fig. 7.6a. The distortion to varying degrees influences the geometry of the real failure surfaces. A few theories of elastic-plastic action have been proposed but are of limited use because of their complexities. Moreover, most materials exhibit a transition in which neither ideal elasticity nor plasticity occur, Fig: 7.6c.

A second shortcoming is in representing the state of stress at failure by a simple expression such as the Coulomb equation. Part of the loads applied to the mass are supported by neutral stress, the remainder by effective stress, and no simple expression of plasticity suffices to represent the total stresses at failure. For example, it was shown in Chapter 3 that for a saturated clay in undrained shear, where $\varphi = 0$, α should be 45 degrees. However, the actual failure plane angle is determined by effective stress, and $\alpha > 45$ deg.

A third shortcoming is the effect of the plastic strains on the state of stress. A loose sand becomes denser and stronger, a sensitive clay becomes weaker. The real state of stress at failure is strain dependent, whereas simple plasticity presumes the strain and stress are independent.

In spite of these shortcomings, plastic theory is a valuable tool for solving real problems.

7:4 Rheology

In many materials including most soils and rocks the response of the mass to load is time-dependent. The time rate of consolidation of soil, Chapter 3, is an example. The analytical treatment of time dependent behavior is termed *rheology*.[7:2]

BASIC ASSUMPTIONS OF RHEOLOGY / Theoretical rheology is an analytical representation of time-dependent response to load in terms of simple mechanical models or analogs, Fig. 7.9. The immediate elastic response that is proportional to load is represented by a spring termed the *Hookean substance*. A rate of movement that is proportional to stress is analogous to viscous flow of a liquid, and is represented by a liquid-filled cylinder with a small orifice through which laminar flow can occur. This is often termed a *dashpot* or *dashpot analog*, or Newtonian *substance*. The failure that occurs at a certain stress level is represented by a *stress link* or *fuse*, termed the *St. Vennant substance*. The characteristics of all of these elements in terms of stress response can be described mathematically.

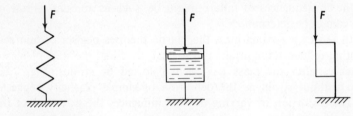

a. Hook or elastic spring – strain is a function of stress

b. Newtonian or viscous dashpot – Rate of strain is a function of stress

c. St. Vennant, yield point or fuse link – Beginning of strain is a function of stress

Figure 7.9 *Rheologic analogs.*

A *rheologic theory* is constructed by combining these simple elements in such a way that they simulate the real behavior of the mass. This requires a clear mathematical representation of observed behavior under simple load condition. The combination of elements to synthesize the real response is found by trial.

SIMPLE RHEOLOGIC MODELS / A number of rheologic models are useful in soil and rock mechanics. The compression of air in soil voids, followed by its flow from the soil can be represented by a spring and dashpot in series, Fig. 7.10a, the *Maxwell model*. The Terzaghi time rate of consolidation theory is a rheologic model consisting of a series of springs and dashpots in parallel, Fig. 7.10b. This is sometimes called a *Kelvin model*.

A combination of the Kelvin and Maxwell models, Fig. 7.10c, is termed the *Burger model*. This might represent the time rate of consolidation of a partially saturated soil with initial plus primary consolidation.

The rheologic models are helpful in approximating the response of soil to complex load variations, where observations of real soil behavior are difficult to measure. The model combination is usually based on simple loading. Therefore, if a more complex load pattern alters the soil, the model may not be a valid representation of the behavior under complex load changes.

SEC. 7:5] SOLUTION OF PROBLEMS IN MASS BEHAVIOR

There is a real danger in utilizing rheologic models in assuming that the analog which represents the time dependent behavior is a representation of the mechanical process that determines that behavior. This is to a great extent true in time-rate of consolidation, but not necessarily valid in other cases.

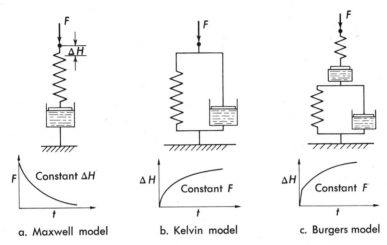

Figure 7.10 *Rheologic models useful in soil and rock mechanics.*

7:5 Solution of Problems in Mass Behavior

CHARACTER OF REAL PROBLEMS / The solution of real engineering problems concerning the response of soil and rock masses to stresses requires an understanding of the character of the materials and the nature of the loads.

Rocks and soils have engineering properties that may be approximated by simple equations; however the inherent errors in these approximations (such as the difference between the true α and $45 + \varphi/2$ in undrained shear of clay) must be evaluated before using the approximation in analysis.

The masses of soil and rock are three-dimensional, are non-homogeneous and seldom isotropic. While they may be approximated by simple forms, such as the semi-infinite homogeneous isotropic elastic body, the effects of the simplification must be evaluated. The localized discontinuities, such as cracks, cavities, and unusually rigid or weak zones also have a profound effect on the response of the mass.[7:5]

The loads are known with varying degrees of accuracy. The dead load due to soil weight and the weight of an engineering structure can be determined within a few per cent. The live loads due to wind, occupancy, vehicles and earthquakes can often be only estimated.

The environment, including the effect of excavation and nearby construction profoundly influences soils and rocks, changing their behavior. Hopefully, these changes can be predicted by laboratory tests; in some cases, however, there are surprises. For example, pore pressures in joints in the rock abutments of the Malpasset Dam in France reduced the strength sufficiently that the mass sheared, taking a portion of the dam with it. Over 300 people were killed when the valley below the dam was scoured clean by the deluge.

There are three types of problems the engineer must solve. The foremost is design. Here the objective is to predict the behavior of the soil or rock. On the basis of that prediction the design is developed either to control the behavior or to adapt to it. Because the design generally influences the soil and rock, the process is iterative; design and analysis alternate until a satisfactory solution is reached. What constitutes a satisfactory solution depends only partly on the accuracy of the prediction of soil or rock behavior. For example, an analysis of foundation bearing capacity might show that a wall footing one foot wide would be adequate. However the cheapest way to build the footing might be to use a back hoe that digs a trench 2 feet wide. A satisfactory prediction of soil bearing capacity in this special case could have been only 50 per cent of the most accurate value.

By way of contrast, a refined analysis of the stability of the shell of a large earth dam made it possible to reduce the slope from $2.5(H)$ to $1(V)$ to $2.2(H)$ to $1(V)$ and save 1,000,000 cu yd of fill.

A second objective of analysis is to determine corrective measures for some shortcoming in the original design. The analysis starts from a known condition and works back to the cause of trouble which is much easier than prediction. A new analysis must then predict the effectiveness of the correction or of a redesign.

The third objective is to develop more realistic analyses for future design, or to check the validity of an old method by comparing the prediction with measured performance. The accuracy is limited only by the data available and the time and resources of the investigator. The value can only be measured in the use of the information in new designs or in correcting old mistakes.

THE GENERAL APPROACH / The approach to any problem is controlled by three factors:

1. The existing theories are not always adequate.
2. Past experience does not include some aspects of the new problem.
3. The time, facilities and money for a complete evaluation of the unknown are not available.

The engineer, therefore, must find a satisfactory answer in spite of these controls. What is satisfactory depends on the price of the solution compared to its value, considering the structure's useful life and the cost of its possible premature end. The design analysis for a temporary construction cofferdam that could be flooded without loss of life in case of an excessive river flow would thus be far less demanding than that for a permanent dam that must supply enough power to recover its cost in 50 years and whose failure could cause loss of life and extensive property damage.

A number of tools are available to the engineer. Specific ones are found in Chapters 8 through 11; the general procedure is outlined below:

1. The problem must be identified and then simplified by dividing it into segments and delineating the significant factors.
2. Appropriate theories are utilized with allowances for their limitations.
3. Physical models and mathematical models are utilized where theories prove inadequate.
4. The results of the theoretical and model studies must be interpreted in the light of experience.
5. The gaps in knowledge must be filled intuitively.
6. The answers must be re-evaluated and revised when observations of actual performance show their inadequacies.

Of these steps, the second and third are capable of mathematical evaluation; the remainder require understanding, initiative, and—courage.

TWO-DIMENSIONAL REPRESENTATION / A common simplification is to reduce the problem to two dimensions. If the mass or the loading is uniform and of great extent in one dimension, the two-dimensional representation or cross-section may depict all the relevant factors. For example, the shear that develops behind a long retaining wall involves no displacements parallel to the face of the wall, a condition termed plane strain. A two-dimensional analysis, including shear strengths determined by plane strain testing, is a realistic evaluation of this problem at all points except the ends. An analysis of foundation bearing capacity by plastic theory is feasible in two dimensions, but has eluded solution in three dimensions. However, empirical correction factors have been developed from model tests to extend the results of the two dimensional analyses to three-dimensional problems.

Other simplifications include representing complex structural arrangements of the mass by a simple equivalent. For example, in seepage analysis, alternate strata of pervious sand and low permeability silt can be represented by a single stratum with a high permeability parallel to the bedding and a lower value perpendicular to it.

In any case it is essential that the effect of the simplification on the results be investigated by comparing the results of the analysis with observations of performance.

NUMERICAL APPROXIMATIONS, MATHEMATICAL MODELS / The high speed digital computer with large storage or memory capacity has made it possible to solve many problems by successive approximations or by extremely large numbers of simultaneous equations. Without the computer the labor and time required would be prohibitive.

Many approximations involve dividing the mass into segments whose response to stress can be simply represented. The response of the mass is the sum of the responses of the segments.

In the lumped mass approximation, each segment is visualized as consisting of a discrete mass, connected with the adjoining segment by springs that represent the elasticity, Fig. 7.11a. Anisotropy and variations in elasticity from point to point can be represented by differences in the spring constants.

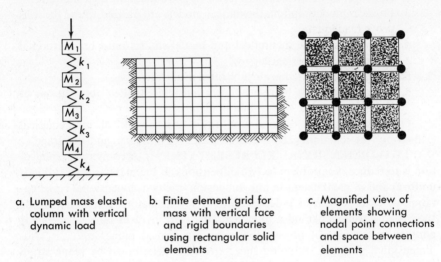

a. Lumped mass elastic column with vertical dynamic load

b. Finite element grid for mass with vertical face and rigid boundaries using rectangular solid elements

c. Magnified view of elements showing nodal point connections and space between elements

Figure 7.11 *Approximations of elastic masses.*

In the *finite element approximation*, (7:6), Fig. 7.11c, the mass is divided into discrete elements each of whose response to load is equivalent to that of the portion of soil it represents. The elements, rectangular or triangular solids, are connected together by frictionless pins at their corners or nodal points. The forces act on each element through the nodal points, and are in static equilibrium. The corresponding strains of each element are compatible with those of the adjoining element because they are connected at the nodal points. The total response of the mass is reduced to a system of simultaneous equations that satisfies equilibrium and compatibility for each element. The response of each element need not be isotropic nor linear so long as the computer can handle the variables.

The finite element approach has made it possible to obtain rigorous solutions to problems that previously could only be approximated crudely. One example is the stress patterns in embankments resting on an elastic foundation, including the effect of building the embankment in layers.[7:8] A second example is the response of an earth or rock slope to horizontal earthquake accelerations.[7:9] While the solution to a specific problem involving non-homogeneous conditions would require a computer, it is also possible to express the results for certain general cases in the form of graphs or charts suitable for routine calculations.

The numerical approximations are capable of solving complex two-dimensional problems in stress and strain as well as three-dimensional problems that are symmetrical about a vertical axis. General solutions to three-dimensional problems are limited by the capacity of the computers available.

Solutions of problems of plasticity can be made by successive approximations. The finite element approach can be used in the transition range between elastic and plastic behavior but special techniques must be employed once there is no unique relation between stress and strain.

Various other numerical approximations such as the method of finite differences are used in solving the differential equations of elasticity for which no direct solution is possible. The details of all these methods of numerical analysis are beyond the scope of this textbook. They are powerful tools for the solution of complex problems. Both their applications and limitations must be clearly understood before the results of such methods can be used.

MODELS / Physical models have been widely used in engineering to solve problems where analysis is impractical or to verify and modify the results of analyses based on approximate theories. Physical models have been of use in both applications in soil and rock engineering. In addition, they can indicate the areas of high and low stress or the shapes of zones of plastic flow that can lead to more realistic theories in the future.

The problem in any model is scale, unless the model has the same dimensions as the prototype. Dimensional analysis is necessary to establish the characteristics of the model that will properly reflect the behavior of the prototype. For models that are geometrically similar to the prototype it can be shown that the dimensionless ratios (known as π terms) must be the same in model and prototype. For simple problems in static plastic and elastic equilibrium,

$$\pi_{(\text{Strength})} = \frac{s}{\gamma L}, \tag{7:10a}$$

$$\pi_{(\text{Rigidity})} = \frac{E}{\gamma L}. \tag{7:10b}$$

In these expressions L is a linear dimension, γ is the unit weight of soil or rock, and s and E are the shear strength and modulus of elasticity. For complex problems, several π terms are required, which are not necessarily compatible. The model, therefore, may not fully represent the behavior of the prototype, and only a full scale model will be realistic.

REFERENCES

7:1 S. P. Timoshenko and J. N. Goodier, *Theory of Elasticity*, 2nd ed., McGraw-Hill Book Co., Inc., New York, 1951.

7:2 N. H. Polakowski and E. J. Ripling, *Strength and Structure of Engineering Materials*, Prentice Hall, Englewood Cliffs, N.J., 1966.

7:3 A. Nadi, *Theory of Flow and Fracture in Solids*, McGraw-Hill Book Co., Inc., New York, 1950.

7:4 W. D. L. Finn, "Application of Limit Plasticity in Soil Mechanics," *Journal of the Soil Mechanics and Foundation Division, Proceedings, ASCE*, **93**, SM5, September 1967, p. 101.

7:5 R. E. Goodman, R. L. Taylor and T. L. Brekke, "A Model for the Mechanics of Rock," *Journal of the Soil Mechanics and Foundation Division, Proceedings, ASCE*, **94**, SM3, May 1968, p. 637.

7:6 O. C. Zienkiewicz, *The Finite Element Method in Structural and Continuum Mechanics*, McGraw-Hill Publishing Co., Ltd., London, 1967.

7:7 C. V. Girijavallabhan and L. C. Reese, "Finite Element Method for Problems in Soil Mechanics," *Journal of Soil Mechanics and Foundation Division, Proceedings, ASCE*, **94**, SM2, March, 1968, p. 473.

7:8 R. W. Clough and R. J. Woodward, "Analysis of Embankment Stresses and Deformations," *Journal of Soil Mechanics and Foundation Division, Proceedings, ASCE*, **93**, SM4, July, 1967, p. 529.

7:9 I. M. Idriss and H. B. Seed, "Response of Earth Banks During Earthquakes," *Journal of Soil Mechanics and Foundation Division, Proceedings, ASCE*, **93**, SM3, May, 1967, p. 61.

PROBLEMS

7:1 Compute the settlement of the clay stratum in example 7:2 if the initial void ratio is 1.25, the compression index is 0.43, and the clay stratum is 20 ft thick.

7:2 After the ground water lowering of example 7:1, the ground water rose until it was 6 ft from the ground surface.
   a. Compute the new effective stress on the surface of the clay stratum.
   b. What has been the total change in effective stress produced by the changing water table?
   c. If the sand has an angle of internal friction of 33 degrees, what has been the percentage change in sand shear strength just

PROBLEMS

above the clay due to both lowering of the water table and raising the water table?

7:3 A mass of clean fine sand 20 ft thick overlying a shale rock has a void ratio of 0.45 and a specific gravity of solids of 2.68. The height of capillary rise is 6 ft. Within the capillary zone the soil is saturated; above it is virtually dry.
   a. Compute the vertical effective stress in the sand at several levels and plot the stress as a functional depth.
   b. The ground water level rises to the ground surface. Compute total, neutral and effective stresses and plot as functions of depth.
   c. The water level falls to a depth of 12 ft. Plot the vertical total, neutral and effective stresses as functions of depth.
   d. Compare the sand strengths for all three conditions at the following levels: (1) ground surface; (2) 6 ft deep; (3) 12 ft deep; (4) 20 ft deep.

7:4 Compute the approximate shear stress in the upper surface of the clay stratum produced by an earthquake whose horizontal acceleration is 0.12 g, in example 7:2.

7:5 A thick clay stratum with a $c = 600$ psf and a unit weight of 100 psf is excavated with a vertical bank. Consider that a small cube of clay in the face of the bank is unrestrained (the lateral stress is zero). How deep could the excavation be made before a state of plastic equilibrium is reached in a clay cube at the bottom of the face?

CHAPTER 8
Problems in Earth Pressure

AN EXCAVATION 33 FT OR 10 M DEEP WAS PLANNED 8 FT OR 2.5 m FROM the corner footing of an existing 5-story building. The bottom of the excavation was to be 22 ft or 7 m below the footing. The contractor drove steel sheet piling along the excavation line and then installed diagonal steel supports at three levels as the earth was removed. The second level of supports was being placed and the excavation was half complete when the nearby steps of the building sagged. Nevertheless, the work continued according to plan. At about the time the excavation bottom was reached, a loud cracking was heard, similar to that of a concrete test cylinder bursting, and the building corner dropped 2 in. or 5 cm and moved toward the excavation one in. or 2.5 cm. An examination of the sheet piling showed it had tilted outward 18 in. or 45 cm at the top. A crescent shaped crack in the original ground surface focused on the bracing nearest the building corner. The sagging steps were within the segment defined by the crack. A second crescent shaped crack was found, nearly concentric with the first, cutting across the ground floor of the building about 20 ft from the sheet piling. The footing was between the crack and the sheet piling. A review of the bracing design disclosed that no allowance had been made for the footing load. The contractor had assumed that the building was pile supported. No consideration had been made to possible deflection of the sheeting; in fact, the bracing was so poorly constructed that considerable movement was inevitable.

Unfortunately the lesson of the effect of inadequate bracing was not learned in this case. Two months after the movement had been arrested by strengthened bracing, the building moved again. The contractor had removed one set of braces to make it easier to lower a concrete bucket into the excavation. The cost of repair to the building was $50,000. An arbitration board required the contractor to pay half for his carelessness. The owner of the damaged building paid the other half for not informing the contractor specifically of the location of that building foundation although the general condition of the specification required protection of any adjoining structures.

SEC. 8:1] THEORY OF EARTH PRESSURE

Real problems in lateral earth pressure involve more than the simple loads produced by soil against a retaining wall. Earth pressure is not a unique property of the soil or rock. Instead it is a function of the material that the retaining structure must support, of the loads that the soil behind the structure must carry, the ground water conditions, and the amount of deflection the retaining structure undergoes.

Engineering structures such as retaining walls, trench and excavation bracing, bulkheads, and cofferdams have a common function—to provide lateral support for a mass of soil. The pressure exerted by the soil on these structures is known as *earth pressure* and must be determined before a satisfactory design can be made.

Some of the earliest theories of soil mechanics dealt with earth pressure on retaining walls. Unfortunately the engineers using these theories have not always realized the significance of the assumptions made in their development. The result has been many failures and a discrediting of soil mechanics by engineers who have had to deal with soils in construction work—an attitude that persists today.

8:1 Theory of Earth Pressure

The general theory of earth pressure can be developed from the stresses in an extremely large, level mass of soil. The total vertical stress in a mass of homogeneous soil at a depth of z is equal to the weight of the soil above. When ground water is present, the vertical stress can be separated into two components: neutral stress and effective stress, as previously described in Chapter 3.

$$\sigma_v = \gamma z, \qquad \text{(3:31a) (8:1a)}$$

$$\sigma_v' = \gamma z - u. \qquad \text{(8:1b)}$$

EARTH PRESSURE AT REST / The stress conditions of an element of soil at depth z in a level mass are shown in Fig. 8.1a. The element can deform vertically under load, but it cannot expand laterally because it is confined by the same soil under the same loading conditions. This is equivalent to the soil being placed against an immovable frictionless wall that maintains the same lateral dimension in the soil regardless of the vertical load. The soil is in a state of *elastic equilibrium,* and the stresses in the lateral direction can be computed from the stress-strain relationships of the soil. The relation between lateral and vertical strain is described by Poisson's ratio (Section 7:2), and for the condition of zero lateral strain the principal stresses are related by

$$\Delta \sigma_x = \left(\frac{\nu}{1-\nu}\right) \Delta \sigma_z. \qquad \text{(7:8b)}$$

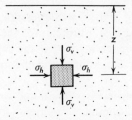
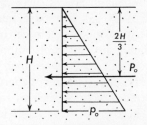
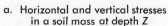

a. Horizontal and vertical stresses in a soil mass at depth Z

b. Pressure distribution and resultant force, P_o, on vertical surface of height, H

Figure 8.1 *Earth pressure at rest.*

The lateral pressure exerted in the at-rest state is given the symbol p_0 and can be computed from the vertical stress σ_v in a dry soil by

$$\sigma_x = \sigma_h = \sigma_v \frac{v}{1-v} = p_0,$$

$$p_0 = \gamma z \left(\frac{v}{1-v}\right) = K_0 \gamma z \quad \text{(dry soil)}. \tag{8:2a}$$

Below the water table the pressure is found from its effective and neutral components:

$$p'_0 = (\gamma z - u)K_0 \quad \text{(wet soil, effective)}, \tag{8:2b}$$

$$p_0 = (\gamma z - u)K_0 + u \quad \text{(wet soil, total)}. \tag{8:2c}$$

K_0 is the *coefficient of earth pressure at rest* and is found from Poisson's ratio. The value of K_0 for saturated clays in undrained loading or quick loading is sometimes also expressed in total stresses that include the neutral stress, and for which Equation 8:2a should be used.

TABLE 8:1 / VALUES OF K_0*

Soil	K_0, effective, Drained	K_0, total, Undrained
Soft clay	0.6	1.0
Hard clay	0.5	0.8
Loose sand, gravel	0.6	
Dense sand, gravel	0.4	

*The total pressure at any depth is the sum of effective and the neutral stresses (except where the undrained total K_0 is used with saturated clay).

SEC. 8:1] THEORY OF EARTH PRESSURE

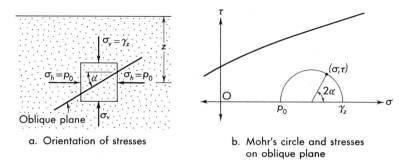

a. Orientation of stresses

b. Mohr's circle and stresses on oblique plane

Figure 8.2 Mohr's circle for earth pressure at rest.

The resultant force per unit of width of the wall, P_0, acting on a wall of height H, can be found by integrating equation 8:2a or from the pressure diagram (Fig. 8.1b). For a dry soil (or a saturated clay in undrained loading), the diagram is triangular and the resultant force is

$$P_0 = \frac{K_0 \gamma H^2}{2}, \qquad (8:3a)$$

and the location of the resultant is at a depth of

$$z = \frac{2H}{3}. \qquad (8:3b)$$

With ground water the effective and neutral pressure diagrams must be computed separately and the magnitude and location of the resultant found by the methods of mechanics.

The stress conditions in the soil mass are far from failure, as can be seen in Fig. 8.2. Stresses in an oblique direction can be computed by Mohr's circle.

ACTIVE EARTH PRESSURE / If the unyielding frictionless vertical wall of the at-rest condition is allowed to move away from the soil, each element of soil adjacent to the wall can expand laterally. The vertical stress remains constant, but the lateral stress or earth pressure is reduced in the same way the stress in a compressed spring becomes less as the spring is allowed to expand. Initially the stress reduction is elastic and proportional to the deformation; but as the difference between the major and minor principal stresses increases with the reduction in lateral stress, the diameter of Mohr's circle grows until the circle touches the rupture envelope. The lateral pressure has reached a *minimum* at this point; the stress conditions are no longer elastic; the soil mass behind the wall is in a state of shear failure or plastic equilibrium; and further movement of the wall will just continue the failure with little change in pressure.

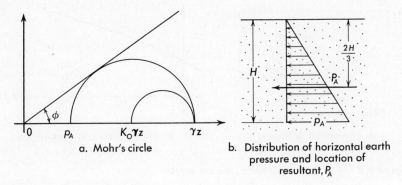

a. Mohr's circle   b. Distribution of horizontal earth pressure and location of resultant, P_A

Figure 8.3 *Active earth pressure in cohesionless soils: sands and gravels.*

The minimum horizontal pressure p_A at any depth z for dry sands and gravels can be found from the Mohr diagram at failure (Fig. 8.3a) and is

$$p_A = \frac{\gamma z}{\tan^2 [45 + (\varphi/2)]}, \tag{8:4a}$$

$$p_A = \gamma z \tan^2 \left(45 - \frac{\varphi}{2}\right). \tag{8:4b}$$

The expression $\tan^2 [45 - (\varphi/2)]$ is often called the *coefficient of active earth pressure* and is given the symbol K_A. The state of shear failure accompanying the minimum earth pressure is called the *active state*. The resultant force P_A per unit of width of wall for the dry sand can be found by integrating the expression for active pressure or from the area of the pressure diagram:

$$P_A = \frac{\gamma H^2 K_A}{2}. \tag{8:5}$$

The action line is through the centroid at a depth of $2H/3$ (Fig. 8.3b).

If the sand is below water, neutral stress must again be considered. The effective active pressure is computed from the effective vertical pressure and K_A. The total is the sum of the effective and the neutral stress:

$$p'_A = (\gamma z - u)K_A, \tag{8:6a}$$

$$p_A = (\gamma z - u)K_A + u. \tag{8:6b}$$

When a dry cohesionless soil is inundated by a rising water table, the effective pressure is reduced to about half its original value. The total pressure, however, is approximately tripled. The location and magnitude of the resultant for a cohesionless soil below water is found by combining the effective and neutral stress diagrams.

SEC. 8:1] THEORY OF EARTH PRESSURE

Example 8:1

Compute the active earth pressure at a depth of 15 ft in a sand whose angle of internal friction is 37° and which weighs 97 lb per cu ft dry and 123 lb per cu ft inundated.

1. The sand is dry throughout.
$p_A = \gamma z \tan^2 [45 - (\varphi/2)],$
$p_A = 97 \times 15 \times (0.5)^2,$
$p_A = 363$ lb psf.

2. The water table is at a depth of 5 ft.
$p'_A = (\gamma z - u) \tan^2 [45 - (\varphi/2)],$
$p'_A = (5 \times 97 + 10 \times 123 - 10 \times 62.4) \times (0.5)^2,$
$p'_A = 1091 \times 0.25 = 273$ psf,
$p_A = \bar{p}'_A + u = 272 + 10 \times 62.4,$
$p_A = 897$ psf.

A similar analysis for saturated clay in undrained loading, using Mohr's circle, (Fig. 8.4a), gives the formula for the active pressure as

$$p_A = \gamma z - 2c. \qquad (8:7)$$

The total force per foot width of wall, P_A, is given by the expression

$$P_A = \frac{\gamma H^2}{2} - 2cH. \qquad (8:8)$$

According to this formula the resultant earth force will be zero when the height of the wall is $4c/\gamma$, even though the soil is in plastic equilibrium. This explains why clay often stands unsupported in high, vertical banks. The pressure diagram (Fig. 8.4b) indicates that the clay is in tension to a depth of $2c/\gamma$. The tension causes vertical cracks and causes the clay to pull away from the wall. The tension part of the pressure diagram disappears upon cracking, leaving a positive pressure against the bottom part of the wall.

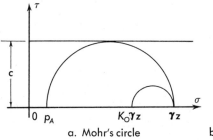
a. Mohr's circle

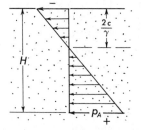
b. Distribution of horizontal earth pressure and location of zone of tension

Figure 8.4 Active earth pressure in saturated cohesive soils: clays in undrained shear.

Thus a short wall up to the depth of $2c/\gamma$ should theoretically support the clay in spite of the cracking. However, water that accumulates in the cracks will add to the horizontal pressure and will require more support. The development of tension cracks explains why vertical cuts in clay sometimes fail without warning, weeks after they were made, and why many of the failures occur during wet weather.

PASSIVE STATE / If, instead of moving away from the soil, the wall moves toward the soil, the pressure against the wall increases. The stress circles increase to the right of the vertical stress γz, which now becomes the minor principal stress. The maximum pressure against the wall is reached as shear failure again occurs in the soil behind the wall.

For dry cohesionless soils the pressure at any depth can be found by Mohr's diagram (Fig. 8.5a) and is

$$p_P = \gamma z \tan^2\left(45 + \frac{\varphi}{2}\right), \qquad (8:9)$$

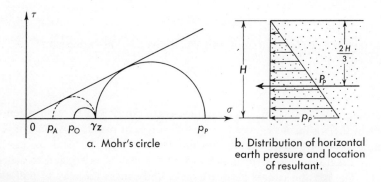

a. Mohr's circle

b. Distribution of horizontal earth pressure and location of resultant.

Figure 8.5 *Passive earth pressure in cohesionless soils: sands and gravels.*

where p_P is the maximum or *passive earth pressure*. The expression $\tan^2[45 + (\varphi/2)]$ is often called the *coefficient of passive earth pressure* and is given the symbol K_P.

The total force per foot width of wall of height H is found from the pressure diagram (Fig. 8.5b):

$$P_P = \frac{\gamma H^2}{2} K_P. \qquad (8:10)$$

The action line is horizontal and is at a depth of $2/3H$.

Below the water table the effect of neutral stress is handled in the same way as for the active state.

For saturated clays in undrained loading, the passive pressure is found by Mohr's circle (Fig. 8.6a) to be

$$p_P = \gamma z + 2c. \qquad (8:11)$$

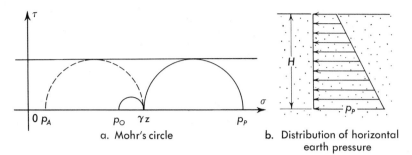

a. Mohr's circle b. Distribution of horizontal earth pressure

Figure 8.6 Passive earth pressure in saturated cohesive soils: clays in undrained shear.

The total force for a unit width of wall is found from the pressure diagram (Fig. 8.6b) and is

$$P_P = \frac{\gamma H^2}{2} + 2cH. \qquad (8:12)$$

Clay.

For soils such as partially saturated clays whose shearing resistance is given by the formula $s = c' + p \tan \varphi'$, the following formulas may be derived with the aid of Mohr's diagrams. For the active state:

$$p_A = \gamma z \tan^2\left(45 - \frac{\varphi'}{2}\right) - 2c' \tan\left(45 - \frac{\varphi'}{2}\right), \qquad (8:13)$$

$$P_A = \frac{\gamma H^2}{2} \tan^2\left(45 - \frac{\varphi'}{2}\right) - 2c'H \tan\left(45 - \frac{\varphi'}{2}\right). \qquad (8:14)$$

For the passive state:

$$p_P = \gamma z \tan^2\left(45 + \frac{\varphi'}{2}\right) + 2c' \tan\left(45 + \frac{\varphi'}{2}\right), \qquad (8:15)$$

$$P_P = \frac{\gamma H^2}{2} \tan^2\left(45 + \frac{\varphi'}{2}\right) + 2c'H \tan\left(45 + \frac{\varphi'}{2}\right). \qquad (8:16)$$

The pressure diagrams for these conditions are similar to those for saturated $c + \varphi$ clays.

The analytical approach to earth pressure is termed the *Rankine method* after that famous Scottish engineer who first applied such reasoning to soil masses.

8:2 Deformation and Boundary Conditions

In both the active and passive states the zones of soil adjacent to the frictionless wall that are in a state of shear failure or plastic equilibrium form plane wedges (Fig. 8.7). Since the angle between a failure plane and the

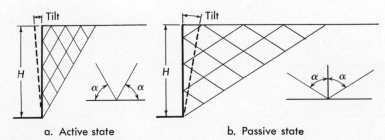

Figure 8.7 *Failure planes and plane wedges in soil behind a frictionless wall.*

major principal plane is $\alpha = 45 + (\varphi/2)$, the wedge is bounded in the active state by a plane making an angle of α with the horizontal, and in the passive state by a plane making an angle of α with the vertical. Within the wedges in both cases are an infinite number of failure planes making angles of α with the major principal plane.

The amount of horizontal movement of any point on the wall necessary to produce either the active or passive state is proportional to the width of the shear zone adjacent to that point. As can be seen from Fig. 8.7, the minimum movement consists of tilting about the base of the wall.[8:1] The amount of tilt is small and depends on the soil rigidity and wall height H, as given in Table 8:2.

TABLE 8:2 / TYPICAL MINIMUM TILT NECESSARY FOR ACTIVE AND PASSIVE STATES

Soil	Active State	Passive State
Dense cohesionless	$0.0005H$	$0.005H$
Loose cohesionless	$0.002H$	$0.01H$
Stiff cohesive	$0.01H$	$0.02H$
Soft cohesive	$0.02H$	$0.04H$

Soft cohesive soils do not remain in either the passive or active condition for long. Slow yield of the soil (often termed *creep*) tends to return the soil mass to the "at rest" state. In the case of walls supporting a soft clay backfill, this means that there will be a continual slow, outward movement of the wall if the wall is designed to support only active pressure.[8:2] Over a period of a few months, however, the change in pressure due to creep is usually negligible.

INCOMPLETE DEFORMATION / The relation between deformation of the wall and earth pressure is shown in Fig. 8.8. The minimum and maximum limits (active and passive states) can be computed by the Rankine theory, and the point of no deformation (at-rest) can be computed by the theory of elasticity. When the deformation is in between these limits, the pressure is also in between but cannot be computed by theoretical methods.

SEC. 8:2] DEFORMATION AND BOUNDARY CONDITIONS   345

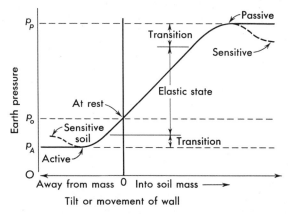

Figure 8.8 Effect of wall movement or tilt on the magnitude of earth pressure.

Instead it must be estimated from the results of pressure measurements on structures correlated with limiting active, passive, or at-rest conditions.

When the deformation is small, the state of stress is elastic and the pressure is proportional to the tilt. When it is large, the stress is in a transition from elastic to plastic conditions and the pressure is no longer proportional to tilt. At very large deformations beyond that necessary for active (or passive) conditions, the pressure may again change toward the at-rest state, owing to a reduction in strength of sensitive soils from shear, as shown in Fig. 8.8.

IRREGULAR DEFORMATION: ARCHING[8:3] / When the movement of the wall is different from the tilting required to establish the active or passive states, both the magnitude and the distribution of the earth pressure are changed. If a section of the wall deflects outward more than the neighboring sections, the soil adjacent to it will tend to follow, as shown in Fig. 8.9a and b. Horizontal shear develops along the boundaries of this

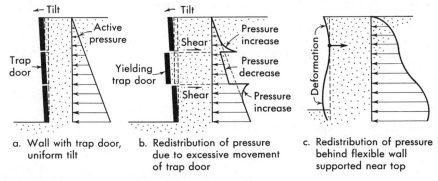

a. Wall with trap door, uniform tilt

b. Redistribution of pressure due to excessive movement of trap door

c. Redistribution of pressure behind flexible wall supported near top

Figure 8.9 Effects of irregular deformation: arching.

section of soil and this restrains it and transfers part of the lateral load it carried to the adjoining soil. The result is a *redistribution of pressure by shear*, sometimes called *arching*, and an irregular pressure distribution. Examples of the effect of arching on earth pressure against typical strutures are shown in Fig. 8.9b and c. The magnitude of the redistribution must be estimated from observations of pressures and deflections of actual structures, since no valid theoretical analyses of arching have been developed.

EFFECT OF WALL FRICTION / The Rankine analysis considers an extensive zone in plastic equilibrium, with the shear pattern not distorted by the engineering structure. This is equivalent to assuming that no shear can develop between the wall and the soil (the *wall is frictionless*). In reality the movement between the wall and the backfill does develop shear or friction, which distorts the shear pattern and the magnitude and direction of the resultant force that acts on the wall (see the example of Fig. 8.13). The error involved is not serious for small walls with smooth faces, but it can be significant for structures higher than 30 ft and with rough faces that induce appreciable shear. Methods for solving such problems are discussed in Section 8:3.

Simple problems of active and passive earth pressure can be adequately solved by the Rankine analysis. For more complex problems either modifications or corrections applied to the Rankine method, or approximate analyses compatible with the real boundary conditions are used. Some of these are discussed in this section. More extensive treatments can be found in the references.

8:3 Computing Earth Pressure

APPROXIMATE ANALYSES OF SLOPING WALLS AND BACK-FILLS / For walls less than 30 ft high a number of useful approximations based on the Rankine analysis have been developed. When the back of the wall is inclined, the force is assumed to act on a vertical plane through the heel of the wall, as shown in Fig. 8.10a. In such cases the weight of the wedge of soil between the vertical plane and the wall is added vectorally to the resultant P_A.

If the backfill slopes at an angle β with the horizontal, the corresponding Rankine formula (derived in the same manner as the formula for a horizontal soil surface) is

$$P_A = \frac{\gamma H^2}{2} \cos \beta \, \frac{\cos \beta - \sqrt{\cos^2 \beta - \cos^2 \varphi}}{\cos \beta + \sqrt{\cos^2 \beta - \cos^2 \varphi}}, \qquad (8:17)$$

and P_A acts parallel to the backfill surface, as shown in Fig. 8.10c, at a depth of

$$z = \frac{2H}{3}, \qquad (8:3b)$$

where H is the height of a vertical plane through the heel of the wall.

SEC. 8:3] COMPUTING EARTH PRESSURE

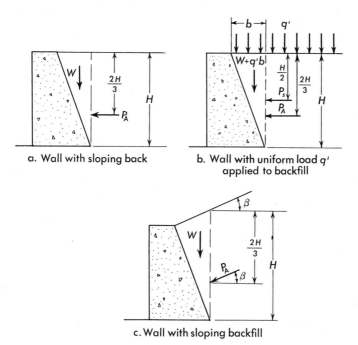

Figure 8.10 *Approximate pressure for walls with cohesionless backfills.*

EFFECT OF SURCHARGE LOADINGS / If a uniform surcharge load of q' acts on the soil behind the wall, as shown in Fig. 8.10b, it produces additional pressure on the wall. In the active state the resultant of this pressure P_s, in pounds per foot width of wall, is

$$P_s = q'H \tan^2\left(45 - \frac{\varphi}{2}\right). \tag{8:18}$$

It acts midway between the top and bottom of the wall.

Sloping surcharges such as piles of materials on a level backfill can be approximated by a uniform surcharge equivalent to the average height of the material within the shear zone.

EFFECT OF CONCENTRATED SURCHARGES / In many real situations loads of limited extent are placed on the backfill. For example, a building foundation, a highway pavement, or railroad is often supported on the backfill close enough to the wall that additional earth pressure is produced. This creates a local distortion of the soil mass and possibly a change in the shear pattern. The distortion is, in effect, a state of elastic equilibrium superimposed on a state of plastic equilibrium. While there are theoretical objections to this concept, large scale model studies show it to be realistic.

According to the Boussinesq analysis of elastic equilibrium, a load of Q at the surface produces a stress increase $\Delta\sigma_h$ at a depth of z and at a horizontal distance of x from a wall and a length of y along the wall, Fig. 8.11a, of:

$$\Delta \sigma_h = \frac{Q}{2\pi} \frac{3x^2z}{R^5} = \frac{0.48Qx^2z}{R^5}, \qquad (8:19a)$$

$$R = \sqrt{x^2 + y^2 + z^2}. \qquad (8:19b)$$

For a continuous load of Q' per unit of length parallel to the wall and at a distance of x from it, Fig. 8.11b, the increase in stress, $\Delta\sigma_h$ uniformly along the wall length is given by

$$\Delta \sigma_h = \frac{0.63 Q' x^2 z}{R^5}, \qquad (8:19c)$$

$$R = \sqrt{x^2 + z^2}. \qquad (8:19d)$$

In either case R is the direct distance from the load to the point on the wall the stress increment acts.

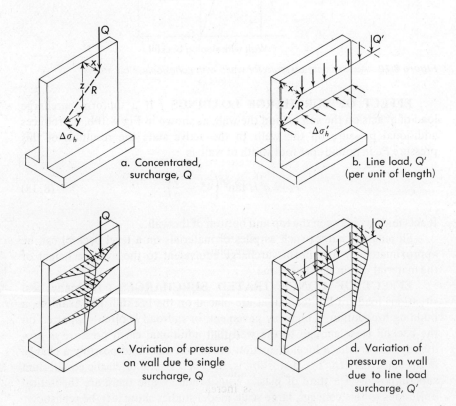

Figure 8.11 Pressures produced by surcharge concentrations on a level backfill.

SEC. 8:3] COMPUTING EARTH PRESSURE

The stress increments above do not consider any restraining effect of the wall on the elastic equilibrium. Tests by Spangler[8] indicate that for the active state these stresses should be increased by a factor of about 1.5. For the at-rest condition they should be increased by a factor of 2.

The pressure increase is not uniform along the wall surface. For the concentrated load it varies in both the y and z directions, Fig. 8.11c. The greatest intensity of increase is opposite the load at a depth below the top of the wall of $z = x/2$ approximately. For a continuous line load of Q' the pressure varies vertically as shown in Fig. 8.11d. The greatest increase is similarly at a depth of $z = x/2$ approximately.

If the surcharge is of great magnitude, the total shear pattern is changed. This effect can be approximated by adding the surcharge load to the backfill weight, utilizing the approximate analysis of plastic equilibrium described in the next section. This approach is also theoretically weak. The real effect is probably between that computed by the Boussinesq or elastic analysis and that found by assuming the surcharge to add to the soil weight in the zone of plastic equilibrium.

COULOMB ANALYSIS / In 1776 the French scientist Coulomb published a theory of earth pressure that includes the effect of wall friction and which can be applied regardless of the slope of the wall or its backfill. He discovered through many observations with dry sand that a retaining wall tilts outward until earth pressure becomes a minimum—the active state. In this condition, the backfill is in a state of shear failure along a series of inclined, parallel, slightly curved surfaces (Fig. 8.12a). The wedge-shaped section of backfill, bounded by the shear surfaces, slides downward and outward as the wall moves outward. Coulomb approximated the shape of the failure wedge by assuming it to be sliding on a plane surface, and derived the active earth pressure from the forces producing equilibrium in the wedge as it commences to move, as shown in Fig. 8.12b.

The weight W of the wedge of earth is computed by the assumed angle of the failure plane η, the soil weight, and the wall and backfill dimensions. Its direction is vertical. The resultant force F of the wedge on the soil is

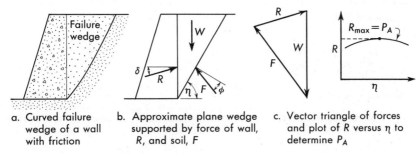

a. Curved failure wedge of a wall with friction

b. Approximate plane wedge supported by force of wall, R, and soil, F

c. Vector triangle of forces and plot of R versus η to determine P_A

Figure 8.12 *Coulomb analysis of active earth pressure.*

inclined at an angle φ with the normal to the shear plane, but its magnitude is unknown. The force of the wall R on the wedge is inclined at the angle of wall friction δ to the normal to the wall. Its magnitude is unknown.

The three forces form a vector triangle (Fig. 8.12c) from which the magnitude of R (and P_A) is obtained graphically. Of course the correct angle of the failure wedge η is unknown. It is found by computing values of R for several assumed values of η and plotting the results graphically. The peak of the curve represents the critical failure plane, and the maximum R is equal to but opposite in direction from P_A. If the soil has cohesion, the cohesive force c along the failure plane is added to the vector polygon. Graphical solutions of the Coulomb analysis and other more exact methods for computing earth pressure are available.[8:4,8:5,8:6]

The value of the angle of wall friction δ can be found by laboratory friction tests. For smooth concrete it is often $\frac{1}{2}\varphi$ to $\frac{2}{3}\varphi$, and for rough stone it is equal to φ.

The Coulomb analysis gives an active earth pressure equal to the Rankine for a frictionless vertical wall and a level backfill. When δ exceeds 0, the resultant pressure computed by the Coulomb method is as much as 10 per cent lower. The location of the resultant is the same.

The analysis can be applied to the case of the sloping backfill or an inclined wall face as easily as for the simple case of the level backfill and vertical wall; only the geometry of the wedge is changed. The effect of a uniform surcharge or a large concentrated surcharge can be estimated by adding the load to that of the soil wedge, but if a concentrated load is beyond the wedge it does not contribute. The plot of R as a function of the wedge angle, η, will include a sharp drop at the angle where the surcharge is no longer within the wedge.

For passive pressure, the curvature of the failure surface increases greatly with increasing values of δ, Fig. 8.13a. Furthermore, a thrust load against the wall in the upward direction could develop a reversed value of δ as shown in Fig. 8.13c and a completely different shear surface. The Coulomb analysis with its assumed straight line shear surface can be seriously in error in computing passive pressure. For the more common case of passive pressure with a downward component of friction between the wall and the soil, the magnitude of the real resultant of passive pressure lies between the

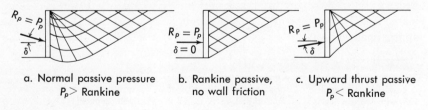

a. Normal passive pressure $P_p >$ Rankine

b. Rankine passive, no wall friction

c. Upward thrust passive $P_p <$ Rankine

Figure 8.13 Shear patterns in passive earth pressure on walls with friction.

Rankine and Coulomb values. More rigorous solutions involving curved failure surfaces should be used for computing passive pressure when δ exceeds about $\frac{1}{4}\varphi$.

LATERAL PRESSURE IN ROCK / Because of the high permanent cohesion, the lateral pressure developed by an intact mass of rock is negligible. According to Equations (8:7) and (8:8) tension would be required to produce shear failure, and the net force will be zero for faces of great height in intact rock of even moderate strength.

Intact rock, however, seldom exists. The pressure exerted by rock is largely controlled by the joints and bedding planes, and water pressure. The lateral pressure exerted by a rock mass can be found by a simplified Coulomb's wedge, Fig. 8.14. The potential shear surface in three dimensions is defined by joints or bedding planes whose angles can be found. The shear characteristics along such surfaces is established by laboratory tests made on large samples that include the potential surface of weakness oriented in the most favorable angle for failure. Large scale field tests for shear on the suspected plane of weakness are more realistic than the laboratory tests and should be made for important structures whose cost justifies the expense.

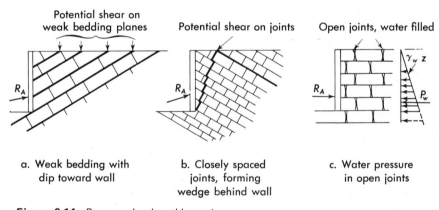

Figure 8.14 Pressure developed by rock masses.

Ground water pressure must be included in the analysis if there is seepage through the joints and bedding planes. Even horizontally bedded rock can develop high lateral pressure from water in joints, Fig. 8.14c. Because of the uncertain water pressures and the variability of strengths along joints and bedding planes, pressure computations for rock masses are approximate at best.

8:4 Retaining Walls

A retaining wall is a permanent, relatively rigid structure of cribbing, masonry, or concrete that supports a mass of soil. It substitutes the steep face of the wall for the gentle natural slope of the earth to provide usable space in highway and railroad cuts, in and around buildings, and in structures below ground level.

The wide use of retaining walls is accompanied by many failures and partial failures because of designs based on rules and formulas that fit only certain limited conditions. For example, the design of walls backfilled with soft clay is often based on analyses that apply only to sand, and the design of walls that support structures that will be cracked by foundation movement is often based on the active earth pressure that requires the wall to tilt outward. The only thing between success and failure in many cases has been an overgenerous safety factor.

REQUIREMENTS FOR DESIGN / A satisfactory retaining wall must meet the following requirements:

1. The wall must be structurally capable of withstanding the earth pressure applied to it.
2. The foundation of the wall must be capable of supporting both the weight of the wall and the force resulting from the earth pressure acting upon it without:
   a. Overturning or soil failure.
   b. Sliding of the wall and foundation.
   c. Undue settlement.

The earth pressure against a retaining wall depends on the deformation conditions or tilt of the wall, the properties of the soil, and the water conditions. For greatest economy, retaining walls are ordinarily designed for active pressure as developed by a dry cohesionless backfill, but if necessary, a design can be developed for any conditions of yield, soil, and water.

Organizations such as state highway departments and railroads who design many retaining walls have developed charts and tables that give earth pressures with a minimum of computation. Nearly all such charts and tables are based on the Rankine formula, with assumed values for the angle of internal friction and the unit weight of the backfill soil.

Some designers use the hydrostatic pressure developed by an imaginary fluid whose unit weight γ_f is termed the *equivalent unit weight*. This is a modification of the Rankine formula, where $\gamma_f = K_A \gamma$.

RETAINING WALL TILT / The earth pressure must be compatible with the wall tilt, which is limited by the rigidity of the wall, the foundation, and any connections with adjoining structures. The tilt for active pressure is given in Table 8:2. As a general rule the designer should allow at least 1 in. of tilt for each 10 ft of wall height ($0.008H$) unless he is certain of the backfill quality and its installation. Ordinarily the tilt stops within a few days after the backfill is placed, although small movements sometimes continue in loose sands subject to vibration and in clays. For isolated straight walls on a soil foundation such tilting is no problem. For curved walls, long walls, or walls of widely varying height, it is necessary to provide joints that will permit movement. These can be sealed after the tilt stops.

SEC. 8:4] RETAINING WALLS

Walls on piles are somewhat restrained, and the earth pressure is probably a little more than the active pressure. The difference is not great, and therefore the active pressure is used for design.

The foundations of walls resting on bedrock cannot deflect. If the wall itself is flexible, like a thin reinforced cantilever, it will probably deflect enough to establish active or near-active conditions. A massive wall, however, cannot tilt and must be designed for at-rest pressure. A cushion of sand between the wall foundation and the rock permits some movement and can result in a lower pressure on the wall.

The tilt is often limited by adjoining structures. If construction of such structures is delayed until after the wall is backfilled, little trouble should develop. Building walls resting on retaining walls often suffer from cracking, owing to retaining wall tilt. In one case, the anchor bolts for columns resting on a 30-ft or 9 m high wall were 2 in. or 5 cm out of line after the wall was backfilled, and in another case a brick wall on top of a retaining wall was split apart by the tilting.

DRAINAGE PROVISIONS / The most important problem of backfill design is keeping the soil dry. Two different methods may be used:
1. Remove water from backfill.
2. Keep water out of backfill.

In all cases the first method should be used, and in some cases both.

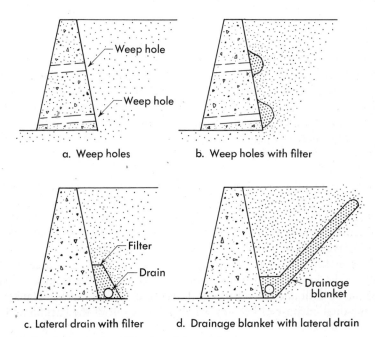

Figure 8.15 *Drains for backfills of retaining walls.*

Removal of water from the backfill is accomplished by drainage, often by the simple expedient of *weep holes* through the wall (Fig. 8.15a). Weep holes should be spaced about 4 to 6 ft or 1.5 to 2 m apart horizontally and vertically and should be at least 4 in. or 10 cm in diameter to permit easy cleaning. If the backfill is coarse sand, a few shovelfuls of peagravel over the inlet to each weep hole will act as a filter to prevent the hole from clogging with sand (Fig. 8.15b). Weep holes have the disadvantage of discharging water at the base of the wall where the foundation pressure is greatest. A better but more expensive drain consists of a 6-in. or 8-in. or 15 to 20 cm perforated pipe parallel to the wall at its base, lying in a filter trench (Fig. 8.15c). Manholes should be provided at the ends of this pipe so it may be cleaned. For soils of lower permeability, such as silty sands and silts, more elaborate provisions are necessary. An inclined drainage blanket (Fig. 8.15d) drains the entire backfill and is simple to construct.

When clays that are likely to swell or soils that are difficult to drain must be used, it may be necessary to take steps to prevent water from entering the backfill. The first step is to locate the sources of water; the second is to divert the water away from the backfill. If surface infiltration is the source of water, the backfill may be paved with a flexible, impervious blanket of asphalt or plastic clay. Surface drains must be provided to remove the water. Special attention should be paid to removing water from the inevitable crack between the top of the wall and the backfill. A small gravel blanket, drained with weep holes through the wall, will be sufficient. If underground seepage is the source of water, interceptor drains can be very effective in preventing water from entering the backfill.

PREVENTING FROST ACTION / In northern climates frost action has caused many retaining walls to move so far that they have become useless. Since stone or concrete is a relatively good conductor of heat, the temperature along the back side of the wall is the same as in the air. If freezing temperatures prevail, if the backfill soil is susceptible to frost action, and if there is a plentiful supply of water, ice lenses will form parallel to the wall and can cause horizontal movements of up to 3 ft or 1 m in a single season.

Frost action can be prevented by substituting a thick blanket of relatively coarse, cohesionless soil such as sand or gravel for the portion of the backfill adjacent to the wall. This blanket should be as thick as the depth of frost penetration in the region. It can be constructed by dumping small loads of sand or gravel against the wall as the backfill is being placed.

Serious frost action can be prevented by cutting off the source of water supply to the ice lenses. A blanket of gravel 8 in. or 20 cm thick is placed below the backfill and provided with filters and a drain. Such a blanket can serve the dual purpose of keeping water pressure out of the backfill and also preventing frost heave.

SEC. 8:4] RETAINING WALLS

BACKFILL MATERIAL / The best backfill is rigid, free-draining, and with a high angle of internal friction so as to develop minimum earth pressure with the least movement. Table 8:3 rates the soils of the Unified Classification for selection.

TABLE 8:3 / RETAINING WALL BACKFILL

GW, SW, GP, SP	Excellent, free-draining backfill.
GM, GC, SM, SC	Good if kept dry but require good drainage. May be subject to some frost action.
ML	Satisfactory if kept dry but requires good drainage. Subject to frost. Neglect any cohesion in design.
CL, MH, OL	Poor. Must be kept dry. Subject to frost. Wall deflection likely to be large and progressive unless at-rest pressure is used.
CH, OH	Should not be used for backfill because of swelling.
Pt	Should not be used.

Artificial materials such as cinders and crushed slag often make good backfill. All the cohesionless backfills are best when well compacted because the higher internal friction angle and the resistance to vibration offset the higher weight.

A wedge-shaped backfill of sand, gravel, or slag at least 50 per cent wider than the failure wedge makes it possible to design the wall for the low pressure of a cohesionless soil even though the remainder of the backfill is clay.

RETAINING WALL DESIGN / The design of a retaining wall is based on the materials available, appearance, the space required, the forces acting, and finally, cost. The materials for wall construction are stone masonry, plain concrete, reinforced concrete, and earth or broken stone. The choice is based partly on appearance. Walls used in conjunction with stone-faced buildings or in residential areas and parks are often of brick or stone masonry. Walls in industrial areas or adjacent to bridges and dams are usually concrete. Cost and availability of materials and labor are important factors in the choice of wall materials. Stone masonry is expensive in the United States and requires skilled workmen; plain concrete is relatively easy to form and requires no steel but may use excessively large quantities of concrete; reinforced concrete is economical for large structures but requires accurate fabrication of the steel and forms and controlled-quality concrete.

Space is an important factor in wall design, since the function of retaining walls is to make more usable, level space than a natural slope will provide.

In addition to having an important bearing on choice of wall materials, appearance often governs the shape of the wall face. Walls should not be designed with vertical faces because the inevitable slight tipping will cause them to lean outward and appear unstable, even though they actually may be quite safe. To give the appearance of stability, it is better to provide the face of the wall with an inward batter of at least 1 (horizontal) to 10 (vertical).

Gravity walls (Figs. 8.12a and 8.15) resist earth pressure by their weight. They are constructed of stone and concrete that can resist compression and shear but no appreciable tension, and so the design is mainly concerned with preventing tension. Tentative dimensions are selected: a top width of 1 to 2 ft (30 to 60 cm) and a bottom width of about 40 per cent of height are typical trial values. Sections are taken through the wall at the base and at one or two intermediate levels. The resultant of all the forces acting above the section, including the resultant of earth pressure, the weight of the wall, and any load acting on the top of the wall, must pass through the middle third of the section to avoid tension.

Two types of reinforced concrete walls are used: *cantilever* (Fig. 8.16a) and *buttress* or *counterfort* (Fig. 8.16b). Cantilever walls are used for heights up to 30 ft or 9 m, and buttress walls are commonly used for heights greater than 25 ft or 8 m. Structurally the cantilever wall is a wide cantilever beam acted upon by pressure that increases uniformly to a maximum at the point of restraint. It is reinforced in the vertical direction to withstand the bending moments and in the horizontal direction to prevent cracking.

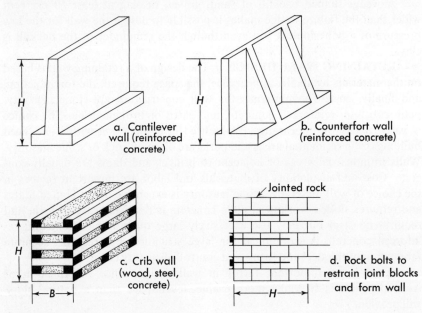

Figure 8.16 Types of retaining walls.

The buttress or counterfort wall consists of a vertical, flat slab supported on two sides by the buttresses and on the bottom by the wall foundation. Ordinarily the support furnished by the wall foundation is neglected and the slab is designed as though it were continuous over the vertical buttress supports. Counterforts are buttresses on the back side of the wall in order to provide more usable space in front. This increases the cost of the wall, since they are in tension and require considerable reinforcing; but on the other hand, a lower wall is usually possible.

The crib retaining wall, (Fig. 8.16c,) consists of a hollow rectangular cribwork of logs, timbers, reinforced concrete beams or steel beams filled with soil or rock. The cribwork can be vertical or tilted toward the backfill for greater stability. Crib walls are relatively cheap and are usually flexible enough to be used where settlement is a serious problem. Structurally the crib is a gravity wall and is designed so that its width is sufficient to keep the resultant within the middle third. In addition the shear at any cross-section must not exceed the shear strength of the soil.

Baskets or containers formed of wire mesh and filled with gravel, termed *gabions*, are a form of gravity wall. These depend on the shear of the fill for internal stability, and their mass to resist the backfill loads.

ROCK SUPPORT / Retaining walls for intact rock masses are not true retaining structures. Instead they are facings to prevent the weathering of the rock, to prevent the local spalling of loose fragments, or to minimize erosion of a soft rock below a dam spillway or hydro plant tailrace. The facing is anchored to the rock by rock bolts (see Section 8:7) or by reinforcing bars grouted into the rock.

Walls to support rock pressures developed in masses with joints and bedding planes can be designed in the same way as walls that support soil, once the resultant force has been computed.

Sometimes the rock mass can be made to support itself as a gravity retaining wall. The rock blocks formed by the joints are tied together by reinforcing rods or rock bolts through the mass and by bearing plates, or beams along the rock face, as shown in Fig. 8.16d. This creates an intact body similar to a crib or gravity wall that resists the pressure from the remainder of the mass by gravity. The design requires a detailed study of the joint pattern so that the blocks are securely restrained to form a unit that cannot separate.

WALL FOUNDATION / Faulty foundations are a frequent cause of retaining wall failures. The combination of vertical and horizontal forces that are supported are not always fully considered in design. Furthermore, it has been pointed out that the deflection of the foundation can be a major factor in the magnitude of the earth pressure acting against the wall. The most economical wall design presumes active earth pressure, which requires movement of the wall. By way of contrast, a rigid wall on an unyielding foundation must resist at-rest pressure.

The design of foundations to resist combined vertical and lateral pressure is considered in Chapters 9 and 10. Generally, a spread footing design requires that the resultant force of the earth pressure, wall weight and foundation weight pass through the middle third of the foundation width.

If the wall is supported on rock or very hard soil, the safety against overturning should be determined. Moments of the earth pressure about the outside corner of the bottom of the foundation, the foundation *toe*, must be resisted by the moment of weight of the wall, foundation and any vertical component of the backfill. The ratio of the resisting to the overturning moments should be at least 1.5 (a safety factor of 1.5 against overturning).

If the foundation is very weak, the weight of the backfill plus the wall can cause a large scale shear failure involving both the wall and the backfill, Fig. 8.17a. If there are external loads on the backfill, such as railroad tracks or structures, both the pressure against the wall as well as the load on the foundation are increased. A *relieving platform* supported on piles, Fig. 8.17b, can carry part of the backfill load. Platforms are particularly useful in wharf construction where the foundation soils are likely to be poor and the loads on the backfill are as high as 1000 lb. per sq ft or 5 tons per sq m. Methods for analyzing the total stability of the mass are discussed in Chapter 11.

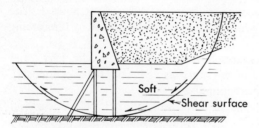
a. Failure of wall, foundation and backfill from shear of soft foundation

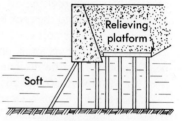
b. Support of backfill by relieving platform on piles

Figure 8.17 Total wall stability.

8:5 Excavation Bracing

In many construction jobs deep excavations must be made before the structure can be built. Often the planning of the excavation is left to the excavation superintendent or to a shovel operator. When expensive excavations or those that may endanger lives or adjacent property are involved, however, the bracing must be designed as is any other important structure.

Since an excavation is an opening in the soil for some specific purpose, the design of the excavation depends on two factors: the nature of the soil and the size of the opening desired. In most cases the primary factor is size, including depth, area, yardage, and the working space required for equipment and the structure within the excavation. The soil strength and its ease of excavation, the depth of ground water and the ease by which it travels through the soil influence both the excavation method and the excavation design.

SEC. 8:5] EXCAVATION BRACING

Open excavations are those that require no bracing to support the soil or to control ground water. The soil is cut to the steepest slope on which it will stand, usually $1\frac{1}{2}$ (horizontal) to 1 (vertical) for sandy soils, and up to vertical slopes for shallow excavations in stiff clay or decomposed rock. The proper slopes are usually determined by trial and error or from past experience in similar soils. The deeper the excavation and the weaker the soil, the flatter are the slopes required. Flat slopes, small size, and great depths all result in considerable excavation beyond that actually required for the structure; therefore open excavations are usually limited to strong soils, large areas, or shallow depths (under 20 ft). For analysis of stability and the slopes required in open cuts see Chapter 11.

CUT BRACING / When it is uneconomical or impossible to use open excavations, bracing must be employed to support the soil. Many methods of bracing have been devised, and some have been even standardized by organizations (such as sewer contractors) who must do considerable trench excavation work. Unfortunately, however, some excavation contractors spend little or no time designing their bracing, and as a result there have been

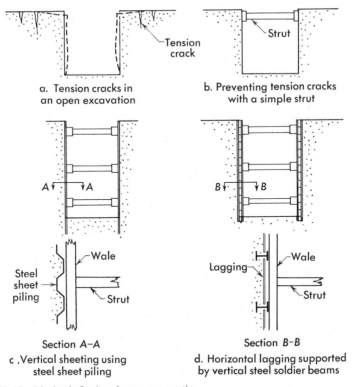

Figure 8.18 *Methods for bracing an excavation.*

numerous failures. Almost every year workmen are crushed to death by the failures of inadequately designed bracing, which occur even in shallow excavations.

The simplest type of bracing is the strut (Fig. 8.18b), a horizontal timber whose ends are jacked against the soil. Struts are commonly applied in shallow cuts in cohesive soils where the soil can stand unbraced for a short time. The struts, applied near the top of the cut, relieve the tension in the soil that occurs above the depth $z = 2c/\gamma$ and prevent the formation of tension cracks (Fig. 8.18a) that would lead to a collapse of the walls of the excavation. If two rows of struts are required, they may be set against vertical timbers known as *soldier beams*.

When more extensive bracing is required, two methods may be employed:
1. Vertical sheeting.
2. Horizontal lagging.
3. Cast-in-place wall.

When the soil is very soft and runny, the time-honored method of vertical sheeting is employed, as shown in Fig. 8.18c. Vertical sheeting (either wood or steel sheet piling) is driven along the line of the excavation before the soil is removed. As excavation proceeds, horizontal members known as *wales* are placed along the inside of the sheeting and braced with struts.

If the depth of excavation is greater than the length of the sheeting, a second row of sheeting may be driven inside the first after excavation has extended near the bottom of the first row.

When the soil is not runny, horizontal lagging (Fig. 8.18d) may be used. If the cut will stand without bracing for a few hours, it is excavated without supports and then horizontal boards known as *lagging* are placed against the soil. These are held in place by vertical soldier beams supported by struts. If the soil requires support at all times but does not run, the soldier beams, consisting of steel W^F sections are driven into the soil, and the lagging is wedged tightly against the soil to prevent any excessive movement. The vertical beams are supported by a wale and strut system in the same manner as vertical sheeting.

Cast-in-place concrete retaining walls can be installed where obstructions prevent pile driving or where the shock and vibration of pile driving are objectionable. One form consists of overlapping concrete cylinders, Fig. 8.19a. These can be constructed in two ways, depending on the strength of the soil and on ground water. In firm clays above the water table a hole 15 in. or 40 cm or more is drilled with a large auger. A reinforcing cage is inserted, and the hole filled with concrete. In less stable soils, the hole is drilled but kept filled with soil or mud during the drilling process. Cement mortar is injected through the hollow stem of the drill from the bottom of the hole up as the drill is withdrawn, so that the mortar replaces the soil and there is never a moment that the walls of the hole are not supported. (The same method applied to foundations is described in Chapter 10). Reinforcing

SEC. 8:5] EXCAVATION BRACING

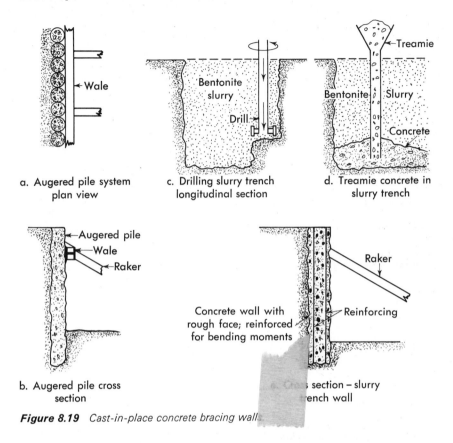

Figure 8.19 Cast-in-place concrete bracing wall.

steel is then set by forcing it down through the mortar; however, accurate placement is impossible. Wales are placed against the cylinders to support the wall as required.

The second wall form uses a trench filled with a slurry of clay and water to support the soil during construction. The *slurry trench* is excavated in short sections, 7 to 20 ft or 21.6 m long, and with a 3 to 5-ft or 1 to 11.5 m wall thickness, by a variety of drilling or digging methods depending on the patented equipment used. The essential feature of the process is that the excavation is kept full of a viscous mixture of water, clay (usually bentonite) and soil that supports the earth by fluid pressure. Gravel and even boulders can be removed by special clamshell buckets and sand can be pumped out by circulation of the mud. After excavation, prefabricated reinforcing steel is lowered through the slurry and positioned by rollers or projections against the walls of the hole. Concrete is treamied into the trench from the bottom up, displacing the slurry which is stored for re-use. Slurry trenches 5 feet thick and nearly 100 ft or 30 m deep have been constructed in a variety of soils. Generally the resulting wall is uniform enough that it can be used as a form for the permanent wall and sometimes as the permanent wall with only a thin facing.

SUPPORT SYSTEM / The bracing for very shallow excavations is supported by using the sheeting, soldier piles or cast-in-place walls as cantilevers. Excavations deeper than 8 to 10 ft or 3 m generally require support systems, Fig. 8.20.

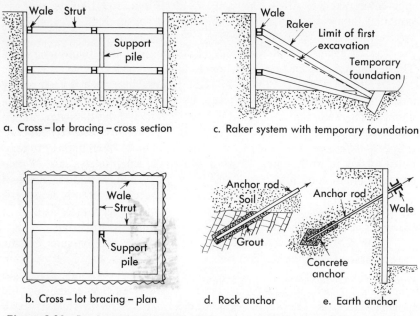

Figure 8.20 Bracing support systems.

Narrow excavations, such as trenches or small building excavations can be supported by horizontal columns or struts, Fig. 8.18b, c, 8.20a, b. If the excavation is wide, the system is tied together in both the vertical and horizontal directions to reduce the slenderness ratio (L/r) of the struts and to minimize buckling of a strut if excavating equipment should strike it. Sometimes the struts and vertical supports are tied together with diagonal members in the vertical plane to form trusses. In this way the upper struts can be used to support construction equipment and the excavation bottom can be free of obstructions.

Cross-lot bracing obstructs the site, and if the excavation width is several times the depth, a system of diagonal supports or *rakers* is convenient, Fig. 8.20c. The excavation is completed to final grade at the center leaving sloping banks to support the sheeting or soldier piles, as shown by the dotted line, Fig. 8.20c. The raker is then set in-place, reacting against a special foundation or against a completed portion of the foundation of the permanent

SEC. 8:5] EXCAVATION BRACING

structure. The next increment of excavation is made to the level of the second wale and raker, and these are installed. If the soil is very weak, and requires a very flat slope for support, the rakers are placed in trenches, leaving the greater part of the bank intact between them. Generally the raker angles are not steeper than 35 deg with the horizontal to minimize the upward component of thrust on the sheet piles or soldier piles.

Anchors or *tiebacks*, Fig. 8.20d, eliminate obstructions in the excavation inherent in rakers or struts. They consist of rods that extend well beyond any potential failure surface into firm undisturbed soil or rock. A number of systems are employed some with high tensile cables grouted into rock and pre-stressed against a wale, and others utilizing ordinary reinforcing steel. The design and construction of anchors is discussed in Chapter 10.

DESIGN OF A BRACING SYSTEM / A bracing system is a temporary structure that usually is removed when the job is completed. Actually it is a dam to keep water and soil from the building site in order that construction can proceed in the dry and consequently is often called a *land cofferdam* or simply a *cofferdam*. The latter term, however, is more often applied to temporary dams in open water. In most situations, since safety, ease of construction, and convenience are the most important considerations and economy of materials is less important, refined, accurate methods of analysis are seldom justified. An understanding of the nature of earth pressure against bracing is necessary, however, for even the most approximate design.

DEFORMATION AND PRESSURE / The earth pressure on the bracing system depends on the type of soil and the amount of deformation or yield of the bracing. Unlike the retaining wall, which is structurally a rigid unit against which the earth is placed after construction is completed, the bracing system is somewhat flexible and is called upon to support earth pressure while it is being constructed. The result is irregular deformation conditions and erratic variations in earth pressure with depth that cannot be calculated by theory alone.

The stresses on an element of intact soil before excavation, Fig. 8.21a, are drastically changed by excavation. The lateral stress, σ_h, is reduced, and the element bulges outward and subsides vertically. The combined effect is bulging in the lower part and subsidence in the upper part of the bank as shown in Fig. 8.21a. The soil is in tension near the surface, because of the downward and outward movement, and tension cracks may form. The crack location is typically between 0.4 and 0.7 times the face height, H, from the top. The first cracks to appear are close to the top; subsequent cracks are progressively further away.

The excavation bracing system restrains the bulge, keeps the crack closed, and minimizes the surface subsidence. If the bracing system is so rigid that no lateral deformation is possible, the earth pressure will be the at-rest, Fig. 8.21c. Usually the system deflects; this alters the total force and the pressure distribution. Excavation to the first support level allows the

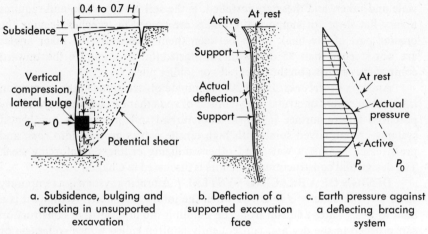

Figure 8.21 Deflection and earth pressure involved in a bracing system.

bracing to tilt, Fig. 8.21b, and the pressure to drop toward the active, Fig. 8.21c. The first support prevents appreciable further deflection at that point. Deeper excavation allows the system to bulge below the support and the earth pressure to drop. The pressure at the support increases correspondingly because the load is transferred by horizontal shear from the bulging zone. At the excavation bottom, the soil is restrained from bulging by the horizontal shear within the mass, and the pressure against the bracing is reduced.

The earth pressure diagram is irregular. The resultant force is somewhat more than the active and the location of the resultant is above the third point associated with the active pressure in a cohesionless soil. Pressures on actual systems computed from strut loads and the bending moments in the vertical sheet piles or soldier piles have generally verified the form of pressure diagram shown in Fig. 8.21c. The actual pressure, however, will vary considerably from one point to another because of the differences in construction sequence and the displacements of the support system.

The general approach to computing the earth pressure for bracing design is to find the resultant of active earth pressure (or at-rest pressure if the system is extremely rigid). The resultant is then increased empirically because the system does not permit the full deflection required for active pressure (or decreased if the at-rest pressure is the basis for design). The resultant is then re-distributed in a simplified pressure diagram, Fig. 8.22 and 8.23. Both the increase of the resultant above the active and the pressure distribution are based on the pressures deduced from measurements on similar full-sized bracing systems.

SEC. 8:5] EXCAVATION BRACING

DESIGN OF BRACING IN SAND / A semiempirical method was developed during subway construction in Berlin, Germany, to approximate measured earth pressures against excavation bracing in dense sand. A trapezoidal pressure diagram (Fig. 8.22a) was found to represent very closely the actual earth pressure and to be compatible with theory.[8:1] The resultant force of the assumed trapezoid is about 28 per cent greater than the active earth pressure. For loose sands a similar pressure diagram (Fig. 8.22b) has been suggested.[8:7] The resultant force in this case is 44 per cent greater than that of active conditions.

Other simplified pressure diagrams have been proposed, including rectangles and triangles. The differences between them illustrate the range in variability of the actual pressures.

For design purposes the trapezoidal pressure is assumed to be acting over each vertical section of the bracing. Strut loads may be calculated by assuming the vertical members to be hinged at each strut except the topmost and to be supported by a concentrated force at the bottom of the cut. Moments in the vertical members can be calculated by assuming them to be acting as simple beams between the assumed hinges. Ordinarily, members subjected only to bending can be designed with 25 per cent higher than usual working stresses, since bracing is temporary and excessive bending can usually be corrected long before an actual failure takes place. On the other hand, a very high factor of safety with respect to strut buckling should be used, since struts are occasionally damaged through careless installation or subsequent construction work and since buckling, unlike bending, occurs suddenly and can precipitate a chain reaction of buckling failures. Steel struts should be designed for the compressive stress allowed by the customary column formulas; and wood struts should be designed with two-thirds the customary compressive stress. Struts must be carefully cross-braced to prevent damage from impact of construction equipment. Strut design should

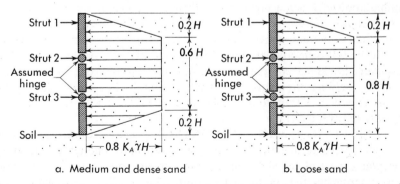

Figure 8.22 Design pressures for bracing in sand. Vertical members assumed to be hinged at each strut (except the uppermost), for computing strut loads. (*After Terzaghi and Peck.*[8:7])

include moments produced by the strut weight. All joints must be reinforced to prevent local failures.

DESIGN OF BRACING IN CLAY / During the construction of the Chicago subway, pressure measurements were made on the excavation bracing in clay, and on the basis of those measurements an empirical design pressure diagram (Fig. 8.23) was developed.[8:8] It is similar to the diagram for

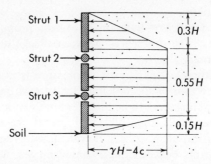

Figure 8.23 Design pressures for bracing in soft to firm clay. Vertical members assumed to be hinged at each strut (except the uppermost) for computing strut loads. (After Terzaghi and Peck.[8:7])

dense sands. The resultant force is 55 per cent greater than that developed in the active state. This pressure diagram applies to medium plastic, inorganic clays such as those found in the Chicago area and should be used in other situations only with caution. For soft clays the at-rest pressure with pressure with $K_0 = 0.6$ to 0.8 should be used. The struts and other members should be designed in the same manner as described for bracing in sands.

DESIGN OF BRACING IN PARTIALLY SATURATED CLAY / The computation of bracing pressures is difficult in partially saturated clay and similar soils exhibiting some cohesion. The maximum pressure would be the at-rest, the minimum, active. In computing the active pressure it is questionable if the full cohesion acts for a long period of time; therefore, the active resultant should be computed neglecting the tension zone. The resultant force lies between the active and at-rest resultants. For design the pressure is distributed in a trapezoid, similar to Fig. 8.22a so that the resultant is equal to the average of the at-rest and active resultants.

DEFLECTION AND ITS CONTROL / The recommended design pressures presume that the bracing system can move enough to mobilize a substantial part of the shearing resistance in the soil, so that the resultant force approaches that of the active state. The necessary movement comes from the deflection of the bracing system: the bending of the sheet piles, soldier piles, lagging and wales, the shortening of struts or rakers, the local

SEC. 8:5] EXCAVATION BRACING

readjustments of structural connections and the displacements of the raker supports.

Typical system movements are often more than are required to establish the design pressure although not uniform enough to develop active pressure distribution. Uncontrolled movements are sometimes as great as 1 to 1½ per cent of the excavation depth. As a result of the bracing movement the surface of soil adjacent to the bracing and even beyond the shear zone stretches, cracking pavements and damaging structures. At the same time the soil surface subsides within the shear zone, as shown in Fig. 8.21a.[8:9,8:10]

The deflection can be reduced by prestressing the bracing of the sheet piles or soldier piles. All movement will not be prevented, but the total movement of the soil mass will be reduced. If the prestress load is equivalent to the recommended design pressures for bracing systems, the mass deflections typically will be ¼ to ½ per cent of the excavation depth, similar to the deflections required for active earth pressure.

It is technically possible (but not always practical) to prevent any movement. In order to do this, the structural members in flexure are designed for a limiting deflection. The system is based on the at-rest earth pressure, and the supporting members are prestressed to that magnitude.

STABILITY OF BOTTOM OF EXCAVATION / When the bottom of the excavation extends into soft clay, there is danger of a failure by bulging upward. The weight of the soil adjacent to the excavation bears on the soil stratum at the level of the bottom of the excavation, and if the bearing capacity of that soil is not great enough to support the weight, a failure will occur. The failure zone can be approximated by drawing a 45° line from one bottom corner of the excavation and connecting it to a circular arc whose center is the opposite bottom corner as in Fig. 8.24. The zone of

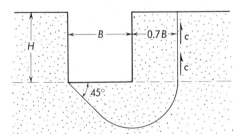

Figure 8:24 Stability of the bottom of an excavation in soft clay.

soil contributing to the failure has a width of 0.7 times the width of the excavation in this case. The downward force of this mass of soil is reduced by shear along the boundary of the mass, so its effective vertical force per linear foot of cut is $Q = 0.7B\gamma H - cH$. The pressure per square foot is

$q = \gamma H - cH/0.7B$. Since the bearing capacity of clay is given approximately by $q_0 = 5c$ (see Chapter 9), the safety factor SF of the bottom of the excavation can be expressed by the relation

$$SF = \frac{5c}{\gamma H - cH/0.7B}. \qquad (8:20)$$

A safety factor of at least 1.5 should be used. If sheeting extends below the bottom of the cut, the effective load is reduced by the shear along both sides of the imbedded portion of the piling.

Excavation bottoms in sand are ordinarily stable so long as the water level inside the excavation is not lower than the ground water level outside. As soon as the level inside is lowered by pumping, an upward seepage of water is created. If the difference in water levels is excessive, the bottom will heave, then become quick, and sand boils will appear. Seepage erosion can cause subsidence of adjoining structures. The analysis of seepage stability is given in Sections 4:2 and 4:3.

LOST GROUND / Both deformation of the bracing and heave of the excavation bottom are accompanied by a subsidence of the soil adjacent to the excavation. This is known as *lost ground*. While in many instances a moderate subsidence is of no consequence, in others even a slight movement of the soil can result in damage to adjacent buildings. In some instances lost ground results from running of sands in a "quick" condition, owing to excessive neutral stresses; in still other cases it may be caused by the slow, plastic creep of clays that are strong enough to stand in open excavations with no bracing at all.

Before any excavation that could cause damage to adjacent structures is begun, a survey should be made to determine the condition of those structures. The location, elevation, and size of all building cracks should be recorded and photographs secured. This information can do much to prevent annoying and expensive lawsuits that often arise during excavation work.

During construction, level readings should be made on points adjacent to the excavation to check the possibility of subsidence that might go unnoticed because of the usual din and confusion within the excavation. The bench mark should be located far enough from the excavation so that it will not subside and produce erratic level readings. A distance of at least five times the depth of the excavation from the excavation should be sufficient.

If subsidence is caused by deformation, it can be reduced by tightening the bracing system or by prestressing it against the soil. If the bottom heaves, it can be prevented by driving the sheeting deeper and by loading the portions of the bottom of the excavation not actually involved in construction with excavation waste or piles of sand. If creep of unbraced soil is the cause, it can be prevented by bracing. If running of sands is the cause, it can be

SEC. 8:6] ANCHORED BULKHEADS

prevented by drainage to relieve the neutral stress or by a water and sandtight bracing system.

8:6 Anchored Bulkheads

An *anchored bulkhead* is a special form of retaining wall of sheet piling that is widely used in waterfront construction. Because it is built from the surface down by driving the sheets, it is adapted to sites where the water level is so high or the soil immediately beneath is so soft that the cost of constructing a retaining wall of masonry or concrete would be prohibitive.

BULKHEAD CONSTRUCTION / The components of an anchored bulkhead are shown in Fig. 8.25a. The wall itself is of interlocking sheet piling. For very low walls, laminated creosoted timbers are occasionally used. Reinforced concrete sheets (often prestressed) 6 to 8 in. or 15 to 20 cm thick and 12 to 30 in. or 30 to 75 cm wide are sometimes employed in salt

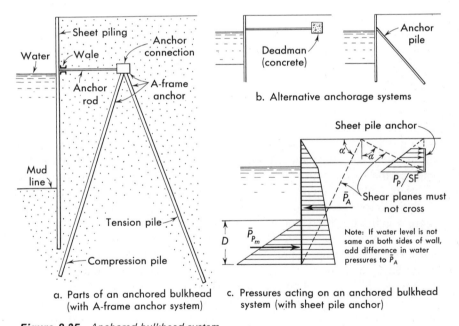

Figure 8.25 *Anchored bulkhead system.*

water locations because of their resistance to corrosion. The concrete must be dense and free of honeycombing in order to protect the reinforcement. Steel sheet piling is the most widely used because of its ease of handling and driving. Protection against corrosion, by coatings or by electrochemical methods (cathodic protection) is desirable, particularly in salt water.

The sheets are driven into the ground to provide lateral support at the bottom. The upper end is supported by the anchor system, consisting of the *wale*, *anchor rod*, and *anchor*. The wale is a continuous beam, usually a pair of channels or a W^F section, that ties the sheets together and carries their load to the anchor rod. Structurally it is simpler to place it on the outside of the wall. However, since it is vulnerable to damage from ships in that position, it is often placed on the inside. The anchor, or tie, rod connects the wale to the anchor. It is threaded for a nut at the bulkhead end or provided with a turnbuckle so its length can be varied and the sheeting aligned after installation. The anchor can take many forms, as shown in Fig. 8.25. The simple concrete *deadman* and the sheet pile deadman are used when the soils are strong enough and there is sufficient space. The A-frame and the single batter pile are employed when space is limited or when the upper soils are weak.

Two general methods of construction are found. When the bulkhead is built in open water and then the fill placed behind it, it is a *fill bulkhead*. When it is constructed in natural ground and then the earth removed from its face, it is a *dredged bulkhead*.

The wale and anchor rod are ordinarily placed as low as possible to minimize bending moments in the sheeting. Their depth is limited by low water; otherwise, their cost of installation becomes excessive.

BULKHEAD DESIGN[8:11,8:12,8:13,8:14] / The forces acting on a bulkhead are shown in Fig. 8.25c. The inner face of the sheeting supports active earth pressure. This is resisted by the wale and anchor near the top and by passive earth pressure distributed along the outside of the sheeting at the bottom. Such a condition is termed *free earth support* because the embedded section of the pile is free to rotate as the unsupported section bulges outward under load. If the sheeting penetrates deeply into a rigid soil, it is fixed against rotation, and the condition is termed *fixed earth support*. The free support analysis described below is satisfactory for many design problems. The more complex fixed earth support analysis is described elsewhere.[8:14]

The pressure acting on the inner face of the bulkhead is essentially the effective active earth pressure. Although the pressure distribution is probably distorted by arching, as shown in Fig. 8.9c, the magnitude and location of the resultant P'_A are probably not changed enough to affect the analysis.

In addition to the earth pressure there may be a difference in water pressure between one side of the sheeting and the other, caused by tidal fluctuations or rainfall infiltration. The unbalanced head can also cause a reduction in the passive pressure in the zone of embedment. Weep holes are sometimes installed to equalize the pressures but can lead to erosion of the backfill.

The depth of embedment is determined by the passive pressure required to support the toe. The effective resultant of passive pressure P'_p is divided by the safety factor, usually 2 to 3, to obtain a *mobilized* or *working* resistance

P'_{pm}. The depth of embedment, D, is found, so that the algebraic sum of the moments of the active resultant and the mobilized passive resultant is zero.

The wale reaction per foot is the difference between the active resultant and the mobilized passive pressures. The wale is designed as a uniformly loaded beam with support at the anchor rods. The anchor rod pull is determined by the accumulated wale load. The rod is designed conservatively because corrosion and physical damage can reduce its capacity.

The moments in the sheeting are determined from the distributed loading (Fig. 8.25c). The working stresses in bending for design of the sheet piling are ordinarily 10 to 20 per cent greater than for other structures, since the real bending moments are less than those computed because of arching.

The deadman anchor is essentially a second wall with passive pressure acting on its face. The working passive pressure is one-third to one-half the maximum. The anchor and the bulkhead must be separated a sufficient distance so that their shear zones do not interfere. The anchor wale is located at the depth of the resultant, which sometimes means a sloping anchor rod. If a concrete deadman is employed, its friction with the soil beneath provides added resistance. The A-frame anchors are designed on the assumption that the heads and tips of the piles are hinged. All anchor systems must be structurally flexible so that rotation and deflection of either the sheeting or the anchor will not develop secondary stresses that lead to failure.

Most bulkhead failures can be traced to inadequate depth of penetration of the sheeting, insufficient anchor resistance and poor structural connections; therefore special attention should be paid to these aspects of design.[8:9]

8:7 Underground Structures

Underground structures, including culverts, water and sewer pipes, tunnels and chambers such as powerhouses, must support both the horizontal and vertical pressures exerted by soil or rock. The analyses are similar to those for horizontal pressure alone: the limiting conditions can be mathematically approximated; the conditions between the limits must be evaluated by empirical adjustments based upon experience.

ELASTIC STATE / The at-rest state in a level mass of soil or rock was described in Section 8:1. The vertical pressures are defined by equations 8:1a and 8:1b. If the underground structure could be built by excavating the opening and immediately replacing the material removed with a perfectly rigid structure of exactly the same dimensions, the mass would remain in the same at-rest state of elastic equilibrium, Figs. 8.1 and 8.2.

Although it is physically possible to construct such a structure, it is far more likely that the mass will deform, and change the stresses in the vicinity of the opening. An approximate analysis for an unlined circular tunnel of radius R at a great depth z was developed by Kerisel, Fig. 8.26, assuming $K_0 = 1$.[8:15] The stresses are expressed in cylindrical coordinates, σ_r, the radial

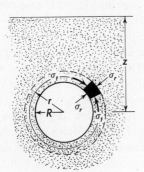 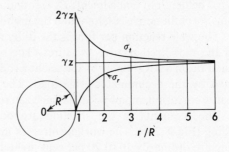

| a. Radial and tangential stress around a tunnel | b. Variation of radial and tangential stresses with distance from tunnel center |

Figure 8.26 Stresses around an unlined tunnel-elastic state.

normal stress and σ_τ, the tangential normal stress, with the origin at the tunnel center.

$$\sigma_r = \gamma z \left(1 - \frac{R^2}{r^2}\right), \qquad (8:21a)$$

$$\sigma_\tau = \gamma z \left(1 + \frac{R^2}{r^2}\right). \qquad (8:21b)$$

At the edge of the opening the radial stress is 0, and the compressive stress in the innermost ring of material surrounding the tunnel is $2\gamma z$. If the tangential stress, σ_τ, is less than the compressive strength, the tunnel will stand unlined with the stresses approximating elastic equilibrium; if not, a state of plastic equilibrium will develop.

More complex cases of elastic equilibrium have been analyzed and can be studied in the references, especially 8:16. In addition, two-dimensional models and the finite element approximation have been utilized in solving problems of complex shapes.

PLASTIC ANALYSIS OF STRESS / An analysis of pressures produced in a state of plastic equilibrium in the vicinity of the underground structure depends on the displacements along the soil-structure interface that define the extent of the shear zone. An approximate solution for both the horizontal and vertical pressures was developed by Terzaghi[8:3] based on the simple shear pattern proposed by Marston and Spangler[8:17] for computing the load on pipes in trenches. Although the real shear patterns are different, depending on the displacements on the soil-structure interface, the computed pressures and the significant factors defined by this approach are useful in the solution of many problems.

The zone of shear in a trench of width B is shown in Fig. 8.27a. The trench is filled with soil above a level surface of a structure at depth H. The

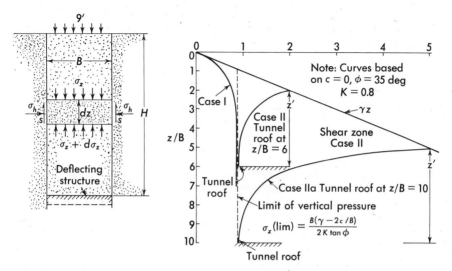

Figure 8.27 Plastic zone in a trench above a deflecting structure and vertical stresses on the structure roof.

structure and the soil above deflect downward, so that shear develops between the soil in the trench and the mass beyond. If the shear stress mobilized is equal to the shear strength of the soil, the equilibrium of a prism of soil at depth z can be described in terms of its weight dW, the vertical earth pressure on its upper and lower surfaces, and the shear strength produced by the lateral earth pressure $K\sigma_z$.

$$dW = \gamma B \, dz$$
$$dW + B\sigma_z = B(\sigma_z + d\sigma_z) + 2s \, dz$$
$$s = c + p \tan \varphi$$
$$\gamma B \, dz + B\sigma_z = B\sigma_z + B d\sigma_z + 2c \, dz + 2(\sigma_z K \tan \varphi) \, dz$$
$$\frac{d\sigma_z}{dz} = \frac{\gamma - 2c}{B} \qquad (\text{if } \varphi = 0).$$

At the upper surface $z = 0$ and the vertical stress is equal to any surcharge, q'. Solving the differential equation with these boundary conditions, the vertical earth pressure σ_z is

$$\sigma_z = \frac{B(\gamma - 2c/B)}{2K \tan \varphi} (1 - e^{-2K(z/B)\tan\varphi}) + q' e^{-2K(z/B)\tan\varphi} \qquad (8:22a)$$

$$\sigma_z = (\gamma - 2c/B)z + q' \qquad (\text{if } \varphi = 0). \qquad (8:22b)$$

The significance of the general equations (8:22a) and (b) is illustrated by Examples 8:1 and 8:2.

In Case I, a shallow trench is backfilled over a deflecting structure, and Case II, a deep trench is first backfilled and then the structure installed by tunneling.

Example 8:1

Case I—Backfill over structure. For Case I, assume a trench of width B and no surcharge. The soil properties are $c = 0$, $\varphi = 35$ deg, and $K = K_0 = 0.8$. The vertical stress is plotted in a dimensionless form, $\dfrac{\sigma_z}{\gamma B}$ and the depth in terms of z/B. The vertical stress, Fig. 8.27b, initially increases with depth at a rate of $\sigma_z = \gamma z$. The rate of increase in σ_z decreases rapidly as the weight of the prism of soil becomes supported by shear along its sides. Eventually at a depth of $z/B = m$ (about 4 in this case), the vertical stress approaches a limit:

$$\sigma_z(\lim) = \frac{B(\gamma - 2c/B)}{2\,K\tan\varphi}. \qquad (8:22c)$$

Below this level, only enough soil shear is mobilized in each increment of depth to support the increased weight, and no additional vertical pressure develops, regardless of depth. Above that level, the full soil shear strength is mobilized.

Example 8:2

Case II—Trench backfilled then tunnel excavated. If a trench should be backfilled, and then a tunnel excavated, the shear mechanism is somewhat different. Before the tunnel is excavated, the vertical stress is γz at all levels. After excavation and deflection of the tunnel roof the shear develops from the bottom up rather than the top down as previously described. The zone of full shear extends upward a distance of $m(z'/B)$ to the point where the vertical stress, σ_z is equal to the initial vertical stress at that level, γz. Above that level the mass is still in elastic equilibrium, $\sigma_z = \gamma z$, and there is no shear mobilized.

The height, $m(z'/B)$ of the shear zone in Case II is not sharply defined and is found by trial. This can be illustrated by Case II utilizing the same soil conditions as the first. Assume that the tunnel roof is at a depth of $z = 6B$. A preliminary estimate of the top of the shear zone can be made from the zone of full shear of Case I where $m = 4$. At that level the vertical pressure can be expressed by $\gamma z = \gamma(6B - 4B)$. This pressure is considered to be q' in Equation (8:22a). Below that level the stresses are computed using Equation (8:22a), considering that $z' = 0$ at the top of the shear zone: $z' = z - 2B$. The vertical pressure decreases rapidly through the shear zone until at $z = 6B$, the σ_z approaches a lower limit, $\sigma_z(\lim)$. This is the same limit as in Case I, and is defined by Equation (8:21c). If the assumed m were too small, the value of σ_z would not approach the limit; if the assumed m were too large, the limit would be approached well above the level of the tunnel. The height of the shear zone in Case II increases somewhat with the tunnel depth. If the tunnel depth should be greater, for example $z/B = 10$ (Case IIa) the limiting vertical stress immediately above the tunnel will be the same as for Case II. The height of the shear zone m, will be $z'/B = 5$; the vertical stress at the upper limit of the shear zone will be larger. If the tunnel depth is less than the height of the zone of full shear as found by Case I, the vertical stress is always less than γz, and the pressure on the tunnel is computed as for Case I.

LATERAL PRESSURE / The simple shear zone for the approximation of the lateral stress is shown in Fig. 8.28. Above the structure, plastic equili-

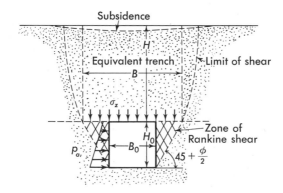

Figure 8.28 Lateral earth pressure on a deflecting tunnel.

brium is established as described in Equations (8:21a) and (b), and the vertical stress σ_z acts on the plane of the top of the structure at a depth of $z = H$. The height of the structure is H_0 and its width B_0. The inward deflection of the tunnel walls produces zones of plastic equilibrium in the mass on both sides, equivalent to the Rankine active state. The angle of the shear surfaces is $\alpha = 45 + \varphi/2$ and the width of each shear zone is $H_0/\tan(45 + \varphi/2)$. The deflection of the shear zone also induces a vertical subsidence adjacent to the subsiding structure surface. The total width of the zone of greatest subsidence, which is equivalent to the trench width B of Equation (8:21) is given by the following:

$$B = B_0 + 2H_0 / \left(\tan(45 + \frac{\varphi}{2})\right). \tag{8:23}$$

This width is used to compute σ_z that acts on both the top of the structure and on the top of the Rankine shear zones as a surcharge. The lateral pressure is computed by the Rankine expressions including surcharge.

The real shear zone extends beyond the limits of the hypothetical trench, as shown in Fig. 8.28. The trench approximation, however, has been useful for solving real problems.

DITCH CONSTRUCTION / The load on a pipe or small culvert laid in a narrow ditch, Fig. 8.29a is a simple case of the application of plastic equilibrium in a vertical direction. Generally the trench backfill is not as well compacted as the surrounding soil, and settles under its own weight. In addition the pipe subsides from its distortion under load and from settlement of the foundation. The movements approximate those assumed by Equations (8:22a) or (b). The total load carried by the pipe per unit of length is $B\sigma_z$—thus the narrower the ditch the smaller the load. Sometimes a small sub-ditch Fig. 8.29b, is used to minimize the pipe load if the walls of the sub-ditch can support the remaining load from above.

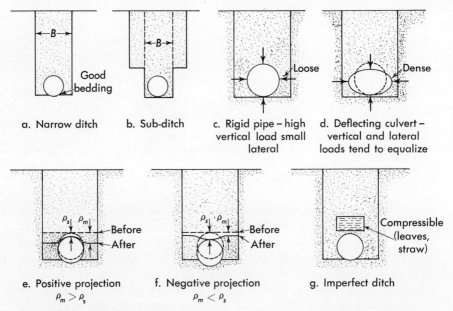

Figure 8.29 Buried conduits.

The lateral pressure depends on the deflection of the conduit as well as the compaction of the soil beside the pipe, if any. If the pipe is rigid and the soil beside it is poorly compacted, the lateral pressure will be negligible and the pipe will carry only the vertical load, Fig. 8:29c. If the soil beside the pipe is compacted and the pipe deflects, the lateral pressure will build up, Fig. 8.29d. In this way a thin flexible pipe can become structurally more stable in a deep ditch than a rigid pipe, provided its walls do not buckle.

SEC. 8:7] UNDERGROUND STRUCTURES

Considerable empirical data on the soil properties to be used in Equation (8:22) have been published.[8:17] Generally they suggest that K for ditches is close to K_A, probably due to lack of support of the ditch wall and the generally inadequate compaction of the backfill which allows enough movement for the active state to develop.

TABLE 8:4 / SOIL PROPERTIES FOR DITCH LOADS

Soil	$K \tan \varphi$
Sand, gravel maximum	0.19
Sand, gravel minimum	0.16
Moist silts, clays	0.13
Saturated clay	0.11

PROJECTION-CUT AND COVER / Culverts in highway fills and shallow tunnels are frequently constructed on the ground surface or in a large excavation, and then the backfill placed around them. This has been termed the *projection condition* in culvert design. Large tunnels or underground structures are similarly constructed in place or prefabricated and lowered into relatively wide excavations. This technique has even been used in underwater tunnels where a trench is excavated in a river bottom and the tunnel sections, built on land, are submerged and then connected in place.

The earth pressure on this form of construction is greatly dependent on the backfill compaction and on the relative deflection of the soil and the structure. The structure, and its foundation deflect downward, an amount ρ_s, while the adjoining mass and its foundation deflect ρ_m. Their difference compared to the mass deflection is termed the settlement ratio R_s.

$$R_s = \frac{\rho_m - \rho_s}{\rho_m}. \qquad (8:24)$$

If it is positive, termed *positive projection*, a plane initially level across the top of the structure is displaced as in Fig. 8.29e. If it is negative, the plane is displaced downward as in Fig. 8.29f. If they deflect equally the value of R_s is 0 and the plane remains plane.

When $R_s = 0$, the vertical stress is σ_z by Equation (8:22) with B equal to the ditch width. The load on the pipe is $D\sigma_z$ and the lateral pressure is $K_0\sigma_z$. When the ratio is sufficiently negative, the condition is essentially that of a narrower ditch whose width is the pipe diameter, D. The vertical pressure is less than for the case of $R_s = 0$. If the ratio is sufficiently positive the pressure will be equal to that of zero projection. However the pipe supports the load of the full width of backfill, $B\sigma_z$. Methods of analysis as well as empirical data relating the pressures on the structure to the rigidity of the structure and its foundation are given in Reference 8:17.

The load on a rigid conduit with a positive projection ratio can be reduced by placing a zone of loose soil or compressible material (leaves, rubbish, pine needles) above the pipe. This creates the shear conditions of Fig. 8.29f and the minimum pressure above the pipe. The permanence of this construction is uncertain, but limited experience with large culverts under deep fills suggests it may be effective for 10 to 20 years, depending on the rate of decomposition.

EARTH TUNNELS / True tunnels are constructed by excavating from below, similar to mining. The method of construction is governed by the ability of the soil to support itself temporarily during construction and on the pressures ultimately developed against the support system. Both are related to the tunnel depth and diameter, to the elastic properties and strength of the soil and to the ground water pressure. With the exception of some stiff clays and partially saturated formations most tunnels in soil require support both during construction and permanently. Above the water table in relatively firm soils, the excavation starts at the tunnel roof and works down in increments, Fig. 8.30a and b. The excavation is made

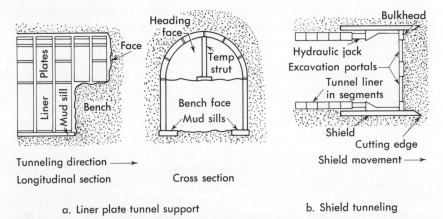

a. Liner plate tunnel support   b. Shield tunneling

Figure 8.30 Earth tunneling methods.

as large as possible and yet depend on the soil to support itself temporarily. That initial increment is then supported by a *liner plate*, a flanged steel plate that can be bolted to similar plates to form a continuous wall, or by *lagging* consisting of wood or steel planks, braced by steel beams. The next increment is then excavated and similarly supported. The process continues until the plates join forming a completed section of tunnel length. Sometimes the liner is stiffened with curved W^F beams or arches.

If the soil is very soft, a cylindrical temporary support called a shield is jacked through the soil, Fig. 8.30b. The face is usually supported by a bulkhead fitted with doors or portals that permit excavation of a limited

SEC. 8:7] UNDERGROUND STRUCTURES 379

portion of the face at one time. The permanent lining is constructed inside the shield as the excavation progresses. The shield is then jacked ahead, using the completed tunnel as the reaction. The lagging construction requires permanent support, ordinarily by a concrete lining. The liner plate temporary support sometimes is adequate for permanent support, but is ordinarily protected from corrosion by a concrete facing. If the liner plate is inadequate, a permanent reinforced concrete lining is necessary.

The elastic at-rest state is present in the undisturbed soil. This is altered by excavation, and if the soil is sufficiently strong, the compressive stress in the hypothetical ring around the tunnel will double [Equation (8:21b)], accompanied by inward deflection. Such a tunnel may not need support. An unlined tunnel through marl at Charleston, S.C. has been supplying raw water from a river diversion upstream for more than 40 years without distress. However, creep at high stress will cause the diameter to gradually reduce. In a few cases such tunnel *squeeze* has reduced the bore to a fraction of its original size. Supports or lining can prevent squeeze. However, because of creep, the pressures in softer clays approach the at-rest, with K_0 between 0.6 and 1.0.

The pressures on tunnels in sands usually require support during construction as well as permanently. Some inward deflection is inevitable with excavation and the shear approaches the conditions described by Fig. 8.28 and Equations (8:22a) and (8:23).

The art of tunneling described in References 8:16 and 8:18, and 8:19 is fascinating. The techniques of excavation must be intimately adapted to the soil and groundwater conditions. Temporary liners, drainage, soil stabilization and internal air pressure to partially balance water pressure are tools utilized by the tunnel builder or *mud hog* in boring through treacherous formations, and are beyond the scope of this text.

ROCK TUNNELS / Rock tunnels include many of the features of tunnels in soil. However, there are a number of significant differences. First the formation may be subject to high residual stresses from tectonic movements or erosion of overburden that obscure the stresses due to the weight of the rock. Second, the joints, bedding planes and shear zones are zones of weakness that destroy the continuity of the mass and focus shear in their direction. Third, the excavation process, often requiring explosives, may disturb the continuity of the mass, adding dynamic stresses and opening closed joints and fissures.

The tunneling operation depends on the hardness and soundness of the rock. Softer materials such as shales, schists and poorly indurated sandstones can be cut by tunnel boring machines if they are strong enough to stand without support until the machine passes. Hard rock must be drilled and blasted in short advances of the face, 10 to 20 feet at a time. If the rock is weak or shattered, temporary support consisting of steel ribs termed *sets* are wedged in place, and often lagging is placed between them to prevent local spalling. Better rock can be reinforced by rock bolts. Permanent lining of concrete is often required to support the pressures which develop or to reduce hydraulic friction in hydro-power tunnels.

The initial state in a level homogeneous rock mass without tectonic stresses is essentially at-rest. Tunneling imposes the increased tangential stress described by Equation (8:21b). Although the theoretical maximum tangential stress, σ_z is $2\gamma_z$, actual measurements indicate the maximum is only $1.5\gamma_z$ or slightly more. Strong, homogeneous rocks can support extremely high unconfined compressive stresses, making lining or other support unnecessary to maintain elastic equilibrium at great depths.

Example 8:3

Compute the maximum tunnel depth in a granite having an unconfined compressive strength of 20,000 lb per sq in. or 1400 Kg per sq cm, weighing 165 lb per cu ft or 2.65 g per cu cm.

$$20,000 \times 144 = 2 \times 165z$$
$$z = 8750 \text{ ft} = 2660 \text{ meters.}$$

If the rock is weakened by joints and other defects or is poorly indurated, plastic conditions develop in the vicinity of the tunnel. In closely jointed or fractured rock such as in a shear zone, the mass resembles a cohesionless soil. The mass deflects inward, with a shear pattern that is approximated by Fig. 8.28, and Equation (8:22a). Even rather closely spaced joints, Fig. 8.31a may resemble the simple model of Fig. 8.28 and the vertical tunnel load can be estimated in that way. The margin of uncertainty is greater, however, because K in a rock with high internal residual stresses is usually indeterminate.

When the joint spacing is less than half the tunnel width, the pressures on a tunnel lining can be estimated from the geometry of the joint blocks above the opening.[8:16,8:18] The shear resistance of the joint surfaces must be estimated or determined by in-place tests. The equilibrium of each system of blocks is evaluated, utilizing the most realistic appraisal of the possible displacements. If the joints are tight and high internal stresses cause high friction across them, possibly no pressures will develop. This would be analogous to a strong arch ring around the tunnel, Fig. 8.31b, that supports the load. More commonly the joint patterns, with some open joints, permit wedge-like masses to drop, or portions of the mass to slide down partially resisted by friction. Typical shear zones and the loads to be resisted are shown in Fig. 8.31c, d and e.

In rocks with high residual internal stresses, local concentrations of tangential stress at the rock surface cause progressive *popping* of wedge-shaped segments of rock. The popping is aggravated in zones of poorer induration or where the rock has been weakened by blasting. Some weak rocks, such as

SEC. 8:7] UNDERGROUND STRUCTURES

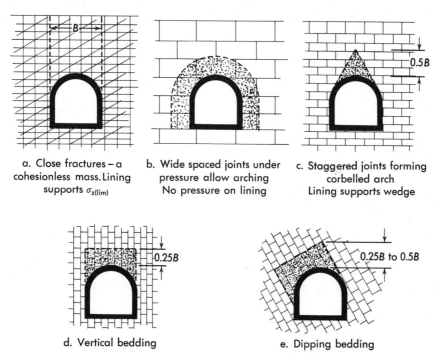

a. Close fractures – a cohesionless mass. Lining supports $\sigma_{z(lim)}$

b. Wide spaced joints under pressure allow arching No pressure on lining

c. Staggered joints forming corbelled arch Lining supports wedge

d. Vertical bedding

e. Dipping bedding

Figure 8.31 *Pressure of rock on tunnel lining.*

shale, creep under load causing squeezing and a reduction of the tunnel bore. A few rocks, such as shale, expand on exposure to air, causing increased pressures, progressive failure or squeeze.

ROCK BOLTS / Rock bolts have revolutionized underground construction, and to a lesser degree surface construction, in rock.[8:19,8:20] They enhance the ability of the rock to support itself and reduce the need for tunnel linings, beams and other supporting members.

A typical rock anchor is shown in Fig. 8.32a, installed in a hole drilled into the rock. The inside end may be fitted with an internal wedge that is forced against the hole bottom by driving the anchor against the rock or with an expansion shell that is forced against the hole walls by screwing the anchor rod into it. Simple anchors also can be grouted into the hole. After the bolt is anchored, it is tensioned by a nut on the face end that squeezes the rock together, in simple precompression.

The rock bolts support the mass by a number of related mechanisms. Essentially they impose a radial pressure, σ_r, or confining stress on the unrestrained surface. This confining stress increases the shear strength of the rock, Fig. 8.32b and develops friction between blocks of jointed rock. If the rock is horizontally bedded, the increased stress on the bedding planes produces friction so that the laminated mass acts as a beam. If the strata are

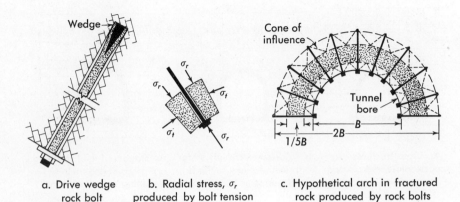

a. Drive wedge rock bolt
b. Radial stress, σ_r, produced by bolt tension
c. Hypothetical arch in fractured rock produced by rock bolts

Figure 8.32 Rock bolting.

badly fractured, the blocks are wedged together to form a continuous arch ring, Fig. 8.32c. Loose rock prisms can be held in place by suspending them from the mass above. Coherent masses are created by tying the joint blocks together.

The bolt length depends on the fracture patterns, the thickness of the beam or arch to be formed, and on the soundness of the rock at the point of wedging. Typical bolts are 4 to 10 feet long, but to form an arch, their length should be approximately half the tunnel width. The spacing and pattern depend on the fractures and the amount of stress to be imparted in the rock.

In closely fractured rock or in formations that spall because of high stress, the bolts serve a dual purpose. They support a continuous protective cover, termed *lagging*, which restrains the blocks that would otherwise drop free and injure workmen. Small steel channels or even heavy wire fencing are used for this purpose.

The installation of a rock bolt system and any auxiliary support such as lagging is largely based on experience. An all-over bolting plan is adopted based on initial studies of the fractures. The final plan is developed in the field when the fracture system and its behavior can be observed directly.

REFERENCES

8:1 K. Terzaghi, "General Wedge Theory of Earth Pressures," *Transactions, ASCE*, **106**, 1941, p. 68.

8:2 K. Terzaghi, "The Mechanics of Shear Failures on Clay Slopes and the Creep of Retaining Walls," *Public Roads*, **10**, December, 1929.

8:3 K. Terzaghi, *Theoretical Soil Mechanics*, John Wiley & Sons, Inc., New York, 1943, p. 48.

8:4 Ibid p. 67.
8:5 J. Brinch Hansen, *Earth Pressure Calculation*, Danish Technical Press, Institution of Danish Civil Engineers, Copenhagen, 1953.
8:6 A. Caquot and J. Kerisel, *Traite de Mechanique des Sols*, 3rd ed., Gauthier-Villars, Paris, 1966.
8:7 K. Terzaghi and R. B. Peck, *Soil Mechanics in Engineering Practice*, 1st ed., John Wiley & Sons, Inc., New York, 1948, pp. 348–350.
8:8 R. B. Peck, "Earth Pressure Measurements, Chicago Subway," *Transactions, ASCE*, **108**, 1943, p. 1008.
8:9 G. B. Sowers and G. F. Sowers, "Bulkhead and Excavation Bracing Failures" *Civil Engineering*, January 1967.
8:10 M. S. Caspe, "Surface Settlement Adjacent to Braced Open Cuts," *Journal of the Soil Development and Foundation Division, Proceedings, ASCE*, **92**, SM4, July, 1966.
8:11 K. Terzaghi, "Anchored Bulkheads," *Transactions, ASCE*, **114**, 1954, p. 1243.
8:12 G. Tschebotarioff, "Design of Flexible Anchored Sheet Pile Bulkheads," 12th International Navigation Congress, Lisbon, 1949.
8:13 *U.S. Steel Sheet Piling*, Carnegie Illinois Steel Corporation, Pittsburgh.
8:14 F. E. Richart, Jr., "Analysis for Sheet Pile Retaining Walls," *Transactions, ASCE*, **122**, 1957, p. 1113.
8:15 J. Kerisel, *Proceedings of the Fourth International Conference on Soil Mechanics and Foundation Engineering*, London, 1957.
8:16 K. Szechy, *The Art of Tunneling*, Akademiai Kiado, Budapest, 1966.
8:17 M. G. Spangler, "Culverts and Conduits," Chapter 11, *Foundation Engineering*, McGraw-Hill Book Co., Inc., New York, 1962.
8:18 K. Terzaghi, "Rock Defects and Loads on Tunnel Supports," *Rock Tunneling with Steel Supports*, Commercial Shearing and Stamping Co., Youngstown, Ohio, 1946.
8:19 L. Obert and W. I. Duval, *Rock Mechanics and the Design of Structures in Rock*, John Wiley & Sons, Inc., New York, 1967.
8:20 H. K. Schmuck, "Theory and Practice of Rock Bolting," *Quarterly of the Colorado School of Mines* (Second Symposium on Rock Mechanics), Golden, Colorado, 1957.

SUGGESTIONS FOR FURTHER STUDY

1. References 8:1, 8:16, 17, 18, 19, 20.
2. M. E. Harr, *Foundations of Theoretical Soil Mechanics*, McGraw-Hill Book Co., Inc., New York, 1966.

PROBLEMS

8:1 A vertical wall 20 ft high has a backfill of sand whose $\varphi = 39°$, that weighs 110 lb per cu ft dry and 131 lb per cu ft saturated.
   a. Compute the active earth pressure diagram and the resultant, assuming the backfill to be dry.
   b. Compute the active pressure, assuming the water table in the backfill rises to the top of the wall. Compare with (a).

8:2 A vertical wall 30 ft high moves outward enough to establish the active state in a dry sand backfill.
   a. Draw the pressure diagram and compute P_A if the sand $\varphi = 37°$ and it weighs 104 lb per cu ft dry.
   b. Compute the pressure and the resultant on the assumption that the wall does not move at all.

8:3 A wall 40 ft high retains sand. In the loose state the sand has a void ratio of 0.67 and a φ of 34°. In the dense state the sand has a void ratio of 0.41 and a φ of 42°. Which would produce the lesser resultant of active pressure, the loose or the dense state? Which would produce the greater resultant of passive pressure? How much difference is there in the resultants for both conditions?

8:4 A wall 25 ft high retains sand weighing 98 lb per cu ft dry and 123 lb per cu ft saturated. The water table is permanently 10 ft below the top of the wall. Estimate φ of 38°.
   a. Compute the effective and total active earth pressure diagrams.
   b. Find the location of the resultant.
   c. How much reduction in overturning moment about the base of the wall would occur if the ground water level could be lowered to the bottom of the wall?

8:5 A vertical wall 25 ft high has a soft clay backfill. The clay weighs 110 lb per cu ft and its strength c is 750 psf.
   a. Compute the at-rest pressure, draw the pressure diagram, and find the resultant.
   b. Compute the active pressure and draw the pressure diagram. Find the resultant, neglecting the tension because of cracks.
   c. How much is the overturning moment caused by earth pressure reduced by allowing the wall to yield enough to establish the active state, neglecting tension?
   d. How much increase in overturning moment would occur if water got in the tension cracks?

8:6 Derive the expressions for active and passive pressures of a partially saturated clay whose shear strength is expressed by $s = c' + p \tan \varphi'$.

8:7 An anchor consisting of a sheet pile wall 10 ft high is embedded in a partially saturated clay whose $\varphi = 19°$, $c = 1800$ psf, and

$\gamma = 112$ lb per cu ft. The top of the wall is at the ground surface.
  a. Compute the passive pressure, draw the pressure diagram for a 1-ft width section of wall, and find the resultant force.
  b. Determine the depth below the top of the wall at which the anchor rod should be attached.

8:8 Compute the active, passive, and at-rest pressures for a sand having a unit weight of 125 lb per cu ft, and $\varphi = 39°$. The sand is placed behind a bridge abutment that is 20 ft high. Draw the pressure diagrams for each condition, and find the resultant.

8:9 A retaining wall 12 ft high supports a level backfill of sand whose $\varphi = 37°$ and $\gamma = 120$ lb per cu ft. The back of the wall has a batter of 1 (horizontal) to 4 (vertical). Trucks park with their rear wheels 3 ft from the wall. Each truck's rear wheels carry 3000 lb, and there is one truck for every 10 ft of wall. Compute the moment of earth pressure about the heel of the wall per foot of wall.

8:10 A retaining wall 25 ft high supports a dry sand fill whose $\varphi = 34°$ and $\gamma = 117$ lb per cu ft. The top of the fill rises on a slope of 3 (horizontal) to 1 (vertical). The back of the wall slopes at an angle of 75° with the horizontal.
  a. Compute the resultant of the active earth pressure and its direction and line of action.
  b. Find the base width required to place the resultant at the third point of the base of a stone gravity wall with a top width of 18 in. and a masonry weight of 155 lb per cu ft.

8:11 A gravity retaining wall 19 ft high supports a level sand backfill. The top of the wall is 2 ft wide and the front face has a batter of 1 (horizontal) to 6 (vertical). Find the base width required so that the resultant of the earth pressure and the weight of the wall will pass through the outside third point of the base. Assume that φ is 35°, the unit weight of the sand is 120 lb per cu ft, and the unit weight of the concrete is 150 lb per cu ft.

8:12 A concrete cantilever retaining wall is 15 ft high and supports a sand backfill that rises on a slope of 2 (horizontal) to 1 (vertical). The sand $\varphi = 32°$ and $\gamma = 113$ lb per cu ft. The wall is 18 in. thick at the top and 30 in. at the base; its back side is vertical. The wall foundation is 2 ft thick and extends 2 ft back of the back face of the wall. The foundation soil is clay, with $c = 1500$ psf. The concrete weighs 150 lb per cu ft.
  a. Compute active earth pressure diagram and the resultant.
  b. Compute width of foundation so that resultant is within middle third.
  c. Check foundation for safety against bearing failure and against sliding. A sand cushion is placed between the foundation and the clay.

8:13 Recompute the resultants of problems 8:10, 8:11, 8:12, using the Coulomb analysis with $\delta = \tfrac{2}{3}\varphi$ and assuming that the resultant meets the wall at a depth of $2H/3$ below the top. Find the base width required and compare with that required for the wall, based on the Rankine analysis.

8:14 An excavation 30 ft deep and 40 ft square is to be made in sand with $\varphi = 40°$ and $\gamma = 122$ lb per cu ft. The bracing system is to consist of horizontal wood lagging supported by vertical 8-in. soldier beams.
   a. Determine the pressure diagram.
   b. Determine thickness of lagging if the soldier beams are 6 ft apart. Wood is southern shortleaf pine (dense structural select).
   c. The uppermost wale is at a depth of 4 ft and the others are spaced 10 ft apart. Design wales and struts, assuming only one strut in each direction at each elevation. Struts to be tied vertically at intersection in center of excavation.

8:15 A long excavation 35 ft deep and 25 ft wide in soft clay is supported by steel sheet piling, The clay $c = 600$ psf, and $\gamma = 110$ lb per cu ft. Three sets of wales are used. The uppermost is 6 ft below the ground and the other two are 12 ft and 24 ft below the first.
   a. Compute the pressure diagram.
   b. Select the steel sheet piling (see manufacturer's catalogs).
   c. Determine the size wales and the strut size and spacing. The struts are braced vertically at their centers.
   d. Check the stability of the bottom of the excavation if the sheet piling extends just below the bottom of the excavation. The clay extends 50 ft below the bottom of the excavation.

8:16 An anchored bulkhead retains sand weighing 120 lb per cu ft saturated and 110 lb per cu ft damp, with an angle of internal friction of 36°. Below the dredge line is sand weighing 130 lb per cu ft, with an angle of internal friction of 39°. Low tide is 26 ft above the dredge line, the tidal fluctuation is 6 ft, and the top of the bulkhead is 15 ft above low tide.
   a. Compute the active and working passive pressure diagrams, using a safety factor of 2 for passive pressure and assuming a maximum difference in water levels of 2 ft, front to back of the wall.
   b. Compute the minimum embedment, assuming the wale to be 2 ft above low tide.
   c. Determine the wale reaction per foot.
   d. Determine the sheet pile section to be used if the maximum bending stress is 24,000 psi.
   e. Determine the wale size and anchor rod diameter if one rod is used for every six sheets.

PROBLEMS

8:17 Design a sheet pile anchor system for problem 8:16, including the sheeting length, the anchor location, the anchor rod length, and the wale size.

8:18 An excavation 75 feet deep, and all above ground water, is to be constructed in a site that does not restrict excavation slopes.
   a. If the excavation is 25 ft square, compare the costs to brace the excavation or to excavate with $2(H)$ to $1(V)$ side slopes. (Use the unit prices for basement excavation, and backfill compaction around structures and the average values for steel sheet piling.)
   b. If the excavation is 200 ft square and the unit prices for earthwork and embankment compaction apply to excavation and later backfilling around the structure, compare the costs of a braced and open excavation.

8:19 A tunnel nominally 15 ft square is driven through a mass of damp sand whose unit weight is 125 psf, angle of internal friction is 33°, and whose capillary cohesion is 200 psf. The tunnel is 60 ft deep. Compute the vertical and horizontal pressures on the tunnel assuming $K = 0.7$.

CHAPTER 9
Foundations

A LARGE RETAIL STORE COMPLEX WAS BEING BUILT OF PRECAST, PREstressed concrete frames supported by individual square footings. After the building frame had been erected, the plumbing contractor excavated a trench adjacent to one of the footings. Suddenly the soil supporting that footing sheared into the trench. The footing and the column dropped, and pulled the prestressed beam attached to the column head down and out. Each adjoining column and beam in succession was pulled out of line. The column-to-beam connections tore loose, the beams fell and with them collapsed 50,000 square feet (5000 sq meters) of roof. Three workmen were crushed to death and several injured, all because of the bearing capacity failure of one small foundation.

A one-story brick building was constructed on a 10 ft or 3 m thick compacted sand fill placed above a 20 ft or 6 m thick stratum of organic clay. A test of the sand exposed in a footing excavation found the fill to be strong and incompressible, confirming the architect's decision to support the building on spread footings. Within a year the building had settled irregularly 6 to 12 in. (15 to 30 cm). Although the building was light, the organic clay consolidated under the fill load, which had been forgotten by all concerned. The settlement was irregular because of the uneven thickness of the clay and erratic lenses of relatively incompressible sand within the stratum.

These two cases illustrate the effect of foundation inadequacy on the behavior of the structure. In each case the footing was structurally sound and did not suffer damage; instead the critical factor in the foundation performance was the soil below—the structural member which was present even before the building was built and the ultimate support of everything above.

Foundation construction is one of the oldest of man's arts. The prehistoric lake dwellers of Europe built their homes on long wood poles driven securely into soft lake bottoms; the ancient Egyptians built their monuments on mats of stone resting on bed rock, and the Bible confirms this by stating that solid rock is more secure than shifting sand. The Babylonians found only deep alluvium in their flood plains between the Tigris and Euphrates, which settled under the weight of their cities. Buildings and walls were supported on mats of masonry, and adjacent parts of structures were provided with sliding connections so that they could settle different amounts without cracking apart. The artisans of the Middle Ages supported their masterpieces on inverted tables of stone, on rafts of timber, or on wood piling, following rules laid down by Roman builders before them. Until the twentieth century, however, the design of foundations was based entirely on past experience, ancient rules, and guesswork. Soil and rock mechanics have given the foundation engineer powerful tools with which he can analyze stresses and strains in the substructure, in the same way he does the superstructure, and formulate a rational design to fit the structure to the capabilities of the soil and rock below.

9:1 Essentials of a Good Foundation

The *foundation* is the supporting part of a structure. The term is usually restricted to the member that transmits the superstructure load to the earth, but in its complete sense it includes the soil and rock below. It is a transition or structural connection whose design depends on the characteristics of both the structure and the soil and rock. A satisfactory foundation must meet three requirements:

1. It must be placed at an adequate depth to prevent frost damage, heave, undermining by scour, or damage from future construction nearby.
2. It must be safe against breaking into the ground.
3. It must not settle enough to disfigure or damage the structure.

These requirements should be considered in the order named. The last two are capable of reasonably accurate determination through methods of soil and rock mechanics, but the first involves consideration of many possibilities, some far beyond the realm of engineering. During the long period of time the ground must support a structure, it may be changed by many manmade and natural forces. These should be carefully evaluated in choosing the location for a structure and particularly in selecting the type of foundation and the minimum depth to which it must extend.

The surface zones of many soils change volume regularly with the seasons. In much of the United States, frost action swells the ground in winter, which means foundations should be placed below the maximum depth of frost penetration. This can be determined by local experience or estimated from Fig. 4.22.

Clay soils, particularly those with high plasticity, shrink large amounts on drying. Such clays and some shales expand when wet. In regions having pronounced wet and dry seasons, the soils close to the ground surface expand

and contract. The outer walls move the most, while the inside, where the soil is protected from the sun and the rain, moves the least. In normally moist regions a prolonged dry spell can cause soil shrinkage and foundation settlement. Accelerated drying and settlement can be brought on by certain types of vegetation that extract moisture from the soil or by boilers and kilns that heat the soil abnormally. In very dry regions the opposite occurs. Added moisture from leaking pipes, irrigation, and even watering lawns can cause a desiccated clay to expand and lift a structure. In most cases the volume change becomes less with increasing depth, and if possible, foundations are placed below the volume-change zone.[9:1] Other methods of handling high volume change are given in Section 9:8.

The scour of river bottoms, especially during periods of flood, has resulted in a number of bridge failures. This occurs in two ways: the normal scour of the river bed and the outsides of bends due to the increased velocity of flow during floods, and the accelerated scour caused by the obstruction offered by the bridge pier to the flow. The first is a characteristic of the stream, and the amount can be estimated by correlating the height of water surface rise during a flood to the increase in depth of the river bottom. In many streams with a sand or gravel bottom a foot rise of the surface is accompanied by a foot or more of scour.[9:2] The accelerated scour can be minimized by good hydraulic design including streamlining the piers and aligning them with the direction of flow.[9:3]

Ice is a serious problem with bridge piers and marine structures such as docks and wharves. Impact of floating ice, carried by river or tidal currents, can cause serious damage. The weight of ice and frozen spray can become great enough to overturn light structures.

All foundations should be designed with allowances for future excavation and construction. Bridge piers must be located to provide clear navigation channels and must be placed deep enough to allow for future dredging of the channel. Construction operations in congested city areas often affect building foundations. The undermining of foundations by deep excavation and subway tunneling can result in the settlement and failure of buildings that have stood safely for years. In such cases the contractors responsible for the damage have been liable, but it is usually cheaper in the long run if such trouble is anticipated when the foundations are designed. Many building codes make a contractor liable for damage to adjacent property only if his excavation is deeper than 10 ft or 3 m.

Ground water is a factor in several ways. First, excavation below the ground water level is expensive and often hazardous because upward seepage loosens sands and tends to create a quick condition, and water standing over exposed clays may soften them. Second, when the ground water level is above the lowest floor, seepage into the structure and hydrostatic uplift become serious problems. Third, changes in the elevation of the water table have caused much trouble. In cities the water level drops because of drainage into sewers and deep excavations or because of pumping for water supply. This may increase building settlements by increasing soil stresses, or it may cause rotting of timbers formerly submerged well below the water level. On the other hand, if the water level rises through flooding, protracted rainfall, or broken water mains, soil strength is decreased and failures may occur. In some cases watertight structures such as empty concrete swimming pools and buried tanks have floated out of their normal locations because of the high water table that normally occurs in late winter and spring.

Underground cavities such as mines, caves, and sewers are hazards to foundations because they sometimes collapse from overload or structural deterioration. Piping or internal erosion of soil into leaky sewers or cavities likewise can cause trouble. If possible, foundations should be moved from these defects or corrective measures taken to make them harmless.

9:2 Stability—Bearing Capacity

The bearing capacity of a soil, often termed its *stability*, is the ability of the soil to carry a load without failure within the soil mass. It is analogous to the ability of a beam to carry a load without breaking. The load-carrying capacity of soil varies not only with its strength but also with the magnitude and distribution of the load. When a load Q is applied to a soil in gradually increasing amounts, the soil deforms, making a load-settlement curve similar to a stress–strain curve. When the critical or failure load, Q_0, is reached, the rate of deformation increases. The load-settlement curve goes through a point of maximum curvature, indicating failure within the soil mass. Different curves (Fig. 9.1) are obtained, depending on the character of the soil that is

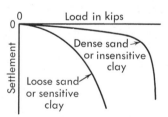

Figure 9.1 Load settlement.

loaded. Dense sand and insensitive clay usually show a sharp sudden failure, whereas loose sand and sensitive clay show a more gradual transition associated with progressive failure.

If the soil is observed during loading by means of a glass sided model or by an excavation adjacent to a full size foundation, it will be seen that there are usually three stages in the development of a foundation failure. First, the soil beneath the foundation is forced downward in a cone or wedge,

Fig. 9.2a. The soil below the wedge is forced downward and outward. Imaginary lines in the soil that were initially vertical now bulge outward like a barrel. Second, the soil around the foundation perimeter pulls away from the foundation, and surfaces of shear propagate outward from the tip of the cone or wedge, Fig. 9.2b. If the soil is very compressible or can endure large strains without plastic flow, the failure is confined to fan-shaped zones of local shear. The foundation will displace downward with little load increase: one form of *bearing capacity* failure. If the soil is more rigid, the shear zone propagates outward until a continuous surface of failure extends to the ground surface and the surface heaves, Fig. 9.2c. This is termed *general shear* failure. The failure can be symmetrical, particularly if rotation is restricted by a column attached to the foundation, or it can tilt as in Fig. 9.2d. Such a *bearing-capacity failure* is not common, but it almost always results in a complete failure of the structure.

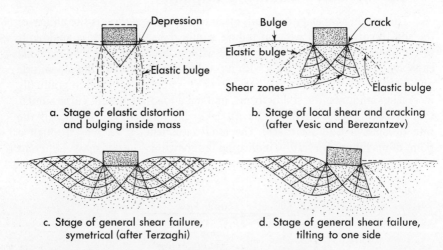

Figure 9.2 *Development of shear failure beneath a foundation.*

No exact mathematical analysis has been derived for analyzing such a failure. A number of approximate methods that have been developed are based on simplified representations of the complex failure surface and of the soil properties.

BEARING-CAPACITY ANALYSIS / A simple and conservative analysis was developed by Bell, extended by Terzaghi, and further modified by the authors. The method approximates the curved failure surfaces with a set of straight lines, as shown in Fig. 9.3.

A foundation having a width of B and an infinite length is assumed, similar to a long wall footing. At the moment of failure the foundation exerts a pressure of q_0, which is the *ultimate bearing capacity*, or simply *bearing*

SEC. 9:2] STABILITY—BEARING CAPACITY

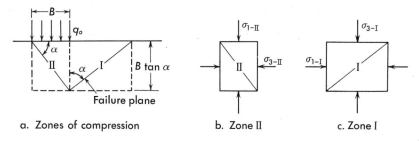

a. Zones of compression   b. Zone II   c. Zone I

Figure 9.3 *Assumed straight-line failure planes and prismatic zones of triaxial compression and shear beneath a uniform load q_0 of width B.*

capacity, of the soil. The soil immediately beneath the foundation is assumed to be in compression similar to a specimen in a triaxial shear test. The major principal stress on this zone, II, is equal to the foundation load q_0 if the weight of the soil beneath the footing is neglected. The minor principal stress on zone II is produced by the resistance of zone I to being compressed. Zone I is like a triaxial shear specimen lying on its side with the major principal stress horizontal. At the moment of foundation failure both zones shear simultaneously, and the minor principal stress on zone II, $\sigma_{3-\text{II}}$, equals the major principal stress on zone I, $\sigma_{1-\text{I}}$.

The minor principal stress on zone I is provided by the average vertical stress caused by the soil's own dead weight and any surcharge q'. The *surcharge* (Fig. 9.4) is any permanent confining pressure above the foundation level such as the weight of a basement floor or the weight of soil above the foundation level:

$$q' = \gamma D_f. \tag{9:1}$$

The height of the failure zone is $B \tan \alpha$, where α is the angle of the failure zone, $\alpha = 45 + (\varphi/2)$. The average minor principal stress due to soil weight is therefore $(\gamma B/2) \tan \alpha$. The total minor principal stress is

$$\sigma_{3-\text{I}} = q' + \frac{\gamma B}{2} \tan \alpha. \tag{9:2a}$$

If the minor principal stress is known, the major principal stress on zone I

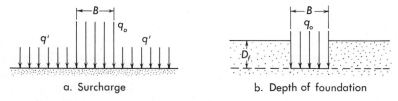

a. Surcharge   b. Depth of foundation

Figure 9.4 *Surcharge and depth of foundation.*

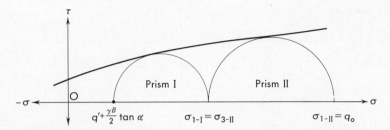

Figure 9.5 *Mohr's circle analysis of bearing capacity based on straight-line failure planes and prismatic zones of triaxial compression and shear.*

can be found graphically by Mohr's circle (Fig. 9.5). This is essentially passive earth pressure and it resists the bulging of zone II. Since this is equal to the minor principal stress on zone II, a second Mohr circle will give the major principal stress on zone II, the ultimate bearing capacity:

$$\sigma_{1-\text{II}} = q_0. \tag{9:2b}$$

The graphical analysis can be used in any soil, regardless of the shape of the Mohr envelope. If the Mohr envelope can be approximated by a straight line of the form

$$s = c' + p \tan \varphi', \tag{3:32b}$$

the ultimate bearing capacity can also be derived analytically from the trigonometry of the Mohr circle (Fig. 9.6):

$$\frac{\sigma_1 - \sigma_3}{2} = \left(\frac{c'}{\tan \varphi'} + \frac{\sigma_1 + \sigma_3}{2}\right) \sin \varphi'$$

$$\sigma_1 = \sigma_3 \left(\frac{1 + \sin \varphi'}{1 - \sin \varphi'}\right) + 2c'\left(\frac{\cos \varphi'}{1 - \sin \varphi'}\right),$$

$$\sigma_1 = \sigma_3 \tan^2 \alpha + 2c' \tan \alpha. \tag{9:2c}$$

$$\sigma_{1-\text{I}} = \left(q' + \frac{\gamma B}{2} \tan \alpha\right) \tan^2 \alpha + 2c' \tan \alpha,$$

$$q_0 = \sigma_{1-\text{II}} = \left[\left(q' + \frac{\gamma B}{2} \tan \alpha\right) \tan^2 \alpha + 2c' \tan \alpha\right] \tan^2 \alpha + 2c' \tan \alpha,$$

$$q_0 = \sigma_{1-\text{II}} = \frac{\gamma B}{2} \tan^5 \alpha + 2c' (\tan \alpha + \tan^3 \alpha) + q' \tan^4 \alpha. \tag{9:2d}$$

This is a general expression for the ultimate bearing capacity for any soil with a straight-line Mohr envelope. It can be used for a cohesionless soil by setting $c' = 0$ and for a saturated clay in undrained shear by setting $\varphi' = 0$, $c = c'$ and $\tan \alpha = 1$.

SEC. 9:2] STABILITY—BEARING CAPACITY

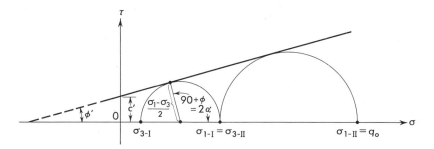

Figure 9.6 *Mohr's circle analysis based on a straight-line Mohr's envelope.*

GENERAL BEARING-CAPACITY EQUATION — TERZAGHI–MEYERHOF / The equation for bearing capacity can be rewritten in a simple form:

$$q_0 = \frac{\gamma B}{2} N_\gamma + c' N_c + q' N_q. \tag{9:3}$$

The symbols N_γ, N_c, and N_q are *bearing-capacity factors* that are functions of the angle of internal friction. The term containing factor N_γ shows the influence of soil weight and foundation width, that of N_c shows the influence of the cohesion, and that of N_q shows the influence of the surcharge. The values of these factors for different values of φ (or φ') are given on Fig. 9.7.

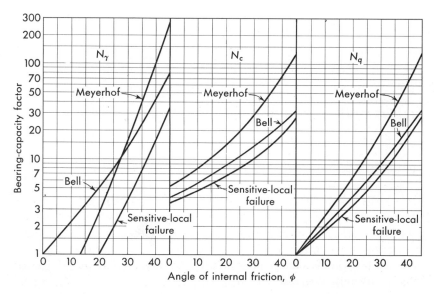

Figure 9.7 *Bearing-capacity factors for general bearing-capacity equation (Adapted from Meyerhof.*[9:5]*)*

This general expression was also derived by Terzaghi[9:4] from a more rigorous analysis of bearing capacity. It is based on approximating the surface of shear by a combination of straight lines and logarithmic spirals. The analysis was later improved by Meyerhof, but the results are expressed in the same form. Meyerhof's values for the bearing-capacity factors are given on Fig. 9.7.

Both the Terzaghi and the Meyerhof analyses assume the development of the full shear surface and complete shear failure. However, loose sands and highly sensitive clays fail by *local* or progressive shear when local cracking develops around the foundation or when the cone or wedge of soil under the foundation forms. Terzaghi suggested an empirical reduction to the bearing-capacity factors for this condition. The reduced Meyerhof factors, which apply to sands having a relative density of less than 30 or to clays with a sensitivity of more than 10, are also shown on Fig. 9.7. The factors of the simplified Bell analysis fall between the limiting values of the more accurate one.

RECTANGULAR AND CIRCULAR FOUNDATIONS / Both the Bell–Terzaghi and the Terzaghi–Meyerhof analyses assume an infinitely long foundation. When the foundation has a limited length, shear develops on surfaces at right angles to those previously described, and the bearing-capacity factors N_c and N_γ are changed. Correction factors to be multiplied by the bearing-capacity factors are given in Table 9:1, where L is the foundation length and B the width.

TABLE 9:1 / CORRECTION FACTORS FOR RECTANGULAR AND CIRCULAR FOUNDATIONS

Shape of Foundation	Correction for N_c	Correction for N_γ
Square	1.25	0.85
Rectangular $L/b = 2$	1.12	0.90
$L/b = 5$	1.05	0.95
Circular*	1.2	0.70

* Use diameter D for width B.

EFFECT OF SOIL PROPERTIES AND FOUNDATION DIMENSIONS / As can be seen by the general equation, the bearing capacity depends on the angle of internal friction φ (or φ'), the soil unit weight γ, the foundation width B, the cohesion c (or c'), and the surcharge q'. The angle of internal friction has the greatest influence because all three factors increase rapidly with only small increases in the angle.

If the angle of internal friction is zero, as for a saturated clay in undrained shear, the first and third terms become very small and only the

SEC. 9:2] STABILITY—BEARING CAPACITY

cohesion contributes materially to the bearing capacity. Thus for all practical purposes in a saturated clay,

$$q_0 = cN_c; \qquad (9:4a)$$
$$q_0 = 5.2c \text{ (for long footings)}; \qquad (9:4b)$$
$$q_0 = 6.5c \text{ (for square footings)}. \qquad (9:4c)$$

Both the first term and the third term in the equation depend on the unit weight of the soil. When the shear zone is above the water table (the bottom of the footing a height of about B above the water), the full soil unit weight is used in computations. When the water table is at the base of the foundation, the submerged unit weight, $\gamma' = \gamma - \gamma_w$, must be used in the first term. The effect is to reduce that part of the bearing by about one-half. If the water table is above the bottom of the footing, the surcharge weight is also affected.

The first term of the equation varies in direct proportion to the foundation width. This means that in cohesionless soils such as sands, the bearing capacity of small foundations is low and that of large foundations is very high. Estimating the bearing capacity of sand by small scale tests can be misleading because the bearing capacity of a full-sized foundation will be much greater. In saturated clays in undrained shear, foundation width has little effect on bearing capacity.

The third term is proportional to the surcharge q'. For a saturated clay where φ is zero and $N_q = 1$, the contribution of surcharge to bearing capacity is small. In a soil with a high angle of internal friction, a small amount of surcharge produces a large amount of bearing capacity.

EFFECT OF ECCENTRICITY / If the load is not applied concentrically, the overturning moment reduces the bearing capacity. According to Meyerhof,[9:6] the eccentrically loaded foundation responds as if it had a reduced width, B':

$$B' = B - 2e. \qquad (9:5)$$

In this expression e is the eccentricity of the resultant of the column load and the foundation weight, Fig. 9.8a. If there is eccentricity in two directions,

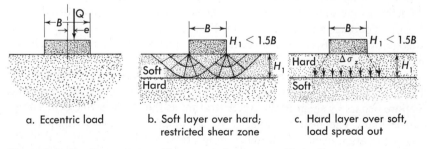

a. Eccentric load b. Soft layer over hard; restricted shear zone c. Hard layer over soft, load spread out

Figure 9.8 Bearing-capacity with eccentric loads or stratified soils.

both the length and width are reduced according to Equation (9:5). The value of q_0 computed from the reduced width is the average, and is used with the reduced width again in computing total capacity, Q.

An older method has been to compute the pressure distribution on the foundation produced by eccentric loading by assuming a linear variation from one side to the other, similar to the stress distribution in an eccentrically loaded column. The maximum pressure is then used in computing the safety factor. This approximation is reasonable for small eccentricities, but the corrected width of Equation (9:5) is more realistic.

EFFECT OF INCLINED LOADING / If the loading is not vertical the shear pattern is altered. The horizontal component of the load increases the lateral stress on the surrounding zone, leaving less resistance to support the lateral stress generated by the vertical component of the load. Meyerhof had proposed corrections to the bearing capacity factors to be used in computing the ultimate capacity under the vertical component of load. These are given in Table 9:2:

TABLE 9:2 / CORRECTIONS FOR INCLINED LOAD[9:6]

| Factor | D_f | \multicolumn{4}{c}{Inclination of Load From Vertical} |
		0	10°	20°	30°
N_γ	0	1.0	0.5	0.2	0
N_γ	B	1.0	0.6	0.4	0.25
N_c	0 to B	1.0	0.8	0.6	0.4

NONHOMOGENEOUS BEARING / If the soil is nonhomogeneous, the analyses are not directly applicable, but reasonable approximations can be made. When there are random variations in c' and φ' or thin repeating sequences of strata with different φ and c parameters within the hypothetical shear zone, Fig. 9.3, the mean of the c and φ can be used. If the range in variation is more than ± 20 per cent of the mean, somewhat higher safety factors should be used in design.

If a weak stratum overlies a strong one, Fig. 9.8b, the shear will be confined to the weaker material and the stronger will not be involved in the failure. The bearing capacity should be computed from the strength of the weaker stratum. Because the shear zone is restricted, the real bearing will exceed the computed value.

If a strong layer overlies a weak stratum, the strong layer spreads the load, reducing the bearing pressure on the weaker material, Fig. 9.8c. Failure occurs by shear in the softer stratum as the stronger one bends down under load. The bearing capacity is computed from the strength of the weaker stratum using a reduced bearing pressure $q = \Delta\sigma_z$ computed by the approximation of Equation (9:6).

9:3 Stress and Settlement

When a load, such as the weight of a structure, is placed on the surface of a soil mass, the soil deflects, resulting in settlement of the structure. This is not a unique property of soils but one shared by all materials. In the same way that the deflection of a beam may be the limiting factor in structural design, the settlement of loads on soil is often the controlling factor in foundation design.

Settlement of the soil produced by loading comes from two sources: the change in void ratio of the soil (or rock), and the distortion or change in shape of the soil immediately beneath the load. The first is termed *compression settlement*; and the second, *distortion settlement* or *contact settlement*.

Both the compression and distortion settlements depend on the stresses produced in the soil by the foundation or other surface loads. By making simplifying assumptions about the physical properties of the soil, the stresses can be computed by the theories of elasticity. The settlements are then found from the stresses, using the physical properties of the soil determined by laboratory tests.

STRESSES DUE TO SOIL WEIGHT / The initial vertical effective stress in a soil mass before a structure is built is approximately the weight of the soil minus the neutral stress. At a depth of z in a homogeneous soil from 7:1b,

$$\sigma'_{z0} = \gamma z - u. \tag{8:1b}$$

If the soil consists of different strata, each with a different unit weight, the vertical stress at any level is equal to the sum of their loads minus the neutral stress.

Changes in the neutral stress can play an important part in the settlement of a structure.

As was described in Section 7:1 lowering of the water table can increase the effective stress and produce settlements comparable to those produced by the weight of a building. Typically lowering the water table 2 to 3 ft is equivalent to the load of a one-story building (see example 7:1). Serious settlements, extending far beyond the building site have been produced by construction drainage and must be included in any settlement evaluation.

STRESSES DUE TO SURFACE LOADS / When a load is applied to the surface of the soil mass, the vertical stress within the mass increases. If the soil were a series of independent columns, the load would be supported by the column immediately beneath it and the others would feel no change. The soil, however, is a coherent mass with the columns of soil interconnected elastically. Load at one point is transferred throughout the mass, spreading laterally with increasing depth.

As a very crude approximation, it can be assumed that load spreads through the soil as though it were supported by a flat-topped pyramid, as shown in Fig. 9.9. The sides of the pyramid are sloped 2 (vertical) to 1

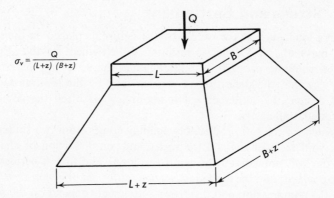

Figure 9.9 Approximate method for computing the average increase in vertical stress at a depth z beneath a rectangular foundation with dimensions L x B. Foundation is assumed to be supported by a pyramid of soil whose sides slope at 2 (vertical) to 1 (horizontal).

(horizontal), which means that the base of the pyramid becomes 1 ft larger in length and width for each foot increase in depth. The average stress increase in the soil at any depth z beneath a foundation whose dimensions are L and B and which has a load of Q and a pressure of q is

$$\Delta\sigma_z = \frac{Q}{(L+z)(B+z)} = \frac{qLB}{(L+z)(B+z)}. \qquad (9:6)$$

This approximation is useful in preliminary studies of settlement. It can be misleading because it fails to show the variation in stress at a uniform depth, and it does not indicate any stresses beyond the pyramid.

A more accurate representation of the stress distribution can be obtained from various theories of elasticity. These show that a load applied to the soil increases the vertical stress throughout the entire mass. The increase is greatest directly under the load, as shown in Fig. 9.10, but extends infinitely far in all directions. As depth increases, the concentration of stress directly beneath the load decreases, but at any depth, if the increases in stress were to be integrated over the area to which they applied, the total force would equal the applied load Q. Near the surface the stress distribution depends on the size of the loaded area and on the contact pressure distribution, but at depths greater than about twice the width of the loaded area the stress distribution is practically independent of the way Q is applied.

Many formulas based on the theory of elasticity have been used to compute soil stresses. They are all similar and differ only in the assumptions made to represent the elastic conditions of the mass and the geometry. One of the most widely used formulas is that published by Boussinesq, a French mathematician, in 1885 and adapted to soil engineering by Jurgenson.[9:7] He assumed a homogeneous, elastic, isotropic mass that extended infinitely

SEC. 9:3] STRESS AND SETTLEMENT

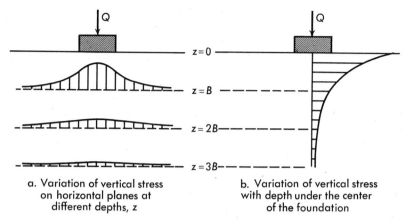

a. Variation of vertical stress on horizontal planes at different depths, z

b. Variation of vertical stress with depth under the center of the foundation

Figure 9.10 Vertical stresses in a soil mass due to a load Q applied to the ground surface by a square foundation of width B.

in all directions below a level surface. A concentrated load of Q is applied to the surface of the mass, and the increase in vertical stress, $\Delta\sigma_z$, at a depth z and at a horizontal distance of r from the point of application of Q is calculated by the formula:

$$\Delta\sigma_z = \frac{3Q}{2\pi} \frac{z^3}{(r^2 + z^2)^{5/2}}. \tag{9:7}$$

Westergaard in 1938 published an analysis that more closely represents the elastic conditions of a stratified soil mass. He assumed a homogeneous, elastic mass reinforced by thin, non-yielding, horizontal sheets of negligible thickness. The formula for the increase in vertical stresses produced by a concentrated surface load on a compressible soil (with Poisson's ratio = 0) is

$$\Delta\sigma_z = \frac{Q}{\pi z^2 [1 + 2(r/z)^2]^{3/2}}. \tag{9:8}$$

Both equations can be used to compute the stress increase caused by a footing if the depth z is greater than about twice the footing width B. For shallower depths the foundation pressure must be integrated over the foundation area to give the stress increase. The results of such integrations are presented in the charts of Fig. 9.11, 9.12, 9.13, and 9.14. The first two give contours of equal stress beneath foundations having widths of B and which exert a uniform pressure of q on the soil surface. The left side of each chart is for an infinitely long foundation and the right side for a square foundation. The depth and horizontal distances are expressed in terms of the foundation width B. The stress contours are expressed in fractions of the foundation pressure q. When the foundation is rectangular, the chart for a square foundation can be used with little error by assuming $B = \sqrt{A}$, where A is the foundation area.

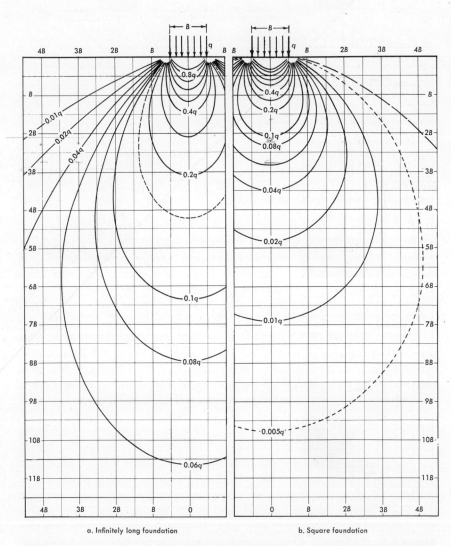

Figure 9.11 Contours of equal vertical stress beneath a foundation in a semi-infinite elastic solid — the Boussinesq analysis. Stresses given as functions of the uniform foundation pressure q; distances and depths given as functions of the footing width B.

SEC. 9:3] STRESS AND SETTLEMENT

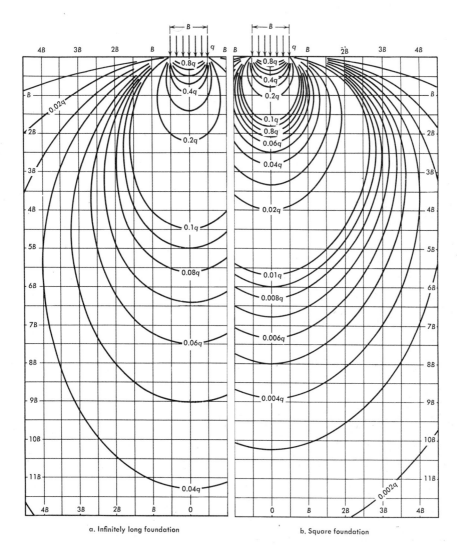

a. Infinitely long foundation

b. Square foundation

Figure 9.12 *Contours of equal vertical stresses beneath a foundation in a semi-infinite homogeneous thinly stratified material — the Westergaard analysis. Stresses given as function of the uniform foundation pressure q; distances and depths given as functions of the footing width B.*

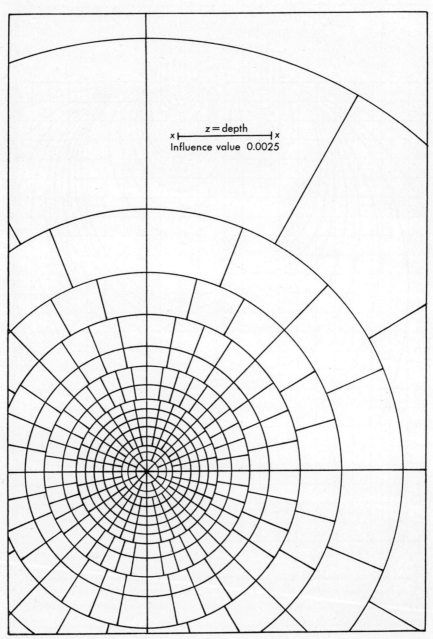

Figure 9.13 Influence chart for computing vertical stresses beneath a uniformly loaded foundations on a semi-infinite homogeneous isotropic elastic soil — the Boussinesq analysis. (Adapted from N. Newmark; chart courtesy of Soil Engineering Laboratory, Georgia Institute of Technology.)

SEC. 9:3] STRESS AND SETTLEMENT 405

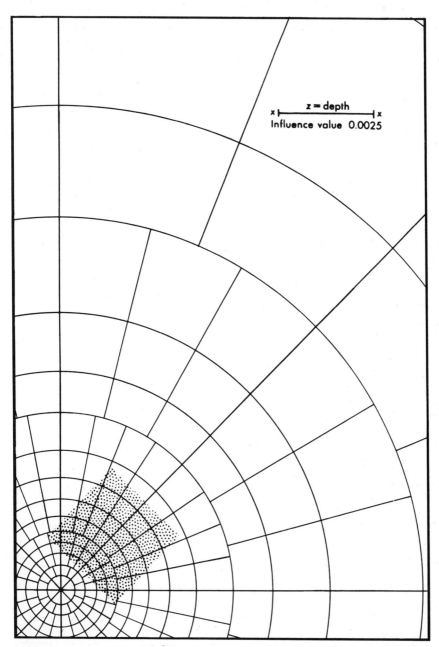

Figure 9.14 Influence chart for computing stresses beneath a uniformly loaded foundation on a semi-infinite homogeneous thinly stratified soil — the Westergaard analysis. (Adapted from N. Newmark; chart courtesy of Soil Engineering Laboratory, Georgia Institute of Technology.)

Figures 9.13 and 9.14 are circular charts originally devised by Newmark.[9:9] The foundation is drawn on tracing paper to such a scale that the depth z, at which the stresses are to be computed, is equal (on the same scale) to the key line x-x on the chart. The paper is placed over the chart so that the point at which the stresses are to be computed is at the circle center. The number of squares covered by the foundation are counted. This number, multiplied by the foundation pressure and the chart influence value 0.0025, gives the stress increase at that depth.

Example 9:1

Compute the stress increase at a depth of 10 ft and at 8 ft from the center of a footing that is 10 ft square and which exerts a stress of 3000 psf on a stratified soil.

1. Using Fig. 9.12, right side: The depth of the point, 10 ft, equals B. The horizontal distance, 8 ft, equals $0.8B$. From the chart, the contour is 0.09.

$$\Delta\sigma_z = 0.09 \times 3000 = 270 \text{ psf.}$$

2. Using Fig. 9.14: The footing (drawn to scale) covers 36 squares.

$$\Delta\sigma_z = 36 \times 0.0025 \times 3000 = 270 \text{ psf.}$$

When several loads act simultaneously, such as the footings of a building, the total stress increase at a point is the sum of the stress increases produced by each load acting independently.

The digital computer makes it simple to compute the stress of a point produced by any combination of loads. Loads distributed over areas are reduced to a number of appropriately spaced concentrated loads producing the same total force, for which Equations (9:7) and (9:8) directly apply. The results can be printed in tabular form for use in settlement analysis or plotted graphically with stress as a function of depth for a given location or stress as a function of location for a given depth, similar to Fig. 9.10.

COMPUTING COMPRESSION SETTLEMENT / The compression settlement for each soil stratum is computed by using the average initial stress, average stress increase, and the stress–void-ratio curve for the soil found by the laboratory consolidation test. The settlements for all the compressible strata are then added to obtain the total for that point.

The average initial effective stress in each stratum is the same as the initial stress at the middle of the stratum because the stress increases in direct proportion to the depth. The average increase in stress, however, is not the same as the stress at the stratum middle because the stress increase-depth relation is not linear. If the stratum is thin and relatively deep, it is sufficient to use the middle stress as the average. If the stratum is thicker than the footing width and if its depth is less than twice the footing width, it should be divided into thinner substrata and the average stresses computed for each.

SEC. 9:3] STRESS AND SETTLEMENT

The initial void ratio corresponding to the initial average effective stress and the change in void ratio caused by the average stress increase are found directly from the stress–void-ratio curve, as shown in Fig. 9.15. If it is certain that the stresses exceed any preconsolidation load, then the compression index can be used to compute the void-ratio change, as given by Equation (3:16). The settlement is computed from the stratum (or substratum) thickness H by

$$\Delta H = \frac{H \Delta e}{1 + e_0}. \tag{3:14}$$

Compression settlement is often a slow process requiring years to develop fully, as shown in Fig. 9.15b. Estimates of the total settlement that will occur in any period can be made from the coefficient of consolidation of the soil by using Equation (3:19) and (3:20). Since different strata are likely to have widely differing rates of consolidation, each stratum must be analyzed separately and the total settlement at any given time found from the sum of their individual settlements.

Consolidation settlement is adapted to a tabular form of computation. The stresses, changes in void ratio, or percentage of settlement, and settlements are computed for each compressible stratum or substratum; the sum of these at each point is the total settlement. The procedure is not immediately adapted to a computer solution because the stress–void-ratio curve or stress–percentage settlement curve generally cannot be represented by a simple mathematical function. The range of virgin compression is the exception; however, few practical problems involve stresses entirely within the virgin range. By approximating the stress–void-ratio cure by short straight lines, computer computation is feasible. Tabular solutions, with the aid of a desk

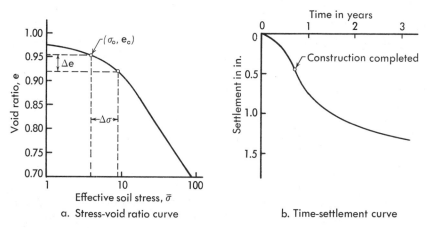

Figure 9.15 Typical soil stress — void ratio curve and time-settlement curve for a structure.

calculator are well suited to settlement analyses using the stress–void-ratio curves.

DISTORTION SETTLEMENT[9:4] / Distortion settlement occurs because of a change in shape of the soil mass rather than because of a change in void ratio. The soil immediately under the foundation deflects downward and bulges laterally like a barrel, permitting the foundation above to settle.

Saturated clays, most rocks, and similar elastic materials behave like a mass of gelatin or rubber when loaded because their modulus of elasticity is nearly constant regardless of the confining stress. If a uniform pressure of q per square foot is applied over their surface, both the loaded area and the adjacent unloaded surface will deform in a sagging profile (Fig. 9.16). This

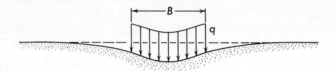

Figure 9.16 Profile of distortion settlement of a uniformly loaded flexible foundation on an elastic solid such as a saturated clay.

is similar to the way in which a bed spring will deform or deflect downward if one sits in its center. The shape of the deformation curve can be computed by methods of the theory of elasticity. For a loaded square area of width B, the settlement ρ of a corner and of the center are given by the formulas:

$$\rho_{cor} = \frac{0.42qB}{E}, \quad (9:9a)$$

$$\rho_{cen} = \frac{0.84qB}{E}. \quad (9:9b)$$

The soil is assumed to have constant volume as the load is applied and to be homogeneous to a depth of at least twice the footing width. Distortion settlement of loaded rectangular areas may be found approximately by the above formulas, by assuming $B = \sqrt{A}$.

Distortion settlement of uniform loads on a noncohesive soil results in a deflection curve that is concave downward (Fig. 9.17). The soil near the edge of the loaded area is unconfined laterally and so is pushed aside by the lateral pressure of the sand nearer the center of the area. The result is that the edges sag through lack of support. The sand at the center is confined by pressure from all sides; therefore it has a higher modulus of elasticity than the sand at the edges, which also means more settlement at the edge than at the center. So far no way has been devised for calculating the shape of the settlement curve. Experiments and observations indicate that the wider the loaded area, the flatter the curve at the center.

SEC. 9:3] STRESS AND SETTLEMENT

Uniform pressures are not common. They occur under large oil tanks where a relatively thin steel bottom rests directly on the soil or on a thin, concrete mat. Approximately uniform pressures occur over a large area when equally spaced columns carry their loads to a continuous, flexible foundation or to wide footings.

The most common way of applying a load to the soil or rock surface is through a relatively rigid foundation of concrete.

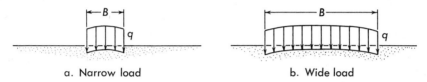

a. Narrow load   b. Wide load

Figure 9.17 Profile of distortion settlement of a uniformly loaded flexible foundation on a cohesionless soil.

The distortion settlement of rigid loads on saturated clays and similar materials whose modulus of elasticity is independent of confining pressure can be computed by the theory of elasticity. The following formula gives the distortion settlement ρ of square foundations of width B on a material with a homogeneous modulus of elasticity E. It can be used for other shapes by assuming $B = \sqrt{A}$, where A is the footing area:

$$\rho = \frac{0.6qB}{E}, \qquad (9:10a)$$

$$\rho = \frac{0.6Q}{E\sqrt{A}}. \qquad (9:10b)$$

This indicates that if the distortion settlement of an area with a load Q is to be reduced by one-half, the area must be increased four times. For equal contact settlement of loaded square areas under different total loads, the average pressure q must vary inversely as the total loads:

$$\frac{q_1}{q_2} = \frac{Q_2}{Q_1}. \qquad (9:11)$$

Computation of the distortion settlement of a rigid foundation on a cohesionless soil is more difficult because the modulus of elasticity varies with the state of stress.[9:10] A crude approximation useful for estimates is to visualize the foundation supported by a prism of soil having the same length and width as the foundation. The prism is divided into short segments and the initial vertical and horizontal stresses computed for each. For each load increment the added horizontal and vertical stress can be found by Boussinesq equations 8:19a and 9:7.

The average vertical and horizontal stress in each segment for each increment of foundation load is then computed. From these the applicable modulus of elasticity (which includes both distortion and void ratio change) is found from test data presented as in Fig. 3.35. The distortion settlement of each segment of the column for each load increment is computed; the sum of the settlements is the total foundation settlement. In a homogeneous sand, most of the settlement occurs within a depth equal to $1.5B$ for square foundations and $2B$ for rectangular foundations. It is sufficient to compute the settlements only for these depths.

CONTACT PRESSURE / The pressure acting between the bottom of a foundation and the soil is the *contact pressure*. It is important in the design of the foundation structure because it determines shear and moment distribution. As shown in Figs. 9.16 and 9.17, the distortion settlement of a foundation that exerts a uniform pressure on the soil is not uniform. The foundation, therefore, must be flexible so that it can conform to the settlement and keep the pressure uniform. If the foundation is rigid, it cannot conform; it finds more support from the points that deflect the least, and it bridges over the areas that deflect the most. As a result there will be a redistribution of pressure, as shown in Fig. 9.18. The pressure will be

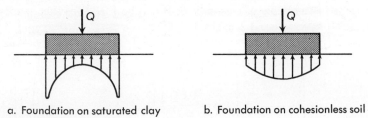

a. Foundation on saturated clay   b. Foundation on cohesionless soil

Figure 9.18 Contact pressure on the base of rigid foundations.

greatest at the outside edges of a foundation on an elastic soil such as a saturated clay and greatest at the center of a foundation on a cohesionless soil. For very wide foundations on a cohesionless soil the pressure is uniform at the center but falls off at the edges where the confinement of the sand is less.

9:4 Settlement Observations[9:11]

Whenever settlement exceeding $\frac{1}{2}$ in. or 1.3 cm is predicted for a structure, careful observations should be made to check the accuracy of the estimates. So many assumptions are made in settlement studies that their accuracy is often poor, and the only way such studies can be improved is by correlating the actual, measured settlements with the predicted.

SEC. 9:4] SETTLEMENT OBSERVATIONS

The first essential to accurate measurements is a stationary bench mark. This should be founded on bedrock if possible. In areas with deep soil such bench marks can be constructed by driving a 4 in. or 10 cm pipe into the soil, through all compressible strata. This pipe is cleaned out and the hole extended into the firm soil and rock strata below by boring with earth drills or a diamond drill (Chapter 6). A 2-in. or 5 cm pipe is placed inside the 4 in. or 10 cm outer casing and securely grouted into the firm stratum or rock. The inner pipe forms the bench mark and the outer pipe acts as a sleeve to insulate the inner pipe from the settling soil. The second essential is permanent reference points at different points on the structure. Readings of settlement should be taken once every few weeks during the construction period and once a year thereafter. Settlement can be read with a good engineer's level, or special water-level devices made for the purpose.

Settlement observations during construction often can give warning of trouble from other sources. Landslides, underground subsidences, and bearing-capacity failures usually begin with slow but gradually increasing settlement rates. Usually the trouble can be corrected before failure takes place if it is caught in time.

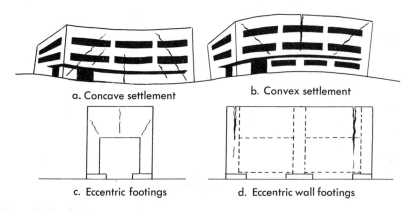

Figure 9.19 Settlement crack patterns.

Excessive settlement usually leads to building cracks and in some cases structural failure. The engineer should recognize the causes of different types of cracking in order to correct them before structural failure results. Uniform settlement will produce no cracking except of water and sewer lines into the structure. Differential settlement can produce cracks, tipping of the structure, or both. The crack pattern depends on whether the center of the building or its edges settle more. The diagrams in Fig. 9.19 illustrate types of failure that occur.

The concave settlement is the usual pattern for a uniformly loaded structure on a compressible soil. The zone of settlement is saucer shaped and extends well beyond the limits of the structure. Nearby buildings may be affected by the zone of settlement and develop new cracks. The convex pattern develops with wall-bearing structures or structures on loose sands.

Cracks often occur when footings are eccentrically loaded. This puts a bending moment into the base of the column or bearing wall and can cause failure. This is particularly true in outside bearing walls that are very close to the property line. The designer is tempted to make the footing wider inside the building than it is outside, and the result can be a crack, as shown in Fig. 9.19c and d.

Tipping is serious in narrow, tall structures such as chimneys and bridge piers. It is likely to occur when the soil compressibility is not uniform. It can also develop when the major cause of the settlement is a heavy load at some distance from the tall structure, as shown in Fig. 9.20. The sagging

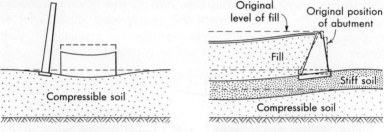

Figure 9.20 *Tilting of structures caused by adjacent heavy loads.*

settlement profile develops beneath the larger load and the tall but lighter structure tilts in that direction. Such settlement caused by the weight of an approach fill for a bridge resulted in the abutment tilting backward against the fill, as shown in Fig. 9.20b.

Not all cracks in structures are caused by settlement. Shrinkage of mortar or of concrete blocks and similar masonry units is a common cause of cracking. Plaster is likely to shrink differently from the wood or masonry base which supports it. Shrinkage cracks are usually vertical and horizontal and are of uniform width or become narrow at both ends. Thermal expansion and contraction are important causes of cracks in exterior walls. Such cracks can be identified by their opening and closing with temperature changes. Vibration, shock, and earthquakes can cause cracking. Usually these have an *x*-pattern at the ends of walls and a ✻, or +, at the middle.

Gage marks on cracks can aid in studying their movement and in their identification. A straight pencil line is drawn across the crack at right angles to it and gage points are set a definite distance apart (such as 2 in. or 5 cm) on each side of the crack. Measurement of the gage distance at regular intervals will show how much and in which direction the crack is moving.

TABLE 9:3 / CAUSES OF SETTLEMENT

Cause	Form of Mechanism		Amount of Settlement	Rate of Settlement
Structural Load	Distortion (Change in shape of soil mass)		Compute by Elastic Theory (Partly included in consolid.)	Instantaneous
Structural Load	Consolidation: Change in Void Ratio under stress	Initial	Stress–void-ratio curve	From time curve
Structural Load	Consolidation: Change in Void Ratio under stress	Primary	Stress–void-ratio curve	Compute from Terzaghi Theory
Structural Load	Consolidation: Change in Void Ratio under stress	Secondary	Compute from log time-settlement	Compute from log time-settlement
Environmental Load	Shrinkage due to drying		Estimate from stress–void ratio or moisture–void ratio and moisture loss limit-shrinkage limit	Equal to rate of drying Seldom can be estimated
Environmental Load	Consolidation due to water table lowering		Compute from stress–void ratio and stress change	Compute from Terzaghi Theory
Load independent (but may be aggravated by load) often environment related, but not dependent	Reorientation of Grains—shock and vibration		Estimate limit from relative density (up to 60–70%)	Erratic, depends on shock, relative density
Load independent (but may be aggravated by load) often environment related, but not dependent	Structural collapse—Loss of Bonding (Saturation, thawing, etc.)		Estimate susceptibility and possibly limiting amount	Begins with environmental change, rate erratic
Load independent (but may be aggravated by load) often environment related, but not dependent	Ravelling, Erosion into openings, cavities		Estimate susceptibility but not amount	Erratic, gradual or catastrophic, often increasing
Load independent (but may be aggravated by load) often environment related, but not dependent	Biochemical Decay		Estimate susceptibility	Erratic, often decreases with time
Load independent (but may be aggravated by load) often environment related, but not dependent	Chemical Attack		Estimate susceptibility	Erratic
Load independent (but may be aggravated by load) often environment related, but not dependent	Mass Collapse—collapse of sewer, mine, cave		Estimate susceptibility	Likely to be catastrophic
Load independent (but may be aggravated by load) often environment related, but not dependent	Mass Distortion, Shear-creep or landslide in slope		Compute susceptibility from stability analysis	Erratic, catastrophic to slow
Load independent (but may be aggravated by load) often environment related, but not dependent	Expansion—Frost, clay expansion, chemical attack (resembles settlement)		Estimate susceptibility sometimes limiting amount	Erratic, increases with wet weather

OTHER CAUSES OF SETTLEMENT / There are many causes of structural settlement, all of which must be considered in design and which should be evaluated in studying the possible effects of settlement on the structure. The major causes, distortion and consolidation, are directly related to the foundation load and are controlled by the foundation design. Consolidation may also be produced by loads induced by a changing environment that increases the effective stress by altering the neutral stress. Other forms of settlement are not directly produced by the structural load although possibly aggravated by it. Neither the amount nor rate of these other causes of settlement can be predicted in advance, although the susceptibility to settlement may be estimated. Table 9:3 summarizes the principal features of each form.

9:5 Allowable Pressure on Soil

After a foundation meets the requirements of location and minimum depth, two conditions remain that must be satisfied: First there must be adequate safety against a failure within the soil mass; and second the settlement of the foundation must not endanger the structure. It is obvious from the methods developed to analyze bearing capacity and settlement that these two conditions are independent of one another. For foundation design, however, it is desirable to know the maximum pressure that can be placed on soil without exceeding either of these two limits. This maximum is known as the *allowable soil pressure*, or *allowable loading* q_a.

PRESUMPTIVE BEARING PRESSURE / The oldest method of determining the allowable foundation pressure is to rely on past experience with similar materials in the region. Most engineers accumulate information on the success of their past designs, and these are used as a basis for future work. In many areas, such as the larger cities, the records of which design pressures were successful and which were not have been assembled and condensed in tabular form. These are called *presumptive bearing pressure* because it is presumed on the basis of past performance that the soil can support such a pressure without a bearing-capacity failure or excessive settlement. Most building codes include such tables, and they are often a helpful guide to local practice.

Table 9:4 gives presumptive bearing pressures based on the author's experiences for simple structures up to four stories.

Unfortunately the use of presumptive bearing pressures often leads to trouble. Most of the tables are based on experiences going back to the nineteenth century and on entirely different types of structures than are built today. The soil and rock characteristics are defined only by a description, and often the most important properties are not mentioned. Sometimes the building code table is a copy of that from some other city and does not reflect local practice at all. For example, one municipal code includes a

SEC. 9:5] ALLOWABLE PRESSURE ON SOIL

TABLE 9:4 / TYPICAL PRESUMPTIVE BEARING PRESSURES*

Very loose sand	Dry 0–1000 psf	Inundated 0–600 psf
Loose sand	Dry 1000–3000	Inundated 600–2000
Firm sand	Dry 3000–6000	Inundated 2000–4000
Dense sand	Dry 6000–12,000	Inundated 4000–8000
Soft clay	0–1500	
Firm clay	1500–2500	
Stiff clay	2500–5000	
Hard clay	5000–10,000	
Layered, laminated, fractured rock	10,000–30,000	
Massive rock, occasional seams	30,000–80,000	
Sound massive rock	80,000–200,000	

* (See Sections 2:9 and 2:11 for meaning of descriptive terms.)

bearing pressure for granite, although there is no igneous rock within 400 miles.

Finally, the table does not reflect the influence of the size and weight of the structure. As a result the use of the table value does not always mean a safe, economical foundation. The engineer has just as much responsibility in determining the allowable foundation pressure for his structure as he has in determining the size of a beam or thickness of a floor slab, and to rely only on experience can be disastrous. It is ironic that most of the foundation failures that the authors have investigated have involved designs that nominally conformed to the local building code.

LOAD TEST / The *load test* or *test plate* method of determining allowable soil pressure (Fig. 9.21) was developed because of the failures of

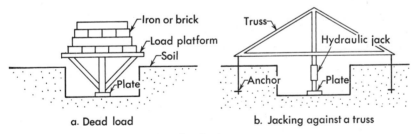

Figure 9.21 Methods for making plate load tests.

the design tables. Essentially the load test is a model test of a foundation. A small plate, usually one ft or 30 cm square or 30 in. or 90 cm in diameter, is placed on the undisturbed soil and is loaded in increments. The results of the test are presented in a load-settlement curve of the test plate. Properly conducted and correctly interpreted, the load test is a valuable *aid* to rational design, but as usually conducted, it is a waste of time and money and often leads to a dangerous sense of false security.

A pit is dug to the level at which it is desired to determine the allowable soil pressure. The pit width should be at least five times the width of the test plate. The minimum size of the test plate is one ft or 30 cm square but larger sizes are desirable. A load is placed on the plate, as shown in Fig. 9.21, by means of a loading platform weighted with pig iron or lead or by jacking with calibrated hydraulic jacks against a beam held down by earth anchors or weights. Loads should be applied in increments of one-fourth the estimated allowable soil pressure and increased until two times the estimated allowable soil pressure is reached for sands and gravels and until 2.5 times the estimated pressure is reached for clays. Settlement under each increment should be read to 0.001 in. or 0.025 mm, and must be referred to a bench mark beyond the limits of the possible settlement profile. Each increment should be maintained constant until the rate of settlement is less than 0.002 in. or 0.05 mm per hr, before adding another increment. The final increment should be maintained at least 4 hr before ending the test. A time-settlement curve (Fig. 9.22a) should be plotted on semilogarithmic coordinates for

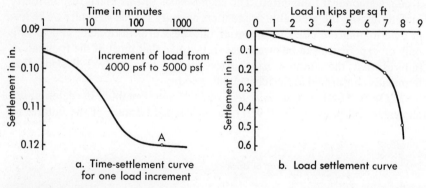

Figure 9.22 *Plate load-test results.*

each increment of load. The semilog plot will show a break into a straight, nearly flat line, point A. This point should be selected as the "ultimate settlement" under each load increment. The loads and their corresponding ultimate settlements should be plotted to form a load-settlement curve (Fig. 9.22b). A definite break or the intersection of tangents to this curve represents the ultimate bearing capacity of a foundation the same *size*, *depth*, and *location* as the test plate.

The load test is useless unless its results are interpreted in terms of a full-size foundation. Interpretation must be based largely on theory because so few field observations have been reliable enough to correlate foundation performance with load test results. Since bearing capacity of footings on a clay soil is independent of the width of the loaded area, the critical pressure determined by the load test is the same for all footing sizes:

$$q_0 \text{ (foundation)} = q_0 \text{ (load test)}. \qquad (9\!:\!12\mathrm{a})$$

In sands and gravels the bearing capacity increases in direct proportion to the width of the loaded area, and so the following correction should be made:

$$q_0 \text{ (foundation)} = q_0 \text{ (load test)} \times \left(\frac{B \text{ foundation}}{B \text{ test plate}}\right). \quad (9\text{:}12\text{b})$$

If the soil is not homogeneous to a depth of at least the width of the proposed foundation, the load test results are meaningless as far as bearing capacity is concerned.

The load test can be used to determine distortion settlement in a homogeneous soil, provided the soil is uniform to a depth of twice the footing width. For saturated clays, the distortion settlement at a given pressure per square foot varies directly with the width of the loaded area, or

$$\rho \text{ (foundation)} = \rho \text{ (test plate)} \times \left(\frac{B \text{ foundation}}{B \text{ plate}}\right). \quad (9\text{:}13\text{a})$$

The load test cannot predict compression settlement. Compression requires time, particularly in the more critical soils such as clays, and the few hours or days allotted to the load test can allow only a negligible part of the total volume change to take place. Furthermore, if the compressible soil extends to any depth, the stresses throughout the stratum resulting from the loaded test plate will be very small. The same pressure applied by a wide foundation will produce much greater stresses in the deep stratum and correspondingly greater settlement.

Example 9:2

Compute stresses in soil stratum 1 ft thick, 10 ft below the ground surface, due to (a) load test, 1 ft × 1 ft, pressure 3000 psf; and (b) foundation, 8 ft × 8 ft, pressure also 3000 psf. Use approximate method for computing stress, formula 9:6.

1. Load plate:

$$\Delta\sigma = \frac{3000 \times 1 \times 1}{(10 + 1)(10 + 1)} = 24.8 \text{ psf.}$$

2. Foundation:

$$\Delta\sigma = \frac{3000 \times 8 \times 8}{(10 + 8)(10 + 8)} = 590 \text{ psf.}$$

In cohesionless soils the compression and distortion settlement occur simultaneously. Most of the settlement in a thick deposit of sand or gravel occurs near the surface because the modulus of elasticity increases and the stress decreases rapidly with increasing depth. There are only limited data that relate the size of the loaded area to settlement in homogeneous cohesionless strata. The following expression is based on observed settlements of structures in natural sand deposits and of test foundations up to 4 ft or 1.3 m width in sands of uniform relative density.[9:12]

$$\rho \text{ (foundation)} = \rho \text{ (test plate)}\left(\frac{B \text{ (foundation)}}{B \text{ (plate)}}\right)^n \quad (9\text{:}13b)$$

The exponent, n, for many sands is between 0.5 and 0.7.

Rules for conducting load tests are included in many building codes and standard specifications. Too often blind observance of these rules leads to trouble because the rules are either based on typical conditions (and no site is typical) or on limited experience. The most important rules are: (1) use as large a load plate as practical; (2) make the tests in both good and bad soils; (3) correlate the results with accurate boring data; and (4) interpret the data with full understanding of the mechanics of bearing capacity and settlement. The load test cannot be the entire answer to the question of allowable foundation pressure because it supplies only a part of the data (bearing capacity and distortion) for a scale model of the proposed structure.

9:6 Rational Procedure for Allowable Pressure and Design

A rational determination of allowable foundation pressure is similar to the design of other parts of the structure. First, a trial pressure is assumed based on experience. Second, the trial pressure is checked by a bearing-capacity analysis to determine the safety against soil failure. Third, if it is safe, an analysis is made of the settlement produced, to see if it is excessive. Fourth, the trial pressure is revised to increase the safety, reduce the settlement, or improve the economy, depending on the analytical results.

In order to do this, the engineer must have accurate data on the soil below and the structure above the foundation. The soil data include the depth and thickness of the soil and rock strata, the level of ground water, and the physical properties of each soil, including its strength and compressibility. If the soil deposit is uniform, the analyses are based on the average properties of each material; if it is variable, the analyses are based on the worst combination of soil properties as determined from the tests.

The data required on the structure are the general features of the structure and its loads. The general features include the use or purpose, the elevations of the lowest floors and particularly basements and pits, the type of structural framing and its sensitivity to deflection, and the possibility of future additions. The load data required are the depth and extent of general excavation and filling and the dead loads and live loads on columns, including the amount of live load that is likely to be continuous. If the lowest floor is supported directly on the ground, its average sustained load is needed.

SEC. 9:6] RATIONAL PROCEDURE FOR ALLOWABLE PRESSURE

The *safety factor* of a foundation, *SF*, is the ratio of the ultimate bearing capacity q_0 to the actual foundation pressure q. The *safe bearing capacity* q_s is the ultimate bearing capacity divided by the minimum permissible safety factor *SFM*:

$$SF = \frac{q_0}{q}, \qquad (9:14a)$$

$$q_s = \frac{q_0}{SFM}. \qquad (9:14b)$$

$$q_a \leq q_s \qquad (9:14c)$$

The allowable foundation pressure q_a cannot exceed the safe bearing capacity, but it often is less.

SAFETY FACTOR / The safety factor required for design depends on how accurately the soil conditions and the structural loads are known and what hazards are involved in a bearing-capacity failure. Any future changes in the site, such as a rising water table or excavation adjacent to a footing that will reduce the surcharge, must be taken into account by the ultimate bearing-capacity equation or else included in the safety factor. For temporary construction work where a failure would be inconvenient but not disastrous, a safety factor of 1.5 is required. For most cases of structural design where there is reasonably accurate data on the soil and loadings, a safety factor of 2.5 is employed with dead load plus full live load. If a large part of the live load is not likely to develop, a minimum safety factor of 2 is permissible. When the conditions are questionable, a safety factor of 4 is sometimes warranted.

PERMISSIBLE SETTLEMENT / The settlement is computed by using the assumed foundation design pressure (not the ultimate bearing capacity). For foundations on soils that settle slowly, such as saturated clays, only the dead load plus any sustained live load is used in the analysis; but for partially saturated clays, silts, and organic soils, which usually settle rapidly, the dead plus total live load is used. In some cases it is necessary to compute the settlement of every column or part of a structure. In many cases it is sufficient to know the settlement of the more critical parts such as footings for delicate machinery, smoke stacks, and the heaviest columns.

The amount of settlement a structure can tolerate depends on its size and construction, and whether it is uniform or non-uniform. If all parts of the structure settle the same amount, the structure will not be damaged. Only access, drainage, and utility connections will be affected, and these can tolerate movements of several inches. Differential settlement causing tilting is important for floors, crane rails, machinery, and tall narrow structures such as stacks, and the limits are established by their operation. Differential settlement causing curvature affects the structure itself and is limited by the flexibility of its construction. Table 9:5 gives the maximum settlements that

TABLE 9:5 / MAXIMUM PERMISSIBLE SETTLEMENT*

Type of Movement	Limiting Factor	Maximum Settlement
Total settlement	Drainage and access	6 in. to 24 in. or 15 to 60 cm
	Probability of differential settlement	
	Masonry walls	1 to 2 in. or 2.5 to 5 cm
	Framed buildings	2 to 4 in. or 5 to 10 cm
Tilting	Towers, stacks	$0.004B$†
	Rolling of trucks, stacking of goods	$0.01S$†
	Crane rails	$0.003S$†
Curvature	Brick walls in buildings	$0.0005S$ to $0.002S$†
	Reinforced concrete building frame	$0.003S$†
	Steel building frame, continuous	$0.002S$†
	Steel building frame, simple	$0.005S$†

* B is base width; S is column spacing.
† Differential settlement in distance B or S.

can be permitted. It is based on both theory and observations of structures that have suffered damage.[9:13, 9:14]

The allowable foundation pressure for each footing must satisfy both the requirements of safety and settlement. For convenience in its determination for each foundation, the results of the bearing capacity and settlement analyses should be expressed graphically, as shown in Fig. 9.23. The safe bearing pressure represents the upper limit without regard to settlement. Curves of equal settlement for different column loads or foundation sizes are also shown. A number of such charts are prepared, each for a different set of soil conditions, different shape of foundation, or for varying ground water and depth of excavation. From these the designer can select the allowable pressure that will keep the total settlement and differential settlement within limits.

REDESIGN ALTERNATIVES[9:15] / When the analyses show that the assumed foundation pressure is not safe or that the settlement will be

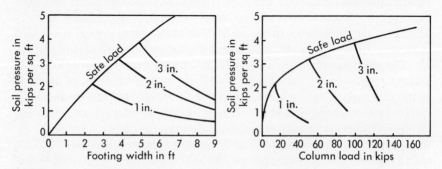

Figure 9.23 Footing design charts showing safe bearing as the upper limit and curves for equal settlement of 1, 2, and 3 in. (2.5, 5, and 7.5 cm).

SEC. 9:6] RATIONAL PROCEDURE FOR ALLOWABLE PRESSURE

excessive, or when they indicate that the safety is so high and settlement so low that the foundation is uneconomical, a redesign is necessary. Ordinarily this is limited to the foundation, but it is often fruitful to extend the redesign to the structure and even to the soil.

The simplest procedure is to change the foundation. By reducing the design pressure, the safety factor against failure is increased. Reducing the pressure is not always effective in reducing settlement, however. If the compressible strata are at a shallow depth below the foundation, the settlement will be reduced almost in proportion to the pressure; but if the compressible layer is far below the foundation, a reduction in pressure may not reduce the settlement materially.

Example 9:3

What is the effect of reducing the foundation pressure from 6000 psf to 3000 psf on the stresses at a depth of 10 ft beneath a column supporting a load of 120,000 lb? A square footing at 6000 psf will be 4.5 ft wide, and at 3000 psf, 6.3 ft wide. From Fig. 9.12 the stress at 10 ft below the 6000 psf footing will be $6000 \times 0.06 = 360$ psf, and beneath the 3000 psf footing it will be $3000 \times 0.12 = 360$ psf. No change.

The limit in size is reached when the foundations touch and form a continuous footing. Such a continuous foundation can bridge over small soft areas, but it cannot reduce the dished-in settlement profile produced by thick compressible strata.

Increasing the foundation depth will increase the bearing capacity in homogeneous soils by increasing the surcharge, particularly if the soils have a high angle of internal friction. If the soil is stratified and becomes stronger with increasing depth, then depth will improve the bearing capacity; but if the soil deposit has a hard crust underlain by softer soil, an increase in depth will reduce the bearing. In most cases, increasing the depth will reduce settlement. However, if the compressible strata are deep, increasing the depth will bring the foundation closer to the source of trouble and will aggravate the settlement. Very deep foundations that transfer the load below the weak or compressible strata are considered in Chapter 10.

Sometimes great benefits can be had by changes in the structure. The column loads can be reduced by reducing the spacing. This helps if the bearing capacity is limited, but it will usually not reduce settlement appreciably because the total structural load is not changed materially. Substituting lightweight construction for conventional forms will reduce the total weight and benefit both the safety against failure and settlement. If the site is sufficiently large, the structure may be spread over a larger area and the concentration of load reduced, which will also benefit safety and settlement. If extensive fill is planned to support the ground floor, its weight is a major factor in settlement. For example, 5 ft or 1.5 m of fill weigh as much as three or four stories of building. Eliminating such fill will reduce settlement materially. Changing the structure by making it so rigid that it will resist distortion is occasionally possible, as shown in Fig. 9.24. The floors and walls can be combined structurally to develop a box girder stiff enough to prevent differential settlement. Introducing trusses into the building frame can do the same thing in steel structures. For very small buildings a continuous mat can be made rigid if its thickness is about one-tenth its span, but the weight of the mat itself becomes a major load on the soil.

Making the structure flexible so that it can conform to the settlement is a simple method of preventing damage from differential movement. Simple structural framing, small wall panels that are not rigidly connected to the floors and columns, masonry walls with low-strength mortar, and ground floors that are reinforced, jointed, and keyed like a pavement will permit the maximum movement with minimum damage. Flexible construction is best adapted to wide, low buildings where the structural framing would ordinarily be light.

Reducing the net load on the soil by excavation (Fig. 9.24b) is a very old method for minimizing settlement. If the weight of the soil excavated equals the structural weight, there will be no increase in the stresses in the soil below and therefore little settlement. Such a design is often called "floating the structure," since the structure appears to be buoyed up by the weight of the soil displaced. The soil, however, is a solid that expands when the load is relieved by excavation and which recompresses when reloaded; therefore there will be some settlement even with a good balance. A perfect balance is impossible because of the variable live load in the structure and the variations in the unit weight of the soil excavated.

If some areas of the structure are heavier than others, the balance can be improved by *differential excavation*: the excavation depth is increased in the areas of greatest load so that the net increase in stress in the compressible

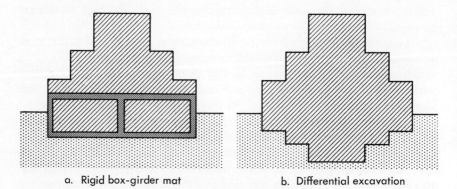

a. Rigid box-girder mat   b. Differential excavation

Figure 9.24 *Reducing differential settlement by changes in the building or excavation.*

strata is relatively uniform across the total breadth of the structure. A changing water table can upset the balance. This method requires a very careful evaluation of the soil and loading conditions in order to be successful.[9:15]

Example 9:4

A 10-story building weighs 150 lb per sq ft, the roof 50 lb per sq ft. Settlement analyses show that consolidation of a clay stratum 35 ft below the ground surface will be excessive due to the building weight. How could excavation reduce the stress?
1. The building load average is $10 \times 150 + 1 \times 50 = 1550$ psf.
2. If the soil weighs 120 lb per cu ft, basement excavation to reduce the net loading to zero would be $1550/120 = 12.5$ ft.

Changing the soil to increase the safety or reduce the settlement includes drainage, densification, altering the soil by an additive, and changing to another site. Drainage improves the soil by reducing the neutral stress. In a cohesionless soil the bearing capacity is often doubled by lowering the water table below the foundation shear zone, but it must be permanent before it can be counted on in design. Reducing the neutral stress in a compressible soil will cause it to consolidate and become stronger and, of course, to settle. If the consolidation can be completed before the structure is built, the bearing capacity will be increased and the settlement reduced. Loss of drainage after construction is complete is not critical in the compressible soil and can be helpful because, as the neutral stress increases, the effective stress decreases.

Preconsolidating a compressible soil by a surcharge load is very effective in increasing strength and reducing settlement. In effect this is another form of drainage, by consolidation, and was discussed in Chapter 4. Stage loading makes it possible to improve the soil by the weight of the structure itself. For example, the soil at a riverside site was too weak to support with safety the full load of a grain elevator. The load was limited to half the capacity for the first year to maintain safety. The soil consolidated and became stronger and could then support the full load.

Densification of loose sands by shock and vibration is effective in increasing bearing capacity and reducing settlement. Altering the soil by injecting cementing agents, changing the chemistry of clays, and fusing it with heat are collectively termed *soil stabilization*. All these methods are discussed in Chapter 5.

Sometimes it is better to move to a different site. It is physically possible to design a satisfactory foundation for any site if enough money can be spent. However, the cost of foundations added to the cost of the property can make cheap real estate very expensive. A complete study of all the economic factors involved is required to determine if special foundations are worth their cost.

9:7 Footings and Mats[9:16, 9:17]

FOOTING FOUNDATIONS / A footing is an enlargement of a column or wall in order to reduce the pressure on the soil to the maximum allowable. Beneath a wall the footing may be continuous, forming a long, rectangular, loaded area, called a *wall footing*. Beneath a column the footing may be any shape, but economy in construction favors the square. Rectangular shapes are often used if clearances prevent squares, or two or more square footings may be joined to form a single, rectangular footing under several columns. Occasionally round or hexagonal footings are used, especially under chimneys or heavy machines, but the economy in materials of such shapes is usually overbalanced by the additional labor required for their construction.

Structurally the footing is a wide beam, acted on by a distributed load (the soil pressure) and supported by a concentrated force (the column). Modern practice calls for reinforced concrete, which is designed by the usual methods. The soil pressure is usually assumed to be acting uniformly over the footing. This is a conservative assumption for footings on sand but may be somewhat unsafe for footings on saturated clays. The soil pressure under a rigid load on clay is high at the outer edges of the loaded area and decreases to about half the average at the center. The pressure under a concrete footing will not have such extreme variation, owing to upward deflection of the footing edges, but it will not be uniform. Therefore it is well to be conservative in the structural design of footings on clay.

To avoid eccentricity, the centroid of the footing should coincide with the centroid of the load coming on it. Ordinarily this is no problem, but along the exterior walls where the property line limits the extent of the footing or where elevator pits, machinery, and utilities obstruct the space, a concentrically loaded footing is impossible. If the eccentricity is small, it is sometimes possible to design the footing and column to absorb the unbalanced moment (Fig. 9.25a). A better method is to combine two adjacent

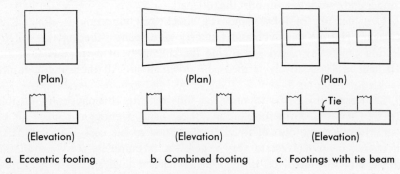

Figure 9.25 *Footing designs to avoid eccentric moments.*

footings into one large footing, as shown in Fig. 9.25b. The footing should be proportioned so that its center of gravity is at the same point as the center of gravity of the two column loads. If the column spacing is large, two adjacent footings can be connected by a small tie beam that absorbs the eccentric moment (Fig. 9.25c).

STRAP FOUNDATIONS / A *strap foundation* is a continuous footing that supports three or more columns in a straight line. It consists of a number of spread footings that have been connected together. Straps are employed for two reasons: to provide structural continuity and to achieve construction economy. When single footings are large and closely spaced in one direction, they can be combined to form a continuous shallow beam. This can bridge over small (less than half the column spacing) weak areas and achieve some economy through structural continuity. The strap foundation is often cheaper to build because the foundation excavation is a continuous trench rather than a series of isolated pits. The strap is designed like a continuous beam with a uniform load on one side and with concentrated supports on the other.

MAT FOUNDATIONS / A *mat* or *raft* is a combined footing supporting more than three columns not in the same line. The mat provides the greatest total foundation area for a given space and the minimum foundation pressure, and therefore the maximum safety against soil failure. If the compressible strata are located at a shallow depth, the mat will minimize the settlement. However, if the compressible strata are deep, it will have little effect and in some cases, because of its weight, can increase the total settlement slightly.

A mat has other advantages. Like the strap it can bridge over small isolated soft areas in the foundation. It provides economy in design and construction by developing structural continuity and by permitting a uniform excavation depth. Cost comparisons between mats and large footings show that when the total area of the spread footings is more than from one-half to two-thirds the building area, a mat may be cheaper. Mats are employed when hydrostatic uplift must be resisted because the weight of the building is used to overcome the upward pressure.

MAT DESIGN PRESSURE / The pressure on the soil-mat interface depends on the stiffness of the structure-foundation system and the deflection and settlement of the soil under that load. If the structural system is very rigid, Fig. 9.26a and b, the pressures will resemble those for a perfectly rigid load; nearly uniform but falling off at the edges in a deep cohesionless soil and higher at the edges and less at the center in an elastic compressible material such as clay. If the structure-foundation is sufficiently flexible so that the total load for any segment of the mat defined by lines drawn midway between the columns is not changed by differential deflection of the structure, then the pressure on that portion of the mat will be equal to the column load divided by the area of the mat segment. For uniform column loads the

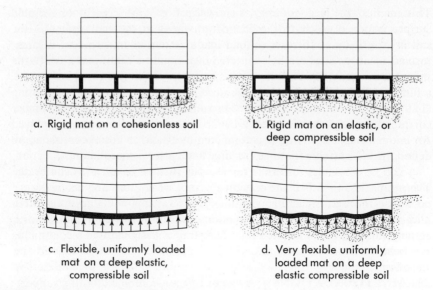

Figure 9.26 *Pressure distribution on mat foundations.*

pressure in this case will be approximately uniform, Fig. 9.26c. If the mat is so flexible that it deflects upward appreciably between columns, the pressure will be greatest at the columns and less between as in Fig. 9.26d. The average pressure on each mat segment will be equal to the column load divided by the area of the corresponding mat segment.

Most mats are neither as rigid as depicted by Fig. 9.26a and b nor so flexible that the load is uniform. For deep sands the over-all pressure is so nearly uniform that this can be assumed for design. For elastic soils the pressure distribution on the mat is statically indeterminant and must be compatible with the stress-deflection characteristics of both the soil mass and the structure. This can be analyzed in two ways: by successive approximations or by representing the soil by a simple mathematical model that can be solved directly.

In the method of successive approximations, a pressure distribution is assumed, similar to Fig. 9.26b. The deflection of the structure and the settlement of the soil are then computed. If the assumed pressure was correct, the two deflection curves should coincide. If not, the assumed pressure is revised and the process is repeated, until the deflections are compatible.

The second method makes use of an imaginary soil parameter, the *coefficient of subgrade reaction*, k_s, previously discussed in pavement evaluation, Chapter 5.

$$k_s = \frac{q}{\rho} \qquad \left(k_s = \frac{\sigma}{\rho}, \text{Equation 5:5} \right). \qquad (9:15)$$

This implies that the settlement or deflection of any segment of the soil surface is proportional to the foundation pressure on that segment. The soil is thus assumed to react to load like a system of independent springs, instead of as an elastic mass. Generally, it assumed that k_s is a constant, independent of B; however, Equations (9:10a), (9:13a) and (b) show it to be a function of B. Therefore, more than one k_s value must be used in analysis: (1) a maximum corresponding to the column spacing and (2) a minimum corresponding to the total mat width. The value of k_s also varies with time. An initial k_s must be found corresponding to elastic or initial consolidation deflection and a final value corresponding to long-term consolidation.

The mat-structure system is represented by an equivalent elastic beam. The moments and deflections are then computed by the theories of beams on elastic foundations, utilizing four possible values for k_s: maximum and minimum and for long-term and short-term settlements. Although the assumption of a coefficient of subgrade reaction is not compatible with real soil behavior, model tests show that the moments and deflections in the structure computed in this manner are reasonably reliable if the structure is relatively flexible.[9:18]

The structural design of a mat is similar to that of an inverted floor. Flat slab designs are used for small mats and have the advantage of unobstructed surface. If additional thickness is required at the columns, it is provided by lowering the bottom of the mat. Beam and slab designs are used for larger mats. To minimize thickness, the beams are wide and shallow. The space between the beams is filled with lightweight concrete to provide a flat floor surface, or a solid mat is used with the beams defined only by their reinforcing steel. Mats can be combined with the basement walls to form an inverted T-beam foundation or with the walls and upper floors to form a box girder (Fig. 9.24a). Such a rigid foundation-structure system can be designed to resist differential settlement.

9:8 Special Problems in Shallow Foundation Design

HYDROSTATIC UPLIFT / Structures below the ground water level are acted on by uplift pressures. If the structure is weak, the pressure can break it and cause a blow-in of a basement floor or collapse of a basement wall. If the structure is strong but light, it may be forced upward or *floated* out of its original position. Uplift is taken care of by drainage or by resisting the upward force. Continuous drainage blankets as described in Chapter 4 are very effective but must be designed with filters to function indefinitely without clogging. If possible the water should be disposed of by gravity because pumps sometimes fail.

The entire weight of the structure can resist uplift if a mat foundation is employed. In addition it is sometimes possible to increase the mat thickness to provide more weight. However, it must be kept in mind that each foot of extra thickness of mat concrete resists about $1\frac{1}{2}$ ft or 45 cm of head because the uplift is also increased by one ft or 30 cm. Anchors, grouted into bed rock or driven into hard soil, can provide uplift resistance. The anchor resistance is limited to the buoyant or effective weight of the soil or rock engaged and to the connection between the anchor and the earth, whichever is the smaller.

Structures designed to resist uplift must be waterproofed to prevent seepage and to minimize dampness. Bituminous membranes applied to the outside are useful if the head is only a few feet. If the heads are greater, the concrete structure itself must be made water tight. There are three essentials: (1) high-quality dense concrete placed without honeycombing; (2) water stops across all construction or expansion joints; (3) masonry waterproofing (usually cement, powdered iron and an agent to rust the iron) applied externally if possible, but internally if necessary to fill the hair cracks and to minimize capillary movement.

If the uplift occurs infrequently or if the value of the structure does not justify elaborate measures to overcome uplift, damage to the structure can be prevented by intentional flooding to balance the uplift pressure. Relief holes with flap valves are sometimes installed in the bottom of sewage treatment tanks so they will flood automatically if they are empty at a time of ground water rise. Flooding in buildings is seldom desirable because of the damage to the contents.

SEVERE VOLUME CHANGE[9:19] / Soils having high volume change were discussed in Section 3:9, and under depth in Section 9:1. In most cases the best way to avoid trouble from high volume change is to place the foundation below the level of severe movement. Movement of the upper soil strata along the column extending upward from the footing can break the column and lift the structure. This can be prevented by placing a layer of weak, isolating material around the column. Mineral wool, vermiculite, and even sawdust have been used for this purpose. When the building is supported by deep footings, any grade beams must be separated from the soil so that they will not be damaged by soil or shale expansion. The space between the beams and the soil should be filled with vermiculite or mineral wool to keep out soil.

In arid regions where swelling is the most serious problem, it is sometimes possible to utilize a bearing pressure that exceeds the swell pressure. This is not always successful because the stress increase produced by the foundation decreases rapidly with increasing depth, while the swell pressure may not.

Small structures can be placed on relatively stiff mats that rise and fall with the volume change but which do not deflect enough to cause trouble. Flexible foundations and structures that can deform without damage can be used if the volume change is not very irregular.

In arid regions special attention must be paid to drains, leaking pipes, and other sources of water that could cause heave. Piping under floor slabs can be placed in concrete troughs so that leakage can be kept out of the soil and so that soil movements will not cause the pipes to leak.

LOESS / Loess soils are ordinarily hard and incompressible from the partial cementing of clay and calcium carbonate. If they become wet, they soften and are very compressible. Surface water must be drained away from foundations on loess and piping must be routed so that leaks will not cause damage.

Similar settlements from structural collapse sometimes occur in the loose soils of arid regions. The settlement can be minimized by prewetting the site, but at the cost of reduced bearing capacity.

COLD-STORAGE STRUCTURES / Frost action beneath cold-storage warehouses is a serious problem even in warm regions. If the temperatures are not very low, replacement of the frost-susceptible soil with a clean coarse sand or gravel is sometimes sufficient. This also provides some insulation. Insulation with cork or foam glass from 4 to 8 in. or 10 to 20 cm thick is effective if the soil is warm. Isolation, by placing the cold area on piers so that air can circulate below is sometimes used for small structures. When the cold is severe, it may be necessary to warm the soil by hot-air ducts or hot-water coils. Methods for analyzing the heat balance in such cases have been published.[9:20]

BOILERS AND FURNACES / Foundations for boilers, furnaces, and hot industrial processes can heat the soil and cause it to lose moisture and settle. The preventive measures are similar to those for cold-storage warehouses: soil replacement, insulation, and isolation.

FOUNDATIONS ON FILL / The ability of fill to support structures safely and with tolerable settlement depends on two factors: (1) the character of the fill and (2) the settlement of the foundation of the fill. A properly compacted fill on a good foundation can be as good or better than virgin soil although some archaic codes prohibit building on any fill. If the fill is uniformly compacted to the minimum standards given in Chapter 5, and if tests of the fill and foundation show them to have the required bearing and incompressibility, foundations on fill can be designed as any other foundation.

Old fills, and fills placed over low areas underlain by compressible or weak strata should be considered unsuitable unless extensive tests show them to be usable or unless the proposed structure can be adapted to low bearing and irregular settlement. Frequently poorly compacted old fills continue to settle for years from secondary consolidation, aggravated by a breakdown of the soil structure by moisture. For example, a one-story masonry building on a 50 year old 50 ft or 15 m deep fill settled nearly 4 in. or 10 cm during the first 6 years after its construction, largely because of the additional weight of 3 ft or 1 m of new well-compacted fill placed to provide a better foundation for the building.

Sanitary land fills and fills above organic debris and rubbish suffer from continuing organic decomposition and physico-chemical breakdown.[9:21] One result is continuing erratic settlement. A more insidious effect is the production of methane and hydrogen sulfide gases that can be explosive and poisonous. A number of explosions, and some cases of illness and death can be traced to gas accumulation in structures built on fills over organic matter. The best approach is to remove the organic matter in the immediate area of the building and replace it with good fill. Venting has sometimes been effective in eliminating the gases. Corrosion of underground piping is a hazard in debris-ladened fill. Anti-corrosion protection or the replacement of the fill in the vicinity of the piping is usually necessary.

LATERALLY LOADED FOUNDATIONS / Foundations that support both vertical and horizontal loads are required for retaining walls, bridge abutments, and many kinds of industrial process machinery. The bearing capacity that supports the vertical component of the load is computed from the reduced bearing capacity factors of Section 9:2. Load inclinations exceeding 30 degrees should not be analyzed by correcting the equations for vertical bearing but must be evaluated in terms of passive earth pressure on a sloping surface which is beyond the scope of this text.

Resistance to the lateral load is provided two ways: (1) sliding resistance of the bottom of the foundation; and (2) passive earth pressure on the face of the foundation opposite to the thrust. The sliding resistance is either equal to the adhesion and friction between the foundation and the soil or the shear resistance of the soil (or rock), whichever is the smaller. Too often there is a thin layer of disturbed soil in the bottom of the footing excavation that acts as a plane of weakness. For maximum sliding resistance the soil surface is made irregular (but not notched as occasionally advocated) and cleaned just before pouring the foundation. To avoid disturbing the soil while placing reinforcing a 4 to 6 in. (10 to 15 cm) working mat of concrete is poured on the freshly exposed soil surface. The steel is then placed on this surface which is left rough so as to bond with the remaining concrete.

If the sliding resistance is inadequate, it can be enhanced by a shear key on the edge of the foundation to which the thrust is directed. This adds to the passive pressure against the foundation face. The design should not rely on full passive pressure because considerable lateral movement is necessary for its development. Ordinarily one-third of the potential passive pressure is assumed to be available.

Excavation adjacent to the foundation can destroy passive pressure. Saturation of the soil can reduce it. Therefore, designs relying on passive pressure should be conservative.

9:9 Foundations on Rock

Generally there has been little concern for the bearing capacity and settlement of foundations on rock. Tradition, going back to the virgin stone platforms of the Great Pyramid of Egypt, as well as the Biblical example of

SEC. 9:9] FOUNDATIONS ON ROCK

the wise man who built his house on rock to withstand rain, floods, and wind, has witnessed the security of bedrock support. Facts, such as the compressive strengths of hard limestone, sandstone and granite that exceed the capacity of good concrete, lull the modern engineer into a sense of false security. Foundations supported by rock do sometimes experience bearing capacity failures and settlement. Therefore, these possibilities must be investigated and any shortcomings minimized by design.

BEARING CAPACITY / The mechanics of bearing capacity failure of homogeneous rock masses of great extent should be similar to that of soil because their Mohr envelopes are similar (Chapter 3). The major consideration in rock shear at low confining stresses is the drop of strength after failure and at low strains, termed brittle failure. From this it is deduced that the bearing capacity of rock is controlled by local shear, accompanied by cracking around the perimeter of the loaded area. This was confirmed by model tests by the authors nearly 30 years ago. For homogeneous rock, Equation (9:3) and the bearing capacity factors for local shear should be used, Fig. 9.7. For ordinary concrete foundations on sound rock, bearing capacity is no limitation, because the rock is stronger than the concrete. For intensely loaded pile tips and the concentrated loads of steel supports in tunnels, bearing capacity of the softer homogeneous rocks such as shale and sandstone can be critical.

If the rock is jointed, the mechanism of potential failure is different, depending on the size of the loaded area, the joint spacing, joint opening, and the location of the load. Three simple possibilities can be analyzed, Fig. 9.27a, b, and c. In the first case where the joint spacing, S, is a fraction B

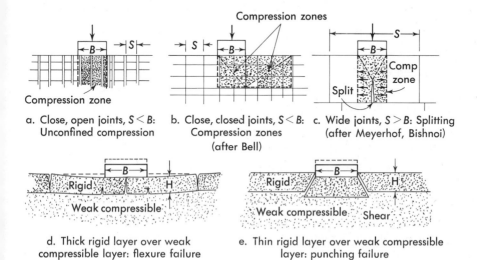

a. Close, open joints, $S < B$: Unconfined compression

b. Close, closed joints, $S < B$: Compression zones (after Bell)

c. Wide joints, $S > B$: Splitting (after Meyerhof, Bishnoi)

d. Thick rigid layer over weak compressible layer: flexure failure

e. Thin rigid layer over weak compressible layer: punching failure

9.27 Bearing capacity failure modes in rock.

and where the joints are open, the foundation is supported by unconfined rock columns. The ultimate bearing capacity approaches the sum of the unconfined compressive strengths of the rock columns. The total capacity is always less than the sum of the column strengths because all do not have the same rigidity, and therefore some will fail before others reach their ultimate capacity.

If the joints are closed so that pressure can be transmitted across them without movement, the shear mechanism is essentially that described by the Bell-Terzaghi analysis, Section 9:2. The bearing capacity can be evaluated graphically by Fig. 9.5. If the joint spacing is much greater than width, $S \gg B$, the mechanism is different. The cone shaped zone that forms below the foundation splits the block of rock formed by the joints. This condition has been analyzed first by Meyerhof,[9:22] and extended by Bishnoi and Sowers.[9:23] The results are approximated by an abbreviation of Equation (9:4a), assuming that the load is centered on the joint block, and little pressure is transmitted across the joints.

$$q = JcN_{cr}. \tag{9:16}$$

Values of N_{cr} derived from models for splitting failure depend on the S/B ratio, and φ. These are given in Fig. 9.28 for circular footings. The values for square footings are 85 per cent of the circular. The factor J considers the effect of the thickness of the upper layer: if $H/B > 8$, $J = 1$; if $H/B = 1$, $J = \frac{1}{2}$ approximately.

When the rock formation consists of an extensive hard seam underlain by a weak compressible stratum, two forms of failure occur, depending on the ratios H/B and S/B, and on the flexural strength of the rock stratum. If H/B is large and the flexural strength is small, the rock failure occurs by

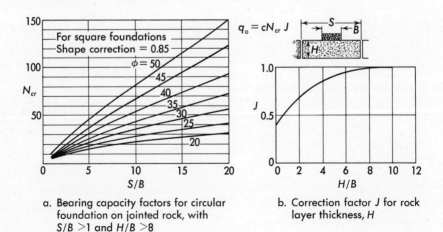

a. Bearing capacity factors for circular foundation on jointed rock, with $S/B > 1$ and $H/B > 8$

b. Correction factor J for rock layer thickness, H

Figure 9.28 *Bearing capacity factors for rock splitting. (After Bishnoi.)*

flexure, Fig. 9.27d. If the *H/B* ratio is small, punching is more likely, Fig. 9.27e.

Neither case has been adequately studied and so only possible methods for analysis are suggested. A foundation over a rock cavity also fails by flexure or punching and could be analyzed approximately by those mechanisms. A major unknown factor in either case is the location of vertical joints and their effect on failure.

SETTLEMENT / Settlement of foundations on most rock formations is controlled by the joints. Settlement of hard rock with closed joints is negligible. If the joints are open or irregular, the observed settlement is comparable to the measured joint separations under the loaded area. Some porous limestones, and weathered shales, poorly indurated sandstones, earth-like rocks such as tuff consolidate similar to soils. The settlement potential can be evaluated by consolidation tests of undisturbed samples.

DESIGN / The design of foundations for rock is similar to foundations resting on soil. However, because much higher pressures are commonly used, and because of unforseen defects, special treatment of individual conditions is usually required. A narrow soft zone beneath the foundation, Fig. 9.29a, will not seriously reduce the bearing of the adjoining rock if the edges of the sound rock do not crumble. The fissure is cleaned out to a depth of twice its width and filled with concrete to support the rock corners above.

An irregular or sloping rock surface, Fig. 9.29b, is no problem if it can be sufficiently cleaned to obtain good bond and if the slope is less than the angle of internal friction of the rock or concrete. Excavation or blasting to level the rock may weaken it and do more harm than good. Resistance to

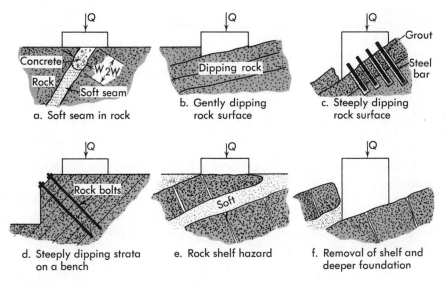

Figure 9.29 Rock foundation problems.

sliding can be generated at little cost by drilling holes in the rock and inserting reinforcing bars as dowels, Fig. 9.29c.

Sloping seams of weakness, Fig. 9.29d and e, that define blocks that can slide easily can be corrected in two ways. The unstable mass can be removed or it can be reinforced by rock bolting.

Cavities and zones of high porosity that could lead to consolidation or bearing failure require grouting to strengthen the mass or to support the crust on which the foundation rests. Foundation design in rock is usually tentative. Ingeneous field decisions are required during construction to adapt the design to the rock formation as it is uncovered.

9:10 Foundations Subject to Vibrations [9:24, 9:25]

Foundations are subject to vibrations from a number of sources, natural and artificial. Earthquakes, wind on tall, narrow buildings and towers, fast flowing water, and wave pounding on marine structures produce both transient and near-continuous vibrations of variable frequencies. Machinery is the most important cause of vibration. Reciprocating engines, compressors, pumps, and oscillating machines are important sources of continuous, low frequency vibrations and are usually the most serious causes of trouble. Electric motors, rotary pumps, and turbines produce continuous high-frequency vibrations. Shock and transient vibrations often are caused by stamping machines, forges, pile drivers, moving vehicles, and blasting. Vibration consists of complex repeating motions and can include both rotation and translation in all three directions. Continuous vibrations usually have a constant frequency determined by the source and complicated by harmonics generated by the structure, while transient vibrations from shocks may have a variable frequency, depending on both the source and its supporting system.

Vibration can occur in different directions, Fig. 9.30a: vertical, longitudinal and transverse linear motion (equivalent to an orthagonal coordinate system), and in three directions of rotation: rocking, pitching and slewing, depending on the source of the vibration and the freedom to move. Many vibrations of structures are a complex combination of all six, which must be unravelled to understand the mechanism of motion and to devise corrective measures.

MECHANISM OF VIBRATIONS / If an impulse of short duration is applied to a body that is supported elastically, it will vibrate at its *natural frequency*, which depends on its mass and elastic properties. For a perfectly elastic body (Fig. 9.30b) whose weight is W and whose resistance to deflection in force per unit of deflection (pounds per foot or grams per centimeter) is K, the natural frequency f_n is given by

$$f_n = \frac{1}{2\pi}\sqrt{\frac{Kg}{W}}. \qquad (9:17)$$

SEC. 9:10] FOUNDATIONS SUBJECT TO VIBRATIONS

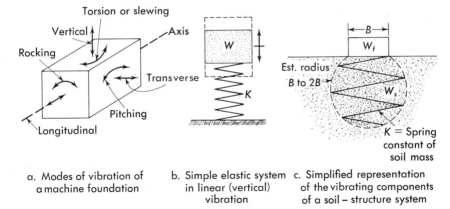

a. Modes of vibration of a machine foundation
b. Simple elastic system in linear (vertical) vibration
c. Simplified representation of the vibrating components of a soil – structure system

Figure 9.30 *Vibrations of a structure.*

This means that the natural frequency increases as the square root of the rigidity and decreases with the square root of the weight of the body. When energy is lost in the process, the vibration is said to be *damped*, and the natural frequency is somewhat less.

The damping is described in terms of the *damping ratio*, C, which is an indication of the amount of vibration energy lost in each cycle: $C = 0$ denotes no loss; $C = 1$ indicates all the impulse energy is dissipated in one cycle of vibration.

The natural frequency of a structural column or beam can be estimated from its weight and rigidity, using Equation 9:17. The natural frequency of a foundation-soil system as shown in Fig. 9.30c is much more complex. The resistance per unit of deflection K can be estimated from the distortion settlement ρ. This depends on both the modulus of elasticity of the soil and the size of the foundation. The weight of the vibrating body, W, is the sum of the weight of the foundation, W_f, and the portion of the soil mass below the foundation which is vibrating, W_s. Therefore the natural frequency of the soil is not a property of the soil alone but also depends on the weight and size of the foundation and the load it carries.

The intensity of the vibration is also a factor because the modulus of elasticity of some soils changes with confining pressure and with the strain. Tests of soil masses with vibrators having masses of from 2000 to 6000 lb or one to 3 tons and with square bases from 2 ft to 3 ft (60 to 90 cm) wide indicate natural frequencies of from 700 vibrations per minute for peat to 1800 vibrations per minute for very dense sand. For heavier and wider foundations the natural frequency would be less. These natural frequencies, unfortunately, are comparable to the vibrations or multiples of the vibrations generated by many reciprocating machines such as pumps and compressors. They are lower than the frequencies generated by turbines and high-speed motors, however.

FORCED VIBRATION-RESONANCE / If a periodic impulse is forced on a simple elastic system at a very low frequency, $f \ll f_n$, the mass and its elastic support will respond together, with the same amplitude of motion in *forced vibration*. The movement is neither *magnified* nor *attenuated* If the impulse is forced on the system at a very high frequency, $f \gg f_n$, the elastic mass cannot respond, because of its inertia and the elasticity of the system absorbs the difference in movement. The vibration effect on the mass is attenuated. These effects are illustrated by the extreme ends of the graph, Fig. 9.31. If the vibration is applied at approximately the natural frequency, $f = f_n$, each reoccurring movement of the source applies additional impetus to the movement of the body. If there is no energy loss or damping, the amplitude of the movement of the elastic system is increased each cycle by the energy imparted by the impulse, and eventually becomes indefinitely large. This is termed *resonance*. With damping, the amplitude of the vibration of the elastic mass increases with each impulse until the energy lost in each cycle of vibration is equal to the energy input of the impulse during that cycle. The ratio of the magnitude of the vibration of the elastic system to that of the impulse is termed the *magnification*. For a continuous source of vibration, the magnification depends on the damping factor and how close the frequency of the impulse matches the natural frequency, as shown on Fig. 9.31. (Although high damping reduces the natural frequency somewhat, this does not change the concept of resonance.)

An impulse at half the natural frequency can cause a magnified vibration by adding energy every other cycle. The magnification is less for a given damping ratio because the energy input is only half that for one impulse per cycle. Similarly, resonance can occur with input frequencies of one-third

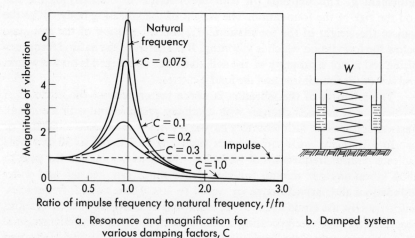

Figure 9.31 *Resonance in a damped system.*

and one-fourth the natural frequency. The magnification for these conditions is seldom great unless the damping ratio is very small.

EFFECT OF VIBRATIONS / Soil vibrations have a number of important effects. First, the vibration can be transmitted to other foundations and to other structures at some distance from the vibration source. These transmitted vibrations can be annoying and even damaging. If some foundation-soil system should be in resonance, severe damage could result. Second, the vibration can cause a reduction in the void ratio of cohesionless soils and result in severe settlement. Ordinarily the settlement will be small if the relative density is greater than 70 per cent, but if the vibration is severe, as in the case of resonance, settlements can occur until the relative density is nearly 90 per cent. Third, vibration in loose, saturated cohesionless soils can bring about a quick condition, loss of strength, and failure. Soils with cohesion are resistant to vibration settlement and are not affected appreciably.

Corrective measures include reducing the vibration of the source, changing the soil-foundation system to prevent resonance, and stabilizing the soil to prevent vibration damage. Vibration can be reduced by installing isolating systems such as spring mounts on the source or by cushioning them on vibration-absorbing materials. It also can be minimized by changing the type of machine, such as substituting rotary for reciprocating compressors. The frequency of resonance is changed by altering one or more of the factors in Equation 9:17. Increasing the size and weight of the foundation will reduce the resonant frequency of the system. Increasing the modulus of elasticity of the soil by densification or stabilization (Chapter 5) will increase it. Changing the speed of the source can be helpful, of course, if it is possible mechanically. In general the natural frequency should be less than half or more than one and one-half the vibration frequency. Stabilizing the soil by injecting a cementing agent or by densification can prevent settlement or loss of strength. This is discussed in Chapter 5.

REFERENCES

9:1 "Theoretical and Practical Treatment of Expansive Clays," *Quarterly Colorado School of Mines*, **54**, 4, October 1959.

9:2 E. W. Lane and W. M. Borland, "River Bed Scour During Floods," *Transactions, ASCE*, **119**, 1954, p. 1072.

9:3 E. M. Laursen and A. Toch, "Scour Around Bridge Piers and Abutments," *Bulletin 4, Iowa Highway Research Board*, Ames, 1956.

9:4 K. Terzaghi, *Theoretical Soil Mechanics*, John Wiley & Sons, Inc., New York, 1943.

9:5 G. G. Meyerhof, "The Influence of Roughness of Base and Ground Water on the Ultimate Bearing Capacity of Foundations," *Geotechnique* **5**, 3, September 1955, p. 227.

9:6 G. G. Meyerhof, "The Bearing Capacity of Footings Under Eccentric and Inclined Loads," *Proceedings of the Third International Conference on Soil Mechanics and Foundation Engineering*, **1**, Zurich, 1953.

9:7 L. Jurgenson, "The Application of Theories of Elasticity and Plasticity to Foundation Problems," *Journal, Boston Society of Civil Engineers*, July 1954.

9:8 H. M. Westergaard, "A Problem of Elasticity Suggested by a Problem of Soil Mechanics: Soft Material Reinforced by Numerous Strong Horizontal Sheets," *Contributions to Mechanics of Solids*, The Macmillan Company, New York, 1938.

9:9 N. M. Newmark, "Influence Charts for Computation of Stresses in Elastic Soils," *Bulletin 38, University of Illinois Engineering Experiment Station*, Urbana, 1942.

9:10 D. J. D'Appolonia, E. E. D'Appolonia and F. Brissette, "Settlement of Spread Footings on Sand," *Journal of the Soil Mechanics and Foundation Division, Proceedings, ASCE*, **94**, SM3, May 1968, p. 637.

9:11 K. Terzaghi, "Settlement of Structures in Europe and Methods of Observation," *Transactions, ASCE*, 1938, p. 1432.

9:12 *Load Test Research*, Law Engineering Testing Co., Atlanta, 1967.

9:13 A. W. Skempton and D. H. McDonald, "The Allowable Settlement of Buildings, *Proceedings of the Inst. of Civil Engineers*, **5**, 3, London, December 1956, p. 727.

9:14 D. E. Polshin and R. A. Tokar, "Maximum Allowable Differential Settlement of Structures," *Proceedings of the Fourth International Conference on Soil Mechanics and Foundation Engineering*, **1**, London, 1957, p. 402.

9:15 A. Casagrande and R. E. Fadum, "Applications of Soil Mechanics in Designing Building Foundations," *Transactions, ASCE*, 1944, p. 383.

9:16 L. S. Goodman and R. H. Karol, *Theory and Practice of Foundation Engineering*, The Macmillan Company, New York, 1968.

9:17 J. Bowles, *Foundation Analysis and Design*, McGraw-Hill Book Co., Inc., New York, 1968.

9:18 A. B. Vesic, "Beams on Elastic Subgrade and Winklers Hypothesis," *Proceedings of the Fifth International Conference on Soil Mechanics and Foundation Engineering*, **1**, Paris, 1961, p. 845.

9:19 G. F. Sowers and C. M. Kennedy, "High Volume Charge Clays of the Southeastern Coastal Plain," *Proceedings of the Third*

Panamerican Conference on Soil Mechanics and Foundation Engineering, **1**, Caracas, 1967.
9:20 W. Ward and E. C. Sewell, "Protection of the Ground From Thermal Effects of Industrial Plant," *Geotechnique*, **2**, 1, June 1950, p. 64.
9:21 G. F. Sowers, "Foundations on Sanitary Land Fill," *J.San.Eng.D.*, *P.ASCE*, **94**, S1, February 1968.
9:22 G. G. Meyerhof, "Bearing Capacity of Rock," *Magazine of Concrete Research*, April 1953.
9:23 B. W. Bishnoi, *Bearing Capacity of Jointed Rock*, A Thesis presented to the Georgia Institute of Technology in partial fulfilment for the Ph.D. in Civil Engineering, 1968.
9:24 R. V. Whitman and F. E. Richart, Jr., "Design Procedures for Dynamically Loaded Foundations," *Journal of the Soil Mechanics and Foundation Division, Proceedings, ASCE*, **93**, SM6, November 1967, p. 169.
9:25 G. P. Tschebotarioff, *Soil Mechanics, Foundations, and Earth Structures*, McGraw-Hill Book Co., Inc., New York, 1951.

SUGGESTIONS FOR FURTHER STUDY

1. G. A. Leonards et al., *Foundation Engineering*, McGraw-Hill Book Co., Inc., New York, 1961.
2. G. F. Sowers, "Shallow Foundations," *ibid*.
3. C. W. Dunham, *Foundations of Structures*, McGraw-Hill Book Co., Inc., New York, 1950.
4. A. S. Vesic (Editor), *Bearing Capacity and Settlement of Foundations, A Symposium*, Dept. of Civil Eng., Duke University, Durham, N.C., 1967.
5. *Proceedings of the International Conference on Soil Mechanics and Foundation Engineering*, London, 1957; Paris, 1961; Montreal, 1965; and Mexico City, 1969.
6. A. B. Carson, *Foundation Construction*, McGraw-Hill Book Co., Inc., New York, 1965.
7. F. E. Richart, Jr., J. R. Hall, Jr., and R. D. Woods, *Vibrations of Soils and Foundations*, Prentice-Hall Inc., Englewood Cliffs, N.J., 1970.

PROBLEMS

9:1 A long footing is 3 ft wide. Its base is 2.5 ft below the ground surface. Find the safe bearing capacity if the soil is a saturated clay having a unit weight of 110 lb per cu ft and a strength c, of

3000 psf. The safety factor is 3. Use Mohr's circle and compare with bearing capacity computed by the general formula, using both the Bell–Terzaghi and Meyerhof factors.

9:2 A square footing is 8 ft wide with its base 4 ft below the ground surface. The soil is saturated clay having a unit weight of 120 lb per cu ft and a cohesion of 4000 psf. Find the safe bearing capacity by the Meyerhof factors if a minimum safety factor of 2.5 is required.

9:3 A long footing 5 ft wide is 3 ft below the surface of a sand weighing 130 lb per cu ft saturated and 110 lb per cu ft dry, and having an angle of internal friction of 37°. Compute the safe bearing capacity for a safety factor of 2.5, using (1) the graphical method; (2) the general equation with the Bell–Terzaghi factors; and (3) the Meyerhof factors. For each case find the capacity (a) the water table 10 ft below the footing; (b) the water table at the base of the footing; and (c) the water table at the ground surface.

9:4 A column carries 200,000 lb. The soil is a dry sand weighing 115 lb per cu ft and having an angle of internal friction of 40°. A minimum safety factor of 2.5 is required, and the Meyerhof factors are to be used in computation.
  a. Find the size of square footing required if it is placed at the ground surface.
  b. Find the size of square footing required if it is placed 3 ft below the ground surface.
  c. Find the size of footing required for (b) if the water table rises to the ground surface, increasing the soil weight to 132 lb per cu ft.

9:5 A steam turbine whose base is 20 ft × 12 ft weighs 2400 kips. It is to be placed on a clay soil with $c = 3000$ psf. Find size of foundation required if the minimum safety factor is 3. The foundation is to be 2 ft below the ground surface.

9:6 A column carries 340 kips. It is to rest on a square footing on sand with $\varphi = 38°$ and $\gamma = 120$ lb per cu ft. The safety factor is 2.5.
  a. Find size of square footing if it is at ground surface.
  b. Find size of square footing if it is 4 ft below the ground surface.
  c. Would it be cheaper to lower the footing as in (b) if the column is 18 in. square and the footing is 2 ft thick than to place it at the ground surface? Concrete costs $40.00 per cu yd in place, and excavation costs $0.95 per cu yd.

9:7 A column carries 475,000 lb to a square footing that rests 3 ft below the surface of a partially saturated clay. If $\varphi' = 15°$, $c' = 1000$ psf, and $\gamma = 113$ lb per cu ft, find the footing size required for a safety factor of 2.5.

PROBLEMS

9:8 A smoke stack foundation 35 ft × 35 ft exerts a pressure of 5000 psf at the surface of a sand that weighs 108 lb per cu ft dry and 125 lb per cu ft saturated. Below the sand at a depth of 30 ft is a clay seam that is 6 ft thick, weighing 105 lb per cu ft saturated.
   a. Construct a diagram showing the variation of vertical stress increase under the center of the foundation as a function of depth.
   b. Construct a similar diagram showing the initial effective stress in the soil as a function of depth. The water table is at a depth of 10 ft.
   c. Construct a diagram showing the increase in stress at the center of the clay stratum (a depth of 33 ft) as a function of the horizontal distance from the footing centerline. What is the average stress increase directly beneath the foundation?

9:9 The clay of problem 9:8 is normally consolidated and has a compression index of 0.63 and a void ratio of 1.42.
   a. Find the settlement caused by the initial load and the average stress increase in the stratum.
   b. Below the clay is more sand. If the coefficient of consolidation is 0.02 sq ft per day, compute the time required for (1) 25 per cent, (2) 50 per cent, and (3) 75 per cent consolidation.

9:10 Find the additional increase in stress and increased settlement of problem 9:9 caused by permanently lowering the water table to a depth of 20 ft. This occurs after the settlement due to the foundation load is completed.

9:11 An elevated water tank weighing 400 tons rests on four square footings 20 ft apart (center to center). The allowable soil pressure is 5000 psf. The soil consists of 25 ft of gravel underlain by 8 ft of clay underlain by more gravel. The water table is below the clay. Both clay and gravel weigh 110 lb per cu ft. The void ratio of the clay is 0.80 and the compression index 0.32.
   a. Compute the average effective stress in the clay before and after construction. Use the Westergaard chart.
   b. Compute the tank settlement.
   c. Compute the tank settlement if the footing area is doubled. How much was the settlement reduced? (Express as a percentage of the original settlement.)

9:12 A monument has a base 40 ft × 60 ft. It weighs 3000 kips. It rests on a stratum of sand 40 ft thick underlain by a stratum of soft clay 5 ft thick. The clay rests on bedrock. The water table is at the ground surface and the γ of the sand is 130 lb per cu ft saturated and the γ of the clay is 110 lb per cu ft.
   a. Compute the average effective stress in the clay before and after construction. Use the Westergaard chart.

b. Compute the settlement of the monument if, for the clay, $e_0 = 1.13$ and $C_c = 0.31$.
c. Compute the time required for 80 per cent of the ultimate settlement to take place if $k = 7 \times 10^{-9}$ cm per sec.
d. Recompute stress and settlement, using approximate method.

9:13 A turbine foundation mat 20 ft × 40 ft carries a load of 1600 psf. The soil consists of 25 ft of clay overlying dense sand. The water table is at the ground surface. The clay has the following characteristics: $c = 2000$ psf, $\gamma = 110$ lb per cu ft, $E = 750,000$ psf, $c_v = 0.0015$ sq ft per day. The stress-void-ratio curve is given in problem 3:9.
a. Is the foundation safe if a minimum factor of safety of 2.5 is necessary?
b. What is the contact settlement?
c. What is the average stress increase in the upper 5-ft substratum of clay, the next 10 ft, and the bottom 10 ft of clay?
d. What is the total compression settlement?
e. How much settlement will take place in one year if a thin layer of sand is placed between the clay and the concrete mat?

9:14 Prepare a report describing the failure or excessive settlement of a structure due to faulty foundations. Include the following items:
a. Soil conditions.
b. Foundation.
c. Description of failure.
d. Cause of failure.
e. Corrective measures if any.

9:15 A building 100 ft × 80 ft of reinforced concrete weighs 700 psf of gross area. Columns are to be spaced 20 ft apart. The soil is deep, dry sand weighing 120 lb per cu ft. The following results were obtained with a load test made at the soil surface, using a 1 ft square plate.

Load (lb)	Settlement (In.)
1000	0.05
2000	0.10
3000	0.15
4000	0.20
5000	0.30
6000	0.60

The footings are to be placed 3 ft below the ground surface. The minimum safety factor is 2.5.
a. Plot load test and get failure q_c.
b. Compute φ

c. Find minimum square-footing sizes.
d. Find footing settlements for (c). (Assume that the settlements at higher pressures can be found by extending the straight-line portion of the load settlement curve and that formula 5:13b applies.)
e. If settlement is excessive what should be done?

9:16 A bridge pier has a total load of 1800 kips. It is located in a flood plain in which the water velocities are nominal. The soil profile consists of:
1. 30 ft of firm sand weighing 125 lb per cu ft, $\varphi = 35$ deg.
2. 20 ft of dense sand weighing 135 lb per cu ft and $\varphi = 42$ degrees.
3. 6 ft of clay, normally consolidated with $e = 1.25$ and $c_c = 0.48$.
4. 100 ft of dense gravelly sand.

Ground water is at a depth of 3 ft. Assume that the designer wishes a spread footing foundation and presumes a depth of 12 ft will provide safety against scour.

a. What is the minimum width of a square foundation required at a depth of 12 ft if the land can flood and if the foundation has an eccentricity of load of 3 ft perpendicular to the roadway. A safety factor of 3 is required.
b. Find the depth if penetration of a steel sheet piling cofferdam required to provide a safety factor of 1.5 against heave if the contractor excavates to the foundation level under water and then pumps the excavation from a ramp so the water level is 2 ft below the excavation bottom. What will be the approximate rate of pumping if the firm sand has a permeability coefficient of 3×10^{-3} per min and the dense sand is relatively impervious. Assume that 100 ft from the excavation the ground water level remains unchanged and that a two-dimensional flow net is a reasonable approximation of the seepage conditions.
c. What stress changes and settlement (or expansion) are produced by (1) excavation followed by (2) dewatering. Assume c_c is the same for both increase and decrease in stress in the clay.
d. How much settlement is produced by the weight of a concrete footing 10 ft thick poured in the excavation, assuming the ground water level regains its level 2 ft below the ground surface.
e. How much stress change and additional settlement takes place when the 1800 kip total load, 1200 kip dead load of the bridge is added to the foundation?

9:17 A chimney weighing 4000 kips is placed on a circular foundation with a bearing pressure of 4000 psf (including the foundation

weight). The soil profile consists of sand from the surface to a depth of 40 ft underlain by soft organic clay 6 ft thick. Under the soft clay is dense sand to a depth of 80 ft, underlain by rock. The upper sand weighs 103 pcf dry and 123 pcf saturated, and has an angle of internal friction of 32 degrees. The soft clay weighs 105 pcf saturated, has a void ratio of 1.5 and a compression index of 0.71. The ground water table before construction is 3 ft below the ground surface and is lowered permanently to 15 ft below the ground surface before foundation construction is begun. The base of the foundation will be 10 ft below the ground surface.
a. Compute the effective stress in the clay before construction.
b. Compute the effective stress increase in the clay caused by lowering the water table, and the settlement ultimately produced by the unwatering.
c. Compute the stress change in the clay caused by excavating the foundation and building the chimney, and the settlement resulting from these operations, assuming that the time interval between excavation and the pouring of concrete is so short that there is no time for decompression of the clay.
d. If the coefficient of consolidation of the clay is 0.02 sq ft per day, how long will it take for 90 per cent of the settlement to develop?
e. Is the foundation safe from bearing capacity failure under static load?
f. Is the foundation safe from bearing capacity failure if it must undergo a wind of 75 mph? The chimney is 20 ft in diameter and 250 ft high.
g. If the stack safety is marginal, what should be done to remedy the situation?
h. If the settlement is considered excessive, what alternate foundations might be considered and why?

ial deposit more than 100 ft thick. The formation was pre-
CHAPTER 10
Deep Foundations

THE SITE FOR A 20-STORY HOSPITAL WAS UNDERLAIN BY A RECENT alluvial deposit more than 100 ft thick. The formation was predominately soft, compressible clay. However, there were several strata of firm to dense sand within the clay; one about 40 ft or 12 m below the surface was about 10 ft or 3 m thick. Because the soft clay had insufficient bearing capacity for spread footings, the decision was made to support the structure on piles.

Timber piles were driven at several locations on the site. None could penetrate deeper than about 45 ft or 14 m below the surface. Several of the piles were load tested and none settled more than $\frac{3}{8}$ in. or about one cm at the design load of 30 tons. On the basis of the tests the structure was supported on more than 10,000 piles with their tips embedded in the dense sand. Within a year after the completion of construction the building had settled in a saucer-shaped depression nearly a foot below its original level at the center. Within 20 years the depression was more than 400 ft or 120 m in diameter and nearly 3 ft or one m deep at the center. Major repairs were necessary to keep the building serviceable. Today the settlement is increasing slowly in the secondary compression range, and although only minor repairs are now necessary, the movement will continue indefinitely.

Deep foundations are employed when the soil or rock strata immediately beneath the structure are not capable of supporting the load with adequate safety or tolerable settlement. Merely extending the level of support to the first hard stratum is not sufficient, although this is the decision that is often reached, as in the case of the hospital foundation. Instead, the deep foundation must be engineered in the same way as the shallow foundation. Like the shallow foundation, the deep foundation including the soil and rock strata below must be safe and free of objectionable settlement.

Two general forms of deep foundation are recognized: *piles* and *piers*. Piles are relatively long, slender shafts that are forced into the ground. Although piles as large as 5 ft or 1.5 m in diameter are sometimes driven, most are less than 2 ft or 60 cm. Piers are larger, constructed by excavation, and usually permit visual examination of the soil or rock on which they rest. In effect they are deep spread footings or mats. A sharp distinction between piles and piers is impossible because some foundations combine features of both.

10:1 Development and Use of Piles

Piles are older than history. The Neolithic inhabitants of Switzerland 12,000 years ago drove wooden poles in the soft bottoms of shallow lakes and on them erected their homes, high above marauding animals and warring neighbors. Similar structures are in use today in jungle areas of Southeast Asia and South America. Venice was built on wood piling in the marshy delta of the Po River to protect early Italians from the invaders from Eastern Europe and at the same time enable them to be close to the sea and their source of livelihood. Venezuela was given its name (meaning little Venice) by the Spanish explorers who found the Indians living in pile-supported huts in lagoons around the shores of Lake Maricaibo. Today, pile foundations serve the same purpose: to make it possible to build homes and maintain industry and commerce in areas where the soil conditions are unfavorable.

USES OF PILES / Piles are used in many ways, as shown in Fig. 10.1. *Bearing piles* that support foundation loads are the most common form. They do this either by transferring the load of the structure through soft strata into stronger, incompressible soils or rock below, or by spreading the load through soft strata that are not capable of supporting the concentrated loading of shallow footings. Bearing piles are used when there is danger of the upper soil strata being scoured away by current or wave action, or when wharves and bridges are built in water.

Tension piles are used to resist upward forces. These are used in structures subject to uplift, such as buildings with basements below the ground water level, aprons of dams, or buried tanks. They are also used to resist overturning of walls and dams and for anchors of guy wires, bulkheads, and towers.

Laterally loaded piles support loads applied perpendicular to the axis of the pile, and are used in foundations subject to horizontal forces such as retaining walls, bridges, dams, and wharves, and as fenders and dolphins for harbor construction. If the lateral loads are great, they can be resisted more effectively by *batter piles* driven at an angle. Frequently a combination of vertical and batter piles is used, as in Fig. 10.1c. Piles are sometimes employed to compact soils or to serve as vertical drains through strata of low permeability. Closely spaced piles and wide thin *sheet piles* that interlock together are used as retaining walls, temporary dams, or seepage cutoffs.

SEC. 10:2] PILE DRIVING

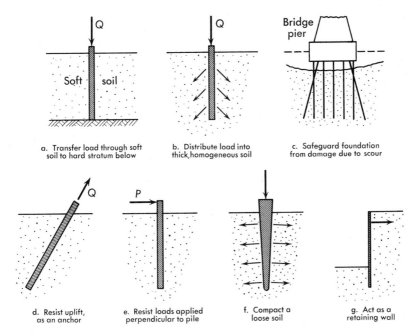

Figure 10.1 Uses of piles.

10:2 Pile Driving

The operation of forcing the pile into the ground is *pile driving*. Like many construction operations it is an art, the success of which is dependent on the skill and ingenuity of the workers. However, also like many construction operations it is increasingly dependent on engineering science for greatest effectiveness. Even more important both the art and engineering mechanics involved in the construction are major factors in the ultimate ability of the pile foundation to fulfill its function. Therefore, the foundation engineer must be ultimately involved in construction and the construction engineer in the design. The oldest method and the one most widely used today is by means of a hammer. Oriental constructors have used a stone block as a hammer for centuries. It is lifted by ropes held taut by laborers arranged in a star pattern around the pile head. The rhythmic pulling and stretching of the ropes throws the stone up in the air and guides the downward blow on the pile head. The Romans used a stone block hoisted by an A-frame derrick with slave or horse power, and guided in its fall by vertical poles.

PILE DRIVING EQUIPMENT / While the simple A-frame, pile-driving rig of the Romans is still in use today (with mechanical power), the more common machine is essentially a crawler-mounted crane (Figs. 10.2

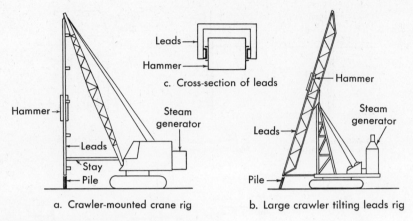

Figure 10.2 *Essential parts of pile-driving rigs.*

and 10.3). Attached to the boom are the *leads*: two parallel steel channels fastened together by U-shaped spacers and stiffened by trussing. These serve as guides for the *hammer*, which is fitted with lugs so as to slide between them. The leads are braced against the crane with a *stay*, which usually is adjustable to permit driving of batter piles. A steam generator or air compressor is required for steam hammers.

The pile is placed between the sides of the leads under the hammer. Lateral support is sometimes provided by sliding guides placed in the leads at the mid-point or quarter-points of the pile.

Some large pile-driving rigs are mounted on I-beam bases that are supported by steel beams and timber cribbing. They are moved by skidding them along the beams or on rollers. Rubber-tired motor crane rigs are used for highway work, and even fork lift trucks mount hammers for working inside buildings. Barge-mounted rigs are available for marine construction and compact rail-mounted rigs for work on tracks. Sometimes small *swinging leads* suspended from cables are used when there is insufficient room for a crane.

The most important feature of the driving rig, from the engineer's point of view, is its ability to guide the pile accurately. It must be rugged and rigid enough to keep the pile and hammer in alignment and plumb in spite of wind, underground obstructions, and the movement of the pile hammer.

PILE HAMMERS / The simplest hammer is the *drop hammer*, a block of cast steel commonly weighing from 500 to 2000 lb. It is raised 5 to 10 ft above the pile by a winch and then released and allowed to drop. The drop hammer is simple but very slow and is used only on small jobs where the contractor must improvise his equipment or where the cost of bringing in heavy driving equipment is not justified.

SEC. 10:2] PILE DRIVING 449

Figure 10.3 Pile-driving rig mounted on crawler treads with leads that can be tilted for driving batter piles. (Courtesy of Raymond International Inc.)

The *single-acting* steam hammer employs a heavy cast-steel block known as the *ram*, a piston, and a cylinder (Fig. 10.4a). Steam or compressed air is introduced into the cylinder to lift the ram 2 or 3 ft or 60 to 90 cm and then is released to allow the ram to fall on the head of the pile. These hammers are simple and rugged, and they deliver a low-velocity blow whose energy is relatively constant in spite of wear, adjustment, or small variations in steam pressure. Their characteristics are given in Table 10:1.

The *double-acting* or *differential-acting* hammer (Fig. 10.4b) employs steam or air pressure to lift the ram and then accelerate it downward. The blows are more rapid, from 95 to 240 blows per minute, thus reducing

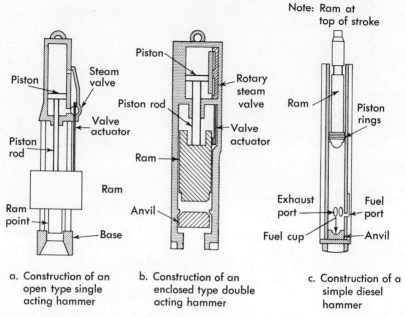

Figure 10.4 *Construction of pile-driving hammers.*

the time required to drive the pile and even making the driving easier in loose sands. They can lose some of their effectiveness with wear or poor valve adjustment. The amount of energy delivered in each blow varies greatly with the steam or air pressure and careful inspection is necessary to ensure that it is constant and the amount specified. If the number of hammer blows per minute is approximately the rated value, as given in Table 10:1, the steam pressure is probably correct.

Steam hammers can operate on both steam and compressed air. Steam operation is more efficient, particularly with circulating steam generators. If the hammer is to be operated underwater, as can be done with enclosed double-acting types, air is required.

Diesel pile hammers are available in an increasing range of sizes. They consist of a solid-bottom cylinder and an enclosed piston ram. The ram is raised upward mechanically and then allowed to fall. Fuel is injected into the cylinder while the hammer drops and is ignited by the heat of the air compressed by the ram. The impact and the explosion forces the cylinder down against the pile and the ram up, to repeat the cycle automatically. The important advantages of the diesel hammers are that they are self-contained, economical, and simple to service. The energy per hammer blow is high, considering the weight of the hammer, but it is developed by a high-velocity blow from a middle weight ram. The biggest disadvantage is

TABLE 10:1 / CHARACTERISTICS OF TYPICAL PILE-DRIVING HAMMERS

Hammer	Type	Wt. Ram (lb)	Stroke (in.)	Energy per blow (ft lb)	Strokes per min
Vulcan 2	Single Acting	3,000	29	7,620	70
Vulcan 1	Single Acting	5,000	36	15,000	60
Vulcan 0	Single Acting	7,500	39	24,375	50
McKiernan-Terry S-3	Single Acting	3,000	36	9,000	65
McKiernan-Terry S-5	Single Acting	5,000	39	16,250	60
McKiernan-Terry S-14	Single Acting	14,000	32	37,500	60
McKiernan-Terry 9B3	Double Acting	1,600	—	8,700	145
McKiernan-Terry 11B3	Double Acting	5,000	—	19,150	95
McKiernan-Terry C-5	Diff Acting	5,000	—	16,000	100
Vulcan 50C	Diff Acting	5,000	—	15,100	120
Vulcan 65C	Diff Acting	6,500	—	19,200	117
Vulcan 80C	Diff Acting	8,000	—	24,450	111
McKiernan-Terry DE20	Diesel	2,000	113*	18,800*	48*
McKiernan-Terry DE40	Diesel	4,000	129*	43,000*	48*
Link Belt 520	Diesel (DA)	5,070	—	30,000*	80*
Raymond 65CH	Hydraulic	6,500	—	19,500	130*

* Maximum energy of maximum stroke and minimum speed.

that the energy per hammer blow varies with the resistance offered by the pile and is extremely difficult to evaluate in the field. In some types of diesel hammers the length of the ram stroke can be observed visually and the available energy approximated by the product of the stroke and weight. In others the hammer energy can be estimated from the air pressure generated in a recoil chamber above the hammer. Because of the variable energy the diesel hammer is best adapted to conditions where controlled energy is not critical, or where it can be closely monitored at critical times.

A double-acting hammer actuated by hydraulic pressure is somewhat faster and lighter than equivalent steam hammers because the operating pressure is much greater. The compact hydraulic pump system is easier to move than the bulky air compressor or steam generator, although the higher pressures do involve more critical mechanical problems. The light double-acting hammer, will develop the same foot-pounds of energy at the instant of contact with the pile head as the heavy ram of a single-acting steam hammer falling 2.5 to 3 ft or 75 to 90 cm. The effects of the two blows, however, are different, owing to the widely different velocities of the falling rams at the instant of striking. Consider the driving of a railroad spike, using first a tack hammer and striking a hard, fast blow. Then use a heavy iron sledge hammer that drops a few inches so as to develop the same amount of energy. The slow, heavy blow drives the spike, whereas the tack hammer bounces. The same difference in effect can be observed in the driving of piles. Experience has shown that the weight of the driving ram should be from one-third to two times that of the pile.

Most pile hammers require the use of *driving heads*, *helmets*, or caps that distribute the force of the hammer blow over the butt of the pile. The head is made of cast steel and contains a renewable wood, fiber or laminated metal and rubber or plastic cushion block on which the hammer strikes. Heads for driving reinforced concrete piles may also provide for a wood cushion between the driving head and the pile.

BEHAVIOR OF THE PILE DURING DRIVING / Pile driving is a fascinating operation that never fails to attract crowds of onlookers. Clouds of steam and the repeated hammering are arresting, but they often obscure what merits much attention from the engineer—the behavior of the pile during driving. In very soft soils the first few blows of the hammer may drive the pile several feet; in fact, the pile may "run" into the ground under the static weight of the hammer. In harder soils, however, each blow of the hammer is accompanied by definite distortion of the pile and consequent losses of energy. If a piece of chalk is held against a pile and is moved with a steady horizontal motion while the driving is progressing, a graph will be traced on the pile that represents the vertical movement of the pile time. A typical example of such a graph is shown in Fig. 10.5. The blow of the

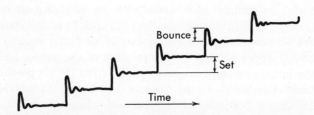

Figure 10.5 Graph of movement of the head of the pile during driving.

hammer produces an initial downward movement of the pile, but this is followed by a partial rebound or *bounce* that represents the temporary elastic compression of the pile and the soil surrounding it. The initial movement minus the bounce is called the *set* and is the net movement of the pile into the soil under one hammer blow. The average set for several hammer blows can be found from the driving resistance, which is the number of blows necessary to drive the pile a specified distance, usually 1 in., 6 in., or 1 ft (2.5, 15, or 30 cm).

When the pile is very long and the driving hard, the pile behavior is more complex. At the instant of impact the top part of the pile moves downward. The section of pile immediately below is compressed elastically and the tip of the pile momentarily remains fixed. The zone of compression travels swiftly down the pile, reaching the tip a fraction of a second after the initial impact. As a result of this compression wave the entire pile does not move downward at any one instant but instead moves in shorter segments.

SEC. 10:2] PILE DRIVING 453

OTHER METHODS OF PILE DRIVING / In cohesionless soils *jetting* can be used to place short, lightly loaded piles in their final position or as an aid to driving long, heavily loaded piles. The jet consists of a $1\frac{1}{2}$- to 2-in. pipe, with a nozzle half that diameter, that is supplied with water at from 150 to 300 psi. It can be used to wash a hole in the sand before driving or can be fastened to the pile, singly or in pairs (or even embedded in concrete piles), so that driving and jetting can proceed simultaneously. Since jetting loosens the soil, it is usually stopped before the pile reaches its final position, and the last few feet of penetration secured only by hammering. If too much water is used, the jetting can loosen previously driven piles. It is of greatest benefit in dense sands and of little help in clays.

Where stiff clays or soft rock must be penetrated at high levels in order to reach the bearing stratum, time and expense can be saved by *preboring*. In dry soils this is done with an auger, and the pile is dropped into the open hole. If the soil is continuously stiff, a concrete pile can be cast in the open hole, forming a *bored pile* (discussed later in the chapter). If the soil contains soft seams, the hole can be made with a rotary well drill and kept open by a slurry of soil and water. The pile is driven through the slurry to bearing in firm strata below.

Spudding is the driving of a heavy steel W^F section into the soil to punch through obstructions or to break up hard seams that could damage or even prevent penetration of small piles. The spud is withdrawn before the pile is driven.

Jacking is employed to drive piles when the vibration of hammering is not permissible or when the head room is too small to permit use of a pile hammer. It is used principally in underpinning where the piles are jacked in short sections, using the existing structure as a reaction.

Vibrators have proved effective in driving piles in silty and sandy soils.[10:3] The vibrators consist of pairs of counterrotating weights oriented to provide an up-and-down motion. Speeds of from 735 to 2500 rpm have been used with vibrators weighing from 26,000 to 30,000 lb, respectively. The 26,000-lb vibrator, driven by a 200-hp electric motor, develops a dynamic force of nearly 350,000 lb. A smaller 100-hp vibrator designed for sheet piling driving and extraction weighs 11,000 lb and delivers 54,000 ft lb per min at speeds of from 700 to 1000 rpm.[10:1]

The giants have been used in synchronized pairs to drive piles 4 ft in diameter and even larger caissons.

The sonic vibrator generates the vibration in resonance with the pile. In this way the impulse of the vibration is in phase with the elastic compression wave that travels down the pile, and the energy is used most effectively in overcoming friction and point resistance. The variable frequency necessary to match the natural frequency of the pile's elastic column is generated by an internal combustion engine drive. The speed of driving is phenomenal in many cases, and the noise and shock nuisance are less than for the impact hammers.

10:3 Pile Capacity

The ability of a pile foundation to support loads without failure or excessive settlement depends on a number of factors: the pile cap, the pile shaft, the transfer of the pile load to the soil, and the soil and underlying rock strata which ultimately support the load. The pile-cap analysis and design is essentially a structural problem and is covered adequately in textbooks on reinforced concrete design. It is rarely a critical problem or a source of trouble. The analysis and design of the pile shaft involves both the pile and the soil. Ordinarily the shaft capacity is dictated by construction needs and is far more than is needed for the ultimate load, but it can be critical with heavily loaded slender piles or when construction difficulties are encountered. The transfer of the pile load to the soil is termed the *pile-bearing capacity*. It is a frequent source of trouble in pile foundations. The ability of the underlying strata to carry the load depends on the combined effect of all the piles acting together. Although the capacity of the underlying strata seldom receives attention, it is a frequent source of trouble in pile foundations.

PILE SHAFT / The pile shaft is a structural column that is fixed at the point and usually restrained at the top. The elastic stability of piles, their resistance against buckling, has been investigated both theoretically and by load tests.[10:4] The buckling of a pile depends on its straightness, length, moment of inertia, and modulus of elasticity, and the elastic resistance of the soil that surrounds it. Both theory and experience demonstrate that the lateral support of the soil is so effective that buckling will occur only in extremely slender piles in very soft clays or in piles that extend through open air or water. Therefore the ordinary pile in sand or soft clay can be designed as though it were fully braced or were a short column. This is substantiated by load tests of 100 ft long piles in soft clay at a Midwest site. H-piles failed above the ground surface at the yield point of their steel, and concrete piles failed by crushing at the compressive strength of their concrete (Fig. 10.23).

The most important consideration in limiting the capacity of the shaft is faulty construction, particularly of connections between two sections of pile. This can lead to deflection of the lower part of the pile and the development of a dog-leg and to a reduction of the pile cross-section and a loss of strength as a short column. A study of dog-legged piles shows that their capacity is not materially reduced, provided the surrounding soil is firm.[10:5] The reduction in column strength can be prevented by careful control of the construction procedures.

EFFECT OF PILE ON SOIL / The stress pattern, settlement and ultimate capacity of the pile foundation depend on the effect of the pile on the soil. The pile, represented by a cylinder of length L and diameter D, Fig. 10.6a, is a discontinuity in the soil mass that either replaces or displaces

SEC. 10:3] PILE CAPACITY

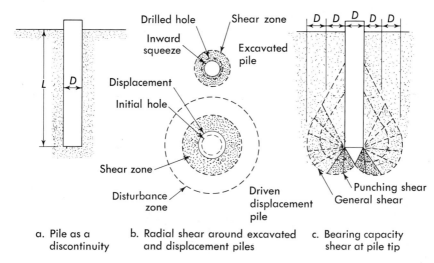

Figure 10.6 *Effect of a pile on the soil during driving.* (*Adapted from Meyerhof,* [10:6] *and Vesic.* [10:7])

soil depending on whether it is installed by excavation like a pier, or by driving.

Excavation disturbs the soil by changing the stress pattern; the soil may squeeze inward, Fig. 10.6b, disrupting the structure of clays and reducing the density of sands. Forcing a pile into the hole or placing wet concrete in the hole may partially force the soil back, creating more disturbance.

Driving the pile creates even greater disturbance. The tip acts as a small footing that accumulates a cone of soil, and punches down, forcing the soil aside in successive bearing capacity failures, Fig. 10.6c. A zone of disturbed or remolded soil is formed around the pile, with a width of from D to $2D$. If the driving is aided by jetting or predrilling a small hole, the disturbed zone is smaller. Within the disturbed zone there is a reduction of cohesive strength in saturated clays and cemented soils. In most cohesionless soils there is an increase in density and angle of internal friction.[10:6] However, in a very dense soil there might be a density reduction in the immediate vicinity of the pile due to shear and a slight local reduction in the angle of internal friction.

The displacement of driven piles has two effects. First there is heave of the ground in saturated clays and dense cohesionless soils. The heave sometimes pushes previously driven piles laterally as much as 1 or 2 ft or 30 to 60 cm, or raises the ground surface an amount equivalent to the volume of soil displaced. Second, high lateral pressures are set up in the soil. The limited data available indicate that the total lateral pressure in saturated clay can be as much as twice the total vertical overburden pressure; and in sands, the effective lateral pressure can be from one-half to four times the vertical effective stress.[10:7] In saturated clays even higher pressures have been indirectly indicated by the collapse of cofferdams and thin-walled open pipe or steel shell piles and the shoving of structures near piles being driven.

The tubular pile collapse, Fig. 10.7, occurred in a group of 36 piles driven 3 ft apart into stiff clay. The displacement of piles driven in clay on one project broke spread footings and raised a wall of an adjoining building 3 in.

Figure 10.7 Collapse of a hollow tubular pile in a large group, driven into stiff clay.

In saturated clays the pressure increase is largely in the neutral stress phase. This is dissipated into the surrounding soil with time, causing the lateral pressure to drop towards its original value, somewhat less than the overburden pressure. The reduction in neutral stress in the clay is accompanied by a regain in strength, which in some cases eventually exceeds the original strength of the undisturbed soil.

Driving piles with hammers produces shock and vibration that are transmitted through the ground to adjoining structures. This can annoy the occupants, and when sufficiently severe, cause physical damage. If there are very loose saturated fine sands present, the vibration may cause temporary liquifaction, loss of bearing capacity and severe damage. This is a rare occurrence, however. More commonly, the vibration in loose sand deposits causes a subsidence of the ground surface in spite of the pile displacement. The subsidence may extend as far as 100 ft from the structure, depending on the pile length and severity of driving, and cause settlement and damage to nearby buildings.

TRANSFER OF LOAD / The pile transfers the load into the soil in two ways as shown in Fig. 10.8:[10:8,10:9] First by the tip in compression, termed *end bearing*, and second, by shear along the surface commonly

SEC. 10:3] PILE CAPACITY

termed *skin friction* (although true friction does not develop in all cases). Piles driven through weak strata until their tips rest on a hard stratum transfer the greater part of their load by end bearing and are sometimes called *end bearing piles*. Piles in homogeneous soils transfer the greater part of their load by skin friction and are thus called *friction piles*. Nearly all piles develop both end bearing and skin friction, however.

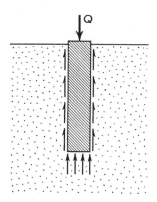

Figure 10.8 *Transfer of load from pile to soil by end bearing and skin friction.*

STRESS FIELD AROUND PILE / The initial stress field around a pile installed by drilling or jetting is probably close to the at-rest; depending on any stress reduction accompanying squeeze and any stress increase produced by displacement.

Upon loading the pile, the stress field is changed by the transfer of the pile load to the soil.

The analysis of stresses produced by a vertical load introduced below the surface of a semi-infinite isotropic elastic solid was developed by Mindlin, Fig. 10.9.[10:10] It is the equivalent of the Boussinesq analysis for surface loads. The vertical stress increment, $\Delta\sigma_z$, produced by a load Q at a depth of L, the pile length is given by the expression

$$\Delta\sigma_z = \frac{Q}{L^2} I_p, \qquad (10:1a)$$

$$I_p = f(z/L; x/L). \qquad (10:1b)$$

Contours of equal vertical stress in terms of I_p for unit depths of $z/L = m$ and load Q are shown on Fig. 10.9. The right half shows the vertical stresses at a radius of $n = x/L$ and $m = z/L$ for end bearing only at depth L; the left half shows the vertical stress for a uniform distribution of Q by skin friction along the length of the pile, Fig. 10.8. The vertical stress computed by the Boussinesq analysis assuming that the pile end bearing is at the surface of an

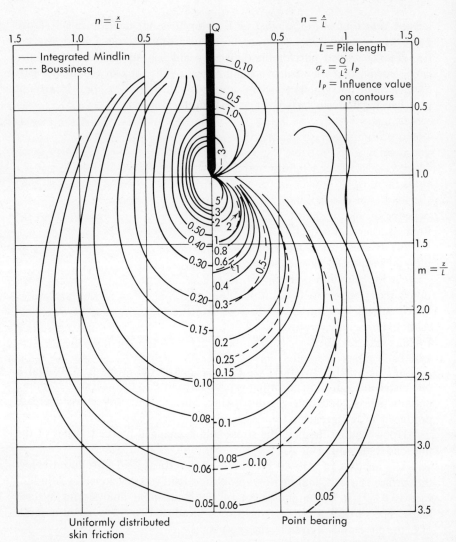

Figure 10.9 Contours of equal stress around and below the point of a pile in a semi-infinite homogeneous elastic solid: The Mindlin analysis. (After O. Grillo. [10:10])

elastic mass is shown by a dotted line on the right half of the diagram for comparison. It shows that the increased stress due to end bearing inside the elastic mass near the pile tip is approximately half of that found by the Boussinesq analysis for surface loads.

Above the pile tip, within a cylindrical zone whose radius is about half the pile length, the end bearing produces a negative $\Delta\sigma_z$, or a reduction in vertical stress in the mass. The radial stresses (in the lateral direction) are similarly influenced by the vertical stress transferred into the soil by the pile.

SEC. 10:3] PILE CAPACITY

Above the point of loading the radial stress is reduced; below the point of loading, it is increased.

The combined effect of end bearing and skin friction on the stress field depends on their relative magnitudes as well as the distribution of skin friction along the pile length.

Estimates of the vertical stress produced by a pile at loads below failure can be made with the use of Fig. 10.9b. Limited field observations of piles in homogeneous materials suggest that for lengths exceeding 20 diameters, the end bearing is about $\frac{1}{4}$ to $\frac{1}{3}$ the total; for shorter piles the proportion carried by end bearing increases in proportion to D/L. If the soil or rock at the pile tip is more rigid than along the shaft, the end bearing would be greater. As the load approaches failure the proportion of load transfer in end bearing depends on the ultimate or limiting shear in end bearing compared to the limiting shear in skin friction.

STRESSES ADJACENT TO A PILE / The vertical stress, σ_{zp} immediately adjacent to an unloaded drilled pile is γz, Fig. 10.10.

As the pile load is increased, there is a reduction in the vertical stress immediately adjacent to the lower part of the pile due to the load transferred in end bearing as can be seen from the tension zone of Fig. 10.9. Although this may be partially offset by the increase in vertical stress caused by load transfer in skin friction above, the net effect for long slender piles will be a stress reduction. In addition the subsidence of the soil mass surrounding the pile produces a vertical stress reduction similar to that in a

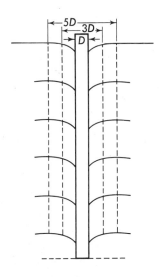
a. Distortion and zone of arching around a pile approaching failure

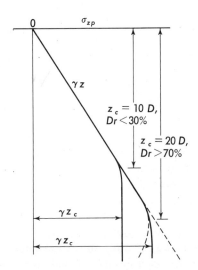
b. Vertical stress, σ_{zp}, adjacent to a loaded pile approaching failure

Figure 10.10 Vertical stress adjacent to a loaded pile. (Adapted from Vesic.[10:7])

backfilled trench, Fig. 8.27. As a result the vertical stress adjacent to a loaded pile is less than γz below a critical depth z_c, Fig. 10.10b. Below the depth of z_c the vertical stress immediately adjacent to a drilled or jetted pile depends on the pile load. At failure, tests indicate that it is approximately γz_c. At smaller loads it is somewhat higher. For driven piles (which have already "failed" successively during driving) the lateral pressure below z_c is apparently close to γz_c, regardless of load. Large scale tests in soils by Vesic[10:7] at the Georgia Institute of Technology and by Kerisel[10:11] in France indicate that the critical depth z_c is a function of relative density. For $D_r < 30\%$, $z_c = 10D$; for $D_r > 70\%$, $z_c = 30D$; for intermediate densities it is approximately proportional to relative density.

The lateral pressure of soil against the pile surface can be expressed by the equation

$$\sigma'_h = K_s \sigma'_{zp}. \qquad (10:2)$$

The earth pressure coefficient K_s depends on the displacement of the pile and the density or compressibility of the soil.

TABLE 10:2 / COEFFICIENT OF LATERAL EARTH PRESSURE IN COHESIONLESS SOIL ADJACENT TO PILE AT FAILURE

Soil	Displacement Condition	K_s
Loose Sand, $D_r < 30\%$	Jetted, Drilled Pile Driven Pile	0.5 to 0.75 0.75 to 1.5 2 to 3
Dense Sand, $D_r > 70\%$	Jetted, Drilled Pile Driven Pile	0.5 to 1 1 to 2 3 to 5

For the jetted or drilled piles the values of K_s increase with load; the maximum occurs at failure.

STATIC ANALYSIS OF BEARING CAPACITY / The ultimate bearing capacity of the pile or pier is the sum of end bearing and skin friction at the instant of maximum load:

$$Q_0 = Q_{EB} + Q_{SF}. \qquad (10:3)$$

The ultimate values for Q_{EBC} and Q_{SFC} can be analyzed separately. Both are based upon the state of stress around the pile (or any deep foundation) and on the shear patterns that develop at failure.

In end bearing the pile tip resembles a deeply buried spread footing. When the pile is loaded, a cone of intact soil adheres to the tip. As the tip penetrates deeper with increasing load, the cone forces the soil aside, shearing

SEC. 10:3] PILE CAPACITY

the mass along a curved surface, Fig. 10.6c. If the soil is loose, compressible or has a low modulus of elasticity, the mass beyond the shear zone compresses or deforms, allowing the cone to penetrate further. This is a form of local shear, similar to that described for shallow foundation, Chapter 9. If the soil or rock is very rigid, the shear zone expands until the total displacement allows the cone to punch downward. Various forms of shear zone have been proposed to evaluate end bearing. Like the results of shallow foundation analyses these can be expressed in the general form

$$q_0 = \frac{B\gamma}{2} N_\gamma + c N_c + q' N_q. \qquad (9:3)$$

For piles where B is small, the first term is often omitted:

$$q_0 = c N_c + q' N_q. \qquad (10:4)$$

Although many different bearing capacity factors have been derived for deep foundations the range that has been verified to some extent by full size test piles is depicted on Fig. 10.11. The lower curves are the Meyerhof

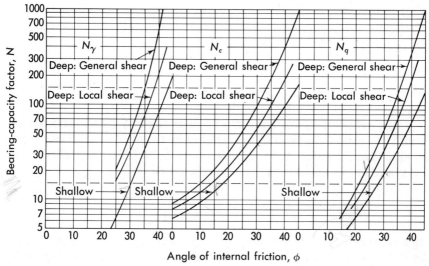

Figure 10.11 Bearing capacity factors for shallow and deep square or cylindrical foundations. (Adapted from Meyerhof [10:6], and Berezantzev.[10:12])

factors for shallow foundations, corrected for a circular or square shape. The upper curves are for general shear failure, adapted from Meyerhof,[10:6] and require the complete development of the shear zone that can only occur in a rigid-plastic solid or a dense sand. The middle curves are adapted from Berezantzev's work in sands; they fit the results of load tests of driven large scale models and full-sized piles.[10:12]

It is difficult to define the correct bearing capacity factors to be used for any situation. The shallow factors apply to end bearing piles or piers resting on hard strata overlain by weaker formations. They also apply to piles embedded in soft clays and loose sands. The highest factors apply only to the harder clays and very dense sands in which the point is embedded to a depth of $10D$. Factors for conditions between these limits can be interpolated with caution. For many real cases, tests show that the middle curves apply.

If the piles are driven into the ground, the appropriate angle of friction is that obtained after driving. In sands the increase is from 2 to 5 degrees above the value before driving according to Meyerhof.[10:6] If the foundation is installed by jetting or drilling, the angle remains virtually unchanged.

The appropriate value of q' at the foundation level depends on the pile length.

$$q' = \gamma z \text{ if } z < z_c, \tag{10:5a}$$

$$q' = \gamma z_c \text{ if } z > z_c. \tag{10:5b}$$

The skin friction that acts along the pile shaft is equal to either the sum of friction plus adhesion on the pile face or to the shear strength of the soil immediately adjacent to the pile, whichever is smaller. If f is the skin friction

$$f = \begin{cases} c' + \sigma'_h \tan \varphi' \\ c_a + \sigma'_h \tan \delta, \end{cases} \tag{10:6a}$$

where c_a is the adhesion and δ is the angle of friction of soil against the pile face.

The values of c_a and $\tan \delta$ can be determined by direct shear tests in which one half of the shear box is replaced by the same material as the pile surface. Load tests on full-sized piles suggest the following values for c_a compared to c of saturated clays in undrained shear

$$c_a = 0.9c \qquad\qquad c < 1 \; ksf \tag{10:6b}$$

$$c_a = 0.9 + 0.3\,(c - 1) \qquad c > 1 \; ksf \tag{10:6c}$$

In these expressions, c and c_a are in kips per sq ft. The lower rate of increase in c_a for the soils in which $c > 1 \; ksf$, appears to be a slight gap that forms around the pile during driving, and possibly due to tensile stresses that develop around the upper end of the pile shaft during loading. There is some evidence that c_a increases slowly with the passage of time until it equals c.

Typical values of $\tan \delta$ are given in Table 10:3, based on limited test data.

The effective lateral stress is computed from Equation (10:3a) and (b) using the coefficients estimated from Table 10:2.

Empirical correlations have been made between the end bearing and skin friction of piles in cohesionless soils and the penetration resistance

TABLE 10:3 / COEFFICIENT OF FRICTION, COHESIONLESS SOILS AGAINST PILES AND SIMILAR STRUCTURES

Material	Coefficient of Friction, $\tan \delta$	δ
Wood	0.4	22°
Rough concrete, cast against soil	$\tan \phi$	ϕ
Smooth, formed concrete	0.3–0.4	17
Clean steel	0.2	11
Rusty steel	0.4	22
Corrugated metal	$\tan \phi$	ϕ

measured during site exploration. The point resistance of the Dutch static cone, Fig. 6.7, in cohesionless sand is approximately equal to the end bearing of a pile in the same material. The skin friction of steel and concrete piles is approximately double the cone friction, for piles with $L/D > 20$. For the Standard Penetration Resistance, N, Meyerhof[10:13] suggests the following:

$$q_0 = 8 N \text{ (in kips per sq ft)} \tag{10:7a}$$

$$f_0 = 0.04 N \text{ (in kips per sq ft)}. \tag{10:7b}$$

Immediately after driving, the soil strength (and adhesion) are in the remolded condition. After a clay soil has had an opportunity to reconsolidate and, in some cases, to harden thixotropically, the adhesion and the strength immediately adjacent to the pile increase and can even exceed the original soil strength. Piles pulled out of clay frequently are covered with a skin of soil several inches thick that adheres tightly to their surface.

The total pile capacity is nominally the sum of the mobilized end bearing and the product of the mobilized skin friction and the surface area of the pile. The ultimate or failure load, Q_0, however, is not necessarily the sum of the ultimate end bearing and the ultimate skin friction. First the end bearing and the skin friction along different sections of the shaft may not be mobilized simultaneously. Consider a pile whose shaft is in a weak, non-rigid soil but whose point rests on a rigid stratum. A relatively small downward movement of the pile would be sufficient to produce bearing-capacity failure, but the same movement would not be great enough to produce skin-friction failure. Therefore only part of the skin friction would be mobilized at the instant of failure. The deflection of the pile shaft under load (which is greatest at the ground surface but less at the point), the different rigidities of different strata in contact with the pile, and the compression of soil beneath the pile point also contribute to the unequal mobilization of end bearing and skin friction. As a result the actual pile capacity can be materially less than the sum of the ultimate values. The difference is aggravated in ultrasensitive soils where failure brings about a loss of strength. For these reasons the skin frictions of weaker strata are generally neglected in analysis.

Generally a driven pile has a higher ultimate capacity than one placed by excavation or jetting, because both skin friction and end bearing reach their ultimate values during driving.

A second cause of difference between the computed and the actual ultimate capacity of piles arises from negative skin friction.[10:14] The stresses introduced into the soils by the pile and by any surface loads such as fill not supported by piles will cause the soils to consolidate. If there is a highly compressible stratum at some level above the pile point, its consolidation will cause the soils above to move downward with respect to the pile. Instead of supporting the pile, these strata, by their downward movement, now add load. This negative skin friction has been great enough to cause failure of pile foundations in a few cases and must be considered in design.

Example 10:1

Compute the ultimate bearing capacity of a 14 in. square reinforced concrete pile 65 ft long driven into a thick stratum of homogeneous, insensitive clay. The shear strength of the clay is given by $s = c = 1250$ psf, and the unit weight of the clay is 113 lb per cu ft. The adhesion is 0.9 times the cohesion. The water table is at the ground surface.

1. End bearing (Fig. 10.11):

$$q_0 = 9 \times 1250 + 65(113 - 62.4),$$
$$q_0 = 11{,}250 + 3{,}280 = 14{,}530 \text{ psf},$$
$$Q_0 = 14{,}530 \times \frac{14 \times 14}{144} = 19{,}800 \text{ lb}.$$

2. Skin friction S:

$$S = 1250 \times \frac{14}{12} \times 4 \times 65 \times 0.9 = 341{,}000 \text{ lb}.$$

3. Total ultimate capacity:

$$Q_0 = 19{,}800 + 341{,}000 = 360{,}800 \text{ lb}.$$

LOAD TESTING / The most reliable method of determining pile capacity for most sites is a load test. Pile load tests are made to determine the ultimate failure load of a pile or group of piles or to determine if the pile or pile group is capable of supporting a load without excessive or continuous settlement.

The bearing capacity of all piles except those driven to rock does not reach the ultimate until after a period of rest. Load tests are not a good indication of performance unless made after this period of adjustment. For piles in permeable soils this period is two or three days, but for piles partly or wholly surrounded by silt or clay it can be more than a month.

SEC. 10:3] PILE CAPACITY

Pile load tests can be made by constructing a loading platform or box on top of the pile or group of piles (Fig. 10.12a) on which the load is applied, using sand, pig iron, concrete blocks, or water. A safer and more easily controlled test uses a large, accurately calibrated hydraulic jack to apply the load (Fig. 10.12b). The resistance above the jack can be secured with a loaded platform or by a beam held down by piles in tension. An added advantage of the jacking method is that the load on the pile can be varied rapidly and cheaply. Settlements are measured by a precision level or preferably by micrometer dial gages mounted on an independent support.

The loads are applied in increments of one-fifth or one-fourth the design load until failure or two times the design load is reached; the load is then reduced to zero by increments. Each load is maintained constant and the settlement is measured at regular intervals until the rate of movement is less than 0.0005 in. per hr. A load final-settlement curve is plotted similar to that for the plate load test.

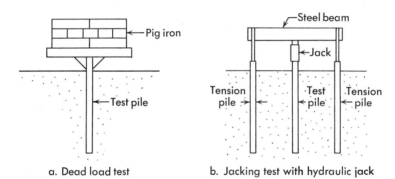

Figure 10.12 *Pile load testing.*

Many different criteria for working load have been proposed, but the best is the same as for any other foundation: the load having an adequate safety factor (1.5 to 2 when a test is made) or the load giving the greatest permissible total settlement (as described in Chapter 5), whichever is the smaller.

SETTLEMENT OF A SINGLE PILE / Settlement of a single isolated pile comes from the elastic shortening of the pile shaft and to some extent from the distortion of the soil around the pile. These can be determined best from a load test. Settlement can be computed from the static analysis of pile capacity by computing the elastic shortening of each section of the pile shaft from that portion of the total load remaining in that section.

The major settlement of all piles except those that are end bearing on rock comes from the consolidation of the underlying soil by stresses developed by the pile group. This is considered in Section 6:4.

TENSION PILES / Tension piles can be analyzed by the static method (with no end bearing) or by tension load tests. The resistance of tension piles with enlarged bases can be determined best by load tests.

10:4 Dynamic Analysis of Pile Capacity

Since the driving of a pile produces successive pile-bearing failures, it should be possible theoretically to develop some relationship between pile capacity and the resistance offered to driving with a hammer. Such *dynamic analyses* of pile capacity, often termed *pile formulas*, have been employed for over a century. While in some cases they have been able to predict pile capacity accurately, in others they have not, and their indiscriminate use has led to both overdesign and failures.

The pile loading and "failure" produced by driving with a hammer occurs in a small fraction of a second, whereas in the structure the load is applied over a period ranging from hours to years. A fixed relation between the dynamic and long-term capacity can exist only in a soil whose shear strength is independent of the rate of loading. This is approximately true in a dry cohesionless soil, and in wet cohesionless soils that are of intermediate density or so coarse grained that shear does not develop appreciable neutral stresses. In clays and in both very loose or dense fine-grained saturated cohesionless soils, the strength depends on the rate of shear; in such soils a dynamic analysis can have no validity.

WAVE ANALYSIS / The dynamic process of pile driving is analogous to the impact of a concentrated mass upon an elastic rod. The rod is partially restrained along the surface by skin friction and at the tip by end bearing. This system can be approximated by a lumped mass elastic model, Fig. 10.13a.[10:16] The distributed mass of the pile is represented by a series of small concentrated masses, W, linked by springs that simulate the longitudinal resistance of the pile. The skin resistance can be represented by a rheologic model of surface restraint that includes friction, elastic distortion, and damping.

When the hammer strikes the pile cap, a force R_c is generated that accelerates the cap (W_c) and compresses the cap. This transfers a force, R_0 to the top segment of the pile, W_1 and causes it to accelerate, slightly after the acceleration of W_c. The compressive force induced in the top of the pile, R_1 induces acceleration in the next segment of pile, W_2. A wave of compression moves down the pile. The vertical force at any instant, t, is equivalent to the spring compression. Diagrams of force throughout the pile length for successive time intervals, t_1, t_2 ..., are shown in Fig. 10.13b. The force wave is partially dissipated in overcoming skin friction on the way down; at the bottom the force remaining overcomes end bearing. In order for the pile to penetrate deeper, the force in the wave must exceed the accumulated

SEC. 10:4] DYNAMIC ANALYSIS OF PILE CAPACITY

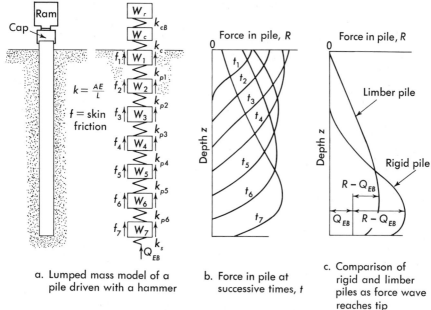

a. Lumped mass model of a pile driven with a hammer

b. Force in pile at successive times, t

c. Comparison of rigid and limber piles as force wave reaches tip

Figure 10.13 *Wave analysis of a driven pile.*

sums of the ultimate skin friction and the ultimate end bearing; if it does not, the pile is said to meet *refusal*.

The shape of the force wave depends on the pile rigidity; a rigid pile (stiff springs) exhibits a sharper force wave with a higher peak than a limber pile. The force overcoming end bearing, $R - Q_{EB}$ is greater for the rigid pile, Fig. 10.13c. The peak force is also a function of hammer energy and efficiency: the higher energy produces a higher force. The force, divided by the cross-sectional area of the pile equals the stress produced during driving. If the maximum stress exceeds the pile strength, the pile will be damaged, a condition known as *overdriving*.

Although a wave analysis of pile driving provides a clear picture of the mechanics of the process, it has limited usefulness in evaluating pile capacity. The energy dissipation by skin friction and the equivalent spring constant and resistance of end bearing are difficult to evaluate under field conditions, and virtually impossible to predict in advance. The calculations are easily made with a digital computer, but are tedious by hand. Generally the analysis is used for diagnosing the causes of unusual driving behavior or as a guide to more efficient choice of equipment or pile.

APPROXIMATE METHODS / Approximate methods of dynamic analysis, the socalled *pile formulas*, have been used for more than a century, and are still useful in predicting the ultimate pile capacity from simple observations of driving resistance.

All the dynamic analyses are based on the transfer of the kinetic energy of the falling pile hammer to the pile and the soil. This accomplishes useful work by forcing the pile into the soil against its dynamic resistance. Energy is wasted in the mechanical friction of the hammer, in the transfer of energy from the hammer to the pile by impact, and in temporary compression of the pile, pile cap (if any), and of the soil. The basic relationship, therefore, will be

$$(R_0 \times s) + \text{losses} = W_r \times h \times (\text{efficiency}) \qquad (10:8)$$

where R_0 is the resistance of the pile to driving; s, the distance it moves into the ground from one hammer blow (the set); W_r, the weight of the pile hammer; and h, the height that the hammer drops. The relation is solved for R_0 which is then assumed to be equal to the capacity of the pile under sustained loading, Q_0.

The major uncertainty in this approach and the basic difference between all the pile formulas is the way in which the energy losses and the mechanical efficiency of the process are computed. The most complete is that of Hiley as described by Chellis.[10:17] The mechanical efficiency of the hammer is expressed by e, a coefficient that ranges from 0.75 for drop hammers operated by a drum winch, or for most steam pile hammers that are not new, to 0.9 for new double-acting hammers and higher for hydraulic hammers. The energy available from the hammer after impact can be approximated by the method of impulse and momentum. This considers the coefficient of restitution, n, which ranges from 0.9 for aluminum–plastic laminate to 0.25 for a hammer striking on the head of a wood pile or a wood cushion block in a pile helmet or cap. In addition it involves the weight of the hammer, W_r, and the pile weight W_p. The available energy after impact is the hammer energy multiplied by

$$\frac{W_r + n^2 W_p}{W_r + W_p}.$$

This shows that as the weight of the pile increases with respect to that of the hammer, the relative inertia increases and there is less energy available for useful work. For long piles this is not strictly valid because the pile moves in a wave rather than as a rigid body.

The energy lost by elastic compression of the pile, any helmet, and the soil can be approximated by assuming a linear increase in the stress acting from 0 to R_0 while the compression develops. The energy loss will therefore be

$$\frac{R_0 c_1}{2} + \frac{R_0 c_2}{2} + \frac{R_0 c_3}{2},$$

where c_1, c_2, and c_3 are respectively the temporary elastic compression of

SEC. 10:4] DYNAMIC ANALYSIS OF PILE CAPACITY 469

the helmet, pile, and soil. The value of $c_2 + c_3$ is the bounce of the pile with each hammer blow (Fig. 10.5) and is easily measured as described before. The value of c_1 must be estimated from the value of R_0 and the shape and material of the helmet. The resulting energy balance and dynamic formula are,

$$\underbrace{R_0 s + R_0\left(\frac{c_1}{2} + \frac{c_2}{2} + \frac{c_3}{2}\right)}_{\text{Work done on pile}} = \underbrace{W_r h e\left(\frac{W_r + n^2 W_p}{W_r + W_p}\right)}_{\text{Energy available to pile}}; \quad (10{:}9a)$$

$$Q_0 = R_0 = \frac{W_r h e}{s + \tfrac{1}{2}(c_1 + c_2 + c_3)}\left(\frac{W_r + n^2 W_p}{W_r + W_p}\right). \quad (10{:}9b)$$

NOTE: The formula is dimensionally homogeneous and both h and s must be in the same units.

For double-acting hammers the rated energy E in the same length units as s is substituted for $W_r h$.

Detailed tables of the constants for use in the Hiley formula have been published.[10:17] Although the values of e, n, and (for long piles) W_p must be estimated, which requires considerable experience, the method is reasonably accurate for piles driven in cohesionless soils. A safety factor of from 2 to 2.5 is ordinarily employed to obtain the safe load.

For long piles and very rigid piles, the Hiley formula is overly conservative because only a fraction of the total pile weight is accelerated at one time, as is demonstrated by the wave analysis. The approximation is more realistic if the moving mass, W_p, is taken to be the weight of the pile cap plus the weight of the uppermost part of the pile. The proper length depends on the pile rigidity and weight per unit of length: for heavy steel mandrels and precast concrete piles it is 30 to 50 ft, or 9 to 15 m.

ENGINEERING NEWS FORMULA / The formula (10:9b), can be simplified by substituting arbitrary constants for the different factors in the equation. The *Engineering News* formula was derived from observations of the driving of wood piles in sand with a free-falling drop hammer. The value of $c_1 + c_2 + c_3$ is assumed to be 2 in. or 5 cm, and both the hammer efficiency and impact factor are assumed to be 1. The result is

$$R_0 = \frac{W_r h}{s + 1}. \quad (10{:}10a)$$

A safety factor of 6 was introduced to make up for any inaccuracies arising from the use of the arbitrary constants. Since the height of fall of drop hammers is usually measured in feet and s is measured in inches, a factor of 12 was added to make it possible to use the mixed units. This reduced to the familiar form of the equation:

$$R_s = \frac{1}{6}R_0 = \frac{W_r(h' \times 12)}{6(s+1)},$$

$$R_s = \frac{2W_r h'}{s+1}. \qquad (10\!:\!10\text{b})$$

In this expression, h' is the hammer drop in feet and s is in inches. R_s is the safe pile load, including the built-in safety factor. The formula was later modified for steam hammers by substituting 0.2 in. for the temporary compression to give

$$R_s = \frac{2W_r h'}{s+0.1}. \qquad (10\!:\!10\text{c})$$

Numerous pile-load tests show that the real safety factor of the *Engineering News* formula averages 2 instead of its apparent 6, and that the safety factor can be as low as $\tfrac{2}{3}$ and as high as 20. For wood piles driven with free-falling drop hammers and for lightly loaded short piles driven with a steam hammer, the *Engineering News* formulas are a crude indication of pile capacity. For other conditions they can be very misleading.

10:5 Pile Groups

Since piles are ordinarily closely spaced beneath footings or foundations, the action of the entire pile group must be considered. This is particularly important when purely "friction" piles are used and when the hard stratum on which the points of end-bearing piles rest is underlain by more compressible soils.

GROUP BEARING CAPACITY / The group bearing capacity is computed by assuming that the piles form a giant foundation whose base is at the level of the pile points and whose length and width are the length and width of the group (Fig. 10.14a).[10:18] The group capacity is the sum of the bearing capacity developed by the base of the "foundation" and the shear developed along the vertical sides of the group "foundation".

The bearing is computed by using the general bearing-capacity Equation (9:3). The factors for deep foundations are used when the pile length is at least ten times the group width and when the soil is homogeneous; in all other cases the factors for shallow foundations are used. The shear around the group perimeter is the soil strength, determined without any increase in lateral pressure from pile displacement, multiplied by the surface area of the group. While model tests show that the actual group capacity is always slightly less than the computed values, the difference is well within a safety factor of 2.

SEC. 10:5] PILE GROUPS

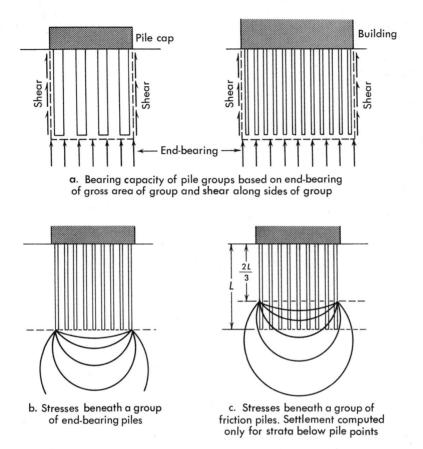

Figure 10.14 *Approximate method for analyzing bearing capacity and settlement of pile groups by assuming a group to act as a single foundation unit.*

PILE EFFICIENCY / The efficiency of the pile group e_g is the ratio of the group capacity, Q_g, to the sums of the capacities of the number of piles, n, in the group:

$$e_g = \frac{Q_g}{n\, Q_0}. \qquad (10:11)$$

Although a number of empirical formulas have been derived for group efficiency, none have been shown to be realistic. Instead the efficiency should be evaluated from the group capacity using the definition, Equation (10:11). The group capacity increases with spacing while the individual capacity of piles in clay does not. A plot of theoretical efficiency vs spacing, Fig. 10.15a shows the group capacity equals the sums of the individual capacities at an optimum spacing, and an efficiency of 1. The optimum

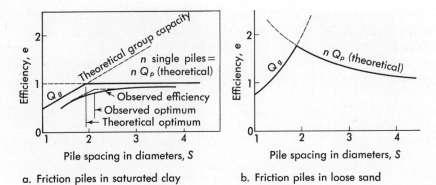

Figure 10.15 *Efficiency of groups of long friction piles.*

spacing S_0 for long piles in clay and group efficiency at the optimum spacing is given by the following:[10:18]

$$S_0 = 1.1 + 0.4n^{0.4}, \qquad (10{:}12a)$$

$$e = 0.5 + \frac{0.4}{(n - 0.9)^{0.1}}. \qquad (10{:}12b)$$

Typically S_0 is 2 to 3 pile diameters center to center. Model tests in clay[10:18] indicate that the actual efficiency at the optimum spacing is somewhat less than 1 (0.85 to 0.9) and increases slowly at greater spacings. For design, utilizing the common safety factor of 2, the error in assuming the real efficiency to be 1 at the optimum spacing is inconsequential.

For piles in cohesionless soil the capacity of the individual piles increases with reduced pile spacing because of increased soil strength due to densification.[10:19] The optimum spacing, Fig. 10.15b, is very small, and has an efficiency greater than 1. Unfortunately it is impossible to drive the piles that close. Practical spacings are 2.5 to 4 diameters, center to center.

GROUP SETTLEMENT / The settlement of a group of piles results from consolidation of the soil strata beneath the pile points. Such settlement will exceed that of an isolated pile that carries the same load as each pile in the group unless the piles are end bearing on rock or on a thick stratum of incompressible soil. The group settlement may be analyzed by again considering the group to represent a giant foundation. When the piles are end bearing, the base of the imaginary footing is assumed to be at the level of the pile tips, as shown in Fig. 10.14b, and the stresses are computed on that basis. When the piles are supported by friction, the stresses beneath the footing are computed by assuming that the entire group load is introduced in the soil at a depth of from one-half to two-thirds the pile length. The load is distributed at this level over the gross area of the group. Settlements of the soil strata beneath the pile points are computed from these

SEC. 10:6] LATERAL LOADS   473

stresses. Such computations are approximate at best and are likely to give settlements that are higher than the observed values. They can indicate when trouble from settlement is likely to occur.

The exact stress distribution for a load of significant width applied well below the surface of an elastic solid has not been solved. As an approximation, the stresses may be computed by the Boussinesq or Westergaard analyses as for surface loads, Fig. 9.11–9.14, and then reduced. The basis for the reduction is the Mindlin analysis which shows that the stresses beneath a point load deep below the surface are about half of those found by Boussinesq.

Example 10:2

Compute the soil stresses due to piles at a point 10 ft below the tips of a group of friction piles 60 ft long. The outer or gross dimensions of the group are 6 ft × 6 ft and the ground load is 500 kips.
1. The group load of 500 kips is assumed to act at a depth of $\frac{2}{3}$ × 60 = 40 ft. The area over which it acts is 6 ft × 6 ft = 36 sq ft.
2. The depth at which the stresses are to be found is 10 + 20 = 30 ft below the point the 500-kip load is applied. This is 30/6 = 5 times the group width or 5b.
3. From Fig. 9.12, surface loading of a thinly stratified mass, the stress at a depth of 5B directly below the loaded area is 0.014q.
4. The stress due to the piles, computed from the chart applicable to surface loading is

$$\Delta\sigma_z = 0.014 \times \frac{500}{36} = 0.194 \text{ kips per sq ft.}$$

5. The real stress increase is probably between 0.100 and 0.194 or about 0.15 kips per sq ft.

10:6 Lateral Loads

VERTICAL PILES / A vertical pile loaded laterally deflects as a partially supported cantilever beam, Fig. 10.16. If the loads are small the resistance of the soil is reasonably elastic. This can be approximated by assuming the soil to react as a series of horizontal springs whose rigidity can be expressed as a coefficient of subgrade k (Chapter 5). The differential equation of the bending beam can be solved for deflections and moments as well as the soil pressure by successive approximations or the finite element approach. Such solutions are available in graphical non-dimensional form for different assumed variations of k. The curves of Fig. 10.16 developed by Reese and Matlock,[10:20] give deflections, moments and soil pressures for a pile of constant stiffness and k increasing linearly with depth: $k = k'z$. The

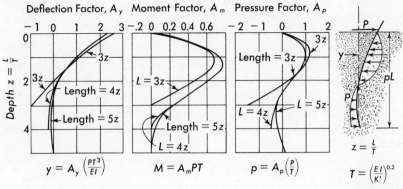

Figure 10.16 Behavior of a laterally loaded vertical pile. (After Reese and Matlock.[10:2])

values of k or k' are best determined from a full sized pile load test of the soil formation. Typical values are given in Table 10:4.

The curves are expressed in terms of the relative stiffness, T in in.:

$$T = \left(\frac{EI}{k'}\right)^{0.2}. \tag{10:13a}$$

where E and I refer to the pile cross-section. The depth is expressed by the dimensionless coefficient Z

$$Z = \frac{L}{T}. \tag{10:13b}$$

If the lateral load is great enough, the soil pressure will exceed the soil strength and the pile will fail. The failure resistance is sometimes computed to be the passive earth pressure against the upper part of the pile. This is unrealistic because the usual theories of passive pressure assume two-dimensional or plane strain shear while the laterally loaded pile will fail in

TABLE 10:4 / COEFFICIENT OF SUBGRADE REACTION FOR LATERALLY LOADED PILES

Soil	k'
Soft clay	1–5 lb per sq in. per in.
Stiff clay	10–20 lb per sq in. per in.
Loose sand	5–10 lb per sq in. per in.
Dense sand	25–50 lb per sq in. per in.

three-dimensional shear at a pressure exceeding the passive. Moreover, the deflection accompanying failure is so great that a structure supported by a laterally loaded pile would be in distress long before failure of the pile is reached.

Typical load test results suggest fully embedded vertical piles can support lateral loads of only 1/10 to 1/5 of their vertical capacity without excessive (more than $\frac{1}{2}$ in.) deflection. If greater rigidity or greater lateral resistance is needed, batter piles are required.

BATTER PILES IN GROUPS / Batter piles combined with vertical piles are the most effective device for resisting horizontal thrusts. Anchors for wharves and bulkheads that combine a vertical pile in tension and a batter pile in compression, as shown in Fig. 10.17, have proved to be compact

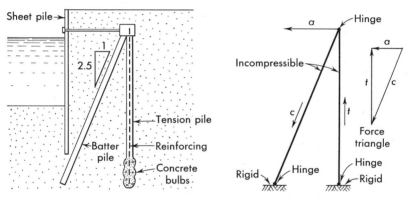

a. Batter and vertical pile used in A-frame anchor for sheet-pile bulkhead

b. Simplified analysis for combined vertical and batter piles

Figure 10.17 Batter and vertical piles in combination.

and economical. Batter piles combined with vertical piles have been utilized to support retaining walls and similar structures that develop horizontal loads. A rational analysis of batter-pile loading is impossible because the problem is statically indeterminate to a high degree. One approximate method assumes the piles to be hinged at their points and at their butts. (Fig. 10:17b)

10:7 Types of Piles and Their Construction

PILE SHAPES / Constructors through the ages have tried and used with varying degrees of success many shapes and types of piling. Each shape has probably been successful under certain conditions. However, the use of a certain type or shape of pile that has proved successful in one job may not meet with success in a different situation. In the United States the establishment of large and well-equipped pile-driving organizations has led to the general use of a few types and shapes of piles.

Four basic shapes are commonly used: first, uniform cross-section throughout the length; second, enlarged base; third, tapered; and fourth, sheet. These are shown in Fig. 10.18.

The uniform-section pile comes in a variety of forms: cylindrical, square, octagonal, fluted, and H-section. The uniform section provides uniform column strength from the point to the butt, and "skin friction" is well distributed over the entire shaft. It is well adapted to splicing and cutting, since each section of the pile is identical.

In order to increase the end bearing and the friction on the lower portion of the pile, different forms of enlarged points have been used. In one form a large, precast point is attached to a cylindrical pile, while in another form a bubble of concrete is forced into the soil at the pile point. Piles of this shape have proved very effective in developing end bearing on firm, cohesive soils and even in loose sands. They are of little value as friction piles and have little advantage over uniform-section piles when used as end bearing on rock.

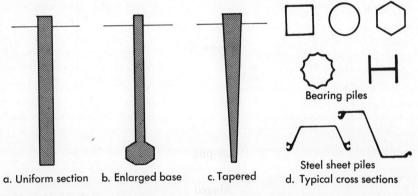

Figure 10.18 Basic pile shapes.

The tapered shape originated with wood piles that conform to the natural shape of the tree. However, the taper has been imitated in concrete and steel in order to permit easier construction. Tapered piles are useful in compacting loose sands because of their wedge action but in other cases may be less effective than uniform-section piles. Both the point bearing and the skin friction on the lower portions of the tapered pile are low, since both the point area and the surface area of the pile are small. The result is that end-bearing, tapered piles require greater lengths than uniform-section piles in order to support a given load. Tapered piles that depend on friction for support may transmit a large portion of their load to the upper, weaker soil strata and produce objectionable settlement.

SEC. 10:7] TYPES OF PILES AND THEIR CONSTRUCTION

Sheet piles are relatively flat and wide in cross-section so that when they are driven side by side they form a wall. Many different forms of wood, concrete, and steel sheet piles have been developed for special purposes such as cofferdams, wharves, retaining walls, and cutoffs. Some have arch-shaped and Z-shaped cross-sections to provide rigidity, and most all types are made to interlock with adjacent sheet piles to form a soiltight wall.

Piles that are hollow have a distinct advantage over those that are not, for it is possible to inspect the entire length of the piles after driving. Piles may deviate from the vertical, develop sharp bending or "dog-legs," or may be damaged from overdriving. Hollow piles may be inspected by dropping a burning flare into them or by reflecting the sun's rays into them with a mirror, but other forms must be assumed satisfactory without any check. Therefore higher safety factors should be used with piles that cannot be inspected. Hollow piles that are driven with open ends and then cleaned out make possible an examination of the soil beneath the pile point. When an open-end pile rests on an irregular rock surface, the rock may be smoothed by drilling; and if the pile is found to be hung on a large boulder above the supporting stratum, the boulder may be drilled or dynamited to permit the pile to reach its full depth.

WOOD PILES / Wood is one of the most commonly used pile materials because it is cheap, readily available, and easy to handle. Some kind of timber suitable for piling will be found available in nearly every section of the world. Spruce, fir, and pine up to 100 ft long; oak and mixed hardwood piles up to 50 ft; southern pine up to 75 ft; and palmetto are commonly used for piling. Untreated timber piles completely embedded in soil below the water level will remain sound and durable indefinitely. When the campanile of St. Mark's in Venice fell in 1902, it was found that the 1000-year-old piles were in such a good state of preservation that they were left in place and used to support the new tower. Sound timber piles that have been water soaked for many years should not be allowed to dry before redriving, since in drying the wood fiber becomes "short" and brittle.

Above the water table untreated timber is subject to decay and damage from termites and other insects. In salt water timber is susceptible to marine borer attack.[10:21] Many types of marine borers are found but most are related to the lobster and crab or to the clam and oyster families. The crab-like limnora destroy the wood from the outside in, leaving the pile as a slender spindle of wood (Fig. 10.19a). The clam-like toredo destroy the wood from the inside out; they enter the pile through a small opening, destroy the inside of the pile, and leave it a hollow shell. Timber piles can be made to last longer through treatment with zinc chloride, copper sulfate, or numerous patented chemicals. Creosote impregnation has proved to be one of the most efficient and long lasting means of protecting timber piles. From 12 to 25 lb of creosote per cubic foot of timber are forced into the wood by vacuum- and pressure-treatment processes. In areas of very severe

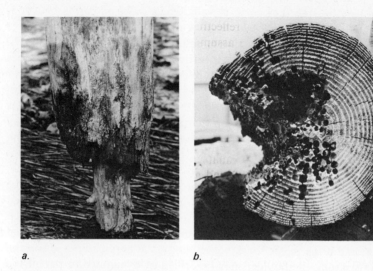

Figure 10.19 Wood pile hazards. a. Limnora attack necking pile at the water line. b. Toredo attack destroying center and one quadrant of pile cross section. c. Splitting from over-driving.

marine borer attack combination treatment of copper arsenate followed by a coal tar–creosote mixture, both applied under heat and pressure are necessary to assure a life of 15 to 25 years in salt water.

Timber piles tend to suffer badly from overdriving. The tops of the piles become "broomed," and the shafts are very likely to split or break, as shown in Fig. 10.19b, when stiff resistance to driving is encountered.

SEC. 10:7] TYPES OF PILES AND THEIR CONSTRUCTION 479

On one lock construction job on the Mississippi River it was necessary to drive several thousand wood piles through an undisclosed stratum of cemented sand. Subsequent excavation disclosed that many of the piles had splintered and broken. It was later found necessary to spud a steel beam through the cemented stratum before driving the timber piles.

Timber piles are ordinarily capable of supporting safely from 15 to 30 tons per pile. Timber piles have been utilized for design loads of 50 tons or more, and the safety of these designs have been demonstrated by load tests. The major problem of such designs is ensuring that the structural quality of the wood piles is uniformly high so that there is no danger of the piles breaking during driving. The very low cost for materials and for driving often makes timber the cheapest pile foundation per ton supported.

PRECAST CONCRETE PILES[10:22] / Precast concrete piles are uniform-section circular, square, or octagonal shafts with sufficient reinforcing to enable them to withstand handling stresses. The smaller sizes are 8 to 12 in. or 20 to 30 cm wide and are usually solid. Larger sizes are solid or are hollow so as to reduce their weight. Prestressing makes it possible to secure adequate strength with relatively thin concrete walls; diameters up to 54 in. or 1.4 m with walls 4 in. thick, similar to concrete pipe, have been used where great stiffness and high bearing capacity are required.

Precast piles are used principally in marine construction and bridges where durability under extreme exposure is important and where the pile extends above the earth as an unsupported column. In the latter case the reinforcing is increased as dictated by the column requirements. Typical lengths of the small, solid piles are 50 to 60 ft or 15 to 18 m, and of the larger hollow piles, up to 200 ft or 60 m. Typical loads for the small piles are 30 to 50 tons, and for the larger piles, over 200 tons.

Two factors limit the use of precast piles. First they are relatively heavy compared with other piles of comparable size. Second it is difficult to cut them off if they prove to be too long and even more difficult to splice them to increase their length.

CAST-IN-PLACE CONCRETE PILES / Concrete piles that are cast in the ground are the most widely used types of piles for 30 to 60 ton loads. These can be divided into two groups: *cased piles* in which a thin metal casing is driven into the ground to serve as a form, and *uncased* piles where the concrete is placed directly against the soil. Many types of each have been developed, and the engineer will find it enlightening to study the catalogs of pile contractors to see the different methods of construction.

The *Raymond Standard* pile (Fig. 10.20a) is one of the earliest cased types. A thin metal shell with an 8-in or 20 cm diam tip and with a taper of 0.4 in. in diameter for each foot of length is driven into the ground on a close-fitting steel core or mandrel. After the mandrel is withdrawn, the tapered hole, supported by the shell, is filled with concrete. This pile is employed for lengths up to 37 ft or 12 m and for loads of 30 to 40 tons.

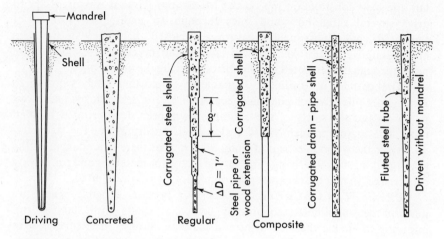

Figure 10.20 *Cast-in-place concrete piles.*

The *Raymond–Step-Taper* pile (Fig. 10.20b) consists of a series of cylindrical, corrugated sheet-metal sections, each 8 ft long and 1 in. in diameter larger than the one below and screwed together to form a continuous tube. The minimum tip diameter is $8\frac{5}{8}$ in. or 22 cm but larger tips up to $13\frac{3}{8}$ in. or 35 cm can be used by starting the pile with the larger cylindrical sections. The pile is driven by a loosely fitting mandrel that drives against the tip and the shoulders or *plow rings* of each larger section. Lengths up to 96 ft or 29 m and loads of from 40 to 75 tons, depending on the tip diameter, are used.

The *Cobi pile* and the *Hercules pile* employ a cylindrical corrugated sheet-metal shell, similar to drainage pipe, from 8 to 21 in. or 20 to 53 cm inside diameter. The shell is closed at the lower end with a flat or cone-shaped boot. It is driven with a cylindrical steel core that expands to grip the inside of the pipe and its corrugations tightly. The Cobi-type core expands by air pressure in a rubber tube, whereas the Hercules type expands by mechanical wedging. Lengths up to 100 ft or 30 m are possible.

The *Union Monotube* consists of a thin-walled, fluted steel tube that is driven into the soil without a mandrel or core. The fluting makes the thin steel shell capable of withstanding the driving stresses without buckling. Monotubes as long as 125 ft or 37 m carrying loads of 30 to 60 tons, are used. They are particularly suited to small jobs because they require no special driving equipment such as the mandrel.

Many variations of the thin shell cast-in-place pile are used. The *Button Bottom*, employs an 18-in. or 45 cm diameter precast concrete pedestal at the bottom of the 12-in. or 30 cm corrugated pipe shell. After the mandrel is

SEC. 10:7] TYPES OF PILES AND THEIR CONSTRUCTION

withdrawn, the corrugated shell is filled with concrete to make a continuous pile. This form develops unusually high end bearing but reduced skin friction because the 18-in. point punched a larger hole than the pile shaft.

The thin shell piles have a number of features in common. Ordinarily they are not reinforced, because they are in compression when supporting vertical loads. Reinforcing can be added before pouring if the piles are to resist tension or flexure. The thin shell is usually not considered a part of that reinforcement, because of possible corrosion. It is easy to cut them off if too long or to increase their length during driving by welding on more shell. They can be inspected after driving to check their straightness. The shell keeps water and soil away from the wet concrete and develops a shaft of uniform quality. The thin shells sometimes are damaged by obstructions that tear them or smooth out their corrugations and reduce their strength, or they may collapse because of the extremely high lateral pressures that develop in stiff clays and dense sands.

The uncased concrete pile formed by a temporary casing, is shown in Fig. 10.21a. A steel pipe or *casing* is first driven into the ground. Soil is kept out of the lower end by a precast concrete plug or a metal pan that is supported by a core. After driving the core is removed and the pile filled with concrete. The core is lowered into the concrete and the casing is pulled out while the core forces the concrete against the soil and prevents it from moving upward with the casing. The *bulb pile* is formed similarly except that the casing is initially only partly filled with concrete. The casing is then raised and the core forced downward by hammering, which forces the concrete out in a bulb (Fig. 10.21b and c). After forming the bulb, the casing is filled with concrete and then pulled out, leaving the bulb or *pedestal pile*. The uncased piles ordinarily are not reinforced. The bulb piles, however, make excellent tension piles but in such applications require reinforcing.

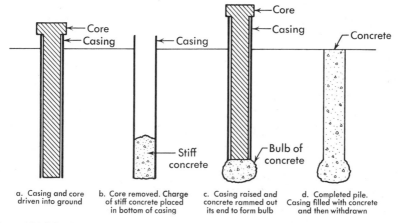

Figure 10.21 Construction of uncased bulb pile.

Uncased piles formed in a casing are suitable for loose sands and firm clays where the lateral pressures developed will not squeeze the unprotected fresh concrete. Lengths of 60 ft or 18 m and loads of from 30 to 75 tons are the usual limits for such piles. Uncased piles require heavy driving rigs and special apparatus for pulling the casing and are economical where the size of the job justifies the initial expense of equipment.

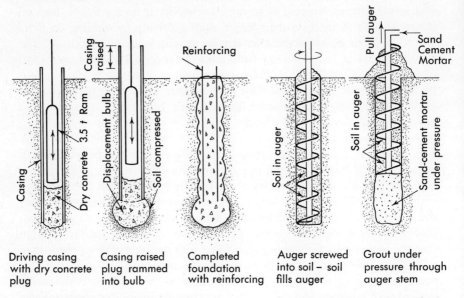

Figure 10.22 Special uncased piles.

The *Franki* uncased concrete pile, Fig. 10.22a, is formed by ramming a charge of dry concrete in the bottom of a 20-in. or 50 cm steel casing so that the concrete grips the walls of the pipe and forms a plug. A 3.5-ton ram falling 10 to 20 ft or 3 to 6 m inside the casing forces the plug into the ground, dragging the casing downward by friction. At the bearing level, the casing is anchored to the driving rig, and the concrete plug is driven out its bottom to form a bulb over 3 ft or 1 m in diameter. The casing is then raised while successive charges of concrete are rammed in place to form a rough shaft above the pedestal. Lengths up to 100 ft or 30 m with capacities of from 100 to 1000 tons are typical. With reinforcement they make excellent tension piles.

A number of forms of augered uncased piles are available. The *Augercast* pile, Fig. 10.22b is drilled with a continuous auger with a hollow central stem. The rate of drilling is such that the auger screws itself into the ground rather than expelling the soil. The hole, therefore, remains full of intact

SEC. 10:7] TYPES OF PILES AND THEIR CONSTRUCTION

soil, until it reaches the bearing stratum. At that point the auger is slowly withdrawn. At the same time sand cement grout is pumped down the hollow stem. The rate of auger pulling is controlled so that there is always a positive pressure on the grout to fill the hole, prevent hole collapse, and to force the grout a few inches into loose sands and gravels. The resulting pile develops both end bearing and skin friction from its irregular shaft, Fig. 10.18c. The process is economical and free of vibrations, an advantage in building additions and underpinning. Lengths of 60 ft or 18 m and diameters of 14 to 18 in. or 35 to 45 cm are commonly used.

Bored piles consisting of augered holes filled with concrete can be used where the soils are firm enough to stand without support. Diameters from 6 in. or 15 cm up and lengths of more than 50 ft or 15 m are commonly constructed. (Diameters larger than 22 in. or 56 cm are considered as piers and will be discussed later.)

It is difficult to measure the quality of the augered piles except by load testing. Test piles should be installed before the design is final. After their strength has been verified by load tests, continuous inspection is essential. If there is any change in construction procedure, new load tests must be made to check the effect of change.

STRUCTURAL STEEL PILES / Structural steel shapes, particularly the H-pile and W^F sections, are widely used for bearing piles, usually when high end bearing is required on soil or rock. The cross-sectional area is small compared with the strength and makes driving much easier through obstructions such as hard, cemented seams, old timbers, and even thin layers of partially weathered rock. The sections can be obtained in pieces and can be easily cut off or spliced. The sections ordinarily driven are 8BP36 to 14BP117, with working loads of from 40 to 150 tons. Wide-flange sections as deep as 36 in. or 90 cm have been driven, and built-up piles from channels and railroad rails are occasionally used. The lengths are limited only by driving; 14-in. H-sections over 300 ft or 90 m long have been installed.

H-piles driven onto rock have demonstrated their capability of supporting loads up to the yield point of the steel. Figure 10.23 shows the local buckling of a 100 ft long H-pile driven through soft clay and then load-tested to 400 tons, approximately the yield point. Apparently the pile bites into the rock and establishes full bearing in spite of irregularities in the surface. In very hard rock, the point end is sometimes reinforced by gusset plates on the web to prevent local buckling. The H-pile penetrates soil with minimum displacement and development of heave and lateral pressure. When H-piles are used in friction, the surface area between the flanges is so great that friction failure occurs by shear parallel to the web across the outer edges of the flanges, and by friction against metal along the outer faces of the flanges.

The structural sections have three disadvantages. First they are

Figure 10.23 *Failure of long H-piles above ground at yield point stress in steel. Piles were driven through 100 ft of soft clay to end bearing on rock.*

relatively flexible and easily deflected or twisted by boulders. In fact a few H-piles have been deflected so far that their points skidded along the bearing stratum instead of biting in. Second the soil packs between the flanges so that the friction area is equal to the rectangle which encompasses the pile rather than to the total area of the pile surface. Third, corrosion reduces the effective cross-section. In most soil a corrosion allowance of 0.05 to 0.1 in. is realistic because a heavy film of rust protects the pile from further attack. In strongly acid soils such as fill and organic matter, and in sea water, corrosion is more serious. Cathodic protection or jacketing with concrete is necessary to prevent deterioration.

STEEL PIPE PILES / Steel pipes filled with concrete make excellent piles. In most cases they are driven with the tip closed by a flat plate or a conical point. The flat plates are cheapest and tend to form a conical point of soil ahead of them during driving. An X-shape of plate welded to the tip helps the pile to break through gravel and cemented layers and to bite into bed rock. Open-ended pipes are used where minimum displacement is essential. The plug of soil that pushes up into the pipe must be removed at intervals to prevent its packing and causing the pile to drive as though it had a closed end.

Both the closed-end and the open-ended pipes are filled with concrete after driving (after cleaning the open pipe, of course). This increases the shaft load capacity because both the strength of the steel and of the concrete contribute to the column strength.

Pipes from 10.75 in. (27 cm) OD × 0.188-in. (4.8 mm) wall to 36 in.

(90 cm) OD × 0.50-in. (13 mm) wall have been driven with capacities of from 50 to more than 200 tons. Lengths are limited by the driving equipment; pipe piles over 200 ft or 60 m long have been installed.

Pipes are light, easy to handle and to drive, and can be cut off and spliced readily. They are stiffer than H-piles and not so likely to deflect when they strike an obstruction. They have the distinct advantage that they can be inspected internally after driving and before concreting.

In driving steel piles, the hammer must strike squarely over the centroid of the section. An off-center or wobbling hammer will "accordion" pipes and batter structural sections, which destroys the effectiveness of the blow. The carbon content of the pile is important, for if it is too high, the pile will split; if too low, the steel will yield. On one job where 100 miles of 10.75-diam by 0.25-in. steel pipe were driven through clays in lengths up to 160 ft, it was found that cold-formed steel with 0.22 per cent carbon and 0.6 per cent manganese drove best.

COMPOSITE PILES / Composite piles are a combination of a steel or timber lower section with a cast-in-place concrete upper section. In this way it is possible to combine the economy of a wood pile below the ground water level with the durability of concrete above water, or to combine the low cost of cast-in-place concrete with the great length or relatively greater driving strength of pipes or H-sections.

The design and construction of the splice between the two sections is the key to a successful composite pile. The head of the lower section must be protected against damage; a tight joint must be maintained to exclude water and soil from the shell; good alignment between the sections must be maintained to prevent dog-legging; and the splice must be as strong as the weakest member it connects.

Two methods are used. In one the lower section is driven its full length. The metal shell, on its driving core, is attached to the lower section, and the whole assembly is driven to the final penetration. The core is then withdrawn and the shell concreted. A second method consists in driving a steel casing first. The core is withdrawn and the lower section is placed in the casing like a projectile in a gun. The lower section is then driven out the casing by the core. The thin steel shell for the cast-in-place section is lowered into the casing and locked on the lower section, after which the casing is withdrawn and the pile concreted.

Composite wood-concrete piles 140 ft or 43 m long and steel pipe–concrete piles 180 ft or 55 m long have been used with loads of up to 30 and 60 tons, respectively.

SAND PILES / Holes rammed into soil and filled with sand or crushed slag for the purpose of compacting and draining a soil are known as *sand* or *wick* piles. They have little structural strength other than that of the compacted sand. They are constructed in the same way as uncased cast-in-place concrete piles but a free draining material is used instead of concrete.

10:8 Design of Pile Foundations

The design of a pile foundation is similar to the design of any other part of a structure. It consists of assuming a design, then checking the proposed design for safety and revising it until it is satisfactory. Several such designs are then compared, and the final one is selected on the basis of cost and time required for construction.

Piles in a foundation may be valueless in some locations, and under some conditions their use actually may be very harmful. For example, a layer of reasonably firm soil over a deep bed of soft soil might act as a natural mat to distribute the load of a shallow-footing foundation. The driving of piles into the firm layer might break it up or remold it. The result would be a concentration of load in the soft soil strata, with excessive settlement likely to take place.

SELECTION OF PILE LENGTH / The selection of the approximate pile length is made from a study of the soil profile and the strength and compressibility of each soil stratum. Such studies may be made by using the methods of pile group analysis discussed in Section 10:5. End-bearing piles must reach a stratum that is capable of supporting the entire foundation without undue settlement or failure, and friction piles must be long enough to distribute the stresses through the soil mass so as to minimize settlement and obtain adequate safety of the entire group of piles.

SELECTION OF POSSIBLE PILE TYPES / The pile type and the material from which it can be made must be carefully chosen to fit:
1. The superimposed load.
2. The amount of time available for the completion of the job.
3. The characteristics of the soil strata through which it penetrates as well as that of the strata to which the load must be transferred.
4. The ground water conditions.
5. The size of the job.
6. The availability of equipment and getting it onto the site.
7. The availability of material for the piles.
8. The building code requirements.

If the structure is a bridge abutment or a wharf, the depth of the water, its velocity, ice condition, and the possibility of marine borers or chemicals in the water attacking the pile material must be given full consideration. Scour is quite likely to take place around new bridge piers and abutments because of the increased water velocity; the piling in such cases should be protected by concrete and the structure should be braced by batter piles.

If the foundation loads are low and scattered, a pile of low cost per foot and per pile may be the most economical. If the loads are high and concentrated within small areas, a pile having a high load-supporting value will probably be the lowest in cost per ton of load. If the load is a single load of over 300 tons, and there are several such load points, some type of pier may be more economical.

SEC. 10:8] DESIGN OF PILE FOUNDATIONS

The shipping of pile-driving equipment is always expensive. The contractor who has his equipment within truck-hauling distance of a job has a marked advantage over the contractor who must load and unload his equipment from railroad cars. A few piles required on a job do not justify the cost of moving heavy, large pile-driving equipment. Light, easily handled piles, which may be driven by equipment that will be used for erecting the superstructure may provide the most economical job. When the job includes several hundred piles 40 ft and longer, the large pile-driving rigs are more economical, since more piles can be driven per working shift and a greater variety of pile types is available.

If a pile foundation job is located in the heavy timber portions of the country, the delivered price of wood piles is low. Therefore the total cost of using many piles at 20 tons each is less than that of using concrete piles at 40 tons each.

Since concrete is generally a part of every construction job, it is available in all parts of the country for piling. Steel may be cheaper in industrial areas but may be more expensive when the job is a long distance from the mill.

The characteristics of handling and shipping the piles may influence the choice of type. The ability to "nest" pile shells one within another makes a more compact load, especially for shipment by water. Light weight with high resistance to rough handling is a decided advantage with steel pipe piles, fluted pile shells, and the corrugated shells used as a part of many cast-in-place piles. Wood piles are also light and easy to handle. Long H-beams and precast concrete piles must be handled in slings to prevent bending or breaking.

The requirements of building codes are becoming more uniform throughout the country. Codes generally specify certain maximum loads allowed per pile and more generally are providing for load tests to determine the allowable maximum loads.

PILE DESIGN LOADS / The design of the pile shaft is governed by stresses produced by driving. During driving the actual or working load on the shaft equals the failure load between the pile and the soil, R_0. The driving stresses can be estimated from Equation (10:9b), regardless of the type of soil because only the dynamic resistance is concerned. The shaft should have a safety factor of at least 1.3 with respect to R_0, which means that the safety factor with respect to the design load, Q_a is larger than for other short columns. The safe load on the pile is governed by the soil-to-pile connection and the group capacity. These are analyzed as described in Sections 10:3, 10:4, and 10:9, and the appropriate safety factor applied depending on the reliability of the analyses and the structural loading data.

SPACING / The final pile spacing is based on the analysis of pile group action. The piles are placed so that the capacity of the pile group acting as a unit is equal to the sum of the capacities of the individual piles.

Greater spacing may be necessary in dence sands and stiff clays to minimize lateral pressures from displacement.

TOLERANCES / It is impossible to construct piles in exactly the required location or angle because they tend to drift out of line when they encounter hard or soft spots in the ground. The designs and specifications should allow for 2 in. of tolerance in the top of small piles in soil and 6 in. (and sometimes more) for piles driven through water. Out of plumbness and doglegging of one or 2 per cent of the pile length generally does not affect pile capacity and should be permitted by the design and specifications. More is usually permitted if load tests show the piles are adequate.

INSPECTION AND RECORDS / No important pile-driving job should be carried out without competent engineering inspection and the keeping of complete records of the driving of every pile. The field reports should record the following:

1. Time, weather, and working conditions.
2. Type and size of hammer, weight of driving ram, weight of driving cap.
3. Actual length of the hammer stroke.
4. The number of blows struck per minute by the hammer.
5. Steam or air pressure at the hammer or boiler. The length of pipe and hose between the boiler and hammer should be noted.
6. Length driven, pile size, etc.
7. Number of blows of continuous driving required to drive the last foot or few inches. The count should not begin until the pile is in motion if there has been a suspension of driving.
8. Rebound or bounce of the pile. This may be observed by inscribing a graph on the surface of the pile with a fixed bar or crayon.
9. Suspension of driving if prior to the final penetration.
10. The condition of the inside of *every* pile shell *immediately* before filling with concrete. (An electric light or flashlight on a line, sunlight reflected by mirrors, a burning ball of oil soaked rags, or other light source may be used.)
11. Heaving of the adjacent ground.
12. Shrinkage of the adjacent ground.
13. Heaving of the piles after driving or heaving of the soft concrete inside the pile shell.

PILE CAP / The load of a wall or column must be transferred to the pile by means of a footing or pile cap. In designing the cap, consideration must be given to the fact that the pile butts may be from 2 to 4 in. or 5 to 10 cm out of their required position. In some cases the piles may be pulled or jacked into position, depending on the rigidity of the pile and the soil, but it is cheaper if the pile cap is designed with allowances for some misalignment. In dock structures the pile must resist horizontal forces and often must resist rotation. In such cases the pile must be anchored to the cap by adequate embedment and in some cases with reinforcing steel. The structural design of the cap is similar to the design of a footing foundation. Care must be taken to see that the footing is rigid enough to transfer the load to the outermost piles in the group.

10:9 Pier Foundations

The pier foundation is a relatively large, deep foundation. Its function is to transfer a foundation load through soft soil to hard soil or rock or to transfer a load through soils that may be scoured away by rivers or tidal currents. The chief differences between piles and piers are size (piles larger than 24 in. or 60 cm in diameter are sometimes called *piers*) and the method of construction. Piles are ordinarily forced into the ground without previous excavation whereas piers usually require soil excavation ahead of or during their construction.

Piers are divided into two classes, *open shafts* or wells, and *caissons*, depending on the method of construction. Open shafts are merely deep excavations that are provided with bracing or lining when needed as the work progresses. A caisson is a box or chamber that excludes water and soil from the excavation. It is usually prefabricated above ground and then sunk to the required level by excavation from within. The word caisson is often applied to any pier but, strictly speaking, it refers only to those that employ the box or chamber that is lowered as excavation proceeds.

The materials employed and the type of structure are dependent upon the load, ground water conditions, depth of load-supporting strata, building code requirements, and the availability of materials and equipment. If the pier is to be in water, the velocity of the water, its maximum depth under scouring conditions, and the effect of ice and debris must be provided for in the design.

BEARING CAPACITY AND SETTLEMENT / A pier is actually a deep footing foundations that is supported by end bearing and by shear or friction along its sides. The end bearing is computed by the general bearing-capacity Equation (9:3), using the appropriate bearing-capacity factors in Fig. 10.11. The skin friction on the pier is either friction plus adhesion or shear strength, whichever is the smaller. In computing friction and shear, no allowance is made for any increase in lateral pressure from displacement because the piers are constructed by excavation.

For piers resting on hard rock, with soil above, skin friction is neglected. For piers entirely in a homogeneous soil, the skin friction in the soil can be included. If the rock is weak, and sufficient end bearing cannot be obtained, the pier can be extended below the rock surface. The skin friction or shear below that level augments the total bearing of the pier.

It is essential that the soil or rock immediately beneath a very heavily loaded pier or caisson be examined because local soft or compressible seams could seriously impair the bearing or settlement. If the caisson is large enough and can be unwatered, one to three 2 in. or 5 cm holes are drilled in the bottom, 1.5B to 3B (pier) deep. These are probed with a hooked rod to find soft seams or cavities. It is also possible to make core borings in the bottom, with the drill rig on the ground surface, particularly if the pier cannot be unwatered. Boring in a deep open shaft is difficult, and only the best workmanship is permissible. It is essential that loose material be removed from the bottom—every inch of loose soil means virtually an inch of settlement.

The settlement of piers is often the governing factor in their design. Settlement can be estimated by the methods described in Chapter 5, by assuming the end of the pier to be a footing. The stresses in the soil below such a "footing" are considerably less than those computed by the Boussinesq or Westergaard methods because those analyses are based on loads at the ground surface. Based on the Mindlin analyses, Fig. 10.9, the stresses beneath a narrow deep end-bearing pier can be as little as half those computed by surface load methods.

OPEN SHAFTS / The simplest form of the open shaft is an open excavation similar to a dug well (Fig. 10.24a and b). Shallow wells in firm soil can be dug by hand. Large power augers are capable of drilling open

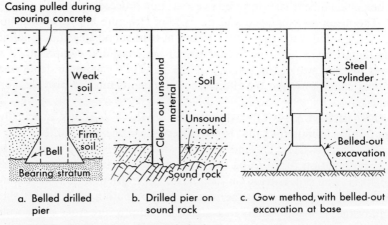

Figure 10.24 *Construction of open shaft piers.*

excavations as large as 10 ft or 3 m in diameter and more than 9 m or 27 ft deep. Special underreaming drills can be used to enlarge or *bell out* the bottom of the shaft to nearly double the shaft diameter. After it is drilled, the shaft is filled with concrete to form the pier.

When the pier must extend below the ground water level, or when the soil is not strong enough to stand without support, some form of bracing is required. The simplest is a metal cylinder that is lowered into the shaft immediately after drilling to hold it open until the concrete can be placed. The cylinder is lifted out as the concrete is poured, because the pressure of the wet concrete is ordinarily capable of supporting the soil and keeping out water. In very soft or wet soils it is sometimes necessary to drill and install lining cylinders successively in short sections, 8 to 16 ft long. The cylinders telescope, forming a tapering shaft known as the *Gow caisson* (Fig. 10.24c). Piers as deep as 100 ft with belled bottoms have been installed by this method, using both hand and auger excavation.

If the pier rests on rock, a man is lowered into the hole to clean the surface so that there is no soil between the rock and the concrete. This is hazardous because of gas and danger of blow ins. If the rock is seamy or the surface is badly weathered, it is sometimes necessary to remove the unsound materials by air hammers or even blasting. A level bottom is unnecessary. If the surface is steep, the pier should be keyed to it by a socket into the rock or by steel dowels placed in drilled holes.

The *Benoto* system makes it possible to excavate pier foundations to great depths through soil which is not easily drilled with augers. The Benoto drill is a specialized crane equipped with a grab bucket similar to a clamshell, but with four blades that can remove boulders and soft rock. A heavy steel churn drill is used to break up boulders and penetrate soft rock and a long cylindrical bucket can bail the soil-water slurry from the hole. The machine is also equipped with a chute to aid in concreting. An auxiliary machine aids in driving casing, when needed, by rotating the steel tube back and forth to reduce driving friction. Ordinarily the casing is removed during concreting and the completed foundation is similar to the drilled pier, Fig. 10.24a and b.

The *Chicago well* is an open shaft lined with vertical wood sheeting held in place by steel hoops inside the shaft. It is dug by hand and the sheeting installed in 4- to 6-ft or 1.2 to 2 m lengths. Shafts as deep as 200 ft (60 m) and 12 ft or 4 m in diameter have been placed by this method.

The wet excavation method is sometimes used in soils that are too soft to permit open excavation without support. A large rotary well drill bores a hole that is the diameter of the finished pier. The hole is kept full of a mixture of clay, water, and heavy minerals that has the same unit weight as the soil and which provides an internal pressure that keeps the hole open. After drilling, a cylindrical steel shell is lowered into the excavation, and then the mud is replaced by clean water. Concrete is tremied through the water to form the pier. Such methods do not permit thorough cleaning of the excavation or an inspection of the stratum on which the pier rests.

The balance of pressure is often augmented by extending a casing a few feet above the ground surface and keeping the mud level at its top. The coarse cuttings can be removed by pumping them up a hollow drill stem, a process termed *reverse circulation*. In this way the high velocity of fluid through the stem can keep the gravel-sized particles in suspension. Cylindrical buckets with bottom valves or internal pistons also are used to suck up the coarse cuttings, but they disturb the hydrostatic balance. Concrete is tremied into the mud-filled hole to concrete the shaft from the bottom up. A casing can be placed after drilling is complete, and the hole flushed out and unwatered for *final cleanup*.

CAISSONS / Three forms of caisson are used in the United States; the caisson pile, the open caisson, and the pneumatic caisson. The caisson pile is a large diameter pipe (24 to 60 in. or 60 to 150 cm in diameter) which is driven with open ends by a very large pile-driving rig. The soil within the pipe is excavated, after which the pipe may be entered for inspection or for cleaning the surface on which the caisson rests. It is then filled with concrete to form the pier.

In the process known as the *drilled-in-caisson* (Fig. 10.25a) the pipe is fitted with a tool-steel cutting edge that can be driven into rock. The soil inside

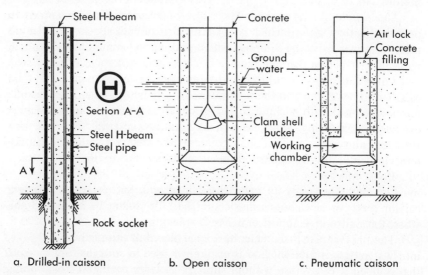

Figure 10.25 *Construction of different types of caissons.*

can be removed with a small-size bucket, by blowing out with a jet of compressed air, by driving a coring tube into the soil and then pulling the tube, or by adding water and churning the soil into a slurry, after which it can be bailed out. After cleaning, a large well drill bit is placed inside the pipe and used to drill a socket in the rock. The pipe is then driven until the cutting edge develops a watertight seal against the rock. The drilling continues until a socket from 2 to 10 ft deep is formed in the solid rock. The caissons may be unwatered and the rock sockets inspected before concreting. In order to increase their column capacity, they may be provided with a steel H-beam core. Drilled-in-caissons have been built with diameters from 24 to 30 in or 60 to 75 cm, as long as 250 ft or 80 m, and with capacities as high as 2000 tons each. They have the advantages that any type of soil may be penetrated and obstructions such as boulders can be drilled out, they can be extended through partially decomposed rock until sound rock is encountered, and visual inspection is ordinarily possible before concreting. They can be placed at a batter of 1 (horizontal) to 6 (vertical) and in this way can be combined with vertical caissons to form A-frames that resist lateral loads.

The open caisson (Fig. 10.25b) is an open box that has a cutting shoe on its lower edge. As the soil is excavated from inside, the box is forced down by weights until it comes to rest on the desired bearing stratum. Open caissons are often used for constructing bridge piers in open water. The caisson is prefabricated on land, floated into position with pontoons, and then lowered into place. No attempt is made to unwater the caisson until it is seated on the bearing stratum. Sometimes it is necessary to seal the bottom with concrete placed under water before unwatering is possible.

The width of the opening or *dredging well* must be great enough to permit the use of a clamshell bucket for excavation (at least 10 ft or 3 m but preferably more). Water jets are sometimes placed in the outside walls to reduce friction during sinking.

Large caissons for bridge piers are made up of a number of small caissons, or *cells*, each with an independent dredging well but all opening into a common bottom chamber. The depth to which open caissons may be extended is limited by skin friction which overcomes the effects of weighting. Boulders and other obstructions that catch under the cutting edge may limit the depth, since it is difficult to remove them.

PNEUMATIC CAISSONS / Compressed air or pneumatic caissons must be used when an acceptable bearing stratum cannot be reached by open caisson methods because of water conditions. The high cost is justified only where the loads to be supported are very high.

The pneumatic caisson (Fig. 10.25c) is like an inverted tumbler lowered into water. It is a box with an open bottom and an airtight roof or cover that is filled with compressed air to keep the water and mud from coming into the box. The lower section is a working chamber in which the excavating is carried on and the pier is constructed. Above the working chamber is the air lock, which permits the workmen and materials to enter and leave without loss of air pressure in the working chamber.

The pneumatic caisson process provides a better means for controlling the sinking of the caisson and makes possible the removal of boulders, logs, and debris from under the cutting shoe. Also the foundation bed upon the rock or bearing strata can be better prepared and inspected.

The workmen have to work under air pressure sufficient to balance the pressure of the surrounding mud and water. This working pressure limits the depth to which pneumatic caissons may be used and greatly slows down the progress of sinking the caisson. There are numerous hazards in the use of the pneumatic caisson and only experienced engineers and contractors should undertake their design and use.

10:10 Anchors

An *anchor* is a special form of deep foundation designed to resist a lateral or upward force. It is used to resist hydrostatic uplift or provide support for anchored bulkheads, excavation bracing, or tied retaining walls.

GENERAL FEATURES / A number of different forms are used: anchor piles and blocks or thrust walls in soil and anchor rods in rock. The anchor blocks and thrust walls are analogous to footings or retaining walls. Their capacity is controlled by bearing capacity or earth pressure against the anchor, and by the resistance of the mass, whichever is smaller. The anchor pile and the anchor rod in rock are analogous to friction piles. Their capacity is controlled by the skin friction or shear developed along the anchor shaft and by the resistance of the mass, whichever of the two is the smaller.

ANCHOR BLOCKS / The mechanisms of shear resistance depends on the depth below the surface, Fig. 10.26. For shallow blocks $z_2 < 4$ $(z_2 - z_1)$. The shear patterns with horizontal loading resemble passive

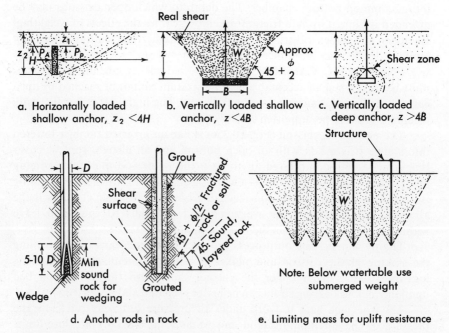

Figure 10.26 Anchors in rock and soil.

SEC. 10:10]   ANCHORS   495

pressure, Figs. 8.5 and 8.6.[10:23] The resultant of earth pressure against the anchor face approaches that of passive pressure for the entire depth z_2, minus the resultant of active earth pressure on the opposite face. (Tension in cohesive soils should be neglected.) For small anchors, with horizontal thrust, wall friction is neglected and the Rankine analysis applies. For anchors with an upward thrust or pull, the wall friction acts upward, and the passive pressure is less than the Rankine value. The Coulomb analysis is an adequate approximation.

For vertical uplift blocks the shear pattern is that of Fig. 10.26b. This can be approximated by a truncated cone or pyramid whose sides slope at $45 + \varphi/2$ with the vertical, using the slow or drained φ. The uplift resistance is equal to the weight of the block and the soil it engages. The vertical component of c (in undrained shear) can be added for dynamic uplift resistance but is neglected for static loading. Below the ground water level the submerged unit weight applies.

If the anchor depth z_2 exceeds about 4 times its height, $z_2 - z_1$, for horizontal loads or 4 times its width B for vertical loads, then the anchor shears the soil like a small inverted deep footing, Fig. 10.26c. Limited data from model tests of square anchors in saturated clay in undrained shear as well as full scale pull out tests [10:24] indicate resistances, r_0, per unit of area perpendicular to the thrust is similar to that for footings.

$$r_0 = q_0 = cN_c + q' N_q. \tag{9:4a}$$

For typical anchors $N_c = 5$ to 7, between the values for shallow and deep foundations. This suggests that for other soils the general bearing capacity, Equation (9:3), with factors for shallow foundations may be a reasonable approximation.

ANCHOR PILES AND RODS / Anchor piles in soil and grouted rods in rock transfer their loads by shear along their surface. The uplift resistance for the anchor pile can be computed in the same way as skin friction for a bearing pile. Of course, end bearing is neglected.

The anchor rod in rock is placed deep in a drilled hole, Fig. 10.26d. If the rock is strong and sound it can be secured by wedges or expansion sleeves like rock bolts. In softer rocks it is held by filling the hole with grout. A number of forms of anchor rods are used: plain bars with an expansion sleeve or an enlargement held by grout, high strength steel tendons similarly anchored, or ordinary reinforcing bars grouted the greater part of their length. The capacity of the anchor is limited by two factors: the pull-out resistance of the bar, and the resistance of the rock mass engaged by the bar.

The pull-out of anchor wedges and expansion sleeves is dependent on the rock, the bolt size, and design. In hard rock, 1 in. (2.5 cm) diameter sleeve bolts have ultimate capacities of 10 to 15 tons. The capacity of grouted bars depends on the shear between the grout and the rock as well as on bond or bearing between the bar and the grout. In soft rock, grout to rock shear should be limited to 30 or 40 psi, in hard rock to 100 to 300 psi. Grouted rods can also be used in coarse sand, gravel and very fractured rock. The grout-to-soil shear is equal to the soil shear at the perimeter of the cylinder of grout.

Both the anchor rod and the pile are limited to the resistance of the mass surrounding the rod. In rock the geometry depends on the joint pattern. For cohesionless soils and closely jointed rock a cone with its angle $45 - \varphi/2$ with the rod is conservative. For widespread rock joints that overlap, a cone angle of 45 degrees is more realistic. For uplift anchors the limit of capacity is the weight of the cone (submerged weight below the water table).

TESTS / Because of the uncertainties in anchor wedging, bond, and the resistance of the soil or rock surrounding the anchor, pull-out tests are essential for safe design. Moreover, it is prudent to proof test all anchors to 1.3 to 1.5 times their working loads, because minor peculiarities in soil or rock structure have a great influence on anchor capacity.

After the overload test the full working load (or a large fraction of it) is retained permanently as a *prestress*. This minimizes anchor deflection, which can be damaging for structures subject to occasional uplift. The prestress also induces compressive stresses in the soil or rock and increases their strength (which is one of the functions of rock bolting).

ANCHOR SYSTEM / The total capacity of a system of anchors is limited by the weight of the portion of the mass that might break loose when all are loaded simultaneously, Fig. 10.26e. For uplift anchors closely spaced the limit is the weight of the soil mass penetrated by all the anchors. For lateral anchors the limit is the passive earth pressure or sliding resistance of the mass. For masses of complex shape, the resistance of the mass is analyzed as for slope stability, Chapter 11.

10:11 Underpinning

Underpinning is the construction of new foundations under existing structures. The underpinning of structures is a highly specialized type of foundation engineering involving construction methods adapted to very limited working space and the handling of soils already under load. The work is done by only a few highly skilled contractors whose many years of experience qualify them for such exacting and critical work.[10:25]

Underpinning is necessary when the foundations of a structure prove incapable of supporting the structure with adequate safety or without undue settlement. Underpinning also is required when changing conditions, such as the construction of a nearby building with a deep basement or the excavation for a subway, make existing foundations inadequate.

Two procedures are commonly used: first, installing new foundations in small pits excavated beneath the existing foundations; and second, installing new foundations adjacent to the old and transferring the load between new and old with steel beams.

The pit method (Fig. 10.27a and c) requires excavating a small hole beneath part of the existing foundation. A new, deep footing is poured in this hole, or pipe piles are forced into the soil by jacking against the existing foundation. The pipes, in sections about 2 ft or 60 cm long, are jacked into the soil and then excavated by small buckets, a steam jet, or augers. The new foundation is placed section by section so that the old foundation is never without some support.

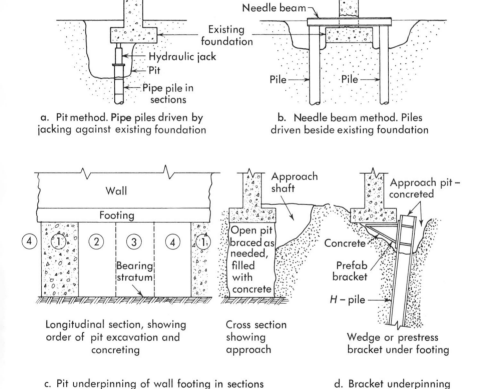

Figure 10.27 *Methods for underpinning an existing foundation with piles. Where the new bearing stratum is close to the surface, deep footings or piers are substituted for the piles.*

The second method (Fig. 10.27b) involves driving piles or constructing a new foundation as close as possible to the old. This is necessary when the old foundation is so small or weak that a pit beneath it is impossible. It is often cheaper than the pit method, for more room in which to work is available. The load is transferred from the old to the new foundation by horizontal *needle beams* which are placed beneath the old footing or through it.

Clamps tightly bolted to a notched concrete or welded to a steel column make it possible to place the needle beam above the footing.

A bracket system, Fig. 10.27d, employs a small drilled pier or pile reinforced to support bending moments. The pile or pier is placed immediately adjacent to the foundation and terminated about 2 ft or 60 cm below it. A bracket of steel or reinforced concrete is then constructed under the old foundation and either wedged or prestressed to pick up the load. Steel members are encased in concrete after prestressing, for corrosion protection.

The transfer of the load from the old to the new foundation is accompanied by some settlement. This may be eliminated by jacking the new foundation against the old until the new carries the load. The settlement is prevented by extending the jack as the new foundation deflects under load. After the new foundation has stopped settling, the jack is replaced by steel wedges which are then encased in concrete. This is known as the *Pretest method* of underpinning.[10:24] This method, developed by Lazarus White in constructing subways in New York, has made it possible to build beneath the largest structures without damaging them and to preserve monumental structures despite the ravages of time or the demands of deep new construction next to old.

REFERENCES

10:1 "Data on Pile Drivers and Extractors," L. B. Foster Co., Pittsburgh, 1968.

10:2 W. S. Housel, "Michigan Study of Pile Driving Hammers," *Journal of the Soil Mechanics and Foundation Division, Proceedings, ASCE*, **91**, SM5, September 1965, p. 37.

10:3 D. D. Barkan, "Foundation Engineering and Drilling by the Vibration Method," *Proceedings of the Fourth International Conference on Soil Mechanics and Foundation Engineering*, **2**, London, 1957, p. 3.

10:4 L. Bjerrum, "Norwegian Experiences with Steel Piles to Rock," *Geotechnique*, **7**, 2, June 1957, p. 73.

10:5 J. D. Parsons and S. D. Wilson, "Safe Loads on Dog-Leg Piles," *Transactions, ASCE*, **121**, 1956, p. 695.

10:6 G. G. Meyerhof, "Compaction of Sands and Bearing Capacity of Piles," *Journal of the Soil Mechanics and Foundations Division, Proceedings, ASCE*, **85**, SM6, December 1959.

10:7 A. S. Vesic. "Ultimate Loads and Settlements of Deep Foundations in Sand," *Symposium on Bearing Capacity and Settlement*

of Foundations, Duke University, Durham, North Carolina, 1967, p. 53.
10:8 L. C. Reese and H. B. Seed, "Pressure Distribution Along Friction Piles," *Proceedings, ASTM*, 1955.
10:9 H. M. Coyle and L. C. Reese, "Load Transfer for Axially Loaded Piles in Clay," *Journal of the Soil Mechanics and Foundation Division, Proceedings, ASCE*, **92**, SM2, March 1966.
10:10 O. Grillo, "Influence Scale and Chart for Computation of Stresses Due to Point Load and Pile Load," *Proceedings of the Second International Conference on Soil Mechanics and Foundation Engineering*, Rotterdam, 1947.
10:11 J. Kerisel, "Deep Foundations—Basic Experimental Facts," *Proceedings of the Conference on Deep Foundations*, Mexico City, 1964, p. 5; also, J. L. Kerisel, "Vertical and Horizontal Bearing Capacity of Deep Foundations in Clay," *Symposium on Bearing Capacity and Settlement of Foundations*, Duke University, Durham, North Carolina, 1967, p. 45.
10:12 V. G. Berezantzev, V. S. Khristoforov, and V. N. Golubkov, "Load Bearing Capacity and Deformation of Piled Foundations," *Proceedings of the Fifth International Conference on Soil Mechanics and Foundation Engineering*, **2**, Paris, 1961, p. 11.
10:13 G. G. Meyerhof, "Penetration Tests and the Bearing Capacity of Cohesionless Soils," *Journal of the Soil Mechanics and Foundation Division, Proceedings, ASCE*, **82**, SM1, (Separate 866, 1956).
10:14 L. Zeevaert, "General Considerations on Problems Related With Pile and Pier Foundations," *Proceedings of the Conference on Deep Foundations*, Mexico City, 1964.
10:15 W. S. Housel, "Pile Load Capacity, Estimates and Test Results," *Journal of the Soil Mechanics and Foundation Division, Proceedings, ASCE*, **92**, SM4, July 1966.
10:16 T. Ramot, "Analysis of Pile Driving by The Wave Equation," *Foundation Facts*, Raymond Concrete Pile Co., New York, **3**, 1, Spring, 1967.
10:17 R. D. Chellis, *Pile Foundations*, McGraw-Hill Book Co., Inc., New York, 2nd Ed, 1961.
10:18 G. F. Sowers, L. Wilson, B. Martin, and M. Fausold, "Model Tests of Friction Pile Groups in Homogeneous Clay," *Proceedings of the Fifth International Conference on Soil Mechanics and Foundation Engineering*, **2**, Paris, 1961, p. 155; **3**, p. 261 and 279.
10:19 H. Kishida and G. G. Meyerhof, "Bearing Capacity of Pile Groups Under Eccentric Load in Sand," *Proceedings of the Sixth International Conference on Soil Mechanics and Foundation Engineering*, **2**, Montreal, 1965, p. 270.

10:20 L. C. Reese, and Hudson Matlock, "Non–Dimensional Analysis For Laterally Loaded Piles With Soil Modulus Assumed Proportional to Depth," *Proceedings of the Eighth Texas Conference on Soil Mechanics and Foundation Engineering*," University of Texas, Austin, 1956.

10:21 R. D. Chellis, "Finding and Fighting Marine Borers," *Engineering News Record*, March 4, March 18, April 1, April 15, 1948.

10:22 *Concrete Piles*, Portland Cement Association, Chicago.

10:23 G. P. Tschebotarioff, "Retaining Structures," Chapter 5, *Foundation Engineering*, McGraw-Hill Book Co., Inc., New York, 1962.

10:24 G. G. Meyerhof and J. I. Adams, "The Ultimate Uplift Capacity of Foundations," *Canadian Geotechnical Journal*, **5**, 4, November 1968.

10:25 E. A. Prentis and L. White, *Underpinning*, 2nd ed., Columbia University Press, New York, 1950.

SUGGESTIONS FOR ADDITIONAL STUDY

1. W. C. Teng, *Foundation Design*, Prentice Hall, Inc., Englewood Cliffs, N.J., 1962.
2. A. B. Carson, *Foundation Construction*, McGraw-Hill Book Co., Inc., New York, 1965.
3. J. H. Thornley, *Foundation Design and Practice*, Columbia University Press, New York, 1959.
4. G. A. Leonards et al., *Foundation Engineering*, McGraw-Hill Book Co., Inc., New York, 1961.

PROBLEMS

10:1 Prepare a table showing the point diameters, point areas, and surface areas of the lower 10 ft of different sizes of concrete, steel, and wood piles. Consult the catalogs of different pile-driving contractors and manufacturers of wood and steel piling.

10:2 A wood pile with an 8-in. diam tip and a 14-in. diam butt 35 ft long is driven into a dry loose sand weighing 107 lb per cu ft; angle of internal friction of 32°. Compute its bearing capacity.

10:3 A 10BP57 steel pile 40 ft long is driven into saturated clay weighing 110 lb per cu ft and having an undisturbed c of 2000 psf and a remolded c of 1000 psf. Compute its skin friction, assuming (1) the skin friction is developed over the entire pile surface and (2) the skin friction is developed along the surface of a rectangle that encloses the pile. Depending on which governs, compute the end bearing for either the gross area of the enclosing rectangle or the net area of the pile. (Use the undisturbed c for end bearing and the remolded c for skin friction.)

10:4 A 14-in. OD pipe pile 150 ft long with a flat end is driven into a deep deposit of clay having the following characteristics:

Depth (ft)	c (psf)	Weight (lb/cu ft)
0–40	1600	120
40–110	500	105
110–140	900	110
140–180	3400	130

Water table is at ground surface.
a. Compute the capacity in kips.
b. If 25 of these piles are to be driven in a group, determine the minimum spacing to ensure that the group capacity will not be less than the sum of the capacities of the individual piles.

10:5 A steam hammer weighing 3000 lb and falling 29 in. is used to drive a precast concrete pile 12 in. square and 40 ft long The total bounce is 0.3 in. per blow, and the driving resistance is 4 blows per inch. The coefficient of restitution is estimated to be 0.40 and the value of c_1 is 0.2 in.
a. Compute the safe capacity by the Hiley formula, using a safety factor of 2.
b. Compute the safe capacity by the *Engineering News* formula.
c. How do they differ and why?

10:6 A steam hammer weighing 1000 lb that falls 3 ft is used to drive wood piling, Compute the dynamic resistance according to the *Engineering News* formula if the pile penetrates $\frac{1}{2}$ in. under each of the last few hammer blows.

10:7 Compute the safe bearing capacity of a steel $10\frac{3}{4}$ OD × $\frac{1}{4}$ in. wall pipe pile 45 ft long driven into sand by a steam hammer with a 5000-lb ram and 36-in. stroke. The "bounce" is measured and found to be 0.3 in., and the net penetration or "set" is 0.4 in. per blow. The hammer efficiency is 75 per cent and the coefficient of restitution of the pile hammer on the pile cap is estimated to be 0.80; c_1 is 0.1 in.
a. Compute the safe load, using the Hiley formula and including a safety factor of 2.
b. Compute the safe load, using the *Engineering News* formula.
c. Which shows the greater safe load and why?

10:8 A pile load test produces the following data. The pile is 10 in. × 10 in., 57-lb H-section.

Load (tons)	Settlement (in.)	Load (tons)	Settlement (in.)
20	$\frac{1}{8}$	80	$\frac{9}{16}$
40	$\frac{1}{4}$	100	$\frac{13}{16}$
60	$\frac{3}{8}$	120	$1\frac{1}{2}$

a. Find the safe pile load, using a safety factor of 2.
b. How much of the above settlement can be attributed to elastic shortening of the pile if the pile is assumed to be end bearing? Assumed length of pile is 30 ft.

10:9 A machine weighing 2500 kips is to be supported on piles. The wood piles at 20 tons must be 40 ft long, and steel pipe piles loaded to 50 tons must be 50 ft long. Which will be cheaper? The wood piles will cost $2.25 per ft driven; the pipe piles, $5.00 per ft.

10:10 An anchor consists of a vertical tension pile and a pile driven on a batter of 1 (horizontal) to 3 (vertical). Both piles are 70 ft long. If the horizontal load is 20 kips, compute the load in each pile. Assume the piles hinged at both ends.

10:11 Prepare a report describing a pile or caisson construction job. Include the following items:
1. Soil profile.
2. Reason for selection of piles or caissons.
3. Equipment.
4. Description of features of piles and their construction.
5. Tests, if any.

10:12 A caisson pile 3 ft in diameter, of concrete with a compressive strength of 6000 psi is drilled 2 ft into a horizontally bedded close jointed sandstone formation. Unconfined compression tests of the intact sandstone found failure at 5000 psi. The angle of the failure plane α was 65 deg.
a. Determine the safe load on the caisson.
b. If the joint spacing were 6 ft by 6 ft, what would the safe load be?

10:13 For the soil profile and loadings described in Problem 9:17, suggest alternate types of deep foundations including the advantages and disadvantages of each. Which would be your choice and why do you think it best?

10:14 A closed end 10.75 in. diameter pipe pile 60 ft long is driven into a homogeneous sand formation 80 ft thick underlain by a compressible clay stratum 10 ft thick. The sand weighs 110 pcf above the water table and 130 pcf below. The angle of internal friction of the sand is 36 degrees and the relative density 62 per cent. The ground water table is at a depth of 13 ft. The clay weighs 107 pcf saturated, has a void ratio of 1.38, and a compression index of 0.63. It is normally consolidated. Below the clay is more sand.

a. Compute the safe bearing capacity of the pile assuming that it is jetted to a depth of 40 ft and then driven.
b. Compute the safe bearing capacity assuming that the pile is driven for its full length.
c. Compute the stress increase below the single pile, and the settlement, assuming that the pile was driven. Pile load is 40 kips.
d. Compute the stress increase due to 9 of these piles driven 3 ft on centers, and the resulting settlement in the clay. Each pile carries 40 kips.

CHAPTER 11
Stability of Earth Masses

A LARGE HOUSING PROJECT WAS PLANNED FOR A GENTLY SLOPING HILLSIDE, cut with several steep gullies. The grading plan required cutting into the hillside for roads and building sites, following the contours. The excavated soil was used to fill the gullies and make more nearly level sites on the downhill side of the road. Culverts were provided in each swale beneath the fill, so the embankments would not impound water in the gullies above the road.

The fills were placed under strict supervision and tests showed that they met the specified density. The buildings were nearing completion when one developed serious cracks. It was located on the down-hill side of the road, and its spread footings were supported on the deepest fill on the site, immediately above an old spring. The contract administrator notified the contractor that the cracking was apparently caused by consolidation of the fill, due to faulty compaction. He ordered the contractor to underpin the building at no expense to the owner. Fortunately nature worked faster than the letters from the administrator. The cracks rapidly opened, and the down-hill segment of the fill, including half of the building, slid down and out 6 to 8 ft.

The cause of the cracking was a landslide, not poor compaction. The virgin soil consisted of alternate seams of stiff clay and sand, horizontally bedded. Water seeped continuously through the sand seams from higher terrain beyond the project. The gullies served as the outlets for the seepage until the fills were placed. The impervious clay of the fill acted as a dam for each sand seam, causing the water pressure to build up until the sand could no longer support the shear stresses imposed by the embankments. The cure was to remove the fill, and place a drainage blanket drain of clean sand on the gully sides to relieve the pressure buildup.

When portions of large earth masses become detached and move, the results are usually spectacular and often disastrous. Tremendous landslides have buried entire cities and dammed up rivers; slides in open cuts have caused the abandonment of canals, highways, and railroads; levees have broken during periods of high water, flooding valuable farm land and driving people from their homes; and earth dams have failed, producing tremendous surges of water that scoured out valleys and left death and destruction behind.

11:1 Analysis of Stability

The safety of an earth mass against failure or movement is termed its *stability*. It must be considered not only in the design of earth structures but also in the repair and correction of failures. The design of open-cut slopes and embankment, levee, and earth dam cross-sections is based primarily on stability studies—unless the project is so small that occasional failures can be tolerated. When failures such as landslides and subsidences do occur, their correction requires stability studies to determine the cause of failure and the best method of preventing future trouble.

CAUSES OF EARTH MOVEMENTS / Earth failures have one feature in common: There is a movement of a large mass of soil along a more or less definite surface, as shown in Fig. 11.1. In most cases the earth mass remains

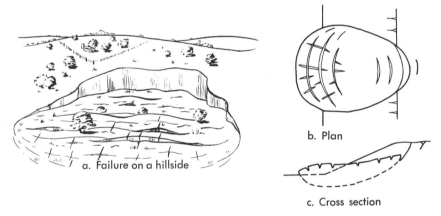

Figure 11.1 *Failure of an earth mass by sliding.*

intact during the first stages of the movement, but finally it becomes distorted and broken up as movement progresses. Some failures occur suddenly with little or no warning, while others take place leisurely after announcing their intentions by slow settlement or by the formation of cracks.

Movements occur when the shear strength of the soil is exceeded by the shear stresses over a relatively continuous surface. Failure at a single point in the mass does not necessarily mean that a soil mass is unstable. Instability results only when shear failure has occurred at enough points to define a surface along which the movement can take place. It is hard to determine the cause of many earth movements. Actually anything that results in a decrease in soil strength or an increase in soil stress contributes to instability and should be considered in both the design of earth structures and in the correction of failures. Table 11:1 will serve as a guide in analyzing for stability.

TABLE 11:1 / CAUSES OF INSTABILITY

Causes of Increased Stresses	Causes of Decreased Strength
1. External loads such as buildings, water, or snow.	1. Swelling of clays by adsorption of water.
2. Increase in unit weight by increased water content.	2. Pore water pressure (neutral stress).
3. Removal of part of mass by excavation.	3. Breakdown of loose or honeycombed soil structure with shock, vibration or seismic activity.
4. Undermining, caused by tunneling, collapse of underground caverns, or seepage erosion.	4. Hair cracking from alternate swelling and shrinking or from tension.
5. Shock, caused by earthquake or blasting.	5. Strain, and progressive failure in sensitive soils.
6. Tension cracks.	6. Thawing of frozen soil or frost lenses.
7. Water pressure in cracks.	7. Deterioration of cementing material.
	8. Loss of capillary tension on drying.

Failure could result from any one or any combination of these factors. Most are independent, but some may be interdependent. For example, progressive strain in loose saturated sands causes a neutral stress increase that leads to a loss of strength and possibly a breakdown or collapse of soil structure. The possible number of combinations of factors leading to stability is staggering: 15!, or 1,307,674,368,000. The effect of water is vital: water pressure or changes in water are involved in 10 of the 15 factors listed.

In cases of failures that result in property damage or loss of life, the engineer is often called upon to determine *the* cause of the failure. In most cases a number of causes exist simultaneously, and so attempting to decide which one finally produced failure is not only difficult but also incorrect. Often the final factor is nothing more than a trigger that set in motion an earth mass that was already on the verge of failure. Calling the final factor *the cause* is like calling the match that lit the fuse that detonated the dynamite that destroyed the building *the* cause of the disaster.

STABILITY OF SLOPES / Among the most common of earth-mass failures are those resulting from unstable slopes. Gravity, in the form of the weight of the soil mass and of any water above it, is the major force tending to produce failure, while the shearing resistance of the soil is the major resisting force. The failure surface has the shape of the bowl of a teaspoon or half an egg sliced lengthwise, with the smaller end at the top of the slope and the wider end at the bottom, as shown in Fig. 11.1.

The failure usually occurs in one of three forms (Fig. 11.2). The *base failure* develops in soft clays and soils with numerous soft seams. The top of the slope drops, leaving a vertical scarp, while the level ground beyond the

SEC. 11:1] ANALYSIS OF STABILITY

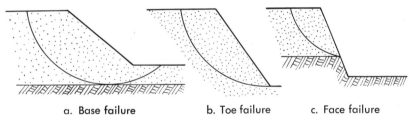

Figure 11.2 Types of failures approximated by circular arcs.

toe of the slope bulges upward. *Toe failures* occur in steeper slopes and in soils having appreciable internal friction. The top of the slope drops, often forming a series of steps, while the soil near the bottom of the slope bulges outward, covering the toe. The *slope* or *face* failures are special cases of toe failures in which hard strata limit the extent of the failure surface.

Other forms occur. If there are zones or surfaces of pronounced weakness, these may define part of the shear surface. If there are large external forces, they will distort the shear surface.

STABILITY BY TRIAL / Stability analysis is a problem in plastic equilibrium. When the mass is on the verge of failure, the forces causing movement have become equal to the resistance of the mass to being moved. A slight increase in forces is sufficient to produce continuing strain, as described in Chapter 7. Because of the irregular geometry of the mass and the complex force system in any real problem, the methods of direct analysis, such as used for earth pressure, are seldom applicable. Instead, a process of trial and revision is the most useful approach to determining the safety factor of a tentative design or the potential failure of an existing slope.

First a potential failure surface is assumed, and the shearing resistance acting along the surface is calculated. The forces acting on the segment of soil bounded by the failure surfaces are determined, and then the safety factor of the segment is calculated as follows:

Safety against rotation

$$SF_m = \frac{\text{resisting moments}}{\text{moments causing failure}}; \qquad (11{:}1a)$$

Safety against translation (straight-line movement)

$$SF_t = \frac{\text{forces opposing motion}}{\text{forces causing motion}}. \qquad (11{:}1b)$$

Theoretically, if a very great number of different segments are assumed, the smallest safety factor found for any will be the actual safety factor of the mass. In practice, however, the smallest safety factor found by analyzing a few, well-chosen, possible failure segments will be sufficiently accurate.

Generally, the forces causing motion are considered to include inertia, gravity and all external loads. Those that resist motion include the soil strength and other forces along the potential surface of movement.

CIRCULAR ARC ANALYSIS / The general method for analysis of the stability of slopes was first suggested by the Swedish engineer K. E. Petterson[11:1] as a result of studies of a landslide in the harbor of Gothenburg. The actual surface is approximated by a segment of a cylinder which in cross-section is an arc of a circle (Fig. 11.3). The overturning moment M_0 per

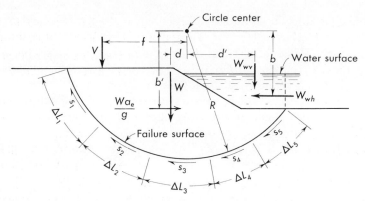

Figure 11.3 *Circular arc analysis of an earth movement.*

foot of width about the circle center is the algebraic sum of the moments due to the weight of the mass W, the horizontal and vertical components of water pressure (if the slope is inundated) acting on the surface of the slope, W_{wh} and W_{wv}, and any other external forces acting on the mass V.

$$M_0 = Wd - W_{wh}b - W_{wv}d' + Vf + \frac{Wa_e b'}{g}. \qquad (11:2a)$$

In this expression, b, b', d, d' and f are the respective moment arms of the centroids of the weights or action lines of the forces about the circle center. The resisting moment is provided by the soil strength. If the shear strength is s along each segment of the arc ΔL, whose radius is R, then the resisting moment for each foot of width is

$$M_r = R \sum s \, \Delta L = R(s_1 \Delta L_1 + s_2 \Delta L_2 + \cdots). \qquad (11:2b)$$

The safety factor of the circular segment is found by

$$SF = \frac{M_r}{M_0}. \qquad (11:3)$$

The basic circular analysis can be applied to any slope and combination of forces where the shear strengths of the soils are independent of the normal

SEC. 11:1] ANALYSIS OF STABILITY

stresses on the failure surface, such as saturated clays in which failure occurs so rapidly that there are no changes in water content or soil strength.

Example 11 : 1

Calculate the safety of the following assumed segment (Fig. 11.4) if the crack, which is 5 ft deep, is filled with water. Arc radius is 32.5 ft.

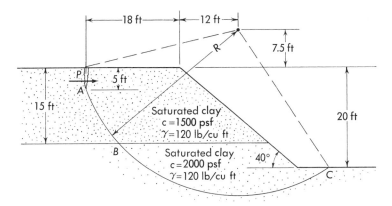

Figure 11.4 Example 11: 1. Circular arc analysis of a slope in a stratified deposit of saturated clays.

1. Divide arc into two segments, AB and BC. Determine length of each.

$$AB = 12 \text{ ft}; BC = 44 \text{ ft}.$$

2. Calculate resisting moment.

$$M_r = 1500 \times 12 \text{ ft} \times 32.5 + 2000 \times 44 \times 32.5 = 3{,}445{,}000 \text{ lb-ft}.$$

3. Calculate weight of segment and find centroid by methods of statics.

$$W = 76{,}000 \text{ lb}; a = 10.2 \text{ ft}.$$

4. The moment caused by the weight is $76{,}000 \times 10.2 = 775{,}000$ lb-ft.

5. The resultant force of water pressure in the crack P is $\dfrac{5 \times 62.4 \times 5}{2} =$ 780 lb per ft. It acts horizontally at a distance of 10.8 from the center of the arc; $M = 8400$ lb-ft.

6. The total moment tending to cause overturning is

$$775{,}000 + 8400 = 783{,}400 \text{ lb-ft}.$$

7. The safety factor SF is given by

$$SF = \frac{3{,}445{,}000}{783{,}400} = 4.4.$$

METHOD OF SLICES / In order to compute the stability of slopes in soils whose strength depends on the normal stress, it is necessary to determine the effective normal stress along the failure surface. A rigorous solution for the normal stress is not available, but the method of slices developed by Fellenius[11:2] has proved to be a workable approximation.[11:3]

The failure zone is divided into vertical slices, as shown in Fig. 11.5. They need not be of equal width, and for convenience in computation, the boundaries should coincide with the intersections of the strata with the circle and with the slope face. In the basic analysis, it is assumed that each slice acts independently of its neighbor: There is no shear developed between them, and the normal pressures on each side of a slice produced by the adjoining slices are equal.

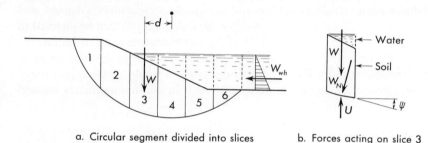

a. Circular segment divided into slices   b. Forces acting on slice 3

Figure 11.5 *Method of slices for circular arc analysis of slopes in soils whose strength depends on the confining pressure.*

The vertical force acting on each slice W is the weight of the soil in the slice plus the weight of water directly above the slice. The weight of any external load on the slice, such as a structure, is also included. The net or effective downward force acting on the curved bottom of the slice is the total weight minus the upward force due to neutral stress, $W' = W - U$. The upward force U is found by multiplying the neutral stress u (computed from the flow net as described in Chapter 4) by the slice width.

If the slice is sufficiently narrow, the curved boundary can be approximated by a straight line that makes an angle of ψ with the horizontal axis. The component of the vertical force normal to the plane, W'_N, is computed by $W'_N = W' \cos \psi$. The shear strength along that segment of arc can be expressed as follows:

$$s = c' + p' \tan \varphi', \qquad (3:32a)$$

$$s = c' + \frac{W'_N}{\Delta L} \tan \varphi'. \qquad (11:4)$$

The total resisting moment for all the arc segments is found as before by Equation 11:2b.

The overturning moment can be found, as previously described, by Equation 11:2a. The moment due to the vertical forces is the algebraic sum of the moment of the total weight W of each slice about the circle center Wd. To this must be added algebraically the total moments due to the horizontal component of water pressure on the slope, and water pressure in cracks.

Many variations and refinements of this basic method have been developed. While none are rigorous, they have proved to be sufficiently accurate for analysis and design. The differences in the more refined methods lie in the assumptions made regarding the shear and normal forces on the sides of the slices.[11:4, 11:5]

The analysis requires trial to a large number of assumed failure surfaces. That which has the smallest safety factor is the most critical surface—the one on which failure is most likely to occur. The analysis is well adapted to tabular or digital computer solution.[11:6] A grid of possible circle centers is defined, and a range of radius values established for each. A computer can be directed to print out all the safety factors or just the minimum one (and its radius) for each circle center. A plot of minimum safety factor for each circle center in the form of contours, Fig. 11.6, can define the location of the most critical circles (more than one circle may be possible in non-homogeneous masses) and the minimum safety factor.

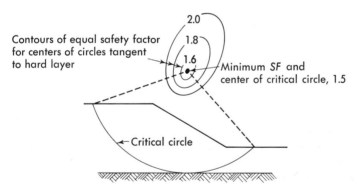

Figure 11.6 *Method for locating center critical circle by contours of equal safety factor.*

EARTHQUAKE STRESSES / The horizontal acceleration of an earthquake, a_e, imposes a transient force of Wa_e/g on the mass. A simple approximation of its effect is to add this inertia force, multiplied by its moment arm, to the overturning moments. The real effect is less critical because the acceleration is momentary. More realistic analyses, such as that proposed by Newmark[11:7] consider the strains produced and their effects.

CRACKING AT THE TOP OF THE SLOPE / As described in Chapter 7, the upper part of a slope in a cohesive soil is in a state of tension. Under continued tension, vertical cracks develop which destroy part of the shear strength and which can contribute to failure if they fill with water. The depth d of the tension cracks can be approximated by

$$d = \frac{2c}{\gamma}. \qquad (11:5)$$

The soil above this level does not contribute to the resisting moment of the failure arc, as shown in Example 11:1. If the crack fills with water, water pressure contributes to the overturning moment.

EFFECT OF SUBMERGENCE AND SEEPAGE / Submergence of a slope has three effects. First the weight of the circular segment is increased by the weight of the water above the slope and the greater soil unit weight, which increases the overturning moment. Second the increase is more than offset by the resisting moment due to horizontal water pressure. Third, neutral stress increases on the failure surface, depending on the seepage flow net that develops, and offsets much of the gain in strength produced by the additional weights of the soil and water. The result is that the submerged slope usually has a higher safety factor than the same slope without submergence.

When the level of submergence is reduced so rapidly that the neutral stress within the slope cannot adjust itself to the new water level, the condition is termed *sudden drawdown*. The helpful moment due to horizontal water pressure is reduced. The weight of the soil and water is also reduced, but the neutral stress is not greatly changed. As a result the safety factor drops sharply, usually below that for the non-submerged condition. This is often the most critical condition in the design of the upstream face of an earth dam.

Seepage through the soil toward the face of a slope is caused by excess neutral stress within the soil mass and results in lower strength and a smaller safety factor compared with that of the same slope without seepage. This condition is often critical in deep excavations, highway and railroad cuts, the downstream face of earth dams, and in natural slopes.

SLOPES IN HOMOGENEOUS, SOFT CLAY / The special case of a uniform slope in a homogeneous, soft clay whose shearing resistance is given by the relation $s = c$ can be solved analytically, and the results are presented in the form of a dimensionless number, m, termed the *stability number*. The stability number depends only on the angle of the slope, β, and on the *depth factor*, n_d, which is the ratio of the depth of a hard, dense stratum measured from the top of the slope to the height of the slope. The height of slope, H_c, at which a failure will occur is given by the relation:

$$H_c = \frac{c}{m\gamma}, \qquad (11:6a)$$

and the safety factor of a slope of height H is given by

$$SF = \frac{c}{mH\gamma}.^* \qquad (11:6b)$$

* This is *not* the same safety factor as found by Equation 11.3 but instead is the safety factor with respect to cohesion.

SEC. 11:1]   ANALYSIS OF STABILITY

A chart (Fig. 11.7a) showing the relation of the stability number to the slope angle and to the depth factor has been prepared from the results of a study by D. W. Taylor.

From the chart it can be seen that toe failures occur for all slopes steeper than 53°. The location of the center of the failure arc can be found from a

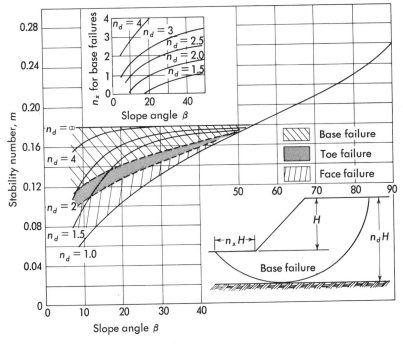

a. Chart for stability number, m

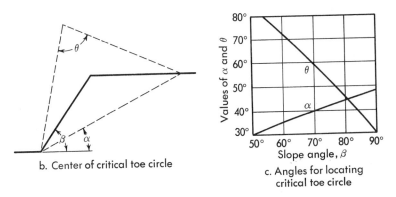

b. Center of critical toe circle

c. Angles for locating critical toe circle

Figure 11.7 Chart for finding the stability of slopes and the location of the critical circle in homogeneous saturated clay soils in undrained shear, $\varphi = 0$. (After D. W. Taylor and W. Fellenius)

chart (Fig. 11.7c) developed by W. Fellenius,[11:2] a Swedish engineer who pioneered in the analysis of slope stability. For slope angles less than 53° three possibilities of failure exist, depending on n_d. When n_d is 3 or more, a base failure will occur whose failure surface is tangent to the hard stratum and whose center is above the mid-point of the slope. For values of n_d between one and 3, a base failure, toe failure, or slope failure may take place, depending on the slope. For values of n_d less than 1, only slope failures can take place. The different possibilities of failure and the stability factors can be determined directly from the chart.

Example 11 : 2

Find the safety factor of a proposed slope of 30° on an embankment 60 ft high with rock 40 ft below the base of the embankment. The soil weighs 120 lb per cu ft and the shear strength, $s = 4700$ psf.

(1.) $n_d = \dfrac{60 + 40}{60} = 1.7.$

(2.) $m = 0.17$ from chart.

(3.) $SF = \dfrac{4700}{0.17 \times 120 \times 60} = 3.9.$

SLOPES IN HOMOGENEOUS, COHESIVE SOIL / Similar analytical studies have been made of the special case of uniform slopes in homogeneous soils whose shearing resistance can be expressed by

$$s = c' + p \tan \varphi' \qquad (3:37)$$

A chart of the stability number for different values of the slope angle β is shown in Fig. 11.8.[11:8] All failures are toe failures unless a hard stratum appears above the toe of the slope. In such cases the height of the slope should be measured from the top of the slope to the level of the hard stratum.

SLOPES IN HOMOGENEOUS, COHESIONLESS SOIL / Instead of failing on a circular surface, sand slopes fail by sliding parallel to the slope. Each sand grain can be considered as a block resting on an inclined plane at the slope angle β. When the slope angle exceeds the angle of friction of sand on sand (the angle of internal friction), the sand grain will slide down the slope. The steepest slope that a sand can attain, therefore, is equal to the angle of internal friction of the sand. Usually this is the minimum value of φ, since near the surface the sand is very likely to be in a loose condition. The *angle of repose* of sand as it forms a pile beneath a funnel from which it is poured is therefore about the same as the angle of internal friction of the sand in a loose condition.

SEC. 11:1] ANALYSIS OF STABILITY

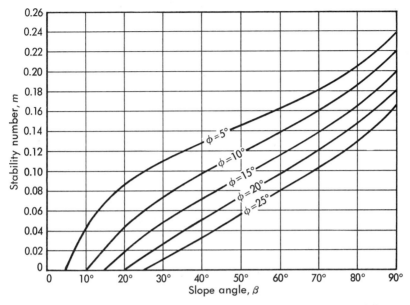

Figure 11.8 *Chart for finding the stability of slopes in homogeneous, partially saturated clays and similar soils. (After D. W. Taylor.)*

STABILITY OF BLOCKS / In many cases the presence of extremely soft strata or strata weakened by neutral stress causes failure to take place by translation or straight-line movement along the plane of weakness instead of by rotation on a circular surface. This often occurs when earth dams are founded on stratified deposits, since the horizontal force of the water acts parallel to the possible planes of failure.

The procedure is essentially one of trial and error. A possible failure plane (usually a weak stratum) is selected, and vertical boundaries ab and cd are arbitrarily assumed (Fig. 11.9). The wedge $abdc$ is acted upon by its own weight W, the force of active earth pressure P_A at the right, and the force of passive earth pressure P_P at the left. Shear along the failure plane bd

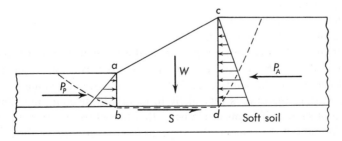

Figure 11.9 *Sliding-block analysis of a slope.*

resists the unbalanced forces acting on the block. If the total shear force is S, then the safety factor SF is given by

$$SF = \frac{S}{P_A - P_P}. \qquad (11:7)$$

In the case of an earth dam, water pressure should be added to the effective active earth pressure causing movement. When thin, horizontal seams of cohesionless soil are present, the neutral stress must be computed in order to determine the effective strength and shearing resistance in the soil.

Stability problems in cracked soils or jointed rocks involve multiple planes of weakness with different orientations. These cannot be analyzed by the simple representation of a two-dimensional cross-section such as Fig. 11.9. Instead, sliding takes place on two intersecting planes in a direction parallel to their intersection. The forces in such a three-dimensional system are most readily handled by vectors.

SHEAR STRENGTH FOR ANALYSIS / The shear strength of a soil varies greatly, depending on the environment and particularly on the degree of saturation, the effective stress and its changes due to neutral stress variations, and the effects of progressive strain. Saturation destroys capillary tension and causes the buildup of neutral stresses under quickly applied loads. Therefore, if the environmental conditions indicate any possibility of soil saturation, the shear strength in that state is employed in analysis.

The effective stress depends on the rate of loading compared with the rate of drainage within the soil. If the loads will be placed on the soil quickly, such as by rapid construction of a large embankment, the undrained strength is used. If the loads are applied slowly, the undrained strength will be safe, but the drained strength with proper consideration of neutral stress will result in more economy. For excavated or natural slopes that are exposed for long periods of time, it is necessary to use the drained strength because the unloading produced by erosion or excavation eventually reduces the effective stress on the soil and thereby the strength.

The consolidated undrained (CU or R) shear test is often a realistic model of shear in slopes where the mass develops equilibrium under static loads but failure is caused by a sudden change in load such as produced a by flood or heavy rainfall. The drained (D, CD, or S) test is best when the pore water pressures can be fully evaluated.

Temporary strength produced by capillary tension is misleading in utilizing the results of undrained or consolidated undrained tests to analyze natural slopes or excavation slopes. While the slope may be temporarily stable, supported by capillary tension, either saturation or drying can eliminate the tension and cause a loss of shear strength.

Progressive failure brings about failure at one point before the adjoining point is highly stressed and a transfer of load from one point of the soil mass to the next. If the soil is sensitive, the average strength on the failure surface is not the maximum soil strength but somewhat less.

Judgment and experience are necessary in interpreting the laboratory test results and selecting the strengths to be used in stability analysis. The best shear strength for design is that found by analyzing actual failures in that same soil. This is particularly true when the problem is to correct a past failure. When a new design must be made for a situation for which there is no record of failures, it may be expedient to make an artificial failure—a full scale model of the proposed slope, embankment, or dam, that can be made to fail under careful observation and control. The results of such a full-scale test, when correlated with the laboratory data for the same soil, furnish the engineer with the effective strength for design of similar structures.

SAFETY FACTORS FOR DESIGN / When existing slopes and embankments have been analyzed for safety, it has been found they have relatively small factors of safety when compared with those of other structures. Although a safety factor of 2 or 2.5 is not uncommon in building design, the same factors applied to embankments would make the cost so high that they could not be constructed. Many earth structures having a *computed* safety factor as low as 0.9 have been proved stable by the test of time. The following table gives the significance of different values of the safety factor for soil masses. The safety factors in Table 11:2 apply to the most

TABLE 11:2 / SIGNIFICANCE OF SAFETY FACTORS FOR DESIGN

Safety Factor	Significance
Less than 1.0	Unsafe
1.0–1.2	Questionable safety
1.3–1.4	Satisfactory for cuts, fills, questionable for dams
1.5 or more	Safe for dams

critical combination of forces, loss of strength, and neutral stresses to which the structure will be subjected. Under ordinary conditions of loading, an earth dam should have a minimum safety factor of 1.5. However, under extraordinary loading conditions, such as a design superflood followed by sudden drawdown, a minimum safety factor of 1.2 to 1.25 is often considered adequate.

The safety factors for use in design include an allowance for the differences between laboratory tests results and the real shear strength of the soil. If the design is based on an analysis of a failure, lower values are acceptable.

11:2 Open Cuts

Open cuts are excavations in which no bracing is used to support the soil. They are used in constructing excavations when hard soil is encountered that needs no support and in highway, railroad, and canal cuts where the cost of long lines of bracing would be great. Cuts less than 30 ft or 9 m deep are ordinarily designed on the basis of experience. Railroad design manuals and standard highway specifications give $1\frac{1}{2}$ (horizontal) to 1 (vertical) as a standard slope for most conditions and 2 to 1 for soft soils.

DEEP CUTS / Deep cuts should be investigated first on the basis of a preliminary soil study and a slope-stability analysis utilizing the drained shear strength. If the soil is a swelling or fissured clay or subject to unusual seepage, further investigation is necessary. The condition of nearby cuts in similar soil should be checked and, if necessary, a trial cut excavated on a slope steep enough to cause failure. The soil strength determined by an analysis of the failure, correlated with the laboratory data on the soils in the cut, should be used to determine the safe slope. In extreme cases where accurate analysis is impossible because of erratic soils, it may be necessary to place bench marks at the top of the finished slopes to warn of any unusual movements that could lead to failure.

IMPROVEMENT OF CUT STABILITY / The stability of a cut can be improved by decreasing the soil stress or by increasing its strength. Soil stress can be reduced in most cases by making the slope flatter. If the sections that require improved stability are short, the slope may be partially supported by a small retaining wall or by cribbing. Water pressure in cracks in cohesive soils can be relieved by surface drains above the slope to intercept water and by horizontal drains driven into the face of the slope.

Piles driven through the potential shear plane can increase the resisting moment slightly. Better still, piles driven near the top of the slope can support part of the potentially sliding mass. Piles are most effective in increasing marginal stability; if the slope is very unsafe, enough piles to be effective will be uneconomical.

Soil strength in cohesionless and slightly cohesive soils can be increased by relieving neutral stresses with surface drains and horizontal drains in the face of the slope. Good drainage has always been the most effective measure for improving slope stability where water is a factor in instability. The strength of cohesive soils is difficult to improve permanently. In some instances large ventilation ducts, driven into the slope, have been able to reduce the soil's water content and increase its strength, but the method is very expensive.

CUTS IN LOESS / True loess, a cemented soil, has high shear strength in spite of its loose structure. It will stand vertically in cuts as deep as 40 to 50 ft or 12 to 15 m. Sloping cuts are stable only until rain falls. The bare, porous soil absorbs the water, which seeps downward rapidly because of the high vertical permeability. The cementing of the soil breaks down in water, and so the slope disintegrates by slumping until it becomes vertical. Vertical cuts (Fig. 11.10) will stand for years with only occasional slumping or scaling along the vertical cleavage planes. Cuts should be made wider than is necessary in order to allow room for the debris that collects.

SEC. 11:3] EMBANKMENTS

Figure 11.10 Highway cut in loess. Note vertical slopes and wide shoulders.

11:3 Embankments

An embankment is an artificial mound of soil used to carry railroads and highways across low spots, or to impound water. Since embankments are constructed of filled-in material, they are often termed *fills*, but this term applies also to other earth construction.

HIGHWAY AND RAILROAD FILLS / Highway and railroad fills are usually designed on the basis of experience unless heights greater than 30 or 40 ft (9 to 12 m) are involved. The standard slopes are usually $1\frac{1}{2}$ (horizontal) to 1 (vertical) or 2 to 1 unless the embankment is subject to flooding. Highways fills are carefully constructed of selected soils compacted to prevent settlement and a rough surface (see Chapter 5), but railroad fills are seldom highly compacted because rough surfaces can be prevented by proper maintenance of the ballast.

HIGH FILLS AND FILLS SUBJECT TO FLOODING / High fills and those subject to flooding require careful analysis and design based on the shear strength and compressibility of soils to be used in the fill construction. The different soils that could be used should be selected and tested as described in Chapter 5, and their shear strengths and other characteristics should be made available to the design engineer. The slopes required by each different soil to provide a safe embankment should be determined by stability analyses. From these data trial designs can be made, using the different soils, and the cost of each design can be estimated. The best soil is the one that gives a satisfactory fill at the lowest cost.

Compaction of soils wetter than optimum can cause a pore pressure buildup during construction. If the soil is very wet and the embankment high, the pressures may be great enough to cause localized slides, ordinarily face failures, in the wet zone. Undrained shear tests with pore pressure measurements can diagnose this possibility in advance and help set limits of moisture, or control the rate of construction to match the rate of drainage.

Fills subject to flooding are especially critical. Railroad fills that have stood the pounding of heavy trains for years often collapse after periods of flooding. They should be designed on the basis of the shear strength, determined after soaking samples of the soil in water as described in Chapter 5. Typical slopes for such fills may be as flat as 3 (horizontal) to 1 (vertical) or even 4 to 1 when made of soils that soften readily on absorbing water. The highway fill in Fig. 11.11 failed because a culvert beneath it clogged with

Figure 11.11 Slide in highway embankment weakened by saturation.

debris so that water backed up against the fill. The soil on the downhill slope was weakened by seepage (the design had not considered flooding) causing the slope to slide in a base failure. Note the trees tilted uphill in the base bulge, which is typical of base failures. Below the slide the trees tilt downhill.

LEVEES / Levees are small, long earth dams that protect low areas of cities and towns, industrial plants, and expensive farmland from flooding during periods of high water. Unlike highway and railroad embankments, settlement is not an important consideration; and unlike earth dams, levees must often be placed on poor foundations. Since levees usually extend for many miles, the cost of borrow materials and of construction is extremely important. Ordinarily a dragline working on top of the completed sections of the levee is used for construction, as shown in Fig. 11.12. It is capable of excavating and placing large volumes of soil quickly and cheaply, but it must utilize the soils found adjacent to the levee. Compaction of the soil interferes with the dragline operating cycle and so is seldom done. Because of poor soils and little compaction, levees in the United States are ordinarily constructed with very flat slopes, such as 5 (horizontal) to 1 (vertical) on the outer slope and 3 to 1 on the inner. The slopes are determined by experience in most cases.

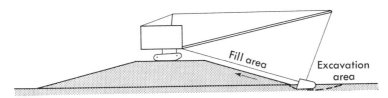

Figure 11.12 Construction of an embankment by a dragline.

High levees, levees in restricted spaces where flat slopes cannot be used, or levees protecting critical areas such as power plants should be designed on the basis of soil tests and stability analyses. In such cases careful soil compaction is required, but since steeper slopes can be used, the saving in soil volume compensates, to some extent, for the added cost of construction.

11:4 Embankment Foundations

Most difficulties with embankments come from faulty foundations, Fig. 11.13. It is not difficult to construct a fill that is strong, free from volume change, and incompressible; but if the soil below it is poor, failure may occur in spite of careful construction. The failure commences below the fill and in some cases can spread into the fill itself, obscuring the actual cause of the trouble.

FILLS ON THICK STRATA OF WEAK SOIL / Fills on deep soils of little strength fail because of inadequate bearing capacity, Fig. 11.13a. The formulas for bearing capacity can be used to analyze such failures if the weak stratum is at least half as thick as the base of the fill is wide. Otherwise the possibility of failure must be determined by trial and error, using the circle method of analysis. If a hard crust overlies the soft soil, its strength should not be relied on to support the load. In one case a levee 40 ft or 12 m high was built on top of a thin crust of hard clay that lay over a thick stratum of soft clay. Twelve hours after the levee was completed, it had sunk until it was only a few feet above the ground surface. The hard clay, which held up the partially completed fill, broke under the full load and allowed the fill to drop. Bulges or mud waves appeared in the ground surface adjacent to the toe of the fill.

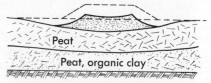

a. Mud waving and embankment subsidence from shear in thick soft clay

b. Embankment settlement from consolidation of compressible soils

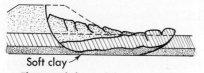
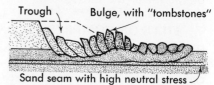

c. Elongated shear slide in a stratum of soft clay

d. Spreading slide in embankment above a thin seam of sand with high neutral stress

Figure 11.13 Embankment foundation problems.

Failures of this type can be prevented in a number of ways. Lightweight fill materials, such as slag, or wide, flat slopes can reduce the stresses beneath the fill to a safe amount. A gravel berm adjacent to the toes of the slope acts as a counterweight to prevent the bulging from taking place and thus may prevent failures. If the soil is normally consolidated, its strength can be improved by consolidation under the weight of the fill. Construction must proceed slowly, however, in order to give the soil time to consolidate. Vertical sand drains or sand piles can decrease the length of the drainage paths and may increase the rate of consolidation.

If the soft stratum is relatively thin (5 to 10 ft thick), it will be cheaper to excavate the soil beneath the fill area and replace it with something better. If the soft soil is from 10 to 20 ft thick, it can be replaced by *displacement*.[11:9] In this method the fill is constructed on the top of the soft soil as high and as steep as possible. In some cases it is expedient to allow it to sink under its own weight, displacing the soft soil in a bearing capacity failure. In other cases, it is feasible to remove the soft soil from beneath the fill by blasting. Dynamite is introduced under the fill, as shown in Fig. 11.14. The innermost charges

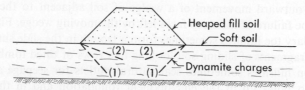

Figure 11.14 Blasting to remove soft soil beneath a fill. Dynamite charges marked (1) are detonated first and are followed by those marked (2) a fraction of a second later.

have a slight delay. The first blast removes the soil from the sides of the proposed fill position, and the second, a few thousandths of a second later, removes the soil beneath the fill and allows it to settle into place. The method has been very successful in many cases but requires trained, experienced personnel. A pile foundation can be used to support fills in especially critical areas, such as dock and harbor installations. On top of the piles is placed a concrete slab, known as a *relieving platform*, which supports part of the weight of the fill (see Fig. 8.17). The method is extremely expensive, however.

FILLS ON COMPRESSIBLE SOILS / In some cases the soil may be strong enough to support the fill without failure but is so compressible that the fill settles badly (Fig. 11.13b). This is particularly true of organic silts and clays and true to an extreme degree in peat. Highways across marsh areas often assume wavy profiles because of irregular settlement.

Excessive settlement due to compression may be minimized by pre-consolidating the soil through slow construction, by the use of sand piles, or by excavating the compressible soil. Any procedures involving consolidation require careful study to determine their effectiveness in each different situation.

FILLS ON THIN STRATA OF SOFT CLAY / Fills on relatively thin strata of soft clay fail by sliding horizontally along a complex failure surface that extends upward through the fill, as is shown in Fig. 11.13c. Failures of this type usually occur during or shortly after construction before the clay stratum has a chance to consolidate under load. The safety against this type of failure can be determined by the sliding-block analysis, using half the unconfined compressive strength of the clay stratum as the shear strength. Increased safety against this type of failure can be provided by a lightweight fill, flat slopes, and slow construction that permits the soil to consolidate and gain strength under the load. Vertical drains into the soft stratum in some cases can accelerate the rate of consolidation. If the stratum is close to the ground surface, it may be most economical to remove it completely.

FILLS ABOVE THIN, COHESIONLESS STRATA SUBJECTED TO NEUTRAL STRESS / When water pressure builds up in thin, cohesionless strata beneath a fill, failure may take place suddenly without any warning.

Although the neutral stress is generally greater under the center of the embankment, so is the confining stress due to the weight of the fill. Near the toe the confining stress is less and $(p-u)$ tan φ may approach zero. Failure starts with outward movement of a wedge of soil adjacent to the toe, followed by toe failure of the steep faces left by the moving wedge, Fig. 11.13d. The secondary toe failures sometimes form a trough in the slide surface with the intact fragments of sliding soil looking like rows of tilted tombstones.

Fills on hillsides sometimes seal natural outlets for seepage and cause pressures to build up in the embankment foundation and in the natural slope above (see Fig. 3:1b). Dams and levees create high pressures by the reservoirs they form. Temporary high water pressures sometimes develop during construction because the weight of the embankment acts on discontinuous silt and fine sand seams from which water cannot drain.

The safety against failure due to water pressure in cohesionless seams is analyzed by the sliding-block method, by considering the neutral stress and its effect on strength along the surface of sliding. This is easily done where the head is known, such as when it is produced by a man-made lake or reservoir. In other cases the head must be estimated from ground water observations or from the weight of the embankment. The safety can be increased by drains that intercept the pervious strata.

11:5 Earth and Rockfill Dams 11:11—11:17

Earth and rockfill dams are special embankments designed to impound water more or less permanently. Dams are the most critical of all engineering structures, for their failure can cause great property damage and loss of life. Earth dams, if not properly designed and constructed, are particularly vulnerable because the very material of which they are constructed can be weakened or disrupted by the water they are supposed to hold. It spite of these dangers, earth dams have proved to be among the most enduring of structures. Dams built in India and Ceylon over 2000 years ago are still storing water for irrigation. The largest structure ever built is the Fort Peck earth dam in Montana, with over 124,000,000 cu yd of embankment. Many earth dams over 400 ft or 120 m high are in use in many parts of the world and earth dams 1000 ft high are a reality.

CRITERIA FOR USE / Dams of earth are employed for a number of reasons. First, materials suitable for earth construction are available at a large proportion of dam sites. Second, earth or broken rock is easily handled, either by hand in remote areas with cheap labor or by great machines. Third, earth and rockfill dams are often suited to sites where the foundations are not strong enough or sufficiently incompressible for masonry dams. Finally, the earth or rockfill dam is frequently cheaper than any other type. Of course there are some disadvantages: Suitable materials are not always available; greater maintenance is usually necessary; and a separate spillway is required.

A successful earth or rockfill dam must satisfy two technical requirements: safety against hydraulic failure and safety against structural failure. These include failure of the foundation and the embankment, acting together as a unit, and any part of either.

Hydraulic failure can be external or internal. Externally the dam must have enough spillway to be safe against overtopping, and both the upstream and downstream faces must be protected against surface erosion. Safety against internal hydraulic or seepage failures was discussed in Chapter 4. The dam and the foundation must be sufficiently impervious that the loss of water is not objectionable, and they must be resistant to seepage erosion or piping. The latter has been an important cause of failure of earth dams, and provisions to prevent seepage erosion are essential features of design.

Structurally the dam and the foundation must support the weight of the embankment and the load of the water under the worst possible combinations of maximum reservoir, seepage forces, changes in reservoir level, and earthquake acceleration. The embankment is analyzed in the same manner as other earth structures but the required safety factors are higher, as given in Table 11.2. Settlement of the embankment can create cracks through which seepage erosion could develop. Ordinarily, settlement of the foundation of 1 per cent of the dam height and settlement in the embankment of 1 or 2 per cent of the height will not be serious, provided there are no sharp differences between adjacent parts of the dam.

COMPONENTS OF AN EARTH DAM / The components of an earth dam are shown in Fig. 11.15a. The basic parts are: (1) the foundation, (2) the cutoff and core, (3) the shell, and (4) the drainage system. The *foundation*, either earth or rock, provides support for the embankment and also may resist seepage beneath the dam. The *core* holds back the water, and if the foundation is pervious, the core is extended downward to form a *cutoff*. The *shell* provides structural support for the core and distributes the loads into the foundation. The internal drains carry away any seepage that passes through the core and cutoff and prevent the buildup of neutral stress in the downstream part of the embankment. The internal drains take many forms, depending on the anticipated seepage; trench drains and sloping drains just downstream from the core, blanket drains between the dam and the downstream foundation, the toe drains at the downstream toe. All must be provided with filters, as described in Chapter 4, to prevent internal erosion and clogging.

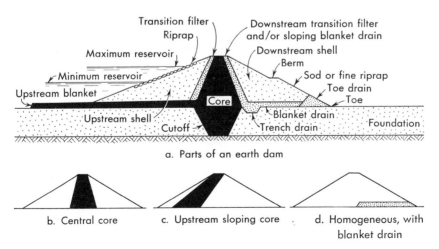

Figure 11.15 Basic earth and rockfill dam cross-sections.

Transition filters are frequently necessary between the core and shell to prevent migration of the fine-grained core into the shell (which is usually coarser). They are sometimes omitted if the grain sizes of the core and shell are not greatly different or if the seepage gradient through the core is small. *Riprap* is required on the upstream face to prevent erosion and wash by waves. Sod or fine riprap is required on the downstream face to prevent rain wash. Berms are often provided to permit access to the face of the dam during construction or for maintenance afterwards. They also help prevent rain wash by breaking the long continuous slope. The parts of a rockfill dam are similar, but the rockfill itself is a drain.

CROSS-SECTION / The three basic cross-sections are the central core, the upstream core, and the homogeneous, as shown in Fig. 11.15c, d and e. The *central core* provides equal support for the core and is most stable during sudden drawdown. It utilizes the minimum amount of core material. The *upstream core* or sloping core is most stable with a full reservoir and provides the cheapest design when there is little or no sudden drawdown. The upstream and central core are common for rockfill dams. The *homogeneous* cross-section results when both the core and shell are made of the same material. Unless a core zone is defined by extensive internal drainage, a homogeneous cross-section requires flatter slopes than a core type and for that reason is seen more frequently in dams less than 50 ft high.

CUTOFF AND CORE DESIGN / Cutoffs and cores are made of earth, steel sheet piling, concrete, or a curtain of grout injected into the soil. Earth is cheap and can be made sufficiently flexible so that it will remain watertight in spite of small movements in the dam or foundation. Almost any soil can be used if its permeability is sufficiently low (less than 10^{-4} cm per sec) and provided it does not develop swelling pressures. Earth is used for cutoffs if it is possible to cut a trench in the foundation with slopes steep enough so that the volume of soil required is not excessive. Earth is always used for the dam core unless no suitable material is present. The minimum thickness of the core or cutoff depends on the soil; clay cores as thin as 5 per cent of the head have been used, but better practice calls for the minimum thickness of clay cores to be 10 to 20 per cent of the head and for silty cores to be 30 to 40 per cent of the head.

A trench filled with a clay slurry can be used in a cutoff where open excavation is impractical because of a high ground water table or where sheet piling is impossible because of boulders. The trench is excavated with a dragline and is maintained full of a highly plastic clay-water slurry. The trench is backfilled with a pasty clay that will consolidate under its own weight. Cutoffs nearly 100 ft or 30 m deep have been made in this way. They are usually placed upstream from the dam toe so as not to weaken the foundation unduly, and are connected to the dam core by an upstream blanket.

Steel sheet piling and grouting are used for cutoffs where it is impractical to excavate for an earth core. Steel sheeting is sometimes used as a core for dams made of sand but is seldom worth the cost otherwise. Concrete is sometimes used for cutoffs but is rarely used today for cores because it cracks under the inevitable movements of the dam.

SHELL DESIGN / The design of the shell consists of selecting the material on the basis of its strength and availability for construction, as outlined in Chapter 5, and then determining the slopes necessary to provide stability for the embankment. The design process is essentially trial and revision: Tentative slopes are selected, the stability is analyzed, and the design is then revised to provide greater economy or greater stability.

Typical upstream slopes range from $2.5(H)$ to $1(V)$ for gravels and sandy gravels to $3.5(H)$ to $1(V)$ for micaceous sandy silts. Typical downstream slopes for the same soils are $2(H)$ to $1(V)$ to $3(H)$ to $1(V)$. A seepage analysis is made of the trial design to determine the flow net and the neutral stresses within the embankment and foundation. Safety against seepage erosion and the amount of leakage through the dam are computed.

Stability analyses are made of both faces of the dam, using the method of slices previously described. The upstream face is usually analyzed for three conditions: full reservoir, sudden drawdown, and reservoir empty before filling. The downstream face is analyzed for full reservoir and minimum tailwater and also for sudden drawdown of tailwater from maximum to minimum if that condition can develop.

Shells of broken rock, both dumped and compacted make excellent dams. Typical downstream slopes are often steeper than $2(H)$ to $1(V)$ even in very high dams. The entire dam shell is a drain, with no pore pressure above the tailwater level. Because of the sharp contrast between the grain sizes of the shell and core, carefully designed transition filters are essential.

If the stability is insufficient, there are several possibilities for improvement. First the slope can be flattened. Frequently only the lower half or third need be changed, making a composite slope that is steep at the top and flatter at the bottom. Berms at the toe of the slope serve the same purpose. Second the soil strengths can be improved by increasing the required density or by using different materials. Third, weak zones in the foundation can be corrected by preconsolidation or by their removal. Fourth the position of the core and cutoff can be shifted. Finally, internal drainage can be designed to reduce neutral stresses in the downstream foundation and shell.

11:6 Earth Movements in Nature

CLASSIFICATION OF NATURAL EARTH MOVEMENTS / Earth movements are commonplace geologic phenomena that are part of the process of mass wasting. Tremendous quantities of fractured or weathered materials are constantly on the move down slopes and into streams where they are carried away to be deposited elsewhere. The impelling force for all these movements is gravity, assisted at times by water pressure, expansion and contraction forces, earthquake shock, and man's interference with nature.

There are four different classes of natural movements:
1. Creep—the slow, relatively steady movement of soil down slopes.
2. Landslides—fairly rapid movements of soil or rock masses in combined horizontal and vertical direction.
3. Subsidences—movement of earth masses vertically downward.
4. Rockfalls—vertical superficial rock movements.

CREEP / Creep is a slow, nearly continuous movement of soil resembling the creep of metals under small stresses or the plastic flow of concrete. It is manifested by the tipping of fence posts and similar rigid objects embedded in the soil. The best indication of creep is the gentle curving of trees, with the convex side point downhill in the direction of movement, Fig. 11.16. (Trees in areas subject to landslides exhibit an abrupt change in trunk tilt that corresponds to each movement.)

Figure 11.16 Tree trunks bowed downhill in a creep zone.

The mechanism of creep is not fully understood. On slopes in which the safety factor is low, the movement is probably true creep (Section 3:18) at stresses close to shear failure. On flatter slopes, seldom less than 4 deg, creep may be the result of alternate shrinking and swelling with seasonal changes in moisture coupled with the continuing downhill force of gravity. Generally creep is confined to the upper 15 to 20 ft of the soil or broken rock mass and it is most rapid close to the ground surface. It is an indication of potential trouble—a quasi-equilibrium state that can be easily upset and turned into a landslide by some engineering construction such as a deep cut or a heavy fill.

Creep cannot be stopped, but its rate of movement can be decreased materially by drainage to increase the strength of the soil and to prevent the periodic swelling and shrinking. In most cases the best method of preventing trouble is to make allowances for it. Bench marks should be set in solid rock or in level areas not subject to movement by creep. Pipe lines should be made with flexible joints when laid in slopes on which creep is taking place. Building foundations must be strong enough and deep enough to resist movement or should be tied together so the entire structure can move. The latter procedure, of course, would be practical only for very small buildings.

LANDSLIDES[11:18] / Natural landslides are difficult to analyze because of their complex nature. The strength of natural deposits is so variable and the number of different forces acting is so great that theoretical studies are at best only indications of what is likely to occur. Most landslides do not occur spontaneously. The slope is usually unstable for years and gives warning of its instability from time to time by slow settlement or the formation of cracks.

Finally, some event takes place that increases the stress or decreases the strength to the point that failure can take place. This event triggers the failure, although it actually may be insignificant by itself. Loud noise has been known to start slides in loose debris in mountainous areas, and the added weight of water due to a hard rainfall is often responsible for the start of slides in humid regions.

Many different ways of classifying slides have been devised. Some systems classify them according to the type of deposits or the appearance of the failure, while other systems classify them according to the forces causing failure or by the force that triggered the movement. Actually each failure should be considered by itself, and any classification should be used for descriptive purposes only.

FLOW SLIDES / If a soil or rock mass should suddenly lose its strength, it will behave like a liquid and flow downhill and spread out over the flat land below. The basic cause of failure is ordinarily neutral stress that builds up until the strength becomes insufficient to support the load. If the material is loose, structural collapse as described in Chapters 1 and 3 contributes to the water pressure increase and loss of strength.

Loose saturated cohesionless soils are particularly vulnerable to sudden flow of a large portion of the mass brought on by localized shear or shock.[11:19]

The local failure produces a pore pressure buildup that causes additional failure and more pore pressure. Pile driving in a shore deposit of loose fine sand triggered a landslide that carried houses and people several hundred feet into a shallow lake. An explosion triggered the flow of a saturated fine grained industrial waste that engulfed the processing plant that produced the waste.

Water pressure buildup from rainfall has produced flow slides in coarse grained cohesionless deposits like talus and rock debris and in closely jointed rock. Heavy rain and snow melt at Gros Ventre Wyoming saturated a fractured conglomerate on a mountainside, and built up enough neutral stress so that several million cubic yards flowed down to destroy a small village and dam up a river in the valley. The lake has remained for 40 years, covering farms and houses. Neutral stress accumulating in fragmented rock from a coal mine waste dump in Wales brought on a flow slide that buried a school and killed more than 100 children in a few seconds.

Severe cyclical loading from earthquakes has caused liquefaction of sand deposits and devastating flow slides. According to Seed,[11:19] the widespread damage that accompanied many past earthquakes was the result of flow slides. One of the disastrous landslides at Ancorage, Alaska in 1963 has been attributed to flow in a sand seam underlying a firm clay. The water in the collapsing voids in a flow slide is forced to the surface in small sand-water geysers that add to the terror of the phenomenon.

FLOW SLIDES IN SENSITIVE CLAY / Flow slides also occur in the ultra sensitive marine clays such as the marine clays of eastern Quebec, Canada. While the undisturbed strength of the clay is moderately high, the remolded strength is extremely low. Failure at an isolated point in the mass, induced by local shear, brings about progressive failure and flow in which large blocks of intact clay float on a stream of viscous liquid remolded clay.

LINEAR SHEAR SLIDES / Linear shear slides occur along well defined plane surfaces. Although crescent-shaped segments of soil break loose in the upper end of the slide zone, and bulges and mud waves occur at the lower end, the failure zone is elongated and most of the movement is linear.

The movements take several forms as shown in Fig. 11.17. The simplest occurs on a sloping plane of weakness such as a seam of weathered shale, Fig. 11.17a. A common variant is a wedge of loose fill whose plane of contact with the virgin soil becomes saturated by seepage from pervious strata. Another common variant is a cohesionless seam dammed up by an impervious fill, Fig. 11.17b. Failure in linear slides can be triggered by excavation that reduces support at the toe, by local softening of the soil or water pressure in pervious strata or joints.

Movement on a horizontal plane can occur by squeezing of a soft clay seam, Fig. 11.17c. The block above moves out as if it were on a belt conveyor. Water pressure in joints aggravates such a movement.

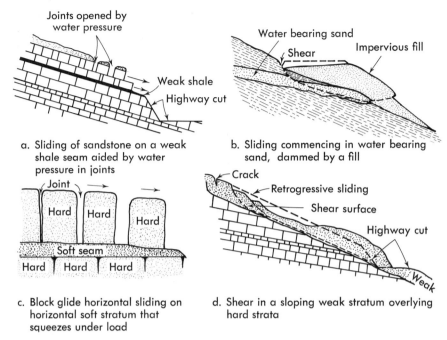

Figure 11.17 *Linear shear slides*

A third form is shear in a stratum of weak material, usually a weathered zone, underlain by a hard boundary. This is a rotational slide that is severely flattened by the solid boundary below, Fig. 11.17d. These slides are common in residual soils overlying rock. They also occur in hard, highly plastic clays that gradually expand and weaken from the ground surface down, creating a blanket of soft soil above a harder unweathered base. Such slides have developed in slopes as flat as $4(H)$ to $1(V)$ or 14 degrees with the horizontal where the clays have been exposed to rainfall by highway cuts.

Progressive failure or *retrogression* is characteristic of many landslides but particularly of linear shear forms. The slide progresses up the hillside, with each successive movement leaving a steep slope that is unstable and which subsequently fails as the earlier slide continues to move down. Progressive failure is also characteristic of soils which lose strength upon continuing shear strain. Even some stiff clays and shales can fail by progressive movement and loss of strength.

Linear slides in weathered rocks including shales are often triggered by the slow expansion or elastic rebound of the material exposed by excavation or earlier sliding. Whole hillsides sometimes begin to move although the materials are strong enough that a conventional stability analysis indicates adequate safety.[11:20]

ROTATIONAL SLIDES / Rotational slides take place in homogeneous soils, particularly clays, and in thick deposits where there are numerous non-continuous planes of weakness. The classic examples of the rotational slides are the deep base failures that occur in the soft clays of the coastal areas of Norway and Sweden. Failure is triggered by undercutting of the toe by dredging or erosion and by external loads on the upper part of the slope.

A major slide occurred in the edge of a valley in a deep deposit of glacial clay. It was caused by filling at the top of the slope to gain a parking area for a manufacturing plant. The underlying clay sheared under the weight of the fill. The bulge at the toe of the slope was 15 ft high. The force exerted by the moving wall of clay against an adjacent railroad trestle was great enough to bend the steel legs of the support towers and to shear their concrete foundation pedestals.

Steep slopes in stiff clays fail in rotational toe slides, Fig. 11.18a. The clay cliffs of the Great Lakes suffer from such movements when they are undercut by wave action. Several successive slides are often seen at the same point, forming narrow, arc-shaped terraces or steps leading down the slope, their surfaces intact and with trees and shrubs growing on them.

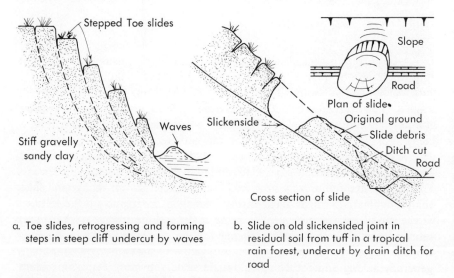

a. Toe slides, retrogressing and forming steps in steep cliff undercut by waves

b. Slide on old slickensided joint in residual soil from tuff in a tropical rain forest, undercut by drain ditch for road

Figure 11.18 Slides in stiff soils, initiated by undercutting.

MODIFIED ROTATIONAL / Modified rotational slides occur when multiple surfaces of weakness distort the shear zone. One typical form occurs in clays with seams or lenses of cohesionless soils. When water pressure (neutral stress) builds up, the strength drops, sometimes so low that shear failure occurs. Failure begins with a horizontal movement, but this leads to an elongated rotational slide that often cuts across the cohesionless seams and allows them to drain. An example of such a slide is shown in Fig. 11.19. It took place after snow melt, spring rainfall, and finally a leaking water pipe augmented the normal ground water pressure in numerous thin sand seams in a thick clay deposit.

Figure 11.19 Landslide caused by water pressure in thin seams of cohesionless silt and fine sand in a thick stratum of clay.

A second form occurs in residual soils that retain the structural defects, such as joints, of the original rock.[11:21] The joints can sustain no tension. Often they are surfaces of advanced weathering that are weaker in shear than the remainder of the mass. Failure may start with a plane tension crack at the top of the slope or a plane bulge at the toe, followed by shear in the remainder of the mass along a curved surface. Such slides are common in tropical regions with deeply weathered rock. The weathered formation is generally strong and therefore will stand safely on steep slopes. However, local failures occur in areas where the joints are favorably oriented, and particularly after equilibrium has been upset by some construction operation such as excavating for a highway, Fig. 11.18b.

CORRECTING LANDSLIDES / Most landslides are caused by a combination of factors. Before any remedial action can be taken, a careful study must be made to determine which factors are the significant ones for that particular situation. The structure of the soil and rock formations and the physical properties of the different materials must be established. The ground water levels and the pressures in cracks and fissures are particularly important.

Correction involves controlling as many of the factors as possible. Often the best method is found by trial. Drainage, flattening the slope, stabilization of the soil by grouting, removing external loads, erosion protection at the toe, and providing support with piling and retaining walls—all have proved to be successful under certain circumstances, but no method is of value unless it fits the specific needs of the particular slide.

SUBSIDENCES / Subsidences are actually vertical earth movements. They are of two types: rapid, caused by undermining or failure of the underlying strata; and slow, caused by consolidation. Rapid subsidence occurs frequently in areas of abandoned mines. Disintegration of old timbering in shallow workings causes caving of the rock above and the formation of a cavity beneath the soil. Sooner or later the soil bridging the cavity will break apart until an intact mass slides vertically downward. The same phenomenon occurs in areas underlaid by cavernous limestone. The countless sinkholes that dot the landscape in many parts of the eastern United States as well as other parts of the world are the results of small subsidences. In a few cases subsidences have been caused by the underground erosion of cohesionless strata by artesian water.

Rapid subsidence can be caused by excavation for sewers, tunnels, and buildings. If more soil is removed from an excavation than the finished volume of the excavation, it indicates that the soil is squeezing into the hole as it is being removed. This phenomenon, known as *lost ground*, is particularly troublesome in soft clays such as are found in Chicago, Detroit, Cleveland, and other cities in glaciated areas. The lost ground results in subsidence of the surrounding ground surface and often causes damage to adjacent buildings. Careful bracing of excavations and checks to determine possible building settlement are required to prevent such troubles.

Slow subsidence caused by consolidation occurs in areas in which the soil stresses increase materially. The Long Beach area of California subsided at a rate of 10 in. or 25 cm per year for the period 1941 to 1945 and is still sinking. The excessive pumping from the many oil wells in the area is reducing the neutral stresses in the oil-bearing rocks and is increasing the effective stresses. The rocks therefore consolidate as the oil is removed and the ground surface sinks correspondingly. Mexico City is also subsiding at a rate of several inches per year. This is caused by the pumping of the city water supply from the sand strata that are interbedded with the soft, volcanic clays beneath the city. The only remedy for such subsidences is to make allowances for them in design of structures or control the withdrawal of water or oil. They cannot be prevented without correcting the causes.

Subsidences may induce stresses in the soil or rock strata that aggravate failure by another unrelated mechanism. The failure of a water reservoir in Baldwin Hills, Los Angeles, is an example. Subsidence over a wide area, blamed on oil well pumping, caused an old fault to re-open. One limb of the fault cut across the earth dike foundation and the reservoir bottom; its movement ruptured the reservoir lining. Water seeping from the leak softened the foundation, aggravating the leak. The reservoir eventually failed by piping followed by collapse of a portion of the dike into the hole eroded by the seepage.

ROCKFALLS / Rockfalls are movements of detached rock fragments down steep slopes. They often occur in cuts in badly jointed rock and in cuts where all the materials loosened by blasting were not removed. Periodic checks should be made on the condition of rock cuts or other steep slopes in rock and all unstable pieces removed. Some roads must be closed during periods of heavy rain or freezing and thawing when water pressure or frost wedging can set loose rock in motion. In some cases it has been practical to anchor loose rocks with rods and cables to prevent their movement.

Rock bolts can be used to prestress jointed formations, and bind them together in a coherent mass. Concrete facings, supported by rock bolts, can prevent the movement of small, loose pieces, and at the same time protect the rock from weathering. The facing is drained to prevent a buildup of neutral stress in pervious seams or joints.

Deep cuts in rock are frequently provided with horizontal berms to catch falling rock and prevent them from tumbling down the full height and gaining enough momentum to create serious damage. Sand cushions on the berms have been effective in minimizing the bouncing of such rocks. Rockfills or fences are used to catch the rocks and keep them from rolling beyond the slope toe and endangering people and structures below. In mountainous regions, avalanche sheds of heavy timber or concrete are sometimes built over railroads and highways to protect them from rockfalls and snow slides. Any preventive or corrective program must start with a complete picture of the joints, fissures and bedding planes upon which the failures focus. Control of the movements on these surfaces is the essence of the design.

REFERENCES

11:1 L. Bjerrum and Nils Flodin, "The Development of Soil Mechanics in Sweden, 1900–1925," *Geotechnique*, **X,** 1, March 1960.

11:2 W. Fellenius, *Erdstatische Berechnungen*, rev. ed., W. Ernst u. Sohn, Berlin, 1939.

11:3 D. W. Taylor, *Fundamentals of Soil Mechanics*, John Wiley & Sons, Inc., New York, 1948.

11:4 R. V. Whitman and P. J. Moore, "Thoughts Concerning the Mechanics of Slope Stability Analysis," *Proceedings of the*

Second Pan American Conference on Soil Mechanics and Foundation Engineering, Brazil, 1963.

11:5 John M. Lowe, III, "Stability Analysis of Embankments," *Journal of the Soil Mechanics and Foundation Division, Proceedings, ASCE,* **93,** SM4, July 1967.

11:6 R. V. Whitman and W. A. Bailey, "Use of Computers in Slope Stability Analysis," *Journal of the Soil Mechanics and Foundation Division, Proceedings, ASCE,* **93,** SM4, July 1967.

11:7 N. Newmark, "Effects of Earthquakes on Dams and Embankments," *Geotechnique,* **15,** September 1965, p. 140.

11:8 D. W. Taylor, "Stability of Earth Slopes," *Journal of the Boston Society of Civil Engineers,* July 1937.

11:9 *Blasters Handbook,* E. I. Dupont de Nemours and Company, Wilmington, Delaware, 1958.

11:10 G. F. Sowers, *Earth and Rockfill Dam Engineering,* Asia Publishing House, Bombay, 1961.

11:11 A. Casagrande, "Notes on the Design of Earth Dams," *Journal of the Boston Society of Civil Engineers,* **37,** 1950.

11:12 R. W. Clough and R. J. Woodward, III, "Analysis of Embankments Stresses and Deformations," *Journal of the Soil Mechanics and Foundation Division, Proceedings, ASCE,* **93,** SM4, July 1967.

11:13 "Problems in the Design and Construction of Earth and Rockfill Dams," *Journal of the Soil Mechanics and Foundation Division, Proceedings, ASCE,* **93,** SM3, May 1967, p. 129.

11:14 "Symposium on Rockfill Dams," *Transactions ASCE,* **125,** Part II, 1960.

11:15 J. L. Sherrard, "Earthquake Considerations in Earth Dam Design," *Journal of the Soil Mechanics and Foundation Division, Proceedings, ASCE,* **93,** SM4, July 1967.

11:16 H. B. Seed, "Earthquake Resistant Design of Dams," *Journal of the Soil Mechanics and Foundation Division, Proceedings, ASCE,* **92,** SM1, January 1966.

11:17 H. B. Seed and A. C. Tory, "Seismic Coefficient for Earth Dam Design," *Journal of the Soil Mechanics and Foundation Division, Proceedings, ASCE,* **92,** May 1966.

11:18 K. Terzaghi, "Mechanism of Landslides," *Application of Geology to Engineering Practice, Berkey Volume,* Geological Society of America, 1950.

11:19 H. B. Seed, "Landslides During Earthquakes Due to Liquefaction," *Journal of the Soil Mechanics and Foundation Division, Proceedings, ASCE,* **94,** SM5, September 1968, p. 1053.

11:20 L. Bjerrum, "Progressive Failure in Slopes of Overconsolidated Plastic Clay and Clay Shales," *Journal of the Soil Mechanics and*

Foundation Division, Proceedings, ASCE, **93**, SM5, September 1967, p. 1.

11:21 B. St. John, G. F. Sowers, and C. E. Weaver, "Landslides on Slickensided Surfaces in Residual Soils," *Proceedings of the Seventh International Conference on Soil Mechanics and Foundation Engineering*, **2,** Mexico City, 1969.

SUGGESTIONS OF FURTHER STUDY

1. *Landslides and Engineering Practice*, Special Report 29, Highway Research Board, Washington, 1958.
2. D. P. Krynine and W. R. Judd, *Principles of Engineering Geology and Geotechnics*, McGraw-Hill Book Co., Inc., New York, 1957.
3. Publications of the Swedish and Norwegian Geotechnical Institutes.
4. W. J. Turnbull, and M. S. Hvorslev, "Special Problem in Slope Stability," *Journal of the Soil Mechanics and Foundation Division, Proceedings, ASCE*, **93,** SM4, July 1967, p. 499.
5. *Proceedings of the International Conferences on Soil Mechanics and Foundation Engineering*, **I,** Cambridge, Mass., 1936; **II,** Rotterdam, 1948; **III,** Zurich, 1953; **IV,** London, 1957; **V,** Paris, 1961; **VI,** Montreal, 1965; **VII,** Mexico City, 1969.
6. *Design of Small Dams*, U.S. Bureau of Reclamation, Denver, 1960.
7. A. W. Bishop, "Use of the Slip Circle in the Stability Analysis of Slopes," *Geotechnique*, **V,** 1, March 1955.
8. "Proceedings of the 1966 Slope Stability Conference, ASCE," *Journal of the Soil Mechanics and Foundation Division, Proceedings, ASCE*, **93,** SM4, July 1967; thirty papers on a wide variety of stability problems.
9. J. L. Sherrard, R. J. Woodward, S. F. Gizienski and W. A. Clevenger, *Earth and Earth-Rock Dams*, John Wiley & Sons, Inc., New York, 1963.
10. D. U. Deere, A. J. Hendron, F. D. Patton, and E. J. Cording, "Design of Surface and Near-Surface Construction in Rock," Chapter 11, *Failure and Breakage of Rock, Eighth Symposium on Rock Mechanics*, Am. Inst. of Mining Metallurgical and Petroleum Engineers, New York, 1967, p. 238. (Also see entire Part III.)

PROBLEMS

11:1 A slope of 2 (horizontal) to 1 (vertical) is cut in homogeneous, saturated clay whose $c = 1100$ psf and whose $\gamma = 112$ lb per cu ft. The cut is 42 ft deep and the clay deposit extends 17 ft below the bottom of the cut. The clay rests on rock.

a. Compute the safety of this slope, using the charts for homogeneous soils.

b. Check, using the circular arc analysis.

11:2 A cut is excavated at an angle of 45°. It is 30 ft deep. The soil profile from the surface down is:

Depth (ft)	Soil	Shear Strength (psf)	Unit Weight (lb/cu ft)
0–10	Stiff clay	$c = 1500$	$\gamma = 118$
10–25	Stiff clay	$c = 1200$	$\gamma = 105$
25–40	Firm clay	$c = 1000$	$\gamma = 112$
40	Shale (rock)		

a. Find the safety factor with respect to base failure, assuming the center of the failure circle to be above the mid-point of the slope.

b. Find the safety with respect to toe failure, assuming the center of the circle to be the same as for the homogeneous case.

c. Check the results, using the stability factor chart and assuming that the effective c is the weighted average c of all the strata.

11:3 A cut at an angle of 65° with the horizontal is to be made in a partially saturated clay with $c' = 500$ psf, $\varphi' = 15°$, and $\gamma = 115$ lb per cu ft. How deep can this cut be made before a minimum safety factor of 1.2 is reached? Use charts for homogeneous soil.

11:4 An excavation 30 ft deep and 65 ft wide at the bottom is to be made in clay with $c = 780$ psf and $\gamma = 110$ lb per cu ft. How wide should the top of the excavation be if the minimum safety factor is 1.3? (Use chart of stability number.)

11:5 An embankment of sand is 40 ft high, 30 ft wide at the top, and has side slopes of $1\frac{1}{2}$ (horizontal) to 1 (vertical). The embankment soil $\varphi = 42°$ and $\gamma = 124$ lb cu ft. The foundation soil consists of clay with $c = 800$ psf. Compute the safety of the embankment against a sliding-block failure along the line of contact of the foundation and the fill.

11:6 A canal is dug in a soil whose characteristics when saturated are $c = 600$ psf, $\varphi = 16°$, and $\gamma = 124$ lb per cu ft saturated. The canal is to be 22 ft deep and the slopes are 2 (horizontal) to 1 (vertical).

a. Compute the safety factor when the canal is full of water. (Use chart.)

b. Compute safety factor if canal is suddenly drained, leaving the soil saturated.

c. Which condition is worse?

HINT: When canal is full of water, the unit weight is reduced by buoyancy, an amount of 62.4 lb per cu ft.

PROBLEMS

11:7 An embankment 75 ft high and 40 ft wide at the top is constructed with a slope of 50°. The soil is partially saturated clay. The uppermost 50 ft of soil has $c' = 2000$ psf, $\varphi' = 15°$, and $\gamma = 114$ lb per cu ft. The remaining 25 ft has $c' = 900$ psf, $\varphi' = 12°$, and $\gamma = 111$ lb per cu ft. Compute the safety factor with respect to toe failure. Use the method of slices. Assume that the center of the critical circle is the same as in homogeneous saturated clays (Fig 9:6c).

11:8 A highway fill 25 ft high, 30 ft wide at the top, with slopes of 1.5 (horizontal) to 1 (vertical) is constructed across an area of soft, compressible soil. The fill weighs 120 lb per cu ft and is well-compacted, homogeneous, sandy clay. Foundation soil has $c = 420$ psf, $\gamma = 106$ lb per cu ft saturated. Compute the safety against a bearing-capacity failure, assuming that the entire fill remains intact.

11:9 An earth dam is 100 ft high and 20 ft wide at the crest. The upstream slope is 3 (horizontal) to 1 (vertical), and the downstream slope is 2.5 to 1. The dam consists of a clay core 20 ft wide at the dam crest and 30 ft wide at the base. The clay core $c = 200$ psf and $\gamma = 100$ lb per cu ft. The remainder of the dam is sand with $\gamma = 121$ lb per cu ft and $\varphi = 41°$. The foundation soil is clay with $c = 2500$ psf, $\gamma = 115$ lb per cu ft. A thin seam of sand with $\varphi = 42°$ extends from the reservoir under the dam at a depth of about 4 ft below the embankment. The seam ends a few feet upstream of the toe. If the maximum head on the dam is 85 ft above the foundation, compute the safety of the downstream half of the dam against sliding. Assume that the pressure in the sand seam is equivalent to the full head.

11:10 Prepare a discussion of a landslide or slope failure from the published description in an engineering journal or magazine. Include the following points:
1. Description of failure;
2. Chain of events leading to failure;
3. Probable cause;
4. Corrective measures, if any.

APPENDIX 1
Unit Costs

ENGINEERING ANALYSIS AND DESIGN CANNOT BE DIVORCED FROM COST of construction. Science and technology have made it possible to do many remarkable things; whether or not they will be of use depends on their ultimate value compared with their cost. The following table will give the student some concept of the cost of soil and foundation work. It is based on typical costs in the United States in 1970 as published in *Engineering News Record* and similar publications.

FOUNDATIONS
Basement excavation in soil	$ 0.60 to $	2.50/cu yd
Basement excavation in rock	5.00 to	30.00/cu yd
Footing excavation in soil	2.00 to	10.00/cu yd
Footing excavation in rock	5.00 to	40.00/cu yd
Footing concrete, including any forming	45.00 to	75.00/cu yd
Drilled pier foundations, including concrete	45.00 to	65.00/cu yd
Wood piles, untreated, including driving	2.00 to	3.00/lin ft
Wood piles, treated, including driving	2.50 to	3.50/lin ft
Concrete 50-ton precast, including driving	7.00 to	10.00/lin ft
Concrete 50-ton cast-in-place, including driving	4.50 to	7.00/lin ft
Steel H-pile, 10 in., 50 ton, including driving	6.00 to	8.00/lin ft
Steel pipe pile, $10\frac{3}{4}$, 50 ton, including driving	6.00 to	8.00/lin ft

EXCAVATION BRACING
Wood sheeting (20 ft deep)	4.00 to	6.00/sq ft
Steel sheeting, or H-beam and lagging, (40 ft deep)	5.00 to	10.00/sq ft

DRAINAGE
Wellpoint installation and operation, 30 days	15.00 to	40.00/lin ft*
Trench excavation in soil	2.00 to	6.00/cu yd

* Of header pipe.

Pumping, 20-ft head	.05 to	1.00/1000 gal
Drain pipe, 8-in. concrete	2.00 to	4.00/lin ft
EARTHWORK (Dams, embankments, large fills)		
Soil excavation, placement, compaction in embankments	0.40 to	0.80/cu yd
Rock excavation, placement, compaction	1.00 to	3.00/cu yd
Backfill, around structures, compacted	2.00 to	5.00/cu yd
SOIL STABILIZATION		
Soil-cement, cement, mixing and compaction	6.00 to	8.00/cu yd soil
Cement grouting, including injection	2.00 to	3.00/sack cement
Chemical grouting, including injection	50.00 to	100.00/cu yd soil
EXPLORATION		
Auger boring	1.50 to	3.00/lin ft
Test boring, including split barrel samples	4.00 to	6.00/lin ft
Diamond-core drilling	8.00 to	12.00/lin ft
Undisturbed sampling	10.00 to	40.00/sample

APPENDIX 2
Age of Geologic Formations

Era	Period	Epoch	Age Range*†
Cenozoic	Quartenary	Recent	0–11,000 years
		Pleistocene	11,000–1 million
	Tertiary	Pliocene	1–13 million years
		Miocene	13–25
		Oligocene	25–36
		Eocene	36–58
		Paleocene	58–63
Mesozoic	Cretaceous	Upper	63–90
		Lower	90–135
	Jurassic		135–181
	Triassic		181–230
Paleozoic	Permian		230–280
	Pennsylvanian ⎫	Carboniferous	280–310
	Mississippian ⎭		310–345
	Devonian		345–405
	Silurian		405–425
	Ordovician		425–500
	Cambrian		500–600
PreCambrian	Proterozoic		600+
	Archeozoic		?

* Kulp, J. L., "Geologic Time Scale," *Science*, **133**, 1961, p. 1105–1114.
† There is considerable controversy in ages in the Paleozoic Era and in the beginning of the Recent and Pleistocene epochs. The Pleistocene may have begun as long ago as 2.5 million years and ended as late as 5000 years ago according to some authorities.

Index

AASHO, classification, 69
    compaction, 210
A-horizon, 61
Active earth pressure, 339
Activity of clay, 29
Adobe, 79, 203
Adsorbed water, 24
AHLVIN, R. G., 250
Air photo interpretation, 263
ALGERMISSEN, S. T., 311
Allowable soil pressure, 414
Alluvial fan, 46
Alluvial soils, 45
Anchored bulkhead, 369
Anisotropy,
    permeability, 93
Anchors, 495
Angularity, 22
Angle of internal friction, 133
Angle of repose, 514
Aquifers, 62
Arching, 345
Area ratio, 247
Artesian water, 63
ASTM, 311
ATTERBERG, 27
Atterberg limits, 27
Auger boring, 268
Auger sample, 269
AVERY, T. E., 311

B-horizon, 61
Backfill (retaining walls), 355
Backfill drainage, 358

Back pressure, 150
BARKAN, D., 498
Base exchange, 26
Base failure, 507
Batter of retaining wall face, 355
Batter pile foundation, 475
Batter piles, 446
Bearing capacity analysis, 391
    of circular and rectangular
        foundations, 346
    of piers, 489
    of piles, 454, 461
    of rock, 431
Bearing capacity equation, 395
Bearing capacity factors, 345, 461
Bearing piles, 446
BEATON, J. L., 250
Bentonite, 79
BEREZANTSEV, V. G., 499
BESKOW, G., 197
BISHNOI, B. W., 439
BISHOP, A. W., 155, 537
Bituminous stabilization, 237
BJERRUM, L., 498, 535, 536
Blanket drain, 186
Blasting for embankment foundation,
    522
Blowout, 96
Boilers, foundations for, 429
Boils, 172
Boring depth, 267
    records, 289
    spacing, 266
Borrow, 203, 220
Boulder, 76
Boussinesq's equation, 401

BOWLES, J., 438
BREKKE, T. L., 324
Buckling of piles, 454
Bulb pile, 481
Bulkhead, anchored, 369
Bureau of Reclamation, 537
Buttress wall, 356

CBR, 249
Caisson, 492
Calcite, 12
Caliche, 79
California bearing ratio, 249
Cantilever wall, 356
Capillary flow, 179
Capillary moisture, 90, 178
Capillary tension, 87, 89, 140
CAQUOT, A., 383
CARSON, A. B., 439, 500
CASAGRANDE, A., 36, 197, 536
CASAGRANDE, L., 197
Cast-in-place concrete piles, 479
Cation exchange, 26
Cavity, 56, 246
CEDERGREN, H. R., 198
Cemented soils, 33, 46, 51
　shear strength, 150
　stabilization by, 234
CHAN, C. K., 250
Changed condition, 257
CHELLIS, R. D., 499, 500
Chemical stabilization, 237, 240
Circular arc analysis, 508
Clay, 13
　active pressure of, 341
　excavation bracing in, 366
　landslides in, 512
　passive pressure of, 342
　shear strength of, 141, 149
　slope stability in, 512
Clay minerals, 13
Clay sizes, 19
Clay soils, 27
CLOUGH, R. W., 324, 536
Coefficient of earth pressure, 338, 340,
　342, 460
Cofferdam, 363
Cohesion, 147
Cohesionless soil,
　active pressure of, 340

density of, 75
elasticity, 132
excavation bracing in, 365
passive pressure in, 342
slides in, 529
slope stability of, 514
strength of, 130
Cohesive soil, 27
Cold storage, foundations for, 429
Color of soils, 75
Column strength of piles, 454, 487
Compaction of soils, 206
　density control, 229
　methods, 222
　requirements, 214, 216
Compaction test, 210
Compactive effort, 209, 206
Composite piles, 485
Composition of soils, 5, 10
　estimating, 75
Compressibility,
　coefficient of, 100
　estimating, 75
　of compacted soil, 213
　of soils and rocks, 101
　of soil layer, 99
　test, 98
Compression index, 101
Conduit, 371
Consolidation, coefficient, 110
　percentage, 109
　rate, 106
　test, 98
Consolidation, secondary,
　time rate, 112
Contact pressure on foundations, 410
COOLING, L. F., 82
CORDING, E. J., 537
Core recovery, 275
Cost, 540
Cost of soil investigation, 259, 541
Coulomb's theory of earth pressure,
　349
Counterfort wall, 356
COYLE, H. M., 499
Cracks, in buildings, 411
　in slopes, 511
Creep, behind walls, 344
　of slopes, 528
　of soils, 152
Creosote, 477
CRISP, R. L., 250
Critical gradient, 173

INDEX 549

Culverts, 377
Cutoff, 175
Cuts (*see also* Excavation), 517
   in loess, 518
Cycles of change, 41

Dam, earth, 337
Damage due to excavations, 368
   due to drainage, 189
   due to frost, 191
   due to settlement, 411
D'APPOLONIA, E., 438
Darcy's law, 91
DE BEER, E., 267
DEERE, D. U., 82, 311, 537
Deformation, due to earth pressure, 343
Deliquescent stabilization, 302
Deltas, 48
Denison sampler, 301
Density, critical, 135
   test, 229
Density, describing and estimating, 75
   of cohesionless soils, 30
   relative, 31
Depth, of boring, 267
   of foundations, 151
Description of soils, writing, 78
Design, of anchored bulkheads, 369
   of earth dams, 337
   of excavation bracing, 365
   of foundations, 177
   of pile foundations, 227
   of retaining walls, 355
Dessication drainage, 190
Diamond bit sizes, 276
Diamond drilling, 274
Diatoms, 48, 800
Differential excavation, 422
Differential settlement, 412, 420
Diffuse double layer, 25
Direct shear test, 127
Displacement method for fills, 333
Disruptive gradients, 175
Distortion settlement, 408
Ditch, 185, 376
Double-acting steam hammer, 449
Dragline, 521
Drainage, 182
   effects of, 189
   of dams, 176, 525

   of retaining wall backfills, 353
Drainage systems, 184
Drainage, vertical sand, 191
Drawdown, 169
Drilled-in-caisson, 492
Drop hammer, 448
Dry density, 207
Dunes, 50
DUNHAM, C. W., 439
DUNCAN, N., 82
Dutch cone, 277
DUVALL, W. L., 83, 383

Earth dams, 525
   design of 526
   sections, 525
   seepage through, 167, 176, 526
Earth movements, causes of, 506, 528
Earth pressure, 337
   active, 339
   at rest, 337
   effects of deformation on, 343
   on bulkheads, 369
   on excavation bracing, 363
   on piles, 460
   on retaining walls, 351
   on tunnels, 372, 375
   passive, 342
Earthquakes, 261
   stresses, 319, 511
Effective size, definition, 20
Effective stress, 94, 135
Elastic equilibrium, 320
Elasticity, modulus of, 122
Electrical resistivity, 286
Electrical surface charge, 25
Electro-osmosis, 183, 190
Embankments, 203
   design of, 519
   foundations of, 521
   materials for, 214
End-bearing pile, 457
Engineering News formula, 469
Equipotential line, 163
Equivalent fluid weight, 352
ERICKSON, H. B., 251
Erosion, seepage, 174
Evaporation drainage, 183
Evaporation bracing, 358
   in clay, 366
   in sand, 365

Excavation methods, 217
Excavation slopes, 518
Exchangeable cation, 25
Expansion, 105
Exploratory investigations, 266, 270
Explosive compaction, 227

Fabric, 28
FADUM, R. E., 438
Failure surfaces, behind walls, 344
   in slopes, 505
Failure theories, 124
Failure wedge, behind excavation
      bracing, 364
   behind walls, 344
FAUSOLD, M., 499
FELD, J., 38
FELLENIUS, W., 535
Fence diagram, 291
Field density tests, 229
Field tests, 305
Fill, 19, 203, 429
   construction 227, 232
   placement, 221
   requirements of, 216
Fills, 519
Filter, 176, 185
Fines, 19
FINN, W. D. L., 324
Flakey grain, 23
FLETCHER, G., 311
Floc, 32
Flocculent structure, 33
FLODIN, N., 535
Flood plain deposits, 47
Flow channel, 165
Flow line, 162
Flow net, 163
Flow slide, 529
Footings, 424
Forchimer's method, 165
Foundations, allowable pressure
      beneath, 415
   bearing capacity of, 391
   depth of, 389
   design charts for, 420
   essential requirements of, 389
   failure of, 391
   of embankments, 521
   of retaining walls, 357
   on rock, 430

   pier, 489
   pile, 446
   settlement of, 406, 411
   stability of, 391
   stresses due to, 399
   types, 424–45
   vibrations of, 434
French drain, 185
Friction, internal, 133
   of pile and soil, 463
   of wall on soil, 349
Friction pile, 457
Frost action, 191
   in retaining wall backfill, 354
   prevention of, 194
   soils susceptible to, 194
Frost heave, 191
Frost line, 192
Frost penetration, 195
Fuller's earth, 80

Geologic age, 542
Geologic information sources, 260
Geologic investigation, 259
Geophysical exploration, 283
GIBBS, H. J., 155, 311
GIRIJAVALLABHAN, C. V., 334
Glacial deposits, 51
GLOSSOP, R., 82
GOLUBKOU, V. N., 499
GOODMAN, L. S., 438
GOODMAN, R. E., 324
Gow method, pier, 491
Grain shape, 21
Grain size, estimating, 76
Grain size chart, 20
Grain size scales, 20
Grain size tests, 19
Gravel, 80
   sizes, 20, 76
Great soil group, 59
GREGG, L. E., 83
GRILLO, O., 494
GRIM, R. E., 36
Ground-water table, 62, 281
Grouting, 243
Gumbo, 80

Halloysite, 15

Index

HANSEN, J. B., 383
Hardpan, 80
HARR, M. E., 198, 388
HAZEN, A., 154
Head, 162
Head loss, 91, 162
Heave, 85, 96, 172
HENDRON, A. J., 537
HENKEL, D. J., 155
Hiley formula, 468
HOLTZ, W. G., 154, 250, 311
Horizontal loads, on piles, 473
H-piles, 493
HOUSEL, W. S., 448
HVORSLEV, M. J., 311
Hydraulic fills, 231
Hydraulic gradient, 92
Hysteresis, 131

Ice layers, 193
IDRISS, I. M., 324
Igneous rock, 42
Illite, 16
Injection stabilization, 244
Intercepting drains, 184

JENKINS, D. S., 83
Jetting piles, 453
JOHN, K. W., 83
JOHNSON, A. W., 197
Joints, 65
JUDD, W. R., 82, 537
JURGENSON, L., 438

Kaolin, 80
Kaolinite, 15
KAROL, R. H., 251, 438
KENNEDY, T. B., 251
KERISEL, J., 383, 449
KISHIDA, H., 499
KO, H. K., 155
KRISTOFOROV., V. S., 399
KRUMBEIN, W. C., 36
KRYNINE, D. P., 537

Laboratory test programs, 294, 308
Lacustrine deposits, 48
Lagging, 359
Lake deposits, 48
LAMBE, T. W., 36, 154, 155, 157, 197, 250
Landform, 265
Landslides, 529
  correction of, 533
LANE, E. W., 437
Laterite, 61
LAURSEN, E. M., 437
Leads, 448
LEGGETT, R. F., 83, 310
LEONARDS, G. A., 250, 439, 500
Levees, 520
Limestone terrain, 54
Line of seepage, 167
Liquefaction, 139
Liquid limit, 27
Liquidity index, 28
LITVINOV, I. M., 251
Load-settlement curve, 391, 415
Load test, 415
  interpretation, 417
  pile, 464
Loam, 80
Loess, 50
  cuts, 518
Loess loam, 51
Lost ground, 368
LOWE, J. M., 536
LUEDER, D. R., 311

Macrostructure, 35
MANSUR, C. I., 197
Marine borers, 478
Marine deposits, 49
Marl, 81
MARTIN, C. B., 499
Mat foundations, 425
MATLOCK, H., 500
Maximum density, 207
Mechanical stabilization, 239
Meniscus radius, 88
Metamorphic rock, 57
MEYERHOF, G. G., 311, 438, 498, 499
Microstructure, 28
MIDDLEBROOKS, T. A., 197
MILLER, E. A., 251

MINDLIN, R. D., 458
Models, 331
Modulus of elasticity, 122
　of cohesionless soils, 132
Mohr's circle, 119
Mohr's rupture theory, 124
Moisture content, optimum, 207
Moisture control of fills, 221, 229
Moisture-density curve, 208
Monotube piles, 480
Montmorillonite, 12
MOORE, P. J., 535
Moraines, 51
Muck, 81
Mud, 81
Muskeg, 59, 81

NADI, A., 324
Needle beam, 498
Neutral stress, 95, 107, 171
　effect on cohesionless soils, 135
　in embankment foundations, 523
　ratio, 96
Newmark chart, 404, 405
NEWMARK, N. M., 430, 536
NICHOLS, H. L., 250

OBERT, L., 83, 383
Observations on settlement, 410
Octahedral stress, 126
Open caisson, 493
Optimum moisture, 208
Organic soils, 18, 48
OSTERBERG, J. O., 197
Over-driving piles, 478
Oxbow lakes, 48

PALAKOWSKI, N. H., 334
PARSONS, J. D., 498
Partly saturated clay, earth
　　pressure, 366
　shear strength, 149
　slope stability, 515
Passive earth pressure, 342
PATTON, F. D., 537
Pavements, 247

Peat, 81
Peat deposits, 81
PECK, R. B., 154, 311, 383
Pedestal piles, 481
Pedogenesis, 58
Pedology, 59
Penetration resistance, 274, 279
Penetration test, 272, 276
Perched water table, 63
Permafrost, 196
Permeability, 91, 93
　tests, 93
PERROTT, W. E., 251
PETTERSON, K. E., 508
Pier foundaiton, 446
　bearing capacity of, 489
　settlement of, 490
Piezometric level, 162, 169, 282
Pile bounce, 452
Pile driver, 448
Pile driving, 447
　records, 488
Pile foundations, cap, 488
　design, 486
　safety factor, 487
Pile groups, bearing capacity of,
　　470
　settlement of, 472
Pile hammer, 448, 451
Piles, definition, 445
　batter, 475
　bearing capacity of, 451
　cast-in-place concrete, 479
　composite, 485
　displacement of, 435
　formulas, 467
　lateral loads, 473
　load tests on, 464
　precast concrete, 479
　shapes of, 476
　spacing of, 472, 487
　steel, 483
　tapered, 476
　tension, 466
　uses of, 446
　wave analysis, 467
　wood, 477
Pipe piles, 484
Piping, 174
Piston sampler, 300
Pit underpinning, 497
Plastic equilibrium, 325
Plastic limit, definition, 29

Plasticity, describing and estimating, 77
Plasticity chart, 72
Plasticity index, 29
Plate load test, 415
Pneumatic caisson, 493
Pneumatic-tired roller, 225
Poisson's ratio, 123
POLSHIN, D. E., 438
Pore pressure coefficient, 145
Porosity, 7
Potential energy, 162
Precast concrete piles, 479
Preconsolidation load, 100, 103, 423
PRENTIS, E. A., 500
Pressure, earth (*see* Earth pressure)
Prestress (*see also* Pretest), 367
Presumptive bearing pressure, 415
Pretest underpinning, 498
Principal stress, 117
PROCTOR, R. R., 206
Proctor test, 210
Projectile pile, 485
PRUGH, B., 250
Public Roads classification, 69
Pumping, frost, 192

Quality control, 230
Quicksand, 81, 173

RAMOT, T., 399
Rankine formulas, 343
Raymond Standard pile, 479
  Step-Taper pile, 480
Recharge, 189
Reconnaissance, 259
REESE, L. C., 324, 499, 500
Refusal, 274
Relative density, 31, 75
Relief hole, 176
Relieving platform, 358, 523
Remolding, 35
  due to pile driving, 455
Repose angle, 514
Residual soil, 43, 53, 56, 57
Residual strength, 135
Resistivity exploration, 286
Retaining walls, 351
  backfill for, 355
  foundations, 357
  tilt, 353
  types, 356
Rheology, 327
RICHART, F. E., 197, 383, 439
Ring shear, 128
ROBESON, F. A., 250
Rock, definition of, 3
  classification, 64
  foundations on, 430
  igneous, 42
  metamorphic, 57
  minerals, 11
  sedimentary, 53
  slides, 86
  strength, 64, 151
Rockfalls, 535
Rock bolt, 381
Rock fill, 217, 228
Rock flour, 81
Rock mechanics, 5
Rollers, rubber-tired, 225
  sheepsfoot, 223
Rotary samplers, 300
Rupture, 124

Safety factor, foundations, 419
  pile foundations, 470, 487
  stability of earth masses, 507, 517
ST. JOHN, B., 537
Sand, 19, 81
  density, 31
  shear strength, 130
Sand piles for compaction, 226
  for drainage, 191
Sandstone, 53
SANGLERAT, G., 311
Saprolite, 44, 58
Saturated clays, shear strength, 141
Saturation, degree of, 6
SCHMUCK, H. K., 383
SCHULTZE, E., 311
SCOTT, R. F., 155
Scour, 309
Secondary compression, 111
  Coefficient of, 112
SEED, H. B., 36, 155, 250, 324, 494, 536
Seepage erosion, 174

Seepage, forces, 171
  control of, 175
  line of, 167
  rate, 165
Seismic exploration, 283
Seismic potential, 260
Sensitivity, 33, 104, 148
Settlement, allowable, 419
  causes, 414
  distortion, 408
  methods for reducing, 420
  of embankments, 523
  of piers, 489
  of pile groups, 472
  of structures, 411
  rate of, 11, 407
Settlement computations, 407
Settlement cracks, 411
Settlement measurements, 411
SEWELL, E. C., 438
Shaking test, 76
Shale, 53
Shear strength, 124
  describing and estimating, 74
  for slope analysis, 516
  of cemented soils, 150
  of compacted clays, 212
  of dry cohesionless soils, 130
  of partly saturated clays, 149
  of saturated cohesive soils, 141
  of wet cohesionless soils, 135
  relation to penetration resistance, 280
Shear tests, 127
  vane, 302
Sheepsfoot roller, 223
Shelby tube samplers, 299
SHERRARD, J. L., 536
Shrinkage, 113
Shrinkage cracks, 114
Shrinkage limit, 28, 113
Sieves, 20
Silt, 81
  sizes, 19
Single-acting steam hammer, 449
Sinkhole, 56
Site inspection, 262
SKEMPTON, A. W., 82, 154, 438
Slaking, 115
Slices, method of, 510
Slickensides, 68
Slope stability, 506
  of cohesionless soils, 514

  of partly saturated clays, 515
  of soft clays, 513
Slurry trench, 361
SMITH, T. W., 197
Soil-cement stabilization, 235
Soil classification, 68
  for drainage, 184
  for fills, 214, 216
  for subgrades, 214
Soil composition, mineral, 11
  phases, 6
Soil definition, 2
Soil deposits, 45
Soil horizon, 60
Soil identification, 72
Soil investigation, costs, 259
  planning, 258
Soil mechanics, 4
Soil profile, 289
Soil sampler, field density, 229
  split barrel, 272
  undisturbed, 297
Soil samples, sizes for testing, 294
Soil stabilization, 233
Soils engineering, 4
Soldier beam, 359
SOWERS, G. F., 154, 155, 251,
  310, 430, 439, 499, 536, 537
Spacing of boring, 267
Spacing of piles, 471
SPANGLER, M. G., 383
Specific gravity, 7
  of solids, 11
Split barrel sampler, 272
Spring, 64
Stability, of blocks, 515
  of bottom of excavation, 367
  of earth masses, 505
  of foundations, 391
  of slopes, 506
Stabilization of soils, 233
STAFFORD, G. V., 197
STOKES, W. L., 82
Stone, 81
Stress, beneath foundations, 400, 457
Stresses, combined, 116
  earthquake, 317
  effective, 94, 317
  neutral, 95
  normal, 116
  octahedral, 126
  principal, 117
  shear, 116

INDEX

Structure, composite, 33
　cohesionless, 30
　cohesive, 33
　crystalline, 35
　dense, 29
　describing, 78
　dispersed, 32
　flocculent, 33
　honeycombed, 31
　loose, 29
　matrix, 34
Subgrade, 203, 247
　construction of, 247
　evaluation, 248
　modulus, 248
Subsidence, 534
Subsidence due to excavation, 367
Sudden drawdown, 512, 527
Sump, 185
Surcharge, 393
　behind retaining walls, 347
Surface tension, 86
Swamp, 64
Swedish circle, 508
Swelling, 114
SWIGER, W. F., 251
SZECHY, K., 383

Talus, 45
Tamping, 223
TAN, T. K., 154
Tapered piles, 476
TAYLOR, D. W., 155, 197, 535
TAYLOR, R. L., 324
Temperature gradient, 192
TENG, W. C., 500
Tension piles, 446, 466
Terminal moraine, 51
Terraces, 46
TERZAGHI, K., 98, 154, 197, 311, 382, 437, 438, 536
Test boring, 269
Test pit, 276, 296
Textural soil classification, 69
Thawing, 192
Thermal osmosis, 181
Thin-walled samplers, 299
THORNLEY, J. H., 500
Till, 51, 81
Time factor, 109
TIMOSHENKO, S. P., 334
TOCH, A., 437

Toe failure, 507
TOKAR, R. A., 438
TORY, A. C., 536
Transported soil, 45
Trench, 372
Triaxial shear test, 128
TSCHEBOTARIOFF, G. P., 383, 439, 500
Tuff, 51, 81
Tundra, 81
Tunnels, 371, 378
TURNBULL, W. J., 250, 537

Uncased piles, 481
Unconfined compression test, 147
Underpinning methods, 496
Undisturbed samples, preparation, 308
　tests on, 294, 309
Undisturbed sampling, chunk, 296
　deep, 295, 297
　definition, 296
　in sand, 301
Undrained shear, 146
Unified Soil classification, 70, 214
Uniformity coefficient, 20
Unit weight, definition, 7
Uplift, 171, 427

Vacuum drainage, 183
Vacuum wellpoint, 188
Vane shear test, 302
Vapor movement, 179
VARNES, D. J., 82
Varved clay, 52, 81
Vertical sand drains, 190
VESIC, A. S., 438, 439, 498
Vibration compaction, 226
Vibration of foundations, 434
Vibroflotation, 226
Void ratio, 6
　of single-grained structures, 28
Voids, 5
Volcanic deposits, 18, 43
Volume change, 115
Volume changes in shear, 135, 138

Wale, 359
WALKER, F. C., 250
WARD, W., 438

WARKENTIN, B. P., 154
Water, adsorbed, 24
Water content, 7
Water pressure (*see* Neutral stress)
Weathering of rock, 10, 44, 96
WEAVER, C. E., 537
Wellpoint, 187
Wells, drawdown of, 169
WESTERGAARD, H. M., 438
Westergaard's formula, 401
Wet caisson, 491
WHITE, L., 498
WHITMAN, R. V., 36, 157, 439, 535, 536
Wick piles, 190
WILLIAMS, N. F., 197
WILSON, S. D., 498
WINTERKORN, H. F., 250
Wood piles, 477
WOODWARD, R. W., 324, 536, 537

Work sheet, 290
WU, T. H., 154

X-ray diffraction, 17

YODER, E. J., 250
YONG, R. Y., 38, 154

ZANGER, C. N., 198
ZEEVAERT, L., 499
Zero air voids, 207
ZIENKIEWICZ, O. C., 324
Zone of freezing, 193

REALIZING
THE OBJECT-ORIENTED
LIFECYCLE

PRENTICE HALL OBJECT-ORIENTED SERIES

C. BAUDOIN AND G. HOLLOWELL
Realizing the Object-Oriented Lifecycle

D. COLEMAN, P. ARNOLD, S. BODOFF,
C. DOLLIN, H. GILCHRIST, F. HAYES
AND P. JEREMAES
*Object-Oriented Development:
The Fusion Method*

S. COOK AND J. DANIELS
Designing Object Systems

B. HENDERSON-SELLERS
A Book of Object-Oriented Knowledge

B. HENDERSON-SELLERS
Book Two of Object-Oriented Knowledge

B. HENDERSON-SELLERS
Object-Oriented Metrics: Measures of Complexity

H. KILOV AND J. ROSS
Information Modelling

P. KRIEF
Prototyping with Objects

K. LANO AND H. HAUGHTON
Object-Oriented Specification Case Studies

J. LINDSKOV KNUDSEN, M. LÖFGREN,
O. LEHRMANN MADSEN AND B. MAGNUSSON
*Object-Oriented Environments:
The Mjølner Approach*

M. LORENZ
*Object-Oriented Software Development:
A Practical Guide*

M. LORENZ
Object-Oriented Software Metrics

D. MANDRIOLI AND B. MEYER (eds)
*Advances in Object-Oriented
Software Engineering*

B. MEYER
Eiffel: The Language

B. MEYER
Reusable Software

B. MEYER
An Object-Oriented Environment

B. MEYER AND J. M. NERSON
Object-Oriented Applications

POMBERGER AND BALSCHEK
*An Object-Oriented Approach in
Software Engineering*

P. J. ROBINSON
Hierarchical Object-Oriented Design

R. SWITZER
Eiffel: An Introduction

K. WALDEN AND J. M. NERSON
Seamless Object-Oriented Software Architecture

R. WIENER
Software Development Using Eiffel

REALIZING THE OBJECT-ORIENTED LIFECYCLE

Claude Baudoin and Glenn Hollowell

Prentice Hall PTR
Upper Saddle River, NJ 07458

Library of Congress Cataloging-in-Publication Data

Editorial/production supervision: Betty Letizia
Acquisitions editor: Paul W. Becker
Editorial assistant: Maureen Diana
Manufacturing Manager: Alexis R. Heydt
Cover design director: Jerry Votta
Cover design: Wanda Espana
Cover art: Eve Chenu

 © 1996 Prentice Hall PTR
Prentice-Hall, Inc.
A Simon & Schuster Company
Upper Saddle River, New Jersey 07458

The publisher offers discounts on this book when ordered in bulk quantities. For more information, contact: Corporate Sales Department, Prentice Hall PTR, One Lake Street, Upper Saddle River, NJ 07458.
Phone: 800-382-3419   Fax: 201-236-7141 E-mail (Internet): corpsales@prenhall.com

All rights reserved. No part of this book may be reproduced, in any form or by any means, without permission in writing from the publisher. All trademarks and registered trademarks are the property of their respective owners.

Printed in the United States of America

10 9 8 7 6 5 4 3 2 1

ISBN 0-13-124454-X

Prentice-Hall International (UK) Limited, *London*
Prentice-Hall of Australia Pty. Limited, *Sydney*
Prentice-Hall Canada Inc., *Toronto*
Prentice-Hall Hispanoamericana, S.A., *Mexico*
Prentice-Hall of India Private Limited, *New Delhi*
Prentice-Hall of Japan, Inc., *Tokyo*
Simon & Schuster Asia Pte. Ltd., *Singapore*
Editora Prentice-Hall do Brasil, Ltda., *Rio de Janeiro*

Trademark Information

The following is a list of known trademarks and registered trademarks used in this publication. Other products or services may be trademarks or registered trademarks of their respective owners. No investigation has been made of common-law trademark rights in any word, because such investigation is impracticable.

Cadre is a registered trademark of Cadre Technologies, Inc.

Cohesion and **VAX** are registered trademarks of Digital Equipment Corporation.

dBase is a registered trademark of Borland International, Inc.

Discover/Education is a trademark of International Business Machines Corporation.

ISIS is a registered trademark of McDonnell Douglas Corporation.

Lotus is a registered trademark of Lotus Development Corp.

MacIntosh is a registered trademark of Apple Computer, Inc.

NewWave is a registered trademark of Hewlett-Packard Company.

NExTStep is a registered trademark of Next Computer, Inc.

OPEN LOOK is a registered trademark of AT&T.

OS/2, OS/400, and **PS/2** are registered trademark of International Business Machines Corporation.

OSF/1 and **OSF/Motif** are registered trademarks of Open Software Foundation, Inc.

PostScript is a registered trademark of Adobe Systems, Inc.

PowerPoint is a registered trademark of Microsoft Corporation.

RS/1 is a registered trademark of Bolt Beranek and Newman, Inc.

SAS is a registered trademark if SAS Institute, Inc.

SEMATECH is a registered trademark of SEMATECH, Inc.

Softbench is a registered trademark of Internetix, Inc.

Software Through Pictures is a registered trademark of Interactive Development Environments, Inc.

UNIX is a registered trademark of Novell, Inc.

WorkStream is a registered trademark of Consilium, Inc.

Table of Contents

Acknowledgements xiii

Foreword xv

Introduction xix

Chapter 1—The Challenges of Prevalent Software Lifecycle Practices

1.1 Criticality of Software in Business 1
1.2 System Development Bottlenecks 4
1.3 Software Quality Issues 11

Chapter 2—Defining a Lifecycle Architecture

2.1 Rationale and Objectives 19
2.2 What is an Architecture? 21
2.3 Why a Lifecycle Architecture? 24
2.4 Principles of a Lifecycle Architecture 27
2.5 Benefits 39

Chapter 3—The Software Process

- **3.1** The Software Process: Introduction and Definitions 43
- **3.2** Software Lifecycle Models 49
- **3.3** The Meta-Lifecycle Model 54
- **3.4** Lifecycle and Objects 74
- **3.5** Defining a Specific Software Process 77
- **3.6** Software Process Enactment 88
- **3.7** Software Process Improvement 91
- **3.8** Conclusion 102

Chapter 4—Development Activities

- **4.1** Overview of the Development Activities 105
- **4.2** Object-Oriented Methods Overview 108
- **4.3** What is a Method? 117
- **4.4** Models Used in Analysis and Design 119
- **4.5** Object-Oriented Analysis 126
- **4.6** Object-Oriented Design 136
- **4.7** How to Select a Method 140
- **4.8** Object-Oriented Programming 146
- **4.9** Testing 150
- **4.10** Software Distribution and Support 154

Chapter 5—Object-Oriented CASE Tools

5.1	The Benefits of Computer-Aided Software Engineering	161
5.2	Examples of Object-Oriented CASE Tools	170
5.3	How to Select an Object-Oriented CASE Tool	180

Chapter 6—Software Development Environments

6.1	Integrating the Development Process	201
6.2	The Dimensions of Environment Integration	203
6.3	Integration Standards	209
6.4	Process Management Facilities	221
6.5	Selecting a Software Development Environment	224

Chapter 7—Project and Deliverables Management

7.1	The Planning of Object-Oriented Projects	235
7.2	Deliverables Management	266
7.3	Reusing Objects	269
7.4	Object-Oriented Metrics	276

Chapter 8—Future Directions for Software Engineering

8.1	Evolution of the Activities and Deliverables	281
8.2	Transformation Engineering	285
8.3	After OO: Are Agents the Next Paradigm Shift?	297
8.4	After OO: Are Formal Methods the Next White Knight?	300
8.5	Other Research Topics	305

Chapter 9—Shaping Object Technology: Standards and Consortia

9.1 The Long Road to Object Technology Acceptance 317
9.2 A Case for Standards (de *facto* or *de jure*) 321
9.3 The Key Players 328
9.4 Recommendations on Standards Participation and Conformance 338

Chapter 10—Object-Oriented Execution Environment Concepts

10.1 Requirements 341
10.2 Architectural Components 343
10.3 Information Bus Technology: Is it Object-Oriented? 354

Chapter 11—Objects in Industry: Case Studies in Computer-Integrated Manufacturing

11.1 What is Computer-Integrated Manufacturing (CIM)? 371
11.2 Microelectronics Manufacturing Science and Technology (MMST) Project 374
11.3 The SEMATECH CIM Application Framework 383
11.4 SEMATECH's SkillView 395
11.5 The Semiconductor Workbench for Integrated Modeling (SWIM) 399

Chapter 12—Transition to Objects

12.1 Risk Management 408
12.2 Models for Transition 418
12.3 Cultural Factors 422
12.4 Planning the Transition 427
12.5 Coexistence With Legacy Systems 431
12.6 Object-Oriented Enterprise Integration 432
12.7 Partnering: A Potential Low-Risk Way of Getting Started 448
12.8 Now It's Your Turn 450

Table of Acronyms 453

Glossary 459

Bibliography 489

Index 501

Acknowledgements

Any serious technical book is a significant project and never solely the work of its authors. This one was no exception. As the person that, in addition to being co-author, initiated and nurtured the book through the SEMATECH and Prentice Hall gestation, I want to acknowledge the other primary members of that team.

Carla Jobe was an outstanding contributor by performing as Project Manager. She was responsible for most of the project planning, design and layout of the book, and all of the execution of the finished "camera-ready" product. Additionally, she demonstrated her talents as a multi-faceted artist (she is also an accomplished Cellist) by producing the 60 graphics found in the book.

Kathleen Allen-Weber performed as independent "Editeur Extraordinaire" for the book. Her skills in French as well as English allowed her to harmonize the authors' styles as few others could have done, and her dedicated participation throughout the multiple editing passes provided significant continuous improvement. Her professional skills were well known in advance, but her attitude, people skills and sense of humor were welcome bonuses in the long and arduous process.

Claude Baudoin was, well, Claude Baudoin—a creative, knowledgeable, articulate, and demanding co-author.

The cover art is a 1989 painting entitled "Stasis & Kinesis" by French-born artist (MFA in Sculpture), Eve Chenu, of Richmond, Virginia. It was inspired by a passage addressing the idea of beauty from James Joyce's "Portrait of the Artist as a Young Man."

Last and never least, heart-felt appreciate goes to Prentice Hall Publisher, Paul Becker, and SEMATECH management, especially Alan Weber and Gary Gettel, for their support and patience.

Glenn Hollowell
Senior Computer Scientist
SEMATECH

Foreword

You may wonder why SEMATECH, a Semiconductor Manufacturing Research Consortium, would sponsor a book on object-oriented software engineering. It's a good question. From a selfless standpoint, the answer is simply that we believe the information presented here is too valuable **not** to share. However, our motivations don't stop there.

This book was a natural outgrowth of SEMATECH's strategic research agenda in manufacturing systems technologies. Given that part of our mission is to look five to ten years into the future for those innovations that can potentially have the most impact on manufacturing competitiveness, object-oriented software has long been an obvious candidate. Modern semiconductor factories depend heavily on their manufacturing information and control systems, so major improvements in capability, flexibility, and/or cost of these systems represent a significant opportunity. However, balancing this opportunity is a natural (and prudent) reluctance on the part of the manufacturing customer base to adopt any new technology without a thorough shakedown (ideally in someone else's factory). Factory managers care most about schedules, application functionality, system performance and reliability, and minimizing risk to their operations—not technology.

Object-oriented software **is** different, and herein lies both the promise and the challenge. To be fully exploited, one must think differently about the problem at hand, and the solutions that the technology enables. This kind of change takes time and resources, and therefore lots of patient management support. By contrast, the principles of good engineering practice are relatively constant, and have been refined over many generations of engineers. And while software engineering is a relatively new discipline, we nevertheless have more than three decades of experience upon which to draw. This book amplifies and projects this experience into an object-oriented technology context, providing a tailored and proven methodology for developing complex, reliable software systems.

The SEMATECH development and review processes that gathered and organized the background information for this book involved a wide variety of experts from SEMATECH member companies, software suppliers, universities, government agencies, consulting companies, and other research consortia; the fact that they are too numerous to name makes us no less grateful for their efforts. The idea to publish this in book form, however, belongs to Claude Baudoin and Glenn Hollowell, and with SEMATECH's support, they enthusiastically undertook the task. It turned out to be a much larger job than anyone anticipated, because much of the technology was (and still is) in a continuous state of flux, especially in the computing infrastructure and development tools areas. Recognizing that this is the case with any new technology, we felt it was better to get the book into the market rather than wait for further evolution of the technology.

SEMATECH believes this book meets a genuine need in the professional software development market, and that it can help to accelerate the adoption of object-oriented technology in mainstream industrial applications. The semiconductor industry has cleared a number of hurdles in the process of evaluating and

Foreword

embracing this technology over the past few years, and we sincerely hope our experience helps you overcome the challenges you will undoubtedly face along the way.

Alan Weber
Manager, Manufacturing Execution Systems Program
SEMATECH Factory Integration

Introduction

Most of us have so thoroughly memorized the traditional business and technology rules of our systems development specialty that we can recite them by rote. These rules represent a conventional wisdom that is both familiar and comfortable, but many are becoming outdated and irrelevant. In the business climate of the late twentieth century, such inertia can be lethal. Companies once so strong in their field that it was impossible to imagine that they could be dethroned are scrambling merely to survive. Alliances unthinkable only a few years ago are commonplace today, all in the name of self-preservation. Companies that were insignificant competitors in the recent past are now dominant. Most companies with the corporate will to excel are deliberately destabilizing their normal business activities just to remain competitive.

The business landscape is rapidly changing

While enhanced customer satisfaction is the primary goal of these changes, information technology is necessarily a primary ingredient of any company seeking to revitalize itself. Like most modern activities, these changes have become enshrined in euphemisms such as "reengineering," "downsizing" and "outsourcing." Regardless of the labels or the underlying principles of these changes, they all have two common goals: to increase systems development productivity and to improve

Individual companies must keep pace

quality. From these emanate many other worthwhile, but ancillary, objectives such as lower cost.

Information technology is a requirement for survival...

What does all this have to do with a book that concentrates on a particular systems technology? Everything. Most experts agree that we are well into the Information Age and will remain there for many years to come. The world is just as predatory in the Information Age as it was in the Stone Age, the Industrial Age, or any other age. The difference now is that the victors and the vanquished will not be determined by either brute strength or pure intellect. Rather, survival will depend on how rapidly and effectively new data can be gathered and abstracted into meaningful information, and how effectively viable concepts can be realized in implementations.

Why an Object-Oriented Lifecycle?

...and object-oriented lifecycle is becoming a key ingredient

Since the dawn of living organisms, survival of the fittest has been a constant rule, and with the emergence of Homo sapiens, it now applies to concepts and institutions as well. We believe that object-oriented concepts have evolved into a "species" of Information Technology (IT) that will soon take its place at the top of the pecking order. Likewise, we believe the application of object-oriented concepts to the entire software lifecycle is paramount to its complete success. Our distant human ancestors found that those who captured, domesticated, and adapted a source of sustenance to fit their needs fared far better in the scheme of things than did the nomadic hunters. So it is with systems developers that adopt up-to-date lifecycle technology over ad hoc or outdated techniques.

This book assumes a basic understanding of object technology and focuses on demonstrating the lifecycle's relevancy

The material in this book does not aim to convince you of the theoretical goodness of object technology. (We leave this up to marketers who have been systematically replacing the word "good" with the phrase "object-oriented" in their brochures.) Its focus is to convince you of the robustness of object technology, persuade you of its coming prevalence, illustrate its

Introduction

practicality, and demonstrate some of the ways it can be implemented to accomplish real-world productivity and quality improvements. If we succeed in these objectives, you will be equipped with the practical knowledge to take advantage of object-oriented technology, and its impending dominance will no longer be perceived as threatening. In pursuit of these objectives, we assume that the reader has a moderate level of knowledge about object-oriented concepts and the general application of analysis and design methods, plus an appreciation of the need for the structure brought about by following a formal lifecycle plan. We also reference recommended source material for those who feel they lack some of these basic concepts.

Already, we see evidence that the object-oriented lifecycle approach in this book can benefit software organizations in different application domains. The immediate gains from implementing object-oriented technology, methods, and tools are:

The benefits of an object-oriented lifecycle

- A better understanding and more accurate specification of requirements;
- Traceability of requirements through analysis and design phases;
- Reduced cost of developing and communicating analysis and design information;
- Ability to "buy versus make" software components from an emerging market in objects;
- Greater consistency in development, as a result of the discipline of the methods;
- Increased productivity through reusable designs, software components, and systems;
- Reduced cost and time to maintain and enhance software yet to be developed;

- Improved motivation and retention of highly skilled computer professionals.

Acknowledgment to SEMATECH

Much of the material in this book has its roots in a body of work produced in SEMATECH's Software Methods and Tools program. These original concepts have since been influenced by other SEMATECH programs and the authors' observations and experience working in the object-oriented software lifecycle.

SEMATECH was formed to help save a declining U.S. manufacturing industry

SEMATECH (for SEmiconductor MAnufacturing TECHnology) is a consortium of ten U.S. semiconductor manufacturing companies and the U.S. government. SEMATECH's purpose is to develop and share generic manufacturing technology among its member companies to increase the competitiveness of the U.S. semiconductor industry. The commercial participants in the consortium include companies whose primary business is the manufacture and sale of semiconductors, those who manufacture semiconductors strictly for use in their own finished products, and those who do both. SEMATECH was formed in 1987 to help stem a disastrous slide of the U.S. semiconductor industry's market share that threatened the total demise of this industry in the U.S. Since then, the U.S. semiconductor industry has reversed this trend and much of the credit is attributed to SEMATECH.

SEMATECH determined CIM is a tool for strategic competitive advantage

SEMATECH's member companies determined early that in order to improve the competitiveness of the U.S. semiconductor manufacturing industry, and to revitalize the domestic equipment industry that supports it, a robust Computer-Integrated Manufacturing (CIM) program was a necessary component of SEMATECH's development strategy. Furthermore, they recognized 1) that the use of computer systems in every facet of semiconductor manufacturing would be essential, and 2) that standalone computerization, islands of automation, or even interfaced systems would not suffice. Therefore, to establish and maintain a competitive edge, semiconductor manufacturers and equipment suppliers would need to develop systems that could be highly integrated within the factory and throughout the enterprise. That level of integration is almost impossible with legacy computer software technology and current development/support methods.

Because of its focus on advancements in CIM, SEMATECH has led the semiconductor industry by example in its endorsement of object technology and in the development of a software lifecycle process. It has accomplished this leadership by also being a good follower. This has been achieved through a strategy of finding the "pockets of excellence" within its member companies and from leading authorities. SEMATECH benefits from having access to most of its member companies' systems organizations and from their people's participation in SEMATECH-led projects. Continuous interaction with domain experts from the various member companies helps assimilate knowledge and build consensus. Finally, concepts are prototyped and implemented in joint projects between SEMATECH and its member companies or their U.S. suppliers, and results are furnished as lessons learned to all other appropriate parties.

SEMATECH and its members enjoy cooperative innovation

SEMATECH undertook the specification of an object technology lifecycle because none existed that covered the required scope. The material in this book specifically describes the SEMATECH approach to that software lifecycle. Although we believe object technology is the best alternative to the current software technology (which is inadequate), many of the fundamental concepts found in this book are not restricted to usage with object technology.

SEMATECH developed a new lifecycle out of necessity

This lifecycle is not specific to the CIM domain

Because SEMATECH is specifically chartered to improve semiconductor manufacturing, the lifecycle architecture is promoted by SEMATECH primarily in the CIM domain, but we have no doubt that this lifecycle architecture is applicable to all software domains. CIM in its fullest meaning is a strategy for integrating information systems technology across a manufacturing enterprise. SEMATECH's philosophy toward the software lifecycle supports the belief that there is an integrated "software lifecycle architecture" that encompasses everything from the requirements gathering phase to the deactivation of obsolete systems.

Scope of the Book

Every organization has a lifecycle — even if it's chaotic

Every software-centric organization (and most organizations are nowadays, whether or not they realize it) has a software process. That process may be so chaotic and unmanaged that it is unrecognizable as a process, but it is a process nevertheless. Creating and maintaining systems in an atmosphere where the organization *controls* the software process rather than *reacting* to the demands of near-random circumstances requires not only good management, but also a well-planned lifecycle architecture. An acceptable software lifecycle involves many different aspects. It requires:

- A set of principles to which the lifecycle architecture adheres;
- A specific lifecycle model;
- A well-defined process;
- The necessary support services;
- Metrics that define success, failure, or something in between;
- A continuous improvement plan.

Introduction

The twelve chapters of this book explore many of the process and technology elements required to achieve this level of discipline.

Chapter 1—Overview of the problems posed by the prevalent software development practices.

Chapter 2—Introduction and justification of a lifecycle architecture and its nine principles.

Part I -
The Challenges of
Systems Development

Chapter 3—Discussion of the history and requirements of a lifecycle model. Introduction of a radically more comprehensive model called the "Meta-Lifecycle Model."

Part II -
The Development
Process

Chapter 4—Overview of object-oriented method concepts and a survey of the unique aspects of each concept. Discussion of the strengths of different object-oriented methods and programming languages for various typical applications.

Chapter 5—Discussion of object-oriented Computer-Aided Software Engineering (CASE) tools and general examples of product use. Presentation of a comprehensive list of evaluation criteria for object-oriented CASE tools.

Chapter 6—Exploration of the contrast between an integrated "software development environment" and a set of individual CASE tools. Presentation of a comprehensive list of evaluation criteria for software development environments.

Chapter 7—Survey of the critical requirements of software development project management, with emphasis on unique aspects of object-oriented projects.

Chapter 8—Examination of the technology factors that will likely influence the evolution of software engineering at the turn of the millennium.

Chapter 9—Discussion of emerging object-oriented technology standards (*de facto* and *de jure*) and directions these are likely to take.

Part III -
Run-time technology

	Chapter 10—Overview of the object-oriented execution environment and the contrast with traditional execution environments.
Part IV - Adoption of Object Technology	**Chapter 11**—Presentation of a series of mini-case studies of object-oriented projects and applications by SEMATECH, SEMATECH member companies, and SEMATECH suppliers.
	Chapter 12—Discussion of challenges, requirements, risks in migrating to an object technology lifecycle environment, enterprise integration, and partnering to lower the startup risk.
Intended Audience	The content of this book will be useful to readers with at least one of the following motivations for reading it:

- Learning the elements of a lifecycle;
- Determining a starting point for organizing the implementation of a software lifecycle architecture suited to object technology;
- Improving the quality and maturity of a software development organization;
- Obtaining a recipe and a checklist for software project management;
- Getting started in selecting and implementing software tools that support the lifecycle of object-oriented software systems.

Target organizations   Appropriate organizational audiences for this book include:

- Medium-sized and large end-user organizations providing computer systems and support. Organizations with a very small number of applications development/support personnel (e.g., under ten) should find the underlying concepts applicable, but probably will have to adapt the specific techniques recommended.

- Suppliers of systems or applications for commercial sales. This includes an increasing number of suppliers of "intelligent" equipment with embedded software applications.
- Consultants/contractors to the above organizations.
- Academic institutions teaching lifecycle concepts.

This book assumes a familiarity with the basic concepts and terminology of Information Technology. However, a glossary of terms and acronyms, and a bibliography are included at the end of the book.

An understanding of some basic concepts helps

The software industry is very dynamic, frequently offering new technologies and new ways to perform old tasks. This constant state of flux means that people must often change their work processes and procedures while remaining committed to the accomplishment of the work at hand. This book is dedicated to making the vision it describes a reality. We recommend that software systems suppliers and support groups begin now applying the software lifecycle process within their organizations and to their projects. These may be projects expressly undertaken to gain experience with various parts of the software lifecycle architecture, such as the introduction of object-oriented methods, or integrated frameworks and tools. Material in this book provides information especially useful for getting started with the practices and tools discussed. However, since systems technology evolves rapidly, the specific references made to methods, tools, and products in this book should not be interpreted as specific endorsements or immutable recommendations.

Our goal is to convince you to improve your lifecycle

Our Systems Developer's Toast to our readers:

> May all your analyses be logical,
>
> May all your designs be defensible,
>
> May all your programs execute predictably,
>
> May all your objects be reusable,
>
> And may all your maintenance be scheduled.

Glenn Hollowell and Claude Baudoin
Austin, Texas and Palo Alto, California
August 1995

Chapter 1

The Challenges of Prevalent Software Lifecycle Practices

Today's business environment is rapidly changing. An out-of-date systems lifecycle technology prevents an organization's software from keeping up with the rate of change and further exacerbates the challenges of business evolution. Industry experts are beginning to realize that meeting the software lifecycle challenges will require unconventional actions to support the current business and technology environment.

Rapid business change requires a complementary software strategy

1.1 The Criticality of Software in Business

In the past, most corporate managers viewed computers and systems as "back office stuff" that had little to do with the real mainstream efforts of turning a profit (other than the necessity of printing the invoices and doing the accounting). "Real managers"—those who were the decision makers—did not themselves use computer terminals, reserving those workstations for clerical people. The idea that the manufacturing, marketing, and management people needed computers to achieve genuine productivity and value-added work seemed merely a necessary indulgence for the technologists. All the same, the upper echelons of management did not want anyone to criticize their company for lagging behind the current technology trends.

Software was not originally considered essential technology

Therefore, they somewhat reluctantly went along with the procurement of additional computer hardware and the development of more software.

High levels of dependence on software

How times have changed! Gone are the days when there was an army of secretaries and administrative support personnel in every office. The wave of computers that swept though corporate offices washed away tremendous numbers of clerical support staff. Modern companies would be extremely noncompetitive if they maintained the level of administrative overhead that was necessary in those noncomputerized, bygone days.

Most companies adapt

Fortunately, most companies successfully made the transition. Computer integration in the business structure of most companies is wide-ranging. Electronic mail, or e-mail, to cite an everyday example, replaces much of the former sea of paper memos. Or consider that ordinary people can now produce in *two hours* glitzy presentations that in the precomputer era took a *person-week* of clerical time. Computer systems track and store more detailed business transactions than would have been possible with hordes of clerical workers.

In today's technical world, the design of mechanical and electronic parts would be unthinkable without computer aids and simulation programs. Complex manufacturing processes are controlled by computers in "lights-out" factories.

Competition is the driving force

How did this transformation take place in so short a period of time? Global competition, the quality movement, corporate downsizing, and numerous other factors played a role. In most companies, acceptance of the proliferation of computer technology was not a matter of choice. They adopted it and used it to improve their effectiveness in an effort to avoid being gobbled up by their competitors.

Today, large companies prosper or suffer based on the availability and quality of their software systems. Regardless of a company's size and industry type, most businesses recognize that computer systems are so integral to daily operations that without them, they could not operate with any semblance of normalcy. For that reason, most medium-to-large companies even have elaborate disaster contingency and recovery plans.

The quality and availability of software systems are a matter of survival

As significant as these past changes may be, there is currently an explosion of business change occurring [Peters 91] that is driving an acceleration of new requirements for software. Functionality requirements are always dissimilar enough that existing, static systems will never suffice. Users' requirements are always strong enough that the system must be more reliable and easier to use. Cost constraint requirements are always severe enough that gains in development productivity must be significant. And lastly, management and competitive requirements are usually demanding enough that there is always pressure to make unrealistic schedule commitments.

Systems must react to continuous business changes

Perhaps only now do we realize the extent of the computer revolution. For the majority of companies, software applications are an essential part of business life, and their timeliness, quality, and capabilities are becoming more important every day. In many cases, they are as central to an organization's strategy and goals as its product line or the quality of its service.

Systems have become essential

The Needs of the Extended Enterprise

In another, related development, most businesses are creating relationships with other organizations that require direct computer-to-computer exchange of data. Electronic Data Interchange (EDI) has become the prevalent data exchange technology, and many businesses are using it to streamline their old, paper-based processes. EDI is one aspect of cooperative systems that extend beyond the walls of a single organization and

Inter-business data exchange is a growing trend

constitute so-called *extended enterprise systems*. The requirements for direct computer-to-computer interaction between businesses are expanding:

- Paperless transactions (purchase orders, invoices, and electronic funds transfer) will become commonplace.

- Suppliers will provide mechanisms for open access by their customers' computers, allowing the retrieval of information about an order's status or a planned defect correction.

- Meanwhile, the Internet and the worldwide web (WWW) services running on it are revolutionizing the concepts of on-line information retrieval.

1.2 System Development Bottlenecks

Two bottlenecks: bad lifecycle models and legacy systems

Computer systems ranging from corporate Information Systems (IS) to plant floor process control computers play a role in improving business productivity. Yet the productivity achieved through computerization is nowhere near its potential for two fundamental reasons: an inadequate software lifecycle model, and the "boat anchor" effect of legacy systems.

Management attitudes have changed about Information Technology

Recognition of the need for increased software productivity and visibility of the magnitude of this problem have finally reached the highest levels of corporate management. Why? The situation is adversely affecting the capability of corporations to routinely conduct their business.

The larger the company, the more complex the systems issues

For many small businesses, the required computer applications may be fairly simple to obtain and maintain. They may be able to run their business with off-the-shelf commercial software or by purchasing turnkey systems from industry-specific Value-Added Resellers (VARs). On the other hand, for the typical medium-to-large organizations, life with fast-moving computer technology is not nearly as straightforward. Most of these have come out of the explosive growth period of computer systems with a hodgepodge of hardware and software systems that range from the old (five to fifteen years) to those just released.

There is a very old saying in the computer software business: "you can make a computer do anything you want, it is SMOP (Simply a Matter of Programming)." This is usually a systems person's sarcastic response to a manager or a user upon being asked to perform a miracle, such as creating overnight a piece of software that no one had thought of requesting ahead of time. What this bitter remark really means is: "You wouldn't ask for a new piece of mechanical engineering or a new circuit board to be designed overnight. Well, guess what? The fact that software is 'soft' does not mean that I can create it by just snapping my fingers!"

The "soft" in software does not mean "easy"

However, because software is directly linked to the competitiveness of a business, yesterday's outlandish request may have become today's vital necessity. Yet even if computer technology is capable of doing several orders of magnitude more today than ten or even five years ago, SMOP is no more realistic today than it was then. There are always cost and resource constraints to consider, as well as some incompressible series of steps required to engineer a product. Lofty development goals must be balanced with the realities of living within resource constraints, all the while performing enhancements and maintaining many old systems.

Inadequate Lifecycle Models

A software lifecycle is a set of procedures for:

Lifecycle definition

- Analyzing a problem;
- Designing a solution;
- Turning the design into an executable software system;

- Maintaining and evolving the system;
- Deactivating the system upon replacement, or when it is no longer useful.

There is a growing demand for systems, not met by current development capabilities

The expansion of computer systems technology demands an ever-expanding capability to develop and support these systems, but without a commensurate increase in cost. The difficulty in fulfilling these conflicting requirements has resulted in a rapidly widening gap between software development and support needs, and available resources. The root cause is the inability of dated lifecycle models to deliver software (Figure 1.1). *Demand* has exceeded the *ability to deploy* systems in most cases, although the point in time that this crossover occurred varies by industry and individual company.

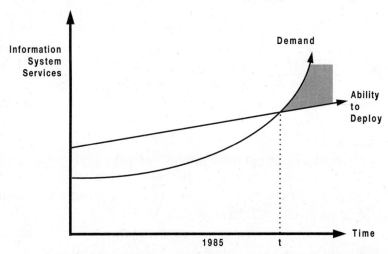

Figure 1.1 – System Demand Has Exceeded Development Capacity

The Challenges of Prevalent Software Lifecycle Practices

A ground swell of emphasis on software engineering has been building over the last few years. This is due to higher quality and reliability expectations from users of software applications, as well as a demand for additional features, functionality, and ease of use. Many of the software engineering approaches that attempt to resolve the myriad of complex lifecycle issues revolve around traditional techniques, specific tools, or a less-than-adequate software lifecycle process modification.

Current software engineering approaches have not bridged this gap

At best, the average organization's software lifecycle model provides the ability to increase software productivity only linearly. A bad lifecycle model invariably keeps an organization from even taking advantage of incremental productivity gains offered by improved development tools. A very bad (ad hoc) lifecycle can result in periods of negative productivity which could include development project failures and extended system outages caused by only minor enhancements or bug fixes.

More specifically, the results of most current lifecycle methods are:

- Rising support costs;
- Lack of interoperability and distributed integration;
- Scarcity of global models for software engineering, maintenance, support, and operations;
- Limited extensibility;
- Data inconsistency and redundancy;
- Difficulty of changing and maintaining complex, monolithic systems.

To respond effectively to the new challenges of global competition, businesses require not only new operating practices, but also new systems development and maintenance lifecycles, as well as the techniques and tools required to accomplish this

The development paradigm must change

shift. In the late 1990s and early twenty-first century, a significant productivity improvement in the software lifecycle of most organizations will be mandatory to remain competitive. Two approaches must be combined to obtain such a positive change: better adherence to the discipline of software development, and a departure from "software handcrafting" methods.

Expand training in lifecycle methods

Typically, even though there is an adequate experience base in a software development organization, the majority of personnel creating or maintaining software does not have specific training in software lifecycle methods and techniques. Consequently, there is not widespread usage of these lifecycle methods, which in turn leads to poor planning and performance of those activities necessary to accomplish a high level of software quality. Furthermore, while most software development organizations systematically perform a requirements analysis to develop software specifications, few hold regular design reviews or software code inspections, or formally review test results.

Deferring the crisis will not help

Even if existing knowledge about software engineering concepts and methods is better disseminated than in the past, this will only result in *raising the slope* of the straight line in Figure 1.1. This will move the intersection point with the demand curve farther to the right, but it will only defer the inevitable crisis. In other words, simple extensions to current practices, methods, and tools are inadequate.

Software engineering is not yet a component-based industrial process

Fundamentally, the issue that remains at the core of the lifecycle inadequacy is that traditional software engineering speeds up the creation of quality handcrafted software—systems that are painstakingly written one line at a time. Only modest improvements will be achieved as long as the activities within the lifecycle are based on the creation of unique "parts" from individual instructions written in a procedural programming language.

Legacy Systems

A progressive organization's strategy often runs into a wall of constraint called "legacy systems." A legacy system is a currently installed system performing a needed function at a tolerable level. A legacy system does not usually follow the current software architectural concepts of the using organization, but for nontechnical business reasons (typically high cost, lack of human resources, or both), it cannot be replaced immediately.

By definition, a legacy system cannot be replaced immediately

Older software technology normally requires more rigid, time-consuming modification procedures. It is a vicious circle that older software systems use technology requiring frequent modification to serve current needs, but as the number of modifications compound, they demand increasing levels of support.

Legacy systems are typically intolerant of change

The resource drain and slow response to needed change often interfere with the desired business strategy and they invariably limit the ability to develop new and more effective systems in a timely fashion. As a result, it is not at all unusual for companies, in order to accomplish the needed effect (Figure 1.2), to find that they must adopt convoluted business procedures to accommodate a system's idiosyncrasies, rather than the other way around.

Legacy systems dictate convoluted business processes

Legacy systems bring another curse. Most current systems in the enterprise have evolved from a collection of independent, monolithic systems composed of islands of capability partially interfaced over the past decades. Even systems of more recent vintage are typically little better because most developers tenaciously cling to the comfortable software lifecycle technology of the past twenty years. Consequently, these island systems are at best interconnected—not integrated—and the resulting environment has little or no modularity. This means that change is expensive in time, cost, labor, and risk.

They bring decades of patches and old technology

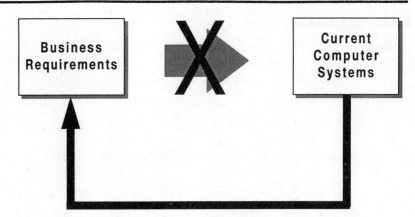

Figure 1.2 – Inflexible Software Often Constrains Business Practices

New lifecycle models must address legacy as well as new systems

On the other hand, it is a business reality that in order to achieve practical acceptance, any new solutions created by new methods and tools must interoperate with old solutions for a long transition period. Therefore, new methods and tools specifically addressing legacy systems integration will achieve quicker and wider acceptance.

Consequences of Obsolete Lifecycle Models and Legacy Systems

Bad lifecycle and inflexible systems add enormous costs

As a result of these two problems, the software community is plagued with an inability to develop correct, reliable software that both meets the consumers' needs, and delivers within predictable dollar, people, and time constraints.

The Challenges of Prevalent Software Lifecycle Practices

In recent years, software quality has become an overwhelming concern. Some of the more dramatic software failures have caused space shuttle mission delays, breaches of security systems, satellites that go astray, and public transportation deaths. More typically, software failures result in losses of hundreds of millions of dollars per year in business. As an example, a major U.S. semiconductor manufacturer found in a 1989 study that just over fifty percent of its factory equipment failures could be attributed to software-related problems. Other examples include a number of capacity-limited semiconductor factories in different companies that are totally dependent on software systems for factory material movement. Tens of thousands of dollars in unrecoverable profit are lost every minute that certain software failures shut down an entire factory's production.

Poor quality has caused spectacular failures

Even though software is not the only engineering discipline with these problems, it is highly visible to the organization because each project is developing a fairly unique solution to an opportunity to fulfill a business need. Unfortunately, the next decade is likely to see these problems worsen before they improve. As early as 1988, Boehm and Papaccio [Boehm 88] and many others pointed out the growing cost of software: in excess of $125 billion per year—and that was seven years ago!

The growing visibility of software

1.3 The Slow Progress of Software Practices

Because of these problems, there has been a considerable effort throughout the world to improve the state of the practice of software development and support. In the U.S., much of this work has revolved around the Capability Maturity Model (CMM) developed by the Software Engineering Institute (SEI). SEI is a division of Carnegie-Mellon University funded by the Department of Defense's (DoD) Advanced Research Projects Agency (ARPA) [SEI 91]. This work defines a process

SEI was formed to improve software practices

maturity framework to help organizations assess the capability of software organizations and improve their software process. The CMM also provides guidance on how organizations can evolve toward a culture of software engineering excellence by organizing these evolutionary process steps into five maturity levels that lay successive foundations for continuous improvement:

- Level 1 – Ad hoc and possibly chaotic,
- Level 2 – Repeatable,
- Level 3 – Defined,
- Level 4 – Managed,
- Level 5 – Optimized.

SEI appraisal results

Through December 1994, SEI accumulated the results of 435 CMM assessments. All of these assessments used the original framework published in 1987 and covered 1,927 projects in 379 organizations of 99 participating companies. Sixteen percent of the sites were outside of the U.S.

The progress has been modest

The results shown in Table 1.1 indicate a small improvement from 1991 to 1994, but this is suspect because only 296 (15%) of the 1,927 project appraisals had been conducted in the original survey. Even in 1994, the results were very unsatisfactory. Additionally, there is no significant difference between U.S. and non-U.S. organizations. See Chapter 3 for more detail on the CMM and its use.

Maturity Level	1991 Organization Profile	1994 Organization Profile
5		0.3%
4		0.6%
3	7%	10%
2	12%	16%
1	81%	73%

Table 1.1 – 1991 & 1994 Organizations Maturity Levels

1.4   Addressing the Challenge

There may be a silver lining to the *software crisis,* for it may be the very stimulus that forces the systems technology community to make dramatic improvements over the next several years. This will not be accomplished by adopting a single new technology or by merely purchasing the latest software tools. There are many solutions proposed to alleviate the software crisis, but the most important component of the solution is the realization that there is *no silver bullet* [Brooks 87]. This means that no one action, no one technique, no one tool, no one change in the development and maintenance of software will alone attain all—or even most—of the objectives.

There is no "silver bullet"

Instead, several approaches are likely to be combined together to solve the problem:

A multifaceted answer to the software crisis

- A well-engineered approach to the software lifecycle, supported by the appropriate techniques and tools;
- The use of objects to break the productivity logjam of instruction-based program handcrafting;
- Application interoperability based on standards;
- A new relationship between the users and suppliers of information systems.

The first two aspects of this solution will be developed extensively in successive chapters of this book. A few more words are appropriate here about the third and fourth components of the approach.

Standards-Based Interoperability

Standards open new possibilities

Standards exert great influence on the ability of organizations to adopt a new model of application development. Emerging as well as established standards provide new capabilities:

- Entire applications, or smaller-size software components, can be purchased and integrated because they meet common standards.
- Legacy applications can be modified, or hidden behind a standards-compliant interface, in such a way that they continue to serve the business needs of their owner without requiring that new applications adhere to obsolete architectures and designs.

External standards are more effective than internal ones

If interoperability and legacy integration were solely a matter of consistency within a single entity, any internal "standard" would do. However, the need to provide integration within the extended enterprise of Section 1.1, and the necessity of finding commercial components ready for use, dictate a more thorough look at broadly accepted external standards.

Interoperability has real business benefits

Interoperability is not some sort of theoretical "Holy Grail" for computer scientists. It has practical and quantifiable benefits that include being able to buy systems components instead of having to make them, and using information technology for uninhibited integration throughout the enterprise. Integration extends beyond the traditional limits of the enterprise to encompass a company's customers and suppliers, making interoperability more challenging and even more crucial.

Evolving Roles of Software Providers and Users

An information systems organization cannot succeed in this integration task with a myopic view of its role in the enterprise. It must maintain linkages and knowledge of the enterprise's overlapping management objectives and operational activities, as shown in Figure 1.3.

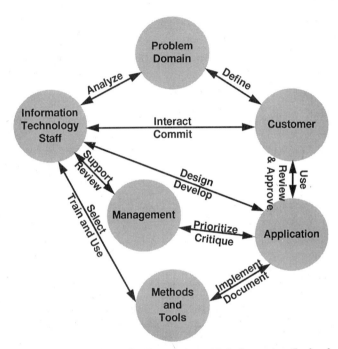

Figure 1.3 - Integrating the Enterprise with Information Technology

Business management, operations, and information technology are represented in this figure by three circles of equal size which overlap in a symmetrical pattern. In order to accomplish the required enterprise integration, each of the overlapping areas in Figure 1.3 represents a discrete subdomain that inherits characteristics from those domains of which it is a part. Figure 1.4 exhibits some of the more significant functions within each of the seven subdomains.

Management and systems activities are interdependent

Realizing the Object-Oriented Lifecycle

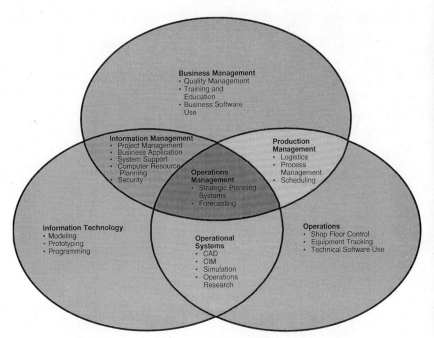

Figure 1.4 - Functions and Roles in the Integrated Enterprise

The role of the information technology providers must change

The way computer-based systems should be developed now and in the foreseeable future implies significant changes for the staff in the business, operations, and information domains. Ultimately, the integrated framework, tools, and repositories that characterize an advanced software lifecycle environment give the operations application users (Operations Domain) and business applications users (Business Management Domain) the capabilities needed to successfully deploy and evolve their own applications. This recasts the applications provider (Information Technology Domain) in the role of software lifecycle architect. For business and operational applications to have the required simplicity, all the complexity currently exposed to the user must be hidden away in underlying systems and software technology (e.g., intelligent agents, multimedia documents, electronic approval, etc.). This implies a significant increase in the sophistication of the system that must be developed and maintained by the staff of the Information Technology Domain in their new roles.

Conclusion

The adoption of object-oriented techniques and standards will not meet the challenges posed by software development needs as long as we remain in a world of ad hoc applications development. However, the various approaches introduced in this section lead us to a concept of an *object-oriented lifecycle* which can significantly improve a systems organization's ability to respond to the business challenges of today and tomorrow.

The object-oriented lifecycle

Chapter 2

Defining a Lifecycle Architecture

2.1 Rationale and Objectives

A particularly elusive challenge in addressing system development involves establishing and maintaining *traceability* between the business goals of the organization and the systems, including their development mechanisms and their underlying technologies. It is a difficult, if not impossible, task to map systems technologies, such as object-oriented techniques and tools, directly to a business goal, such as the reduction of manufacturing cycle time. Yet without a clear link between customer-related business goals and the systems that support them, the development strategies and the technical decisions cannot be justified. For these reasons, system development funding is frequently jeopardized.

Systems must be linked to business goals

One possible strategy for establishing this necessary linkage consists of the following steps, depicted in Figure 2.1:

How to establish these links

- Inspect and analyze the key **business goals** of the specific customer base.

- Translate these business goals into **goals and success factors for systems**.

- Derive a set of **system characteristics** that demonstrably supports the achievement of these goals and success factors.

- Link the selection of **run-time technologies** and **development processes** to these goals and success factors.

- Evaluate the **design** of systems (and the execution of the systems development plans) against these criteria.

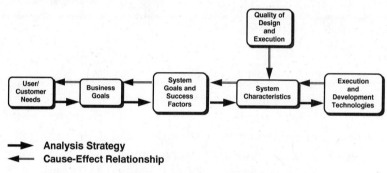

Fig. 2.1 – Linking System Characteristics to Business Goals

Align all stake-holders behind the goals and solutions

The primary purpose of a **system architecture** is to create a consensus regarding the above steps among the various stakeholders involved in creating or evolving a system or set of systems. This might include the development staff, the decision

Defining a Lifecycle Architecture

makers who hold the purse strings, and any subcontractors or consultants involved in the project. A common set of goals, system characteristics, technologies, and processes can be developed, refined, and adopted *a single time* through a consensus process, then applied uniformly by the various parties involved. Without this explicit definition, each in-house or external development group must independently follow the above steps with each project. Over time, these groups are liable to arrive at different and incompatible definitions and decisions, with little or no learning transferred from project to project or shared from group to group. Worse yet, a disciplined development approach may never be adopted at all, which jeopardizes the organization's chances of putting in place the adequate techniques to create quality systems.

2.2 What Is An Architecture?

The word "architecture" has become fashionable in recent years in the information systems community, but it is often misused. Computer manufacturers deserve much of this blame because they have used the term casually with the hope of conferring respectability to their products. For instance, two competing network protocols are IBM Corporation's *System Network Architecture* (SNA), and Digital Equipment Corporation's *Digital Network Architecture* (DNA). In each case, one could argue that a suite of protocols and products is not tantamount to an architecture.

The word "architecture" has been misused

To understand what an architecture is, one might begin with the dictionary, and the domain in which the word earned its meaning, namely, the architecture of buildings. Both *Webster's Ninth New Collegiate Dictionary* and the *American Heritage Dictionary of the English Language* agree that the definition of "architecture" includes:

The original definition of "architecture"

- The art and science of designing and erecting buildings;

- A style and method of design and construction;
- A structure (i.e., a unifying or coherent form, or an orderly arrangement).

Architecture is about structure, not details

If we consider the meaning of "architecture" in the world of modern buildings, we further recognize that only the major stylistic and structural features fall within the definition of architecture, while most details do not. An architecture is a *structure* and a *set of guidelines* intended to ensure the following results:

Five goals of an architecture

- *Functionality*: the building can perform its function. For example, rooms are connected appropriately for their planned use (e.g., there is a loading dock if it is a manufacturing plant, etc.).

- *Robustness*: the building has the necessary foundations to be stable. It can support the weight of the equipment and people it may contain; it can resist earthquakes or hurricanes if it is in the corresponding danger zones, etc.

- *Performance or capacity*: the building has sufficient corridors, elevators, staircases, restrooms, ventilation ducts, etc., to meet the demand on these various facilities.

- *Regulatory compliance*: the building conforms to all applicable codes and laws (e.g., zoning restrictions and access for people with disabilities in the U.S.).

- *Elegance*: the building is pleasant to inhabit, to work in, or to walk by. This not only attracts buyers or tenants, but also enhances the reputation of the architect. Note that this is the most visible part of the architect's job, but it would be of no importance at all if the building were unusable due to violations of any of the previous four points.

What architecture is not

On the other hand, architecture is *not* the selection of specific plumbing fixtures or carpet colors nor the execution of the construction details. The building contractor must follow the architect's guidelines for construction materials in order to ensure that the robustness criterion is met (and perhaps the outside elegance aspect as well) but beyond these, he has wide latitude in the choice of building methods and many of the other details.

The architecture of systems

If we apply these concepts to the world of systems, we can describe the scope of architecture using exactly the same words:

- *Functionality* means that the systems can perform the function for which they were designed (e.g., order entry).

- *Robustness* covers such aspects as the absence of bugs as well as the recoverability of databases in the event of computer failure.

- *Performance and capacity* include various factors such as response time, the volume of information that can be handled, etc.

- *Regulatory compliance* includes meeting customer specifications (e.g., using the Ada language in the U.S. defense community, or offering an interface that conforms to a recognized standard in a particular industry).

- *Elegance* might translate into user friendliness or the ease of integration with other systems.

The importance of a common structure

The word "architecture" is also used in general to denote the *common structure or appearance* of a class of buildings, as in "neoclassic" or "Byzantine" architecture. If we apply this concept to systems, we find an analogy with such concepts as "client-server" or "mainframe architecture." Whereas the common appearance of buildings principally creates a pleasing aesthetic effect, the common structure or appearance of systems has very practical consequences. Systems that share a common structure can be more easily interconnected; systems that share a common appearance allow users to move from one to the other without significant retraining.

Summary of the definition

We can summarize our analogy about building architectures by proposing a concise list of points that comprise a systems architecture:

- A set of principles for the orderly creation of systems and applications;
- Definitions of terms and concepts (to establish a common vocabulary);
- Guidelines for creating detailed specifications and for selecting system solutions.

Some important principles will be listed in Section 2.4 below. The remainder of the book covers the definitions and guidelines in depth.

2.3 Why A *Lifecycle* Architecture?

Architecture is not limited to the execution components

Many who use the word "architecture" in the computer and systems industry implicitly limit it to the execution domain, that is, the set of elements that constitute the run-time environment of the software. For instance, people talk about:

- *Network architectures:* how the various components of a distributed system are connected together, and how the various communication protocols are stacked on top of one another;
- *Computer architectures:* how the various elements of the computer itself (CPU, memory, bus, floating point unit, etc.) are sized and interconnected;
- *Object Management Architecture:* how a system can be constructed from a set of objects that communicate with

one another through the use of an object request broker (we will describe this particular architecture in more detail in Section 10.2).

However, this approach ignores the question of how the system is derived and developed. And since the development effort is itself a complex undertaking, it surely can benefit from some of the organizational ideas presented above. Specifically, let us consider the elements of the earlier definition of an architecture, and see how they might apply to the *system development process* rather than to the running system itself:

"Architecture" also applies to the development of a system

- *Functionality.* In the early days of computing, the word "development" was synonymous with "programming." People often started coding without first analyzing what the system was to do or without explicitly designing how it would do it. Even when the importance of analysis and design was recognized, these were still manual activities of limited scope, documented with common tools such as word processors. There was little need to organize the relationship between the components of this limited environment. Today however, software development requires a number of interrelated components in order to analyze, design, program, test, and maintain a product. But whenever a collection of individually selected tools has been introduced without regard for how they would be integrated into a consistent environment, the cost is higher than the benefits. In other words, the structure of the environment affects whether or not it can do the job.

The structure of the environment affects the ability to do the work

- *Robustness.* The development process itself often breaks down. Software projects are frequently late, or the quality of the resulting systems is unacceptable to the customer. This indicates that development is made up of multiple components that need to be put together in a more organized fashion.

- *Performance.* There is now evidence that some analysis and design methods, and some programming languages, allow projects to complete faster and with fewer quality accidents than others.

Standards or customer constraints must be followed

- *Compliance.* The choice of development techniques and tools is rarely left to the project leader's discretion. Often an organization has a set of standards, either out of a technological belief in the superiority of certain languages or methods, or because it relies on economies of scale by purchasing the same tool for all its system development staff. In some cases, powerful customers, such as government agencies, require from their contractors compliance with their particular standards (e.g., most of the software written for the U.S. Department of Defense must be written in Ada).

The environment should be easy to work with

- *Elegance.* Although many project leaders fail to pay enough attention (and many programmers pay too much attention) to this point, it is important for the set of capabilities used during development to have a "nice feel." This maximizes the productivity and the morale of the developers. No one can continue doing quality work with tools that feel awkward, such as multiple tools that have a completely different "look and feel," or tools that rely too much on textual input as opposed to a more creative, graphical expression of ideas.

The "lifecycle architecture" covers development as well as execution

These five points above show that the same concerns underlying an architect's work also apply to the elements (methods, tools, etc.) that constitute the developer's environment. For this reason, we propose to extend the traditional definitions of an architecture to *encompass the development phase as well as the execution phase.* We call the result of this effort a **lifecycle architecture**.

2.4 Principles of a Lifecycle Architecture

The principles we propose here may still be relatively new in many circles, but they are already well established in areas where the quality of the resulting systems is especially critical. In military control systems, for example, fault-tolerant software is an absolute requirement, even though it is practically impossible to perform realistic tests of the system. Most of the concepts in this and subsequent chapters have been pioneered by the U.S. Department of Defense and Federally Funded Research and Development Centers (FFRDC). We will repeatedly mention two government-funded initiatives: the STARS program (Software Technology for Adaptable, Reliable Systems) and the Software Engineering Institute (SEI) at Carnegie-Mellon University.

These principles are not new in high-reliability domains

In spite of their current low level of adoption, the implementation of these principles rarely requires additional research work. In most cases, it is possible with existing methods and tools. There is one possible exception: advanced computer-aided software engineering concepts such as **process enactment** (see Section 3.6 and Section 6.4) or **transformation engineering** (see Section 8.2). These still require more research and more technology transfer.

They can be applied with existing methods and tools

Principle #1: An Architecture is a Long-Term Roadmap

The effort needed to develop an architecture (for an organization or for an entire industry) can only be justified if this investment reduces the development cost for *many* projects. Therefore it must cover the long term by taking into account the future use of emerging technologies. This requires two things:

Amortize the cost over many projects

- That the architecture be inherently flexible; and

- That mechanisms be put in place to revise it periodically as technology evolves.

1.	An Architecture is a Long Term Roadmap
2.	One Architecture
3.	Specifications, Not Products — Do Not Choose the Carpet Color!
4.	"**U**nderstand, **S**implify, **A**utomate" Principle
5.	Systems = Development + Execution
6.	Adopt Applicable Standards — Do Not Conform to Anticonformism
7.	Include "Legacy Applications" — Evolution, Not Revolution
8.	Work Through Partnerships and Consortia — There is Strength in Numbers
9.	Empower the Users — Aim for Model-Driven Systems

Figure 2.2 – Nine Principles of a Lifecycle Architecture

The architecture must evolve to survive

Given that computer technology evolves rapidly, and that it is impossible for any company or group to forecast perfectly what will emerge over an extensive period of time, it is important for the architecture itself to be able to evolve, or it will become obsolete. To achieve this, *an architecture proposal must therefore include the mechanisms of its own evolution.* One approach is to put in place a **review board** which periodically assesses an organization's prescribed architecture to determine whether or not it is still adequate, and how to make modifications, if necessary. Clearly, the evolution must not remain on paper. Any change must be transferred to the development groups to which it applies. Otherwise, they will either continue to follow the prior guidelines without realizing that these are

Defining a Lifecycle Architecture

obsolete, or worse, they will meet the new needs with their own separate—and divergent—guidelines.

Principle #2: *One* Architecture

Again it is primarily a matter of *return on investment* to conclude that a single architecture should apply to a broad range of systems. Systems that are interconnected, as is the case with business and manufacturing systems in most industrial environments, should definitely share a single architecture. Otherwise, the interconnection of these systems may be impossible to realize cost-effectively.

Apply a single architecture to multiple systems

Indeed, one of the purposes of developing an architecture within a company should be to **integrate the enterprise.** This permits the sharing of data and platforms, as well as development skills between separate "islands of automation." Developing an architecture to address business systems without worrying about the interface with manufacturing systems would be like designing a hospital without paying attention to the location of the streets from which the ambulances enter!

This allows easy interconnection

Principle #3: Specifications, Not Products (Do Not Choose the Carpet Color!)

Architectures sometimes become obsolete because they are confused with *sets of detailed standards*. Often, when a Director of Information Systems says: "We need a common architecture," he is really asking: "Should we use TCP/IP or SNA?" or "Which database should we select?" etc. Few companies completely define the architectural framework before plunging into product selections. Consequently, the list of selected products proves to be inadequate, or worse, incompatible.

Do not confuse architecture with standards

The truth is that *multiple products can coexist within an architecture if they all meet the same architectural specifications.* For instance, most large companies have multiple electronic mail systems in place. For the most part, this is transparent to

An architecture does not impose a single product choice

the users, provided there is a common protocol (such as X.400 or the simple mail transport protocol, smtp) defined in the architecture and supported by all these systems. With this prerequisite, architectural compliance does not imply selecting a single product. It may still be preferable to converge on a single product for reasons of cost (such as availability of larger discounts) and of efficiency (such as simpler support and training), but this is not mandated by the architecture.

Principle #4 (The "USA" Principle): Understand, Simplify, Automate

Do not blindly automate current human procedures

A successful architecture should maximize the systems' responsiveness to business needs. It should not be purely technology-driven, which sometimes comes from a "systems for systems' sake" mentality. On the other hand, it should not consist of blindly automating the current manual procedures. We propose to achieve this careful balance through the "Understand, Simplify, Automate" approach depicted in Figure 2.3. Using this principle, an organization may produce or procure new systems or applications, or it may streamline business processes and *eliminate* the need for a system, as explained below. The current buzzword for this principle is "Business Process Reengineering."

Examine and restate the problem

For example, one business need may be to lower the cost of inventory. To address this, the **Understand** phase would consist of collecting, organizing, and analyzing information so that all concerned understand the problem environment. One might examine the components of cost, and whether or not inventory costs are the actual problem or merely a symptom resulting from some other more fundamental business issue (e.g., excessive manufacturing cycle time). This understanding is used to restate precisely the problem to be solved.

Defining a Lifecycle Architecture

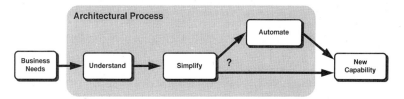

Figure 2.3 – "Understand, Simplify, Automate" Principle

The **Simplify** phase involves careful analysis of the description of the "as is" practice, and the elimination of all non-value-added steps. One can then determine what *organizational* changes can be effected to improve the business process, regardless of whether this is a manual or an automated process. In our inventory cost example, the analyst may discover that the company's procurement policies can be improved. He may find that despite the fact that some suppliers have the ability to engage in just-in-time (JIT) delivery contracts, the company still reorders materials in larger batches than needed, based on an older model of procurement.

Eliminate unnecessary steps

In many service organizations, the Simplify phase may consist of eliminating paperwork, for instance, by requiring management signatures only above certain threshold levels, or by

Change business rules when applicable

combining information from several requisition forms into a single one.

Do not skip the simplification phase

In other cases, while the existing process may be found to be simple and adequate, excessive manual processing or obsolete systems may detract from its efficiency. In this instance, no simplification may be needed, but one should still not skip uncritically from the **Understand** phase to the **Automate** phase. A bona fide attempt at simplification should first be made, or one may end up replacing manual chaos with automated chaos!

To automate or not to automate?

At that point, the organization has a decision to make, symbolically represented by the question mark on Figure 2.3:

- A new, simpler (and manual) process might be sufficient to improve inventory costs, in which case the **Automate** phase is not needed at all; or

- The new process may require computer automation in order to function. For example, to eliminate the paperwork from the order entry process, a just-in-time reorder and delivery system may require an on-line connection between the customer and its suppliers. In this case, the automation needs of the new process can now be analyzed, an application or system can be designed or purchased, and the new process and its supporting software can be introduced at the same time.

Principle #5 (The Lifecycle Principle): Systems = Development + Execution

Most authors focus on build-time or run-time components only

This is simply a restatement of the "Lifecycle Architecture" definition introduced earlier in this chapter. We reiterate it here because it is one of the fundamental principles on which this book is based. Most books, papers, or articles on software and systems engineering, or on object technology, focus on only one of two aspects:

Defining a Lifecycle Architecture

- The **build-time** components, including analysis and design methods, and the Computer-Aided Software Engineering (CASE) tools that support them; or

- The **run-time** components, including services like database management systems, distributed computing environments, object request brokers, or message buses that are used by the executable software.

Although focusing on one or the other is entirely appropriate in a technical paper that does not attempt to cover the complete spectrum of systems development, a successful architecture must not only address both aspects as equally important, but also the many links they have to one another. For instance, using a database management system at run-time implies that information models should be developed during build-time. Moreover, it is useful to have tools to automate the translation of the model into the Data Definition Language (DDL) statements that create the actual database.

Both are important. They are related...

This principle of the architecture says that both development and execution must be considered, but they must *not* be confused. Let us examine the pros and cons. In many cases, the two environments are not necessarily different. For example, a network of UNIX workstations is generally recognized today as one of the best platforms for software development. It may also be quite suitable to support the execution of the final product (e.g., the management of a telecommunications network), but there are also good reasons why the two environments may need to be different:

...but they do not have to be identical

- Some of the features needed in the development environment (high-resolution screen, mouse, significant disk space, etc.) may not be needed in the execution environment. This is especially true if some of the run-time components interact only with machines, not with humans. These features may not even be available in the execution environment (there is still a lot of software being developed for embedded 16-bit microprocessors with no disk and limited random-access memory).

- Conversely, the run-time environment may require costly options for safety and security (e.g., redundant computers, replicated disks and databases, etc.) whose cost would not be justified in the development environment.

- Some external constraints that apply in one environment may not make sense in the other. For example, almost all software developed for the U.S. Department of Defense must be written in the Ada language, but even the DoD does not require that the CASE tools used to create the software be themselves written in Ada.

The development and execution environments can be addressed separately

Therefore, the architecture should *allow*, but not *require*, the two environments (development and execution) to use the same technologies. Since there may be a difference, and since "divide and conquer" is a time-honored way to address complex problems, much of this book will address the two areas separately. But one should remember that both should be considered necessary and complementary components of the architecture.

**Principle #6: Adopt Applicable Standards
(Do Not Conform to Anti-Establishment Trends)**

Acknowledge and incorporate standards in information technology

A successful architecture should take into account the standardization activities that occur in official organizations and in the industry at large, should incorporate these standards whenever applicable, and should make room for incorporating emerging standards that have not yet reached maturity. There are two main categories of reasons why standards are important:

- Applications rarely exist in isolation. Obedience to applicable standards allows multiple applications to

Defining a Lifecycle Architecture

interoperate in three distinct ways: sharing information, coexisting within the same graphical user interface so that the user can effortlessly move among them, and interacting directly with one another when needed.

- Users rarely interact with a single application, especially given the number of once-mundane office tasks (e.g., writing, calculating, drawing) that have become computer-aided. Standards allow users to migrate from one application to another without having to learn everything from scratch, and without losing the information they created with a previous application.

We therefore recommend that an organization permanently keep an eye open for the appearance and evolution of standards in both the development and execution domains. Throughout this book, we will mention some of the emerging standards that we find most promising. There are two purposes for doing so:

Monitor the standards bodies

- To *adopt* these standards, when possible, in one's own developments;
- To *identify which suppliers* have a roadmap for their commercial offerings that promises to converge with applicable standards.

Principle #7: Include "Legacy Applications" – Evolution, Not Revolution

In order to be accepted, most architectures must include the possibility of supporting migration from existing systems, rather than having to throw them away completely to introduce new capabilities. During this transition period, systems will include a mixture of new and old components and will require appropriate interfaces between them. Eventually, as more components are re-architected or replaced by new ones, the old components and their interfaces can be retired.

Support a smooth migration from legacy systems

Principle #8: Work Through Partnerships and Consortia – There Is Strength In Numbers

Recruit partners

An architecture will be more successful, and the risk inherent in any change will be lowered, if the architecture is adopted and pursued in concert by all the elements of the *extended enterprise*. This includes all the components of an organization as well as its main external business partners (customers and suppliers—including system and application suppliers). This suggests that partnerships and consortia are useful vehicles for the development and adoption of architectures and, in noncompetitive areas, of systems that comply with them. This is true even if the most immediately visible effect of working with others is the initially slow process of building consensus on the definition of those areas open to cooperation and those that must remain in each organization's competitive domain.

Benefits of working with others

For the typical "end-user organization" (i.e., a company or agency that is not a supplier of computer hardware or software), the main benefits of working within a consortium or partnership are:

- Access to the knowledge of computer systems that other members of the consortium already possess. This knowledge may be greater because the other participants are bigger, or because their role as suppliers makes them more dependent on state-of-the-art expertise.

- Instead of being saddled with the indefinite support of the basic technology on which a proprietary architecture may rely, the organization might choose (for a price) to offload this support to the suppliers of a technology that many others would also use.

Defining a Lifecycle Architecture

Principle #9: Empower the Users – Aim for Model-Driven Systems

Strive for user-friendly system representations

A successful architecture must reduce the conceptual gap between the users' understanding of their activity and the way in which the applications that support or automate this activity are described. The ideal solution is the one shown in Figure 2.3 where a system or application is completely described through a set of **models**. These models should be intuitive enough for the end user—with minimal training—to understand. Moreover, the subsequent transformation of a set of models into a running application should be 100% automated. At that point, the end user could create or maintain an application simply by creating or updating the models to reflect the new business requirements, and by pressing an "install" button to place the new or revised application on line.

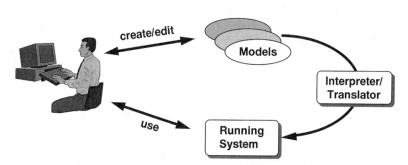

Figure 2.4 – Model-Driven Computing

A long-term vision and two reasons why it should not be perceived as threatening

This vision is clearly not achievable in the next few years, but it is a valuable goal towards which to strive. Many systems professionals understandably fear that the emergence of such a model-driven capability would threaten their livelihood by making them redundant. But this is unlikely for the following reasons:

- Unless it becomes possible to generate complex applications from user-level models, the amount of software to be developed and maintained will continue to expand faster than the number of people with the requisite level of training. Access to previously untapped sources of expertise, for instance, in the ex-Soviet Union and in southern Asia, will delay but not eliminate the crisis. Ultimately, the need will outgrow the supply.

- Model-driven computing is already happening in some limited domains. The best example is probably provided by spreadsheet programs like Lotus 1-2-3 or Microsoft Excel. In this case, a person with moderate financial analysis skills, but no software skills whatsoever, effectively creates a working model of the relationship between certain numbers every time he enters a formula in a cell rather than simply a number. When the user saves the spreadsheet, recalls it later, and modifies some of the inputs or formulas to obtain a new output, he is in effect using and modifying a model.[1]

---

[1] In the technical domain, another example of model-driven computing is provided by LabView, from National Instruments. In the business systems arena, a product called Edify, from Edify Corporation, provides another good example. It facilitates the building of interfaces between disparate systems, or between data and voice systems, using a flow chart of the task as its programming paradigm.

2.5 Benefits

The main benefit of an architecture-driven process is to provide a framework for a more effective response to rapidly changing business and technical requirements. The response is more effective because:

How an architecture helps respond to new needs

- A developer faced with a new problem is already in possession of a set of guidelines within which a solution can be designed.

- The emergence of a new technology can be handled for the foreseeable future through revisions to the architecture, rather than creating a challenge for every new project. As a result, the organization is less likely to find itself lagging years behind new technology (e.g., using pre-relational database technology in the mid-1990s).

- Any modern architecture promotes a modular approach (which is more effective than a monolithic one) to the design and evolution of systems.

- An architecture provides guidelines and tools for the integration of "legacy" applications within the complete system. This allows the organization to exploit intelligently its past investments (until they can be phased out in a practical and economically acceptable manner), without constraining its future to past selections of proprietary solutions.

- The architecture does not just define the end state of the product, but because it is a lifecycle architecture, it also describes the activities needed to create the product, and the components of the environment necessary to support these activities. As CASE tools become increasingly sophisticated, the combination of formally described requirements and designs with automated tools will finally allow portions of the development work (significantly, the most tedious and error-prone ones) to become automatic.

An architecture is also one criterion to select suppliers with compatible approaches

This improved capability to meet new and changing needs applies to external suppliers as well as to the internal development groups of a company or organization. Indeed, the lifecycle architecture can be the basis for a better and more stable definition of the cooperation between a customer and his system and software suppliers. It also allows the customer to select his suppliers according to whether or not their products and their development processes meet the customer's architectural requirements and thus, ultimately, his quality requirements and his business needs.

Model-driven applications will not eliminate jobs

The architecture assumes that the development of systems will increasingly consist of automated translation (or interpretation) of explicit models of the requirements. Does this mean that we envision, or even promote, the disappearance of the traditional Information Systems organization and of the programmer as a profession? The answer is both yes and no:

Development jobs will move from writing custom applications . . .

- As users are trained to describe their needs directly in an appropriate modeling notation to a development facility, the need for traditional programming work on user applications will diminish considerably. Think again of spreadsheet programs. Their emergence has suddenly rendered useless a large number of financial utilities, written redundantly all over the world. For example, there is no longer a need for a mainframe-based loan amortization program, painstakingly written in COBOL, when this calculation can be done by a single built-in function in Excel.

. . . to creating modeling environments

- At the same time, sophisticated development skills will continue to be needed in order to program the "primitives" used by such systems. In the above example, someone had to program the amortization calculation only once, but he had to do it with 100% accuracy, good performance, and enough generality to satisfy the needs of millions of unrelated users! And someone had to program all the other facilities included in the spreadsheet program to manipulate tables and charts, "cut and paste" information, etc.

In summary, we do envision a shift from traditional development. Today, requirements are typically "thrown over the wall" by users to the Information Systems (IS) staff. Then some time later, a completed application is thrown back over the same wall. A new mode of cooperation will emerge where the separation of tasks is very different. Users will provide the expression of their needs directly in a modeling notation, and they will continue to refine their applications through successive prototypes under their direct control. The IS staff will provide the *environment* that allows this to happen, including the tools and the basic libraries of elementary functions from which larger applications can be built. The low-level programming role may indeed lose its importance or even disappear, but there will be a great demand for highly skilled IS professionals trained to handle the needs of end users who are in the driver's seat.

Chapter 3

The Software Process

3.1 The Software Process: Introduction and Definitions

By "software process," we mean the **order and manner in which the activities of the software lifecycle are executed.** In the rest of this chapter, we will describe in more detail what this means, but for the moment let us point out that we did not use the phrase "software *development* lifecycle." Had we included the word "development," we could have inadvertently given the impression that the early stages of a project (feasibility study, requirements gathering) or the late stages of its lifecycle (maintenance and retirement of a software system) are not subject to the same process concepts as the more central stages. This would have been misleading, since we intend to define the notion of process as something that encompasses the entire lifecycle of a system, from cradle to grave.

The software process covers the whole lifecycle, not just development

Such a broad definition of the software process is relatively new. Historically, the first two decades of computer history were marked by a narrow understanding that equated "software" with "programming." In other words, the only thing

The concept of "process" was more limited in the past

that was considered part of the software process (although this phrase was never used) was coding, compiling, linking, and testing.

The software lifecycle emerged when analysis and design were recognized as crucial

During the 1960s, and definitely by the 1970s, it was clear that there were other activities related to software. It had become clear that the main reasons for *quality* problems were ill-defined requirements or bad designs, and that maintenance was becoming the "black hole" where *productivity* was being lost. In fact, after this technology trend resulted in the separation of analysis and design from programming, it became obvious that the coding/compiling/linking/testing activities were, compared to the rest, neither as critical nor as time-consuming as was earlier believed. This led to the concept of a **software lifecycle;** models of the lifecycle were created, as well as methods to deal with analysis and design in a rational manner. At the same time, a whole industry in software tools emerged to support the various steps in the lifecycle.

The process is more than a series of technical steps

Although this was a significant step forward, a dimension was still missing. By defining software engineering purely as a set of *technical* activities, the originators of the lifecycle concept had not gone far enough. Events in the late 1980s and early 1990s revealed (with perfect hindsight, of course) that there is more to software engineering than a sequence of activities. Some of the missing management-oriented concerns include:

- Defining precisely who does what and when;
- Defining the precise entry and exit points of each activity;
- Placing development activities in the broader context of project management;
- Staffing several crucial new roles properly;
- Defining metrics and performing quality measurements throughout development, not just at the end.

The Software Process   45

It is only when adding these concerns that one moves from the concept of a software lifecycle to that of a **software process**. The following balance sheet summarizes this discussion:

The process balance sheet

	Coding
+	Compiling
+	Linking
+	Testing
=	**Programming**
+	Analysis
+	Design
+	Maintenance
=	**Software Lifecycle**
+	Management Issues
=	**Software Process**

The complete process incorporates and addresses different types of concerns:

The process addresses three concerns

- **Technical issues:** The selection and use of CASE tools and programming languages in support of a given process.

- **Managerial issues:** Project management, process definition and education, continuous improvement, etc.

- **Hybrid technical/managerial issues:** For instance, configuration management may initially proceed from a management concern, but this also has definite technical implications.

Across this spectrum, from technical to managerial, we propose to enumerate in this chapter the elements of the software process that can be supported by components of a computer-aided environment, as well as those that drive the selection and introduction of such an environment. In this sense, Chapter 3 sets the stage for Chapter 6, which will deal specifically with the software engineering environment.

The Big Picture

Software development is part of a larger context ...

The software process should be considered as a part of a larger whole, which is the **enterprise process**. The development of software is not an end in itself, regardless of whether this is software intended for internal use or for external sale:

- If the software is being developed for internal use, the software process should be embedded within a management process leading to the decision that a certain capability will be developed rather than acquired. That more global process should also periodically audit whether the situation that warranted an internal development continues to exist, or if a migration to off-the-shelf software is preferable in order to lower the maintenance costs.

- If the software is developed to sell as a product (or if it is bundled into a complete system which is sold), then the software development process is part of a larger **product development process.** This in turn is part of the overall process of the enterprise (including Marketing, Research and Development, Engineering, Sales, Service, and Finance).

... but this book focuses on the software aspects

In either case, the software development process is really a subprocess of the enterprise process. This book focuses on the software process. It does not address the surrounding levels, but assumes that they are already in place or are being developed separately.

Terminology

Twelve important terms are defined

A number of precisely chosen terms will be introduced in this chapter and used throughout the rest of this book. The word "process" has been defined in the introduction, but will be repeated here for the sake of completeness. In addition, all these words are included again in the glossary at the end of the book.

The Software Process

- **Process:** the order and manner in which the activities of the software lifecycle are executed.

- **Lifecycle:** the set of phases that a software product undergoes from conception to retirement, but not necessarily in a strict sequential order.

- **Phase:** a portion of the lifecycle defined by a discrete set of activities, milestones, and deliverables.

- **Method:** a set of activities, techniques, deliverable definitions (including textual and graphical representations), and rules that can be applied in a repeatable manner to a given phase of the software lifecycle

    Do not confuse "method" and "methodology"

- **Methodology:** the discussion or comparison of methods (the suffix "-logy" comes from the ancient Greek *logos*, meaning "discourse"); it should therefore not be used in lieu of the word "method."

- **Technique:** a precise way of executing a task (e.g., analyze written requirements, find objects, perform a code review, etc.).

- **Deliverable:** a concrete output of a phase of the lifecycle that is to be delivered directly to a customer or provided as input to a subsequent phase of the lifecycle.

- **Model:** an abstract representation of the logical essence of something, constructed for the purpose of gaining understanding or insight into this "something." A model allows one to focus on those characteristics of interest, while removing details or constraints not directly useful to the activity at hand.

- **Notation:** a textual or graphical convention for the representation of a deliverable (e.g., for the diagramming of a model).

- **Rule:** an objective criterion that can be used to decide whether or not a certain representation of a system is complete or correct.

- **Review:** a milestone that frequently provides a "go/no go" decision on whether or not an activity was completed correctly, and whether or not certain deliverables are accepted.

- **Tool:** an application program that generates or manipulates models, prototypes, programs, or other representations of the software process or product.

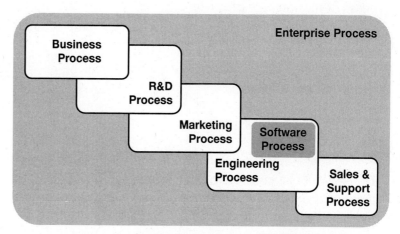

Figure 3.1 – A Hierarchy of Processes for the Enterprise

3.2 Software Lifecycle Models

Is there a common, standard, widely accepted, universally applicable model of what constitutes the software development lifecycle? No. The "right" model depends on too many factors to allow a "one size fits all" lifecycle. These factors include the type of software being developed, the users for which it is intended, the profile and level of training of its developers, the way the software is being distributed, etc. In this section, we will examine *several* models in order to promote an understanding of the components of the lifecycle, and to explain how lifecycle models have become more sophisticated over time. We will then generalize this discussion in Section 3.3, and in Section 3.5, provide guidance to defining *your own* software lifecycle.

There is not a single model of software development

The Waterfall Model

Originally, software lifecycle definitions were linear in nature. In other words, development passes through several sequential stages to reach eventual completion. This type of lifecycle has the following general characteristics:

A sequential model

- The various steps in the process are accomplished by people with different job definitions: "system analysts" worry about requirements and design, while "programmers" perform coding through testing.

- After they have been interviewed to gather requirements, end users of the software have little or no involvement in the process, even though it may take many months until the software is delivered and they become involved again in testing, accepting, and using it.

- CASE (Computer-Aided Software Engineering) tools are used to generate graphical or textual deliverables (analysis models, design models, code, executable programs), but the translation from one representation to the next is mostly done by people; some design-to-code translation may be performed automatically.

A classical representation of this model is pictured in Figure 3.2 and is described in more detail in [Royce 70]. This model has been called the "waterfall model" because the activities flow downward from step to step.

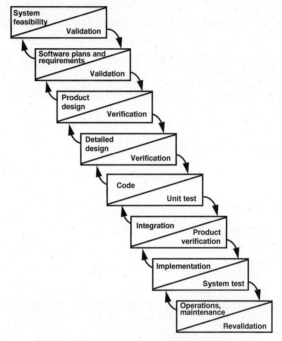

Figure 3.2 - The Waterfall Model

Some flexibility has been added to the list of activities

Many variants were developed over the years on the basis of this initial model. One change consists of varying the exact list of stages in order to adapt them to a different domain or to simplify them for small organizations. Here are two examples of variations on the original waterfall model:

Example 1	Example 2
• Requirements Analysis • External Design • Internal Design • Detailed Design • Physical Implementation • Operation • Maintenance and Support • Retirement	• Requirements • High-level design • Low-level design • Development • Release • Problem Management

In the first example, design has been split into several phases because the domain in question requires very careful selection of technical solutions through several successive levels of detail. In this case, "detailed design" includes what is usually called programming or coding, while "physical implementation" refers to the installation and acceptance phases.

Example 1: several design phases

In the second example, the lifecycle is simplified. All enhancements to the existing software are accomplished through an iteration of the same complete process. Therefore, it is not necessary to add an explicit maintenance phase. On the other hand, the release process is crucial for quality reasons, so it deserves attention as a separate phase.

Example 2: a simplified lifecycle

The waterfall model is also often modified in order to allow different policies for **backtracking.** The original lifecycle model only permitted backtracking to the previous step. This soon proved too limiting. For instance, assume that during operation, a user discovers that the software does not behave as expected, and that the developer realizes that a requirement has been misunderstood. There is really no point in backtracking all the way, one step at a time, to the requirements analysis stage. It is faster and just as effective to go back to the requirements immediately. Then, one can move forward again, making design and implementation changes, performing new tests, and installing the corrected version.

Backtracking one step at a time is too limiting

The waterfall model was very helpful in the early evolution of software engineering

The waterfall model, with or without modifications, was used for an increasing proportion of software projects during the 1970s. Its principal merit was to force development teams to formalize the transition between phases through formal documents and reviews. This brought a sorely needed level of discipline to the "art" of programming. Moreover, the waterfall model is well suited to the subcontracting of software efforts. This is because the requirements are supposed to be defined once, reviewed, and agreed upon *before* design starts. Thus in theory, it is possible to give these requirements to a subcontractor who will deliver a complete system that meets the requirements specification.

It has since proven too rigid; it hinders rapid response

However, as competitive pressures and productivity requirements demanded shorter "concept-to-market" cycles, the waterfall model imposed its systematic, linear, and somewhat plodding view on the development process. Merely obtaining a change in the *title* of a report could take three months (and this only after submitting the necessary forms—with signatures—and holding several reviews) when religiously applying this model. Although this is sometimes a joke told at the expense of Management Information Systems (MIS) departments, it is also often unfortunately true. In other words, this model limited the ability of an organization to follow the rapid changes in both the user requirements and the available technology. In addition, it did not allow the users to recognize an inadequate understanding of the requirements until late in the process.

The Spiral Model

To address these limitations, a new model was needed that:

- Allows iteration (trying different successive versions of a design);

- Supports parallelism (concurrent development of different parts of the system by different people);
- Automates the production of certain deliverables to reduce cycle times;
- Supports prototyping to show early and incremental results to the users.

Barry Boehm's "spiral model," proposed in [Boehm 88], has significantly influenced the current thinking by responding to these needs. As shown in Figure 3.3, the spiral model is a life-cycle design approach that builds the product incrementally.

Boehm's approach is incremental

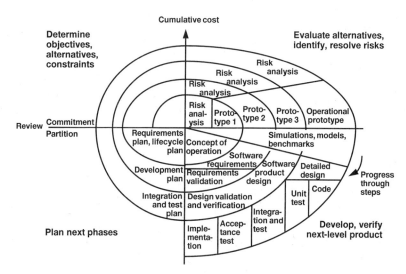

Figure 3.3 - The Spiral Model

During each increment, represented by the successive turns of the spiral, the traditional development activities (requirements, design, coding, testing) represent only one aspect (the bottom

Traditional development is only one of four categories

right quadrant of the diagram) among four. Within this quadrant, the component activities resemble those of the waterfall model, although they are partitioned differently and their definitions are enhanced. But more importantly, the model adds three other aspects neglected in previous models, that have to do with planning, determination of objectives, and risk analysis (and reduction) through **prototyping.**

The spiral model lowers the risk of developing the wrong thing

Starting near the center, the development team goes through these successive turns, iterating between analysis, prototyping, validation, and planning. At each turn, more of the application's requirements are identified and addressed. Instead of consuming resources for a long time before anything usable begins to appear, increasingly functional versions of the system are created incrementally, and thus the cost of the project increases only in proportion to the increasing completeness of the deliverables.

It is similar to concurrent engineering and compatible with objects

In the software domain, the spiral model appears to be the closest analog to the concept of "concurrent engineering" introduced in the late 1990s in the hardware domain. This approach was intended to perform design and manufacturing in parallel. It also appears to be quite compatible with the iterative development approach supported by the use of object-oriented techniques.

3.3 The Meta-Lifecycle Model

A Customizable Set of Lifecycle Components

A more flexible model of the lifecycle is needed to accommodate evolution

While both the waterfall and the spiral models each have significant strengths, neither is universally applicable. Other models exist, and new ones (or at least new variants of the existing ones) are periodically proposed. Others will undoubtedly be developed. It is unlikely that a "model to end all models" will ever be found, if only because software processes must adapt to a changing technology (new languages, improved tools, the emergence of reusable component libraries, etc.). However, in examining various lifecycle models, it appears that most share certain common features. Thus a logical way to accommodate both the present and future needs of the software development community is to create a **metamodel** (i.e., a "model about models"). The metamodel contains all the common elements that the various specific models may contain, but without prescribing the way in which these elements should be assembled. This assembly can then be performed to suit the needs of a particular user or industry.

We call this "build-to-suit" approach to the software process the **Meta-Lifecycle (MLC) Model.** It is a complex of technical *activities, components,* and *critical success factors* that together provide a comprehensive set of building blocks for lifecycles. (Although we just said "comprehensive," note that pure management processes such as marketing surveys, project initiation, customer education, or product retirement are outside the scope of the model. It could conceivably be extended to cover these additional aspects of the software business). This is not a prescriptive lifecycle model, but a "framework" that an organization can use to design its own lifecycle model, according to its specific needs. The Meta-Lifecycle Model is graphically depicted in Figure 3.4.

Introducing the MLC

The Meta-Lifecycle consists of four major parts. We will examine each in turn. These descriptions are fairly long, and often restate well-known principles of software engineering. But they are meant to point out key considerations and raise sufficient questions for the reader to make intelligent choices between alternatives.

The MLC has four parts

Meta-Lifecycle Components – Technical Components and Support Components

The *layers* of the model represent the technical components and support components that comprise the activities and

The layers represent the components

artifacts of the development process. Technical components are central to the development process. Without them, the process would make no sense because there would be no product! Support components facilitate the execution of the process. Without them, the software product *could* be generated, but with lower productivity and quality.

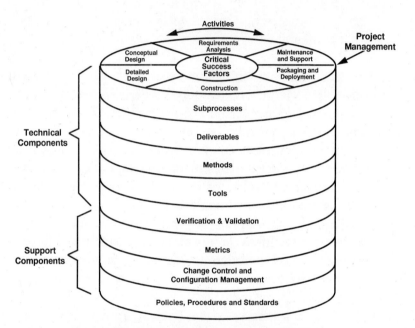

Figure 3.4 – The Meta-Lifecycle Model

Critical Success Factors

The core represents factors for success

The **inner core** of the model represents intangible critical success factors that affect the quality of *all* lifecycle activities or artifacts, as well as the product itself. These factors include **reuse, traceability, process integration, and continuous improvement.** By constantly striving to incorporate these factors into their products and processes, software organizations will produce better products and increase customer satisfaction.

a) **Reuse** is often falsely seen as dealing principally with code. In reality, it applies equally well to most of the intermediate products of the software development process, including requirements specifications and designs. The various paragraphs about reusable components in the subsequent section on "Activities" will provide more detail.

Reuse applies to more than code

b) **Traceability** establishes relationships between artifacts produced in different stages of the lifecycle, as well as relationships between those in the same phase. This is an important concern because large, integrated systems tend to evolve over time in a way that obscures these relationships. This is best illustrated by an example.

Various artifacts must be explicitly linked

Suppose that during the early development of a product, a study was conducted to determine the best choice of a data storage mechanism for a directory of electronic mail addresses (this scenario is based on a real case). A reasonable option, *a priori,* was to use a database management system for all data access. However, tests performed on a first version of the software showed that while the response time was acceptable for directory updates (which were rarely done one record at a time, and were only performed by specialized personnel who usually could tolerate the longer response times), the response time for a query—especially a query based on a partial name (e.g., "Baud*")—was unacceptable for the casual user who might need to invoke this program many times a day. In the end, the operating system on which the software needed to run had a very efficient string searching utility that could be applied to a plain text file. Hence a working solution consisted of doing the updates directly into the database, as originally planned (which allowed concurrent updates, checkpointing, backup, and all the other safety and security features usually associated with Database Management System, or DBMS, technology), and periodically creating a simple column-formatted report (in batch mode) to execute the retrievals

An example of traceability between a performance requirement and a design decision

through the text search command supplied by the operating system.

Lack of traceability may lead to the mistaken reversal of this design decision

The solution appeared to be a good one. It provided safety during updates, performance during queries, and a minimal delay in having access to new or changed records. However, suppose that the reason for using the string search on a report instead of a database query was never documented anywhere. Months or years later, someone may need to perform a different type of query or to add a new field to the database. In the process of updating the design and code of the original application, this person may run into difficulties with the current design, or he may simply *think* he can improve this strange and unexplained design by getting rid of the intermediate file and reverting to using the DBMS query capability. As a result, either the query performance will again become unacceptable (but the new version may now be deployed anyway as the only way to provide access to the new feature), or the same backtracking and redesign will need to be repeated.

This form of traceability can be achieved through documentation

What could have been done instead? Clearly, the design documents, or the code itself, could have contained information about the reasons an intermediate file was used for query purposes. From a performance viewpoint, such information would have alerted the person or team in charge of the product upgrade that eliminating the file was risky. This could have directly led to a better design, or if a design could not be found that preserved the performance of the product while offering the new functionality, to a decision to abandon the upgrade effort, all this at a fraction of the cost that might be incurred without that knowledge.

There are other forms of traceability: defects to code ...

This is just one example of traceability, for which some authors have coined the term "design rationale documentation." In this case, the main benefit is increased *maintainability,* especially in terms of the *productivity* of the organization doing the maintenance. However, there are other applications of this concept. For instance, one can link bugs, as recorded in the sort of defect tracking database that is likely to exist for a large product, to the specific section of code containing the bug and that has since been fixed. When modifying the same section of code again, each developer can be asked, as a matter of routine procedure, to find out whether or not there is any linkage between that section of code and previously detected bugs, and to examine his own code for the possibility that he might introduce the same bug again. An example of such a traceability entry is as follows: "Do not use variable `salary` here, but use `old_salary` because the statistic being computed is wrong if we use raises that have not yet been approved." In this case, it is the *quality* of the software that directly benefits.

Another frequent example is the linkage between test cases and modules (test case T1 exercises function `update_salary`), which saves test time and improves the rate of bug detection, thus having an effect on both productivity and quality.

... or test cases to code

It is important, and challenging, to maintain traceability during changes. For instance, problems found during design may necessitate a review of the requirements. In this case, the iteration back through requirements analysis should be explicit, and a revised specification should be published. Otherwise, the traceability between the corrected, detailed design and the now-obsolete requirements specification would be lost.

Traceability can be lost when changes are made without updating affected documents

c) **Process integration** considers the entire software development context to decide how best to execute each task, instead of selecting each component independently. Integration typically addresses the compatibility between analysis and design methods, tools, project support environments, and the like. Many of these concerns will be addressed in later chapters of this book.

There must be compatibility between the various development activitities

The end of an activity can be formalized, and measurements can then be made

Integration also concerns the processes themselves. Clearly, development activities and their products should properly flow from one to the next in the defined sequence (e.g., an output of analysis should be useful and effective as an input to design). The handoff between different phases should be materialized by **milestones, checkpoints,** or **reviews** that are well understood by the development staff. Moreover, these should add value to the whole process, rather than causing delays or acting as pure bureaucratic rubber-stamping exercises. The results of each activity should provide **estimators** of the size or difficulty of later efforts. For instance, the number of object classes defined during object-oriented analysis should become, at least over time, a predictor of the size of the delivered software.

Quality is a constant improvement effort

d) **Continuous Improvement.** This is the linchpin of all modern quality programs. Quality is not something you can achieve and then put aside, making no further effort. It is something you must strive constantly to improve over previous levels, because your competition is also hard at work to produce better products more profitably.

Use metrics, but periodically refocus on new improvement areas

Metrics are one example of how the continuous improvement principle has yielded unexpected but very positive consequences. Metrics are used to drive an organization to achieve a certain level of measurable proficiency. Once the target level is reached, there is a tendency to keep measuring the same things and to derive continued—but useless—satisfaction in monitoring compliance. Instead, what must be done is to find new metrics that measure something else on which a more significant improvement is needed. The constant evolution of software methods and tools ensures that there will be no lack of additional things to measure in the future.

The Software Process

Meta-Lifecycle Activities

We finally come to the **wedges** depicted in Figure 3.4, which represent different types of activities that are performed as the development progresses through the lifecycle. Depending on the particular lifecycle to be realized, the activities may be performed sequentially, iteratively, or not at all. Indeed, not every project necessarily requires all activities, but when designing a new lifecycle, all should be examined for inclusion. The circular nature of the model's cross section is intended to suggest that an iterative lifecycle is *possible*. Although it is not mandated by the MLC, it may be appropriate in many domains. In this case, the analysis of the requirements for a new version may immediately follow, and in many cases will overlap, the end of the lifecycle for the previous version.

Wedges are the activities performed to develop software

Several of the activities contained in the MLC will be discussed in more detail in future chapters, particularly in terms of the methods and tools that can be used to support them. However, we will examine here some of the **subprocesses**, other **technical components, support components,** and **critical success factors** that merit attention when planning the contents of each of these activities.

A list of what is important during each activity

a) Requirements analysis

Subprocesses:

- *Establishing and cultivating a relationship with the customer*, or with a Marketing organization that has the authority and competence to speak for the customer.

Cultivate good sources

- *Identifying reusable requirements.* Although reuse is a critical success factor, identifying reusable requirements is a subprocess. One of the most underutilized productivity improvements in software consists of finding elements of previous requirements analyses that may apply to the problem at

Reuse previous requirements analyses

hand. In other words, the concept of reuse is not limited to code. Reusability will, of course, be difficult if the requirements are monolithic, or if all the functionality is described in one area while all the information access requirements are described in another. This is why object-oriented requirements analysis (just like object-oriented design and object-oriented programming) inherently lends itself better to reuse.

Get information by looking at existing products

– *Reverse engineering of requirements from an existing product.* This is typically used as an alternative to reuse, particularly in the absence of clearly documented requirements. There are two variants of this practice. Where there is no surviving requirements specification, one might extract requirements from the organization's own older product. On the other hand, in order to create a list of functions to be emulated and/or surpassed, one might want to extract requirements from a competitor's product. In both cases, the reverse engineering subprocess requires either extensive reading of the user documentation, where available, or extensive use of the product itself, if the documentation does not exist or cannot be obtained. Without documentation or on-line help, it may take considerable effort by extremely creative people to explore all the nooks and crannies of the software, as well as to document all the limitations of these functions. Finding these limitations is important to make the new product better than the old one, and not just an imitation of it.

Simplified prototyping as a proof of concept

– *Prototyping as part of requirements analysis.* What should be prototyped during this phase, and how? It is relatively easy to design user interface mock-ups and have them revised by a future user. But designing a complete "proof-of-concept" prototype that *behaves* very much like the eventual software might seem as difficult as producing the real thing. In reality, one can use randomly generated numbers instead of calculating real outputs. Another technique is to substitute arbitrary constants for data that will ultimately be read from a database. Finally, error conditions can usually be ignored in a prototype, except when the main purpose of the software is precisely to handle errors (e.g., an alarm system). Where more extensive or more accurate prototypes are desired, one must determine whether or not they should be created through the same techniques that will be used for the final development. If so, then the prototype becomes the "first turn of the spiral" in Barry Boehm's spiral model. If not, it might be realized with different, perhaps non-production-worthy tools, purposely to avoid the temptation of embellishing the prototype into the final product. There is often a lure (due to budget pressures and late schedules) to "productize" a throw-away prototype and perhaps release it prematurely as the real thing.

Other Technical Components:

- *Deciding which methods, notations, and tools should be used to obtain, organize, and document the requirements. This includes the subprocesses required to iterate the requirements when a review concludes that there are errors or gaps in a first pass. It also includes deciding how the requirements should be documented so they are most clearly understood during design.*   *Organize the requirements gathering and analysis work*

- *Verification and validation (V&V) of requirements. This can be accomplished through reviews, prototyping, or both. It should include prioritizing the*   *Review and prioritize the requirements*

requirements in terms of their relative importance and in terms of the estimated cost of meeting them. This allows the rational allocation of the organization's finite resources to achieve the highest user satisfaction within the available budget.

Support Components:

Predicting the downstream effort

– *Metrics about requirements.* The most important metric is one that provides a forecast of the size and complexity of the resulting design. For instance, the total number of screens and fields (or presently, "buttons" and pull-down menu items) may be a good metric for a transaction-oriented package. For a real-time equipment control program, the appropriate variables may include the number of commands that can be sent, the number of data points that can be measured, and the frequency at which these measurements or commands need to be performed.

Requirements should also be under configuration control...

– *Placing the successive versions of the requirements under configuration management (CM).* With old style, monolithic requirements, CM meant, at best, a simple version control system. Furthermore, few organizations cared to formally reissue a revised requirements document after the requirements had changed. More often, the changes were captured in a simple memorandum which served as the starting point for incremental design and programming changes. However, with larger systems, this is a dangerously unprofessional practice in that it impedes the organization's ability to correctly evaluate its designs and to perform the appropriate tests of the software. Therefore, some form of version control is a must, as is the concomitant discipline to periodically reissue the requirements.

In the past, requirements specifications or models were monolithic. They consisted of either a single document, or a single file or database created by a CASE tool. This simplified configuration management, for instead of calculating the differences (or "deltas") between versions, the CM tool could store each version in its entirety. Furthermore, there was little need to provide a "merge tool" to combine separately modified versions into a single one.

... but this requires modularity

However, object-oriented requirements add a new twist to CM. The requirements may now be much more modular, including a smaller, separate specification for each object that has been identified and analyzed. It is now possible to revise only one section and to leave the others alone, assuming that the interactions between these objects are left unchanged, or simply to add a new object without changing previously existing ones. Because of this, the requirements now have configurations, just like the code. Sets of versions of change-controlled objects are *themselves* (the sets) subject to change. It is therefore possible to place requirements documents under full configuration management, and to benefit from the types of services provided by a comprehensive CM system.

Object-oriented requirements are more likely to be modular and lend themselves to CM

Critical Success Factors:

- *Integration with project management and planning activities.* Every time a certain group of requirements has been elicited and understood, it should be incorporated into the project plan, since the amount of work required to design or implement the required features can now be evaluated on a more rational basis.

Each requirement corresponds to a future task

Some components and factors are the same as those of requirements analysis

b) **Conceptual Design** (sometimes known under other names, such as **global design, architectural design** or **high-level design**) requires subprocesses and other components that are often similar to those introduced above for requirements analysis. Therefore, the following description lists only the required elements and indicates meaningful differences from the previous discussion.

Subprocesses:

When is it appropriate to expose outsiders to the design activity?

— *Customer relationships, if applicable.* If the organization is developing software that will be mass-marketed, it probably wants to limit external exposure of the product architecture to a few consultants bound by strict nondisclosure agreements. If, however, the software is being developed for a specific customer, or for a small number of customers of which one or two have a special partnership agreement with the supplier, then it can be entirely appropriate to have customer representatives review a conceptual design and comment on it. Note that the most effective customer representatives for a design review may not be the same people from whom one might best gather requirements. The reviewers may be computer professionals from the customer's own information systems division rather than end users.

Reverse engineering of a design can be the enemy of necessary innovation

— *Reverse engineering.* It is very unlikely that one can reuse the design of a product from another organization, since this design is probably unavailable. In general, this subprocess of conceptual design will only be added in the case of a new version of an old product. We offer one word of caution about this, however. Reverse engineering an old design to create a new one will perpetuate old architectural choices which may no longer be appropriate. For instance, an architectural design choice from the past could be the use of the UNIX file system to store information for which a relational DBMS was too slow at that time. In the mid-1990s, using an object-oriented DBMS may be a much better choice, since it provides better performance as well as the promise of a more extensible system (not to mention the elegance of directly mapping an object-oriented design to the class schema of an object-oriented database management system, or OODBMS).

- *Prototyping.* Here, the goal of prototyping is less likely to be ensuring that the software will "do the right thing," but that it will do it with the right quantitative characteristics (speed and volume) and that the design is feasible (i.e., that it is in fact implementable using *some* software technique). This is a "reality check" about the organization's ability to create software that meets the requirements.

    Verify the feasibility and performance of the design

- *Iterative refinement of the design.* Based on customer or peer reviews of the design, and on the knowledge gained from prototypes, the conceptual design may undergo successive revisions. If this is very likely to happen, it is preferable to leave some details out during the first pass. Otherwise, these details may represent wasted effort.

    Make successive revisions

Other Technical Components:

- *Defining the deliverables* of conceptual design (e.g., block diagrams and definitions of the functionality or behavior of its elements) so that they can serve as inputs to the next phase.

    Deliverables must be useful for detailed design

- Methods, techniques, and tools.

    Organize the work . . .

... and review it at the end—especially for quantitative factors

- *Design verification and validation (V & V).* At this level, the aspects to be reviewed are somewhat different from the verification and validation of requirements. In particular, quantitative factors are now likely to be introduced. Will the proposed architecture support the required performance or data volumes? Will it lead to a volume of software beyond the size and complexity that the organization, given its resources, can create and maintain? Does the high-level design allow or require the use of off-the-shelf packages, such as a DBMS, a report writer, or a graphical user interface builder? From the productivity viewpoint, an affirmative answer to this last question is good because it represents low-level designs and code that do not need to be developed again. On the other hand, from the perspective of the licensing cost of the product, the same answer may be bad, especially if it is intended as a mass-marketed product that is likely to be price-sensitive.

Support Components:

Design metrics can correlate with the magnitude of the requirements or with the coding effort

- *Metrics collection.* Because we are now in the second phase of the lifecycle, design metrics can be used either to shed light on the previous phase, or to predict aspects of the following ones. "Backward" design metrics can be correlated to characteristics of the requirements. For instance, over time, it can be very helpful to develop correlations between the number of individual requirements and the total number of design features included in the architecture. Rules might also be derived to identify those requirements that create a need for specific architectural components. "Forward" design metrics may predict the size and complexity of the code to be developed during the construction activity.

The Software Process

- Configuration management.

Critical Success Factors:

- Integration with project management and planning activities.
- *Reuse of design models.* Again, object-oriented approaches should make the design eminently more reusable than with previous approaches.

c) **Detailed design** is the next step. It is also known by several other names, although "low-level design" is clearly not favored because of the pejorative connotation of "low-level." In terms of designing the software process, detailed design is similar to conceptual design.

Process-wise, detailed design is similar to the previous step

d) **Construction** is a more generic name than "programming" and is more appropriate in the context of promoting the reuse of existing components. Besides, "programming" can have too much of an implication of working from scratch and building programs one statement at a time.

Assembling the actual components

Subprocesses:

- *Coding* is usually more clearly defined than the creation of analysis and design artifacts. This is simply because coding is an older discipline than the other activities. The emergence of new tools has created quite a few potential new subprocesses. For example, a programmer may be expected to compile his code first with a **profiler** in order to analyze performance, and then with a **static** or **dynamic analyzer** to detect quality problems, and finally, without any of these code "instrumenters" to prepare the final object module.

Coding now includes the use of quality analysis tools

- *Populating reuse libraries* with any newly created classes and methods makes them part of the pool of reusable code for future efforts.

Create reusable code

Other Technical Components:

- *Verification and validation of the programs* through a number of techniques and tools. This includes code reviews, unit tests, performance analysis, and other language-specific procedures (such as testing for memory leaks in C programs).

Support Components:

Traditional code metrics, as well as reuse metrics

- *Code metrics.* This is the time to collect metrics about code size, number of defects found during inspections, ratio of reused methods to new methods created, etc.

Version control, branching, and merging support

- *Source code control.* Usually a subset of configuration management, this phase involves special techniques because code can be manipulated at a finer level of granularity than analyses and designs. For instance, a version control tool will actually compare successive versions of a code module line by line, in order to store only the incremental differences between versions. This is invisible to the user. On the other hand, because human decisions are required when the files contain overlapping changes, the complementary process of merging two or more versions of a program into a single file (which inherits all the changes made in either version) is visible to the programmer.

The process of producing a release depends on the size of the customer base

e) **Packaging and Deployment (or Release).** Initially, these seem to be quite different, depending on the customer base. For example, the "production" or distribution mechanisms are likely to be significantly more elaborate for a mass-produced package such as a word processor or a spreadsheet package than for a company-specific internal product. On the other hand, in the case of a mass-produced package, there are so many customers and their relationships with the supplier are so distant (the customer typically buys from a software reseller), that it is not practical to set up a formal customer acceptance procedure. In spite of these differences, there are some common subprocesses that should be included in all cases.

Subprocesses:

- *Formal authorization of the release* by a quality assurance group and by the management of the organization that produces the software;

Release authorization is a formal step

- *Repeatable production and shipment* of the media and documentation from a common, well-controlled master copy.

f) **Maintenance and Support.** These are key to continued customer satisfaction; therefore, they include several very specific subprocesses and metrics.

Customer satisfaction depends on the support process

Subprocesses:

- *Receiving and acknowledging defect reports and enhancement requests* is the first key task. Today, this tends to imply the use of specific tools that track all reported incidents and requests by customer/requester, severity, version number, configuration, etc.

Defect tracking

- *Prioritization of the reports and requests* is also important so that the software group does not work first on what it finds easiest and most exciting, but rather on the changes that will likely have the greatest positive effect on the users. It is also important to distinguish between those enhancements that fit within the company or product strategy, and those that would require a significant amount of resources while making the product lose its focus.

A rational process to prioritize maintenance work

Continued tracking and measurement

— *Tracking the disposition of the requested changes* is required to be able to respond to additional inquiries about the same problems. This is also needed to provide quality metrics about the software organization. For example, does it meet its own forecasts for implementing the changes?

Support Components:

Metrics help identify root causes of problems or help communicate quality improvements

— *Defining comprehensive metrics* (number of newly opened and closed change requests per month, etc.) is often necessary to help management allocate the appropriate amount of resources to maintenance. Metrics, if properly exploited, also lead to detecting repetitive problems which should be addressed through root-cause analysis rather than by merely implementing a symptom-based solution. Finally, metrics can help create or restore customer confidence by showing correlations between the introduction of new processes, methods, or tools, and a decrease in the rate of new defect introduction.

Project Management

The mortar between the other MLC pieces is project management

The **mortar** that holds the different parts of the model together is project management, that is, estimating, planning, budgeting, reallocating resources (when necessary) between the technical and support components, ensuring that the critical success factors are addressed, and tracking the individual tasks that comprise the scope of a project.

More details in Chapter 7

Project tracking is typically done at different levels of granularity, depending on the activity. During requirements analysis, it is hard to determine what percentage of the requirements have been identified and documented. One probably cannot know in advance exactly how many there are. But after a detailed design has been created, the rate of progress during programming can be judged more accurately, and tasks can be tracked precisely at the person-day level of granularity. The same applies to testing, assuming that tests have been defined during the requirements phase. The rate of progress should be communicated regularly to management and/or to the customer. Project management will be discussed in more detail in Chapter 7.

How to Use the Meta-Lifecycle

The MLC model does not preclude, and in fact encourages, devising the lifecycle that is most appropriate for a particular project. The only constraint dictated by this metamodel is that all its components should be considered when an organization decides to tailor its own custom lifecycle. The MLC thus provides a frame of reference and a shopping list for the tasks within a project. It is important to note, however, that this is a "strongly suggested" list; individual components should not be rejected without thorough justification.

You can or should tailor the MLC to your needs

How does Figure 3.4 differ from Figures 3.2 or 3.3, apart from their obvious differences in graphical appearance? The waterfall model and the spiral model are both **process models**, but the meta-lifecycle model is a **component model**. It dictates what tasks (= components) should be addressed in a software project, but it does not say what the exact order and repetition of such tasks should be.

The MLC gives you that flexibility

The waterfall model and the spiral model are both *instances* of the MLC, but they are only two possible instances. This shows that the MLC is inherently more flexible than any specific lifecycle or process model, since it can accommodate the evolution of our understanding of what constitutes a good process. For instance, if reuse becomes a well-defined technique in the future, and specific steps are outlined to perform it correctly, this will require a change in any particular

The MLC is more resilient to change than fixed lifecycle models

process model that attempts to include reuse, but it should not require a change in the MLC itself.

3.4 Lifecycle and Objects

Objects impact the lifecycle

The introduction of object-oriented technology creates an interesting challenge with respect to the definition of a software lifecycle.

There is positive impact...

On the one hand, the emergence of objects should facilitate the definition of, and obedience to, a clear lifecycle model for software. There are two reasons for this:

- Object technology facilitates and promotes reuse, which in turn goes hand in hand with having good requirements and high-level designs before plunging into implementation.

- Because objects implement a clear distinction between their requirements specification and their implementation (thanks to encapsulation), it is easier to construct a good high-level design of a system (i.e., the "interface definitions" of the objects) prior to starting any implementation.

...but also certain challenges due to reuse across applications

On the other hand, an object-oriented system is built from a much looser collection of entities than was the case with structured systems. In older technology, there might exist a hierarchical relationship between applications and some of the subroutine libraries procured from a different source or even developed by the same organization. However, the amount of reuse *across applications* is usually modest because the subroutines represent only a small part of a complete application, and the lack of inheritance limits the impact that a change in one component can have on the others. With objects, this is no longer true. They constitute a very dynamic and interdependent collection of entities, and if applications are indeed designed in a rigorous object-oriented way, reused objects can constitute a large majority of the application's code.

The Software Process   75

What happens in terms of the lifecycle of these objects versus the lifecycle of the applications they form? Clearly, some precautions must be taken.

The lifecycle of objects versus the lifecycle of applications

- Each application still has its own specific requirements and design. It also has some specific "glue" code used to combine lower-level objects into the complete application. Therefore, it continues to have its own lifecycle.

- At the same time, each individual object may undergo *implementation changes* without affecting the functionality of the applications that use them. Therefore, each object also has a lifecycle! For instance, a lack of performance in a scheduler object may result in a complete redesign, even though the interface definition remains the same. If the new implementation is correct, when the redesigned object is placed into service, and especially if linkage between objects is dynamic, every application that invokes it immediately inherits the increased performance, with no other change in functionality.

Based on these two paragraphs, one might conclude that nothing is simpler than just decoupling the two levels. A part of the organization, functioning as the "components shop," can devote itself to creating new objects and providing new and better implementations of old ones (without changing their definition in order to preserve the applications that use them). Another part of the organization functions as an "assembly shop" that creates complete applications from existing objects.

Each of the two can be executed by a different group . . .

Functionally speaking, this is likely to work. However, from the viewpoint of schedules, and therefore of customer satisfaction and competitive viability, this is not sufficient. There must be a synchronization mechanism between the lifecycles of the "using" and "used" objects. Otherwise there is no assurance that the "used" objects will be ready when the "using" object

. . . but synchronization is necessary

needs to be tested and released. Furthermore, if a lower-level object has been modified and contains a bug, the testing of a higher-level object can suddenly go awry.

In other words, the problem with an object-oriented approach is that now an application no longer has a single lifecycle. It has one top-level lifecycle and a multiplicity of embedded lifecycles for each of its objects. This requires that there be a synchronization mechanism between all these interdependent lifecycles.

As is often the case, there are two approaches to solving this quandary: a management approach and a technical one.

Strengthen the planning and review process

- The management approach consists of taking this situation into account in the process definitions, and introducing in the process the necessary planning and review sessions such that the "components" group and the "assembly" group work in concert. For example, the "components" group can publish a schedule (established with input from the "assembly" group) that clearly defines an iterative lifecycle containing the following phases:

    - A phase during which requirements for the creation of new objects and changes in the implementation of existing objects can be input;

    - A design and implementation phase for these objects, during which new requirements cannot be input, unless on a tightly controlled exception basis;

    - A phase during which the new or modified objects are available for testing;

    - A production phase during which the objects in question are available for full usage; this terminates, on an object-by-object basis, upon the release of the next version.

The Software Process

- The technical approach consists of using an object-oriented development environment that includes an *object-level* configuration management capability. For both pragmatic and theoretical reasons, such systems are still fairly rare. In practice, the market for configuration management tools is still a world of statically compiled and linked pieces of code. In theory, since dynamic binding is one of the core principles of object orientation, it is difficult to see how a new version of a "used" object could be created without impacting one of its "using" objects.

Strengthen the configuration management support

At this point in the evolution of the technology then, the best approach is to pay attention to both aspects. When selecting an object-oriented development environment, ask a lot of questions about the configuration management of objects, and at the same time, define a software lifecycle and a process that take into account the specific challenges of the parallel evolution of a family of interdependent objects.

3.5 Defining a Specific Software Process

The objective of this section is not to propose a specific process, but to discuss the techniques used to define or model one. Software development work must be designed, and this should be done *consciously, explicitly, and thoughtfully*.

Techniques to define a process model

- **Consciously.** In many cases, the organization of the work is the result of previous practice or of unquestioned assumptions. For instance, the distinction between analysts and designers is an old practice developed largely in MIS groups. It is based on the assumption that two categories of people with two different sets of skills are needed to handle the earlier tasks of the lifecycle (analysis, high-level design) and the later ones (coding, testing, maintenance). This distinction has often become impractical as the lines between

Challenge previous assumptions

the activities have blurred, and as rapid prototyping techniques and tools are being considered. When designing a work process, one should challenge previous assumptions rather than keeping a traditional organization solely because of the weight of tradition, or to preserve a hierarchy of job grades and titles.

Formally document the process (see notations a few pages down)

- **Explicitly.** The process that is designed should be formally documented so that it can survive management changes and can be reviewed periodically (e.g., once a year). This will force the organization to consider the various elements of the process definition:

    – Which elements of the meta-lifecycle will be included in the process?

    – In what order?

    – What are the roles and responsibilities that must be assigned or created in the group?

    – What are the reviews and decision points that must be scheduled in the project plan?

It takes effort and commitment to define a process

- **Thoughtfully.** The process definition may take a lot of time at first (although process definitions for similar or derived projects in the same organization will hopefully inherit many aspects from the first definition effort). It may require a series of meetings, a significant commitment of an organization's best people, and several drafts of documents containing the following complementary parts:

    – A graphical representation of the overall process;

    – A textual description of the intended content of each task;

    – A project plan template showing how the activities will be organized in time;

    – An organization chart showing who will fulfill which role and who will report to whom.

The Software Process

Criteria for a Good Process Definition

Lifecycle processes must be designed to ensure several important properties.

- *Interfaces between software lifecycle processes and other processes of the enterprise are defined and managed.* For example, if the purpose of the software organization is to develop and support a control system for a succession of machines, how are the software requirements developed in conjunction with the hardware engineering group? If the software is directly visible to the customer of the product, does the process include analyzing the end users' requirements, rather than just the requirements of the machine designers (who may care more for the integrity of their hardware than for the human factors of the user interface)?

    A software process definition must include the linkage to other processes

- *Nontechnical processes are supported.* These include customer relationships, project management, planning, etc. For instance, if the organization handles the support of a mass-marketed product, will there be a hot line for customer calls? And how will customer complaints or enhancement requests be fed and prioritized into the rest of the product lifecycle? Some sort of review board may be required to filter all the input and to assign an "objective priority" to all the requests. Will the organization (perhaps through a Marketing Department) create its own roadmap for the evolution of the product, or will it largely execute a customer-driven enhancement process?

    It must include nontechnical activities ...

- *All lifecycle activities are supported.* For this, see the proposed activities at the top of the MLC diagram on Figure 3.4.

    ... and everything listed in the MLC

- *Support components are included.* Also see the MLC for this.

- *The critical success factors outlined in the MLC are built into the process model* of the key activities and supported by the appropriate members of the organization.

Defining a process will fail if there is not enough buy-in from the contributors

- More generally, *organization commitment* at all levels is present. No lifecycle process will be successful without buy-in from most of the practitioners involved. A lifecycle model that is developed in isolation and whose implementation is dictated by management is destined to fail. Every organization has some process to achieve buy-in on challenging or controversial issues. Implementing a new lifecycle process qualifies as such an issue, and therefore requires significant consensus building. If the organization is converting from another regimented lifecycle model, the task will only require convincing people that the new way is better than the old. But if it involves a significant increase in control over the group's work habits, adoption becomes much more difficult. One of the leading arguments against adoption is that the new process will stifle the creativity of the "free spirits" in the group. To get the buy-in necessary for a reasonable chance of success, a significant amount of training on lifecycle concepts and methods will, at minimum, be required. This will provide the concerned users with a better understanding of the theory, benefits, and mechanics of organized software processes. Analyzing the causes of past quality or schedule problems may also lead the team to its own realization that the situation requires changes.

Management buy-in is also critical

On the other hand, one should not imply that only the "troops" need training and may resist change. In many organizations, individual developers or small teams identify process deficiencies and suggest changes. However, they may be rebuffed when they request management support to provide the necessary resources (whether human or financial) to implement the changes. Sometimes, it is because they have not prepared a coherent justification for their request. More often, it is because management, in its relative ignorance of the concepts and economic consequences of process-driven software development, believes that the developers "just want new toys" or "ought to work harder." Clearly, without management commitment to implement better processes, meaningful change is unlikely.

Task Definitions

When creating an organization-specific model, the first task that confronts the software manager or "software process coordinator" is to decide which activities will be included in the process. The MLC provides only guidelines. The number of separate activities may depend on the size of the organization, its existing structure, the type of software being created, and the nature of its customer base. Any one set of fixed rules is unlikely to meet everyone's needs. Instead, we recommend the following pragmatic approach.

How to define which activities should be included in a custom process model

a) Start by listing the MLC's recommended activities.

Seven steps

b) Examine several of the other lifecycle models proposed in the literature (e.g., the waterfall model or the spiral model) to select some additional activities or some different names that may be more relevant to the organization. If you decide to use the spiral model, it is important to think about how many "turns" of the spiral may constitute the right level of compromise between productivity and completeness.

c) Brainstorm the definition of each activity for the specific organization in question, using the contents of Section 3.3 as an input, as well as other literature and the experience of the participants.

d) From this information, decide which activities may need to be separated into multiple stages.

e) Decide whether there are activities that can be safely eliminated or combined into a single task.

f) Write up the results and submit them to a broad review.

g) Iterate steps c–f until the list is stable.

Describing activities using the ETVX method

As part of this definition work, each activity must be described. Just using a name or a text description is not enough because words are inherently ambiguous. One method that has had some success is called **ETVX**, standing for "**E**ntry, **T**asks, **V**erification, e**X**it." Using this method, each activity has four characteristics:

When to start

- **Entry.** What conditions must be true before it makes sense to begin the activity in question? Entry criteria will usually include the availability of certain documents or deliverables from a previous phase. However, other entry criteria may consist of *events*, such as the fact that certain reviews were held, or that management has authorized the expenditure of resources on the new activity.

What to do

- **Tasks.** These are what we called "subprocesses" in the previous section of this chapter. For instance, the construction activity might include the following tasks: identification of reusable code, programming, cataloguing of new reusable code, and integration. Each of these tasks would in turn be the object of a definition in terms of:
    - Specific methods to be used to perform each task;
    - Tools to be used;
    - Documentation to be accumulated;
    - Guidelines about how much effort to devote to each task; etc.

What to check

- **Verification.** This consists of additional steps to be performed during that activity. These steps verify the correctness of the tasks included in the activity, instead of directly contributing to the production of the deliverables. In the case of construction, verification might consist of a combination of unit testing, code reviews, integration testing, test coverage analysis, and performance profiling. Note that these tasks do not necessarily follow the productive tasks, but may be interleaved with them. For example, unit testing and code reviews should occur immediately after the programming of each module, while the rest of the verifications require that at least some integration take place first.

- **Exit.** These are the criteria that allow the project manager to pronounce the activity complete. Because the exit criteria of one activity will typically form the bulk of the entry criteria of one or more other activities, this is the gating factor that allows the process to move forward. Again, some exit criteria simply consist of the existence of one or more deliverables, while others require a formal management decision. A nice way to combine the two concepts is to require that documents (e.g., requirements specifications) be signed off by appropriate levels of management. In this manner, one avoids taking the mere existence of an analysis document as sufficient proof that it meets its goals. Instead, the sign-off procedure will ensure, if applied critically, that the document has been reviewed and that any noted improvements have been processed, allowing its formal acceptance.

How to close that step and move on

Graphical Representations

As the old adage says, "a picture is worth a thousand words" (a famous, if unattributed, variation specifically suited to the world of computers is that "a picture is worth 1,024 words"). When it comes to presenting, explaining, teaching, or debating a process definition, a graphical representation of the process is often much more suitable than pages of dry text.

Graphics are more expressive than narrative text

ISPW work on process notations

Many graphical notations have been used to describe software processes. Indeed, a whole conference (the International Software Process Workshop, or ISPW) is devoted to the definition of software processes. Each of its participants is essentially a proponent of a different description of processes, often with at least some graphical component. Without going into a complete review of this new field, it is worth mentioning three different notations, each of which was initially developed for a somewhat different purpose:

Rehabilitating an old notation

- **Flow Charts.** Flow charts are perhaps the oldest notation used in software. For analysis and design purposes, they have lost some of their luster because of their low level of abstraction and their tendency, when used without any rules of "well-formedness," to promote the creation of "spaghetti code." However, they can still be useful in some cases. Figure 3.5 is an example of a high-level process definition using this notation. Note how the decision boxes shown in the chart correspond to exit criteria. They control the iteration of some task, or groups of tasks, until an acceptability criterion is met.

Using actigrams

- **SADT diagrams.** SADT stands for Structured Analysis and Design Technique. SADT diagrams come in two flavors: *datagrams* (in which the nodes are data elements, and the arrows denote controls) and *actigrams* (in which the nodes are actions, and the arrows denote information). In the case of software process modeling, only actigrams are typically used. Figure 3.6 is a simplified version of a process definition actigram found in [Freeman 87]. In this notation:

The Software Process

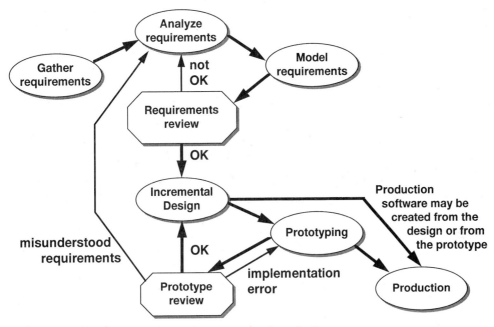

Figure 3.5 – Flow Chart Representation of a Process

- Each box is a function or task.

- An arrow coming from the left represents an *input* (e.g., a deliverable from a previous task).

- An arrow leaving on the right represents an *output* (e.g., a deliverable from this task).

- An arrow coming from below a box represents a *mechanism* (e.g., a human or computer resource) that can be applied to the task.

- An arrow coming from above a box represents a *control* (e.g., the decision to perform the task, or the policies and procedures that apply to it).

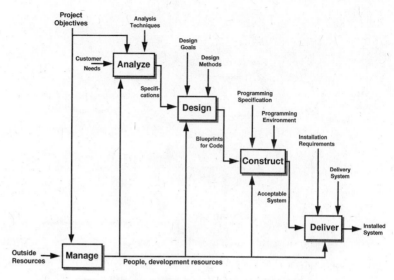

Figure 3.6 – SADT Representation of a Process

A more complete notation that captures the ownership of each task

- **Team Process Charts.** This is a lesser-known but very interesting technique that provides a clear representation of the responsibility of various participants in the process. The technique has its origin in the "quality teams" or "quality circles" first instituted by Japanese companies, then by American ones, in order to improve manufacturing processes. One of the major limitations of the other two notations presented is that they fail to completely resolve who owns a task or decision. This should typically be captured in a separate document, but often ends up being omitted altogether. Team process charts avoid this problem because a task does not exist unless it is allocated to a specific group.

A description of team process charts

Figure 3.7 provides an example roughly equivalent to that of Figure 3.5. Each vertical band corresponds to the tasks

The Software Process

allocated to a specific person or group (some people use horizontal bands instead, essentially rotating the whole diagram 90° so that progress consists of moving from left to right rather than from top to bottom). The names of the people or groups in question are listed at the top of the columns, and as shown in the figure, some notion of organizational hierarchy can be included. There are five kinds of shapes in this diagram: rectangles denote *tasks*, rounded rectangles denote *meetings*, diamonds represent *decisions*, circles represent *milestones*, and dog-eared pages represent documents, or in general, any tangible deliverable (including code). When a task or meeting belongs to several people or groups, the corresponding icon is extended to span the relevant columns.

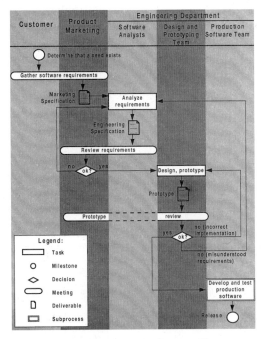

Figure 3.7 – Team Process Chart

More Information on Process Definition

Additional sources

There are several good sources of information that should be consulted about process definition guidelines:

STARS

- One is the **Software Technology for Adaptable, Reliable Systems (STARS)** program of the U.S. Department of Defense. It is, of course, readily accessible to military organizations and their contractors, but since the STARS program aims to elevate the process capability of the whole U.S. software supplier chain, its information is also available to many other interested organizations.

ESPRIT

- For European companies and organizations, work on process models has been performed with European Community funding through the **European Strategic Programme for Research in Information Technology (ESPRIT)**.

Sandia Labs

- In the U.S. semiconductor industry, **Sandia Laboratories** (Albuquerque, N.M.) in particular has worked on definitions of the task **attributes** that should be listed in a process model.

ISPW

- There have been multiple occurrences of an **International Software Process Workshop (ISPW)** devoted entirely to this topic. One of the ISPWs defined a sample problem and challenged its members to propose solutions (i.e., process definitions) using the formalism of their choice. These solutions have been published in subsequent workshops and use a wide variety of concepts and notations with more or less formal (i.e., mathematical) content.

3.6 Software Process Enactment

Guaranteeing that the process is followed

About 1990, an increased understanding of the importance of a good software process combined with the emergence of powerful Integrated Project Support Environments (IPSEs) to yield a new concept: **software process enactment.** One of the dictionary definitions of the verb "enact" is "to ordain, decree." This new buzzword thus refers to the ability to guarantee that the process is followed, rather than relying solely on the developer's willingness to follow a definition that exists only on paper.

Although this might make one think of Orwell's "big brother is watching you," there are different degrees of process enactment. Participants in the STARS program (see Glossary) have done significant work to define *five degrees* and how they relate to the capabilities of an IPSE:

Five degrees of enactment

1) *Manual.* Based on written documentation, the developer only follows the process at his own discretion. The CASE environment or tools have no knowledge of the existence of a process. Enforcement of the process is also left to human oversight (e.g., through reviews).

Levels 1 and 2 are voluntary

2) Same as 1), but with *on-line help* provided by the environment in the form of process manuals and forms. The developer is still able to use or ignore what he wants.

3) Same as 2), but in addition, the environment contains *aids*, such as shell scripts, to automate the correct sequencing of some of the tasks, and to translate data formats between tools. For instance, if the process says that each module must be compiled with some performance profiling tool (in order to allow an analysis of where the program spends its time), the scripts can take care of this so that the user does not forget it. Another example might be to automatically invoke the text editor on any module that compiles with errors, and if the module compiles correctly, to invoke the configuration management package's check-in tool.

Levels 3 and 4 introduce some automation, at the user's discretion

Note that this capability is typically included in IPSEs that are sold independently from specific CASE tools, since the supplier of the IPSE needs to include some shell-level customization capability to adapt to an indefinite number of CASE tools. Chapter 6 will include more details on this topic.

4) Same as 3), with the addition of *interactive,* but *passive,* enactment by the environment. This means that the IPSE collects metrics, presents forms to the user, spawns ancillary tasks, etc., but strictly in response to user requests to accomplish such tasks. Although the IPSE now provides more help and contains a fairly complete representation of the process in order to do so, the user still sits in the driver's seat.

At level 5, the environment controls what the developer can and cannot do

5) This is the last, definitely "big brother"-ish stage, more benignly called *active enactment.* At this stage, the IPSE completely drives the process, keeping track of which activities have been performed, and releasing work units to the user as demanded by the process rules. In its complete form, this active enactment means that the user comes to work in the morning and sees on the workstation screen a series of icons corresponding to tasks that are allowed. Tasks that cannot be performed because some preconditions have not been met (e.g., developing a piece of code corresponding to an unapproved design) are not displayed at all; and there is no way for the user to sidestep the environment and start them.

Work in progress

Formalizing these levels has not only served to stir up a healthy discussion of which level of discipline is desirable or acceptable in a software organization, but it has also triggered the creation of some prototypes of process-based environments. For instance, the Software Process Management System (SPMS) was first developed at Lockheed in 1990-91 (under the aegis of STARS) to provide a demonstration of process support and enactment. It was later enhanced through further work at Science Applications International Corporation (SAIC). This interesting project is described in [Krasner 92].

3.7 Software Process Improvement

The emergence of the software process as an area of concern is still a recent trend, and most organizations have not yet achieved a mature definition of their processes. The Software Engineering Institute (SEI) at Carnegie-Mellon University has collected information about the best practices followed around the world, and organized them into a comprehensive and general-purpose approach to software process improvement, the **Capability Maturity Model (CMM)**.

The Capability Maturity Model

The SEI CMM

According to the CMM, there are five levels of maturity in a software organization's capabilities. These are defined in terms of processes (see Figure 3.8):

Five levels of maturity

- Level 1 is officially called the "initial" level. It is less charitably dubbed the "chaotic" level. An organization at this level may well be able to produce software, but there is no predictability from one project to the next. About eighty percent of the world's software organizations are at this level.

The initial level

- Level 2 is the "repeatable" level. The organization has learned to be consistent from one project to the next. This does not mean that the organization is particularly efficient, but rather that it has codified some of its practices and can execute them in a comparable manner from project to project. To progress from Level 1 to Level 2, an organization needs to pay attention to a certain number of **Key Process Areas (KPAs)**:

The repeatable level

- Managing requirements;
- Project planning;
- Project tracking and oversight;
- Subcontract management (if applicable);
- Software quality assurance (including quality metrics);
- Software configuration management.

The defined level

- Level 3 is the "defined" level. The process is characterized and fairly well understood by all participants (including managers and developers). To progress from Level 2 to Level 3, the key process areas are:
  - Adoption of a process focus by the organization;
  - Definition of a software process for the organization;
  - A training program to teach the process;
  - Integrated software management;
  - Software product engineering;
  - Intergroup coordination;
  - Peer reviews.

The managed level

- Level 4 is the "managed" level. Not only are all the activities defined, but their execution is controlled and measured. This is achieved through the following key process areas:
  - Software quality management,
  - Quantitative process management.

The Software Process

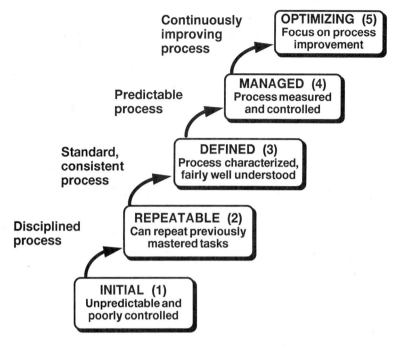

Figure 3.8 – The SEI Capability Maturity Model

- Level 5 is the "optimizing" level. There is no room left for a quantum jump in capability, since an organization at this level has implemented all the techniques and methods known to enable quality software development. However, the organization routinely practices further improvement by continuously refining its procedures, metrics, etc. The key process areas to reach this level are:

    The optimizing level

    — Process change management,

    — Technology change management,

    — Defect *prevention* (as opposed to defect correction only).

Continuous improvement requires setting new goals from time to time

Continuous improvement requires an ongoing focus on monitoring and analyzing quality levels for the purpose of identifying and acting upon new opportunities. For instance, metrics that have reached such high and stable levels as to provide essentially no new information should be dropped. They should be replaced by other measurements on the basis of which new, meaningful plans can be made.

Benefits and Costs of SPI

A complete approach based on improvement cycles

We call **Software Process Improvement (SPI)** the complete approach that consists of identifying an organization's current software process capability, putting in place an improvement plan, and conducting and monitoring the improvements. This approach follows principles that have been used to construct many other quality improvement programs, principles first stated in the "improvement cycles" of Walter A. Shewhart and W. Edwards Deming (see [Shewhart 31] and [Deming 86]). By applying this SPI approach, an organization will:

- Improve the quality of its software products;
- Improve the accuracy of its cost and schedule estimates;
- Increase customer satisfaction;
- Increase productivity and reduce the overall cost of development.

Greater maturity can cause an apparent decrease in productivity

The last point is worth a comment. As an organization moves into higher levels of the SEI CMM model, total productivity is often perceived to diminish because of all the time spent in collecting metrics, writing documentation, performing project or deliverable reviews, waiting for a step to be signed off before starting the next one, etc. There are even rumors about a division of a major computer company that had been rated at SEI Level 5 (one of only a handful of companies worldwide to have reached this level), but which now produces only one line of code per developer per day on average. What should be made of such stories?

First, the trade-off between quality and productivity is often couched in the wrong terms. Preserving productivity first, when quality is insufficient, simply amounts to producing more bugs per day than your competitors! Just as "97% fat free" is a misleading label when applied to a type of food that normally contains only three percent fat anyway, claiming a productivity of 50 lines per developer per day is misleading if 45 of them have to be thrown away later because they correspond to the wrong design.

But is productivity with insufficient quality real?

Secondly, the metric of lines-per-person-per-day is obviously biased toward software processes in which coding still represents the bulk of the effort. There are two reasons why this is the wrong metric:

Metrics are changing

- Since errors in analyses and designs are more costly to repair than errors in the code itself, creating a correct analysis and a correct design has a much bigger impact on the overall process.

- Reuse (especially the reuse of objects) considerably decreases the total amount of code that needs to be written.

Thirdly, the level of process definition required by an organization depends on its size, and the nature and criticality of its products. The above-mentioned division of a nameless company provides extremely critical software, including some used in military command and control systems. In other words, one line per person per day may be quite acceptable in this case, provided that it is *at least 99.99%* probable that each one of these lines is correct! Indeed, the SEI model has been produced with funding from the U.S. Department of Defense, largely as a reaction to the level of process immaturity observed

Compromises may be necessary depending on the application domain

in software contractors to the military. This is a domain in which high costs can be tolerated (for critical elements), provided such costs guarantee quality. In other domains, a different estimate of spending limits might apply. Ultimately, in a competitive arena without life-and-death impact, the decision rationale comes down to whether it makes *economic* sense to spend more in order to achieve a greater level of capability.

The Process of SPI

What does SPI include?

Carrying out a Software Process Improvement involves the multidisciplined activities of assessing, analyzing, planning, implementing, and monitoring changes to the organization's software practices. This section profiles the SPI approach in terms of:

- Who gets involved and the various roles of the participants;
- The range of technical activities the program covers;
- A sketch of the generic work plan for the program;
- When and how the approach is customized to the organization using it.

A simplistic view: measure, improve

An intuitive approach to SPI might simply consist of an iteration between two steps: measure and improve. Furthermore, a single person or group would be tasked to perform these two steps. However, this is far from sufficient, and this limited approach would either fail or provide little improvement.

Begin and end each cycle with management commitment

The SPI process (SPI is itself a process since it has steps that must be executed in some organized manner) begins and ends with *management*. This means that from the outset, management must commit to SPI and be involved. This will occur if management values the results of the process and recognizes its impact in terms of lower costs, improved competitive position, suitable return on investment, etc. Management's role in the process is to set directions and priorities, to establish policies and standards, to charter new functions, to invest the people in these functions with the appropriate authority, and to provide the required resources.

At the center of the SPI process is its **champion** (or evangelist, to use a different metaphor) within the organization. It is this person's or group's role to understand the current software process, and to identify and "sell" throughout the organization (including management and the developers) various alternatives for improving it.

The role of the champion

The primary focus of the SPI process is the organization's software practitioners—the software-related units plus the organizational context in which they operate. Their role is to examine the current approach to software development, production, and support in light of others' experience, to collect and analyze data about results achieved with the current approach, and to help select—and then implement—changes and additions to the current practices. Note that the SPI approach does not prescribe specific methods or tools to develop or maintain software; rather, it is committed to improve the results achieved with the current methods and technologies. The primary concern is whether or not the overall process is complete, well defined, and suitably integrated, and whether or not all participants in the process have a common understanding of it. This being said, SPI also establishes a reliable approach to determining when and how to change or replace any method or tool used.

The role of the developers

An improvement cycle, as defined by Shewhart in 1931 and since used by Deming, is a four-step process consisting of a *measurement* step, followed by *analysis, change,* and *implementation* steps. This has been further refined by SEI and by SEMATECH into the six-step process depicted in Figure 3.9.

SEMATECH's six-step process

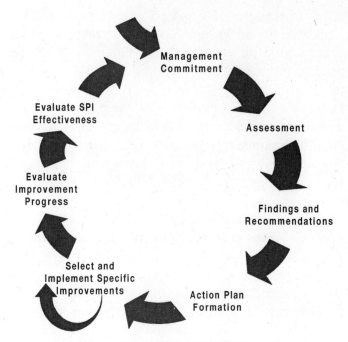

Figure 3.9 – The Six-Step SPI Process

Start at the top

1. *Obtain management commitment* and approval of the program. This may seem hard to achieve, since the lack of hard data regarding the maturity of the organization may make this a typical "chicken-and-egg" dilemma. In such a case, the best practical advice is to start with a modest effort, one that does not require the commitment of great resources. Proceed through the rest of the cycle, then wait for the second "turn" (around the improvement cycle) to become more ambitious, now that solid evidence of the need for improvement has been collected.

Perform an objective, broad assessment

2. *Assess the maturity of the current software process* in terms of the SEI maturity model. The SPI process deals not only with the development of software, but

also with the following functions, which are integral parts of the software process:

- Project planning and management;
- Collecting and using metrics;
- Inspecting results during all stages of the process;
- Prerelease and regression testing;
- Documenting (both for users and for developers/maintainers);
- Managing the product's configuration and releases.

3. *Identify what improvements* the software process needs, and those actions that will achieve the improvements. Some representative examples of improvements are: *Look for possible solutions*

- Peer reviews of programs prior to testing;
- A steering committee that guides the change process;
- Controls over product modifications and version releases;
- Definition, collection, and use of various metrics.

4. *Select change actions* identified in Step 3 and implement each in turn. The participating organization determines which improvements will yield the greatest benefit, the schedule for implementing them, quantitative objectives for each improvement, and the resource level needed in each case.   *Prioritize changes and act*

5. *Measure results* of each change implemented and adjust the changes until improvement objectives are achieved.   *Close the loop*

Evaluate and continue

6. *Assess progress* and adjust the SPI process itself. Repeat Steps 4-6 and include Steps 3, 2, and 1 as needed.

Figure 3.10 shows how the various participants—managers, software practitioners, and SPI champions—are involved in each of these six steps.

Step	Participant		
	General Management	Software Process Change Agents	Software Practitioners
Step 1: Get Management Commitment and Approval	Recognize needs and opportunities. Approve and commit to the SPI process. Charter a steering committee.	Establish need. Get approval of and commitment to SPI process from management and practitioners.	Understand the SPI process objectives. Recognize the commitment of management to the SPI process.
Step 2: Assess the Current Software Process	Respond to SPI process briefings. Subscribe to the assessment results.	Assess present software process (involves selected practitioners).	Respond to issues and questions as required.
Step 3: Create an Action Plan Based on Assessment Findings	Establish standards and policies. Approve the strategic plan of improvement actions. Charter a group responsible for executing the action plan.	Develop a strategic improvement plan that may include a process model and description, standards and policies, a strategic plan of action, and metrics to assess progress.	Provide input to the strategic improvement plan and priorities.
Step 4: Select and Implement Improvement Action	Monitor implementation of the plan. Ensure continued resource commitment.	Tactically plan and schedule improvements from the action plan; set objectives.	Pilot-test process improvement/metrics that require it. Implement planned improvements.
Step 5: Assess Results Achieved and Adjust as Necessary	Monitor and evaluate progress.	Analyze and report progress/problems.	Measure results.
Step 6: Assess/Adjust the Improvement Process. Repeat.	Provide motivation.	Reassess and adjust improvement process.	Suggest adjustments.

Figure 3.10 – Roles and Responsibilities in the SPI Process

Tailoring the Approach

The SPI process may need to be tailored

A "cookie-cutter" approach to software process improvement is unlikely to work, given the wide disparity in business domains, sizes, and maturity of the organizations involved. Instead, it is important to recognize that the SPI approach can be tailored without losing its usefulness. The SPI process seeks to ensure that the needs of each participating organization are met in the most effective manner possible.

- The SPI approach is based on SEI's generic work for the software industry, but several organizations throughout the world are now applying this work in different contexts. Various consulting groups have been qualified by SEI to apply the CMM, and by taking into account the idiosyncrasies of their respective domains of activity, these consultants improve the applicability of the CMM.

    There is some domain-specific help available

- The very first step in implementing the generic approach within an organization is to rework it to fit that particular organization. Both managers and software practitioners participate in making the required adjustments, thereby developing a customized form of the approach to which all the participants subscribe.

- When applying Step 4, the organization has great latitude in deciding which changes are the most likely to be effective and economically feasible in its specific context.

Motorola is one example of a company in which the general SPI concept has been successfully applied and customized. The **Quality Systems Review (QSR)** is Motorola's general approach to assessing itself and its suppliers with respect to quality practices. Motorola conducts about sixty QSRs per year, with several teams of trained assessors who spend a significant part of their time applying the process. The QSR is divided into ten "subsystems," with Subsystem 10 devoted to software. In turn, this subsystem is divided into 12 "elements," reproduced in Figure 3.11. These elements essentially map directly to the areas of focus defined in the SEI CMM as enabling an organization to progress from Level 1 to Level 2 (which, although it might seem a modest goal, would place the organization in the top twenty percent of the industry with respect to software process maturity). And finally, in order to address

Motorola's QSR: a solid process improvement effort

each of the 12 elements, both the assessors and the organization being assessed have to answer several questions and compile evidence to support each answer. There are a total of 99 questions among the 12 elements of Subsystem 10, which thoroughly cover the components and quality of the assessee's software process.

1.	Is an approved, documented process used to guide the development and maintenance of all software that impacts Total Customer Satisfaction?
2.	Are software project planning and control mechanisms in place and followed?
3.	Is software developed as part of a total system using a phased development approach, intermediate deliverables and review and approval based on entry and exit criteria?
4.	Is software developed in support of documented (formal, written, approved, updated and available) requirements, with conformance to these requirements verified?
5.	Is software developed and maintained under documented plans for configuration management and change control, including installation and customer configuration?
6.	Is software developed using proper tools and documented, approved procedures for security and information recovery, including disaster protection?
7.	Does software undergo system/acceptance testing by individuals or organizations not directly involved in the design or implementation of the product being tested? Does testing reflect customer usage?
8.	Are there established goals for software quality including six sigma performance as the overall goal? Do the measurement systems provide tracking of progress towards these goals as well as highlight quality issues from the customer perspective?
9.	Does the quality assurance organization act as a customer advocate in software matters by assuring conformance to customer requirements and specifications and proper execution of the approved development process?
10.	Is there a mechanism used to ensure continuous software development process improvement?
11.	Is there a capability improvement program in place for all software organizations, including deployment and assessment of training?
12.	Is the process used by subcontractors under control and is conformance to requirements of subcontracted software verified?

Figure 3.11 – The Twelve Elements of Motorola's Quality Systems Review for Software

3.8 Conclusion

Software process definition and improvement are key to success

Software process is a fundamental concept, as the length of this chapter indicates. Attempts to create sophisticated software development environments without the underpinnings of a well-defined process have failed to meet quality and productivity expectations. In reaction to these negative experiences, organizations like SEI have spearheaded the effort to make process definition and process improvement the cornerstones of software development.

The concepts of process definition and process improvement are now ready to move from the research centers and government-sponsored think tanks to the product development departments of the "real world." This needs to be a deliberate evolution, based on several well-defined steps:

They are ready for "real-world" application

- Education about process concepts, lifecycle models, and their impact;
- Design of a process that suits the organization's needs;
- Process enactment at the desired level through the proper methods and tools;
- The establishment of a continuous software process improvement program.

After taking these steps, the organization can proceed to select and use the technical components of its development and execution environments. The framework for success is there!

Chapter 4

Development Activities

In Chapter 3, we talked about the software process models in general. The emphasis was not on the contents of the specific steps of the process, but on how the process should be defined and how it should evolve. It is now time to become more specific about the activities that constitute the development and delivery of object-oriented software. Some of these activities will be specific to an object-oriented context, while others will not.

Getting more specific about the steps

4.1 Overview of the Development Activities

Although the exact number of phases depends on the specific model selected by an organization, there are generally three main creative phases: analysis, design, and construction. In addition, there are some non-value-added but critical phases, like validation testing and packaging. We will follow this generic model in this chapter. The reader will have to map these phases to the particular lifecycle model and terminology used in a specific organization.

We are using a generic list of activities

What are the deliverables and artifacts?

A good way to describe a development activity is to focus on its **deliverables**: what does it produce? More specifically, we will talk here about **artifacts**, which are those visible deliverables of an activity that constitute some representation of the system being developed (as opposed to a report from a technical review, which is a *deliverable* but not an *artifact* because it is not a part of the product itself). Artifacts of the software development process include, but are not limited to:

- Graphical and textual representations of the requirements;
- Graphical and textual representations of the design;
- Source and object code.

Table 4.1 provides a summary of the analysis, design, and construction activities, listing detailed activities they may include and artifacts they may produce.

Analysis versus Design

There is a fundamental difference between analysis and design

There is an extensive (and almost religious) debate surrounding the difference between analysis (requirements) and design. In general, the distinction between those two concepts is smaller when using the object-oriented approach than in the structured approach. However, we believe that the use of common concepts (objects, classification, messages, etc.) across the two phases has caused some experts to overlook a fundamental difference between them.

The meaning of "analysis"

The Greek root for the word "analyze" means to "take apart." It is the opposite of "synthesis," which is to "put together." In chemical analysis, one separates the elementary components of a complex substance in order to determine its composition. In the same manner, the analysis of a set of user requirements consists of "taking them apart" in order to discover their constituents. This activity remains solidly within the **problem**

Lifecycle Phase	Activities	Artifacts
Requirements Analysis	• Obtain reference documentation, including existing models and specifications of previous or related systems, if any. • Interview domain experts and other sources of requirements. • Create usage "scenarios." • Produce information, state, and process (functional) models of the problem domain. • Investigate technology constraints (choice of operating system, database, etc.). • Produce high-level project plan. • Produce high-level test plan. • Review and approve all deliverables.	• Requirement specification document, consisting of: (a) Textual requirements, including statements of needs, scenarios, glossaries, etc. (b) Analysis models, as specified in the selected method (see Section 4.3) • User manual (initial draft).
Design	• Develop architecture for the solution. • Produce information, state, and process (functional) models of the solution domain. • Prototype various solutions to select one that will meet the requirements, or review the proposed design to ensure that it will meet the requirements. • Produce database designs, if applicable. • Identify reusable components. • Produce detailed project plan. • Produce detailed test procedures. • Review and approve all deliverables.	• Design models, as specified in the selected method (see Section 4.4). • Revision of user manual in view of design decisions, as applicable. • Design rationale. • Database schemas. • User interface mock-ups (screen designs). • Detailed project plan. • Test plan.
Construction	• Create source code (manually or via tools). • Create and catalog new reusable components. • Process source code through compiler, linker, and static quality analysis tools. • Perform unit tests. • Integrate complete product or system. • Review and approve all deliverables.	• Source code. • Executable code. • Configuration management repository entries. • Reusable code catalog entries.
Validation	• Compute metrics on code. • Produce *golden test results* (basis for future regression test result comparison). • Document and correct problems found during validation. • Approve product for delivery.	• Corrected source and executable code. • Defect reports and enhancement requests. • Documentation of known defects and limitations.

Table 4.1 – Development Activities and Produced Artifacts

Analysis looks only at the problem domain

domain: no new objects should be invented just because a computer-based solution is being considered. Only objects and activities that preexisted the computer system are examined. When one switches to creating new objects, messages, etc., used to build the system (the actual set of artifacts depends on the design method), a fundamental shift occurs. We are now in the **solution domain**, applying creative skills to generate new constructs that the problem space did not imply.

A clear distinction helps make analysis models reusable

The importance of this distinction lies in the reusability of the analysis models. By avoiding the confusion between analysis and design, one can unequivocally state that the analysis model remains entirely valid even if fundamentally different options are selected for implementation at a different time or in a different organization. The allocation of manual and automated tasks in, for example, a material handling system could thus be reevaluated in the future, resulting in a completely different system design. Yet the basic analysis of the problem is the same, and provided the analysis-design distinction has been maintained, it can be reused in both generations.

4.2 Object-Oriented Methods Overview

Before Objects

Previous methods come in two flavors

One way to evaluate object-oriented methods, and to examine their benefits, is to contrast them with their predecessors. Two classes of methods had established a prevalent position during the 1970s and 1980s: **functional decomposition,** and **structured analysis and design:**

- "The underlying strategy of **functional decomposition** consists of selecting the processing steps and substeps anticipated for a new system, using previous experience as a guide. The focus is on *what* processing the new system

Development Activities

requires. The analysis then specifies the processing and functional interfaces." [Coad 90]

- The data flow approach, which is one of the main forms of **structured analysis and design,** involves the mapping of the "real world" into data flows and data stores. "This mapping requires the analyst (and, more significantly, the customer) to follow the flow of data whenever looking at the real world, and map that flow into subsequent analysis and specification." [Coad 90]

Thanks to their earlier appearance, functional or structured methods (which we will call *traditional* methods) are more common and more generally accepted than object-oriented ones. They are well developed, with a fairly stable set of graphical notations and many sources of training, consulting, and automated tool support. Therefore, if maturity were the primary or sole criterion used to select a method, these traditional analysis and design methods would remain the most acceptable ones.

The traditional methods have the advantage of stability ...

However, these methods are proving to be insufficient, especially in technical domains like real-time control systems, computer-aided design (CAD), computer-integrated manufacturing (CIM), and telecommunications. Reasons for their shortcomings include:

... but they are proving to be inadequate

- Functional decomposition only looks at the specification of mappings between inputs and outputs. It can be successfully used for analysis that deals with the static coupling of serial processes, which is typical of batch-oriented Management Information Systems (MIS) applications. But because it concentrates on separately analyzing each activity step and not on how these steps interact in time, functional decomposition is not sufficient to specify completely a real-time application.

- As for structured analysis, "following the flow of data" is not a basic method people use to organize and manage complexity when examining a problem space.

The problem is separation of data and function

In both cases, the analyst or designer concentrates separately on the issues of *function* and *information* contents, without considering the interaction of these aspects (ripple effect) on an application. This leads to many difficulties:

- Adding, deleting, or modifying data elements can require modification of every part of the program that accesses this data. And because there is no limitation on how many parts of the program are exposed to the data structures, this impact can be extensive.

- For the same reason (many parts of the program can touch the same data elements), an error on one part of the program can damage data in ways that are very difficult to identify.

- The resulting program is not modular; it consists of a number of subroutines and data definitions having complex links with one another. Adding or modifying features cannot be done by discarding or adding a single module in an incremental manner. Adding new functionality requires changes to the data structures, and vice versa.

- It is also virtually impossible to isolate a subsystem for testing purposes; everything depends on everything else.

- Interaction with the customer or user during analysis is difficult. Users will not cleanly separate information from processes when describing what they want the system to do.

The Concept of Objects

The idea that a system can be viewed instead as a population of *interacting objects,* each of which is an atomic bundle of data and functionality, is providing an increasingly attractive alternative for the development of complex systems. This is a radical departure from prior methods, and one that requires a completely different approach to understanding systems. In fact, this is actually a much more natural way to consider systems, unless one's thinking has been steeped for too long in the traditional methods. In that case, some "unlearning" of the old ways is necessary before one can appreciate the true elegance and simplicity of the object-oriented approach.

View a system as a collection of interacting objects

A basic understanding of the object-oriented approach is necessary to understand the rest of this chapter. Numerous books (e.g., [Cox 86] or [Taylor 91]) have been written to communicate this information; and the application of object-oriented concepts to analysis and design—not just to programming—has fostered the emergence of more books on the topic. Here we will provide only a generalization of the basic ideas, and refer the reader to the bibliography for more detailed sources.

A summary of the basic, language-independent ideas

- An **object** is defined as some bundle of data (also known as the **attributes** of the object) and operations (also known as its **methods**) that can act on that data. Objects communicate with one another by sending messages which request that the receiving object execute one of its defined operations.

Combine data and operations

- **Encapsulation** means that the operations form a protective, isolating "capsule" or layer around the object's contents (its attributes or the detailed definition of its methods). Other objects are not permitted to know or directly access the internal representation of an object, thus providing extensive decoupling within a system. This is also called **information hiding.**

Protect the object's contents

Classes are abstractions of objects

- Objects are grouped into abstractions called **classes**. All objects of the same class have the same operations and the same list of attributes, with the individual objects having different sets of attribute values.

Class hierarchy

- Classes are organized hierarchically, which permits **inheritance**. All objects of a "subclass" automatically possess the attributes and methods of the "superclass," unless these are explicitly redefined in the subclass.

There are other, language-dependent characteristics

There are several other common characteristics of object-oriented systems. Often, they are consequences of implementation choices made in object-oriented programming languages like C++ or Smalltalk rather than inherent aspects of the approach. **Dynamic binding** is such a characteristic. It is the ability to send a message to an object without knowing in advance which particular method will be invoked. This is necessary since an object may belong to different subclasses of the same class, and it is not known at compile time which particular method will be executed. This provides considerable flexibility, for instance, in allowing new objects and classes to be added to an existing application. But it is not as essential to the object concept as the above characteristics.

A Brief History

Simula 67

In the 1960s, O.J. Dahl and K. Nygaard of the Norwegian Computing Center (Oslo University) developed a language called Simula 67. As the name indicates, the original purpose was to create simulation programs. They deduced that in the world of simulation, where "real-world" objects interact and their individual attributes (e.g., state, speed, and position) are modified in response to events "received" by this object, a new concept of modularity was much more productive than the previous methods. They therefore developed a special language to express this new concept efficiently.

After this pioneering effort, the concept of objects remained limited to a few interested communities. In the 1970s, David Parnas defined the concept of information hiding as a method of decomposing software into a structure that provided better tolerance to change. This is a refinement of an idea of modularity that can be traced back to many sources. By the mid-1970s, the concept of **abstract data types** was defined simultaneously in several papers and books in Europe and in the U.S., but without yet being generalized to the current concept of objects. Also in the 1970s, Alan Kay et al., within the Learning Research Group of the Xerox Palo Alto Research Center, developed Smalltalk, an object-oriented programming language that did much to spread the initial popularity of object-oriented techniques. Smalltalk followers were a small group of devotees during the 1980s. Then, as low-cost versions of the programming environment became available on personal computers, and as the concept of objects entered the mainstream, the language gained more and more disciples.

Slow evolution during the 1970s and early 1980s

Suddenly, in the late 1980s, two phenomena occurred. First, several other object-oriented programming languages emerged. They included Eiffel, Objective C, C++, and others. Of those, C++ became the most popular, both because of its roots in the C language and because of the support of AT&T Bell Laboratories. Secondly, object orientation ceased to be associated solely with object-oriented programming (OOP). The notions of object-oriented analysis and design were abstracted from the original OOP concepts and applied, even in cases when the implementation language was procedural (Pascal, C, Ada, or COBOL).

Later changes increased industry acceptance

Advantages

Taking an object-oriented approach to the analysis, design, and construction of an application or system yields the following benefits:

Several benefits of OO methods

Better mapping to the "real world"

- Because the concepts captured by the analysis correspond directly to the "real-world" objects the users understand, the analysis takes less effort and is more easily verified by the users. This is especially true for application domains in which the user manipulates real "things" (e.g., parts on a manufacturing floor, freight cars in a railway control system, sensors and actuators in a control system). Analysis and design techniques such as "finding the objects" can easily be understood by users who have little knowledge of systems analysis; therefore, less translation of the needs is required.

Reuse of analysis and design

- By using inheritance during analysis and design, and not just during programming, one can save time by reusing fragments of prior analyses and designs. Reuse is done at the class level, which is more beneficial than reusing only procedures or data definitions.

Mapping to object languages

- The encapsulation and inheritance properties of the analysis and design can translate directly into the features of an object-oriented programming language; there is less "conceptual distance" between the three levels of abstraction than with traditional approaches.

Separation of interface and implementation

- Thanks to encapsulation, the designs of the various objects are independent of one another. As long as an object's communication interface does not change, its internal design has no impact on other objects.

Challenges

The methods' instability is an obstacle

At the same time, the lower degree of maturity of the object-oriented approach makes the benefits of this approach harder to realize. Moreover, there is a confusing array of methods being proposed, as shown in [OMG 94]. The industry has not yet settled on which are the best practices, notations, and rules—or even on common definitions of widely used terms such as "class" and "object." Object-oriented languages are by nature

Development Activities 115

more precisely defined, but they do not yet have the blessing of national or international standards committees. In spite of their benefits, many professionals remain uncomfortable with the idea of switching from structured to object-oriented methods and languages.

The questions that potential new users must resolve for their organization in order to adopt the new paradigm include:

Questions new users should ask

- How difficult will it be to learn the new methods?
- After a reasonable learning curve, will all the participants in the effort be able to master them?
- How effective will the method be in bringing about a new generation of quality systems or software in a timely manner?
- How should one interpret the conflicting evidence presented in the popular technical forums? (In other words, is the glass half empty or half full?) How can one avoid falling into the traps encountered by others?

Later in the book, we will devote an entire chapter to the issue of migration to object-oriented technologies and advanced software development processes; therefore, we will not belabor the same points here. In summary, we will show that the adoption issue can be resolved through a combination of:

Summary of adoption challenges

- Careful selection of the analysis and design methods to be applied, based on recent information (since the field is still changing) and on criteria that include ease of use and the availability of training, support, and consulting.

Selecting methods with care

- Healthy skepticism with respect to the snake oil salespeople. Object-oriented technology is sometimes oversold. One cannot expect reuse to provide benefits during the first attempts to use object-oriented techniques in a new domain

Beware of hype

since at that point, there is nothing to reuse. Yet this caveat is often omitted.

Pilot projects and partnerships

- Adoption of a transition strategy based on pilot projects and on working, whenever possible, with more experienced partners.

Project and risk management

- Applying good project and risk management principles, just as one should on any kind of project. In almost all cases of failure of an object-oriented project, it turns out that basic principles of transition to a new technology were violated, or that unrealistic expectations were set.

Analyze, then track, costs and improvements

Ultimately, as is the case with any new technology insertion, the costs of switching (including retraining and retooling) must be carefully analyzed and weighed against the expected benefits, and sufficient **process metrics** must be put in place so that serious deviations can be detected and corrected before serious damage ensues. In some rare cases, if a pilot project has not borne out the promises of the new approach, staying with or returning to structured methods (while the dust continues to settle in the object-oriented world) may be a rational decision.

Analysis versus Design Methods: Same or Different?

The same methods can serve both analysis and design

Although we believe that there is a significant difference between analysis and design, as explained earlier in this chapter, there is little distinction between object-oriented analysis and design *methods*. The same techniques and notations can typically be used in both contexts. Whether one produces an analysis or a design depends more on the context, the procedure followed, and the semantics of what is modeled, than on the method or the symbols it uses. Most authors claim that their methods span both activities equally, although the titles of their books and the contents of their examples usually place the emphasis on one or the other.

4.3 What is a Method?

A software development method ensures a disciplined approach to software development. Methods provide rules and approaches for identifying and modeling various aspects of the problem and solution spaces. They also provide a precise notation for describing and diagramming various models. They are independent of computer-based tools. One may benefit from a method without the support of computers; indeed, many models are conceived on the back of a paper napkin in the company cafeteria. Of course, computer-aided software engineering (CASE) provides much help in using the methods beyond that early stage. Although methods in general, and object-oriented ones in particular, should prove useful across the complete software lifecycle, at this point, they really exist only for *analysis and design*. Extension of the methods to encompass implementation and maintenance is certainly needed, but this is not yet the subject of significant work since at that stage, developers go back to relying on textual code.

Methods provide rules, approaches, and a notation

The emergence of a new method (there are at least twenty object-oriented analysis and design methods in use at this time) usually starts with the publication of a book or technical paper in which a **methodologist** proposes:

New methods are usually proposed by "experts"

- An approach for how to model the "real world" or a computer system;
- A precise notation, or language, used to describe the model;
- Rules or guidelines for performing the analysis of the problem or the synthesis of the solution.

Experts may recommend using different methods for the various phases of the lifecycle. Others claim that their method applies across all the analysis and design phases. In practice, most methods are stronger for certain steps or for certain

Some methods apply preferentially to one step or the other

classes of problems, and no method has gained universal acceptance as equally suited for both requirements analysis and design.

Experts continue to evolve and combine methods

This situation will improve for two reasons. First, methodologists constantly monitor how their methods are applied and they continue to improve them, sometimes by borrowing concepts from one another. Secondly, some independent experts, whom we could call **method assemblers**, combine the best aspects of several methods and apply them to meet a customer's needs. These are usually highly paid system consultants who intervene for a limited amount of time during the crucial stages of a project. Sometimes they report their findings in journals or conference papers, and thus influence the methodologists, their peers, and the industry at large.

We will review methods and selection criteria; we do not propose a specific method

Having introduced methods in general, in the next four sections we will discuss the components of analysis and design methods, review some of the most popular ones, and offer a list of criteria that readers can apply in their own method selection process. We do not propose a specific method, for two reasons. First, as much as the reader might wish to get a firm recommendation, this is neither easy nor desirable. No two problems or organizations are alike, and the field of object-oriented analysis and design is an evolving one. Each potential "customer" of a method should therefore perform at least *some* review of the field from his unique perspective. Eventually, the market will freely decide which methods deserve preeminence and which ones should fade into obscurity. In addition, there have been several reports in which users claimed success with object-oriented methods because they had blended several complementary aspects of different methods in order to suit the needs of their particular domain. They acted as their own "method assemblers" in the sense used above. To exercise this freedom wisely, it is of course necessary to become familiar with (or at least to read about) several different methods.

The happiest users combine aspects from several methods

[OMG 94] and [OMG 94a] are good sources of information with which to begin. Another requirement is to have a flexible CASE tool (see Chapter 5) that supports different methods, or even allows an expert user to customize or blend the different notations (rather than implement a single method strictly by the book). There are a few such tools on the market now.

Finally, we will add a few words about how methods are named. Each method is usually given a name by its author, often the same as the title of the book or paper in which it was first published. Since these names are long and confusingly similar ("object-oriented systems analysis," "object-oriented requirements analysis," etc.), the prevalent tendency of the industry is to refer to a method by the last name of its main author(s) (e.g., "Shlaer-Mellor" or "Booch"). Sometimes, in the presence of two "vintages" of the method (usually captured in two successive versions of an author's book), the year can be appended to remove ambiguities. Thus "Booch 86" was significantly updated in "Booch 91." After a while, however, the old version is forgotten and "Booch" refers clearly to the 1991 update. Note also that this practice is sometimes controversial with coauthors who lose name recognition in the process. The best case in point is the Object Modeling Technique (OMT). Although it was the result of a collaboration between several authors, in most bibliographies the work is cited as "James Rumbaugh et al.," and most people say "I use Rumbaugh" instead of "I use OMT."

Methods are named after their main author(s)

4.4 Models Used in Analysis and Design

Three different types of models (i.e., abstract representations of a system) commonly appear in many of the methods we will examine. By way of introduction, we will first describe each of these models and their relationships. These models allow the analyst or designer to understand, document, explain, and

Three types of models are common

review a system or application, thus removing much of the ambiguity and obscurity that prevail when a system is either described solely in English, or when it is coded without such models being first developed. These models are also absolutely critical for enterprise integration or system integration (i.e., to plan and execute the logical connection between complementary applications).

The Information or Object Model

Definition of the objects

This model contains the definition of objects in the system (or in the "real world," in the analysis phase). This *static* model typically includes:

- Names of the *classes* of objects (e.g., "elevator car," "call button");

- *Attributes* of the objects (e.g., "capacity");

- Sometimes the list of *services* or methods offered by each class of objects (e.g., "start," "stop," "press");

- *Relationships* between objects and classes; this includes not only inheritance relationships between classes, but also "part of" and other relationships (e.g., each elevator car contains one button per floor served; "up/down/floor" buttons are three types of buttons with some common characteristics, etc.).

The information model looks like an ERA model, but OO makes it different

In some methods, the combination of object classes, attributes, and relationships resembles closely the "entity-relationship-attribute" (ERA) model of Chen [Chen 77] and others. Note however, that if the method is truly object-oriented, the notion of class hierarchy and the presence of the services rendered by the class provides a significant departure from classical ERA models. Whereas the attributes are key to an ERA model because it permits the unlimited access to the *contents* of the entities from any procedure in the system, the purpose of listing the attributes in the object model is much more secondary. The attributes list helps plan the internal design of the object, or it documents the nature and purpose of the object to the reader of the model, but it has nothing whatsoever to do with how other objects will interact with this one. Only the description of the semantics of the services is needed to document this.

A number of experts believe that an information model is *not* an inherent part of object-oriented analysis. In their view, all that is needed to specify the objects is their behavior or their interaction. Because objects are encapsulated, they say that the data they contain is immaterial at this stage. Specifying the attributes of a class of objects is a part of design. Yet there are some attributes that are readily apparent at analysis time. In an airline reservation system, for example, the flight number and the name of a passenger are pieces of information that are clearly relevant and important, regardless of how a computerized system will be designed. In that case, capturing this information in the model is not only useful, but it can help the analyst ask some relevant questions of his customers (e.g., "What happens if a different model of airplane needs to be substituted for the original one?" or "How does one keep track of two passengers on the same flight who are both named Mary Smith?"). On the other hand, some details are clearly not relevant at the analysis stage, such as the number of characters required to encode a flight number or a name, or perhaps the plane's registration number.

Is the information model really a part of object analysis?

The Behavior or State Model

Most methods recognize some form of state model as a way to describe the behavior of the objects in the system (some methodologists would argue, however, that this is not the only way to capture behavior). A state model shows at minimum:

The state model describes the behavior of objects

- The legal or stable conditions, or *states*, in which a system or object may exist (e.g., "going up," "going down," and "stopped");

- The *transitions* that are allowed between states (e.g., an elevator cannot switch from "going up" to "going down" without passing through the "stopped" state);

- The *events* that cause an object to change states (e.g., "passenger presses button"), and perhaps the events triggered by a change of state (e.g., "button lights up").

State models describes the lifecycle of the objects

State models are often said to depict the *lifecycles* of the objects because they show the transitions between states. Indeed, because the system has a natural "rest" state when nothing is happening, many state models do look like a cycle. Successive transitions tend to return the system periodically to this state. In the elevator example, this may be what one would find in the middle of the night in an office building, when all the cars are stopped, with their doors closed, on the first floor.

Different forms: Moore and Mealy models

Beyond this, opinions differ about the exact structure and complexity of state models and on the exact notation used. A traditional representation consists of **state transition diagrams (STD)**, directed graphs in which vertices represent states, and edges (shown as arrows) indicate transitions. Events are usually annotations added to the transitions arrows. There are two principal forms of state models, corresponding to two types of finite automata proposed in the 1950s by E.F. Moore and G.H. Mealy (for a description of the "Moore machines" and "Mealy machines," including a proof that they are equivalent, see [Hopcroft 79]). In the **Moore model**, *actions* performed by the system or object when it enters a given state are also indicated. In the **Mealy model**, actions are likewise indicated, but they are associated with the transitions rather than with the states. Thus arrival into the same state along different routes may trigger different actions, an often more realistic situation (e.g., the

Development Activities

same call button will not be turned off when the elevator stops on its way up or on its way down).

A variety of notations are used in real-time systems to indicate time dependency. Consider the kind of state transitions that occur on the basis of the passing of time, rather than upon receiving an event from another part of the system. Whether or not an explicit *timer* object is required in the model to issue the necessary events (e.g., starting to close the elevator doors) may differ from method to method.

The modeling of time dependencies

When there is a need to express a *hierarchy of states* (e.g., "moving up" and "moving down" might be two substates of a parent state called "running"), or to express *simultaneity* between the state changes of several objects which are sensitive to the same event, more complex notations can be used, such as David Harel's **statecharts** (see [Harel 87]).

Hierarchical state diagrams

The Function or Process Model

In the past, the one area upon which traditional or structured techniques have focused is the modeling of the processes or functions of the system, that is, how actions are performed. Indeed, it is because processes and data were modeled completely separately that difficulties arose in the first place. Because the functions that can be executed by an object-oriented system are split into smaller chunks corresponding to each method of each object, this has also been called "methods modeling."

The process model is similar to the functional descriptions of traditional methods

How process models are represented differs widely across methods. Some methods have no such model at all. Others focus on describing the *syntax* of the method invocation and the *semantics*—but not the design—of the methods (e.g., what parameters are passed in and out, what logical assertions one can make about the state of the object before and after execution, and what the relationship is between the input and

Diverse approaches and notations are used

output parameter values). Other methods revert more to the functional decomposition model and actually show a flow chart or pseudo-code description of what happens when the method is executed—which, of course, constitutes a design of the method. Since the methods of an object have full access to its attributes (we are now within the object), **data flow diagrams (DFD)** may be applicable, and indeed, are used by some methodologists. They allow the task to be decomposed into finer actions that read or write different attributes, which then take the place of the *stores* in traditional DFDs.

Relationship Between the Various Models

Events tie object models together

The various models that capture the requirements or design of a system are interrelated and must, of course, exhibit cohesiveness. One way in which they do so is that a process model shows operations that consist of sending an event to another object. This event will in turn trigger a transition in that object's state model, which will cause another process to be executed. All these interactions between objects can be made explicit in an **object communication model** (or **object interaction model**), that shows which events can be sent by which objects to which other ones.

The object communication model should not be an afterthought

Some methods build this object communication model early on, *a priori*, as part of the early analysis process. Later on, one can verify that the state and process models are consistent with the preexisting communication model. Other methods only build the communication model mechanically and *a posteriori*, purely as a way to document the inter-object communication from the fully developed set of state and process models. The first approach may seem preferable, because it allows some cross-checking between models. In the second approach, by contrast, the models will automatically be consistent.

Development Activities

Another distinction between various forms of communication models is whether they show only that some interaction exists, or whether they specify the name or type of the information being exchanged. Some methods specify the data being exchanged directly on the interaction diagram, while others capture this information in text templates. Still others omit it completely.

Is the data being exchanged also modeled?

Other Requirements

The use of graphical notations has become so pervasive in object-oriented analysis and design that people tend to forget that there are other types of information worth capturing, even though these do not lend themselves well to graphics.

OOA&D may also include non-graphical models

Some methodologists provide templates or forms to describe such requirements of the system as reliability, portability, performance, serviceability, documentation, training, etc. It should be noted that for many systems, these requirements are stated either very vaguely, or very specifically, but in an arbitrary way. It is not always clear what the merits of 99.7% availability are, as opposed to 97% or 99.97%, but the costs of achieving the higher figures may be quite significant. Methods that propose to capture this information often use a form of Tom Gilb's *quality templates*, described in [Gilb 88] (Chapter 19, especially, page 361).

Another type of requirement best captured in text concerns "system engineering" (i.e., what does it take to make the system exist in its environment?). This can take the form of imposed standards for user interface "look and feel," communication protocols, database query languages, etc.

Document other constraints

4.5 Object-Oriented Analysis

There are many object analysis methods

There are more object-oriented analysis methods in existence than most users would care to consider. In [OMG 94], *twenty-one* different methods are briefly described. Yet not all the methods in existence are included in the OMG survey, largely because this compilation work is dependent on the willingness of experts to participate in the survey, and of course, on the ability of the editors to identify the survey recipients in the first place.

Details are available in OMG books

Since [OMG 94] and [OMG 94a] provide descriptions and comparison tables about a number of methods, we do not re-state the same information in detail here. Instead, the following paragraphs provide qualitative commentary on some of the notable aspects of a few well-known methods.

Shlaer-Mellor

The Shlaer-Mellor method uses information, state, and process models

Sally Shlaer's and Steve Mellor's "Object-Oriented Requirements Analysis" is described in [Shlaer 88] and [Shlaer 91]. The method is taught by the authors' company, Project Technology Inc. of Berkeley, California. The method conducts requirements analysis in three main phases corresponding to three types of models: *information models, state models,* and *process models*. There is one information model for the whole system, one state model for each object that has a lifecycle, and one process model to describe the actions executed upon arrival in each state.

Information models are very close to ERA models

The information modeling stage parallels classical Entity-Relationship-Attribute methods and produces an **Information Structure Diagram (ISD)**. Figure 4.1 shows a simplified ISD for an elevator control system. Shlaer and Mellor give specific attention to the notion of *associative objects* (i.e., objects which carry attributes pertaining to the relationship between two or

Development Activities 127

more other objects. This is the "DestinationButton" object in the figure).

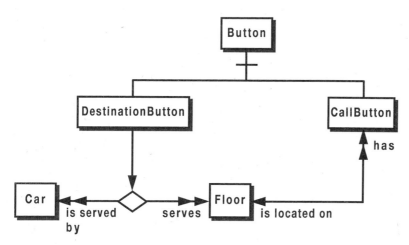

Figure 4.1 – Shlaer-Mellor Information Structure Diagram

At this stage, the Shlaer-Mellor method insists that "many-to-many" relationships (e.g., between projects and departments in a matrixed organization) must not be directly represented, and that an additional object must be included in the model to break such associations into several simpler ("one-to-many") relationships. Their attribute modeling also requires the analyst to distinguish between object identifiers (i.e., keys) and *referential attributes* (i.e., foreign keys). Since these steps correspond to the design and normalization constraints of relational databases, they have been controversial aspects of the method. These concepts are not necessarily applicable in an object-oriented system, where one is not limited to tabular data representations. Moreover, they also appear to be a legitimate part

Some rules are similar to relational database normalization, and should be part of design, not analysis

of a design activity, not an analysis one. Fortunately, many users of the method have elected to skip this step during initial analysis and have found that it can be relegated to a later design stage without affecting the usefulness of the rest of the information model.

State models are linked to events sent by an object to another

State modeling uses traditional (flat) STDs. In state models, transitions are caused by the reception of *events* emitted by other objects. This is the closest the method comes to the concept of messages between objects. An object communication model is abstracted at the end from the set of state models. In effect, instead of asking up front "Given the nature of the problem, what *should* the interactions between objects be?" Shlaer and Mellor ask: "What do the interactions between objects *turn out* to be?"

Process modeling uses both text and graphics (DFD)

The modeling of the actions required at each transition between states consists of two parts: a description in a Program Description Language (PDL)-like notation of the actions performed upon entry in each state (i.e., a Moore model); and a separate process model for each state. Process models use DFDs in which the data stores must correspond to the objects contained in the information model.

Shlaer-Mellor offers an easier transition from structured methods, but is not purely object-oriented

In general, Shlaer-Mellor appears to provide a moderate paradigm shift from structured analysis and design methods. Because notations and concepts from prior methods are used (including ERA-like information models, STDs and DFDs), the Shlaer-Mellor method seems easier to adopt by organizations in transition, but is open to the criticism that it ignores or postpones some of the key aspects of object orientation, such as information hiding (e.g., the processes described by process models are not restricted from accessing the contents of multiple objects).

Coad-Yourdon

This method, simply called "Object-Oriented Analysis," is proposed in a very readable book, [Coad 90]. Peter Coad and Edward Yourdon provide extensive treatment of the information model, which is described in **layers.** This is visually equivalent to the classical presentation trick of superimposing multiple transparencies on top of one another such that different subsets of the information can be combined at will. Each layer carries part of the information being modeled (e.g., the attributes or the relationships), but shows the objects in the same place so that the superimposition "works." The analyst can thus focus on one or more specific aspects of the model while maintaining its overall consistency. Here is a brief review of the various layers and their usage:

Coad's information model is layered, providing multiple views

- The **object layer** shows only the existence and name of each object.

- The **subject layer** shows the grouping of objects in clusters in order to manage the complexity of larger models.

- Other layers complete the information model by containing the attributes of the objects and the relationships between them.

- The **service layer** describes the services (i.e., the interactions between the objects).

Proponents and users of the method claim that one of its most attractive features is the capability to visualize all aspects of the model on the same diagram. Critics claim that only toy problems can be represented in this fashion—real systems being too complex to draw or read in this format. Part of the problem is that there are only two levels in the hierarchy, subjects and objects. As for the model of dynamic behavior, its treatment in the book is incomplete. Object lifecycles and process descriptions are only briefly addressed.

The layers may be hard to use on a large system

Object Behavior Analysis (OBA)

ParcPlace's method does not use many graphical notations

OBA is provided by ParcPlace Systems of Sunnyvale, California, a supplier of object-oriented programming environments. It was originally described in [Gibson 90], received further treatment in [Rubin 92], and as of this writing, a book on this subject is in preparation. In the meantime, the method is taught by ParcPlace consultants. OBA is unique among analysis methods in that it does not rely much on graphical notations. Most of the artifacts of the method are lists and tables which can be created using a word processor or a spreadsheet package. In addition, ParcPlace Systems supplies a tool, MethodWorks, supporting the OBA method.

OBA is split into five basic steps:

System-level goals and requirements

- **Set the context.** This is a scoping and planning stage, with emphasis on the goals and objectives of the system under study, and a look at the existing system, if any. Having system-level goals and requirements is considered an important check to determine later if you have developed the correct system and, more practically, to know when the analysis phase is complete. In spite of the extra work involved, performing an analysis of the "as is" system is justified by the insight it provides into improvements that are necessary but may not have been previously expressed.

Behavior analysis through activity scripts

- **Understand the application.** As its name indicates, OBA begins by analyzing the *behaviors* of the system, using **activity scripts** to describe what happens when a certain behavior occurs. The focus is on understanding how multiple interacting objects cooperate to achieve the required behavior of the entire system. A behavior corresponds to one precise succession of events—this is similar to what other methodologists or users have called "scenarios." If multiple outcomes are possible, such as "normal takeoff" and "aborted takeoff," then two different scripts are written. It

Development Activities 131

is also noteworthy that scripts refer to specific object *instances*, not to object *classes;* the abstraction of classes is done later in OBA. Table 4.2 shows a typical script related to the elevator control system example.

Agent	Action	Recipient	Result
Passenger	Presses button	"Up" button on 1st floor	Stop on 1st floor is scheduled for one of the elevators
Elevator	Arrives at floor	First floor	Doors open
Passenger	Presses button	"4" button inside elevator	Stop on 4th floor is scheduled
Elevator door	Closes		Elevator starts up
Elevator	Reaches floor	4th floor	Elevator stops, door opens, stop is removed from schedule
Elevator door	Closes		Elevator is idle

A script for an elevator control system

Table 4.2 – An OBA Script Example

- **Identify objects.** Objects and their features (i.e., attributes) are derived from the scripts by looking at their participants and the roles they play.

Find the objects from the scripts ...

- **Organize and classify objects.** Glossaries of objects, object features, and behaviors are created for traceability purposes. Associations are identified, and class hierarchies are derived from the identification of common behaviors (*not* common attributes). At this point, all the static information about objects is known, and is captured in **object modeling cards.**

... and create an object model

- **Determine the dynamic model.** Flow diagrams and state change diagrams are created for objects that need them, although OBA does not recommend a specific notation or method to derive these artifacts.

State and process models

The OBA method also contains a number of recommendations for the **architectural design** phase that immediately follows analysis.

Object Modeling Technique (OMT), a.k.a. Rumbaugh

OMT is the best recognized method

OMT, described in [Rumbaugh 91], was originally developed at General Electric's Advanced Concepts Center, now a division of Martin-Marietta. Since its publication, it has achieved a high level of recognition in the industry. Like several other methods, OMT relies on three basic types of models to capture the information, dynamic, and functional requirements of the system.

The state model uses Harel's statecharts, capturing much more information than flat state transition diagrams

The information model is similar to the ISD of Shlaer-Mellor. To illustrate the notational differences, Figure 4.2 shows the OMT information model equivalent to the ISD of Figure 4.1. In a significant improvement over most other methods, OMT uses **Harel statecharts** to model the lifecycle of objects. Flat state transition diagrams are insufficient to model complex systems with nested and concurrent states. Statecharts were invented by Dr. David Harel, of the Weizmann Institute of Technology in Israel, to allow analysts to represent a hierarchy of states, as well as the situation in which multiple objects may change states simultaneously as the result of the same event. This is, of course, an oversimplification. The Harel notation, described in [Harel 87] and [Harel 88], is very rich and offers many more capabilities. A simple Harel statechart for the elevator car object, which shows only a subset of all the capabilities of the notation, appears in Figure 4.3. Some CASE tools allow statecharts to be drawn, and at least one tool, Statemate from i-Logix, can create an animated simulation of the behavior of the system being designed.

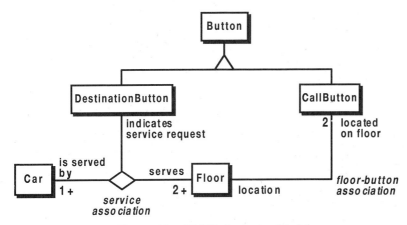

Figure 4.2 – OMT Information Model

Class Responsibility and Collaboration – Wirfs-Brock et al.

In spite of the title of their book, "Designing Object-Oriented Software," Wirfs-Brock, Wilkerson, and Wiener propose, in [Wirfs-Brock 90], a method that covers requirements analysis as well as design.

The method is based on perceiving objects as "cooperating and collaborating agents." This cooperation depends on well-defined responsibilities, captured in a *"contract"* definition. The method uses an artifact called **class responsibility and collaboration (CRC) cards** to explore these relationships during analysis and design. The cards can be used in a "role play" in which several people represent the various objects. Through their interaction, they explore and reveal the required interactions between objects. Concurrency is supported in terms of contracts, responsibilities, and messages.

In Wirfs-Brock's method, objects cooperate through contracts, described by CRC cards

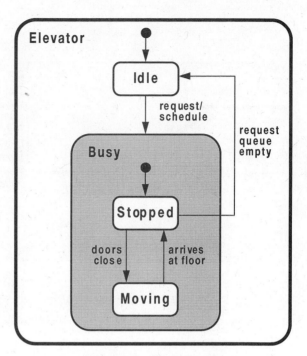

Figure 4.3 – Harel Statechart

Colbert's Combination of Methods

Colbert is a "method assembler" who consulted on military systems

In [Colbert 89], Ed Colbert proposed to combine a certain number of techniques and notations together. He is thus part of a group of method assemblers who synthesize new methods by choosing the best from others. Colbert's approach has mostly been applied by its author to U.S. military projects. Because of this, it exhibits some dependencies on the constructs of the Ada language (packages and tasks) and may not be as language-independent as one would like. Moreover, the fact that there is no book or company behind the method makes it harder to adopt than some of the others previously mentioned.

An interesting aspect of Colbert's method is that he first determines how the system under study interacts with the rest of the world, and then captures this information in a **Context Object Interaction Diagram (OID)**. This is similar to the context diagram of structured analysis methods and gives in one place a clear view of the system's behavior. In other methods, one must read the set of all state models to understand this external behavior. Once the system is split into multiple objects, that is, as soon as architectural design begins, lower-level OIDs can be created. This approach contrasts with the Shlaer-Mellor method, in which the object communication model was an *ex post facto* artifact.

He first models the system's interaction with its context

Objectory

Ivar Jacobson originally developed this method at Ericsson in Sweden. Since then, it has slowly emerged into the mainstream, promoted by Objective Systems (Jacobson's company) and by his book, [Jacobson 92].

Objectory comes from Sweden

Objectory distinguishes between the *requirements model* and the *analysis model*. The requirements model consists of **use cases**, which are similar to the notion of scenarios present in other methods, or the scripts of OBA. A use case is formally defined as "a special sequence of transactions in a dialog between a user and the system." Use cases have recently attracted a lot of attention, and other methodologists seem to be introducing them (or variants of them) as an initial step of the analysis. Figure 4.4 provides an example of a use case for the elevator application. All the other artifacts of this comprehensive method, starting with the analysis model, are derived from the use cases. The analysis model contains three types of objects: **interface objects, entity objects,** and **control objects.**

Objectory introduces "use cases" to capture the requirements before analyzing them

A use case for the elevator system

> **Basic Course**
>
> *SimpleTrip* is triggered by the arrival of a *Passenger* at a floor where no elevators are waiting. The *Passenger* presses the *CallButton* in the direction of desired travel. After some time, an *Elevator* arrives, announced by a chime and the lighted direction arrow above the doors. The *Passenger* boards the *Elevator* and presses the *DestinationButton* corresponding to the desired floor. The doors close, the *Elevator* travels to that floor and stops, the doors open, and the *Passenger* exits.
>
> **Alternative Course**
>
> If other *Passengers* have issued requests from outside or inside the *Elevator*, the *Elevator* may stop at other floors between those at which the *Passenger* enters and exits.

Figure 4.4 – A "Use Case" Example

A disciplined method, perhaps harder to adopt than others

The Objectory method is very organized and requires a disciplined approach to all stages of the development lifecycle. It is not limited to analysis and design, but includes construction, testing, project management, component management, etc. It is often presented in a way that implies a massive commitment by an organization to a specific process, rather than simply to the adoption of an analysis or design technique. Although this shows a positive attention to process needs, it is not conducive to the "creeping adoption" often required in somewhat conservative or risk-averse organizations. This approach also seems more in harmony with the culture of large, centralized companies than with the "pilot project" approach typically preferred in the American business culture.

4.6 Object-Oriented Design

Object analysis and design use similar notations

Most object-oriented methodologists agree that analysis and design are closer to one another in the object-oriented approach than in previous approaches. This is often described as a narrower *semantic gap* between analysis and design. For this reason, similar notations are often used during analysis and the early stages of design. During design, however, it becomes

Development Activities

necessary to express concepts that are nonexistent in analysis, such as the types of the attributes of a class, or the logic of its methods.

Generally speaking, there is also agreement that design can be separated in two phases. The first, called "architectural design," "conceptual design," or "high-level design," deals with the decomposition of the system into subsystems, or large, complex objects. It also solves larger issues of system engineering, such as which communication or data management methods to use. Alternatives are considered, and the rationale for specific selections should be documented explicitly to guide future maintainers who might otherwise waste time reconsidering the same alternatives. This is not common practice, but it should be.

Design consists of a high-level phase ...

The second phase is called "detailed design" or "low-level design." In this phase, attributes and methods are specified at the level of the individual objects. This is also where reuse is the most likely to occur, as it is possible to steer the design so that lower-level objects correspond exactly to those available in existing object libraries. Conversely, this stage is also where one determines which components are not yet available and need to be bought or built. Finally, this is where some trade-offs can be made, such as space versus performance.

... and a detailed phase

Here are several methods that distinguish themselves by the extra attention they bring to design activities and models.

Booch

Grady Booch's Object-Oriented Design is described in [Booch 91]. Although his earlier work predominantly addressed systems written in Ada, the author has taken pains to convince the reader that his method is language-independent. To this effect, several examples are treated in different languages (C++, Smalltalk, Object Pascal, CLOS) at the end of the book. But

Booch's method was initially aimed at systems written in Ada

the method retains an Ada-like flavor in the sense that real-time command and control systems seem to be the best domain for the method. Indeed, the state transition and timing diagrams provide significant support for real-time needs. Nor has the language dependence completely disappeared. During lectures or tutorials, the author himself resorts to Ada or C++ concepts and syntax to clarify some of the subtle distinctions between similar notations in his diagrams.

Distributed systems design

The Booch method devotes more effort than most to describing the allocation of processes to processors, which is important for the design of distributed systems.

The graphical notation is awkward at first

The notations used by Booch are nontraditional and somewhat difficult to master. Objects and classes are drawn as clouds. Some of the conceptual distinctions are captured by shadows, dotted lines, little boxes containing letters in white on a black background, etc. All these notations are hard to use by hand when discussing a system's structure on a white board or on a notepad, and they are also difficult for CASE tool suppliers to implement in a legible manner.

Buhr

Buhr focuses on real-time systems

Dr. R.J.A. Buhr's method, described in [Buhr 90], is targeted specifically at the design of real-time systems in which complex threads of control exist. In such systems, multiple processes execute in parallel and require synchronization through rendezvous mechanisms. In Buhr's approach, these interactions are studied by making distinctions between several kinds of agents:

- **Actors:** processes that send messages to request services from other agents.
- **Reactors:** processes that wait for messages requesting services from them, but do nothing on their own.

- **Gates:** processes that provide a synchronization mechanism between other agents. A message queue is an example of a gate; it provides communication between a message-sender process and a message-receptor process.

Buhr's method offers a graphical notation, called **machine charts**, that corresponds directly to the constructs of the Ada language: packages, tasks, monitors, etc. These charts facilitate the generation of Ada code. Accordingly, the method has been mostly used in command and control systems for military applications.

Machine charts abstract Ada constructs

Buhr's method seems to provide powerful analysis capabilities for the synchronization aspects of real-time systems. His "event scenarios" and "visit scenarios" represent in detail the possible successions of interactions between objects. He also uses a variety of icons to denote waiting places, time-outs, periodic interactions, "balking" connections, etc. However, its applicability to industrial projects would require at least three changes:

A rich set of real-time analysis concepts and notations

- Additional coverage of the information and process aspects, as opposed to the method's exclusive focus on the real-time threads of control of the system;
- Some way to represent the system in a hierarchical manner in order to manage the complexity of very large systems;
- Adaptation to languages other than Ada.

More importantly, it is not clear that the method can really call itself object-oriented. The fact that it is very much related to the Ada language, and that Ada itself was only "object-based" but not "object-oriented" until 1995, leads many experts to state that the method is *not* object-oriented.

Is Buhr's method really object oriented?

Hierarchical Object-Oriented Design (HOOD)

HOOD includes many semantic checks

The **HOOD** method was developed and used by the European Space Agency. It uses a single type of diagram, and more than 300 syntactic and semantic rules have been developed to automate the checking of these diagrams by computer. This is a rare example of the pursuit of completeness and correctness in the method.

It is also very Ada-focused

HOOD is very focused on implementation in the Ada language. It may also have suffered from having been prematurely frozen at a time when it might have benefited from Darwinian evolution. This was done to enable the developers of CASE tools to aim at a stable target.

Object-Oriented Structed Design (OOSD)

Wasserman's method maps an object approach to structured design notations

OOSD is described in [Wasserman 90]. Anthony Wasserman reuses several of the notations of structured design (hence the name of the method), but after an initial step of object identification. The notation is fairly complex in terms of the number of graphical "widgets," but this was intended from the start to be supported by a CASE tool, namely Software Through Pictures, which is supplied by Dr. Wasserman's company, IDE.

4.7 How to Select a Method

A selection effort is required

An organization that decides to move to an object-oriented approach to system development should execute its own selection process. This allows the organization to take into account its specific needs and culture, as well as the most current information.

We will offer some criteria

This section proposes a list of criteria which the ideal object-oriented analysis and/or design method would meet. These criteria can be used not only to select one of the existing methods for a new application, but also to periodically review

Development Activities

whether or not the method currently in use has in fact met reasonable expectations.

We will not attempt to judge how each of the methods described in Sections 4.5 and 4.6 meet these criteria. Because we may not be able to do this fairly, this would be a disservice to some methodologists and to potential users who might overlook a method that has improved since the publication of this book; or who may have specific requirements that we would not have considered when doing our own performance rating. Moreover, it will be helpful to the potential user to go through the exercise of surveying the field, tailoring the selection criteria, judging how each method meets the criteria, etc., all the way to completing a final selection. Finally, this is a tremendous opportunity to learn about the field and to develop a carefully considered approach to object-oriented analysis and design.

It would be unfair to recommend a specific method

The list of 19 criteria below is split into two parts. The first 6 properties should be mandatory for every user of methods. The last 13 properties may be strongly desired, but may not be "showstoppers" for every organization.

A. MANDATORY CRITERIA

1. Support an object-oriented approach

This seems obvious, but some methods that claim to be object-oriented do not distinguish clearly between object and class, or do not respect the encapsulation of attributes within the object (e.g., by requiring that an object directly access the state of another). Other methods do not allow inheritance to be modeled, provide no guidance on how to use it, or do not exploit it to provide reuse. Polymorphism is another characteristic which some applications require (e.g., sending the request for its balance to an object of type "account," even though there may be several types of account with different calculation methods).

Encapsulation, inheritance, polymorphism

Adherence to the principles of object orientation must be thoroughly verified.

2. Suitability to the application domain

Support for concepts important to the user

Some domains require specific characteristics. For instance, telecommunications software and operating systems have stringent needs to represent concurrent activities and their synchronization, or to specify the performance required (message round-trip delay, etc.). The method must have the necessary constructs to meet the domain's specific needs. Unfortunately, the only good way to evaluate this is to conduct a small but fairly complete analysis or design. Otherwise, it is possible to overlook some subtle holes in the method.

3. Completeness

Full representation of requirements and/or design

At either the analysis or design level, the method must support a complete understanding of the application or system. It should not just statically model the objects, but it should also cover their dynamic (behavioral) and functional aspects. In particular, it should be possible to proceed to design the implementation of the object services without having to "step outside" of the method.

4. Continuity from analysis to design

Consistency in models and notation, whether or not there is a single method

There are several ways to interpret this criterion. If the customer is intent on choosing a single method that applies across both phases, then the method itself is likely to provide a relatively smooth transition from one phase to the other. The notations should already be consistent, and the models output by the analysis phase should be exactly what is needed to start designing. If the customer decides instead to select separate methods for analysis and design (perhaps after trying the first approach and failing to find a single satisfactory method), the same consistency is desirable but is much harder to ensure. Again, in this case,

Development Activities

a small but thorough pilot example must be developed to determine how easy the transition can be.

5. Documentation and Assistance

The method must be fully supported by a professional quality training program. Existence of CASE tools that support the method is not a substitute for training. A catalog containing only public courses offered a few times a year is not sufficient because this does not allow "just-in-time training." The method should be documented in a widely available textbook of high publication quality. This not only ensures that it is easier to learn, but it also increases the probability that the method will be used by other customers and by consultants.

Training and a textbook

6. Scalability

The method must be suitable to the construction of large models. The potential user should estimate how many classes, states, methods, etc., he is likely to encounter in a typical application, and should consult references who can attest to having successfully applied the method to a problem of similar size and complexity.

Applicability to large systems

B. DESIRED CRITERIA

7. Consulting Services

There should be a body of consultants with verifiable references which the proponent of the method can recommend. If the only available consultant on the method is its author, then there is too high a risk that he will not be available to assist if a significant problem requires outside help.

Availability of trained consultants

8. Presence of Rules

The method should not merely contain guidelines about developing and documenting models, but should also provide rules that can be used to audit a model and

Rules to determine model consistency and completeness

determine, in a systematic manner, whether or not it is complete and self-consistent. This will ensure that the models provide an accurate understanding of the application.

9. Maturity

Extent of prior use

The method should already have been used in multiple applications. Of course, someone has to be the first one to use anything. Depending on the criticality of an application, one may or may not be able to afford to be the guinea pig. Some level of maturity will also ensure that the rate of change is down to a tolerable level.

10. CASE Tool Support

Existence of a commercial tool

The method should be fully supported by a commercially distributed and supported CASE tool. Furthermore, at least one such CASE tool should meet the criteria laid out in Chapter 5.

11. Traceability

Linkage between requirements and design

The method should support tracing an element of the design back to the original requirements that this design element satisfies. This implies two things: (1) there is a way to capture this information, (2) there is a way to follow the link back from the design to the requirements.

12. Downscalability

Applicability to small projects

The method should be applicable to small projects. It should not systematically involve expenses, overhead, or lead times that make it prohibitive for small projects (e.g., those that require two to three people for less than six months).

13. Reuse

Management of reusable analysis and design components

The method should support the reuse of analysis and design fragments. One elementary mechanism for reuse consists of a glossary of objects and behaviors that can be

Development Activities

searched (perhaps using as simple a tool as a word processor).

14. Compatibility with Prototyping

The method should define its various elements in such a way that one can create some early prototypes. In particular, it should be possible to flesh out the behavior of some objects earlier than others. It should also be possible to use the same method to design stubs that can exercise some objects when most of the rest of the system is still very sketchy.

Support for incremental prototype building

15. Ongoing Improvement

The proponent(s) of the method should be actively pursuing improvements through trial use and/or a research program. Methods that are prematurely frozen are guaranteed to become obsolete as object-oriented technology evolves and as the complexity of systems continues to increase.

Active efforts to evolve the method

16. Independence from Programming Languages

The method should not make any assumptions about the implementation language or be limited by it. The fact that some object-oriented languages do not support multiple inheritance, for example, is not a reason to exclude this characteristic from the analysis or even the conceptual design stage. Mapping to a single-inheritance language can be done at the very end of design or at implementation time.

Compatibility with multiple languages

17. Modularity Support (Coupling and Cohesion)

A good principle to increase the modularity and the maintainability of a system is to decrease the coupling between modules and to increase their internal cohesion, so as to localize changes. Thanks to information hiding and encapsulation, object orientation already decreases the coupling between objects. However, the method

Guidelines to decrease coupling between objects

should provide additional guidance (possibly in the form of metrics for the number of methods or attributes per object, etc.) to further enhance the (de)coupling and cohesion attributes of the system.

18. Metrics

Guidelines to evaluate the quality of an analysis or design

Beyond the specific use of metrics just mentioned, other metrics should be proposed within the method in order to evaluate the "goodness" of an analysis or design, and to provide a basis for improvement. Any metrics coming from the structured analysis and design days are probably inapplicable, since inheritance and reuse can render them obsolete, irrelevant, or even misleading.

19. Communication Model

Early modeling of object communication

Based on the discussion in Section 4.4, the method should include some form of communication model—preferably one that is constructed early in the analysis or design—so that it can periodically be checked for consistency with other behavioral models.

4.8 Object-Oriented Programming

"Construction" is a better term than "programming"

The "construction" or programming stage, placed in the broader context of the software development lifecycle, consists of transforming the detailed design into a working system. Especially in an object-oriented approach, the term "construction" is better than "programming" or "coding" because the reuse and assembly of preexisting components should increasingly replace the development of entire systems from scratch. In other words, construction may include three activities:

- Extracting components from a reuse library;
- Acquiring commercial, off-the-shelf (COTS) components;
- Building new required components.

Object-Oriented Programming Languages

Invoking reused or purchased components, or building new components, requires the use of a programming language. Because the semantic gap is smaller than if one chose C, Pascal, or COBOL, **object-oriented programming languages** (OOPLs) are the natural choice for implementation of an object-oriented design. They directly support the notions of classes, inheritance, information hiding, polymorphism, and dynamic binding. Note that OOPLs are closer to the roots of object technology than analysis and design methods, and have been in existence much longer. Over the years, they have continued to evolve by adding new, useful concepts like multiple inheritance. Many Smalltalk advocates will disagree with this last statement. Rather than judging the Smalltalk language incomplete, they consider the concept of multiple inheritance unnecessary since their language does not support this specific notion.

Object-oriented programming languages directly support the concepts of an object-oriented design

There are fewer OOPLs than object-oriented methods, partially because writing a paper or a book is not tantamount to inventing a language. A programming environment (compiler, linker, debugger, and miscellaneous other tools) is also needed. Until recently, there were no "native compilers" for OOPLs. One needed to first precompile the source code into a procedural language equivalent, such as C for C++, then perform the actual compilation step. This was not only painful (because it introduced an extra step and an extra file), but it also impeded debugging (since an execution error would be reported against the intermediate source code). Tracing the error back to the original code was difficult. In recent years, compilers that directly process the OOPL source and produce the necessary symbol table for the system's debugger have appeared, so this problem is in large part remedied.

The choice of OOPL is more limited, especially if a native compiler is desired

Browsers are more important than with procedural languages

The most powerful programming environments include tools such as class browsers and version control software. By contrast, procedural languages are usually supported by "point" tools (compiler, debugger, etc.) and delegate other functions to generic, language-independent tools. This is partially due to a holistic view of programming activities that facilitates reuse. But it is also a natural consequence of the extreme degree of fragmentation to which some languages lead. In Smalltalk, for instance, it is common to have methods consist of fewer than ten lines on average, which means that a sizable system may consist of thousands of classes. If the class browser were not intimately integrated with the rest of the programming environment, it would be impossible to keep track of the "signature" of all these methods in order to invoke them correctly.

Past adoption barriers

In the past, several factors hindered the adoption of OOPLs. The first, of course, was the lack of familiarity with object concepts. In addition, some languages were only supported by esoteric platforms. The Mesa language, for example, was only available on Xerox workstations. Not only did the developer need such a workstation, but the final application was also limited to that platform, since a specific run-time environment was required to execute it. This problem became a moot point after the world converged on essentially two standards: UNIX and Microsoft Windows. At that point, however, most Object-Oriented (OO) languages continued to require a comprehensive run-time environment, sometimes simply because the language could not be compiled but had to be interpreted. One of the reasons why C++ overtook its rivals is that it was the first language to remove this dependency and to allow applications to run without a large or expensive run-time environment. However, such freedom typically comes at the expense of less support for object-oriented debugging during execution.

Development Activities

Here is a list of some object-oriented programming languages that have achieved significant use:

Some OOPLs

- We must start by mentioning the "granddaddy" of all object-oriented languages, **Simula**. Although no longer in use, Simula was invented in 1967 for the specific purpose of writing simulation programs. It included many of the current concepts of object orientation and was in many ways superior to some of the languages that survived it.

  Simula 67 started it all

- Among the C-based languages, the most successful by far has been **C++**. The language, invented at AT&T Bell Laboratories, is described in [Stroustrup 90]. Its success is partially based on the mistaken belief that "it is easy to learn C++ because it is a superset of C." People who believe this tend to become terrible C++ programmers because they do not follow a strict object-oriented approach. C++ lets them circumvent encapsulation and mix procedural code with object-oriented code, so they continue to use the undisciplined constructs permitted by C. For this reason, courses that advertise and teach C++ as "a better C" or "an extension of C" should, for the most part, be avoided. Courses entitled "object-oriented design using C++" are safer, provided the course description reveals that the title is an accurate indication of its content. When used correctly, C++ is a powerful language that implements all the concepts of object orientation, including:

  C++ is the market winner

  – Multiple inheritance (which Smalltalk, among others, does not permit);

  – The ability to override method definitions in subclasses;

  – Virtual classes: classes which do not have instances themselves (only their subclasses being allowed to have instances).

Objective C was used for the NeXT operating system

- Another C-based language is **Objective C.** The operating system of the NeXT computer was written in Objective C. The language was not widely used on other systems, but an agreement between NeXT and Sun Microsystems has led Sun to support Objective C in its DOE (Distributed Objects Everywhere) product (released in 1995).

Smalltalk is going through a revival

- **Smalltalk,** invented in the early '70s, follows a completely different approach in that it does not attempt to be a superset of a previous language, like C++. After being ignored for a long time, Smalltalk was first rediscovered as a good language to teach object-oriented programming precisely because of its total adherence to pure OO concepts. By 1992, it began to experience a revival as a "serious" language in which one could write real applications. The appearance on the market of inexpensive Smalltalk environments for PCs, Macintoshes and UNIX workstations was a major factor in this renewed interest. (For a description of the language, see [Goldberg 83]).

Eiffel implements disciplined OO programming

- **Eiffel** resulted from the work of Bertrand Meyer, described in [Meyer 88]. It implements a very rigorous way to define the "contract" between a class and its clients.

OO versions of LISP and other languages

- **CLOS** (Common LISP Object System) is the best-known object-oriented version of the LISP language.

- Object-oriented versions of other common procedural languages have appeared, such as **Object Pascal** and **Object COBOL.** They follow the same principle as C++ (that is, add object orientation while retaining compatibility with the base language) in order to emulate its success, but so far, to no avail.

4.9 Testing

Testing is one form of validation

After construction (including the compilation and linking, also known as the "building" of the system), the components of the

Development Activities

system must be tested. Testing is one of the best-known verification and validation (V&V) activities, which ensure the quality and the integrity of the work products. However, V&V is not limited to testing, but also includes reviews, audits, etc.

There are at least five kinds of testing that can be performed on a computer system or application: unit testing, integration testing, alpha testing, beta testing, and acceptance testing. One does not always apply all five kinds of tests to a system, because the last two rarely coexist. But it is not uncommon to run four of the five.

Five different kinds of testing

Unit Testing

The scope of unit testing depends on what one's concept of a unit is in a specific systems context. In object-oriented systems, however, the logical interpretation of the unit concept is the *class*. Each class should be tested for conformance to its requirements or design specifications, which implies that each of its methods should be invoked at least once—and probably multiple times if different outcomes are possible. For example, the method "close_elevator_door" should be tested in the case where the elevator door is already closed, in the case when it is open and can close properly, and in the case when it is open but an obstacle prevents it from closing. Unit testing will mostly find errors in detailed design and in coding. Unit testing typically requires the creation of **stubs,** which are pieces of code that can simulate the environment with which the tested class would normally interact. The stubs perform the following functions:

In an object-oriented system, a unit is a class

Unit testing stubs need to be written

- Set up the proper initial conditions, perhaps by first calling other methods of the class;
- Invoke the method according to the test specification;
- Generate any necessary responses to method calls issued by the class under test;

- Verify that the reply to the call (if any) contains the correct information;
- Report on the correct or incorrect execution of the method.

Unit testing is time-consuming because of the extensive creation of stubs that may be required. On the other hand, a failed unit test is normally much easier to analyze and correct, since the problem is limited to the single class under test.

Integration Testing

Integration testing verifies that the units work together

Integration testing is performed when the complete system has been built and is for the first time being put through its paces by the developers. The purpose is to find malfunctions that result from the interaction between the components, rather than from the internal malfunction of a specific component. Unless unit testing has not been performed well, integration tests are more likely to find design errors than coding errors. There is no longer a need for stubs, although it may be necessary to create simulated inputs if the real environment for execution is not available.

Developers rarely think of all the cases to test

Integration testing is typically performed by the development team. Although the developers are best able to fix any problems they find, they are also the least able to imagine the variety of situations the system may face. They naturally put the system through the situations they imagined when they designed it, and not necessarily the situations that the requirements were intended to reflect; hence the need for the third type of testing.

Acceptance Testing

Customer acceptance testing requires an agreed-upon script

Acceptance testing is performed by an external customer of the system (or of a product that requires the presence of the system to accomplish its function) as a step in accepting delivery of the product. There is often a financial and/or legal consequence. If

Development Activities

acceptance testing fails, the product will be rejected, resulting in nonpayment or even penalties. Therefore, the "script" the customer will use to perform acceptance testing should be very clearly defined, and reviewed ahead of time by both parties for accuracy and fairness. It must not contain any new capabilities that were absent from the stated requirements, or failure is likely.

Alpha Testing

Alpha testing is a form of acceptance testing in the sense that it consists of assessing the execution and performance of the product from the same perspective as an external customer, even though it is conducted by an internal, but independent team. There is no legal implication in this case. The alpha team's goal should be to ensure that all possible fair acceptance tests have been run against the system so that a customer's final acceptance test would pass. For this reason, the alpha test team must exercise all possible legitimate functions of the software without being influenced to leave some parts of it untouched. This is why the test team should be separate and independent from the development team.

An internal but independent group performs alpha testing

Beta Testing

Beta testing is performed by a select group of external customers who are contracted to be guinea pigs and to provide early feedback before a full rollout of a product. For beta testing to be effective, there must be an exchange of services and resources between the supplier and the customer. The supplier must provide software, training, support, and documentation; the customer must provide time from its people, and typically, computer resources. The benefit to the supplier is clear: find the problems early, before full deployment. The benefits to the customer are less clear, which is why it is often difficult to find a willing "beta test site" (unless the customer is interested in the early availability of software that works reasonably well, or unless the supplier offers a reduced price for the final product

Beta testing requires an explicit cooperative agreement, but does not apply to all situations

in exchange for the customer's resources invested in the beta test). Beta testing is often performed on products that are produced for a mass market, such as personal productivity software (word processing systems, etc.). On the other hand, acceptance testing by external customers presumes a small number of key customers, since it is clearly impractical on a large scale.

Different types of tests detect errors from different sources

Acceptance testing, alpha testing, and beta testing all tend to test the software against its requirements, and will therefore find errors of any type, regardless of the phase of the lifecycle in which they were introduced. Note that if unit testing and integration testing could be done exhaustively, these "customer-oriented" tests would only find errors introduced during requirements analysis. These errors are differences between what the customer intended, what he wrote, and what the analyst understood. However, since there is typically an infinity of possible test cases, lower-level errors left undetected by the previous tests are also found.

4.10 Software Distribution and Support

Packaging and Deployment

Transfer the system to its target environment

The goal of software distribution is to successfully transfer the system from the development environment to the execution (or target) environment. These may actually be the same, as in the case of an application developed for internal use. However, for commercial software, they are normally different. When this is the case, distributing the software implies that it is transported between the two environments via some medium.

Steps included in packaging

Packaging the software for distribution involves the following steps:

Development Activities 155

- Gathering the files that need to be included in the distribution. If the work has been planned correctly, this may be unnecessary. All the files may be in a well-defined directory. Selecting at the last moment which files to ship and which ones to exclude is both time-consuming and error prone. If for example, some configuration files are omitted, they can render the software useless. On the other hand, including source code or other elements from the development environment can also have negative effects. At best, it clutters the distribution medium. At worst, it breaches the confidentiality of the supplier's design.

    Creating a master set of files

- Creating or updating the release notes with the latest information available. Release notes should, in particular, contain a description of any known bugs that were consciously left in the release because they were considered too minor to delay the release date. Information on "work-around" procedures should be included.

    Release notes

- Preparing installation procedures or scripts. The user-friendly, customizable "installer" package one often finds on the first floppy disk of a personal computer distribution kit typically performs a complex sequence of tasks, including finding and removing previous versions. In a UNIX environment, where users are assumed to be more computer literate, today's installation procedures are often very skimpy, consisting of minimal shell scripts or simply paper instructions. Suppliers of UNIX-based software would be well advised to emulate the ease of installation provided by their PC or Macintosh counterparts.

    Installation scripts

- Creating one copy of the distribution medium, and verifying that it can be installed correctly. This is a necessary precaution before mass production. The test installation should be performed on a "clean" system to avoid the situation in which the installation works at the supplier site because of some particular leftover files or environment conditions, but does not work on the customer's differently

    Create and verify the master copy of the distribution medium

configured system. One should not only verify that the installation succeeds, but also that the software itself runs correctly. Often a missing file will not cause the installation to abort, but it will prevent a successful execution.

Duplication
- Duplicating the distribution medium (typically cartridge tape, floppy disk, or CD-ROM) after the test installation has proved successful. As part of this procedure, unique license keys may be generated and written in specific files (or patched into the executable program) in order to provide license-based access control.

Assembly and distribution
- Physically putting together the medium and the documentation (along with additional enclosures such as warranty cards, license agreements, etc.) and shipping them.

Deployment differs across types of markets

Deployment takes place at the far end of the supply chain. Depending on the type of software, it may be performed directly by the customer or by agents of the supplier (field personnel, resellers, etc.). The larger the distribution volume is, the more likely it is that the installation will be performed directly by the end user. Deployment includes:

Preparation
- Preparing the target environment to receive the software. This may involve configuring a new system, or it may consist of removing an old version and making sure there is enough disk space available. Some prerequisite software or hardware may also be installed at this point.

Installation
- Installing the software itself, using the computer call or human procedures provided in the installation kit.

Configuration
- Configuring the software and possibly customizing it. Most network-oriented software will require that some network resources (e.g., the location of a database server) be identified. Some environment variables need to be set to specify default directories for files manipulated by the application.

Development Activities

- Documentation may need to be extracted from the distribution medium and placed on line. With the emergence of new documentation support capabilities such as web browsers and other hypermedia capabilities, installing the documentation is becoming more complicated (but also more powerful) than simply removing the cellophane wrapper from a set of manuals. — *Placing documentation on line*

- Performing the customer's acceptance procedures, if applicable. — *Acceptance*

- Training the users and commencing use of the product. — *Training and start-up*

Packaging and deployment end with a transition out of the development lifecycle and into the rest of the product's life.

Maintenance and Support

Most systems are developed for use over a long period of time—often a few years, sometimes many more. Usage of the product will typically generate a stream of support activities. Some of these will in turn result in maintenance (i.e., modifications to respond to unanticipated problems). Both support and maintenance involve collecting and classifying information and responding as necessary. Table 4.4 shows five types of events and the typical responses they trigger. — *Software may not wear out, but it requires support*

Providing support can be quite an elaborate operation, depending on the number of customers and the complexity of the application. A customer, whether internal or external, is naturally very sensitive to the quality of the communication with the support staff. The support hours are a traditional issue, particularly in international situations. In a more mundane way, the features of the communication system matter. For example, the automatic routing of phone calls to the proper number after hours and e-mail access are not insignificant. — *Good support is crucial and may be hard to provide*

Problem tracking

A database of problem reports is usually a necessity. Note that it serves several functions:

- Liaison between the support function and the development staff, who can directly find the description of the problems in the system and report when these have been fixed;
- An expertise repository for the support staff, who can look for similar problems and solution descriptions when fielding a customer call;
- A historical database from which product quality metrics and problem frequency analysis can be performed.

Five types of events and the support tasks they generate

Event	Response
Questions	Answers. If the same questions occur too often, this may indicate an unfriendly design, or documentation that is insufficient or inaccurate. In these cases, the support staff must be able to generate change requests internally.
Problem report	Change request, unless it can immediately be determined that the problem stems from incorrect usage, in which case, advice can be provided. Even in this case, however, statistics about "pilot errors" should be kept because they may again point to documentation problems.
Enhancement request	Change request. Contrary to problem reports, these requests may be reviewed by Marketing in order to be prioritized with respect to the strategic direction for the product evolution.
Environment change (e.g., a new version of the operating system or of the hardware)	Change request. This may take a high priority because a customer may conceivably be left without a workable product if there is an incompatibility.
License expiration	Collect maintenance fee or license renewal fee, provide new license key if necessary.

Table 4.4 – Support Events and Responses

Support events that result in change requests trigger maintenance. Maintenance is really another form of development, except that it is typically incremental and smaller in scope and effort than the initial development. Nonetheless, it may consist of a complete application of the development lifecycle, especially for enhancements, which may become complete projects requiring teams of significant size. A requirements specification is often needed, followed by conceptual and detailed design, coding, testing, and distribution. This also happens for the repair of bugs that come from a missing or misunderstood requirement. If a bug comes from an error in design or in coding, the complexity of the task is reduced because a smaller subset of the lifecycle has to be applied.

Maintenance is one form of development; it may be as complex as development itself

The completion of an enhancement or bug repair (a maintenance action) requires a new form of testing, called **regression testing**. The goal of regression testing is to verify that the software has not regressed (i.e., does not fail in places where it used to work). This is necessary because repairing one bug may inadvertently cause another bug (previously fixed in the same area of the code) to reappear. Over successive releases of the software, regression testing leads to the accumulation of a **regression test suite**, a set of test cases that verify that old bugs have not resurfaced. It is part of the nature of regression test suites to grow almost continuously. It is very difficult to ever remove a test from a regression test suite without incurring a risk. This is just one more good reason to avoid, through reviews and other techniques, introducing the bugs in the first place!

Regression testing prevents old bugs from coming back

Chapter 5

Object-Oriented CASE Tools

5.1 The Benefits of Computer-Aided Software Engineering

Computer-aided software engineering (CASE) consists of using computers to automate or aid portions of the software development process. This is a simple concept: those who create software should be able to use software to make their own work easier. This is just like the cobbler's children who, in theory, should have the best shoes in the world. In reality, this is seldom the case, and likewise, the software engineer's computer tools may not be as sophisticated as those provided to hardware design engineers or molecular biologists.

Types of Tools

What can CASE tools do for the software professional? Entire books, such as [McClure 89], have been devoted to this topic. At minimum, CASE tools *assist* in creating the software deliverables, *keep track* of them, *automate* some of the transformations, (e.g., source code into object code) and *verify* certain properties of the system. Let us review these aspects in more detail.

Tools can assist, keep track, automate, or verify

Editors for various notations

- **Assistance in creating deliverables.** In the same way that a word processing system assisted us in creating this book, allowing us to type and edit text, a CASE tool can be an elaborate editor for the various notations required during software development. This includes the graphic notations required by various modeling techniques. General drawing tools such as MacDraw and Powerpoint can be used to create simple graphics but cannot maintain the connectivity needed between icons when they are moved around. Most of these tools are inadequate for drawing some of the more esoteric shapes required by certain methods (e.g., Booch clouds). Text editors also fall into this tool category and may include any number of additional features to assist software developers. For instance, they may automatically use different fonts and colors to display the different syntactic elements of source code (e.g., variables names in red, function names in green, etc.).

Tools that organize the work

- **Keeping track.** Once a certain number of models, source code modules, test cases, and documentation files have been created, it is sometimes difficult to keep track of them, especially for a team of several developers. "Keeping track" may be as simple as allowing powerful browsing through directories. However, the real bonus of using CASE tools is in the area of change control and configuration management. This consists of recording successive changes, and controlling which versions of which artifacts are combined to construct a given version of the complete system.

Compilers and other transformation tools

- **Automation.** The most common form of automation is compilation, that is, the generation of object code from source code. Other transformations may be partially automated. "Partially" is the key word; some claim it is possible to convert analysis diagrams directly into code, but at this time, such is an illusion (see Chapter 8 for further discussion of transformation engineering).

- **Verification.** Many different categories of tools fit this general description, because there are many characteristics of a system that may be subject to verification. Of course, the ideal CASE tool, which would take a program, analyze it, and report on the location and nature of every error, has not yet been invented. Nevertheless, there are things that can be checked. Some tools verify that a running program does not "leak" memory locations (i.e., that it does not endlessly allocate memory and then fail to release it). Others measure precisely where a program spends its time. Yet others track which instructions have been executed during a series of tests, and which ones have not.

    Checkers, profilers, etc.

- **Prototyping tools.** These could fall into the "assistance" category, and prototypes are also valuable as a "verification" tool. Because of this multiplicity of functions, they deserve a separate category in this list. One can use them to ascertain the feasibility or the performance of a certain design, or perhaps to check that the requirements for the user interface of an application have been correctly understood.

    Prototyping tools are hard to categorize

The Benefits of CASE Tools

The benefits of CASE tools fall primarily into two categories: increases in productivity and increases in quality. Both result in cost savings and/or shorter time-to-market. In addition, CASE tools create or solidify a common culture and process within an organization.

CASE tools increase productivity and quality

CASE tools improve the *productivity* of software developers because they allow certain tasks to be performed faster. For example, creating a Harel statechart using pen and paper is possible, and is often done as a first step during a discussion or a design brainstorming meeting. Ignore, for the moment, the fact that word processors and presentation tools have accustomed us to a flawless rendering of text and drawings, and that

CASE tools accelerate the creation and modification of deliverables

we have therefore become increasingly intolerant of the imperfections of handwritten text and drawings. It nonetheless remains true that if this statechart needed to be redrawn on paper every time a minor change were made, it would be quite time-consuming. Instead, a CASE tool provides the proper palette of prefabricated icons corresponding to the chosen notation, stores the completed drawing in a file, and makes it possible to perform a small change and print the new version in a matter of minutes. In addition to these "model editors" (of which text editors are one special case), tools that generate part of the source code also save time.

CASE tools prevent or help correct mistakes

CASE tools increase the *quality* of the software produced because they either prevent mistakes from being made, or they help identify and correct those that are made:

- In the area of *prevention,* tools that generate code templates from design models prevent simple typing errors that can occur when all the code is created manually. Static analysis tools detect suspicious constructs *before* they have a chance to impact a running system. Change control or configuration management tools prevent confusion between versions; hence, they prevent the user from selecting and/or using the wrong components.

- In the area of *correction,* debugging tools, profilers, memory leak detectors and other static or dynamic analysis tools help pinpoint problems. They are also used after correction to verify that the problem has been resolved. Defect tracking databases and tools allow developers to more easily retrieve the information submitted with the complaint, and to prioritize their activities so that the more important defects are given the greatest attention.

Object-Oriented Case Tools

Table 5.1 summarizes some of the relationships between types of tools and their impact. Note that this is based only on common categories of tools and is not intended to be an exhaustive list of CASE tools.

Summary of the benefits of typical CASE tools

Type of tool	Productivity impact	Quality impact
Analysis and design model editors	Reduce creation time Save redrawing time	Result is easier to read Tools check correctness
Code template generators	Save coding time	Templates are correct by construction
Static analysis tools		Suspicious code is identified before causing errors
Configuration management	Permits developers to work in parallel, merge later	Confusion between versions is prevented
Debugger, memory leak detector, etc.	Eliminate need to add special statements to code	Cause of errors can be more easily identified
Profiler	Focuses effort quickly	Better performance

Table 5.1 – Impact of CASE tools

Higher productivity obviously means lower costs to develop the same amount of deliverables. These savings can be directly measured across successive projects, assuming the size of the deliverables can be compared from one project to another (which means that there has not been a significant change of development paradigm in the meantime). Rather than consuming fewer resources to do the same work in the same amount of time, an organization may instead be able to reduce the actual time-to-market of a product. This can result in dramatic gains in market share by beating the competition, but is usually impossible to quantify in revenue terms.

Impact of productivity

Higher quality results in higher customer satisfaction, which can help keep existing customers and even win new ones. Many competitive benchmarks are not won by a company as much as they are lost by its opponents because the software

Impact of quality

crashes in the middle of a test run. In this case, a simple bug in the software can cause the sudden evaporation of millions of dollars in potential orders. Quality can also mean cost savings in situations involving mission-critical software (e.g., launching a satellite and correctly placing it in orbit, or reliably controlling a nuclear plant for years). Needless to say, the difference between success and failure in such visible cases can also greatly influence the evolution, positive or negative, of an individual's career.

Another benefit: a common process and culture

Finally, CASE tools foster a common process and culture because they impose common ways of performing tasks. The idea is not to transform the software engineering department into a regimented, creativity-free organization, but rather to add a certain level of discipline exactly where it is most needed. For instance, a version control system may require that a one-line reason for a change be entered when a new version of a file is checked in. Or it may require the "electronic signature" of at least one other developer. Such simple steps progressively establish a new discipline and foster a new level of responsibility which translates as: "I do not change my code just because I want to, I change it because there is a reason that also matters to the rest of my team." The requirement to follow some common rules and practices when producing software means that when one developer is unavailable, others can at least maneuver their way through their colleague's work products. Moreover, the use of CASE tools means that successive projects are more likely to repeat the same procedures successfully, rather than being totally at the mercy of the many differences that exist between various developers' skills and preferences.

The Costs of CASE Tools

There are multiple components to the cost of CASE tools

In the process of adopting CASE tools, one must not only consider the benefits of CASE tools, but also their costs and any known limitations. When comparing tools, people often focus on the "sticker price" (i.e., the initial cost of a single license of

Object-Oriented Case Tools 167

the tool). There are, however, several other important cost components of CASE tools. For CASE tools that run on high-performance workstations, the purchase price of a license may be several thousand dollars per copy, which constitutes the majority of the total cost to the user. For tools that run on personal computers, the price may be only several hundred dollars, in which case it becomes more likely that other components of the price will dominate. To build a complete "cost of ownership" analysis, all the following components should be considered.

- **License price.** This is not always simple to determine. If the tool is available on several different platforms, the price may depend on which one is chosen. There are also a variety of licensing schemes in existence, but few suppliers offer all of them:

    Even the license price is not a simple concept

    - *Single-user licenses* are permanently allocated to specific users, who are registered by their account name (or "login id").

    - *Floating licenses* can be temporarily "grabbed" from a server by a user when needed, and are relinquished (for use by someone else) when no longer needed. Calculating how many floating licenses are needed for a certain population of users is a difficult task. It depends on how often and how long the license is needed. For example, a compiler may be used in brief bursts of activity, while a configuration management system may be in constant use by each developer throughout the day.

        Floating licenses are a useful trend

    - *Site licenses* allow any users located in a particular building or center at a single site to use the product whenever needed.

    - *Corporate licenses* extend this concept to an entire company, regardless of location.

Competitive discounts: when the time comes to bargain!

In addition, suppliers typically offer volume discounts for large quantities of single-user or floating licenses, and there is always the possibility of competitive discounts. This means that a supplier will offer a special price (usually with strings attached, such as a higher quantity or the purchase of add-ons) for a limited time in order to win an order.

CASE tools may require additional resources

- **Infrastructure.** A certain amount of supporting hardware and software may be necessary to run the CASE tool(s). Because they store a lot of "metadata" which allows them to keep track of the structure of the deliverables they help create, CASE tools tend to consume huge amounts of disk space. A CASE tool may also require some networked resources, such as a database that serves as an information repository. This database may need to reside on a separate computer for safety or performance reasons. For instance, the workstations or PCs on which a new tool is installed may have been sufficient for other tasks, but may suddenly run out of capacity if the tool is a "CPU hog." Since some tool suppliers hesitate to disclose all the infrastructure enhancements that may be required, these are costs that are hard to estimate. Yet they must be included in the total cost.

Additional annual fees

- **Maintenance cost.** The right to obtain free upgrades when new versions are released, and the access to direct support when the tool malfunctions or explanations are required, are provided for a yearly fee that ranges anywhere between ten and twenty percent of the initial license price. If the same tool is used for several years, it is therefore quite possible that the total amount paid to the supplier over time will be twice the "sticker price" of the tool.

Users must be trained, which costs more than just money

- **Training.** Each user needs some formal training, and for more complex tools, this may take several days (typically two to five). If the number of users is large enough, a special in-house training class is more cost-effective than sending everyone to a public class—although it has the side effect of shutting down the development group for several days. Sometimes, this initial training is included with the purchase. Alternately, "training credits" are sometimes provided so that a certain number of users can be sent to public classes at no extra cost. Moreover, future recruits will need to be sent to a public class when they join the organization (this can be predicted based on expected growth or on the average attrition rate).

- **Internal support.** It is not practical to assume that paying for maintenance will solve all the support needs. Internal expertise must be developed, and some of it should be centralized in one or two in-house experts who will be so designated. Otherwise, everyone's productivity suffers when a problem arises. Some tools will require the equivalent of one-tenth of a person to support them. Other tools may require a full-time "guru" position, especially if they serve large groups. Two good predictors of the difficulty of support are:

    Do not leave the tool users without support

    - Whether or not the CASE tool uses a database;
    - Whether or not it provides some distributed functionality over a network (i.e., it is not used in isolation by each individual user).

The Process Context

Since we have just considered an important factor, namely cost, that can weigh against adopting or selecting CASE tools, we need to mention another important obstacle. *Tools should not be selected without a vision of the process they are to support.* The entire motivation behind the Software Engineering Institute's Capability Maturity Model (SEI CMM) was that the introduction of CASE tools in the early 1980s—without an understanding of software process concepts—had *not* improved the average capability of organizations to produce software (see

Tools should be introduced within the context of a well-defined process

Section 3.1). A tool that produces nice graphical models of the requirements, for instance, is useless if no one uses these models as an input for design (i.e., if the organization's process to link analysis and design is not defined or followed).

This introduction can be incremental

This does not mean, however, that the vision needs to be complete before *any* tool can be selected. Instead, a spiral process model (similar to the one depicted in Figure 3.3) can be applied: determine a first approach to the software engineering process, define some tasks, buy the tools that support those tasks, train the users, consolidate the implementation of those tools; then go back to the first step and iterate through increasingly complete definitions of the process. At each iteration, the organization has a consistent (if somewhat imprecise) process definition and is able to provide its members with the tools that support it.

5.2 Examples of Object-Oriented CASE Tools

Narrowing the scope to OO CASE tools

This is not a book about software engineering in general, but about applying lifecycle concepts and object-oriented concepts in the same environment. We will therefore limit our discussion of specific types of CASE tools to those that support an object-oriented approach. This will exclude several very important categories of tools that do not depend on the particular approach chosen, such as configuration management tools or project planning tools.

Commercial disclaimer

In order to illustrate the concepts presented here, it will be necessary to cite specific names of tools and companies. This does not mean that these are the only offerings on the market, or that other products are inferior, but only that these are the ones with which we are the most familiar.

Object-Oriented Analysis and Design Tools ("Upper CASE" Tools)

The purpose of these tools is to let the analyst or designer create, store, and edit the various types of models dictated by the object-oriented method of choice. Most methods include some graphical representation of the requirements (see Section 4.4). This is where the focus of the CASE industry has been. Several trends have emerged regarding this category of tools:

Support for analysis and design methods

- Companies that have been supplying comprehensive CASE tools in support of structured analysis and design, like Cadre Technologies with its Teamwork tool and Interactive Development Environments with Software Through Pictures (StP), have been adding support for object-oriented methods in the same framework. For example, Teamwork supports the Shlaer-Mellor method, and StP supports the Object Modeling Technique (OMT). In some cases, the resulting tool benefits from the robustness of the preexisting tool. In other cases, however, the fact that objects are not the supplier's initial approach is still evident in the way the tool behaves or in the low level of support provided for the object-oriented (OO) methods.

Traditional CASE tools with creeping OO features

- Several smaller companies have recognized that OO methods are still in flux, and that the most satisfied users are those who have combined some carefully selected components from several methods. To support this approach, they chose to create highly flexible CASE tools that support multiple notations and allow the user to customize them. Examples include Protosoft by Paradigm Plus, and ObjectMaker by Mark V Systems. Note, however, that there is a price to pay for such flexibility. These tools are usually less mature and robust than the ones created by older, more established companies; support may be harder to obtain, especially once a method has been customized; and the long-term viability of the supplier is sometimes in question.

Flexible CASE tools supporting multiple or even custom notations

Tools created by methodologists

- Some methodologists have decided that they are themselves in an ideal position to create a tool to support their method. Their customers can benefit from the fact that when a new concept is introduced in the method, the author's own tool is likely to be the first one on the market supporting the new features required. On the other hand, one must usually sacrifice the ability to mix and match multiple notations. A well-known example of this kind of tool is ROSE by Rational, Grady Booch's company. Another is the MethodWorks tool provided by ParcPlace Systems in support of the Object Behavior Analysis (OBA) method (the first version of this tool, dubbed OBAganizer—for "OBA organizer"—was developed by SEMATECH in 1991–92).

Graphical User Interface (GUI) Builders

GUI builders have a harder job with third-party objects

Although GUI builders are not specific to object-oriented methods per se, object orientation brings a particular twist to the design of user interfaces. Traditional GUI builders are "screen painters" that allow the user to lay out the appearance of the screen. In some cases (e.g., fourth generation languages like Uniface or SmartStar), each field is directly associated with a database field, and the run-time support system ensures that the screen is updated when the data changes and vice versa. In a more functionally oriented system, the programmer must issue specific function calls to display data or to capture text typed in by the user. In an object-oriented system, however, the display is not merely composed of fields, but of objects whose attributes can be set or read through their methods. The GUI builder needs to understand this in order to create the necessary messages that cause the various objects to display themselves. The GUI builder must therefore access the definition of the objects contained in the selected user interface (UI) class library. Most tools still do a poor job of this. If the GUI builder is supplied by the same company as the UI class library, then there is hope, but only if the designer does not want to add some additional classes bought from another source.

Object-Oriented Case Tools

Because of this situation, developers have found that GUI builders are of limited use in an object-oriented application. Sometimes they can be used to create an early mock-up of the user interface, purely to have it reviewed by potential users or to allow the documentation effort to start early. However, when it comes to writing the actual system, it is just as productive to call the class library methods directly from the application objects. This is not true if the class library is at too low a level. In that case, writing an intermediate layer of user interface objects first may be a very wise investment of time because it enhances the organization's ability to reuse software.

This sometimes limits their usefulness

An unconventional but interesting way to build a user interface prototype is to employ a completely different environment than the one on which the actual development will take place, but one that offers sophisticated graphical classes. For instance, a system simulation tool like i-Logix or a framework like Apple's MacApp allows rapid prototyping. The benefits of this approach may be such that that it does not matter that the prototype has to be discarded because of platform differences.

Frameworks can make GUI builders less attractive

Browsers

Browsers are search and visualization tools designed to help the developer locate classes. Simple browsers may allow a user to navigate a hierarchy of names or to search for a class name by providing some string of characters contained in it. More complicated browsers may retain descriptions of the functionality of a class, or they may allow the user to search for a class on the basis of attribute types, method names, or even method signatures.

Help in locating classes

Programming Language Support Tools ("Lower CASE" Tools)

The tools needed to support programming in an object-oriented language are not inherently different from those used for procedural languages, provided the language is compiled. There

Compilation versus interpretation

are differences when the language is (or can be) interpreted, as is the case with Smalltalk or CLOS. Here are some other considerations that distinguish these tools from their more traditional ancestors:

Precompilers were a simple solution ...

- **Precompilers.** Initially, and especially for languages derived from C (such as C++ and Objective C), compilation was a two-step process. The original code was first translated into straight C by a precompiler; then using any regular C compiler, the result was compiled into object code. This two-step process considerably simplified the work of the compiler writer, since the source and target languages were somewhat similar in syntax and in level of semantic expression. However, it leads to difficulties during the debugging process, as explained below.

... but made debugging harder

- **Debuggers.** Debuggers for any language should have about the same capabilities: setting various checkpoints; executing in single-step mode or until the next checkpoint is reached; allowing the user to examine data values; and showing which methods are invoked and what the argument values are. In the case of precompiled object-oriented languages, since quite a few extra instructions may have been inserted to take care of object initialization, dynamic binding, etc., the information displayed about the source code is harder to trace back to the original code. Therefore, debuggers that are written specifically to handle the object-oriented language, and that work in concert with compiler options, perform a much more useful task.

Message traces are very useful

Since the basic principle of object-oriented systems is that objects interact with one another through messages that invoke methods, a basic mode of operation of an object-oriented debugger is to display a trace of successive message invocations. If methods are kept small and if the rules of encapsulation are obeyed, most errors in coding or

Object-Oriented Case Tools   175

low-level design can easily be found by examining this trace.

- **IDL compilers.** An increasingly popular way to specify the interface to a class is the Interface Definition Language (IDL), a *de facto* standard of the Object Management Group (OMG – see Chapter 9). This language is especially useful when designing systems that distribute objects across an **object request broker** (ORB). The IDL code must be compiled in order to generate the stubs that allow the broker to invoke the objects. Since the Common Object Request Broker Architecture (CORBA) specification is still evolving, it is important to have an IDL compiler that supports the version required by the user (e.g., version 2.0 if interoperability across different ORBs is needed). The IDL compiler may perform another task (which makes it useful even if an ORB is not used), namely, it can create a template (e.g., in C++) for the class definition. A designer can then specify the interfaces to all the desired classes in IDL, and obtain correct and consistent templates generated for all of them. The class implementation code is later entered by hand.

    IDL specifications can lead to object-oriented class templates

- **Syntax-driven editors.** For twenty years or more, researchers have tried to create text editors that automatically guarantee the syntactic correctness of a program. These tools have had limited success for several reasons. Some are too constraining because they prevent the programmer from entering an incomplete statement or expression, which he might want to do as a means of capturing his thoughts before he fully understands the details of the implementation. Others are cumbersome because the language constructs must be selected from a menu. This can take longer than typing the equivalent text, such as "if then else." In general, changing the program once it has been entered is difficult. Replacing "if <condition> then <statement>" with "while <condition> do <statement>" (i.e., changing

    A long-standing but difficult search for more productivity in editing code

the structure while retaining the code fragments it contains so that they do not have to be retyped) is usually hard to do. Structured editors are often unable to deal with code that has been generated or modified using another editor. Finally, these editors require extensive computing resources to achieve decent performance, because they must maintain complex data structures in memory rather than simple sequences of characters.

"Smart" editors are rare, but customizable editors exist

For all these reasons, really "smart" editors are rarely in use today. Instead, customizable editors like emacs (from the Free Software Foundation) have progressively been enhanced with macros that help the programmer visualize the structure of his program. Some of these macros are also available in "freeware" or "shareware" form. When the program is loaded in the editor, variables and function names might be displayed in a different font or using a different color from elements such as keywords of the language.

Look for OO language support

Since these tools provide productivity improvements, they should be used for object-oriented languages, too. Clearly, an editor that understands how to enhance the display of a C program may not be able to do a very useful job on a C++ source file, so it is important to search for tools that deal specifically with the language being used.

Integrated environments have roots in old languages

- **Integrated programming environments.** The inventors of some object-oriented languages took a proactive approach to solving the issue of providing language-aware tools (editors and debuggers) in addition to the compiler by providing a completely integrated, language-specific support environment. This follows a tradition established with early interpreted languages such as LISP and in a more modest way, BASIC.

They are related to interpreted languages

There is a good reason why integrated environments are associated with interpreted languages. When a language is

Object-Oriented Case Tools 177

compiled, the resulting executable (after linking) is a standalone piece of machine code that contains no trace of the original source code. It can be copied onto another machine where none of the development tools exist, and it can be executed on it. Because of the dramatic differences between the two forms of the program, an integrated programming facility covering the entire lifecycle does not make much sense. It is much more important to have a common editor and a common debugger (see Figure 5.1). On the other hand, one develops *and* executes a program written in an interpreted language from within the interpreter. This mode of development provides a natural framework in which all sorts of additional language-specific tools such as browsers, "pretty-printers," etc., can be added (see Figure 5.2).

Since the predominant interpreted object-oriented language is Smalltalk, it is not surprising that powerful integrated environments were developed for it. More recently, companies such as Microsoft and Borland have followed this trend with their Visual Basic and C++ products.

Smalltalk, Visual Basic, C++

Integrated CASE tool environments are described in the next chapter. Since the programming environments we are talking about here focus on coding only, they are less complete in terms of lifecycle coverage. They do not aim at providing interoperability with tools from other suppliers. They act more like a single, complex tool providing multiple functions (somewhat like a Swiss army knife) than like a family of compatible tools. Yet they can provide significant increases in productivity, and if the language selected by the organization has such a capability, they should not be neglected. There are even organizations that select a language such as Smalltalk, in preference to C++, *because* of the existence of such integrated programming environments.

Programming environments are not complete (in the sense used in Chapter 6), but they are still quite useful

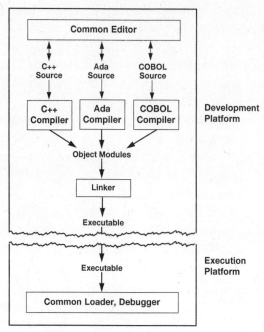

Figure 5.1 - Relationship Between Tools and Compiled Languages

Reconstruct the structure of an undocumented program

- **Reverse engineering tools.** These tools can parse a program, reconstruct its structure, and display it, using different techniques. One form of documentation may be an invocation graph, showing which classes or specific methods invoke other classes or methods. They may also reconstruct a classification tree by extracting the inheritance information from the class declarations. Since these diagrams tend to be very large and complex for a real system, the tool may offer various display and navigation techniques instead of presenting a fixed format. For instance, a graph may be displayed only to a certain distance from the node of interest. In the case of a calling tree, only a few levels may be shown, either above or below the node of interest, etc. Some tools go further and reconstruct diagrams conforming to well-known analysis and design methods, such as OMT or Booch. This is often only partially successful, since all the necessary relationships are rarely apparent from the code.

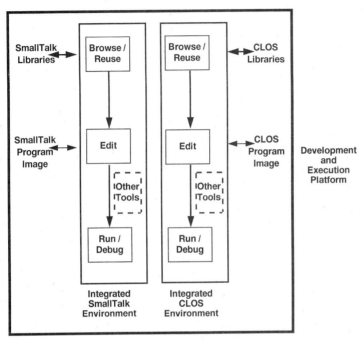

Figure 5.2 - Relationship Between Integrated Environments and Interpreted Languages

Reverse engineering tools sometimes go further than merely displaying the structure of the software; they directly help work on it. The programmer may be able to specify, for example: "replace all declarations of 20-character strings with 80-character strings" or "replace all calls to foo(x,y) with bar(y,x,0)." Because these actions require some pattern matching, they cannot be easily accomplished with a "dumb" text editor. Several capabilities of that nature exist, for instance, in Software Emancipation Technology's Discover.

Reverse engineering tools can help the maintainer

The market for reengineering of OO programs is not here yet.

Reverse engineering tools for object-oriented languages are still in their infancy. Because the market potential is proportional to the amount of code in existence, the priority has been given to procedural languages. The market for COBOL is therefore probably ten times the size of the market for C, which is in turn ten times the market for C++. As object-oriented languages proliferate, so will the need for reverse engineering tools adapted to them. After all, as was the case for C or COBOL, and in spite of the inherent modularity of objects, it is still possible to write cryptic, hard-to-maintain Smalltalk or C++ code.

5.3 How to Select an Object-Oriented CASE Tool

A pragmatic section focused on analysis and design tools

This section contains information to help an organization evaluate and select CASE tools, especially tools in support of analysis and design methods. It includes a list of features that are generally desirable in such CASE tools, with further explanations about those features. Note that slightly different criteria may be applicable to other tools, such as compilers, debuggers, configuration management, and project management tools. We have chosen to limit ourselves to analysis and design tools because this is the area where object orientation requires the most specific support. The same may be said of compilers, too, but choosing a compiler is a somewhat simpler affair.

It is useful to assess the present, not just to select new tools

One way to use this section is for the reader to rate his current environment according to whether or not it supplies the features in question. If too few of these criteria are met, perhaps it is time to consider improving the "toolbox" currently in use. Another way to use the information presented here, assuming that the decision to acquire new tools has been made, is to transform this list into a *request for proposal* that can be submitted to would-be suppliers. After suppliers reply and their products are examined (on paper or through hands-on

Object-Oriented Case Tools

trials), the list can become an *evaluation matrix* to rate and rank the various proposals.

What we *cannot* provide, because of the differing nature of each organization, is a recommendation for the weights that should be applied to each criterion. If one could easily find a tool that meets all the expressed requirements, there would be no issue with the relative importance of criteria. However, this is not the case; each tool is likely to have some of the desirable features and lack others. Some features are more important than others; therefore, one cannot simply say "tool A is better than B because it meets 13 criteria as opposed to 9." For this reason, any rating system used to compare tools should embody some notion of relative importance (e.g., critical, important, optional, indifferent). But one person's "optional" feature may be someone else's "showstopper," meaning that a tool that does not provide this characteristic is automatically excluded from consideration.

The importance of the criteria varies according to the organization

Selection Criteria

We propose and explain below seventeen selection criteria. They are summarized in Figure 5.3.

Seventeen criteria

1) **Support for the desired object-oriented methods.**

   The tool must support either the method or methods selected by the organization. A tool that supports additional mainstream methods (see Chapter 4) should be rated higher than a tool that supports only one method, because this gives the users the ability to change methods later without having to throw the tool away.

   Flexibility is a plus

   The support must be complete and must conform to a very recent version of the method. Object-oriented methods continue to evolve. If a tool supplier stays in touch with the author of the method, and if the tool design is sufficiently flexible to allow the easy introduction or modification of

   Beware of incomplete conformance

diagrams and other notations, it should not be hard to keep the tool in close alignment with the method definition.

Selection Criteria

1	Support for the desired object-oriented methods
2	Not just a pretty drawing tool
3	Generation capabilities
4	Compatibility with customer's lifecycle definition
5	Compatibility with change control and configuration management tools
6	Traceability
7	Repository structure
8	Shared access
9	Import/export
10	Reuse support
11	Platforms
12	Customization and flexibility
13	Ease of use
14	Integration with other tools
15	Support, consulting, and training
16	Costs
17	Continued development

Figure 5.3 – CASE Tool Section Criteria

Some tools were incompletely adapted to new methods

The tool should respect the terminology and iconography of the method it supports. If it does not, it probably means that the tool was created to support another method, and has been "tweaked" to adapt to the one under consideration by the new customer. It is also an indication that too many things are hard-wired in the tool and cannot be easily changed.

2) **Not just a pretty drawing tool!**

A CASE tool must do more than pictures

Some tools are just diagram editors. Their only claim to being CASE tools is that they maintain the connectivity between the various icons that constitute the method's notation, and that their palette of icons contains ready-to-use symbols such as Booch clouds. In reality, a CASE tool should be able to do two more things:

- Prevent the creation of semantically meaningless diagrams;

- Analyze the set of descriptions of a system to produce value-added information.

An example of the first requirement is that a tool which helps draw object-oriented information models should prevent the user from creating a circular graph. Another example is that in a state diagram, there should be a path leading from the start state to every other state, as well as a path leading from every state to one of the end states. In the first example, circularity does not make sense even as an intermediate stage; therefore, it should be strictly prevented. By contrast, in the second case, it is clearly necessary to relax this rule while the diagram is being constructed. It should, however, be applied when the user asks that his work be saved.

Semantic rules

The second requirement is met when the tool is capable of producing reports showing the list of defined classes, attributes, states, etc. It is also met, *ipso facto*, if the tool contains a code generator.

Production of reports

3) **Generation Capabilities.**

Depending on the exact function of the tool and on the method it supports, there are various other artifacts of the software development lifecycle that the tool can produce, either completely or in part (allowing manual completion). These artifacts may include:

Code generation and more

- Code or code templates,
- Documentation,
- Database schemata,
- Test cases and their drivers,

– Prototypes.

4) **Compatibility with customer's lifecycle definition.**

A tool must not constrain the software process

The tool in and of itself should not impose a particular lifecycle or process definition. An organization may start with a somewhat linear, waterfall-like model, and decide to move later to a model that allows more parallel and iterative development between various teams. There is no valid reason why the tool should favor or hinder one choice over another, but the way the tool works could impose certain constraints. This is more likely to occur with tools that constitute a complete, "cradle-to-grave" environment. Such tools are often designed by people who have a clear vision of the whole lifecycle, but who may be imposing their particular philosophies through the tool's policies.

5) **Compatibility with change control and configuration management tools.**

The artifacts produced by the tool must be amenable to CM

Suppose that a CASE tool for an object-oriented analysis method is capable of creating and maintaining information models, state models, process models, code templates, and other documentation artifacts. If all these artifacts are stored in a single, monolithic entity (e.g., a file or database) that constitutes only one item for a configuration management system, then it is impossible to distinguish which changes have been made to which pieces of the whole model. One may be able to find out that one programmer created a new version at 5:43 p.m. on January 14, and that a second programmer made another change at 8:14 a.m. the next day, but that is the extent of the information available.

Instead, one should be able to track successive revisions of each separate artifact, and which combinations of these different artifact versions constitute consistent representations of the complete system. Maybe the first programmer's change consisted of adding the information model for a new object, and what the second did the next morning was

Object-Oriented Case Tools

to add the state diagram for that object. These two changes, taken together, define a valid new configuration. The CASE tool might supply its own facility to create, label, and manage configurations of items, or could even allow the user to apply a configuration management system of his choice (which is possible if each artifact is stored in a separate file accessible by the operating system).

6) **Traceability.**

The tool should maintain as much automatic traceability between artifacts as possible, and allow the user to supplement this information and exploit any information that has been captured.

Traceability may be partially manual

One way of providing traceability is through *links* between items. For instance, the user of a modeling tool should be able to "double-click" (or perform some other easy action) on a class icon in an information model and be taken automatically to the state model for the objects of this class. Some links might be created automatically, while others may be indicated by the user and simply recorded by the tool.

Links that can be followed easily

Another way to provide traceability is to systematically allow the user to attach *annotations* to each item. For example, when changing the data type of an attribute in an object model, the analyst would be able to record in an annotation which part of the requirements specification was not met by the previous version, or which user mentioned in a review meeting that a different type of information was required.

Annotations

Note that an analysis tool capable of storing textual documents, and that supports hypertext links, will automatically allow better traceability of requirements than a tool that does not. An entire requirements specification can be loaded in the tool at the outset, and the various parts of the analysis model can be linked with the words or sentences that suggested them.

Structuring of textual requirements

7) **Repository structure.**

Semantics, not just graphics

The tool should store the models it manipulates in a repository that contains semantic data—not just graphics. For the most part, this is a consequence of requirements **b**, **c**, and **f**, but it is important enough to specify it explicitly.

The repository allows each datum to be captured and stored only once

A true repository, like a database, is organized according to an explicit schema. This implies that common information is stored only once and that the relationships between models the method specifies are expressed directly in the repository structure. For example, the fact that each object in an information model may have one state model is part of the schema definition. Therefore, the name of the object in the information model and the name of the object in the header of the state model are not just character strings that happen to be identical. Instead, they are two instances of the same data item; if the analyst modifies one instance, the other name automatically changes, too.

8) **Shared Access.**

Access by multiple users, with good locking mechanisms

Models should be accessible by several users from their respective workstations or personal computers (assuming the tool is licensed accordingly). A user should be able to check out one part of the model (e.g., a single state diagram), work on it, and check it back in, without preventing other users from working on other parts of the model. Read-only access to components checked out by others may be provided, as long as the user is informed that the component being examined may be undergoing modification by someone else. In this case, the tool should be able to identify who has checked out the component. One useful feature would be to allow users to be wait-listed for a checked-out component, and to be notified (perhaps by an on-screen alert or by an e-mail message) when the component in question has been checked back in.

Object-Oriented Case Tools

The scope of this distributed capability is obviously limited by the physical network—managing collections of objects that are dispersed among users of intermittently connected laptops is not yet a well-mastered process. But if the tool exists on multiple platforms, it should not depend on the homogeneity of the network. In other words, the model should be shareable, in the fashion described above, between users of PCs, UNIX workstations, and Macintoshes, if these are the platforms on which the tool runs.

The network may be heterogeneous

9) Import/export.

The tool should provide the ability to import and export data and graphics according to well-recognized standards. For instance:

Text and graphic interchange

- Model semantics should be imported or exported using the IEEE CASE Data Interchange Format (CDIF).

- Textual reports should be exported in Standard Generalized Markup Language (SGML) or in Microsoft's Rich Text Format (RTF).

- Graphics should be exported using Tag Image File Format (TIFF) or PostScript.

The check-out/check-in capability and the import/export capability may be combined to extract a portion of a repository, move it to another computer, modify it there, and return the modified component back into the original repository. This enables the "nomadic computing" mode of operation that is emerging in the 1990s with the proliferation of laptop computers.

Nomadic computing

10) Reuse support.

A tool can support reuse in two manners. First, it should allow a user to browse existing analysis and design models and to extract fragments to include in other models. Apart from a "cut-and-paste" capability that preserves semantic

Finding objects to reuse

information as well as text or graphics, this should take the form of a library facility in which it is easy to find *qualified* components by name or according to certain properties.

Toward domain-specific class templates?

CASE tools could provide another significant form of reuse if they included templates for commonly found objects. Just as office productivity tools (e.g., word processors, presentation graphics editors, and spreadsheet programs) now offer extensive galleries of clip art and canned formats, a CASE tool might offer general templates for common business objects (e.g., purchase orders, personnel organization, accounts payable), as well as objects found in domains such as manufacturing industries, banking, telecommunications, health care, etc. It may be difficult to define these objects in a generic enough way to make them usable to large numbers of users. Yet the need clearly exists, and the payoff would be great. This is an appropriate challenge for the various market-specific Special Interest Groups (SIGs) of the OMG, which will be further discussed in Chapter 9. These groups could define generic models for their domains, and tool suppliers could implement and redistribute them.

11) **Platforms.**

In practice, the computer platforms in place are a real constraint

Tools should, of course, be selected according to whether or not they run on the customer organization's computer platform(s) of choice. Although in theory, one should worry first about what functionality is needed, and *then* provide the computers needed to support it, in practice, the existence of an installed base makes it too costly or practically impossible to select a CASE tool first and worry later about the platform on which it will run. For the purpose of selecting software, "platform" includes the hardware, the operating system, the user interface standard (e.g., Windows or Motif), and the network protocol stack (e.g., Ethernet with TCP/IP and the Network File System, or NFS). It may be particularly important, for reasons of cost or flexibility, to be able to run the tool on a low-end workstation or on a personal computer. The amount of disk space and memory required by the tool should be examined carefully, although it has recently become possible to purchase large amounts of both for reasonable prices.

If the tool runs on multiple platforms (e.g., PCs and UNIX workstations), one should make sure that all versions have the same capabilities. Some tools formally had PC versions that were significantly less capable than their UNIX counterparts. Furthermore, it should be possible to mix and match multiple platforms on the same network and share repositories, provided of course, that the necessary protocol emulations are available on all machines.

PC versions may be limited in functionality

12) Customization and flexibility.

The tool should use an open, flexible architecture that allows extensions by the customer, given a modest amount of additional training. If the customization process is too arcane for the average user, then it should at least be a service that the supplier provides without charging months of consulting fees. The kinds of elements that can be customized should include:

Flexibility to adapt the tool to the customer's need

- The icons used in the graphics models;
- The contents of the menus (e.g., to add new composite commands);
- The rules used to check the syntax of the models;
- The rules used to generate code, reports, or documentation.

Once the notation, menus, or rules are customized, it should be possible to *lock* these options and deliver to users a version of the tool that implements these changes but which they cannot in turn modify. This makes it possible to define and enforce corporate standards.

The customized features can become a corporate standard

13) Ease of Use.

WIMP (windows, icons, mouse, pull-down menus)

The tool should offer a "what you see is what you get" (WYSIWYG) user interface for all the notations it supports. It should have a modern "look and feel" that is compatible with the standard user interface managers of the corresponding platforms (X Windows, Microsoft Windows or the Macintosh Finder). This ensures the ability to "cut-and-paste" text and graphics across applications, and it minimizes the duration of the learning curve for new users. There should be on-line help, and preferably, it should be context-sensitive, meaning that specific help is provided about the actions that the user can perform or is performing at the point where help is invoked.

14) Integration with other tools.

Support for an integration standard

It is unrealistic to expect that any tool that one selects can satisfy changing needs indefinitely by itself. As needs evolve, different tools may need to be added to the system. This will lead us to the topic of Chapter 6, "Software Development Environments." With regard to the selection of an individual tool, it is important to assess whether or not the tool supports a *de jure* CASE integration standard, such as the Portable Tool Common Environment (PCTE), or even a proprietary standard such as H-P's SoftBench or Sun's Tooltalk (see Chapter 6).

15) Support, consulting, and training.

Multiple course arrangements

The supplier should offer training classes, including classes at the customer site, if so desired. This is often preferable for larger groups that may be impacted by the initial deployment of the tool.

Advantages of purchasing ongoing support

Various forms of support should be provided as part of the support of maintenance agreement:

- Free copies of new revisions of the software;

- Quick response time on patches for major problems that render the tool unusable;
- A telephone hot line with sufficient coverage for the customer's normal work hours;
- Electronic mail access to the support group.

In addition, the supplier should have sufficient consulting staff on hand, and should maintain a list of qualified third-party consultants who know the supplier's product, so that it is possible for the customer to obtain consulting (for a fee) to review and improve the use of the tool or to tutor its users.

Consulting

16) Costs.

There is probably no such thing as an absolute measure of the cost figure that is acceptable or unacceptable for a specific tool. In practice, the cost can be a determining factor in reaching a decision between two or more tools that provide comparable benefits. Of course, all the costs mentioned in Section 5.1 should be considered and transformed into an expected *total cost of ownership* for a period of three to five years (or for whatever the expected time of use might be).

Determine the total cost of ownership

17) Continued development.

Because new capabilities need to be added continuously, the ability of the supplier to continue to develop the tool should be assessed. The new requirements include:

Tools must evolve

- Evolution of methods whose new features must be supported;
- Changing operating systems and other computer technologies;
- Adhering to new standards for data interchange or tool integration.

Selection Process

How to apply these criteria to perform the selection

How does one use the above criteria (or a modified list that takes into account an organization's particular needs) to select a product? This section provides a few guidelines to make the process more objective and rigorous. Note that these guidelines could apply to the selection of almost anything, including methods, project management software, database management systems, etc. We chose to include it in this chapter because this is the first time we encounter the issue of selecting a commercial product.

a) **Assemble a task force.**

Include multiple points of view from the beginning

From the start, it is better to have multiple points of view represented in the selection process, rather than waiting until the end. Otherwise, any of the "stakeholders" in the study (e.g., future tool users) may disagree with the final recommendation and criticize the selection process later on, claiming that the person charged with the selection ignored some of the important criteria. This can be avoided if the selection is performed by a cross section of people representing the various groups affected by the selection. Service groups (e.g., the Management Information Systems or MIS department) that may be affected by the selection because it is they who will have to provide resources and services related to the new software should also be involved. People who are invited to join the task force, but decline because of their workload or other priorities, are less likely to obstruct the adoption process than if they are not asked at all.

Organize the work thoroughly

Someone in the task force, or a separate assistant who is very organized and who has access to word processing and spreadsheet software, should be designated as the task force's secretary. His role is to document the list of criteria, the weights assigned to them, tabulate the ratings agreed upon by the task force, retain documentation from the suppliers as well as notes from task force discussions, and in general, organize the work so that all the members have constant access to all the information gathered during the evaluation process.

b) **Create a list of selection criteria.**

The task force must first agree on the selection criteria. The list provided earlier can be a starting point. The task force may have been specifically told to exclude business issues such as price, letting higher management deal with them; or it may have been instructed to consider these as well. The task force should ask itself whether there are other factors, including "political" ones (alliances with specific companies, prior familiarity with certain products, etc.) that are valid selection criteria.

Develop a customized list of selection criteria

The list is often more clear if it is organized hierarchically. For instance, it is possible to decompose "cost" into "initial cost," "annual maintenance percentage," "flexibility of licensing scheme," "training rates," etc.

Structure the list hierarchically

A frequent mistake is to limit the definition of each criterion to short phrases of two to four words. This invariably leads to ambiguity later on, and can be used to bias the rating according to the subjective feelings of various task force members. Instead, a paragraph should be written to explain each criterion in more detail. The whole task force should review these definitions and agree to them before proceeding. This is a somewhat tedious part of the process, but one that is well worth the effort. An example of the definition of "flexibility of licensing scheme" may be as follows:

Add descriptions of every criterion

> *Flexibility of licensing includes the ability to choose between fixed licenses for named users; floating licenses based on a number of simultaneous, nonspecific users; site licenses allowing unlimited usage at each*

An example: "licensing scheme" criterion

geographical facility; and a corporate license for unlimited use throughout the company. Suppliers offering a greater variety of arrangements will be rated higher. In addition, consider the ease of moving licenses between users in the case of fixed licenses or, for floating licenses, the ability to temporarily exceed the user quota and to pay within a certain grace period, instead of being denied the excess usage.

Communicate this list to suppliers

Once the criteria are defined and written, and the suppliers are contacted in order to schedule the evaluation of their products, these will usually ask to see the list of selection criteria. In general, there is no reason to deny this request. In some cases, marginal suppliers will take themselves out of the race when they realize how serious the evaluation process is and that they do not meet a majority of the selection criteria.

c) **Assign weights to the criteria.**

Four proposed levels of criticality

The relative importance of the criteria should not be assigned at the end, after the products have been studied, because they can be manipulated to obtain a predetermined result. Instead, they should be discussed and settled in advance. On the other hand, it does not add much value to create a very complex scale. The distinctions shown in Table 5.2 have proved to be sufficient.

Through consensus, assign each criterion a weight

Establishing the weights is the first task that requires creating a common result out of multiple opinions. This can be done through some form of consensus building. In other words, discuss the issues until everyone agrees or until each member is willing to accept the prevalent opinion because all other ideas have at least been heard and considered fairly. This can also be done by having each member of the task force vote separately for his preferred ratings, and then selecting the majority or median opinion, but this tends to leave the overruled members more frustrated than if there has been open discussion.

"Showstopper"	A product that does not meet this criterion cannot be selected, regardless of any number of other features.
Important	Meeting this criterion has significant impact on the usefulness of the product, but it is not a "showstopper."
Desirable	This feature may not have a high impact if absent, but it has known benefits to the organization if present.
Optional	An "interesting" feature that may be useful someday.

Table 5.2 – Criteria Importance

d) **Agree on the evaluation scale and calculations.**

The various candidates are going to be examined and rated against each evaluation criterion in the next step; therefore, an evaluation scale must be decided in advance. Scales with a large number of different values (e.g., 0 to 10) are common, but they are prone to a lot of subjective bias. The difference between a "7" and an "8" is often no more than the evaluator's personal preference, and can lead to protracted discussions for a very minor impact on the overall rating. A more objective scale, perhaps not completely devoid of interpretation problems but still easier to apply, is the following one:

A simplified rating scale with only four degrees is easier to use

0 = The product considered does not meet the criterion.

1 = The product partially meets the criterion.

2 = The product fully meets the criterion as specified.

3 = The product exceeds the specification for this criterion.

Questions to determine the rating

This rating scale is easier to use because in a discussion about which rating should be given to a product with respect to a certain criterion, one can ask questions with fairly straightforward answers:

- Is there any feature at all of this product that meets the criterion? If so, the rating should be at least 1, and if not, it should be 0.

- Is there any feature required to meet this criterion that is missing from the product? If so, the rating can only be *at most* 1, but if not, it should be 2.

- Is there any feature that pertains to this criterion but goes beyond the requirements defined by the task force? If so, and if the answer to the previous question is no, then the product should get a rating of 3.

"Showstopper" elimination, and the use of numerical values for weighted averages

Once the rating scale is agreed upon, the way in which ratings and importance are to be combined can be discussed and agreed upon. "Showstoppers" are the easy part: a product that does not meet one or more of the "showstopper" criteria should be eliminated. After that, if multiple products remain in competition, some form of weighted average can be computed from the individual ratings. Note that two minor qualities, or even a larger number, do not make up for a large shortcoming. Therefore, the weights applied to the importance levels defined in Table 5.2 should not form an arithmetic progression such as (4, 3, 2, 1). Instead, a more geometric progression should be used, such as (10, 5, 2, 1). This means that a product will get five more points in its overall score if it exceeds rather than meets an "important" criterion, but it will only get one more point if it exceeds rather than meets an "optional" criterion.

At this point, all aspects of the rating procedure have been defined. Our recommendation then is to create a spreadsheet containing all the necessary calculations that will lead to a final score for each product. The details of this process are trivial for even a casual user of popular spreadsheet programs.

Use a spreadsheet

e) **Perform the detailed rating.**

Finally, after settling all the details, the candidates can be evaluated and rated individually against each criterion. This can constitute a significant amount of work. For example, if there are 50 individual criteria (taking into account the hierarchical decomposition of the top-level list) and 8 products to evaluate, each member of the task force is potentially expected to assign 400 individual ratings. This means that 400 mini-debates could arise in task force meetings about whether a product meets, exceeds, or falls short of a given criterion. In practice, not all task force members will rate all products against all criteria (for reasons of interest, competence, or time available) and forming a consensus opinion may not after all require meeting for dozens of hours. Debate can also be cut short by consensus of the group. Then the task force members who have formed an opinion vote, and an average of their ratings is computed.

Rate each candidate against each criterion

f) **Calculate final scores.**

This is the easy part. It consists of entering all the numbers in the spreadsheet and seeing what comes out at the bottom.

Obtain the results

g) **Review for sanity.**

Just selecting the product that comes out ahead of the others seems rational, but quite often the result of the computation will seem counterintuitive. For instance, the product that heads the list might exhibit some limitations that the task force members feel restrict its appeal. Conversely, an elegant and attractive product may have fared rather poorly.

A critical review of the outcome – last chance to correct weights and ratings (but not arbitrarily)

If this happens, the task force should critique its own work and understand how this occurred. It may be that the first product to be examined looked good at the time, but that exposure to additional candidates progressively jaded the task force, who became more critical in their ratings as time passed. Or it could be that with more knowledge about what the market has to offer, the task force wishes to define some criteria differently. In either case, it is now a simple matter to change certain ratings or weights. One can see immediately what the new results are. Of course, one should refrain from changing the numbers simply to justify an *a priori* selection. This is why having a diverse task force is such a key requirement.

h) **Report the results and obtain a decision.**

An outline for the presentation

Once the task force has converged on a final recommendation, it should present its results to the authority that created it, or that has the power to enact the recommendation (i.e., approve the purchase and deployment of the tool). There is almost a standard outline for such presentations:

- Background: the history and current status of CASE tool usage, and why a new tool is needed;

- Selection process: who the members of the task force are, and what steps they followed (a summary of this section);

- Products considered: how the list was made up and who responded;

- Evaluation procedure: how it was done (through visits to the supplier, hands-on evaluations, calling references, reading "buyers' guides" in computer magazines, etc.);

- Selection criteria and relative weights: "showstoppers" must be described in some detail and other criteria may be summarized;

- Eliminated products: products that failed to meet at least one "showstopper;"

- Ratings: numerical results for the remaining products;

- Business aspects: vital statistics about the suppliers of the top-rated products (size of the company, sales, profits, years in business, location of company offices with respect to where the customer may need support, etc.);

- Costs: what the total cost of ownership of the product is expected to be over the next three to five years;

- Conclusion: a summary of the principal factors for selecting a certain product, and a request for approval of the decision; or if there are multiple, completely equivalent products, a presentation of this fact and a request for guidance on how to proceed.

After the selection process is complete (a process which itself may require more than one iteration), two postmortem activities may be needed:

Postmortem activities

- A complete report on the selection may be written. This follows the same outline as the presentation, but in a more "literary" format and with some of the details that were perhaps only verbalized during the management presentation. This report is not always needed; one should consider whether or not there is a constituency that might benefit from it. Otherwise, the presentation material is sufficient documentation.

Document the study in a report, if needed

- Suppliers considered during the selection process may want a debriefing session. It is only fair to grant such a request, but the confidentiality of the ratings is a matter of company policy. Therefore, the response to this request should be defined in advance. *Under no circumstances* should one supplier be allowed access to the task force ratings assigned to their competitors.

Supplier debriefings

Selection takes time, but it is a key decision

Selecting a new CASE tool, or for that matter any new computer or software technology, is a momentous decision that can have consequences for the capability of the organization for years to come. Therefore, it should be done deliberately and thoughtfully. It is often surprising how much time and effort (by people who have another "day job") must be devoted to this task. Yet the consequences of an ill-advised decision make it important to perform the right steps, and in the right order. The long lists of criteria and evaluation steps in this section may help the reader avoid some mistakes and achieve more efficient and more thorough evaluations.

Chapter 6

Software Development Environments

6.1 Integrating the Development Process

Computer-Aided Software Engineering (CASE) tools, which we looked at in Chapter 5, provide support for specific activities of the software lifecycle. They do not, however, provide any linkage between activities, nor are they automatically compatible with one another. Tool X might be the best one for an organization's design activities, and tool Y might be best for user interface prototyping. But if X and Y are selected separately, there is no guarantee they will work well with one another, or even that they can be used on the same workstation without difficulty.

Tools that are individually useful cannot always work together well

Since the activities of the lifecycle need to be inscribed in a well-defined process, as seen in Chapter 3, the tools that support these activities also need to fit into a common structure which ensures that, taken together, they support this complete process well. We can therefore define the notion of *environment* by contrast with the role of individual CASE tools:

A common structure is needed

A CASE tool provides automated support for an *activity* of the lifecycle; a **Software Development Environment (SDE)** provides automated support for the software *process*.

A real source of productivity

Such environments are receiving increasing attention from software developers and managers because they improve the usability and productivity of the individual CASE tools that they integrate. Once integrated within a complete and consistent environment, CASE tools can finally eliminate the redundancy and inconsistency often found among the elements of a software project. Note that there are several equivalent phrases to designate this concept of environment:

Three equivalent names

- "Software Development Environment" (SDE);
- "Software Engineering Environment" (SEE) (a phrase used almost exclusively in the U.S. Department of Defense and related projects and conferences);
- "Integrated Project Support Environment" (IPSE) (a phrase more often used in commercial circles).

Additional facilities, not just tools

To provide integrated support for the entire process, one needs to add several facilities to the individual tools. The most important addition consists of **integration facilities** that allow multiple tools to be used in a compatible or integrated manner. Building on these facilities, **process management** capabilities can also be provided.

How objects impact environments

Integrated software development environments assume additional importance when an object-oriented approach to software development is used. Since all the tools manipulate a common concept of objects, a well-managed and shareable **object repository** becomes a key capability. But in addition to providing access to objects, integrated environments may themselves use object technology, too. Their components often communicate through messages which can be transported by an object request broker (ORB).

6.2 The Dimensions of Environment Integration

A software engineering environment is made up of three main components:

- A *set of CASE tools*, which provides the required functionality to the developer (as described in the previous chapter);

- A *repository*, which stores all the artifacts created by the tools (analysis and design models, code, test programs, plans, etc.);

- A *framework*, which provides the mechanisms and interfaces to connect the other pieces together.

Three components: tools, repository, and framework

The role of the framework is to provide a common set of integration facilities and utilities so that the different tools can work together, even though they may be provided by different suppliers. This requires, however, that the framework and the tools adhere to a common standard. There are three uses for the integration facilities, and they are referred to as the three dimensions of integration:

The framework provides three dimensions of integration

- **Data integration.** In order for the tools to provide a seamless environment to manipulate the software under development, data integration services allow the tools to access a common repository of information. Each tool requires a different view of the information, but different views of the same objects should not require that the information be duplicated. Duplication not only wastes space, but more importantly, it requires periodic translation of data between tools. If the translation is not performed

Common access to data

every time the data has been modified by one tool, another tool may work on obsolete information. Moreover, if each tool has its separate storage mechanism, simultaneous access to the same subset of the information (e.g., an object model) by multiple users cannot be performed safely. Figure 6.1 summarizes the possibilities and limitations of both scenarios.

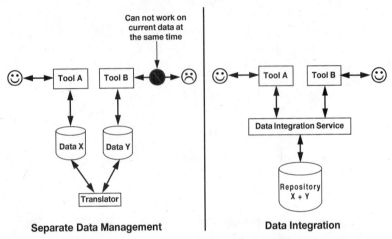

Figure 6.1 – Data Integration versus Separate Data Management

Files are not enough: finer granularity is needed

The file services of an operating system have long provided a basic form of data integration; that is, a text editor can read and write a source code module, and a compiler can read the same file. However, CASE data integration requires more powerful capabilities, akin to those of a database management system (DBMS). For instance, the definition of a class should be shared between an object-oriented analysis or design tool, a class library browser, and a prototype generator. This typically requires that the data services work at a lower level than that of an entire file; thus the operating system services are no longer sufficient, and a finer level of granularity is required. The data integration service can be implemented conceptually by an object query language—whether or not the underlying technology is actually an object-oriented database management system (OODBMS) is a separate question. Data integration can only succeed if a common information model or schema has been defined, including at minimum those data entities used by more than one tool. When this is not the case, and each tool continues to use its own separate objects—even though the actual storage mechanism is common—one sometimes speaks of "shallow integration."

- **Presentation integration.** In order to make synergistic use of various tools, the user must be able to see multiple windows on the same screen, each presenting information from a different tool. To avoid constant changes in the mode of interaction with the computer, all the tools should be consistent in their "look and feel." Not only should they all use the same up-to-date "WIMP" (Windows, Icons, Mouse, Pop-up Menus) approach, but they should also perform the same functions in similar ways. If "Alt-D" means "duplicate" in one tool and "delete" in another, serious confusion is guaranteed. Presentation integration also allows the user to "drag-and-drop" information between tools, so that names of objects never have to be typed more than once, even if they are required in multiple places.

Common user interface management

Achieving presentation integration may simply require that the various tools use the same graphic client-server protocol (e.g., X11R5) and follow the same style standard (e.g., Motif from the Open Software Foundation, or OSF). The framework may optionally provide additional interaction primitives that the individual tools can invoke to enhance their user interface.

Inter-tool communication mechanisms

- **Control integration.** This allows the individual CASE tools to interact with one another directly. A typical example is the interaction between a compiler and a text editor. When the compiler encounters a syntax error on line 27 of a source code module, it can invoke the user's preferred editor and "tell" it to open the source file and position the cursor on line 27, highlighting the part of the line that caused the error. Once the user has modified the file and saved his changes, the editor may signal back to the framework (or directly to the compiler, depending on how the services are designed) that the compilation can be attempted again. Such a level of integration can considerably speed up the iteration between tasks. It does tend, however, to make a user less careful about his work since it is so easy to "just try again" after a small change. Technically, tool integration may be as simple as launching a separate program, using common operating system capabilities, or it may be based on an object-oriented message-passing scheme.

The ECMA reference model, a.k.a. the "toaster model"

A model of how these three forms of integration should be organized, and how the CASE tools should relate to the framework, was developed by members of the European Computer Manufacturers Association (ECMA), under the leadership of Hewlett-Packard's research laboratory in Bristol, England. This reference model was published in [ECMA 90] and is the *de jure* standard in this area. It was adopted, with some modifications, by the National Institute of Standards and Technology (NIST) in the U.S., which effectively made it an international standard. The ECMA/NIST reference model contains a diagram, reproduced in Figure 6.2, that depicts the relationships between the three forms of integration. English readers will recognize the typical kitchen toaster, which is why this diagram is known in the industry as the "toaster diagram." Instead of bread slices, one inserts CASE tools into this "toaster." By extension, the entire model has become known as the

Software Development Environments

"toaster model," even though the toaster diagram is only one small portion of it.

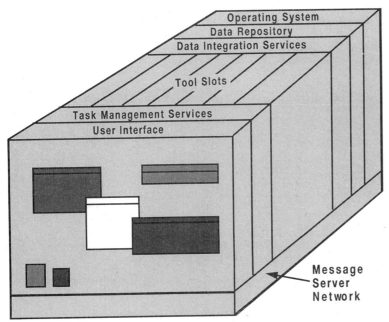

Figure 6.2 – The "Toaster Model"

In this model, one is able to distinguish the various integration services mentioned earlier, as well as a few new elements:

- **Task management services** provide the correspondence between user actions, captured by the user interface management system, and the actions that the individual tools need to perform. For instance, if the integrated environment displays a common menu of possible process steps, clicking on one of the menu items must be transformed into the invocation of the appropriate CASE tool, which has been **registered** in the environment as the agent for this task.

Mapping of user actions to the tools that respond to them

Data storage

- **The repository** is the actual information storage capability which the tools share via the data integration services layer. It can be a relational or object-oriented database, a file system, or a virtual file system (a concept used in such operating systems as UNIX, Mach and OSF/1).

CASE tools may be encapsulated to fit within the framework's slots

- **The tool slots** represent more than the actual CASE tools themselves. Since some CASE tools may not have been designed to work in the context of the control, data, and presentation services provided by the framework, an additional set of interfaces may be added through "encapsulation." A **capsule**, or **wrapper**, is a program that meets the interface specifications of the framework and is engineered to invoke the encapsulated tool according to the protocol it is able to recognize. A good example is the encapsulation of a UNIX command that can accept a parameter (e.g., the name of a file) and several switches. Such commands were designed for a non-graphical user interface, but may still constitute the appropriate mode of interaction with a configuration management system (as an example). A good wrapper for this command would present the user with a dialog box that allows "point-and-click" selection of a file, and lets him press "radio buttons" to select the desired switches. In addition, there may be two buttons called "Execute" and "Cancel." When the user presses "Execute," the wrapper verifies that a file name has been selected, constructs the proper UNIX shell command line, submits it to the command interpreter, tests whether or not the action succeeded, traps any error messages, and displays a pop-up alert indicating success or failure. Thanks to the encapsulation, the user is now provided with a consistent user interface, even though the tool in question had none of the interaction capabilities found in other, WIMP-oriented CASE tools.

6.3 Integration Standards

Initially, the integration of multiple CASE tools was done in a proprietary manner by various suppliers, such as Cadre Technologies and Interactive Development Environments. As the need for a supplier-independent mechanism was recognized in the 1980s, a number of standardization efforts were undertaken with varying degrees of success. We will mention several of these efforts below, and focus on the one most widely supported to date, namely, the European-born Portable Common Tool Environment (PCTE). Much more information on technologies available for integration can be found in [Brown 94].

From proprietary integrated environments to standards-based ones

Presentation Integration

Presentation integration is usually achieved, provided the various tools conform to a common user interface management standard. Fortunately, these standards have now been in place for a number of years, and this dimension of integration is therefore better satisfied than the other two. In the UNIX world, the two standards that jointly permit effective integration are:

Today's UI standards support presentation integration

- **X Windows** (from the Massachusetts Institute of Technology), a client-server protocol and toolkit for the exchange of graphical user interface (GUI) information between the "client" (an application) and the "server" (the workstation display subsystem). In addition, the X/Open Portability Guide (XPG) provides guidelines to write applications based on X Windows.

- **Motif** (from the Open Software Foundation), a user interface "look and feel" standard and window manager that has progressively supplanted proprietary GUI managers such as Sun's Openlook.

"Drag-and-drop" adds considerable productivity

One of the effective ways in which presentation integration manifests itself is the ability to "drag-and-drop" information between two windows that are managed by different tools. Suppose that a browser displays in one window a list of the methods of a class, and that the user is editing source code in another window. In this case, "drag-and-drop" means that the user can click on the name of a method, drag the name to a location within the editor window, and release the mouse button. The name dragged over to the window is automatically inserted in the target location, thus avoiding the effort to retype as well as the risk of a typing error.

Data integration

IBM's AD/Cycle: a data integration precursor with limitations

An early effort at data integration in a CASE environment was IBM's AD/Cycle product ("AD" stands for Application Development). AD/Cycle provided a repository of CASE data that could be used by multiple tools. The repository did not really provide a specific service interface for data integration; that is, access to the data simply consisted of issuing Structured Query Language (SQL) queries since it was handled by the DB2 relational database management system. However, the idea of having a common repository, and of defining enough of a common information schema to push multiple suppliers toward sharing the same information between independently written tools, was innovative. Ultimately, several factors conspired against this particular implementation. First and foremost, relational databases are not well suited to the complex relationships that exist between the data elements manipulated by CASE tools. Moreover, DB2 runs only on IBM mainframes and clones, a model of computing that peaked in the 1970s and has since lost ground. The need for graphical user interfaces pushed CASE tools toward PCs and UNIX workstations even more rapidly than other applications, which made a mainframe-based environment obsolete.

Software Development Environments

The only *de jure* standard effort in this area is the Information Resource Dictionary System (IRDS). This was developed and approved by the American National Standards Institute (ANSI) accredited standards committee X3H4 in 1988. There is a BSI/ISO (British Standards Institute/International Standards Organization) version, but it has differences with the ANSI version. ANSI IRDS is subdivided into six parts:

The IRDS standard

- X3H4.1: reference model for IRDS;
- X3H4.2: service interface to external software and extensions for object-oriented tools;
- X3H4.3: interchange file format for import/export of CASE data;
- X3H4.4: local naming conventions and name administration;
- X3H4.5: integration of an Information Resource Dictionary (IRD) schema from an external source into a local IRD schema;
- X3H4.6: extensions to a distributed, heterogeneous environment.

ANSI IRDS consists of six parts

Data integration must not be confused with data interchange. Again, Figure 6.1 illustrates the difference. On the left-hand side, there is data interchange between the two tools' private repositories, while on the right-hand side, there is data integration because a single repository is shared. X3H4.3 only addresses interchange as a secondary aspect of integration, because it may be necessary to extract the contents of a complete repository in order to move the data elsewhere. The Electronic Industries Association's (EIA) CASE Data Interchange Format (CDIF) is strictly, as its name indicates, an interchange format; it provides no capability to integrate CASE tools within a common framework. The same situation exists with the IEEE Software Template Language (P1175).

Integration is more powerful than interchange

Control Integration

BMS and CORBA as communication mechanisms

An early effort at control integration was the Broadcast Message Server (BMS), which became a component of Hewlett-Packard's SoftBench product around 1990. In more recent years, the emergence of the Object Management Group's (OMG) Common Object Request Broker Architecture (CORBA) as the protocol of choice for distributed object communication (see Section 10.2) has encouraged developers to choose it for inter-tool communication. However, the machinery required by CORBA to ensure delivery of messages across objects located on different platforms has been detrimental to its performance within a single platform, which is where most of the interaction between CASE tools used by one individual occurs. Suppliers are working on lightweight implementations of CORBA for non-distributed applications, a direction that promises interesting improvements.

Tooltalk

Sun Microsystems developed a control integration capability called Tooltalk. Tooltalk is a distributed service—the tools can be located anywhere on a network. It offers both procedural and object-oriented interfaces to the tools. Sun has promoted Tooltalk as an industry standard by making it available to the Common Open Software Environment (COSE) consortium that includes Sun, IBM, H-P, UNIVEL, USL, and SCO. In spite of this effort, Tooltalk has essentially remained a Sun-only product and its acceptance has been limited to users who have no heterogeneity or portability concerns. Sun's position is that Tooltalk and PCTE are complementary and should be merged.

U.S.–based Framework Standardization Efforts

DoD efforts to foster U.S. integration standards

The U.S. Department of Defense (DoD) has played a leading role in trying to foster the emergence of integrated environment standards. This is for good reason; incompatibilities between CASE data and tools used by its numerous subcontractors have caused significant loss of time and information in the development of military systems. The degree to which this has resulted in malfunctions and accidents—and the cost to the taxpayer—may never really be known. DoD-led efforts included:

- The **Common APSE Interface Set - A (CAIS-A)**, a "nested acronym" in which APSE in turn means Ada Programming Support Environment, was a late-1980s effort to create a common set of services for Ada-oriented CASE tools.

- **Arcadia** was a consortium, financed by the Defense Advanced Research Projects Agency (DARPA—now ARPA), to define environment architectures.

The DoD focused on Ada-oriented tools

Efforts by the DoD have met with limited acceptance within the military contractor community, and have had absolutely no *direct* impact in other areas. The main reason is that these efforts have revolved around the Ada language, which is not used much outside of this community. On the other hand, the funding of these projects by the U.S. Government has resulted in significant progress in the understanding of the needs and concepts in question. Therefore, any success enjoyed by other groups is going to be partially due to the efforts of this community.

Little direct acceptance, but a significant influence on other efforts

One commercial U.S. effort, worth at least a historical note, seemed at first to have some potential for success. ATIS (originally meaning Atherton Tool Integration Standard) was the result of Atherton Technologies' effort to create a neutral framework for tool integration, the Software Backplane. Digital Equipment Corporation then decided to adopt ATIS as the integration core of its COHESION software development environment. In the process, Digital not only changed ATIS to mean A Tool Integration Standard, but it also changed the specification, resulting in two incompatible definitions. By 1991, the battle was joined between ATIS and PCTE, although some people were trying to create a compromise architecture in

ATIS was an early commercial contender

which one of the two interfaces would emulate the other. ATIS might still have prevailed and obtained significant results if COHESION had emerged more rapidly, but it did not. With the simultaneous emergence of PCTE as a strong contender, interest in ATIS waned, even within Digital. Therefore, ATIS is no longer a player on the CASE integration scene.

Europe-based Efforts

Other significant results have come from European consortia:

ECMA's PCTE
- We already mentioned the fact that the open *reference model* for tool integration was developed under the auspices of the European Computer Manufacturers Association (ECMA). Another ECMA technical group then developed the PCTE specification, which will be examined in more detail below.

ESPRIT's Sphinx
- The Sphinx project, funded by the European Strategic Programme for Research in Information Technology (ESPRIT), was an early effort to define a method for the encapsulation of CASE tools.

PCIS: NATO's effort at a common interface specification

Once PCTE emerged as a strong contender from Europe, an effort was launched to attempt to merge the PCTE and CAIS-A interface specifications. This effort, started by a North Atlantic Treaty Organization (NATO) working group on Ada Programming Support Environments (APSE), was called the Portable Common Interface Set (PCIS). It concluded in 1994, achieving significant progress in the definition of process management capabilities, including a type/class model of process information, a description of operations written in the Interface Definition Language (IDL) of OMG, and bindings for Ada and C. The adoption of the ECMA reference model—albeit with some changes—by NIST signaled in practice the end of an independent commitment to non-ECMA-based integration efforts by the world community.

Software Development Environments

The Portable Common Tool Environment (PCTE)

PCTE, also known originally as the ECMA-149 standard, is a specification rather than a product, which is appropriate for a standard. The purpose of the standard is to provide an integrated set of environment services (for details, see [Wakeman 93]):

A more detailed look at PCTE

- An Object Management System (OMS) which is the same as the data integration "backplane" of the toaster model. Much to the chagrin of the purists of object technology who would, of course, have preferred an object model, the PCTE OMS is based on an Entity-Relationship-Attribute (ERA) model.

  Object management

- Schema management services, which allow new entities, relationships, and attributes to be declared. In addition, in PCTE, each tool uses only a subset of the information that can be represented in the OMS. This partial view of the repository structure is what constitutes a schema. The schema management services allow the administrator to create *schema definition sets,* which in turn are combined to create a schema for a given tool.

  Schema management

- A number of operating system-like services, including process management (meaning the management of computer process execution and not the management of the software process), interprocess communication, notification, security, concurrency control, and integrity control, which allow multiple users to share the resources of the environment.

  Services to complete the environment

The last characteristic of PCTE means that it has sometimes been presented as a complete computing environment, and not just a facility to support CASE tool integration. This view has not prevailed, however, largely because of the preeminence of the UNIX operating system in the world targeted by PCTE. Conversely, PCTE is sometimes described only as a repository

PCTE is neither a complete computing environment nor just a repository

management system. However, this is too narrow a view, and may be tainted by familiarity with narrower standards such as IRDS. Instead of these two extreme views, PCTE has established itself as a standard for the construction of CASE environments.

American acceptance was slow in coming

Since PCTE originated in Europe, its acceptance in North America was somewhat delayed by the "not invented here" (NIH) syndrome. At the time, Digital Equipment was proposing ATIS as an alternative, and it seemed the two continents would evolve incompatible standards. The adoption of the ECMA reference model by NIST changed this. So did the fact that two of the three prime contractors (IBM, and a Unisys division that has since become Paramax) in the Department of Defense's STARS program (mentioned in Section 3.5) selected PCTE-based approaches to prototype the STARS Software Engineering Environment. In the early 1990s, a number of U.S. companies formed the North American PCTE Initiative (NAPI) in order to promote the use of PCTE. This group then began to explore the relationship of PCTE to object orientation, and the potential overlap between the PCTE model of control integration and OMG's CORBA (Section 10.2).

NAPI and the OMG PCTE SIG

The outcome of this process was a decision made in 1994 by NAPI members to place themselves under the umbrella of the OMG; NAPI then became the **PCTE Special Interest Group (SIG)** of OMG. Its OMG affiliation has allowed the PCTE SIG to gain an international stature and to again involve the European creators of PCTE in its work. As of this writing, the PCTE SIG is working on two directions for PCTE evolution:

Two directions for PCTE evolution: fine granularity...

- **Fine granularity:** evolving the model of the PCTE repository so that an entity in the PCTE object management system can be a single *fine-grained* object, such as a single class or attribute in an object model, or a state in a behavioral model. This has serious consequences in terms of the

Software Development Environments

number of objects that need to be manipulated efficiently, but it is key to effective integration of object-oriented CASE tools. Work on fine granularity in PCTE has benefited from early work done by William Harrison and his colleagues at IBM in Toronto.

- **Object orientation:** providing an object-oriented interface to the PCTE services, and changing the repository model from an entity-relationship one to an object-oriented one, or at least superimposing one of the views on top of the other, if feasible.

... and object orientation

Separately, ECMA submitted PCTE to the International Standards Organization's (ISO) "fast track" adoption procedure, allowing PCTE to become an ISO standard during 1994.

ISO PCTE standard

Implementations of PCTE

In this book, we have refrained from mentioning specific products, both because of the fast-changing nature of the products themselves and because we wish to remain objective. But since there are few examples of PCTE implementations and of PCTE-based environments, and because these have not yet seen wide adoption in the market, we will make an exception in this case and mention several PCTE-related products that appear to embody the most sophisticated principles of integration to date.

Commercial disclaimer

Since PCTE is a standard for the *interfaces* between CASE tools and the rest of the environment, one must clearly distinguish between two types of products: implementations of PCTE (libraries and utilities that implement the standard) and PCTE-based environments (complete SEEs making use of a PCTE implementation). Several implementations of PCTE are:

Two types of PCTE-related products

Five examples of PCTE implementation

- **Emeraude PCTE**, a product of the French company Transtar (a joint subsidiary of Groupe Bull and SYSECA, the software supplier division of the electronics company Thomson-CSF). Through an agreement with Sun Microsystems, a version called Emeraude TCI (Tool Control Integration) was released on Sun workstations in 1994. TCI represents the first attempt at the unification of PCTE and Tooltalk, and is also an introduction of object orientation (specifically, the notion of methods) in the PCTE OMS.

- **PORTOS**, a multi-platform product of Electronic Data Systems (EDS), which uses an object-oriented database management system (DMBS) under the OMS service.

- **IBM PCTE**, from IBM, which also uses an object-oriented DBMS.

- **Heuristix**, a product of Heuristix Systems that implements most—but not all—of the PCTE specification and uses the normal UNIX file system.

- A toolkit from the British company IPSYS.

Software Engineering Environments Based on PCTE

Seven examples of environments based on PCTE

- **AIX CASE**, a complete environment developed by IBM's laboratory in Toronto. The basic technology was the result of a 1991 agreement with Hewlett-Packard, through which H-P's SoftBench product was first ported to the AIX operating system, and then enhanced by connections with CASE tools supplied by companies having partnership agreements with IBM. AIX CASE was initially released in May 1992, running in a homogeneous network of IBM workstations. Since then, IBM has developed AIX CASE client software running on Sun and H-P workstations.

IBM added data integration to Softbench

Since SoftBench does not include data integration, IBM added this capability in the form of a repository (developed using the Versant object-oriented database management

Software Development Environments

system) with a PCTE interface. Furthermore, IBM has been developing a version of AIX CASE running on its OS/2 operating system, with a view toward unifying its AIX CASE and mainframe-based AD/Cycle environments via access to a common repository.

AIX CASE includes the following main components: SDE Workbench/6000, which rather than a tool, is the software development environment itself, allowing the user to display his various objects and tools on the desktop; SDE Integrator/6000, the tool which allows a "super-user" to encapsulate a CASE tool within the environment; and CMVC/6000, a version control and configuration management tool. This latter component transparently uses one of three underlying version control tools: the public-domain SCCS, RCS, or the commercial tool PVCS from Intersolv. CMVC/6000 runs in client-server mode, with a configuration management server resident on one machine per network.

AIX CASE is packaged as separate components

- **COHESIONworX,** an environment from Digital Equipment running on their version of UNIX. COHESIONworX is conceptually derived from DEC's earlier COHESION product, which used ATIS, but has been reimplemented on top of EDS' PORTOS.

Digital's environment

- **Entreprise II**, a product from EII Software, another French offspring of Thomson and SYSECA. Entreprise II started as a project funded by the French Ministry of Defense in 1989. It includes a number of so-called "horizontal services" such as project management, documentation, configuration management, reuse management, communication tools, etc. Note that SYSECA is one of the investors in Transtar, so there is more than mere coincidence in the fact that Entreprise II uses PCTE.

Focus on horizontal services in Entreprise II

- **PCTE Workbench**, a toolkit product from Vista Technologies that stresses the ability to create complex relationships between tools and hypermedia documents.

- An environment created by Paramax, a division of Unisys, as a submission to the U.S. Department of Defense's Software Technology for Adaptable, Reliable Systems (STARS) program.

Process management emphasis

- **Life*FLOW** and **Life*LINK** are two products announced in 1995 by Computer Resources International, a Danish company. Both are based on the ISO version of PCTE; Life*FLOW supports a variety of process management functions including process modeling and enactment, while Life*LINK provides "traceability management."

- **Integrated System Development (ISD)**, a product from Groupe Bull.

A structure that allows multiple suppliers to focus on their respective strengths

The importance of these products (especially the more complete framework-based environments such as Entreprise II, AIX CASE or COHESIONworX) is that they demonstrate the ability to integrate CASE tools bought from multiple suppliers (mostly American), thanks to the use of an open standard. Moreover, these products allow the seamless addition of new tools into the environment at a later date. Many tool suppliers have finally realized that it is better to do one thing correctly than many things poorly. The concept of CASE tool integration is allowing them to capitalize on their strength without shortchanging their users. The SEE IT alliance, created in 1995 by ten U.S. and European companies, embodies this principle. Its goal is to "deliver integrated solutions" for software engineering teams worldwide. It contains companies with broad-ranging software environment offerings (Digital, Cadre, IDE), companies selling framework technology (EDS), and companies that sell specific tools to support one or more of the functions of software development (Atria, i-Logix, Interleaf, Rational).

6.4 Process Management Facilities

In Chapter 3, we described how an organization's *software process* should be modeled explicitly in order to provide discipline and repeatability to the development activities. When an organization uses a set of unintegrated CASE tools, it is impossible to use the computer to manage the process, since each tool only knows about a single task. But when an organization uses an integrated SEE, it becomes possible to capture the process definition and implement it through the environment. Whether or not this is actually done depends on two things: the product's capabilities, and the user's willingness to use them. Products that include extensive process definition and enforcement features are called "process-driven environments."

Integration makes process-oriented services possible

Here is a list of the extra services provided by such an environment:

- **Process modeling.** This service allows a "super-user" to enter in the system the definition of the process followed by the organization. This can be done graphically, using one of the notations presented in Section 3.5. Other environments may use some form of rule or constraint language, or a tabular representation of the activities, the transitions allowed between them, and the exit criteria that allow an activity to be viewed as complete. The environment may come with some predefined process templates which customers can tailor to fit their needs.

  Capture the process definition

- **Role and user registration.** This service allows managers to define categories of users for the purpose of access control (see below). Individual users are then entered in the system and assigned one or more roles. This makes it

  Define the privileges of each user

unnecessary to manage the user privileges in detail at the individual's level. Once a new hire, for example, has been assigned the role of "developer," he automatically and consistently inherits all the privileges and constraints associated with the unique definition of the developer role.

Restrict access to prevent accidents

- **Access control.** Access to different data, tools, and procedures is restricted to some user roles in order to avoid inadvertent modification of artifacts or to enforce process rules. For instance, only a few people may be allowed to create new branches in the configuration management tree, while everyone is allowed to use more basic configuration management, such as checking files in and out.

Instantiate the process model into a project plan

- **Project planning.** Project planning software tends to exist outside of the development environment, but providers of integrated environments, as well as users, have begun to realize that a project plan is an *instance of a process model*, with schedule and resource information added. Therefore, some environments contain an integrated project planning and tracking capability. In this case, a project plan template is created from the process definition, ensuring that none of the tasks and milestones included in the process definition are omitted from the project plan. Since process models can contain iterations, but project plans must have a specific number of tasks in them, the user must specify early on how many iterations are expected to occur. For instance, one needs to know how many times a prototype will have to be modified after being reviewed. A good process-based project planner would allow the project manager to simply say at a later date, "add another iteration here." This would automatically alter the structure of the plan. Then instead of having to manually perform the dependency changes introduced by the new iteration, the manager would only have to fill in the durations and resources applied to the new tasks.

Software Development Environments

- **Process enactment.** This service consists of enforcing the process rules by authorizing, denying, or suggesting specific activities on the basis of the process definition, the state of the various artifacts (e.g., modified, approved, etc.), and the roles of the users. The sophistication of this service depends on the capability of the product, but also on the level of enactment desired by the organization (if the product can support different levels). These levels were defined in Section 3.6. One relatively mild form of enactment, which is also fairly easy to implement, is to send alarms by electronic mail to project managers and other members of the team when certain events occur (e.g., checking a file into the configuration management system or compiling a new module without error).

    Enforce the process rules according to the selected level of rigor

- **Process state and history services.** Keeping track of the state of the process (what deliverables have been completed, what approvals are pending, etc.) and keeping a log of the process history can be invaluable for audit purposes, perhaps to find later the root cause of any error. This information can also be used to improve the process, for instance, by detecting which transitions seem to be problematic. For example, if a requirement specification seems to always require two or three cycles before it is approved, perhaps one should determine whether or not adding some earlier validation would help.

    Keep information for audit and measurement purposes

    Based on the accumulation of historical information, a *metrics service* (which is a particular kind of history service) can provide various interesting correlations between effort, quality, complexity, and other measurements of the characteristics of the artifacts being produced.

PCTE does not support process enactment very well, but will evolve in that direction

At this point in time, one of the limitations of PCTE is that it does not include very strong facilities to implement process enactment. Although the technical details are too arcane for this book, suffice it to say that PCTE is not equipped to intercept or "trap" the calls made by CASE tools to system facilities to verify that process policy rules are not violated by a particular action taken by a tool. A "wrapper" for a tool, for example, can perform some verifications, but only before or after invoking the tool it wraps. It cannot control what the tool or its user will do once the tool is active. And should a policy violation be detected, and the tool's actions result in a change in the contents of the repository, it may be much too late to reverse these changes once the tool completes its execution. It is likely, however, that the results of the PCIS effort mentioned earlier will be one input into the next round of revisions of the PCTE standard, resulting in stronger facilities for process enactment.

6.5 Selecting a Software Development Environment

First Things First

Usually, separate CASE tools are introduced first

What is the right time to select a software development environment? Until recently, most organizations had no integrated environment, and usually, no one who knew what such an environment was. When users are aware of the benefits of integration between CASE tools, it is usually because salespeople from a commercial environment supplier have convinced them of these benefits in order to sell their product. The fact that a standards-compliant framework allows the creation of an *open* integrated environment is obviously not something these salespeople will point out. For these reasons, and because the cost justification of an integrated environment is harder to present than that of individual tools, the environment is usually the last thing to be introduced.

Software Development Environments

The reader may therefore be tempted to reverse the order and choose an environment first, then populate it with tools. This may not work either, for two reasons:

Conversely, selecting the environment first may not work

- Some environments severely restrict the list of tools that can be plugged into it, or offer no facility to easily integrate third-party tools.
- Some environments tend to promote a certain view of the software process, for instance, because they include (or exclude) the capability to monitor or enforce the application of the process.

Because of these constraints, we propose in Figure 6.3 a more logical order for the selection of the complete development environment:

A recommended selection order

- First, select the *process* the organization wants to follow. This determines the phases (analysis, design, etc.) the organization wishes to follow, and the degree of process support desired.

Select the process

- Second, select the *methods* used during each phase. For instance, select specific object-oriented analysis and design methods, and an object-oriented programming language.

Select the methods for each phase

- Third and fourth, but simultaneously, select a *set of tools* that supports the selected methods and an *environment* that both supports these tools (while being open to the subsequent addition of other tools) and offers the required degree of process support, if any.

Select the tools and the environment simultaneously

In an organization that has nothing in place, this rigorous selection order is possible. In many cases, of course, this is not practical. There may be a large installed base of disparate tools that need to be integrated because they cannot be replaced. This may be for economic reasons (the cost of the new tools), for legacy reasons (existing repositories of information would

In practice, this order may not be possible

become useless if the tools that created them were abandoned), or for productivity reasons (to avoid a new learning curve). They may even be the best tools for the job, in which case there is no valid reason to replace them. In this case, the above figure may be hard to follow if the tools do not lend themselves easily to integration. It may be necessary to choose an environment that has extremely powerful integration capabilities (e.g., a script language and a GUI builder to encapsulate tools) so that the existing tools can fit within the environment.

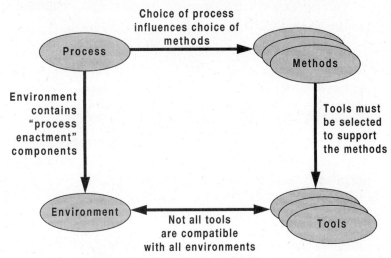

Figure 6.3 – Process, Methods, Tools and Environment: a Selection Order

Selection Criteria

Requirements for the framework as well as the integrated environment

This section mixes requirements for integration framework technology and for environments built on frameworks, because the end user does not really care about the difference between the two. Only the supplier of environments cares specifically about the selection criteria for frameworks. He should not disregard, however, the criteria for a good environment because his prospective customers might use them to judge his product.

a) **Degree of integration.** The framework (or environment) should provide all three dimensions of integration (data, control, presentation), and this integration should fit our earlier description of the needs for cooperation between multiple CASE tools.

Cover the three dimensions of integration

b) **"Look and feel."** The framework should support a unified "look and feel" for the various tools integrated in the environment. "Cut-and-paste" (through a textual and graphical clipboard) and "drag-and-drop" should be provided *between tool windows*.

Presentation standards

c) **Object orientation of the desktop.** Most modern environments running on a graphic workstation present the user with a "desktop" metaphor; that is, the screen represents a work area populated with icons depicting various entities. If one can select such entities (which may represent completely different artifacts of the software development process) and apply to them the same "message" (open, print, duplicate, etc.) with results that depend on the type of entity, then the desktop is said to be object-oriented. To take an example unrelated to CASE tools, the Apple Macintosh desktop is object-oriented. Documents have a type and "know" which application created them. "Double-clicking" on a document always means "open it," whether the document is a data file or a program. By contrast, the Microsoft Windows 3.1 desktop is not object-oriented. One can see only programs, not documents. The Windows File Manager mitigates this shortcoming, but entities (files) are not really typed. Their type is deduced (often correctly, sometimes incorrectly) from the file name suffix.

An object-oriented visual metaphor

Access to fine-grained objects

d) **Repository Granularity.** All CASE tools do not deal with objects at the same level of granularity. An analysis and design tool may want to store each icon in a graphical model as an object in the object management system, while a compiler may have a source code file as its atomic entity, even if it contains hundred of lines in violation of good modularity principles. The choice between coarse and fine granularity may also be a performance compromise. If a CASE tool needs to read hundreds of tiny objects through the data integration interface before it can display an analysis model, it may be much slower than if it is able to read one large object containing the whole model. An environment should offer fine granularity facilities to the tools that wish to take advantage of them and can do so with good performance.

Compatibility with a broad list of tools

e) **Availability of Tools.** The selection of the integrated environment should not force the user to choose from a narrow list of tools because they are the only ones integrated within this environment. Preference should definitely be given to environments that are accompanied by a long and diverse list of integrated tools, or to environments whose suppliers can demonstrate that integrating a new tool is a very small effort (see next criterion). The list of CASE tool functions in Chapter 5 may provide a good checklist to investigate the completeness of the tool suites available within a given environment.

Availability of management services, especially CM

It is also important to evaluate the availability within the environment of common services such as configuration management (CM) and compound document management. Because CM applies to the entities manipulated by the CASE tools (textual requirements, graphical models, code, etc.), and framework technology provides a new way to store

Software Development Environments

these artifacts in an object management system, a file-based CM program may no longer work properly within an integrated environment. Instead, the CM program must be written so that in the process of accessing the objects in order to manage their successive versions, it exploits the framework's data integration interfaces. The same is true of a document management facility, in which *hyperlinks* can be established between different documents. Since hyperlinks now need to refer to objects, not files, a compound document management program must either be cognizant of the data integration interfaces, or it must be directly provided as part of the environment (rather than as a separate product).

f) **Ease of adding tools.** No organization can know in advance the list of tools it will need for years to come. Adding a new tool (from a different supplier) in the environment may require some careful analysis of that tool's data model and its possible interactions with other tools. But once this is done, a reasonably well-trained software engineer with good command of the operating system's command language should be able to perform the actual "plugging in" of the tool within a few days. The documentation for the environment should specifically describe how to do this. If it does not, the prospective buyer should become suspicious. This may mean that adding a new tool is difficult, or that the environment supplier has never really tried it (yes, this can happen), or that he wants to discourage his customers from extending the environment in this manner because he also sells tools.

Ability to customize the environment by plugging in additional tools

The environment should contain facilities to create a simple graphical user interface around a non-graphical tool or command language script. It should be possible to create menus or buttons for the various options that are normally selected on a command line, list boxes for

Graphical wrappers around command line-oriented tools

the selection of input file names, pop-up windows to display results and messages, etc.

Support for process definition and user roles

g) **Process Support.** The environment should support the definition of a software lifecycle process. In particular, it should allow the environment administrator to define the privileges of the users in terms of *roles*. Instead of specifying that a certain user is allowed to create, delete, or rename schemas in the object management system, one can specify separately the following three elements:

- The role called "schema manager;"
- The privileges "schema managers" have, such as the ability to create, delete, or rename schemas;
- The assignment of this particular user to the role "schema manager."

Depending on the degree of process enactment desired by the user, the availability of other process support facilities may be an important selection criterion. A complete environment should offer the ability to define explicitly a process, using a notation similar to one of those in Section 3.5. And it should offer the ability to attach verification or notification tasks to the start or completion of a process step.

Services to manage the repository

h) **Repository Facilities.** The fact that a framework provides a data integration interface (e.g., the PCTE OMS interface) does not automatically imply that an environment constructed from this framework takes advantage of it to offer information management services. Yet without these "out-of-the-box" services, the user of the environment, in order to manipulate the contents of the repository, must buy additional products, or worse, create his own additional modules. Here are some of the functions needed in the environment:

Software Development Environments

- A **browser** to navigate within the repository's structure.

- An **on-line backup** service to create a consistent backup image, for safeguarding purposes, even when the repository is actively being queried and modified.

- A **repository recovery service** to fix inconsistencies in the structure of the repository, should it have been corrupted by a bug in one of the tools or in the environment itself. If the repository is managed by a third-party DBMS, this product may supply the recovery utility.

  A repair capaability is important

- A **schema management** facility to allow new object classes or data types to be added to the structure of the repository without shutting down the system.

- **Import/export tools** to permit the conversion of the repository contents (or part of it) into a format that can be transported to other platforms or processed by other utilities.

i) **Administration.** In the lifecycle of the environment, several events occur that require a series of administration tools. The richness of this set of tools is an important indication of how easy it will be to "live" with a given product. Some of the functions to be supported are:

Tools that help install, configure, and maintain the environment

- **Start** and **stop** the environment. Starting means that users can "connect" to it; stopping may mean a controlled "phase down" in which new users are first prevented from logging onto the environment and current users receive messages asking them to save their work and exit. This allows a controlled stopping process in advance of scheduled maintenance operations.

- **Relocate** or **replicate** resources around the network.

- **Register** new tools so that they can connect to the environment, and so that other tools are able to use control integration mechanisms to invoke them.

- **Register** new users (assigning them a role that automatically defines their privileges) and delete old users.

- Maintain an **audit trail** of all the important actions that affect the system, and in case of problems, permit the controlled access to this audit trail.

Interactive performance can affect user motivation

j) **Performance.** Using an integrated environment can only be helpful to the user if he is motivated to use it. However, this motivation can rapidly disappear if he faces a stiff penalty in terms of performance. If, for instance, it takes much longer to begin editing a source code module within the environment because of the access to an object management system than it takes to invoke the text editor on a file, users will circumvent the system and render it useless.

Location transparency for both data and tools

k) **Ability to be distributed.** It should be possible to view a collection of networked workstations, which may not all run a complete suite of CASE tools, as a complete and location-transparent environment. In other words, the user of the environment should be unaware that some of the documents or tools are located on another machine. For all intents and purposes, the user should believe that the entire environment (tools and data) is on his machine. Two compromises may have to be accepted:

Software Development Environments

- Availability of a given tool may be restricted by the terms of the user's license (e.g., a floating license may be in use by another user).

- Performance may decline when accessing tools and data across the network.

l) **Portability or Interoperability.** The framework or environment selected should be ported (or portable, but ported is better) to the various platforms of interest to the customer. These may include one or more flavors of UNIX and Windows NT. Then, assuming there are versions in existence on multiple operating systems, one may want these versions to interoperate so that they provide a seamless environment in an organization using different platforms. This may allow the object repository to be shared across platforms. For instance, in a mixed PC and UNIX environment, it may be possible to have a full-fledged implementation of the environment, including the repository part, residing on UNIX, while a "leaner" version, without a repository, runs on the PCs. In such a scheme, the PC-based CASE tools run locally, which permits high graphical performance. But when they request access to objects from their local framework, it transparently obtains them from its peer on the UNIX server. Data integrity may be better ensured by this system; moreover, issues of data replication are avoided. This is the same concept as client-server database management systems, but with the framework providing the "glue" between the local CASE tools and the local or remote information repository.

Availability on the computer platforms of interest to the customer, and the ability to network together versions on different platforms

m) **Conformance to Standards.** The integration mechanisms provided in the environment (the framework) should follow a recognized standard. Given the dramatic technology changes of the early 1990s in this domain, this really means that the framework should

PCTE or CORBA

use PCTE and/or CORBA. Other bases for integration may provide a fine technical solution within the scope of a specific vendor's products, but do not ensure that the investment in this environment will survive beyond the specific platform and tools under immediate consideration.

Services required from the supplier

n) **Support – Consulting – Training – Documentation – Costs – Supplier Profile – Active R&D.** These criteria are defined in the same manner as for CASE tools (see Chapter 5).

Chapter 7

Project and Deliverables Management

7.1 The Planning of Object-Oriented Projects

Many aspects of software project management are independent of the approaches and techniques used in the conduct of the project. Some well-known general principles apply to all projects, for example:

Project management is largely unaffected by object technology

"To fail to plan is to plan to fail."

Whether or not a project uses object technology, the computing domain is still immature in its level of planning discipline, especially compared to other domains such as mechanical or electrical engineering. The results, in terms of planning accuracy and delivering quality products on time and for the budgeted cost, are therefore often dismal. This is captured with some humor in *Golub's First and Second Laws of Computerdom:*

Software project planning is still immature

> "A *carelessly* planned project takes three times longer to complete than expected; a *carefully* planned project takes only twice as long."

> "Fuzzy project objectives are used to hide the embarrassment of imprecise estimates."

This chapter focuses on OO projects

There are a number of good principles and techniques of project management applicable to any software or system project, and that should be employed in object-oriented systems as well. There are many books and management courses on this topic. The purpose of this chapter is to address those aspects of project management specific to object-oriented work.

Object technology spells both good and bad news to the project manager:

Separation into tasks is easier

- On the positive side, the *decomposition* of the product into objects that have well-defined interfaces can be used to more easily separate the project into multiple, relatively independent tasks than was the case with previous techniques.

There is less guidance available

- On the negative side, the *novelty* of object technology means that there is little experience in the project management of complex object-oriented projects, and few metrics that can be applied to estimate the effort required by an object-oriented development. One solution is to seek relevant case studies. Another would be to employ industry consultants who have acquired experience elsewhere and who can provide an estimate.

The delayed appearance of code makes progress harder to measure

- Also on the negative side, spending the proper level of effort in analysis and design, looking for reusable components, creating the right framework or foundation classes, etc., may lead to the late appearance (compared to other approaches) of the actual application. Certainly, measuring the progress of the development by counting lines of code is likely to be disappointing for quite a while, until suddenly the pieces start to be assembled and the application begins to emerge. Yet impatient managers share with little children a marked propensity to ask "Are we there yet?" immediately after the trip has started and to continue asking until the journey's end. Moreover, they may be very hard to satisfy, even if the answers to that question are correct. The solution to this problem is not obvious, but once a good work breakdown structure has been created (our next topic), it becomes possible to count how many specifications have been approved and how many designs have been completed and approved. One can then transform this information into a progress metric.

New Work Breakdown Structures for Object-Oriented Projects

Planning usually consists of the following phases:

- First, create a *Work Breakdown Structure (WBS)* of the project. In other words, decompose the project hierarchically into tasks until each piece is small enough to assign a cost or duration to it.   *Project decomposition: the WBS*

- Next, *estimate the effort* required for each task, the resources that can execute it, whether or not the task can be performed by multiple resources in parallel, and what dependencies exist—in other words, which other tasks must be completed before the task in question can begin.   *Effort estimation*

- Create a viable *schedule* from the above information, providing enough time to perform all the tasks, allocating each resource up to a percentage that does not exceed its stated availability, and respecting the interdependencies between tasks.   *Scheduling*

Other plan elements

- Document other risks, costs, and prerequisites, such as procurement of material resources or software tools, in order to construct a *complete plan* (not just a schedule).

Launching

- Review the plan, and upon approval, *launch* it officially.

Structured projects break down the work by lifecycle phases

A good work breakdown structure is therefore a crucial starting point for the planning effort. In systems that use structured analysis and design, the high-level work breakdown structure is driven entirely by the process model, typically a waterfall one. The subdivision of each phase by component of the application or system would then occur only at the lower levels. The hierarchy of tasks would often look like Figure 7.1.

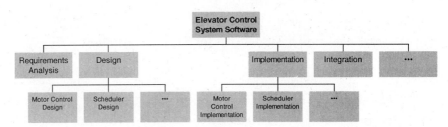

Figure 7.1 – Work Breakdown Structure for a Functional Application

Project and Deliverables Management

In an object-oriented system, the work breakdown structure is much more likely to be inspired by the *product structure*. This is because an object-oriented approach leads to the early identification of interfaces, behind which the implementation of each object is hidden. Each class can therefore be independently analyzed, designed, implemented, and unit tested—there is no need to synchronize the entire project on the basis of the life-cycle phases.

OO projects break down the work according to class structure

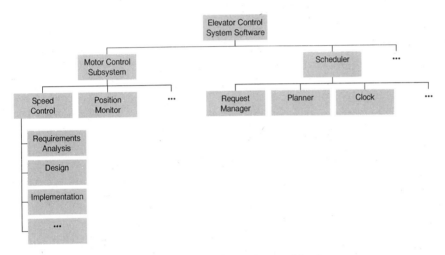

Figure 7.2 - WBS for an Object-Oriented Project

Based on the same example as Figure 7.1, Figure 7.2 shows how the top level of the WBS identifies coarse-grained components that constitute the application or system. At this point, it is quite useful to distinguish between application-specific components and components of a framework—reusable components that other applications may share in the future. Further hierarchical decomposition of each component may occur,

An example of product-driven hierarchical decomposition

leading to individual object classes. Finally, the phases of the selected software process model can be used to perform the last level of decomposition.

The schedule is derived from the WBS

Based on a product-driven WBS, a Gantt chart for an object-oriented project may look like Figure 7.3. As usual, the horizontal axis represents the time line. Vertical dashed lines have been added to indicate task dependencies, which normally appear in a Program Evaluation and Review Technique (PERT) diagram but not in a Gantt chart. Whenever the right edge of the box representing a task aligns with the left edge of another task, as shown by the vertical line, it means in this figure (and in subsequent ones) that the first task is a predecessor of the second one.

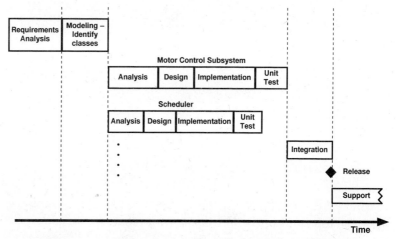

Figure 7.3 – Object-Oriented Project Schedule, First Version

Project and Deliverables Management 241

In this first version of a schedule, which we will refine later, there is no distinction between framework classes and application classes. Note that this is very similar to the chart that would be generated for a functionally designed system, with three notable exceptions:

At this stage, there are few differences from a traditional project

- The modeling of the system into a well-defined list of classes is a necessary step before the work can be partitioned correctly between multiple resources.

- Because these interface specifications become the *contract* between the developers of the different classes, the interfaces between classes must be reviewed and approved through a formal mechanism before being assigned to separate individuals or subteams.

- Since object classes represent smaller chunks of functionality of the system under development, the number of parallel tracks for analysis, design, and development is likely to be high.

This last point can easily become an obstacle to the manageability of the project. For example, what if you have a hundred classes to develop, and the team consists of twelve developers? The combinations of person-to-task assignments are numerous, yet many do not make sense in terms of skills and consistency. In this case, it is much better to develop the project plan according to an intermediate level of decomposition in the WBS. This intermediate level is made up of what Bertrand Meyer calls *clusters* (see [Meyer 95]). These are similar to Peter Coad's concept of *subjects* (see [Coad 90]). Figure 7.4 shows what a project schedule looks like when it is organized by clusters. The shape of the polygons describing each task (note that these are presented somewhat differently in [Meyer 95], and that we took some liberty with the notation in order to make the various figures in this section easier to compare) are meant to indicate that design, implementation, validation, and a

An intermediate level of decomposition is useful: introducing clusters

new step called "generalization," are not entirely sequential but can overlap. In addition, Meyer points out that these tasks do not succeed each other sequentially. Instead, the development of each cluster follows its own iterative life cycle: right-to-left arrows indicating rework are implied between each of the tasks and its predecessors.

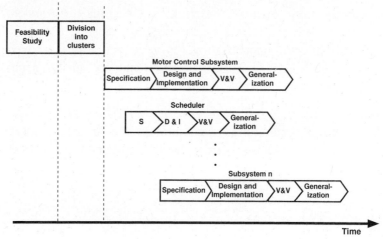

Figure 7.4 – Object-Oriented Project Schedule Using Clusters

This approach indicates, better than the initial one, the existence of independent development lifecycles for groups of objects. It may, however, lack two refinements:

There should be a distinction between application and framework classes

- Not all clusters or components are created equal. In a "real-world" project, there is a strong distinction between *application classes* and *framework classes*. Here we use the word "framework" in the sense given to it by the object-oriented community: a library of reusable classes that provide a foundation upon which to build applications. This is a different sense from the Computer-Aided Software Engineering (CASE) environment notion of a framework discussed in Chapter 6, although one can conceivably create something that is a framework in both senses of the term.

Application and framework classes are indistinguishable in technical terms, but they differ in terms of their intent and their reusability requirements:

These have different characteristics

- **Application classes** are designed to serve a specific purpose. They will only be tested for this purpose, and their developers have only one goal to meet: satisfying the needs of the application's customers. Reusability may be a good idea, but it is not a specified goal of this part of the development; therefore, it will definitely have a lower priority than other requirements.

- **Framework classes** are not only part of the support infrastructure for this project but also for a whole application domain, possibly for the whole corporation. Therefore, they must not only meet the needs of their client classes, but they must also lend themselves to a variety of future uses, some of which are not yet well understood. In other words, they must be designed with generality and reusability in mind.

- There is no clear convergence of the lifecycles of each cluster on a common deliverable. While continuous improvement is a good principle, perfectionism could lead to an indefinite refinement of the clusters, without the ultimate completion of the integrated application.

Each subproject must come to a close

Figure 7.5 depicts a hybrid or composite project schedule model. This model combines a clear beginning and end for the application, with a distinction between the development of

A composite project schedule addresses these diverse needs

application classes and the creation or enhancement of a framework or reusable classes. The initial analysis of the application leads to the identification of two types of needs: some application classes need to be developed to satisfy unique needs; and a framework needs to be created—or an existing one needs to be enhanced—in order to capture common functionality that will be required in application after application. Work on those application classes requiring no new framework features may start immediately. Other application classes can be worked on as soon as the interfaces and semantics of the new or extended framework classes are defined. Thanks to encapsulation, they do not have to wait for the *implementation* of the framework changes. However, those application classes that depend on framework changes cannot be fully tested until the underlying framework classes have been completed. This delay is shown by the horizontal line between the implementation and the unit test of these classes.

The organization is split along the line between application and framework

In such a situation, the project organization often reflects the distinction between the application tasks and the framework tasks (depicted by the heavy horizontal line in the middle of Figure 7.5). The development of the application is assigned to a software team that reports to the manager of the entire product or system of which the application is part. On the other hand, the framework development and evolution are assigned to a separate team that acts as an internal supplier of components and reports to a "software technology manager" or some other similar position. Because the lifecycles of the application and of the components are so dependent on one another, the two teams (more than two if several applications are under development simultaneously and use the same framework) must communicate very well and frequently.

Project and Deliverables Management

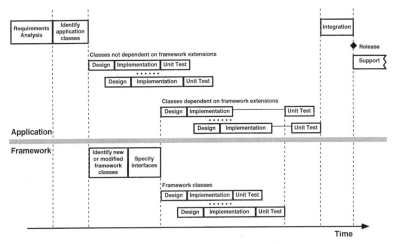

Figure 7.5 – Composite Project Schedule Model

Use of the Spiral Model – Development by Increments

Although the spiral model seen in Chapter 3 was not developed specifically for object-oriented projects, it applies well to them because different groups of classes (e.g., the clusters seen in Meyer's model) can be developed during different turns of the spiral.

The spiral model is useful for OO projects

One reason to use the spiral model is that the exact list of classes required by the application is not always known in advance. The other reasons are the same as in the original presentation of the spiral model, that is, to allow the periodic release of new functionality (thus proving to the organization and its customers that the project is making progress) and to avoid a lot of wasted effort if a mid-course correction is required.

Execute an OO project in small, successive increments

In case studies reported by industry consultant Jon Hopkins, of Rational, it is best to plan and execute a large object-oriented project in increments that each take ten to twelve weeks to complete. These increments are not necessarily planned around a fixed number of classes, but around a number of use cases (see the description of use cases toward the end of Section 4.5), perhaps four or five per increment. In small projects, or for the development of framework classes, he has observed increments as short as four weeks, each consisting of implementing about two use cases only. At the end of each increment, enough system integration is performed to support a demonstration to key users. Each of these demonstrations (accompanied by reminders that the application is not yet complete) yields feedback about the quality, functionality, and style of the application. This feedback helps replan the next stage as necessary. For example, perhaps the priority order of the next use cases needs to be changed, or perhaps some improvements in performance or robustness need to be made in the next increment before it makes sense to add more use cases.

Use cases define the contents of each increment

Project planning thus adds another reason to employ use cases; that is, apart from methodological principles, they make it possible to decide in advance exactly which user functionality will be available during which phase of the project. A list of classes is less directly comprehensible to the customer or to higher levels of management.

Additional Tasks

There are tasks other than the production of classes

When creating the work breakdown structure for an object-oriented project, the project manager must take into account the existence of new tasks that this approach implies. Although these tasks require additional resources, they lead to the higher overall efficiency of the approach. Omit them, and you will find yourself among the crowd of disappointed cynics at the next object-oriented conference! Such tasks might include:

- There must be an **architecture** task that parallels the rest of the project, and it must be staffed by one of the best people available. The architect has multiple roles: to design the overall structure of the application or of a group of applications within a domain; to make sure that this structure is documented, taught, and understood by the development team or teams; to select the framework and software products required to implement the architecture; and to continuously review all specifications and designs for conformance to the architecture. In most cases, only one architect is needed, and he is not the project manager, simply because the skills required are different (vision versus organization). In very large projects, there may be a team of architects, but there should still be one chief architect who controls the consistency of the entire edifice. In very small projects, the architect may be the same as the project manager, or he may be one of the developers.

  Architecture-related tasks

- The **qualification and selection** of components (including commercial frameworks, object-oriented database management systems, object request brokers, etc.) is a separate task. In a small-to-medium-sized project, this may be performed by the architect. In a large project, it may involve different people because of the amount of work required. The architect may define the selection criteria, after which the component selection engineer (or team) gathers information and performs the selection. In contrast to the architect's work, this task ends, or at least decreases in scope, once the development has started. Note that some residual allocation of resources is required throughout the project because the components selected will probably need to be upgraded or will need some troubleshooting over the life of the project.

  Qualification and selection of component technology

- **Reuse** within the project or across projects also requires a separate activity. Newly created classes need to be examined for their reuse potential, generalized if necessary, and

  Reuse-related tasks

captured in some classification scheme that makes them easily accessible by others. Conversely, individual project members may need help in locating components they require. A *reuse library* function is what we are describing. This will be separate from other functions in an organization that is conducting many projects or at least one very large project.

Training is a prerequisite to many other tasks, and must come neither too late nor too early

- **Training** is a significant task, not only because of the time and money it may consume in an organization that uses an object-oriented approach for the first time, but also because it is a prerequisite to many of the other tasks; therefore, it delays their initiation. Training should be provided both on the approach and on the specific methods, tools, and components selected for the project(s) in question. If the nature of the project leads to the creation and maintenance of a "homegrown" framework, it in turn needs to be taught to new project members. Even if the rest of the training is bought from outside organizations, organizing and facilitating the training require effort. It needs to be planned at the appropriate time. This should obviously not occur too late, but if it happens too early, the students will forget some of the lessons because they are unable to begin applying them immediately.

A rigorous curriculum should be developed

It helps to design and publish a curriculum that indicates not only the list of those courses every person on the project should attend, but also the order in which they should be taken (in case the courses are offered several times, which happens in a large organization). Figure 7.6 is an example of such a curriculum (the specific products, method, and language mentioned here by name merely illustrate how specific the course descriptions must be, and are not intended to promote these as superior to their competitors). "XYZ" is the organization's own framework; it can be taught only to students who know both object-oriented design and C++. On the other hand, not all project

Project and Deliverables Management

participants need to know the details of how to use an object request broker or an object-oriented database, so these two courses are not compulsory for everyone.

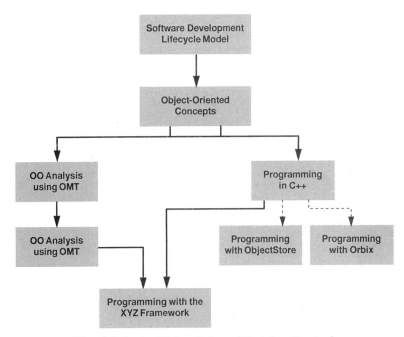

Figure 7.6 – An Object-Oriented Training Curriculum

A well-designed object-oriented training curriculum may require from four to seven weeks of time from each of the participants. This time will probably not be devoted to training in one continuous series of courses; therefore, the total elapsed time from beginning to end will be closer to three or four months! This is clearly not something that can be overlooked in the project plan. Nor can the costs be ignored. An investment on the order of $5,000 per person is typical, and even this assumes that the organization takes advantage of the economies of scale of on-site classes, and the availability of local universities or community colleges for the more common courses, such as those on object-oriented programming languages. It is

The expenditure of time and money is significant but necessary

also a good idea to include some short "refresher" courses in everyone's schedule, and to realize that while the more novice team members will lose some efficiency at the beginning due to their lack of experience, the more experienced engineers will also lose some efficiency, because they will often be interrupted for consultation by the novices! Until the whole organization has come up the object-oriented learning curve, the project manager should include these efficiency factors when converting theoretical person-months to calendar months by specific developers.

Effort Estimation

Task effort estimation is the next step

Once a satisfactory WBS is available, the key factor that will determine the quality of the project plan is the accuracy of the estimates for each task contained in it. Unfortunately, this is where the software industry in general, and its object-oriented flavor in particular, are still in their infancy.

COCOMO does not apply well to OO projects

Under prior approaches (structured analysis and design, functional decomposition, etc.), a few models for software project estimation have been developed and employed since the 1970s. One is the COCOMO (COnstructive COst MOdel) of Barry Boehm, described extensively in [Boehm 81]. Unfortunately, his method and others rely on an *a priori* estimate of the number of lines of code contained in the final product. Although it is not a foregone conclusion, let us assume for a moment that this is a valid method for non-object-oriented projects. Even so, the method would be inherently flawed for object-oriented developments because it ignores two differences:

Initially, the size of the code is not known

- Classes are strictly specified by their interfaces. An estimate of the number of lines of code may be provided by a developer on the basis of experience, but this requires thinking about the *design* of the class, so it violates one of the principles of the approach.

- In an object-oriented system, much new functionality is created by enhancing the underlying framework, or by subclassing or generalizing existing application classes. Less functionality is created by straight new code development—this is the very definition of reuse. A few judicious changes can represent significant functionality addition. Conversely, some apparently modest new needs may lead to a serious reorganization of some parts of the framework. For instance, a change in the class structure may be necessary in order to take advantage of newly recognized commonality between previously unconnected parts of the class hierarchy. This means that "amount of functionality" created or added does not have a simple, linear relationship with the effort required to write code.

Reuse makes the code size less relevant

Another technique, *function point analysis,* avoids the first of these pitfalls but not the second one. Function point analysis, invented by Allan J. Albrecht of IBM, in 1979, consists of counting elementary "atoms" of functionality in an application. A function point can consist of:

Function point analysis is another approach

- Reading or writing one piece of data from/into a file or database;
- Displaying one field on a screen;
- Requesting one simple input from the user;
- Reading the state of a sensor; etc.

The theory behind function points is that the effort required to implement one function point is about the same across all applications, and that a function point count can therefore provide an effort estimate without knowing anything beyond the end user's requirements.

It is less dependent on the design ...

... but it still does not fit OO projects perfectly

This method avoids the premature design syndrome of COCOMO, but it still falls apart when an object-oriented approach is used. For example, the addition of numerous new user interaction functions may be the result of a simple change in a basic framework class. Conversely, a very small change may be an indicator of a new dimension of generality needed in the framework—something that may consume several person-months to accomplish despite the fact that it was triggered by a single additional function point, or even no change in the count at all.

A class-based effort estimation method is needed

A better level of effort metric for object-oriented systems would consist of counting those new and/or modified classes that analysis determines are necessary, and to assign them a complexity level based on the number of their methods. It is possible that the number of arguments of a method is also an inherent indicator of the effort needed to implement it. One interesting aspect of this approach is that it does not require any premature understanding of the design of the classes; only their interface definitions are required.

Use cases are the unit of functionality

Finally, as mentioned earlier, *use cases* provide a simple way to count how many functions are required in a system, and therefore, how complex it is. They could provide a useful estimating method if several assumptions were verified:

- Each use case would have to represent roughly the same amount of effort, which is a highly debatable proposition.

- Previous data on the effort by use case would have to be available, which means that the organization has performed at least parts of some significant projects, and has recorded the amount of effort spent per use case.

Project and Deliverables Management

Even with this change, we are still faced with a basic dilemma, namely, that management usually demands project plans *before* the analysis has been conducted—at least to the extent necessary to enumerate classes and specify their interfaces, or to analyze the user requirements into a set of use cases. In other words, the estimation of the effort must be possible, at least under some relaxed margins of error, before any of the information required by constructive cost models or by serious correlation models is available. While this management requirement is often irritating to the professional who wants to provide a good estimate (and wants to avoid the consequences of initially underestimating the effort), it is a legitimate request in order to understand the feasibility of the project and to allocate resources judiciously. Clearly, in this common situation, the preliminary plan must be based on a much less rational basis, and it must be revised once a specification or an architectural design is available. The following guidelines are then useful to construct an initial estimate:

A rigorous estimation process does not provide the early estimates often required

- Use the experience of people who "have been there before." Since there is no time to gain a more solid footing, it is sufficient to rely on "guesstimates," but it is important to discuss why people make these guesses, to challenge them from various perspectives (e.g., "yes, but last time we didn't have to port to a new platform at the same time" or "we spent a lot of time in training at the beginning of the last project, and we won't have to do that this time"), and to gain consensus. Assembling a balanced team of experienced people to participate in this exercise is useful for obtaining quality estimates. Moreover, the discussion will often yield a better awareness of new tasks, dependencies, or opportunities to reuse other work rather than developing a new product/system from scratch.

Use input from several experienced people

Use code size only when it is actually known

- Use "lines-of-code" metrics only when estimating the effort needed to reengineer an existing system. In this case, the number of lines of the existing system is known, and prior measurements may exist on similar projects.

Extrapolate from projects whose level of effort is well documented

- Use code size as a relative measure only. Do not say: *"We have to redesign this 50,000 line application, and our productivity is 100 lines of code per engineer per day, so this is going to take 25 person-months,"* but say instead: *"We (or another team using the same technology) redesigned application X, which was 25,000 lines of code, using 3 people for 5 months. Based on size only, we should therefore need 30 person-months this time."* (Then look at whether everything can be expected to "scale" accordingly. Perhaps the 15 person-months used for application X included some training that does not need to be repeated for the new application, or is not necessary at all for those people who came from the previous project and are already trained).

Analysis is about thirty percent of the effort

- A thorough object-oriented analysis and design phase will consume about thirty percent of the total effort. This is significantly more than in older methods, which is good. The reason is that the models and the specifications that result from the analysis phase are more complete and more directly usable to create a good design. This rule of thumb can be used in two ways:

    - If someone can estimate, perhaps based on prior experience, the total effort for a part of the development, then a reasonable estimate of the analysis effort can be derived by applying thirty percent to it.

    - If the analysis has already been performed, and has created complete object models (not just information models) of the system, then dividing the effort expended thus far by 0.3 will provide an estimate of the total effort.

Project and Deliverables Management

- It is always useful to apply several methods independently and to compare the results. Often the estimates are actually fairly close, which may mean less than a factor of two apart. In this case, an intermediate number or the maximum of the various estimates can be used for the purpose of an early estimate.

    Try several ways to estimate

- When widely conflicting opinions remain about the effort needed to perform certain tasks, remember that in the history of software project planning, the pessimists have been right more often than the optimists. This is why taking the maximum rather than the average may be the right decision.

    Err on the safe side

- If the total of the individual task estimates looks too high, there will often be a lot of pressure to revise the estimates down, or to lean toward the optimists among the group debating the estimates. Reducing the estimates of tasks that can be pared down is acceptable. Often, it is possible to deliver a working application with fewer bells and whistles, and to add back new features later. Reducing the estimates of tasks that are the incompressible core of the system, however, should be fiercely fought.

    Do not reduce the estimates without good reason

- Always use a strong disclaimer about the validity of any schedule derived without having first performed the conceptual design; and flatly refuse to provide any schedule at all if there is no clear statement of the requirements! Feasible plans are not generated through "what if" scenarios concocted in the coffee room.

    Set expectations about the validity of early plans

- On the first couple of projects in a given domain, count reuse as an extra burden and not as a source of savings. It takes extra effort to make classes more general, or to organize their relationships (inheritance structure, etc.) in order to build a reusable framework rather than a single application made up of non-reusable classes. The savings will come when the framework exists and has been

    Reuse does not save work on the first few projects

validated, meaning that it has actually been used for several applications.

Get outside information

- If there is no in-house experience, try to interview people in other organizations that have worked on similar projects.

Lifecycle Synchronization

The distinction between framework and application influences the planning

We mentioned earlier, when presenting the composite project schedule model of Figure 7.5, that the lifecycles of the applications and of the framework components depend on one another. Because this interdependence must be taken into account when planning a project, this subsection addresses it in a little more detail.

Two lifecycles are dependent on one another

The requirements for the creation of a framework or its extension are only known as a consequence of the requirements placed upon the application. Conversely, the application cannot be fully integrated and tested until the components have completed their own development lifecycle. This interdependency is depicted in Figure 7.7.

Lifecycles must be synchronized with care

Synchronizing these lifecycles so that the necessary components are delivered "just in time" to the teams performing application development is a tricky exercise. It requires not only attention to this situation during planning, but also excellent communication between the various parts of the organization as well as good project tracking techniques and tools. This might include frequent review meetings, "score boards" showing the progress of each component against its schedule, etc. This interdependency of activities is one form of the discipline known as "concurrent engineering."

Project and Deliverables Management

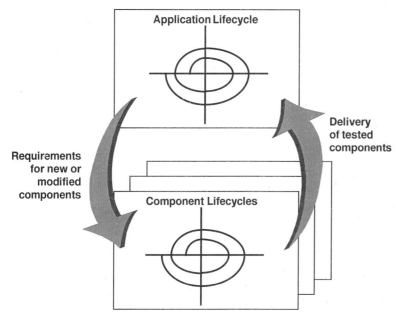

Figure 7.7 – Application/Component Lifecycle Dependencies

Specifying the Interfaces Early

A defining property of object-oriented systems is *encapsulation*, or *information hiding*. This characteristic greatly facilitates the planning of object-oriented projects. Since the users of a certain class C depend only on its specification to be able to invoke it correctly, and do not need to know its implementation, specifying the interfaces early allows the implementation of any class C' that is a "client" of C to proceed in parallel with the implementation of the class C.

Encapsulation facilitates parallel implementation

Object-oriented design methods (such as Booch and others) allow the designer of a class to specify its interface, that is, the messages it can send and receive, and the parameters of these messages. Furthermore, the Object Management Group (OMG) has specified the Interface Definition Language (IDL) in order to provide an unambiguous syntax for the definition of

Specify the interfaces early, using an OO design method or IDL

class interfaces. IDL can be directly transformed (sometimes by automatic means) into templates in implementation languages such as C++. Some users of the object-oriented approach have introduced into their development lifecycle an additional specification document, intermediate between the requirements specification and the high-level (or conceptual) design document. This *Interface Specification* document, a mixture of diagrams from an object-oriented design method and IDL definitions, establishes the *protocol* used to invoke the classes that constitute the framework and/or the application. Adding this task (and the milestone that consists of approving the document) in the project plan results in better parallelism between the application side and the framework component side of the plan.

Subcontracting Objects

The decoupling of implementations allows objects to be subcontracted

This decoupling between the application and the framework, in terms of their design and implementation, offers another advantage. As soon as the Interface Specification document is completed, it is possible to subcontract certain parts of the whole project. As long as frequent reviews occur (e.g., once every two weeks), the development work can occur in different organizations and at separate sites until the complete application begins to be integrated. Frequent communication must also be maintained so that requests for specification changes can be reviewed rapidly, and formally approved or rejected. If they are approved, they may lead to "change orders" with financial consequences. Therefore, the contract or purchase order to the subcontractor (if it is based on a fixed price rather than a cost per hour) must include provisions to fund extra work necessitated by the customer's change of specifications.

First, try to locate commercial components

Clearly, before plunging into a development effort, an organization should first try to find the objects it needs on the market. In the current state of the market for object-oriented components, however, only simple and very generic objects can be found. This includes objects such as graphical user interface (GUI) components, "collection" objects (stacks, queues, etc.), facilities to manipulate compound documents, etc. Domain-specific components, such as objects representing semiconductor wafers or seismic traces, for example, are not for sale. This will change over time, but change occurs slowly. In the meantime, the alternative to developing the objects one requires is to enlist the skills of other organizations that may have more trained people, more robust processes, or more relevant experience. Thus, subcontracting occupies an intermediate place between custom, on-site development (the traditional but expensive way of doing things) and the assembly of commercial, off-the-shelf components (the ultimate goal).

The Role of Prototyping in the Project Plan

The role of prototyping in the development of applications, including object-oriented ones, is the subject of conflicting approaches by various authors and consultants. Opinions range from viewing prototypes as a natural and mandatory part of the lifecycle, to seeing this approach as a dangerous fallacy that spawns many immature systems.

The exact role of prototyping is controversial

As on many other occasions, extreme views are probably impractical or ill-advised, and the best practices lie somewhere in between. Here are three potential ways, with some of their pros and cons, to use prototyping. In each case, the project plan must reflect the presence and resourcing of the corresponding tasks.

Prototyping has several potential uses

- **Prototyping as a way to validate user requirements.** In this approach, prototypes are implemented to provide users with a mock-up of the appearance or behavior of the actual application. The purpose is to have the users review the prototype and decide early on whether or not they wish to amend their requirements. The most frequent kind of prototypes used in this manner are *user interface prototypes*.

User interface prototypes allow users to review the requirements

By using a GUI builder, or by writing some simple code to display graphical widgets (without programming all the actions that need to occur in response to user inputs in the real application), one can create a representation of the various screens with which the user will eventually interact. Rapid iteration can take place. In other words, user feedback may allow new versions of the prototype to be presented every few days at most. After a couple of weeks, if the users are satisfied that the user interface is what they want, this prototyping phase can formally be brought to a close. The prototype is then discarded, but the final appearance of the user interface is used as an input to design the actual system.

This is easy to do once there is a framework

It is an interesting characteristic of object-oriented systems that such prototypes can easily be created once a good framework is in place. A "good" framework, in this sense, is one that contains a hierarchy of user interface classes, with very powerful ones at the top of the hierarchy. This means that a few lines of code can cause a rather complex screen to be composed, and some limited but useful behavior to be exhibited. For instance, in a process control application, it may be possible to display a text entry window to allow a user to type a value, and a dial that shows in analog fashion the value typed by the user.

Some tools add behavior to the user interface

Some modeling tools allow the *prototyping of the system's behavior*. For example, the i-Logix tool takes as input a representation of a system's specification, including Harel statecharts, and creates an executable model that exhibits the specified behavior. The states of the system can be associated with user interface widgets to provide a more realistic prototype. Typical examples discussed in the supplier's literature and in presentations include the prototyping of airplane cockpits, control systems for nuclear power plants or oil refineries, etc.

Project and Deliverables Management 261

- **Prototyping as a way to validate key design decisions.** At design time, the decisions made are unlikely to affect the visible and qualitative behavior of the system (since these are now imposed by the requirements), but they may impact things such as the amount of space required to store data or the performance of the system. At this point, prototypes can help determine the degree of acceptability of different decisions. For instance, in a banking application, the number of customer accounts that can be read and written per second is an interesting performance measure when designing the program that posts the monthly interest. Different storage mechanisms (object-oriented database, relational database, flat files, etc.) and different ways to represent the credit and debit transactions can be tried by prototyping. In this example, the prototype has no need to read or write meaningful records and no need to make correct calculations. It must only read and update records in a manner similar to what the real application will do. The fact that the prototype uses fake information and does not perform real calculations makes it easier to write. Moreover, it makes it easier to ensure that the prototype will be discarded at the end of the exercise. The reason it is so important to discard these prototypes rather than having them included as part of the final application, is that typically, they are developed without any of the solid practices required by the actual development (design reviews, documentation, exhaustive testing, etc.).

  Prototypes can verify performance or other design choices

- **Prototyping as incremental development of the actual application.** This is a fundamentally different choice from the previous two, and a controversial one. In this approach, each prototype is a real system that implements a subset of the application's use cases. It corresponds to one turn of the spiral model of software development. Each successive prototype is a superset of the previous one, and the n-th prototype is presumed to constitute the final system ready

  Prototypes that lead to the final application are the contentious ones

to be released. Among the dangers of this method, the following ones seem to be the most serious:

It is hard to predict completion
- The number of actual iterations necessary is impossible to know in advance. Therefore, project planning is especially difficult.

It is easy to overestimate the accomplishments
- The appearance of successive prototypes will lead managers and users to incorrectly believe that the development is near completion. Arbitrary decisions to "release what we have now" may be made if the project approaches the date or the total spending level originally planned. These decisions ignore the fact that several aspects of the application do not yet exist, or have not yet been tested or documented properly.

It is also easy to criticize the lack of certain features
- Conversely, the fact that something that looks and behaves like the actual application has significant gaps in it will make many users and managers uneasy. They may start listing missing features and demanding urgent corrective action, ignoring the fact that these interim gaps are intentional. For example, if what is currently being reviewed is Prototype Number 4, and validation of the user's password is planned for Prototype Number 7, the plan for Prototype Number 5 may suddenly be thrown up in the air because management demands that this "defect" be "fixed" for the next customer demonstration!

Prototyping is not the same thing as development
In summary, although there are respectable opinions to the contrary, it appears that while prototyping can be a constructive component of the analysis and design phases, it is too different in nature from the development of a robust application to be the only method used to develop one.

Balancing and Completing the Plan

A work breakdown structure and a set of estimates do not automatically produce a plan. There are two reasons: first, a plan is not just a schedule, and secondly, the initial schedule that "falls out" of the planning process may not meet the customer's deadlines. The issues that arise are absolutely not unique to object-oriented projects, so for the sake of completeness, we will only summarize them here and refer the reader to books or courses on project management for further discussion.

Complete planning requires more than scheduling

Let us first address the viability of the schedule. There are three variables related to the overall execution of a project: *functionality, schedule,* and *cost*. In order to achieve an earlier completion date, one possibility is to add resources, assuming that they are available and that it is not already too late to add them (the famous "mythical man-month" problem raised by Fred Brooks in [Brooks 75]). The total expense is likely to be higher, because even if the total number of person-months is the same, the extra people on the project require additional training, software tools, workstations, etc. If, on the other hand, budget limitations are the main concern, then it may be possible to stretch the project over a longer period of time, using money from different budget years for successive phases. But if both the cost and the schedule are constrained, then the only choice is to cut down on the functionality, which may reduce the size of some tasks, and will remove others from the project plan entirely. Working with a project plan is a little like trying to compress a balloon or push down on a spring. If you apply pressure in one area, it tends to pop up in another place. It is not possible, at any rate, to arbitrarily fix all three variables (cost, schedule, and functionality) and still expect the project to succeed. Achieving a viable schedule is a balancing act, and may require many iterations using different sets of assumptions. A good scheduling tool is very helpful here because it can automate the construction of the schedule once a set of assumptions has been made.

Trade-offs must be made between functionality, schedule, and cost

Document the additional elements of the plan

Once the project manager has a schedule, he should worry about the other elements of his plan. Although organizations have different practices, and a written plan document is not always required, it is a good idea to write down and review the following additional elements:

Material resources

- What material resources are assumed? These may include equipment, tools, supplies, etc.

Help needed from other groups

- Which other departments or groups, over which the project manager has no direct control, are expected to cooperate at different times during the execution of the project? For instance, Marketing may need to review specifications, Training may need to schedule new customer courses, Sales may need to be briefed about a new product, and Documentation may have to plan the duplication and shipping of new manuals. If this is not described anywhere, chances are that the lack of coordination will impair the successful completion of the project.

Risks and how to reduce them

- What risks are still present in the project, and how can they be mitigated or overcome if they become a reality? Is there a key person on the project whose sudden resignation or illness would cause a major problem? In that case, can someone else be assigned to the project as a "shadow contributor" to ensure some measure of recovery? Is there an assumption that funding will be obtained from some external source, and what are the alternatives if these funds vanish? For software that needs to be tested in conjunction with hardware, what if the machine in question is not working at the time it is needed? Is there another one somewhere that can be secured? These are just a few of the risks and contingencies that should be raised explicitly in the project plan and reviewed with management before starting execution.

Project and Deliverables Management

Project Tracking

Once the project starts, it needs to be tracked regularly. The purpose of the tracking is to monitor compliance with the plan, and to take corrective actions as soon as the reality becomes different, which almost always happens. Again, this is not limited to object-oriented projects, so here are some brief principles to apply.

The purpose of tracking

First, a plan is a useful document, but actual events are stronger than any plan. Things will go bump in the night—an engineer goes on jury duty, the network breaks down for half a day, or the customer requests a late change to the specification. Even if a buffer has been built in the schedule for such events, actual occurrences will invariably be different from the original assumptions. (Note that if these buffers were truly estimated well, it would mean that the reality would not be worse than the plan, but better.) Even in that case, tracking the plan is still very important. No plan is perfect, and the work can be re-planned periodically to provide new estimates of completion dates, provided this is done through an agreed-upon procedure.

React rapidly to unforeseen events in the course of the project

A regular forum to review and update the plan is desirable. In many organizations, a weekly status meeting—however brief—is the best way to track progress. During this review, each person is asked to provide a brief status of all the tasks he is working on. Some tasks may have been completed as planned—or even early—since the last review. Others are late for some reason, which means that one of several decisions must be made. The team can enter a new estimated date of completion in the plan and see what impact this will have on the entire project; or they can add resources to that task, perhaps by temporarily moving someone else whose work is not on the critical path or who is ahead of his own schedule. As a result, a new version of the plan is created, which is then distributed to all the project members so that they have their new "marching orders" until the next meeting—or until another

Conduct a weekly tracking meeting

unexpected event occurs. An alternative to meetings, which engineers dislike so much, is for them to individually report their progress to the project manager. This seems more productive but has significant drawbacks. In the presence of a problem, for example, it limits the useful discussion of alternatives and recovery options. Moreover, it generally contributes to isolating people too much from one another, thus preventing the team from sharing hints and tricks.

7.2 Deliverables Management

Configuration management is impacted by object technology

Just as the purpose of Section 7.1 was not to reproduce general advice on project planning but to highlight differences in object-oriented projects, this section does not address configuration management in general. There is information on this subject elsewhere, including some excellent product manuals and some valuable (if expensive) comparison studies from consulting organizations. Instead, we again ask ourselves the question: what differences does an object-oriented approach bring to this activity?

Inheritance makes dependencies more complex

Inheritance is the key feature of an object-oriented approach that impacts configuration management. Because the inheritance tree for the application may be very complex, and may include both "homegrown" classes and classes purchased as part of a commercial library, a small change made very high in the hierarchy can have great consequences over the behavior or correctness of many parts of the system. As a result, the lack of a robust configuration management and application building scheme can lead to a constantly brittle environment in which nothing ever works completely. The team comes to work one morning (at the point in the project when an incomplete but usable application is supposed to exist), only to discover that a small change made by one of them has caused the entire "nightly build" of the application to fail, or that basic parts of the application no longer work. "Not to worry," says the project manager, "we will just fix it and redo the build, and we will soon be working again." Except that in the meantime, another small change is made in another class without informing the users of that class, and so the next build has other major flaws, and so on.

There are ways to avoid this syndrome of constant instability, but they require certain tools and a lot of planning and discipline. Clearly, there must be a configuration management system in which successive versions of all the files (IDL definitions, source code, object code, etc.) are stored. This system must have the ability to execute "make" procedures that understand the dependencies between the configuration items so that only those parts invalidated by a change are rebuilt when requested. This tool, properly used, ensures that all the changes are correctly incorporated. But it does not mean that a running application exists at all times.

Planning and discipline can ensure stability

To achieve this next level of robustness, which is required when testing begins (and which may cover most of the project duration if some form of incremental development or spiral process is used), the following practices help:

Steps to maintain a productive development baseline

- Distinguish between a *test baseline* and a *development baseline*. At any point in time, the test baseline is used to detect defects. In contrast, the development baseline is the one in which ongoing functionality enhancements are made, and the one in which the defects detected in the test baseline are corrected. The test baseline corresponds to some intermediate change levels of all the objects involved in the application, and does *not* change on a day-to-day basis. The development baseline corresponds, by definition, to the last change level of all the objects, and *does* change on a daily basis.

Isolate a test baseline

- Groups that develop downstream applications, users who are previewing the system, or quality assurance teams testing the application, use only the test baseline.

***Periodically freeze
and test the
development baseline***

- At some point, the current test baseline outlives its usefulness. The new defects detected in it are ones that have already been fixed in the development baseline, and users need to move on to testing the next batch of features implemented since this baseline was created. At this point, "freeze" the development baseline by prohibiting all enhancements. If a few developers are actively developing new features, they should place any new code in their own private directories, and if necessary, create their own "builds" that do not impact their colleagues. From now on, the only activity allowed in the development baseline is to fix bugs and to rebuild it repeatedly. Iterate this process until the development baseline meets the following criteria: it has no catastrophic defects; regression tests show that bugs previously eliminated (when constructing the current test baseline) have not reappeared; and significant new functionality appears to be working.

***Validate it from a
user perspective, then
promote it with care***

- When this occurs, have a willing user or the quality assurance team briefly exercise the development baseline (e.g., for a day) to validate the development team's opinion that it is ready. If they confirm that this development baseline should be "promoted" to become the new test baseline, then do the following:

    – Save a copy of the current test baseline so that it can be restored easily if necessary.

    – Replace the test baseline with a *static version* of the development baseline (i.e., one that will not be dynamically modified by changes made later by the development team. For instance, there should be no links back to the development directories).

    – Reopen the development baseline for the next wave of modifications.

    – Announce the availability of the new test baseline to the other teams interested in it, and provide documentation of its new features so that they can begin using or testing it.

- Ask for feedback and be ready to back out the change if an unexpected severe problem occurs.

With these precautions, the high level of interdependencies between components that is typical of a medium-to-large object-oriented application can be managed to ensure uninterrupted progress.

This approach protects the developers' productivity

7.3 Reusing Objects

Using something that already exists "as is" or with some modifications, instead of recreating it, is clearly a source of savings. This concept of *reuse* has been introduced in many disciplines, from the construction industry (with prefabricated building parts, or the reuse of designs in housing developments) to integrated circuit design (with the concept of standard cells introduced in the early 1980s).

Reuse is common in other disciplines

Benefits of Reuse

Reuse is not just a source of cost savings. It has several other important benefits:

Reuse has multiple benefits

- **Time-to-market.** Directly associated with the cost savings is the reduction in the time-to-market of applications that reuse existing components. In an increasingly competitive world, creating an application or computer system faster can result in a tremendous market advantage, and can make the difference between meeting or missing the customer's needs.

- **Quality.** Using components that have already been tested extensively by its suppliers and/or by other users leads to quality increases.

- **Ease of learning.** Repeatedly using the same basic components in different applications reduces the learning curve of the developers.

Early reuse efforts focused on code, but object orientation allows more extensive reuse

While earlier efforts to obtain reuse focused strictly on software code (e.g., the emergence of commercial FORTRAN subroutine libraries in the 1970s), reuse should include all potential artifacts of the software development lifecycle, including specifications and designs. In addition, objects are inherently more reusable than subroutines. This is because the bundling of data with the methods that operate on it creates a more useful entity for reuse purposes than a subroutine. For instance, the object "stack" that contains the methods "push," "pop," "last," and "is-empty," can be reused without further ado (perhaps the type of the objects in the stack needs to be specialized by passing some form of type descriptor when the stack is created). In a structured programming approach, each of the four methods would be a separate subroutine. The stack data itself would have to be explicitly located in such a way that the four subroutines have equal access to it. It may actually have to be declared by the user of the subroutines in his own code. In other words, the stack is *not* ready to be used in this case. As the warning goes: "some assembly required—batteries not included."

Reuse Roles and Responsibilities

Reuse must be organized

Reuse is not just a technical activity, nor is it one that happens by itself after software engineers read a good book on "realizing the object-oriented lifecycle." Several parts of the organization must cooperate and organize various activities in order for reuse to occur.

Management support is required to pay for the cost of reuse

- **Management** must actively support a reuse-based process. Ask a manager whether he supports reuse, and he will probably say "yes, of course!" But bring him a purchase requisition for $50,000 to buy a license to incorporate a commercial library of classes in each copy of your new application, and his answer may waver somewhat. And tell him to forget measuring productivity according to the number of lines of code written per person per month, and he may begin to wonder if reuse is not a conspiracy led by programmers looking for ways to avoid accountability.

In other words, management support implies at least the acceptance of new performance metrics that are unfortunately still immature. It also implies the ability to value the purchase of components, at a *tangible* cost, in exchange for the promise of productivity gains and reduced time-to-market that bring with them only *intangible* benefits. Managers must also support potentially significant retraining costs and be willing to attend some appropriate training classes themselves.

The benefits are less tangible than the costs

Finally, one must realize that in order for reuse to be effective, there must be something to reuse. Most organizations can select some components from the market, but must create the reusable components that correspond to their specific domains or embody their trade secrets. This takes time and consumes resources. Therefore, the net effect of reuse may be negative during the first project, neutral during the next few projects, and only begins to be positive after that. Over time, there is a very gratifying snowball effect with reuse, but in the beginning, patience is necessary. In providing this patient support at first, management is badly served by those people (including not only salesmen for external products but also internal champions eager to allay fears at all costs) who overpromise and therefore set themselves up to underdeliver.

It takes time to populate the reuse library in order to obtain savings

This is not to say that management should shut up and ignore any warnings once an object-oriented approach has been put in place. Normal management oversight is still a healthy practice, and the proponents of the approach should indeed be constantly challenged to measure their results and provide justifications for their rate of progress. However, this oversight must be made from an informed perspective.

Developers must be taught to attempt reuse before writing a new class

- The **software developers** whose job it is to reuse instead of creating everything from scratch are the key participants whose daily actions will make or break the reuse program. They must be trained, they must be provided with the tools necessary to search for reusable components (e.g., browsers for class libraries), and they must be convinced that their performance will no longer be measured solely by the amount of new code they create. They must shed the last bit of a "not invented here" attitude, and they must be motivated to create successful reusable components themselves.

Specific functions take care of the administration aspects of reuse

- The **Reuse Manager** and/or **Reuse Librarian** are new functions required to manage reusable components. As the subject of the last sentence indicates, this may be one or two roles depending on the size of the organization and the complexity of its projects. In a small group, this role may not even occupy one full-time position, and may instead be fulfilled on a part-time basis by one of the developers. The role consists of making sure that the components selected for inclusion in a reuse library are *identified, qualified, and catalogued*. The reuse librarian must be given the necessary tools to organize and document the library, and (in the case of internally generated reusable components) maintain a set of test cases that can be used to validate new versions of these components. Needless to say, the library must be under configuration management. The reuse manager should be in charge of defining and implementing the Software Reuse Plan as defined below. He also manages the resources (people, tools, disk space, etc.) devoted to the maintenance of the reuse library.

Actively look for reuseable components

If possible (the size of the organization may be a limitation here), the reuse manager should be proactive, not reactive. Instead of waiting for components to be submitted by willing developers, he should review the structure and

Project and Deliverables Management

content of applications under development in order to identify common needs that are being solved separately, or those "at risk" of such dispersed actions. He should then intervene to suggest the development of a more generic reusable component.

The Software Reuse Plan

An organization that wants to initiate a serious software reuse effort should develop a comprehensive Software Reuse Plan. This plan, which in most cases will be developed by an ad hoc task force or committee, begins by identifying the organization's *goals* for reuse. For instance, here are the four goals contained in a Software Reuse Plan defined at SEMATECH in 1991:

Start by defining the reuse goals

- G1: Increase the productivity of the software development group.
- G2: Increase the quality of the software products.
- G3: Improve the job satisfaction of the software developers.
- G4: Meet and foster the adoption and evolution of industry standards.

A set of **objectives** can then be defined. If the reuse plan agreed upon and implemented meets these objectives, the previous goals will be met. The objectives should be clearly traced back to the goals, which is why the goals were numbered in the above list. Here are the nine objectives contained in the same plan:

Refine the goals into a list of objectives

- O1: Decrease the number of new software artifacts produced by reusing existing artifacts in their current form or with some controlled modification (this objective supports goals G1, G2).

- O2: Create a situation in which it takes less time to reuse an existing component than to create an equivalent one from scratch (G1).

- O3: Assemble a large enough selection of reusable components so that the added cost of failing to find some components is absorbed by the savings made from finding many others (G1).

- O4: Reduce the total amount of testing required by reusing components already tested (G1, G2, G3).

- O5: Reduce errors by using components that have been reused several times in different contexts, and thus, having had greater exposure, are more reliable (G2, G4).

- O6: Modify the appraisal and reward system to focus on "functionality delivered per unit of time" rather than the amount of code produced (G1, G3).

- O7: Provide a mechanism for the authors of reusable components to be acknowledged in the reuse library in a way that is visible to the users of those components (G3).

- O8: Identify, characterize, and offer to the rest of the industry successfully reused components (G4).

- O9: Include all types of software artifacts in the scope of the reuse plan (G1, G2, G3, G4).

Define strategies to meet the objectives and maintain traceability

The next step in this process is to define *strategies* to meet these goals and objectives. Each strategy can be decomposed into lower-level items, which gets the organization closer to an implementation plan. Again, traceability to the objectives should be maintained so that the relative importance of each strategy is easier to assess. For instance:

- S1: Purchase, or if necessary create, a software library tool to allow reuse of any type of software artifacts (this strategy supports objectives O1, O2, O9).

Project and Deliverables Management   275

- S2: Create a library catalog and populate its contents (O1, O2, O3, O4, O5, O9).

- S3: Specify organization structures and processes for introducing software reuse into the organization (O6, O7).

- S4: Define a measurement program to assess the value of individual reusable components, the overall quality of the library, and the quality of the reuse processes (O4, O5, O6, O8).

- S5: Define a software library marketing approach (O8).

The Reuse Process

In order for something to be reused, it must have been created, located, and incorporated into something else. To reuse commercially available components, one needs only to locate and incorporate them. If the components come from a "software component factory" within the using organization, then all three steps (creating, locating, and incorporating) need to be performed. This occurs repeatedly as new components are required.

Reuse is a three-step process

The model in Figure 7.7 is generally adequate to illustrate this incremental reuse process. Sometimes an intermediate step is added between the two cycles. It consists of a "lookup-tailor-verify-release" cycle that consumes the base components that come from the class library development depicted at the bottom of the figure, and supplies the components needed by the application developers. In many cases, this refinement is not needed and may even obscure the model.

The component lifecycle feeds the application development lifecycle

Figure 7.8 presents the interactions between the various roles and tools involved in a reuse plan. In this Figure, "Identification/Qualification" indicates an activity included in the roles of the Reuse Manager and/or Reuse Librarians elaborated earlier. It consists of finding and acquiring suitable

How the various reuse activities relate

reusable components, or specifying to the library developers which new components should be created. It also includes performing the necessary level of quality assurance to ensure that the content of the library is of high quality.

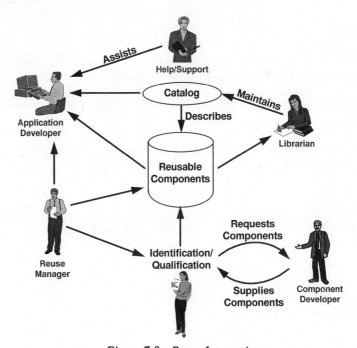

Figure 7.8 – Reuse Interactions

One way to "find" components is to extract from existing applications pieces of functionality that are likely to be needed elsewhere, and to generalize them before other versions appear elsewhere. Clearly, very few organizations today have the experience and the capacity necessary to engage in such a level of proactive reuse.

7.4 Object-Oriented Metrics

Quantitative measurements serve several purposes

A **software metric** defines a means to measure some attribute of a software development process, project, or product. Metrics provide quantitative ways to assess quality and to verify the

Project and Deliverables Management

impact of corrective actions undertaken after each assessment. In his chapter on software process management in [Pressman 92], Pressman presents four reasons for measuring software:

- To indicate the quality of the product;
- To assess the productivity of the people who produce the product;
- To assess the benefits (in terms of productivity and quality) derived from new software engineering methods and tools;
- To help justify requests for new tools or additional training.

But metrics should also provide ways to predict the cost or effort required by a new project. Lloyd Mosemann, Deputy Assistant Secretary of the U.S. Air Force, said in his closing address to the 1993 Software Technology Conference[1]: "More important than the absolute cost of something, more important than whether something takes a year or ten years to accomplish, is the *dependability or predictability of those projections* and, frankly, as a professional community, we have not [predicted] very well when we would deliver capabilities or at what cost."

They should help predict effort and cost, but the success rate is low

Many metrics exist for software products in general. *Process metrics* may include the number of design reviews or code reviews held before a component "passes" the review and goes to the next stage. *Project metrics* include productivity measurements such as the now-infamous number of lines of code produced per programmer per day, or the performance against schedule and cost objectives. *Product metrics* include total code size, but also quality measurements such as the total number of defects found after a product is released, that is, those left undetected by the testing process.

There are process, project, and product metrics

---

[1] As quoted in a special edition of Crosstalk, a newsletter published by the U.S. Air Force Software Technology Support Center

Object orientation invalidates many past metrics, but new metrics are still immature

When it comes to object-oriented software, some of the metrics developed under previous approaches clearly do not apply. Yet what to replace them with is not always clear. Several papers have been published since 1990 on object-oriented product metrics, but these metrics are still very tentative. The proposed measurements are seldom accompanied by convincing arguments that they can indeed lead to improved software. If it is true that "one cannot improve what one does not measure," it should be equally true that it is useless to measure something if one does not know what to do with the information. For instance, the *depth of the inheritance tree* (in the class hierarchy) is proposed as a product metric in [Chidamber 91]. As the depth of the tree increases, each class may inherit more attributes and methods in an unclear way, which makes the design error-prone. However, a very shallow tree is probably equally undesirable, since it means that inheritance is not taken advantage of at all. After all, inheritance is a form of reuse and a fundamental aspect of object orientation; therefore, some inheritance must be good. The question is: how much?

The authors, at this stage of their research, were not able to provide a definite figure of merit. The right answer is probably dependent on the size of the application (its total number of classes) and on whether or not part of the tree constitutes a class library that is developed and tested separately.

We measure what we can, not what we need

In this sense, many of the proposed metrics for object-oriented software irresistibly evoke the image of the person who is looking for his car keys under a lamppost—not because this is where he lost them, but because this is where the light is.

Use cases may be the basis for OO productivity metrics

Regarding project metrics, such as those dealing with productivity, the valid measurements to replace the "lines-of-code" metric are not yet fully understood. The discussion of effort estimation in the middle of Section 7.1 applies here, too.

Estimation and measurement are two facets of the same thing. As we mentioned earlier, the *number of use cases* is a tempting measure of functionality. The problem is that there are "big" use cases and "small" use cases and no clear way to quantify the magnitude of a use case. Counting the number of steps it contains sounds like a rational measure of complexity, but since human languages and thought processes are not rigorous, this is not a reliable indicator. Does the phrase "the elevator travels to that floor and stops" taken from Figure 4.5, for example, constitute one step or two? Different analysts will give different answers.

Object-oriented metrics are receiving some attention in both the academic and industrial communities; therefore, there is hope that advances will occur more rapidly now. The academics must resist the temptation to rely only on features of the languages (classes, number of attributes per method, etc.) because the industry needs metrics that can be applied early in the development process, long before the design is done. At the same time, the industry must resist the tendency to use the old metrics from previous approaches. Moreover, it may have to make a significant investment—in both time and money—to foster the development of the metrics it needs, and to validate them on multiple projects of different sizes. In March 1995, the End User Special Interest Group (SIG) of the OMG took a step in this direction, deciding to create a working group on metrics. We can hope that these are the first signs of an emerging industrial awareness of this important topic.

Progress in research may converge with better awareness in industry to arrive at useful OO metrics

Chapter 8

Future Directions For Software Engineering

Since its inception, software engineering has evolved continuously, and it is *still* evolving. Of course, in the beginning, it was not even called "software engineering." Over a period of about fifty years, a variety of techniques and methods emerged. Some of these never gained a significant market share. Others came to occupy a narrow niche in the whole range of activities, but did not conquer any ground outside that. Still others had their heyday, then retreated before the advance of alternative techniques. For instance, flowcharting is no longer considered an effective way to describe the low-level design of software. Assembly programming is still used, but only to obtain speed and size improvements in critical pieces of operating system code or real-time control applications.

The techniques and methods of software engineering have continuously evolved

8.1 Evolution of the Activities and Deliverables

The Meta-Lifecycle model described in Chapter 3 consists of several elements that may be assembled in various ways to define and execute specific lifecycles for individual projects. As technology changes, lifecycle designs must respond appropriately to these changes in order to support the development of

Lifecycle definitions will change to meet new needs

increasingly complex applications. In addition, the process designs must remain "in sync" with the available and emerging methods and tools, as well as with the evolving skills of the software development work force. This section presents some of the key evolutionary concepts for those lifecycle models.

Increased Focus on Early Stages

More emphasis on analysis and conceptual design will save money on later phases

Due to some technical advances, but mostly to a better understanding of the importance of solid requirements, the development focus shows promise of shifting emphasis to the early phases of the lifecycle. Today, because of current or prior shortcomings in good lifecycle practices, a typical software department that budgets most of its resources for software construction may actually sink most of these resources into maintenance and support. Many of these organizations are now beginning to work more on requirements analysis and conceptual design.

Managers must overcome their urge to see code produced

Initially, this trend will be disconcerting to managers who judge the progress of their projects by the volume of code produced. In a few more years, however, enough data will accumulate about the benefits of "front-loading" the lifecycle effort to lay these concerns to rest. Spending more time up front has several benefits:

- Instead of plunging into the unnecessary development of homegrown components, more time is available to find and apply reusable components.
- There is more time to evaluate several design alternatives.
- When new code *is* developed, it is much more often "correct by construction."

Increasing Use of Prototyping and Iteration

Prototype early, prototype often

Prototyping will become an integrated part of the lifecycle and will foster a tighter coupling among various lifecycle activities.

Prototyping will occur earlier and more often, thanks to tools such as graphical user interface (GUI) builders and a resurgence of interpreted languages such as Smalltalk. More kinds of prototyping will also be planned as part of the lifecycle. This will include not only GUI prototypes, but also prototypes to assess in advance the likely performance of a system, and other metrics.

Finally, Reuse!

Reuse has been touted as the main benefit of object orientation, but it has been hampered by two factors:

Reuse has been slow to appear in either one of its two forms

- By definition, reuse *within an organization* requires that some projects first populate a library of reusable components before other projects can find them.

- Reuse of *commercially available components* requires that a new industry establish itself. This means, however, that it must sort out a number of technical, commercial, and legal problems inherent in creating this new market.

Thus the advent of reuse is largely a matter of maturity, not a matter of technology. The concept of reuse is simple, and is practiced to a certain extent in other forms, such as subroutine libraries. But object technology makes it more attainable and effective. Unfortunately, it will take time to have a major noticeable effect on the state of the software industry. It is bound to happen, however, as more and more reusable objects are developed.

With time, this situation will improve

Reuse has already started to occur at low levels in the "food chain" of software. A good example is graphic widgets and class libraries of low-level design objects (queues, stacks, lists, trees, etc.). But in order for libraries of higher-level reusable components to emerge, industry-wide groups must become active. For instance, reusable domain-specific objects (e.g., for

Graphical class libraries are a start

manufacturing, accounting, Computer-Aided Design or CAD) can only be created successfully if groups of users and/or suppliers are willing to work together in this domain. Organizations such as the Object Management Group's (OMG) Special Interest Groups (SIGs) may play an important role by defining such reusable objects as Common Facilities (see Section 10.2). The Business Objects Management SIG (BOMSIG), for example, could specify a *reusable general ledger,* while the Manufacturing SIG might specify a *reusable recipe manager* that is compliant with OMG's architecture. Multiple suppliers could then provide conformant implementations of the specification, while competing on the basis of performance or other distinguishing features.

An electronic market for reusable components

Reusable components will become accessible from on-line libraries over the National Information Infrastructure (NII), better known as the "Information Superhighway." Assuming licensing issues for such a novel situation have been successfully resolved, users may be able to test them across the network on a trial basis, and then be billed automatically if they choose to retrieve the components. These components will probably include not only code, but also specifications, designs, and test plans that will be compatible with Computer Aided Software Engineering (CASE) tools and repositories.

Increased Formalism of Deliverables

A streamlining of the object analysis and design field

By the end of 1993, there were over twenty distinct object-oriented analysis and design methods, as documented in [OMG 94]. This means that in the next few years, there is likely to be a "shakeout" in this area. Several different methods may be required to cover all the different domains, but their number need not exceed twenty. As a few methodologists begin to establish their preeminence in the field, they will merge some popular features from other methods into theirs. Already the attitude of the main "gurus" of object analysis and design is changing in the direction of more cooperation. In 1993, finding common ground was largely considered an effort by a "thought police" composed of a few misguided methodologists and users. By 1994, however, many authors of methods, aware of the overlap between the twenty-plus object analysis and design methods available, began to delineate an area of cooperation around common core concepts, terminology, and notations. At the same time, some of the less common concepts should probably be subjected to more experimentation before they can be generalized.

This coordinated evolution of methods will lead to several positive consequences for their users:

The shakeout in the methods world will help users

- More complete and precise notations, including common graphical notations to represent common concepts;

- Specific and increasingly rigorous rules to ensure the correctness and completeness of these models;

- Standard representations of these common models in a tool-independent format, permitting the exchange of models between users of different tools;

- More *transformation engineering* tools (see the next section for more details) to decrease the number or difficulty of manual translations between artifacts.

8.2 Transformation Engineering

Transformation engineering is the discipline that enables automatic and human-assisted conversion between lifecycle phases. For instance, code generation and reverse engineering are transformations that require processing between the design and code phases of the lifecycle. Where opportunities exist, these processes will be automated, specifically where productivity and quality benefits ensue. Figure 8.1 shows some of these opportunities.

Automatic or computer-aided conversion of artifacts ...

...between or within phases

Transformation engineering includes transformations *between* life cycle phases as well as *within* phases.

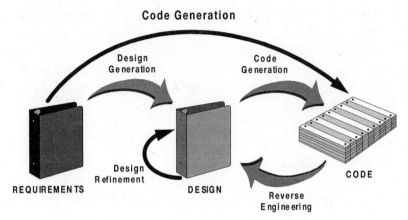

Figure 8.1 – Opportunities for Transformation Engineering

Background

There are few tools to assist with transformation

While the substance and quality of tools supporting work products of each stage of development have improved markedly, little attention has been paid to the development of tools to assist in the conversion of work products between one activity and the next.

This process requires a clear definition of the possible transformations

In order for transformation engineering to be effective, a sufficiently detailed model of the transformations must exist. With a complete definition of the artifacts of the development process, we can start modeling the transformations within and between process phases. Some of these transformations have a higher benefit/cost ratio than others. For these, viable products might emerge soon (based on pent-up demand), provided that customer inputs are integrated into the product requirement.

But it is also likely that some transformations, such as code generation, will continue to require human intervention for quite some time. Today, when a supplier claims to support 100% code generation, it can only mean one of two things:

Some human intervention will remain necessary

- Either the developer has been asked to augment his requirement or design specification with very detailed information, reaching the same low level of abstraction as code;
- Or the domain being addressed is narrow enough that all the potentially needed code has been captured in a reuse library.

Transformation engineering is a development paradigm. It envisions that a user:

A vision of complete analysis-to-product derivation...

- Develop a high-level understanding and specification of what a system (program) is to do during requirements analysis;
- *Transform* analyzed requirements into a conceptual design;
- *Transform* or otherwise evolve the conceptual design into a detailed design;
- *Transform* the detailed design into an efficient program.

Throughout this sequence of transformations, the correctness of the representation must be preserved. Assuming that the transformation process is systematically reproducible, it is possible to perform maintenance on the high-level specifications instead of on the generated code.

...that is compatible with maintenance activities

Expected Benefits

The greatest benefits are in projects that program "in the small"

A benefits analysis appears in [Barstow 87], based on Boehm's Constructive Cost Model (COCOMO, described in [Boehm 81]). This study cites productivity improvements of a factor of 3 for small projects and 1.3 for larger projects. An interesting result that emerges from this analysis is that the greatest benefits are achieved by transformation engineering when an organization employs "programming in the small," which consists of creating software from small components, namely, individual lines of code. This is still the prevalent development approach in the industrial and commercial community.

Organizations that program "in the large" are less impacted

By contrast, organizations that produce massive software (millions of lines of code), such as military or aerospace contractors, have often converted their methods to "programming in the large" or "megaprogramming," a concept described in [Tracz 90] and [Boehm 92]. This means that they assemble software from larger components (but not necessarily object-oriented ones). In this case, the reuse of larger, prefabricated components results, according to Barstow, in smaller percentage benefits. But since the effort level is so high and the total cost of systems may be in the billions of dollars, even a smaller improvement in percentage terms is well justified.

Reference Model

A formalized reference model is needed as a basis for product requirements

A sufficiently detailed model of software engineering transformations is needed for transformation engineering to be effective. This provides a uniform descriptive mechanism for transformations, and a basis to compare examples of each engineering artifact and transformation. Once the artifacts and transformations are well understood, the requirements for transformation engineering products (such as code generation tools and reverse engineering tools) can be written. In other words, the reference model establishes the communication framework to pursue transformation technology.

In a 1990 study performed on behalf of SEMATECH, CASE technology expert Herman Fischer presented such a reference model using *four layers*. The model is intended to serve as a basis for further work to articulate requirements for transformation engineering products.

Fischer defines four layers

- The top layer describes the four layers themselves; therefore, it is called the *meta-layer*.
- The next layer contains *description mechanisms*. At this level, each different software development paradigm may require different components. The model proposed by Fischer addresses the case of an object-oriented paradigm.
- The third layer is the *instance layer*. It provides a schema or type graph and is the most relevant layer for the description of work products and their transformations. In this layer, a particular transformation support environment would be described.
- The bottom layer represents instances of transformations being used on actual projects.

State of the Research

Most of the research activity on software transformations has been focused on helping capture informally expressed requirements and designs, and converting them to *formal* specifications. Formal specifications can then be transformed into various executable programs through automatic and human-assisted means. Formality offers the potential to introduce automatic, repeatable, and mechanized support into the development process.

From informal requirements to formal specifications

Research goals have been twofold:

- **Comprehension:** to improve the understanding of applications, programs, and algorithms. Researchers in this area

Understand existing code

have concentrated on semi-automated, manually assisted transformations from programs to other notations.

- **Construction:** to improve the process of programming. This has focused on transforming formal specifications into programs written in imperative languages. Three approaches have been used:

    Transform specifications into code

    - *Extended compilation.* This is a "batch" process that accepts a very high-level source language—usually a formal textual specification of an application, as well as textual advice from a user—and produces an executable program. Examples of extended compilation languages are SETL, RAPTS and TAMPR. The Ada community should be familiar with SETL (Set Language), because it was used to construct the first group of Ada compilers in the 1979-82 timeframe. It came from a long-running project of the Courant Institute, and has served as a platform for much transformational research. SETL combines executable programs with the syntax and semantics of set theory. RAPTS (Rutgers Abstract Program Transformation System) performs source-to-source transformations on SETL programs; very high-level specifications can be transformed into efficient code. TAMPR (Transformation-Assisted Multiple Program Realization) provides a number of language extensions for FORTRAN, optimizations, and miscellaneous support. This system has been used to automatically convert programs from FORTRAN to Pascal.

    Examples of very high-level specification languages

    Use of expert advice, which can be retained and reapplied later

    - *Metaprogramming.* In this case, the developer provides extensive interactive guidance during the transformation from a textual specification into a lower-level language. The level of formality of the specification language becomes less important than the form of the advice guiding the transformation.

Since the advice is not provided in a static manner as in extended compilation, it has to be retained somehow. This allows the transformation to be repeated later without asking the user the same questions all over again! Research has concentrated on the structure of the transformation developments, but also on how to automate the reapplication of advice when the transformation has to be performed again but with modified specifications.

- *Synthesis.* Knowledge-based reasoning, with or without interactive user guidance, is used to construct a program from specifications written in an even less formal, possibly English-like language. Here, the focus is to avoid limiting the specification language in any way, while automating the transformation as much as possible. Synthesis is similar to metaprogramming, except that deep reasoning replaces the skilled interactive advisor to navigate the space of possible development alternatives.

Expert programming system

State of the Practice of Code Generation

When there is a disconnect between two levels of languages, complete generation of code from languages is used to simplify programming. One language may be an expressive, higher-level one suited to the domain of the application. The other language may be a more arcane or low-level one required for compilation. For instance, many products generate COBOL source code from fourth generation languages (4GL), which are better suited to describe database manipulation and interactive data querying by the user. Similarly, there are domain-specific languages to describe communications networks, system configurations, simulation models, or the real-time interactions that exist in embedded systems. This is a popular application of the code generation concept because it is a stable field.

Most current code generation works between two levels of programming languages

Preprocessors are a trivial example

Code generation is also used to *change paradigms*. Most C++ compilers are really two-part programs: a preprocessor that translates the C++ code into straight procedural C, followed by a transparent invocation of the regular C compiler.

Generation from a design is not yet available

These prior examples concerned language-to-language translators. However, the complete generation of code from a more abstract design notation is much harder. No products have been identified—other than formal languages like GIST[1]—that take diagrams of semantic design specifications and generate complete programs. In fact, the industry learned some hard lessons with respect to code generation. We will describe below three main approaches, each of which has some limitations.

There are products addressing narrow domains . . .

(1) **Narrow Domain.** In some cases, the application can be represented entirely through a fairly simple language like a 4GL or a visual programming language. This happens when the range of "things" to be expressed is well defined and is very much the same across all such applications: query and display of text and numbers from a database, report generation, simple calculations, and data manipulation (e.g., sorting), etc. In those cases, since the details of screen and database interaction do not have to be designed or coded again, a complete application can be generated automatically from the higher-level language, with tremendous productivity and reliability gains.

. . . but there are still limitations

Even in this class of code generation, there are clearly some limitations. The 4GL or visual programming language cannot express all the necessary actions; therefore, the products offer some "3GL callability" feature. The developer can write the more complex parts of the application in a third

---

[1] GIST is a formal specification language developed at the University of Southern California's Information Sciences Institute in Marina Del Rey. A number of tools to process GIST and to transform it into executable programs in the LISP language were developed at ISI in the 1980s.

generation programming language, and call these 3GL modules from the 4GL, or vice versa.

(2) **Template-Only Generation.** Another approach is to keep a completely general domain but to generate only a top-level structure or template for the application. Typically, headers, function calls, and parameter or variable declarations can be generated from any design notation that captures the structure and interactions of a number of functions or objects. Each template must then be expanded manually to include the actual processing logic of the program, usually written in a 3GL.

Other products create a skeleton only

This is helpful in that it creates a uniform style across all modules. Moreover, it avoids simple errors such as calling the wrong function, or parameter list mismatches. But it obviously does not go far enough. The initial output of the generation stage is too incomplete to be more than marginally useful. Furthermore, once the templates have been populated with real code, how does one maintain the software? In most cases, it is impossible to edit the design diagrams and regenerate the code templates in such a way to preserve the previously filled-in detailed code. Instead, once the templates have been filled in and possibly reorganized by programmers, it becomes easier and more natural to maintain the textual representation (the final source code). This means that the code generation is applicable only *once* during the life of each module.

This helps initially, but does not extend to maintenance

(3) **Stretching Claims.** In this third case, the code generation claims to be complete, but only because we use a rich, large, and detailed specification language to describe the application to be generated. Upon further examination, it always turns out that the specification language is itself a *complete programming language*. It has a different syntax from the target programming language, but the same low level of abstraction. As a result, the person who captures

Some claim complete generation because the source language is as detailed as the target one

the design specification in this language is burdened with a new and complex language which is hardly less difficult to use than C or COBOL. Since this language captures every aspect of the application, the application can be maintained in the new language, but productivity remains at approximately the same level. Moreover, debugging can be a problem since errors are typically reported in terms of the generated language instructions, not in terms of the specification language.

Mixing these approaches may yield better results

Although progress has been made with each of these approaches (especially the first two), it is quite likely that a comprehensive solution will involve elements of all three. For instance, the third approach can actually be improved upon by capturing some recurrent aspects of low-level design, such as how to manipulate common data structures (lists, stacks, queues) and how to perform common operations (sorting, applying the same function to every record read from a file, etc.). This is a blend of the "narrow domain" and "stretching claims" approaches described above. The code that is actually generated from a higher-level description corresponds to common computer science algorithms found across diverse applications, while the code that is specific to the logic of an application is described in a language equivalent in complexity to the final programming language.

This requires both more research and more productization

CASE tools for analysis and design are evolving rapidly, and there are already many that provide limited code generation capabilities. Such tools generally fall into the category of template generators described above. Customers would clearly prefer complete code generation that leverages reuse from analysis, design, and class libraries. To achieve this, not only must the "state of the practice" get closer to the "state of the art," but the state of the art itself may need to further advance.

Future Directions for Software Engineering

State of the Practice of Reverse Engineering

Reverse engineering (in software) is a relatively new field, and the term is used loosely by salespeople to describe many products that do *no* engineering. Reverse engineering is definitely more than disassembling an executable module into assembly language source code! Strictly speaking, **reverse engineering** would imply the discovery of the design from the targeted code, or the discovery of the requirements from a design.

Reverse engineering recreates a design

Several companies have products that can discover the low-level design from code. The code is read, analyzed, and transformed into *structure charts* that represent the calling hierarchy of functions or objects. These devices are quite useful for several purposes:

Structure chart generators help understand existing code

- Documenting existing software in order to make it more readable to the development staff (especially to any new members of the team);

- Creating a structure that can be navigated graphically, thus easing the search for bugs or enabling "cleaner" introduction of new features;

- Satisfying the organization's needs for documentation.

The more elaborate reverse engineering products not only create a graphical representation of the code, but also an underlying comprehensive data structure, which is stored in a repository. The repository typically contains three types of information, shown in Figure 8.2:

The resulting structures may be stored in a repository

- A structural representation of the code (i.e., the functions or modules and their calling relationships).

- The layout of the diagrams. Users typically need to edit the diagrams created by the tool, perhaps for better readability. By storing the screen coordinates of each icon in the

The repository may also contain graphical information

diagram, the repository retains the user's layout preferences from one session to the next.

It also stores the code itself

- The code fragments corresponding to every structural item described in the design representation. By storing these code fragments, the repository contains *all* the information that was in the code. Code can be regenerated (this is logically called "forward engineering") and should be identical to the original version, except for pretty-printing differences of no semantic import.

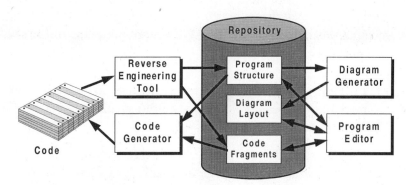

Figure 8.2 – Repository-Based Reverse Engineering

Such a reverse engineering tool supports maintenance

The developer can then perform two kinds of changes. *Structure changes* can be made in one of the graphical views supported by the tool. *Detailed changes* can be made by editing the code fragments that are attached to the "leaves" of the structure. Either way, the repository still contains a complete representation of the program, and the complete source code

Future Directions for Software Engineering   297

can be generated from the repository contents. In addition, the graphical environment contains browsing aids that include search tools to find all definitions and references of a given variable, all variables of a given type, etc. This means that even after maintenance changes are performed, the developer does not have to revert to maintaining the code using only the traditional text-oriented editors.

Since approximately 1993, a few such tools have come into being. One could cite for instance Discover, from Software Emancipation Technology, or Crev and Graphgen from Mark V Systems. (Once again, these names are provided as proofs of existence, not as endorsements or complete lists.)

A few examples

8.3 After OO: Are Agents the Next Paradigm Shift?

In the early days of object technology, some experts said it was nothing more than a different way to describe the techniques they had been using all along in structured analysis and design. It has now become almost universally accepted that the effective use of object technology requires a totally different mind set. Conventional programming languages and analysis/design method usage can be manipulated to obtain the end results of object technology, but effective object technology implementation requires a new tool set that empowers the object-oriented mind set.

A new mind set

Agent technology invokes heated discussions among software people in the same way object technology once did. Are agents just "smart" objects containing goal-seeking methods, or do agents represent a fundamental software technology concept that cannot be represented as a different kind of object? Agent technology proponents claim it is as different from object technology as object technology is from structured technology. Who is right? The free market of technology concepts will have its say, and in a few years, the decision will be obvious.

Agents may be an incremental evolution of objects or they may be a revolution

For now, we will provide a brief summary of the proponents' view of agent technology.

What is a Software Agent?

Two views of agents: simple or complex

The most widely accepted definition of an agent in a direct sense is "that which acts" [Parunak 94]. But within that context, there are two prevalent views of agents. In one view, agents are simple entities that know only a few simple rules, but by cooperating with other similar agents, they are able to perform complex tasks very effectively. In the alternate view, agents are complex enough to exhibit some human-like qualities.

Simple Agents

Simple agents are powerful when many cooperate

The systems most often used to describe *simple but effective agents* are found in insect species, whose individual members have a very low intelligence, but who are able to act collectively to accomplish complex tasks. Ant colonies demonstrate an entire behavior system [Steels 91] that is the result of collective effort. It efficiently constructs networks of paths between food sources and their nests; it sorts larvae, eggs, and cocoons; and it puts food in the nest [Deneubourg 91]. Each of these tasks overcomes formidable obstacles, but no "super ant" computes the network or the sort algorithm to be executed. Likewise, termites build extremely complex mounds [Kugler 90] without master plans; and a single nest of wasps can perform a highly integrated set of tasks without a complex management hierarchy. In all these cases, the individual performs its functions by following four to six very simple rules.

Complex Agents

Complex agents approach tasks through human-like, high-level concepts

Work performed at Stanford in connection with Hewlett-Packard's New Wave concepts [Shoham 93] is often cited as a prevalent vision of advanced agent technology. In this work, an agent is "an entity whose state is viewed as consisting of *mental components* such as beliefs, capabilities, choices, and

Future Directions for Software Engineering

commitments." In Shoham's agent architecture, an agent has a database that contains these capabilities, beliefs, and commitments. This means that it can perform actions for a client. That client can be a human or another software agent. The actions it performs are based on its beliefs—not a set of hard-and-fast action rules. Shoham's agents make choices, enter into commitments, and interact autonomously with their environment based on what the author calls their "mentalistic" state.

"Real-World" Software Agents in the Near-to-Intermediate Future

The business enterprise is becoming so complex that its tasks and processes overwhelm the capability of even skilled human participants. There are just too many events one must react to, and too many things to track to leave much time to conduct business in a proactive manner. For a number of years the software engineering community has been proposing to solve this problem through greater integration of people and software systems, [Hollowell 90], hoping to achieve parity between them. *Personal software agents* will become commonplace in the early twenty-first century. They will act on behalf of project leaders, managers, and technician or clerks, automating some of the tasks that would otherwise overwhelm them by their number or complexity. Fairly soon, an individual or an organization that does not interact with a significant number of software agents will not remain competitive.

Agents will soon gain a practical role as assistants to the knowledge worker

Software agents in businesses will go far beyond today's analytical software practices of pointing out problems after they have occurred. Agents will perform many monitoring tasks and respond productively both in routine situations and upon recognizing out-of-control symptoms. An agent might monitor a series of machines for excessive bearing wear and notify the preventative maintenance scheduling agent or, depending on the severity, shut the machine down and notify a human maintenance technician. Another agent might monitor inventory levels for potentially unfavorable trends and take preemptive

Agents can monitor tasks, respond to alarms, filter information, and make routine decisions

actions. Yet another agent might monitor a person's incoming electronic mail to filter out the trivial and prioritize the remainder. Ultimately, software agents will measure the effectiveness of their own actions and improve the decision-making process in future instances.

8.4 After OO: Are Formal Methods the Next White Knight?

The surfer's metaphor: choosing the right "wave"

In every technical domain—be it aerospace, transportation, materials, or others—technology evolves in successive waves. At any given point in time, it is difficult to pick the right wave in which to invest effort and money. This has been referred to as the "surfer's metaphor." If you choose a wave that has already peaked, you will not get much out of it, and you may find yourself quickly beached. But if you swim farther out and choose a wave that is just forming, you may exhaust yourself trying to catch it, without being sure in advance how much promise it holds. It may be the wave of the century, or it may collapse early. By choosing the object-oriented wave at this point in time, we believe that we (and you) will in most cases be making the right compromises:

Is OO the right "wave?" We think OO is the right "wave"

- The technology has already shown clear promise and encouraging results in a number of different domains.

- It is not so far out that we will become exhausted in its pursuit. Users can already find products on the market, experts, books, training classes, and properly educated students to recruit.

- Its most active years are still ahead, so we are not jumping on a paradigm that is already beyond its prime.

Alternatively to the object-oriented approach, there is interest in developing and using other emerging technologies that appear to have significant potential. These include pattern languages, genetic algorithms, evolutionary programs, and agents. We will leave those for other books and/or other authors, and will address in more detail the area of formal methods.

Formal methods are one alternative on the horizon

Formal methods rely on well-formed mathematical and logical expressions of a system's requirements or design specifications. These expressions are captured in a highly symbolic language, rather than through the diagrams common to object-oriented models (or structured analysis and design models, for that matter). From this unambiguous representation of a specification, various properties of the system can be guaranteed using theorem-proving techniques, and various transformations can be obtained in a rigorous manner. Clearly, the proofs and transformations require automated tools in order to minimize the risk of error. Even so, the correctness of the proof or of the transformed program may still be suspect, given the possibility of defects in the tools.

They use mathematical constructs and reasoning to verify or transform specifications

Formal methods have been more popular in Europe than in the United States, perhaps because the educational systems of Europe stress a more complete basis in mathematics in all technical disciplines. The best known formal methods are Z (pronounced the British or French way, that is, "zed") and VDM. For more information on these, see respectively [Spivey 89] and [Jones 87]. In the U.S., two notable and rare centers of interest in formal methods are the Microelectronics and Computer Technology Corporation (MCC) in Austin, Texas, and X3J21, the accredited standards committee on formal methods established by the American National Standards Institute (ANSI) in 1994.

Formal methods are a predominantly European phenomenon

They help in highly critical applications, or to verify the core of a complex system

Formal methods are demanded, and their complexity is justified, in high-reliability applications, usually in "orange book" applications related to national security. They are also used when the severity of the possible consequences of a program error (e.g., nuclear power plant control) justifies expending a significant effort to understand and prove the correctness of a program. On occasion, formal methods have also been accepted in Europe and Japan for some non-military, non-high-risk applications (e.g., the specification of communication protocol drivers or the kernel of a transaction processing system). In these cases, the justification was that although the part of the system being verified was a small portion of the total size of the project, this part would be executed millions or billions of times over the life of the product. The economic consequences of a defect located in the core of such a system can be devastating. For example, in a well-publicized case, a software defect in a supplier's switching equipment crippled AT&T's entire North American network for a day in 1991. The ability to virtually guarantee a zero-defect design or implementation therefore made this approach economically viable.

There is not yet an intersection with object approaches

In the current state of the art, formal methods and object-oriented methods are completely disjoint and rely on quite exclusive approaches. The latter uses a somewhat informal process, yielding possibly ambiguous interpretations, but maintains a notation that is understandable by the system's users. The other removes ambiguity, but couches the specifications in a language that only computer scientists (and not even most of them . . .) and mathematicians can understand. For instance, Figure 8.3 contains an example, taken from Spivey, of a Z specification. This concerns a computer application to maintain and query a telephone directory.

Obstacles to broad adoption

In the opinion of the authors, formal methods will not see general adoption in the near future, at least in the United States, due to the following limitations:

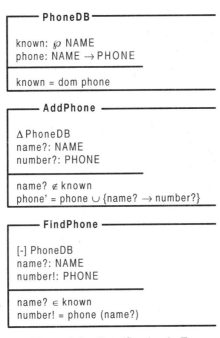

Figure 8.3 – Specification in Z

- They require a level of understanding of mathematics and logic that is currently not prevalent among the software development community in the U.S. There is no indication that a significantly higher proficiency in mathematical logic will be developed any time soon in the educational system.

- There are not enough experts, training classes, or consultants available to support their use.

- The level of experience in formal methods depends a lot on the problem domain. Although there has been research on using Z to specify user interfaces, most applied work is limited to operating system security and data communications.

More use will come with time, but slowly

The last two limitations—and perhaps the first one, too—could be overcome with time. In other words, they are issues of maturity rather than of inherent technical difficulty. Therefore, the applicability of formal methods should definitely be reexamined from time to time. Moreover, even with the current drawbacks, formal methods can be useful for proofs of correctness of small but highly critical parts of a system, as in the examples mentioned above.

Future convergence with objects is possible

Like two geopolitical archenemies, will object-oriented and formal methods ever go through *détente* or even *rapprochement*? This is quite possible, based on the same techniques described in the latter part of Section 8.2. Namely, when the graphical models on which object-oriented methods rely are captured in repositories containing not only a graphical representation but the semantics of the drawings, then it may become possible to enrich the diagrams. More complete annotations such as the cardinality of relationships or the precise types of attributes can be added to the graphical form and captured in the repository. These are precisely some of the types of information that formal specification languages like Z or VDM capture. At that point, it may be possible to envision a convergent path that preserves the rigor of formal methods while offering the readability of object-oriented representations.

Z++ is a first effort in that direction

K. Lano and H. Haughton of Applied Information Engineering, Lloyd's Register, in London, have attempted such a blend of paradigms with the Z++ method. This is described succinctly in [OMG 94]. Z++ is a combination of several notations:

- An object-oriented extension to the Z language;
- Object models borrowed from the Object Modeling Technique (OMT);
- Data flow diagrams, state diagrams, and "entity life histories" borrowed from another method called Structured Systems Analysis and Design Method (SSADM).

Future Directions for Software Engineering

Whether or not Lano's and Haughton's effort represents the beginning of the convergence between object-oriented techniques and formal methods is not yet clear, since Z++ has few followers at this point. A successful evolution in that direction, whether based on Z++ or on another formalism, would undoubtedly represent a watershed in the history of software engineering.

Toward a unified paradigm?

8.5 Other Research Topics

This section has multiple purposes. First, it points out which areas of software engineering have not yet reached the practical usability stage. In that sense, this section is a disclaimer or a *caveat lector*. Another possible use is to indicate what areas of development are likely to see some significant evolution by the end of the millennium. And finally, if you are a student or a thesis adviser in search of a Ph.D. topic, these are areas you may want to consider.

Other less developed ideas

Object-Oriented versus Data-Driven Functional Approaches

Object-oriented methods have often been presented and discussed in opposition to data-driven and functional methods. Some experts have claimed that object orientation actually represents a Darwinian evolution of the concept of data-driven methods, that is, an object is nothing more than a set of data with a well-defined manipulation interface. This is a somewhat dubious claim. One of the fundamental principles of data-driven methods was to create a *public* data representation that all components of the software could manipulate through a query language. To write the query language, one needs to know the names of the data fields and their representation. This is quite the opposite of information hiding.

Object orientation exists in opposition to data-driven approaches...

With functional methods, the chasm is even more obvious. Yet there are domains, such as numerical analysis, where functional approaches still claim a significant following. These are

...or to functional methods

typically domains where the identification of objects is difficult or arbitrary.

A well-defined combined approach might be beneficial

One of the open questions is whether or not there is room for pacific coexistence or even some form of symbiosis between functional and object-oriented approaches. And if there is cooperation, how can it manifest itself in an orderly, methodological approach, so that the practitioner in the field can benefit by applying each as it is needed without confusion? One possibility is to decompose a system into a set of objects, and then apply a functional approach to the design of the methods of a class of objects. But since objects could be further decomposed into smaller-grain objects, this is an ambiguous statement. It only becomes reliable if one can define a criterion for stopping the object-oriented decomposition and switching to functional design. This is definitely an area for further research.

Real-Time (or Time-Dependent) Object-Oriented Software

Real-time systems are a good domain for objects

The real-time aspects of a system, at first glance, are ideal for the application of object-oriented concepts. Real-time systems often consist of a collection of components (sensors, controllers, actuators, etc.) that communicate with one another. Therefore the message-passing paradigm of object technology applies.

Methods rarely do justice to them

However, the picture becomes less clear when it comes to what specific methods to apply. In most methods, real-time factors are not treated in a manner consistent with the rest of the concepts. Two examples will illustrate these difficulties:

Future Directions for Software Engineering

- The notion of a "timer" object is often added during the requirements analysis stage (e.g., in Shlaer-Mellor and in others). The occurrence of time-dependent events can then be represented by messages sent by the timer object to other objects. Yet if one takes the view that requirements analysis defines those objects that are present in the "real world", independently of the existence of the system under design, then the timer object has no place in it. It is a human creation and therefore, it should be part of the design, not the analysis.

    Design artifacts are introduced prematurely to fill gaps in analysis

- In methods based on "scenarios" or "scripts," such as Object Behavior Analysis (OBA), the possible occurrence of events at different points in time causes a tremendous combinatorial explosion of the number of scripts to consider. It is the same with simple error situations, which tend to be numerous in real-time systems, because the passing of time becomes in itself a cause of errors in such systems.

    Time complicates scenario-based analysis

In general, the guidelines and rules for efficient and correct analysis of real-time systems using object-oriented techniques are not as well defined as for non-real-time ones.

More work is needed

Metrics

The metrics that have been defined thus far for object-oriented systems (e.g., in [Chidamber 91]), are primitive. At the same time, the world has become more attuned to the axiom "you can only improve what you can measure." Therefore, more importance is likely to be placed on metrics than in the past.

Object-oriented metrics are currently primitive

Many tout object technology as the solution to the software productivity crisis because it will allow large systems to be built from reusable, qualified components. Managers who have heard these and similar claims before will demand proof of added productivity. Short of witnessing a manifest impact on

They are needed to prove the benefits of objects

the company's bottom line, the best way to assess whether or not the technology is fulfilling its promises is to measure the productivity of the process and the quality of the products. Yet this is practically impossible with the current state of the metrics for object-oriented software.

Old metrics are obsolete

There are nevertheless a few clear principles: do not measure lines of code, since the software will no longer be built by writing individual lines; and look at reuse from both ends of the "food chain." For example, how many reusable components does one developer produce, and how many previously written components has he incorporated during the course of a project? But these metrics suffer from several shortcomings:

New ones are few, ill-defined, and hard to measure

- They are, at best, insufficient.
- They are open to ambiguities unless stated much more precisely than in the previous few sentences.
- Tools are needed to automate the measurement process. Otherwise, it is unlikely that the data will be correctly and completely measured through a manual process.

Model-Driven Computing

Putting the user in the driver's seat

In Chapter 2, we talked about an architectural principle called "model-driven computing." According to this principle, and to the extent possible, the end user should be put in the driver's seat, as was explained through the spreadsheet metaphor (Section 2.4, Principle # 9).

Make more systems behave like a spreadsheet

A spreadsheet is, in its simplest form, a two- or three-dimensional set of numbers and textual labels. But most spreadsheets contain more. Some of the cells contain formulas, often as simple as =SUM(C1:C12), which means "show in this cell the sum of cells C1 to C12," and sometimes much more complicated. Any spreadsheet that contains at least one formula becomes a *de facto* program. If you change the number in cell C3, the cell containing the total of C1 through C12 is immediately recomputed and displayed. Yet the accountant or manager who wrote the formula =SUM(C1:C12) or who used a "point-and-click" method to indicate the desired computation, may or may not have known the required syntax and would probably not call himself a programmer.

What happened here? The spreadsheet program is a *modeling tool*. Its user, by entering twelve specific numbers and adding the formula to calculate the total, has done more than obtaining the total of these particular numbers. He has created a model of all future similar computations to come. Indeed, the individual numbers can now be blanked out, and the spreadsheet can be saved. When retrieved and filled again later, the total will automatically be computed. Clearly, the spreadsheet application itself, as delivered by Microsoft, Lotus, or another supplier, represents an infinity of possible programs. Tens of millions of spreadsheets now reside on millions of PCs worldwide, and it is likely that no two of these (unless copied from one another) represent exactly the same program! And yet a single application program allows this.

It is both an actual program and a tool to write many programs

Let's now think of another domain such as discrete manufacturing. What if a supplier delivered a piece of software that a customer could use to model the operation of his factory, so that the resulting model would subsequently become a working controller for all the equipment in it? The user would enter the capacities of the various machines, their expected mean times between failures, the number of parts that need to be produced, and the "routings" that constitute valid flows between manufacturing steps. And by entering some formulas, perhaps not much more complicated than =SUM(C1:C12), he would in effect tell the program "now make my factory run according to this spec."

An equivalent concept in an industrial application

It is not an impossible dream

We are clearly far from this becoming a reality, but there is no reason to consider it an impossible dream, either. In 1960, the notion that millions of people would eventually have, on a $1,000 computer, a program making many custom, mainframe-based accounting programs obsolete, would have sounded ludicrous. Yet it happened. So it is probably not too early to call for a similar revolution in other domains, whether manufacturing, health-care monitoring systems, telecommunications switching systems, or others.

Design Rationale Repository

Retain information from the design process

Another area that could benefit from a focused research effort is that of design rationale repositories. Such research would define what sort of information should be retained during a design effort to guide the people who may maintain the same system in the future. It would also address how this information should be stored so it can easily be found and exploited.

During design, information is generated about alternatives

During the design effort, many possible solutions are considered, usually only in the designer's mind, and for a very short period of time. Some solutions are rejected immediately, perhaps based on the designer's experience or out of subjective preferences. Others are considered for some time, and may be discussed at length; yet they end up being rejected without being tested, rightly or wrongly, on the basis of a certain balance of advantages and drawbacks. As this winnowing process continues, one or more solutions may be explored. This is when prototyping may be used. This may result in more solutions being rejected as impractical (for example, because of insufficient performance). Finally, a single preferred solution emerges and is implemented in the final design.

This information is usually lost later, at great cost during maintenance

Several months or years later, a change must be made. A designer (not necessarily the same person who did the original design) considers the new requirements, the existing code, and in rare cases, the documentation created to describe the design originally selected. But there is almost never a document that says "do not use a particular feature because it is too slow." So the process restarts from the beginning. The person in charge of designing the change may find that the current design does not lend itself easily to the requested change. Other potential designs are then considered without any information being available to alert the designer that this had been unsuccessfully attempted before (for example, the result was too slow or had some other "showstopper" characteristic). So the designer has to go through all the trial and error process once again from the beginning in order to rediscover which solutions are feasible and which ones are not. This is just as efficient as the exploration of the world would have been in the 16th century if each ship had only carried a captain and a crew of sailors, but no cartographer to create a map of what had been explored along the way. In this respect, it is possible to place the current practice of software engineering on a par with the methods of the Middle Ages!

Let's take for granted that capturing the design rationale would indeed benefit the maintenance of a software package or system. Then there is the issue of how to store this information so that it can be easily navigated. Some sort of graph sounds useful. The nodes of the graph, however, are likely to be labeled with English descriptions of the potential decisions, something that is known to lead to ambiguities and misunderstanding. It also seems important to capture in some manner whether the various solutions are complementary or exclusive, as well as the interdependencies between choices made in different parts of the graph. For instance, one design decision might be to represent a set of data in an array or a linked list. Another decision might be to perform a sequential versus a binary search. However, the two decisions are not independent of each other, since a binary search cannot be performed on a linked list.

Representing all the design decisions is a complex problem

Can design decisions be reused on a different system?

Finally, an open issue is whether this design rationale information can be used in the context of a *different* design, not just the design of a modification of the same program or system. This implies a giant generalization step, which again might be hindered by the use of natural language, and therefore may require the use of some formal language in order to lend itself to a reliable search and retrieval mechanism.

Feasible Software Process Enactment

We lack experience with the stricter levels of process enactment

In Section 3.6, we described the five levels of process enactment defined by the Software Technology for Adaptable, Reliable Systems (STARS) program. The higher the level of enactment, the more constraining it is on the people involved. What is possible, and what is desirable? One can safely assume that as long as the environment brings additional help to the developers without restricting their freedom, this is an improvement. This is the case until one reaches the "interactive passive enactment" model (level 4) mentioned in Section 3.6. But the "active enactment" model (level 5) is an entirely new concept in which the environment drives its user, rather than the other way around.

A very constraining environment may not work, but a semi-flexible one with controlled exceptions may work well

There is a distinct possibility that experimentation will establish that this much mechanistic supervision is too frustrating to software professionals, and that their productivity goes down rather than up. Then a more subtle topic for further research would be to determine whether something in the intent of this level of enactment can be preserved in a less intrusive way. For instance, can work be allowed to happen out of a method's prescribed order within reasonable limits? An example would be to let a developer proceed with coding, even if the design has not been signed off, in order to gain time if the design is approved with no more than minor modifications. However, if too much code is developed without the design being approved, it may be necessary to start issuing warnings about this condition. In the meantime, the developers may be more productive and feel less frustrated if the system lets them take the responsibility for their actions. In most cases, the work they would do out of sequence is useful. This form of enactment would better mimic the way human organizations function; therefore, it would be more natural to all the people involved. On the other hand, it is more difficult to implement, since it requires even more understanding of the process by the tool.

Object-Oriented CASE Tool Interoperability

There are a number of reasons why an organization would want its CASE information to be exchanged or shared between different CASE tools. Here are perhaps the two most frequent reasons:

Two typical needs to transfer data between tools

- A better CASE tool may appear, or the current one may cease to be supported.
- It may be necessary to share an analysis or design with a customer or a subcontractor who uses a different tool.

Even in the world of structured analysis and design tools, these problems have not been addressed particularly well. The Electronic Industry Association's (EIA) **CASE Data Interchange Format (CDIF)**, defined in [EIA 91], provides a fairly well accepted, neutral format for exchange between tools. But it does not always carry all the desired information between all the tools that claim to support it. And this is true despite the fact that the concepts to be supported have been fairly stable for ten years.

EIA CDIF provides only partial support, and only for structured methods

In the object-oriented world, the situation is much worse. The leading methodologists do not yet agree on many of the fundamentals, such as when to use the words "object," "class," and "instance;" whether or not state models should be part of an analysis model; and whether or not "aggregation" is a distinct form of relationship between objects. It should therefore be no

There is no common semantic model of OO methods

surprise that there is no canonical representation of object-oriented analysis and design models.

The OMG attempted a model

In 1992-93, the **Object Analysis & Design Special Interest Group (OA&D SIG)** of the OMG made a first attempt to define a *reference model* of the concepts of object-oriented analysis and design. This work, which is included in [OMG 94a], was adamantly rejected in 1993 by a broad group of object-oriented method gurus. They claimed that such a model constituted "premature standardization" and in private, called the OA&D SIG the "thought police" of object analysis and design. Yet this leaves the interchange problem intact. If one needs to change tools or share models with a business partner, there is no automated mechanism to do this today. The necessary steps to correct this situation could include:

Define the concepts common to multiple methods

- Define a set of common concepts of object modeling—and perhaps a set of optional concepts—that some, but not all, methods include. In terms of the OMG's Object Model Architecture (OMA), this would constitute a *profile* of the OMA object model.

Create a metamodel

- Create a *metamodel* of these concepts (that is, an abstract model of the models to be exchanged). For instance, a state model consists of a set of states, transitions between states, events attached to the transitions, etc. Furthermore, in a Harel statechart, there are additional relationships and attributes, such as hierarchy and orthogonality between states, and the notion of a "memory" of the last substate visited in a given state. The metamodel would capture all these elements.

Define the exchange format

- Define a neutral format, probably as an extension of CDIF, that would encode all the information contained in the metamodel. Alternately, a group within the International Standards Organization (ISO IEC/JTC1/SC7/WG11) is defining a new standard called **SEDDI**, for Software Engineering Data Description and Interchange. The scope of

SEDDI is broader than that of CDIF and has as much emphasis on standard schemas (data descriptions) as data interchange. Extensions to the scope of this effort are progressively being added, including support for Portable Common Tool Environment (PCTE) schemas, and for the IDEF0 and IDEF1X analysis languages.

- Prove that this works by creating at least one "writer" that can read a proprietary tool's database and write the neutral format, and one "reader" that can do the opposite for another tool, and verify that a reasonably complete model can thus be extracted and reloaded in the other tool. A tougher test is to translate the same model *back and forth* between two tools, and check that one ends up with a reasonable equivalent of the original model.

Validate the exchange process

This effort seems like an ideal opportunity for cooperation between an industry consortium (to define the metamodel and format), a university or government research laboratory (to perform the early proof-of-concept work), and specific companies which would then implement translators for their tools. Initially, most CASE companies might resist investing in a capability that threatens to make the users of their tools less dependent on them. The electronic CAD industry went through the same phase in the mid-1980s. Eventually, suppliers realized that if they did not adopt a neutral standard like the Electronic Data Interchange Format (EDIF), they were also preventing the reverse migration, that is, from a competing tool to theirs. They also saw that when customers demanded an export/import capability, the only alternative might be to adopt their competitor's proprietary formats, which was even worse. So they joined the club, albeit reluctantly at first. The CASE industry will likely follow the same scenario.

This will benefit the suppliers, as well as the users

Chapter 9

Shaping Object Technology: Standards and Consortia

9.1 The Long Road to Object Technology Acceptance

Object technology is more than twenty-five years old. Nygaard and Dahl developed the Simula language that defined all the important concepts of object technology in 1967—four years before Dennis Ritchie created C and fourteen years before the introduction of the IBM PC. The Xerox PARC team, led by Alan Kay, developed the Smalltalk language in 1972. Commercial versions of CLOS (Common LISP Object System), Object Pascal, Objective C, Eiffel, and C++ came to market in the early-to-mid-1980s. Why has it taken object technology so long to become a significant influence in the mainstream systems community? There is a combination of specific conditions that account for object technology's slow rate of acceptance.

Why the long gestation for object technology?

Lack of Business Orientation

The first reason for object technology's lack of commercial progress is that for most of the time since its creation, the corporate systems power structure has viewed object technology as research oriented and "not ready for prime time." The fact

Lack of penetration in management systems

that many early object technology enthusiasts disdained the mainstream corporate development environment, which is primarily COBOL-centric, caused this belief to linger longer than it might have otherwise. As reluctant as some may be to admit it, until the COBOL development community accepts object technology, it will never be an unqualified success. Until then, it will remain just another "techie toy" to the people who control the purse strings. COBOL applications development still constitutes more than eighty percent of total systems development. When object technology penetrates the mainstream Management Information Systems (MIS) development world, it will have its first serious opportunity for near-universal acceptance. One can verify this hypothesis by recalling the emergence of the structured technology paradigm.

Object technology didn't speak the language of business: COBOL

Object-oriented analysis and design are beginning to have a significant following in the corporate environment, but most companies will not fully embrace it until it can be implemented in a language that does not require conversion of the masses of COBOL programmers to what is for them an unconventional language. An Object COBOL standard will undoubtedly help the MIS community accept object technology. However, if the object technology marketplace had learned from the precedent of structured technology's adoption, this could have been "sooner" rather than "later."

Waiting for the Structured Paradigm to Fail

Problems with structured methods were not fully recognized until recently

A second major obstacle has been the reluctance of the mainstream software development community to consider a new paradigm—regardless of its merits. The structured technology paradigm itself did not reach its high-water mark until the mid-1980s, so it was the late 1980s before anyone realized that a structured approach could not provide the productivity increases required to keep up with system development demand.

The mid-1980s saw the first documented results of large-scale usage of structured approaches, which fell far short of their promised potential. Only then was the management of mainstream systems development organizations willing to consider a new paradigm. Their key prerequisite for adoption, however, will be the existence of an object lifecycle that is sufficiently robust. This must, of course, include front-to-back (analysis-to-maintenance) object-oriented lifecycle tools that can replace the structured technology tools embedded in their current lifecycle process, or their involvement with object technology will be little more than casual experimentation.

Object technology adoption is subject to its robustness and breadth

Overcoming the Hype Legacy

The third big reason for slow acceptance of object technology is the skepticism surrounding early object technology hype. In the early-to-mid-1980s, the software industry did a "hard sell" on technologies that did not deliver on their promises. The worst offender was the artificial intelligence (AI) movement, which subsequently failed in the eyes of corporate management. Computer-Aided Software Engineering (CASE) tools were also oversold at first because they were falsely touted as a substitute for human intelligence rather than as an aid to human developers. Moreover, CASE tools introduced in a process vacuum did not achieve significant productivity gains.

The legacy of overhyped technology and tools

With Fourth Generation Languages (4GLs), the situation is more complex. 4GLs allowed smaller teams of developers to rapidly create applications combining the use of a database and a user-friendly human interface. In that respect, they were successful at creating more agile system development capabilities. However, they did not solve the basic problem of structured systems, which is that data is defined and stored without any associated behaviors. What this means is that all too often, 4GLs have enabled the proliferation of overlapping applications that all manage the same data differently. For example, when the MIS department was the bottleneck, there was

4GLs delivered localized improvements

probably only one program available to query the accounts payable balance by department. Since the 4GL "revolution," however, there is often one such program in each department, making database administration a nightmare.

MIS management has become wary of false promises

Previous disappointments with these overhyped technologies, together with the unfulfilled promises of structured technology, created an inherent distrust by mainstream systems management of any new systems technology. The more serious players in the object technology movement are doing their best to avoid exaggerating its benefits. But as with all movements, there are zealots who feel no qualms about tossing around highly suspect claims without regard for future consequences. As a result, corporate MIS, the largest of all software development segments, continues to depend primarily on structured technology while waiting for object technology to measure up to its requirements.

General acceptance will be a slow process

Most of the object technology entrepreneurs who want to make money and stay in business for the long term seem to have learned a few lessons from the AI debacle. The overcommitters are still out there, but they tend to be outweighed by pragmatists who recognize there will not be a revolution in corporate MIS. Smalltalk will gain more acceptance, but it will never overthrow the bastions of COBOL. C++ is already well received in technical and commercial computing. In the corporate environment, object-oriented analysis and design are coming into their own, and object-oriented COBOL is not far behind. The combination of all of these will bring genuine acceptance of object technology.

A Lack of Standards

Few object technology standards exist

The fourth significant reason for slow adoption is the fact that at this point in time, object technology has only a few standards, and these are sporadic. Are software standards really important? Almost everyone about to adopt a technology that will cost his organization hundreds of thousands (or millions) of dollars to implement prefers to adopt a technology based on a standard—*de facto* or *de jure*.

9.2 A Case for Standards (*De Facto* or *De Jure*)

Standards Legitimize Emerging Technology

Standards lower the risks involved in adopting a new technology. If a specific product or company proves not to be viable, substitutes can be found that conform to the same specification, making the substitution less painful. This form of investment protection fosters acceptance of new technology, legitimizing its use and providing stability in the market.

Standards foster acceptance by lowering the risk

The Object Database Management Group (ODMG), a consortium of several object technology database product vendors, readily admits that creating legitimacy for their technology is their major rationale for banding together to develop a standard for object databases. Moreover, they needed to build confidence in their approach because of the entrenchment of relational/database technology in corporate organizations. Their products use an object approach to database technology (as opposed to using traditional relational database technology with object extensions). These companies are small when measured against the vendors of relational database products; and they have to combat the reluctance of most customers to make major commitments to proprietary products from small companies with unproven viability. The ODMG aims to overcome this fear by allowing its members to offer better investment protection, through standards, to their customers. Moreover, a standard—particularly one that is sponsored by a group with open membership—provides the opportunity for an organization to have some influence over the future direction of the standard.

An example: ODMG's effort to legitimize object databases

On the other hand, the vendors of relational database technology are not standing idly by, waiting to let object technology database products usurp their market share. They, with their well-established, accredited Structured Query Language standards committee, are already developing object extensions to SQL in an effort to put an object technology spin on their traditional relational database products. Furthermore, these two vendor communities are discussing compromises that would serve the purposes of both.

Standards Are Not Detrimental to Innovation

Standards offer a baseline for decisions...

Standards do not thwart innovation. On the contrary, if a standard exists in their product's domain, most vendors support it, but they also increase their product's functionality beyond the standard to increase its appeal. Standards are also good benchmarks for user comparisons. Unlike the chaos of an ad hoc market with no standards, a product that adheres to a base standard, and then adds value beyond the standard, makes it easier for consumers to recognize and evaluate the product's unique features. This allows the user of that product to make a conscious choice between using only the standard features for portability or interoperability, and using the product's unique capabilities for their added value.

...and usually expand a market

Most companies in a growing software technology field, especially one that is newly emerging, recognize that standards almost always expand a market for all of the participants. In addition, market leaders in a non-standardized domain have natural advantages to exploit as a technology moves into its standardization phase. A company generally loses its market leadership through ineptitude—not by the adoption of standards in their market segment. This point is illustrated quite well by Sun, Digital, and IBM. Digital's operating system, VMS, and IBM's multiple operating systems for mainframe computers are proprietary, while Sun's consists of a *de facto* standard (UNIX) plus extensions. Of the three, it is clearly Sun that made the right decision.

De jure Standards

There are two major categories of recognized standards: *de jure* and *de facto*. There are further subdivisions within each of the two categories. True *de jure* standards carry the authority of law, or at least of government regulation. A familiar example is the set of building codes enforced by local governments. These are standards rather than simply government regulations because a body of experts in the various fields (electrical, plumbing, environmental, and such) came together and developed a consensus standard that was then offered to government entities for adoption and enforcement.

True **de jure** *standards are legally binding*

There are also some *de jure* standards important enough to be international in scope. In that case, government representatives from different countries negotiate and define the standards by treaties between signatory governments. The International Consultative Committee for Telephony and Telegraphy (CCITT) is a prime example of an organization setting *de jure* standards. This is a treaty organization addressing technical, operational, and tariff standards in the areas of facsimile, telegraph, and telecommunications. Its members are countries, represented by their governments (for example, the U.S. State Department represents the United States).

Some have international stature

Many standards called *de jure* standards are actually *accredited* standards where compliance is generally voluntary. At a cursory glance, these standards might appear to be *de jure* because government agencies influence much of their development process. Specifically in the United States, representatives from numerous government agencies are often members of standards committees because they are users of the standards as much as their fellow committee members, who are mostly technologists from product/service suppliers, user organizations,

Accredited standards do not require compliance, unless they are specified in a contract

and universities. However, these standards are only *de jure* when compliance is specifically required in a purchase contract. For example, the National Institute of Standards and Technology (NIST) often participates in accredited standards activities, but usually as coordinator, facilitator, and promoter of an open consensus process. Occasionally, standards promoted by NIST become part of Federal Information Processing Standards (FIPS), which give them real economic weight in that they can bar, at least in theory, noncompliant products from U.S. government purchases.

De facto Standards

De facto *standards are not official*

A *de facto* standard is a specification used by the public like a standard, but not having any official (accredited or *de jure*) recognition as a standard. There are two large subsets of *de facto* standards: *proprietary specifications* controlled by an individual organization, and *consortium specifications*.

Some are proprietary specifications

De facto standards controlled by a single organization are the most heterogeneous of any of the standards categories because each one is a special case. Typically, they come about through promotion outside official sanctioning channels; they then become generally accepted in their sphere of technology.

They usually evolve from a successful product

This most often happens as a result of a company's product becoming so popular that others use it, interface to it, or emulate it. The use of a proprietary specification as an unofficial standard is generally encouraged by its creator in order to place the burden of interoperability on its competitors. *De facto* proprietary standards often take on a life of their own and grow to outlive and become more important than the originating product. Two examples of such *de facto* proprietary standards are the IBM PC hardware bus architecture, and Ashton-Tate's (now Borland's) DBase file format and language syntax.

Another source of proprietary standards is the academic and government-sponsored research community. Standards evolving this way are not proprietary in the same sense as a standard coming from a commercial product, but the end result is usually the same. Positive recognition and all its associated benefits usually motivate noncommercial researchers. What could be greater affirmation than having the results of one's work put into practice by large numbers of people? Generally, that means distribution of valuable knowledge and often a product of some value at little or no cost. The most appropriate example of this is the origination of object-oriented programming (Simula 67) from the research team of Nygaard and Dahl in Norway. Others might include the Transmission Control Protocol/Internet Protocol (TCP/IP) communications protocol, Pascal programming language, and Berkeley Software Distribution (BSD) UNIX operating system extensions.

Business, government, and academic standards

The Consortia Pseudo-Standards Phenomenon

The creation of consortia organized specifically to establish standards in a certain field of technology, especially software, is a recent but growing phenomenon. In the typical case, a consortium results from an alliance of otherwise fierce competitors that agree to cooperate for the specific purpose of developing technical standards in a field of mutual interest. Standards developed by such a consortium are a breed of "pseudo-official-but-still-*de-facto*" standards.

Consortia create software specifications

Software standards consortia form and flourish for several reasons, but they usually grow out of a need to accomplish objectives that are not being fulfilled by the *de facto* standards communities, and to meet them faster than the *de jure* standards community can. By contrast, the accredited standards body's requirements for openness necessitate a lengthy review process. Table 9.1 compares the typical development of *de jure* standards and consortium specifications. The data by no means condemns the *de jure* standards process; open membership has

Consortia move faster than standards organizations

its benefits. But it explains why in some cases the consortium process may be more appropriate.

	Consortium	*De jure*
Time to Develop a Standard	1 to 3 years	5 to 7 years
Membership	Usually limited to paying members	Open
Cost to Participate	Thousands to tens of thousands of dollars	Hundreds of dollars
Reaction to Market	Close to market needs	Either lags or anticipates market needs

Table 9.1 – Compared Developments of Consortium Specifications and de jure *Standards*

Why should it take five to seven years to develop a *de jure* standard when a consortium can accomplish the same thing in one to three years, working on the same technology?

Standards organizations have certain constraints

First, the consortium's structure is typically a not-for-profit corporation. As such, it has the authority to operate by a set of self-generated bylaws and the right to ignore the opinions of nonmembers. On the other hand, the *de jure* committee must adhere to a very strict set of regulations that ensure a fair opportunity for *anyone* who wants to challenge or provide input to a proposed standard. There are requirements about meeting locations, advance meeting schedules, agendas, discussion and voting procedures, and many other requirements intended to ensure an absolutely open process. These include stringent guidelines for the resolution of disagreement on a proposed standard. For example, a draft standard must be held open for months for a public review and comment period. Any negative vote on a ballot can cause a delay and may require additional effort to reach consensus. This can lead to a whole new ballot. Obviously, this also significantly prolongs the process.

There are other significant differences between a consortium and accredited standards committees in the membership qualifications and costs. Because of its private corporate status, a consortium's Board of Directors can set its membership qualifications, voting privileges, and fees at whatever levels it deems appropriate. On the other hand, the accredited standards committee must accept all interested parties willing to pay the yearly fee of (at most) a few hundred dollars. Payment of the fee and regular meeting attendance are the only qualifications for full participation and voting privileges.

Consortia establish their own rules

De jure Anticipatory Standards (The "Yet Another Theoretically Better Architecture" Syndrome)

Besides the slow pace of the process, the most frequent criticism of *de jure* standards groups is the lack of timely response to market needs. *De jure* standards committees must therefore compensate for the unavoidably lengthy process by trying to anticipate the market. This ensures that standards proposed are not outdated even as they are issued. However, some anticipatory standards completely miss the mark, and years of work are wasted, while others are so far ahead of the market that it takes many years to achieve adoption in the marketplace. The most distinguished example of this last situation is the famous seven-layer ISO-OSI (International Standards Organization – Open Systems Interconnect) communications protocol architecture.

Standards try to compensate for slowness by anticipating market needs

Standards Summary

In summary, a consortium can be very responsive to technology market forces, but its standards are only as influential as its membership list. On the other hand, although the accredited standards committee must adhere to a slow and extremely constrained process that ensures openness and consensus, its standards have the national and international recognition that goes with accreditation. Some consortia and accredited standards committees are forging relationships that

New relationships between consortia and accredited standards

may lead to consortium specifications being submitted for consideration in the accredited standards process (sometimes using a "fast track" procedure to speed up adoption). If these relationships receive appropriate attention, there is real benefit for both sides. It is more likely, however, that both consortia and *de jure* standards committees will continue to thrive and coexist well into the future.

9.3 The Key Players

Adding object technology to existing standards

There are many accredited standards organizations, commercial software suppliers, and consortia trying to influence the direction of object technology, each in a specific domain of interest. The definition of standards can do much to remove some of the obstacles against broader adoption. The most notable and visible object-oriented standards are object extensions to several of the existing programming languages—C to C++, Pascal to Object Pascal, Ada to Ada '95, etc. Now it is becoming fashionable in the *de jure* standards community to add object extensions to every conceivable standard. The problem is that there are major inconsistencies between the object standards being adopted and that there are no "object police" to monitor and control them. There are also significant gaps and overlaps of coverage as each of these standards committees acts as an independent entity guarding its special interests.

OMG and RM-ODP are two key players

Because object technology is still very much in the definition stage, nearly every organization is trying to influence its direction. But even well-known experts in the field cannot agree on exact definitions of terminology and concepts. A few organizations are emerging as the primary influencers of object technology standards. Although most of the organizations are very domain specific (database, language, etc.), two in particular are addressing the larger issues of defining interoperability, services, and a common object model. Of the consortia, the Object Management Group (OMG) has the broadest mission and scope, and is the clear leader. In the accredited standards area, excluding programming languages, the international standards committee ISO/IEC JTC1/SC21/WG7 on the basic Reference Model of Open Distributed Processing, commonly known as RM-ODP, is emerging as the dominant force.

We will now take a detailed look at OMG as an organization and at the ODMG, an organization that grew out of OMG. Then we will consider several other influences on object technology, the RM-ODP standard, and harmonization attempts.

The Object Management Group (OMG)

OMG is a consortium dedicated specifically to the object technology software domain and it is very successful in that endeavor. This group, which was organized in 1989 and grew to over five hundred members by early 1995, appears to have the muscle to drive comprehensive *de facto* standards through the development of broadly accepted specifications.

OMG is the most influential software consortium

OMG has a large menu of stated purposes, but even most participants in OMG's proceedings normally describe the organization by its activities in establishing object technical standards. Most of OMG's stated purposes revolve around technical issues, but all have their roots in basic business needs. OMG's management strongly holds a position that object technology is nothing more than a better set of tools for attacking the business issue of making system developers more effective. These better tools should allow the developers to concentrate on fashioning systems that better achieve their customers' business objectives, rather than primarily concentrating on overcoming software technology constraints. At the highest level, OMG has three stated purposes: establishing an open object technology framework, adopting widely accepted object technology specifications, and acting as a technology and marketing center.

OMG's success is due to its market-driven objectives

First objective: define an object technology environment

This first objective is to facilitate the development of an open environment or framework that resides between proprietary operating systems services and the applications. This framework provides an opportunity for applications to achieve portability and interoperability across a wide array of underlying hardware, operating systems, and communications networks. The targeted configurations range from a single CPU workstation to a complex distributed multi-vendor network of computers.

Basic business competitiveness demands continual improvement in cost-effective usage of existing and potential hardware and software. This competitive requirement is behind OMG's desire to bring an industry-wide common framework into existence.

An object management architecture

In recent years, "framework" has become an overused word. In this case, a framework takes on the classical dictionary meaning of a "basic structure, arrangement, or system." The OMG framework is a set of skeletal specifications of OMG's Object Model and Reference Model (see [OMG 92]). The Object Model defines the paradigm of computation used by OMG-compliant applications. The Reference Model identifies the components, interfaces, and protocols that compose OMG's Object Management Architecture (OMA). Neither the Object Model nor the Reference Model provides detailed structures, specific interfaces, or exact protocols. Their purpose is to influence the basic design choices of technology that ultimately are proposed for adoption, and to provide a roadmap for assembling technology that meets OMG's technical objectives.

Second objective: specifications that foster reuse

OMG's second major purpose is to establish specifications such that compliant applications can be built largely from reusable code. Furthermore, OMG would like to see those application parts built from scratch find their way into reuse libraries for incorporation in subsequent applications. When one analyzes how much unique logic actually goes into most new computer applications, the true impact of reuse becomes much more apparent. Only an extraordinarily unique system development project includes twenty percent or more genuinely new logic. Most applications development projects consist largely of regenerating existing logic with a few new twists. Ten percent new logic is a real ground-breaker application; five percent or less is the norm.

Previously, object technology was largely relegated to the systems level or the more technical classes of applications. However, there are high expectations that OMG's adopted specifications will also include the capability for encapsulating large portions of non-object-oriented systems, or entire ones, within object technology shells. The aim of this endeavor is to lengthen the life span of these legacy systems by upgrading them with some object technology traits without performing major reengineering. The most prevalent expected extensions are the addition of more complex interfaces (such as graphical user interfaces, and interfaces for communicating with pure object-oriented systems). The adoption of appropriate object technology specifications, plus facilities to extend the life of legacy systems, will help legitimize object technology for enterprise-wide applications. A large part of Section 10.2 will be devoted to describing the multiple components of OMG's technical approach.

Extend the lifespan of existing systems through object technology "wrapping"

OMG's third major purpose is to act as a technology and marketing center for object technology. In OMG's case, what do "marketing center" and "technology center" mean?

Third objective: a market and technology center

- The best description of OMG's **marketing program** is object technology evangelism—in the most positive sense of the term. OMG's management spends a considerable amount of time giving presentations all over the world. First and foremost, these presentations are sales pitches for object technology and, almost secondarily, descriptions of

OMG's marketing program

OMG's work. OMG also has an arrangement with a training firm to teach courses on its technical specifications.

OMG's technology center

- The **technology center** role consists of coordinating with other organizations that perform related work or wish to incorporate OMG standards into their own adoption process. Two such organizations are the Open Software Foundation (OSF) and International Standards Organization (ISO).

Promoting adoption through publications and conferences

Another portion of the technology and marketing center effort is to keep those already "sold" on object technology continuously informed of the progress of both OMG and the global object technology community. This includes the publication of a very polished, informative, bimonthly newsletter entitled *First Class*. OMG also has an arrangement with a technical publishing house to commercially publish its specifications and relevant reports for widespread distribution. Finally, OMG co-sponsors yearly Object World conferences in the U.S. (one on the east coast and one on the west coast), Europe, and the Far East. These highly professional events draw an increasing attendance of interested object technology users.

Object Database Management Group (ODMG)

ODMG is an OMG spin-off

The Object Database Management Group was organized in 1992 as an outgrowth of OMG. Object database supplier representatives had been meeting as an OMG Special Interest Group (SIG), but felt that OMG's general membership did not give high enough priority to their technology. Consequently, five of the database suppliers who were members of OMG formed an independent consortium (with very stringent membership rules) that focuses entirely on object database specifications. They were careful to keep their specifications compliant with OMG specifications, and in 1994, ODMG commercially published a family of object database specifications that was well received.

ODMG remains a very active organization and is working on follow-on enhancements to its specifications. It has increased its membership significantly, all the while maintaining very strict requirements.

X/Open

X/Open is a worldwide organization whose charter is to promote open systems. It is supported by most of the world's large user organizations and suppliers of software and information systems. It promotes the practical implementation of open systems through its Common Application Environment (CAE), which enhances application programs portability at the source code level through the definition of Application Programming Interfaces (API).

X/Open promotes open systems

X/Open itself does not develop its own specifications for object technology, but it endorses and jointly publishes OMG's Common Object Request Broker Architecture (CORBA) specification. X/Open is also developing a test suite to verify the compliance of supplier software with OMG specifications.

X/Open cooperates with OMG

Open Software Foundation (OSF)

The OSF is a large consortium whose membership size rivals OMG's. It was originally formed by a number of leading computer manufacturers to develop an open operating system architecture able to succeed UNIX. It has since gained a large following of all types of software and hardware developers as well as end users. It produced its OSF/1 operating system by adopting specifications from suppliers through competitive selection, and performing additional integration with its own technical staff. OSF's most widely used specifications, derived through the same process, are its Motif graphical user interface and its Distributed Computing Environment (DCE).

The Open Software Foundation develops technology

A much different consortium model than OMG

Early on, OMG and OSF were rivals. They have very different models of how new technology emerges (OSF develops technical deliverables with its own staff, while OMG adopts proposed specifications), and each claimed vehemently to have the only viable model. Nowadays, OSF is a member of OMG and has endorsed CORBA. Conversely, OMG has adopted certain portions of the DCE specification for interoperability between object request brokers.

Industry-Specific Consortia

Some industry-specific consortia influence object technology

Several vertical industry-specific consortia are influential endorsers and users of object technology standards and specifications. Some develop software for their member companies, others develop common industry software specifications, and still others simply evaluate and make recommendations. Some focus entirely on software, while software is only of partial interest to the rest. However, all of these consortia are similar in the influence they exert on their membership through their actions and recommendations. Two such consortia are SEMATECH (SEmiconductor MAnufacturing TECHnology) and Petrotechnical Open Software Corporation (POSC). Both SEMATECH and POSC are strong advocates of object technology. SEMATECH is an active member of OMG, and POSC has official liaison status with OMG.

Early Adopters

Hardships are a typical attribute of pioneering

Early adopters of object technology that were willing to attest to the productivity improvement potential of this approach had a large impact on its popularization. Some of the most visible were Apple, Borland, and Mentor Graphics. As with all pioneers, they endured certain hardships before breaking through to achieve the productivity advantages of object technology. Typical problems included schedule delays, project cost overruns, and questionable performance. At least, those that followed the example of these early adopters benefited from the lessons they learned.

"Big Name" Suppliers

Almost all of the major brand names in the computing field (e.g., Digital, H-P, IBM, Oracle, Sun, etc.) have become enthusiastic supporters of object technology. This does not mean of course that *all* their products have become object oriented. But it does mean that they have one or more centers developing object technology, that they have a plan to incorporate it in their products, and that they support the activity of other influencers. Their common ground for working on object technology standards is OMG. They are often vehemently vocal in defending conflicting positions at first. But they usually overcome their differences to create an OMG specification, adopt it by consensus, and support it with compliant products.

Most well-known software suppliers support common object standards

The major anomaly so far has been Microsoft. Even though Microsoft is an OMG member and appears to be making a few accommodations to OMG technology concepts, they (because of their dominant market position) are largely charting their own object technology course, based on Object Linking and Embedding (OLE) and the Component Object Model (COM). The OMG is in the process of defining a specification for "interworking" between its specifications and Microsoft's (note that the word "interoperability" is being carefully avoided), but Microsoft's participation in OMG's open process has thus far been lukewarm.

A notable exception

Influence of Government Funding

Government agency funding of private industry and university research projects involving object technology has played a significant role in seeding the technology in industry. This has occurred in several industrialized countries where that type of government policy exists. Two such instances in the U.S. are the Microelectronics Manufacturing Science and Technology (MMST) program, and the Open Object Data Base (Open ODB) specifications project. The MMST program (covered as a case study in Chapter 11) and Open ODB are individual

Government sponsored R&D is a major technology contributor

projects funded by the U.S. Air Force and the Advanced Research Projects Agency (ARPA). A third instance of government funding is NIST's Advanced Technology Programs (ATP). NIST funds dozens of software projects (totaling hundreds of millions of dollars) through programs such as the Software Components ATP or the Technology for the Integration of Manufacturing Applications (TIMA) ATP.

An Accredited Object Standard: RM-ODP

RM-ODP is the dominant international object standard

The Reference Model for Open Distributed Processing (RM-ODP) is a complex suite of ISO standards that will become known as X.900 (pronounced "X dot 900"). The ambitious objective of RM-ODP is to provide a structure for standardizing the full scope of interoperability requirements between distributed systems. The approach to accomplish this goal includes the definition of rules, semantics, and structure for five different hardware- and software-neutral viewpoint languages. The scope of RM-ODP is so large that it has been divided into four distinct parts, each of which constitutes a separate standard. The four parts are:

The scope of RM-ODP is divided into four parts

1) **Overview and Guide to Use of RM-ODP.** It explains the key elements and concepts of the other three parts, and how to integrate them together.

2) **Descriptive Model.** This part contains RM-ODP's guidelines for how to specify ODP systems using object-oriented concepts. It defines the abstract characteristics and a framework for describing the more concrete viewpoint languages in Part 3.

3) **Prescriptive Model.** This part contains the concepts and rules of the five different languages that support specific viewpoints. These "languages" are not languages for implementation, but they define a specification for developing other standards that will govern the interoperability of systems in the various RM-ODP

viewpoint domains. In that sense, RM-ODP is a meta-model. It describes the functions and justification for each of the viewpoint languages, and the relationship of each language to the other languages.

4) **Architectural Semantics.** This part interprets Part 2 and provides the formal architectural semantics. It maps the underlying basic concepts defined in Part 2 to formal methods, but leaves the formal definition of the standard's higher level concepts to be indirectly interpreted.

The compliance of distributed systems to Parts 2, 3, and 4 can be measured.

Harmonizing Object Technology Standards

Almost all of the standards committees in charge of computer languages, along with a number of others, are adding object technology extensions to existing standards. This, together with the *de facto* object standards proliferation, creates an especially chaotic situation in object technology standards. In response, the Accredited Standards Committee, a U.S. standards governing body, established the Object Information Management Technical Committee (X3H7) to *harmonize* the other accredited standards committees' application of object technology extensions. However, X3H7 has no authority to police the other committees' use of object technology or to prescribe any rules. It can only create convergence through the individual and collective influence of its members. Because each standards committee has its own agenda, relevant to its unique domain, this approach has met with only limited success.

X3H7 was chartered to harmonize the work of multiple committees

The OMG is also using the influence of its large and overlapping membership in the accredited standards organizations to push its specifications into the *de jure* standards process. OMG is initiating or accepting requests for liaison with many other

OMG may become the focus of convergence

object technology standards groups, and may become the "object police" simply because there is no other organization with comparable credentials, object technology focus, breadth of membership, and accomplishments to date. The relatively rapid adoption of an OMG specification, compared to the *de jure* standards process, gives that specification a strong position by the time a standard gets drafted.

9.4 Recommendations on Standards Participation and Conformance

Participate in Standards Activities

Participation provides the opportunity to interact with technology providers

If a user organization can afford the expense of membership, the required travel expenses, and the personnel time to allocate, there is no question that it should participate in standards activities. Most suppliers of any size have to participate in standards groups relevant to their product technology to protect their interests. By participating in the same activities, the end-user organizations are able to interact with a variety of different suppliers, which creates multiple benefits for the users and the suppliers alike.

It also helps to forecast emerging technology

Participation gives the technology user a much better ability to predict which technologies will be emerging and make better choices. Discussions of the merits of different proposals often provide insights into the suppliers' technology roadmaps, even if they have not yet been officially disclosed. This interaction can also lead to the user gaining knowledge about technology that is declining. He can therefore avoid making expensive commitments to that particular technology. Such knowledge allows users to better plan, rather than react to technology surprises as they occur.

Standards working groups provide an extremely good avenue for influencing the direction of the marketplace. Suppliers are usually quite receptive to user input in these settings. In general, a user organization that participates has leverage far beyond a user's normal influence with product developers.

Participation can influence future products...

In addition, participation in standards activities frequently increases the user organization's stature and credibility in the technical community. Standards group members are more visible than others to the technical personnel of suppliers, other users, academia, and government.

...and enhance the participants' stature

Monitor Standards Activities

If an organization cannot afford to participate, it can gain some useful insight by regularly monitoring the relevant standards groups' activities. There are several ways to do this. An organization or individual can join a U.S. accredited standards committee for $300 per year and receive all of its hard-copy document and e-mail distribution (regardless of whether or not any meetings are attended). Some consortia provide the same service by purchasing a low-cost (relative to normal membership) subscriber membership. There are also accredited standards groups and consortia that allow receipt of e-mail distribution without the purchase of a membership.

Monitor standards activities if you cannot participate

Adopt and Support Relevant Standards in the Software Lifecycle

Clearly, relevant industry standards should be followed wherever possible in the software lifecycle. However, the object technology *de jure* and consortia *de facto* standards are often on the leading edge, and relevant standards are not always matched by available products. Several actions can be taken to ensure the most reasonable adherence to existing and emerging standards needed in the software lifecycle.

Define a reasonable degree of adherence to standards

Whenever practical purchase products that adhere to standards	There is no prescribed set of standards that fits every organization's needs. Every organization has unique lifecycle requirements, and those lifecycle requirements lead to choices of procedures, methods, and tools. Some of those choices may be made based on the applicability of a standard to the organization's requirements, which in turn leads to a choice of certain procedures, methods, or tools. Conversely, a tool that meets an organization's requirements may adhere to a particular standard, which dictates that other choices must also follow the same standards to interoperate.
Support suppliers that adhere to standards	Whatever relevant standards are chosen, users should encourage, cajole, and demand conformance to those standards from suppliers. If the supplier will not cooperate, other suppliers should be seriously considered before committing to a proprietary tool that will not interoperate with other components of the environment.

Chapter 10

Object-Oriented Execution Environment Concepts

Computer systems and their requirements are generally separated into a development environment and an execution environment. In this categorization, the development environment concerns itself with supporting an application's development and maintenance throughout its lifecycle, while the execution environment provides the underlying services necessary to support effectively the execution of that application. In the context of execution environment support functions, even the software tools used in the development environment are considered applications. One can further subdivide the execution environment into run-time and administration/management requirements. While the run-time portion provides supports for application execution, the administrative/management portion supports the installation and configuration of applications.

Execution environment provides run-time and system administration services

10.1 Requirements

Although the development and the execution environments each have distinct requirements, all object-oriented computing elements must support the integration of *both* environments in order to achieve the three commonly accepted goals for distributed object technology systems:

Development and execution environments must be integrated

- Portability,
- Interoperability,
- Dynamic reconfiguration.

Scalability and extensibility are two additional attributes that require consideration in the implementation of distributed object technology environments.

An organization must build its environment from its unique requirements

Because each implementation is unique, there are no recommendations on object-oriented execution environment products, or even types of products, presented here. Furthermore, object technology is evolving so quickly that before this book is even printed, the best product at this time may be surpassed by a new product introduction. Any organization engaged in a selection process should therefore define carefully its own requirements, evaluate the available products and their capabilities based on those requirements, and continue to monitor emerging technologies after a selection is made and a purchase order is issued.

Object technology must support client-server architecture to remain up-to-date

Almost all new object technology projects of any significant size developed for organizational use (rather than those used by isolated individuals) follow a distributed computing paradigm often referred to as client-server architecture. This is a major technology trend that is likely to prevail for many years to come. That means that an effective object-oriented computing environment implementation must satisfy the requirements for distributed object-oriented applications. To meet this objective and to achieve application portability and interoperability, the execution environment implementation should be based on industry standards.

Object-Oriented Execution Environment Concepts

The underlying operating system that supports an object-oriented execution environment does not have to be object-oriented itself.[1] Placing this type of restriction on the operating system platform choices for the execution environment is not necessary or even desirable. In the current state of available execution technology, it is typical to implement a distributed object-oriented execution environment by layering object-oriented technology products on top of a traditional operating system such as UNIX. Object-oriented languages, object request brokers, and object databases are three types of the many products currently available that can be integrated to implement a customized execution environment. To provide the best possible integration of products in the execution environment, we strongly recommend that conformance to *de jure* or *de facto* standards be one of most rigorous criteria in the selection of all elements of both the object-oriented technologies and underlying system infrastructure.

Object technology components can be implemented on top of traditional architecture

10.2 Architectural Components

An object-oriented execution environment consists of all the technology components required to support the development and execution of object-oriented applications. For discussion purposes, we divide the object-oriented execution environment into six major elements:

Six components of an execution environment

- Application objects,
- Object-oriented languages (or more specifically, the runtime component of the language support environment—typically a library of objects),
- Object-oriented databases,
- Object request brokers,

---

[1] One of the few truly object-oriented operating systems is NeXTstep. Another example is Apple Computer's MacOS.

- Object services,
- Common (object) facilities.

Three components are specific to object technology

The first three elements (application components, language support, and databases) are also part of non-object-oriented systems. The last three categories are specific to object-oriented systems, and have been established and promoted by the Object Management Group (OMG). For lack of an alternative categorization of object-oriented technology, these three groups, as well as the OMG's criteria for their individual elements, are used in the ensuing discussions.

Application Objects

Objects are different in execution and development environments

The definition of an application object in the execution environment is not the same as the typical definition of an object in the development environment. Actually, there are many and varied definitions of an object in the development environment. According to Rumbaugh et al. [Manola 95], development environment object definitions range from "everything in Smalltalk is an object" to "a discrete, distinguishable entity that quantizes data." Even though these definitions are not common, they both characterize a type of object commonly known as a "fine-grained" object, one that contains "atomic" data and methods. Examples of fine-grained objects include the following: a word rather than an entire document, a symbol rather than an entire schematic, a state rather than an entire analysis model, etc.

Execution objects are "coarse grained"

Application objects in the execution environment are slightly less diverse. OMG defines them as software that "corresponds to the notion of a traditional application." Regardless of whether or not this is a precise enough definition, it has the general connotation of a "coarse-grained" object, one that is also a self-contained executable entity. The hairsplitting conceptual problem with defining it as being the same as a "traditional application" is that traditional applications do not use message facilities (such as the Object Request Broker, discussed below).

Object-Oriented Languages

Early object-oriented languages paid a significant penalty, in the form of slower run-time execution, for their higher development productivity. Today this is no longer an appreciable factor. Object languages, like traditional languages, may be fully compiled, with little or no run-time overhead to use features such as messaging and inheritance. Depending on conditions, compiled object programs may run anywhere from twenty percent slower to as much as ten percent faster (thanks to certain efficiencies introduced in the approach) compared to traditional compiled procedural programs. Even when an object language is measurably slower than a traditional language, current processor speeds are so fast that the slower application execution is not perceivable to the user. However, since the performance of the software directly translates into throughput numbers for an industrial process, real-time control of processing equipment may remain an issue. For the most part, though, run-time performance of object languages is no longer a consideration, provided the older-style, interpreted object languages are avoided.

Object languages have overcome their "run-time penalty"

Parallel hardware architectures for concurrent computing constitute a powerful emerging technology and object-oriented languages are more naturally suited to them than traditional procedural ones. In principle, each object is an independent entity, responsible for its own state and behavior. Objects can function naturally in a concurrent environment, whereas procedural programs must be artificially and carefully broken up to achieve concurrency. The application of object languages to parallel computers promises orders of magnitude improvement in run-time performance, depending, of course, on the number of parallel processors.

Object technology offers advantages in advanced hardware and software environments

Object–Oriented Databases

Non object-oriented databases have disadvantages

In practice, most objects contain data that must be preserved from one execution of an application to the next. Although it is possible to decompose objects and store their data in a relational database or in flat files, doing so is very slow and presents some challenges with respect to modern system architectures. Flat files cannot easily be accessed across platforms with both security and concurrency. Relational databases segregate the data from the applications operating on them, and require either a special embedded language (such as SQL) or calls to non-portable subroutine libraries to access the data. A more natural solution is to store objects in their native format, that is, in a database specifically designed to contain objects.

Pure object databases offer large performance improvements in many instances

Object-oriented databases are far faster than relational databases at storing and retrieving complex information. The relationships between objects are represented directly in the database and can be "navigated" using references, instead of calculating *joins* based on data values in a relational database. Douglas Barry is Executive Director of the the Object Database Management Group (ODMG), and has a consulting practice specializing in the selection of object database products for specific applications. Barry states that "nearly every one of my clients performs an application-specific benchmark using one or more object databases against one of the well-known relational database products. These benchmarks have attempted to simulate the target production environments which generally have hundreds of workstations and hundreds of gigabytes of data. For accessing and manipulating complex data, those clients have achieved one hundred to one thousand times performance using an object database product."

Object database products vary greatly in their implementation techniques. The differences generally fall into the following categories:

- *Hybrid relational/object or pure object-oriented.* Hybrid relational/object databases primarily target the large population familiar with relational technology and the Structured Query Language (SQL) database programming language, as well as those who must interface their object-oriented applications with legacy relational database applications. Typically, hybrid relational/object databases are still based on data relationship tables where objects must be expressed in terms of tuples (the set of values contained in one row of a table). Due to the normalization rules of relational databases, an object's attributes may be split between multiples tables. On the other hand, pure object-oriented databases are just that—architected specifically to store object data and the methods that act on that data (i.e., without artificially breaking an object apart). The hybrid relational/object and the pure object database product supplier communities have separate standards. These standards, discussed in Chapter 9, are the emerging object-SQL international standard ODMG consortium standard, respectively.

  Hybrid databases offer a familiar starting point for many organizations

- *Languages supported and their bindings.* To use a hybrid relational/object database in an application, the developer typically embeds SQL constructs within his code. On the other hand, access to pure object-oriented databases is primarily done in Smalltalk or C++, only occasionally in more esoteric languages (a LISP variant, a Smalltalk-like one, etc.). A significant differentiating factor between the pure object-oriented databases is the binding of the language to the database function. In other words, how transparent or integrated are the database commands with the native language? Does the programmer have to "drop out" of the native programming language to make database calls or are they a natural extension of the programming language? Moreover, some object-oriented databases support object data for a single language, while others support multiple languages.

  Language support is typically more transparent for pure object databases than for hybrids

Data and methods storage differs

- *Separate or combined storage of data and methods.* Most databases that use Smalltalk as their base programming language allow for storage of an object's data and methods together in a common database, while most databases using the C++ language require object methods to be stored in separate files outside the database.

Different storage models perform better on different types of applications

When selecting an object-oriented database, an organization must take these differences into consideration. Some models are a better fit than others, given the choices that may already have been made for other components of an object-oriented execution environment. Some products are designed and tuned specifically to store objects used in particular types of applications, such as Computer-Aided Design (CAD).

Object Request Brokers

The OMG adopted the Common Object Request Broker Architecture (CORBA)

In standalone applications programmed in a single language, objects typically execute in a single computer's memory space and communicate directly by sending messages to each other. However, a common communication method is required for applications to interoperate across multiple machines, languages, and operating systems. The OMG consortium introduced the concept of an Object Request Broker (ORB) to fulfill that requirement. OMG members developed and adopted a particular specification, called the Common Object Request Broker Architecture (CORBA), to foster the emergence of several competing yet compatible commercial implementations. According to this specification, an ORB can accept requests for services from client objects independently of their source language, and pass these requests on to the appropriate server objects anywhere within its networking domain. Likewise, responses to these messages are automatically routed back to the requester through the ORB.

The ORB connects three types of objects

In the OMG's model of distributed computing, the ORB acts as a switchboard between three types of objects: Application Objects (described earlier), Object Services, and Common

Object-Oriented Execution Environment Concepts

Facilities (both of which we address below). This overall **Object Management Architecture** (OMA) is depicted in Figure 10.1.

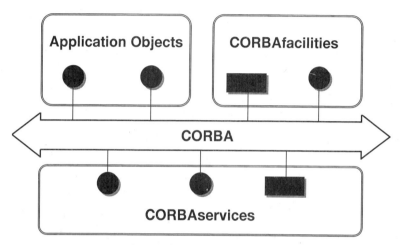

Figure 10.1 – Object Management Architecture

In some regards, the ORB acts as a high-level remote procedure call (RPC) mechanism. In fact, some ORB implementations use RPC as their low-level message send/receive protocol. Unlike RPCs, however, the ORB provides for location-independent requests. This enables many desirable distributed computing characteristics, including dynamic relocation of computing resources based on loading and other factors.

ORB acts like a location-independent RPC

An ORB is generally used to interconnect multi-object components in a distributed, often heterogeneous environment. For example, one collection of objects might provide automation control for machines in a factory. It can interact with another

ORBs provide applications interoperability across multiple computers

collection of objects located on another computer system that manages the scheduling of material to and from those same machines. The automation controller could be written in C++ and reside on an embedded computer inside the machines, while the scheduler could be written in Smalltalk and be resident on a workstation somewhere other than the factory floor. The ORB allows these objects to communicate with minimal overhead as if they resided on the same machine.

"Out-of-the-box" interoperability between different suppliers' ORBs

CORBA version 1.0 did not guarantee that two different suppliers' implementations were compatible. Objects could communicate between different computers running the same ORB product, but not necessarily between computers running different ORB implementations. In contrast, CORBA version 2.0, adopted in 1994, provides a specification sufficiently rich to guarantee "out-of-the-box" interoperability [OMG 95] between different conformant ORB suppliers. This is achieved by specifying several flavors of Inter-ORB Protocols (IOP) through which one ORB can relay object invocations to another ORB.

Object Services

Object services provide basic support to ORBs and higher-level services (Common Facilities)

Object Services (now officially called CORBAservices) are defined by OMG as the low-level applications support functions. They are interfaces and semantics that are commonly needed to build well-formed applications in a distributed object environment. In the context of OMG's specifications, Object Services exist only to provide services to the CORBA Object Request Broker and higher-level services called Common Facilities [OMG 92]. OMG developed a roadmap in 1992 for defining, prioritizing, and adopting specific service specifications. These specifications, like all object specifications conforming to the OMG's Object Management Architecture, are written in the **Interface Definition Language (IDL)**. The process of specifying all the services currently identified will likely continue at least into 1998. This roadmap

Object-Oriented Execution Environment Concepts

and the length of the complete definition process are dictated by two factors: a priority order established by OMG members, and the fact that some specifications are prerequisites for others. OMG is also adopting certain services defined by other consortia and accredited standards organizations, or working with them toward a common specification.

The services specified on OMG's Object Services Roadmap include:

Services the OMG Roadmap has defined to date

- *Lifecycle* - standardized interfaces for creating, deleting, and moving objects;

- *Naming* - a standard interface for finding objects (mapping names to unique object identifiers);

- *Persistence* - long-term, potentially fault-tolerant object storage;

- *Event notification* - specification and delivery of event notices;

- *Security (Access Control)* - access control to objects and methods, and end-to-end security control (i.e., access attributes, interfaces, etc.);

- *Relationships* - standard specification of relationships between objects;

- *Transactions* - ability to commit or roll back groups of object requests;

- *Concurrency Control* - standardized interfaces for controlling object execution to enable correctness in the presence of concurrency;

- *Externalization* - standardized interfaces and storage formats for objects to be delivered in portable, external form (e.g., to portable media) and to be imported from this form into another execution environment;

- *Data Interchange* - standardized formats for sharing information;
- *Licensing* - a standardized interface for controlling software licensing in a distributed environment;
- *Trading* - an interface to post the availability of a service, and to search for an object that provides a given service;
- *Query* - a standardized method to find objects via type, attribute, and relationship information (similar in concept to the SQL "select" statement, but with the necessary changes to support objects);
- *Change Management* - support for configuration and versioning of objects;
- *Properties* - standardized access to additional object characteristics beyond the object's interface defined in IDL.

Common (Object) Facilities

Common facilities are high-level object services optionally provided by a supplier

The Common Facilities (or CORBAfacilities) are defined by OMG's Object Management Architecture as high-level object services that reside logically between the low-level basic Object Services and Application Objects. Whereas Object Services are so basic that they must be present in *all* implementations that comply with OMG's specifications, Common Facilities are higher-level service specifications included at the supplier's discretion.

Common facilities are classified into horizontal and vertical ones

There are no strict criteria to determine if a certain capability constitutes an Object Service or a Common Facility. The general rule is that while a Common Facility is at a higher level of functionality than an Object Service, it is only relevant to some application domains or categories. To make this definition clearer, Common Facilities are classified into *horizontal* and *vertical* categories. Common Facilities in the vertical category address the needs of applications developed

Object-Oriented Execution Environment Concepts

for a specialized market. These may consist of a product category such as CAD, financial applications, etc., or a specific type of industry such as manufacturing, health care, telecommunications, etc. The horizontal Common Facilities use the same criteria of being a high-level optional service, but they are defined by their functionality rather than by a market segment, and are generally applicable across a broad range of applications. For example, an *editable text object* may be used in all kinds of applications, from a multimedia authoring environment to an on-line manufacturing specification management system. This would be a typical example of a horizontal facility.

Since the horizontal Common Facilities are useful in a wide range of applications, OMG found it convenient to further subdivide them into the following categories:

Common facilities groupings

- *User Interface* - includes general rendering management, presentation of compound documents, work management, scripting for task and process automation, etc.;

- *Information Management* - includes compound storage and information access management;

- *Systems Management* - provides common operating systems interfaces for managing applications and computing resources;

- *Task Management* - supports user task management through functions such as workflows, rules, communications, active agents, etc.

The development of CORBAfacilities is the least mature of all the OMG specification activities. This work began in 1994 and the first function targeted is a Compound Document Management facility that includes support for document features associated with traditional paper documents, plus multimedia, version control, and other collection and reporting aspects that

OMG has defined sixty candidate common facilities to be specified

have come into existence with the advent of computer-based visual presentation of text and data. Along with Compound Document Management, OMG has identified sixty candidate facilities for subsequent interface definition work. The following is a representative list of those:

- On-line help,
- Error reporting,
- Electronic mail,
- Tutorials and computer-based training,
- Intelligent agents,
- User preferences and profiles,
- Alarm services.

10.3 Information Bus Technology – Is It Object Oriented?

The Information Bus Concept

An information bus offers flexible connectivity

Hardware buses have been used in computer architectures for many years. They enable a computer system to accommodate such changes as the introduction of new peripheral devices or increased memory capacity that were not foreseen during the initial design. Their main initial benefit is therefore flexibility or interoperability, not cost/performance figures. Indirectly, they may have a positive effect on cost if bus interfaces can be mass-produced once a bus architecture has become a *de facto* standard. An **information bus, or message bus,** offers the software architect the same advantage, namely, the ability to connect various components and to accommodate easily unforeseen requirements for change.

Object-Oriented Execution Environment Concepts

As long as computer systems were either monolithic, grouped in rigid hierarchies, or all based on the same supplier and operating system, designing the communication between multiple applications was a relatively simple problem. However, none of these assumptions is true any more. A typical information system is now comprised of multiple, diverse computers that all need to communicate directly (peer-to-peer) with one another. In many systems, programs or processes communicate directly with one another using common protocols such as Transmission Control Protocol/Internet Protocol (TCP/IP). Each program must know the name or address of every other program with which it needs to share information. To add a new program that communicates with the preexisting ones requires that each of the interacting programs be modified, rebuilt, tested, and then reinstalled. Software maintenance of this kind is time-consuming and a hindrance to system evolution.

Communication between computer systems has become a key need

Using traditional communication protocols like TCP/IP, each pair of communicating processes needs its own communication channel. Both senders and receivers have to manage the list of other processes they are communicating with, and the logic of how to sequence their sending and receiving actions. The developer of an application has to worry about network and operating system features of not only the platform for which he is writing, but also those of the other machines involved. This combinatorial explosion issue is illustrated in Figure 10.2.

Traditional protocols burden the applications with too much housekeeping and with compatibility issues

A message bus provides an attractive alternative without requiring a complete paradigm shift. A bus provides a reliable mechanism for programs to communicate with one another by broadcasting and receiving to/from the bus, rather than talking to each other individually. As a result, each process has only one link—with the bus—and its developer has only one operating system-independent protocol to worry about. One such concept was proposed by the automotive industry in the early 1980s in the form of the Manufacturing Automation Protocol (MAP). Although this particular protocol has had little success

A message bus mediates the data exchange between possibly disparate systems

(compared to its rival Ethernet) as a standard, the basic idea is recognizable in today's information bus products. The concept is illustrated in Figure 10.3.

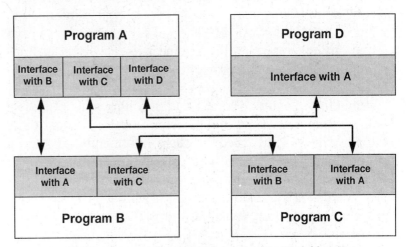

Figure 10.2 – Traditional Communication Model between Multiple Programs

Message buses have certain common features

Although information buses come in various forms, the term typically applies to a communication package that provides the following basic capabilities:

Reliability

- Reliable delivery of messages is ensured without the developer of the application having to worry about it. Often, several levels of reliability and performance can be selected according to the needs of the applications.

Multicasting

- A message sent on the bus can be received by several other programs or processes.

Object-Oriented Execution Environment Concepts

- New "sender" or "listener" processes can be added to an existing system without interrupting the system or reprogramming its individual components—unless they are "interested" in communicating with the new components.

    Dynamic reconfiguration

- As with a hardware bus, a process on an information bus generally does not need to know where it, or its correspondents, reside on the bus. Nor does it care how many other processes are sending it messages, or receiving its messages.

    Location independence

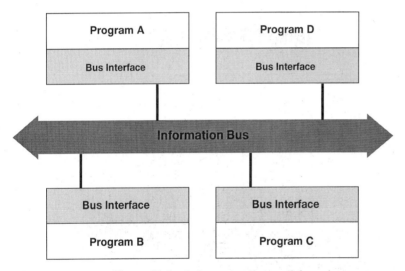

Figure 10.3 – Information Bus model

Implementation

Two different network communication techniques may be used to construct an information bus: broadcast and point-to-point communications. Broadcasting requires only one message to reach several receiving programs. It has the advantage of

Two transmission schemes: broadcast or point-to-point

providing very high information throughput when many programs receive the same message. Moreover, broadcasting scales very well to large networks. As more and more computers are added with programs interested in receiving the same information, the time it takes for each program to receive this message remains essentially constant. Additionally, the amount of network traffic is the same for one receiver as for many.

Broadcasting requires precautions to avoid performance and overflow problems

The main drawback of broadcasting is that each potential recipient process or computer must listen to every message, and "look" at its contents sufficiently to decide whether or not it is interested in it. This can potentially swamp slower computers. Because it is difficult for a sending program to limit its transmission rate to match a slower (or busier) recipient, sustained transmission periods and excessive speed can cause the bus to discard messages. This can be avoided by complex exception mechanisms that take care of message queue overflow. Broadcasting may also complicate the design of the underlying network, since messages need to be delivered to a number of points that are not completely defined in advance and that may vary with time.

An alternative is "publish/subscribe" with defined message types

There is a way to maintain the benefits of broadcasting for the message's sender without causing the recipients to suffer its drawbacks. In the most elaborate information buses, a set of **message types** is defined. Processes can *register* with a bus management authority what types of message they can send and those they want to receive. Then this same *bus manager* component determines the routing of each message based on its type and the registration information. With this scheme, a process only receives messages of those types in which it is interested. By analogy with the distribution of periodicals, this scheme is called "publish and subscribe." The characteristics of this mechanism are that data exchange is unilateral (data is sent but nothing comes back during the same transaction), and that it consists of multicast messages (the same information is sent to all the recipients). This mode is well suited to information distribution applications.

In an information bus application, it is often the case that the recipient of a message wants to continue to communicate privately with its sender. This is true even if the initial message was a broadcast one. Therefore, an information bus also often supports a simpler "request and reply" mechanism. In contrast to the other mechanism, the characteristics of this one are that the interaction is bilateral (a request is sent and a reply is provided as part of the same transaction), and that it consists of point-to-point messages. Figure 10.4 shows the contrast between the publish/subscribe and request/reply mechanisms.

Request/reply mode

1. Publish/Subscribe

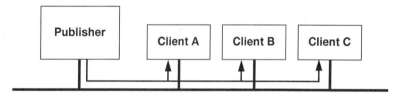

2. Request/Reply

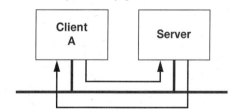

Figure 10.4 – Publish/Subscribe and Request/Reply Communication Mechanisms (Courtesy: Teknekron Software Systems)

Reliability is achieved through message queues

To provide reliable delivery, the information bus includes one or more **message queues** to hold the messages in transit. These queues may be centralized, but are usually distributed in order to provide additional fault tolerance. Various exception processing, journaling, and redundancy features can be applied to the message queues and the manager process itself to ensure fail-safe operation and reliable delivery of messages. It is also usually important that successive messages sent by one process to another be delivered in the exact order in which they were sent.

Usage Examples and Benefits

Four industrial uses of a message bus

Table 10.1 indicates four types of industrial problems that led to the adoption of message bus technology in different industries. In all four cases, the selected product was the same (Digital Equipment's DECmessageQ) but the specific choice is less relevant than the generic examples.

Message buses can help exchange information between manufacturing stations

In a manufacturing application like Work in Process (WIP) tracking, in which uninterrupted operation is especially important, an information bus can bring a useful element of flexibility. Consider a semiconductor factory (or any other discrete manufacturing environment) in which every device capable of sensing wafer position can broadcast its readings on the bus, and every process that may depend on this information can "listen in" and act on messages of interest. The following capabilities are now available:

They add flexibility

- New sensors can be introduced, removed, brought on line, or taken off line without affecting other sensing equipment or tracking programs.

- New processes needing to know wafer position information (e.g., an expert system for scheduling, or a factory throughput analysis program) can be introduced without modifying either the sensing equipment or other tracking programs.

Customer	Business Problem	Solution
ARMCO/ Advanced Materials Company	Increase operational efficiency. Improve productivity. Raise the level of customer satisfaction.	Implement message bus interface in current Computer Integrated Manufacturing (CIM) applications: sensor control, inventory control, production and process reporting, product scheduling, shipping order management, cost analysis.
General Electric Aircraft Engine	Improve factory management by developing bridge to new application.	Install message bus with a "queue adapter" for a legacy computer platform to share data between old and new factory management applications.
Quaker Oats	Reduce cycle time. Reduce inventory.	Implement message bus to automatically move data among their inventory control applications (finished goods, warehouse shipping, damaged goods, quality assurance station, shipping status).
National Semiconductor	Enterprise integration: reduce cycle times and improve the manufacturing cycle by tying shop floor control with MIS applications.	Implement message bus to notify Manufacturing Information Systems (MIS) system of every "move" of WIP in the factory (also see Section 12.6).

Table 10.1 – Industrial Problems Addressed by Message Bus Technology (Courtesy: Digital Equipment Corporation)

New capabilities can be built faster

In such an environment, a message bus speeds up the ability to collect and analyze data produced by various dispersed applications. One result may be improved process control through near-real-time statistical analysis of product, process, and equipment information.

The stock exchange is another target area for message buses

Another good example is a trading system for securities. Consider a network of workstations at which traders enter the buy or sell orders received from clients over the phone. Each order, such as "buy 250 shares of XYZ at $59 7/8" or "sell 1,000 shares of ABC at the current market rate," can be broadcast over the message bus. Any other similar process on the network can listen to these bids and match them against its own list of pending orders. If another trading process finds a match, it can respond with a message addressed directly to the sender of the original bid. This sender may receive several responses to its initial request and after a predetermined delay, it can choose the best response (for example, it is possible that different customers want to sell XYZ at 58 1/2 and at 59, so it is better for the buyer to wait until the lower offer comes in before concluding the deal). The deal will conclude with at least one more "private" message in each direction, meaning "I have accepted your bid" and "I acknowledge that I have sold you this stock," together with some account or transaction numbers for future reference. In this case, note how the information bus imitates the concept of the trading floor or "pit" (i.e., a forum where all the offers and requests can be brought together and reconciled).

The technology was dormant after the MAP effort

These examples illustrate how the information bus concept has propagated beyond the manufacturing domain for which the MAP protocol was originally invented. As a matter of fact, after being dormant for the rest of the 1980s, this technology saw a sudden flurry of activity circa 1991-92. This is attributable to several factors:

Object-Oriented Execution Environment Concepts

- The need for peer-to-peer communication between multiple computers increased as more and more activities became computerized.

- During the same time frame, object technology was not yet ready for commercial use.

Product Examples

In this section, we will describe the characteristics (some common, some unique) of four different products in this domain:

Four examples of products

- The *Teknekron Information Bus (TIB)* from Teknekron Software Systems;

TIB

- *ISIS*, a product supplied by the ISIS division of Stratus;

ISIS

- *messageEXPRESS*, from Momentum Software Corporation.

messageEXPRESS

- The *Framework-Based Environment (FBE)*, a product from Digital Equipment Corporation based on CORBA.

FBE

The intent of these descriptions is to illustrate concepts, not to promote these particular products at the expense of others. Like many others, this is an evolving market in which new products appear and old products evolve, sometimes leapfrogging one another.

TIB and ISIS provide interesting comparisons of the actual messaging schemes. TIB distributes messages to the bus using a broadcast mechanism as described above. The underlying protocol and network is Ethernet. ISIS, on the other hand, originally simulated a broadcast by sending each message to every interested program. This technique alleviates the usual disadvantages of broadcasting, but at the expense of scalability and throughput. Both TIB and ISIS make use of a more fundamental communications protocol called UDP, which

Comparison of TIB and ISIS protocol usage

generally provides the lowest overhead and delay in intermachine communication on a UNIX platform. To obtain this low overhead, UDP is an intentionally unreliable datagram protocol. Since a more complete and reliable protocol like TCP would still be insufficient for a message bus, the reliability required by the information bus is built back on top of UDP as part of the message bus product itself. This only adds a delay of a few milliseconds to the time needed to deliver a message using UDP. ISIS also now supports IP multicast, a protocol-level broadcast, as opposed to TIB's use of a hardware broadcast capability. This should give the two products equivalent capabilities in this respect.

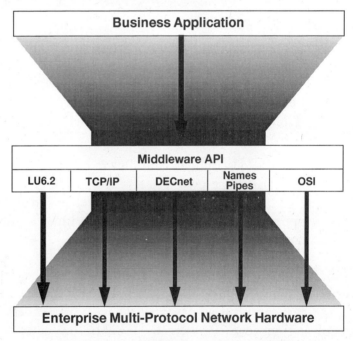

Figure 10.5 – Message Bus Example: EXPRESS
(Courtesy: Momentum Software Corporation)

Object-Oriented Execution Environment Concepts

Momentum's EXPRESS product takes the form of a "small" application programming interface (API) which offers several types of application-level capabilities: message delivery, data query/update, file transfer, and remote procedure calling. The API contains only five distinct "verbs," with the additional information provided as parameters to these calls. The same API is implemented on many platforms, including IBM's MVS, DOS/VSE and OS/400, Digital's VMS, and UNIX. This is a strategy deliberately aimed at providing the ability to write applications on "older" operating systems while providing a high degree of portability, since the API calls are the same on all the supported platforms. This concept is illustrated in Figure 10.5.

EXPRESS uses a simple API but aims at many platforms

In Digital Equipment Corporation's FBE, special attention has been given to the integration of legacy applications (i.e., applications written without this integration in mind, and often using older technologies). Let us examine how the architecture enables such integration. The FBE *integration framework* does not really know about applications, but about *clients* and *servers* for specific types of services (e.g., the placement of a purchase order). An application written with FBE, like A or B in Figure 10.6, can be either a client or a server, and contains a layer of code that supports this definition by issuing or accepting CORBA-compliant messages.

FBE pays attention to legacy applications

A more complex application that performs several tasks, like application C, can be mapped to *multiple* clients and/or servers. The application "exports" several interfaces (a client interface and two server interfaces in the example). A legacy application like D is not directly modified to use the CORBA interface, but a *wrapper* has been written to encapsulate the application and translate between the object-oriented paradigm of FBE and the traditional invocation mechanisms supported by D. For instance, the wrapper may need to take the data contained in a message it has received via FBE, place this data in a file or database, and invoke application D through an operating system call. D, which is rigidly programmed to find its data in this file or database, can perform its normal task without being perturbed by the existence of FBE, since its normal expected environment has been created for it. Upon exit, the wrapper (which may have been "polling" periodically to detect the end of D's execution) may recover some output data from another file or database, which it then translates into a message sent back via MRM.

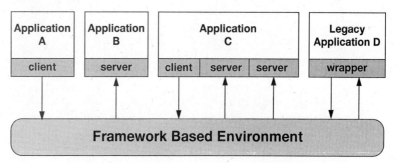

Figure 10.6 – Digital's Framework-Based Environment (Courtesy: Digital Equipment Corporation)

FBE is based on ObjectBroker

FBE is based on Digital's implementation of the CORBA specification, ObjectBroker. A previous version of FBE (circa 1992), the Manufacturing Resource Manager (MRM), was based on DECmessageQ, a pre-CORBA message bus product.

Object-Oriented Execution Environment Concepts 367

Information Buses versus Object Request Brokers

Based on the above descriptions of the technology and some of its implementations, it appears that information buses can be a key simplifying element in the move from a hierarchical architecture to a peer-to-peer system. However, with the emergence of object technology, one might ask whether information buses are a complete alternative to object request brokers or a dangerous distraction.

Are information buses a useful alternative to ORBs?

As is often the case, the real answer probably lies somewhere between these two extremes. Here are some arguments in this debate:

- Because they use a message-based paradigm for application-to-application communication, information buses bring their adopters conceptually much closer to the OMA than the technologies they replace.

  The concept is close to the OMA

- A significant benefit of information buses, in terms of adaptability, is that multiple products are available now from different companies. In some cases, these products have significantly more usage references available today than CORBA implementations.

  Products exist now

- On the one hand, information buses carry *information* between "objects" rather than *requests* to an object to perform a specific task. They do not hide information very well, and unless stringent security is available, it is easy to write a "sniffer" client that can read all the messages sent by other objects and use the information they contain. There is typically no concept of classification or inheritance of messages or objects. Additionally, the access to the information bus is typically provided as subroutine calls from procedural languages (C or COBOL are the most common).

  Buses only carry data, not method invocations

They help with legacy software

- On the other hand, the benefit of this hybrid technology is that it facilitates the wrapping of legacy applications.

They do not interoperate

- Most message buses are not CORBA-compliant, and since there is no common standard to follow, there is no interoperability across different suppliers. Nor has any of them become prevalent enough in the market to become a *de facto* standard, which would force its competitors to provide interoperability features with its own product. Some suppliers, but not all, have stated their intent to offer a CORBA interface in the future. On the other hand, the proprietary nature of the interface definition has enabled some suppliers (such as Horizon with EXPRESS) to offer their product on a variety of platforms in an interoperable way, while CORBA will only reach the interoperability stage when products implementing version 2.0 appear, starting in 1995.

Information Bus Selection Criteria

Technical criteria for message bus products

Based on the previous discussion, and assuming that an information bus capability is desired, what criteria should be used to select a product? Apart from the usual business criteria (robustness of the company, references, license costs, and the sheer technical ability of the product to meet its user's needs), we propose here a list of additional criteria to consider:

Platform Criteria

Existence, portability, and interoperability

– Availability of the product on the various platforms of interest, including those the customer *may* use in the future;

– Portability across platforms (i.e., the calling conventions for the bus interface need to be the same on all supported platforms);

Object-Oriented Execution Environment Concepts

- Interoperability across platforms (i.e., the ability to create a distributed application based on a heterogeneous mix of systems).

Levels of Service

- Ability to request guaranteed delivery of a message.   *Reliability ...*

- Existence of an exception mechanism to inform the requester when a message has *not* been delivered. (This sounds contradictory with the previous criterion, but since there can always be a catastrophic failure of the network or of the requested servers, in reality, both features are needed.)

- Ability to request a low-overhead, less robust means of message delivery for noncritical communication.   *... under user control*

- If the application demands it, ability to specify a maximum delivery delay, and what action to take if that deadline cannot be met.

Security

- Ability of the message sender to limit access to the message contents to a list of approved clients;   *Access control and authentication*

- Ability of the message recipient to authenticate its sender.

Ease of Administration

- Ease with which new servers, new clients, and new message types (or new queues, as applicable) can be registered in the system;   *Ease of reconfiguration and system tuning*

- Ability to perform these changes, or move applications and queues around the system, without stopping and restarting it;

- Existence of monitoring facilities (e.g., to debug problems or to tune the system by finding out which parts of the system have become bottlenecks).

Message Contents

Flexibility of message content and format

- Type of data allowed in the messages. (It needs to be flexible enough to meet the application's needs, but it should be of a sufficiently high level so that the application does not have to do all the raw formatting work.)

- Maximum length of a message, if any. (Some systems have no limit at all; others have a limit that can be changed, but the procedure for doing this may require taking the system down and restarting it with a different parameter in order to rebuild the message queue repository.)

Standards Compliance

Current or planned compliance, especially with CORBA

- Degree to which the system uses well-defined standards for the lower-level protocols and for the encoding of the data contained in the messages;

- Degree of actual activity (not just statements of intent) toward compliance with current and future versions of CORBA.

A carefully selected message bus can be a good interim step

In conclusion, let us say that information bus technology is a step in the right direction, if ORBs are not applicable or available, but adopters should go through a careful selection process and should make sure that the supplier has defined a clear path of evolution to CORBA interoperability compliance.

Chapter 11

Objects in Industry: Case Studies in Computer-Integrated Manufacturing

One reason for the Computer-Integrated Manufacturing (CIM) focus of the following case studies is that CIM is a crucial component of an enterprise's software capability. A second and equally valid reason is that both authors have devoted large portions of their careers to this information technology discipline and feel it provides excellent examples of the application of an object-oriented lifecycle. In particular, the authors each spent over three years at the SEMATECH consortium conducting applied research that addressed a broad spectrum of CIM-related issues. One or other of the authors participated in most of the projects that are being presented as case studies—either at SEMATECH or at their respective SEMATECH member companies.

CIM is an important domain in Information Technology

11.1 What is Computer-Integrated Manufacturing (CIM)?

In a narrow sense, CIM is strictly limited to a set of applications that perform factory floor operational functions. In its

CIM performs factory floor functions, and interacts with most other enterprise applications

broader definition, CIM not only performs these factory floor functions, but it also carries out many higher-level functions associated with manufacturing and interacts with most of the other functions of the enterprise. The latter definition has increasingly become the preferred and most widely accepted view. In businesses with a manufacturing component, CIM influences the operation of the enterprise from factory floor equipment control to enterprise-wide functions through information systems. In this broader sense, CIM supports the relationships between functions such as:

- *Financial* – cost accounting, accounts payable;
- *Design* – Computer-Aided Design (CAD), preproduction simulation;
- *Factory operations* – resource rationalization, scheduling, Work-In-Process (WIP) tracking;
- *Logistics* – inventory control, shipping orders;
- *Equipment control* – material movement robotics, processing automation.

CIM is an integral part of a manufacturing enterprise

Since the ultimate goal of a manufacturing enterprise is to deliver the most competitive and highest-quality product to the marketplace in the least amount of time and at the lowest possible cost, a successful organization strategy is to design activities and systems around the overall *product process*. This process encompasses product requirements definition, product design (in some manner), process design, sales, production, delivery, and service. It involves every group, activity, and employee of the organization.

Manufacturing jobs became transportable

One must consider the nature of international competition to understand the increased impetus behind CIM in recent years. The 1970s and 1980s saw a marked shift toward a service-oriented economy in most highly developed countries. At that time, the common wisdom in boardrooms and Wall Street firms was that a strong domestic capability was unnecessary, and indeed, too costly in view of lower offshore production costs. The stark reality of this fallacy became apparent to most people in the late 1980s as millions of higher-paying manufacturing jobs moved to other countries.

Manufacturing jobs became more "transportable" because international transportation became faster and cheaper, large corporations became increasingly multinational, and tariffs were lowered. Labor-intensive manufacturing in lesser-developed countries became more attractive as their low-cost labor forces gained skills and increased their capability to build high-quality products. These trends were exacerbated when countries such as Japan, Taiwan, and South Korea established national policies to target and promote market dominance in specific high-profit manufacturing industries. With the awareness that a strong manufacturing base was necessary to support a strong economy and to meet standard of living expectations, companies, industries, countries, and even groups of countries began programs to revitalize, modernize, and join forces to develop technology-based manufacturing improvement programs.

Domestic manufacturing plants are important for the economy

Those manufacturing enterprises displaying the most agility to meet the rapidly changing demands of the global marketplace have emerged as the most successful in this era of highly competitive worldwide manufacturing. Because computerization has been the key to improving flexibility in many other aspects of the enterprise, and because manufacturing is typically the least computerized part of most enterprises, much of the manufacturing improvement effort is centered around CIM.

CIM is key to manufacturing flexibility

This new emphasis represents a significant opportunity to apply the concepts of the object-oriented lifecycle, for two main reasons:

Object technology is especially well suited to CIM ...

- The manufacturing domain lends itself more intuitively to object-oriented analysis and design than previous foci of computerization. This is because the manufacturing world is full of real objects that perform actions on one another—a direct match to the concepts of object orientation.

... and mature enough to support the stringent requirements of manufacturing

- The technology has reached the necessary level of maturity just in time to meet the demands of this domain. This was arguably not the case in the 1980s, when the focus was on CAD systems or client-server executive information systems (EIS). Object orientation is the right "technology wave" (to use the "surfer's metaphor" introduced in Section 8.3) for CIM Systems in the 1990s, and probably beyond.

The case studies that follow represent pioneering efforts by SEMATECH and some of its member companies to improve their manufacturing enterprise flexibility and productivity through the use of object-oriented CIM technology.

11.2 Microelectronics Manufacturing Science and Technology (MMST) Project

MMST had extremely aggressive manufacturing goals ...

The Texas Instruments Microelectronics Manufacturing Science and Technology (MMST) project is a celebrated success story of an early object-oriented CIM system [McClatchy 91] [Taylor 92] [TI 92] [Harmon 93]. The MMST project goal was to develop and demonstrate a "next generation" flexible semiconductor manufacturing factory with associated equipment, processes, sensors, and a tightly integrated distributed CIM system (see Figure 11.1).

... requiring new software and hardware approaches

The primary achievements of the MMST project were: 1) the development of a radical new manufacturing technology that allows step-function reductions in the capital cost of building a semiconductor factory (from $500 million to $100 million), and 2) reducing typical factory production cycle times (from 60 to 6 days) while maintaining or improving upon traditional factory quality. A key element in the success of this project was an equally radical approach in the CIM system to monitor and control every facet of this innovative factory.

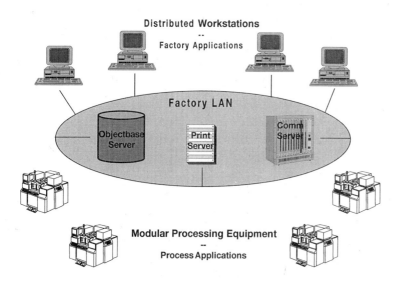

Figure 11.1 – MMST Architecture

The project was executed with shared funding from Texas Instruments, the U.S. Air Force Wright Laboratory and the Defense Advanced Research Projects Agency (DARPA). It lasted from 1988 to mid-1993. The project was not intended to develop a production factory, but its government sponsors did require an extended production environment demonstration. Even though the project has run its course and is no longer active, the technology developed and the lessons learned have spawned follow-on research and production projects throughout the semiconductor industry and beyond.

The initial five-year R&D project generated much follow-on development

The $90 million project used non-traditional software strategies

When the success of a very aggressive, $90 million project is dependent on a software application whose budget is only a small portion of the total, one would expect a narrowly focused system based on a well-tested architecture. Quite the contrary, the approach chosen was very nontraditional. Its strategies were:

- Use object-oriented technology;
- Design for configurability and scalability to an extreme;
- Buy and integrate commercial products wherever possible;
- Conform to existing standards wherever possible;
- Position the system to migrate to emerging standards;
- Target the system for an expected 20-year life span.

MMST followed a "no compromise" object approach

The MMST CIM system did not just give lip service to object technology. It was a no-compromise total immersion in object technology from requirements analysis through implementation and maintenance.

Smalltalk chosen over C++ in 1988!

As far as the development environment was concerned, the MMST team had to start by selecting complementary analysis and design methods at a time when most of these were still in flux and had not yet been documented in books. It also performed a language and programming environment selection, taking the bold step of choosing Smalltalk over C++ in order to force a clean break from past technology.

Commercial UNIX-based environment used throughout

The system's execution environment architecture consisted of off-the-shelf commercial hardware and UNIX-based operating systems supporting distributed object servers, user workstations, and network connectivity to both unique MMST equipment and commercial manufacturing equipment. A key novelty of MMST (see Figure 11.1) is that from the computer systems architecture viewpoint, its Advanced Vacuum processor (a new modular manufacturing station) is completely identical to an operator or engineer workstation. In other words, it has the same CPU, the same operating system, the same Smalltalk-based run-time environment, and the same peer-to-peer ability to send and receive object requests.

The software system supporting the applications was made up of several commercial object technology products consolidated into an infrastructure, including a commercial object-oriented database for persistent object storage and several object-oriented commercial products. Additionally, the project team had to develop some of the infrastructure functionality itself because of the lack of suitable commercial object-oriented products. (The distributed object-oriented execution environment for this project was constructed in the 1989-1991 time frame.) One such capability developed in-house was an equivalent subset of today's Object Management Group (OMG) Object Request Broker (ORB) (before the CORBA 1.0 specification existed). Some needed capabilities were developed internally as extensions to the commercial products being used at the time, and they were subsequently absorbed into the commercial products by the suppliers.

Much object infrastructure had to be developed in-house

In summary, the MMST CIM system was object-oriented, radical, and visionary, but those things in and of themselves did not make it a "good" system. What were the real reasons for its success?

Characteristics that led to success

- It was highly integrated so that it provided a seamless application capability.

- There was little duplication of data even though it was distributed.

- It was pervasive in that all the appropriate capabilities to meet the specific needs of each type of user in the factory were available in one system.

- It was adaptable to rapid change (agile).

The MMST CIM Vision

Innovative in attacking real manufacturing issues

As mentioned above, the MMST CIM system has had notable publication exposure documenting its object-oriented architecture, code reuse, and other technical aspects. The area most neglected in prior references is the significance of the vision to go beyond the functionality of the ordinary CIM systems. More visionary than its exhaustive use of object-oriented technology, however, was the innovative way in which the MMST CIM system used the natural qualities of object-oriented technology to attack "real-world" issues in the manufacturing enterprise. This case study emphasizes that significant point of view.

The prototype MMST CIM system controlled a demonstration factory

Anyone who has "been around the block" a few times in information technology knows that most prototypes have a lot of bubble gum, bailing wire, and developers hand-holding the code to make it work. Prototypes are meant to demonstrate that concepts are workable and to provide a learning experience where risks can be taken in an environment that allows failures before they might be fatal. Prototypes are not meant to be production-worthy (see Section 7.1) and this one was no exception. On the other hand, the MMST CIM prototype was robust enough that it controlled a demonstration factory, running real material through real processes, and making functional integrated circuits. Not every vision for the system was fulfilled, but most were. The greatest benefit was the learning process rather than the resolution of every program bug. The MMST CIM visions and their importance are described in the following scenarios.

Production Planning and Scheduling Scenario

Factory planning and scheduling is a priority for CIM

How does the MMST system know what material to release into the factory and when to start it? Who controls the detailed movement of material once it starts in the factory?

Historically, semiconductor factories, especially those with a diverse product mix, are notorious for large work-in-process inventories and long cycle times. Much academic and industrial research work has gone into improving semiconductor factory planning and scheduling capabilities with techniques as diverse as just-in-time (JIT), operations research (OR), and artificial intelligence (AI). Improving the productivity of a semiconductor factory by only one percent has an extremely large payback. Companies must run their highest technology factories 24 hours per day, 7 days per week to maximize their return on capital investments that often exceed one billion dollars per factory. The potential to use the outcome of the MMST prototype to improve the scheduling of dozens of factories drew significant attention to this aspect of the application.

The **production planning application** controls the material (wafers) released to the factory based on outstanding orders, product mix, and the committed ship date of each order. The planning system provides the human factory planner a comprehensive "what if" simulation capability for determining the potential for re-prioritized intervention in extraordinary circumstances. Furthermore, based on historical precedent and future production levels and product mixes, the planning system predicts future bottlenecks in the factory.

Material release to the factory

After the production planning application releases material to the factory, the **factory scheduling application** controls the movement of the material through the factory and all aspects of dispatching factory resources—both equipment and personnel—to process the material. Rather than allowing material to be pushed into the factory before it is possible to process it, the scheduler controls the material starts and processing stations throughout the factory using a *pull* scheduling philosophy to balance the factory workload and to minimize WIP and cycle time. The scheduling system also tracks performance and reports status on any material or customer order in the factory.

Balancing and controlling factory resources

The CIM scheduler provides automatic alarms to appropriate factory personnel in case "behind schedule" or "out-of-balance" conditions occur in the factory. Additionally, the scheduler is always subject to manual override in extraordinary cases.

Cooperative applications

As products complete processing in the factory, they return to the responsibility of the CIM system's planner. It tracks the factory's actual-to-plan performance, and archives performance history for use as benchmark data to help generate future plans.

Process Engineering Scenario

Machine-embedded computers provide self-monitoring

The CIM system provides highly innovative process control and engineering capabilities. Embedded process computers in processing machines provide a significant amount of self-monitoring and control that include machine performance, Statistical Quality Control (SQC), and dynamic process control with *in-situ* sensors.

Intelligent agents monitor material and machines

Teaming machines with embedded intelligence and the CIM system results in truly futuristic potential. MMST CIM applies the idea of agents deployed into specific machines to watch for events, perform sampling tasks, or analyze and report results. Imagine an engineer about to go home at the end of the day, dispatching a software agent to monitor a certain set of parameters on a suspect machine. When he arrives the next morning, his software agent has performed its analysis and reported its results to the engineer by electronic mail.

Possibility to create processing specifications with object technology concepts

The CIM system also offers the process engineer a powerful set of capabilities for building and maintaining specifications that far exceeds existing capabilities in the semiconductor industry. Manufacturing process step specifications, like other things in the CIM system, are objects. Creation of new processing specifications takes place by obtaining process steps from a library of previously developed processes and joining them in the appropriate order (*reuse*). If a process taken from a library is not precisely what is needed, any modifications required to create a new process step specification are easy to make (*inheritance*). All of these qualities are natural by-products of an object-oriented approach to performing tasks. Besides the obvious benefits provided by the generic qualities of object technology, other innovations include the ability for several engineers to safely work on the same specifications at the same time (concurrent engineering), complete electronic lifecycle maintenance (creation, signoff, activation, revision, deactivation, and history), and most importantly, the ability to write machine-independent specifications.

Traditionally, the creation of semiconductor processing specifications occurs by writing textual style documents describing the work performed by a specific model of a supplier's machine. This method requires that separate specifications be generated with actual machine settings for each new supplier and machine model capable of performing a process. The CIM system's *effects-level specification* capability removes these requirements. It allows engineers to specify the results (*effects*) that an operation is to produce. Each MMST processing machine has the intelligence to determine if it can accomplish the desired effect and to compute the appropriate machine-dependent setting. The *effects* concept and appropriately intelligent machines can provide an order of magnitude reduction in the number of specifications and revisions required, a reduction in the complexity of specifications, and greatly increased quality and life expectancy of a specification.

Effects-level specifications

Equipment Engineering Scenario

Support for equipment engineering is a cooperative function shared by MMST machines and the CIM system. The intelligence built into a machine's embedded control computer allows it to keep self-generated statistics of its own performance in material processing and utilization of each of its processing

Cooperative computing between machine-embedded and CIM software

chambers. Each machine has enough persistent storage to maintain a moderate amount of history. The CIM system is capable of selecting specific parametric data collected by MMST machines and moving it through the network to long-term archives for future analysis.

Any workstation is a "window on the factory"

Along with data collection, users define alarms in a machine's on-board controller. In an alarm situation, the machine sends alerts to a user at any terminal on the network. Also, with the proper authorization, a user may dynamically connect his CIM system terminal to any machine in the factory with an embedded computer—just as if he were standing at the machine and using its attached terminal. This capability is invaluable to an equipment maintainer who is then able to make a virtual tour of the factory, checking on every machine, without leaving his desk.

Machines register themselves with the CIM system

Additionally, when an MMST CIM-conformant machine is connected to the factory network, it announces itself to the CIM system. This startup process includes registering its processing capabilities with the CIM system so that the factory scheduler is able to use its available resources. The CIM system and machines collaborate on many other intelligent functions such as built-in on-line documentation, self-scheduled preventive maintenance, error detection and fault isolation.

Factory Management Scenario

User-defined reporting

For the manager or supervisor, the MMST CIM system provides immediate access to performance data—performance by machine, operation, or person, and consolidations of just about any imaginable grouping of objects that makes sense. All performance data is current as of the moment of the inquiry. There are no cryptic codes required to build an inquiry; therefore, there is nothing to remember—or to forget. Specific or summary inquiries are constructed by manipulating icons that represent familiar objects. Personal preferences in reporting data become the rule rather than the exception. With a click of the pointing device, tabular data can become graphical representations, or vice versa. Specified reports may be generated at predefined times, current data compared to historical data, and many types of filtering accomplished. An object-oriented spreadsheet is available. The cells of this spreadsheet program may contain operating objects such as machines that obtain data from objects in other cells in real-time. Once built, inquiries are available for repeated use. After discovering particularly enlightening data, attaching annotations directly to the screen display and transmitting it to another user for action is an easy matter.

These user-specified analytical and custom reporting capabilities are not just available to people in managerial and supervisory functions, but to all the users of the system. The only limiting criterion is their user profile, which controls access to specific data for security purposes.

Information access defined by user profile

In conclusion, the MMST CIM project created a precedent for the vision of world-class CIM systems that will last well beyond the end of the century. This extraordinary application vision may be specific to CIM, but it can be a model for object-oriented applications in many different functional areas.

MMST can become a model for other domains

11.3 The SEMATECH CIM Application Framework

Late in 1991, SEMATECH's Manufacturing Systems Development thrust began to define and develop semiconductor manufacturing systems strategies to support the complexity, volume, and flexibility requirements for the industry's future manufacturing needs. This led to an evaluation of relevant research and industry activity that included the MMST program described in the prior case study.

MMST was one facet of a broader SEMATECH effort

Generalize MMST into an open architecture

While the MMST CIM system embodied key elements of future semiconductor manufacturing systems requirements, it was targeted to fulfill the requirements of a specific project. Although modular, it was not designed as an open architecture; moreover, it depended on a number of proprietary technologies. Despite these limitations, it was by far the most progressive technology available at the time. SEMATECH and Texas Instruments began a joint feasibility study and detailed planning effort to determine its applicability for generalization and extension. Successful completion of the study led to a project that began in 1992 and resulted in the creation of an open object-oriented architecture that would become SEMATECH's CIM Application Framework.

Why a Framework?

User requirements

This project focus was developed because SEMATECH's member companies agree on several key points:

- The need for viable commercial alternatives to internally developed solutions for at least the required basic capabilities;

- A desire to combine applications from multiple suppliers (internal and external) into integrated, evolvable systems;

- A recognition of open distributed systems technologies and the object-oriented paradigm as a major trend in software systems;

- The requirement that any solution must include mechanisms to integrate existing ("legacy") systems.

CIM application framework was the response

SEMATECH's response to these common requirements was to focus on an open application framework architecture standard that enables and promotes a robust multi-supplier "plug and play" CIM software market.

Objects in Industry: Case Studies in CIM

Goals of the SEMATECH CIM Framework

The SEMATECH CIM Framework is a reusable, domain-specific, implementation-independent system design specification. Since it is based on the object-oriented paradigm, the specification includes information and behavior models as well as communications interface definitions. Moreover, it describes contracted services for all essential components of a manufacturing execution system. Although the design is heavily object-oriented, a variety (or combination) of technologies may be chosen for specific implementations. In other words, the design does not limit implementers to any particular choice of computer hardware, operating system, information bus, database, or programming language.

Implementation-independent specification

Figure 11.2. portrays the heart of the framework as a set of manufacturing abstractions and services that are typically embodied in applications and delivered on distributed computer platforms. This allows developers of conformant applications and equipment control systems to fulfill the continuing customer requests for interoperability between complementary CIM software products. Future versions of the CIM Framework specification will extend this even further by supporting an open market for embedded equipment control software and application functions that are unknown today.

Interoperability between CIM applications

The notion of a framework is not new, even though this CIM Framework is unique in its application scope. The PC DOS desktop computer operating system is an example of an extremely successful "framework" that is analogous to what the CIM Framework hopes to become in the manufacturing systems CIM domain. The interface to the systems services layer described in the CIM Framework specification has some similarity to an environment like Microsoft Windows, which offers prepackaged user interface building capabilities, as well as a common object linking and embedding facility. Imagine how different the PC application software market would be without

Similar to a "Microsoft Windows" environment for CIM-specific applications

these standards. How much would you have to pay for Lotus 1-2-3 or Excel if they had to be reimplemented with each generation of microprocessor, or each release of DOS, or for multiple suppliers' integration shells?

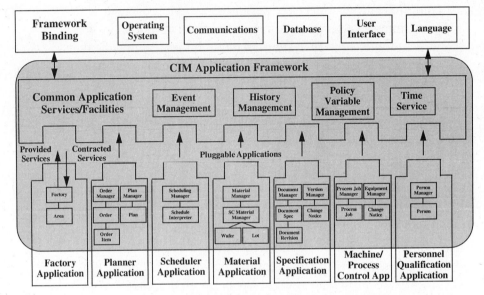

Figure 11.2 – SEMATECH CIM Framework Architecture

Applications developers don't reinvent the CIM infrastructure for each new application

The goal of the CIM Framework is to make it equally easy for innovative software suppliers to provide a variety of reliable, cost-effective, manufacturing-specific applications, and to reap appropriate benefits. The CIM Framework enables them to concentrate on the needed manufacturing end-user functionality by providing them with open interoperability specifications for the target market. As a result, the benefits of the CIM Framework are:

- Reduced development and support cost;
- Lower barriers to entry for new suppliers;

Objects in Industry: Case Studies in CIM

- Ability to evolve systems incrementally and rapidly;
- Ability to introduce new technology while maintaining interoperability;
- Enabling "make versus buy" decisions at the system component level;
- Enabling deeper levels of equipment/factory integration.

Multiple SEMATECH member companies and suppliers are in the process of qualitatively and quantitatively verifying these benefits through development and implementations of conformant systems.

Current efforts will validate benefits

Architectural Principles of the CIM Framework

The CIM Framework serves as a guideline for the development of pluggable applications. The framework specification defines an abstract model that consists of a set of generic abstractions, services, and protocols. It is not the intent of the architecture to define all of the applications that can or should be supplied. Nor is the architecture intended to impose any specific physical realization of the CIM Framework. SEMATECH believes that the CIM Framework can support many diverse implementations—including centralized and distributed architectures—using numerous system technologies.

The CIM Framework is generic enough for diverse implementations

The design of the CIM Framework allows conformant applications to bring to the factory user the benefits of recent key trends in computer systems:

CIM Framework features

- The framework supports the object-oriented principles of object abstraction, encapsulation, inheritance, and polymorphism. Applications should be capable of being distributed as needed and their services available to multiple clients.

Distributed object-oriented peer-to-peer architecture

Fully integrated, dynamically bound applications	• Applications should consist of a set of externally visible objects with well-defined services. External applications needing these services should deal directly with the objects that provide them. Additionally, applications should be dynamically bound to the greatest extent possible. Dynamic binding makes the system more resilient and adaptive to instantaneous changes in factory status or configuration.
Intelligent, proactive object model	• Objects should be given as much intelligence as is reasonable and they should take on a proactive role in improving the state of their domain. Active intelligent objects promote greater responsiveness.
Event-driven object interaction	• Objects should be automatically notified whenever an aspect of another object of interest changes. This eliminates the need for polling for object status and reduces overhead, which increases system responsiveness. In addition, the objects to be notified should be both dynamically bound and transparent to the object that has changed.
Universal object communication with transparent object location	• Any object in the system should be able to communicate with any other object without knowledge of or regard for its location. This enables applications to be written and moved without affecting application code.
Direct object manipulation	• User interfaces should be object oriented and users should interact with the system following the paradigm of direct object manipulation. Icons on the screen should represent familiar "real-world" objects with which a user can interact. Relationships between objects should also be graphically represented so that the user can intuitively navigate from one object to the other visible objects it references.
Generic high-level abstractions	• At the highest level in the CIM Framework class hierarchy, the abstractions should be generic across application domains. For example, while lower-level classes might be specific to semiconductor manufacturing, higher-level classes should be applicable to manufacturing in general.

- Applications should be portable to any node in the system with user interfaces that can be locally or remotely located. This promotes flexibility by separating locality of use from locality of execution.

    Access to any application from any workstation

- Applications should be portable to the greatest extent possible to enable transparent exploitation of advances in computer hardware and system software.

    Any platform, operating system, or network

- The framework itself and framework-conformant implementations should incorporate appropriate industry standards. Standards simplify the requirements, reduce development time and effort, facilitate portability, and enhance diffusion of the technology into the marketplace.

    Standards wherever possible

    The first implementation of CIM Framework was not OMG-compliant because most of the CIM Framework specification was developed prior to the OMG specifications. However, SEMATECH has long been active in OMG activities, and the CIM Framework anticipated the emerging OMG specifications wherever possible and practical. For these reasons, the CIM Framework has always been visualized within the context of the OMG's Object Management Architecture (OMA). Figure 11.3 shows how the various CIM Framework elements fit within the OMA partitions of ORB, Object Services, Common Facilities, and Application Objects (see Section 10.2).

    The first implementation was built prior to OMG architecture specification

Specification versus Implementation

At the heart of the CIM Framework are the core component objects that collectively represent the minimum required to run a semiconductor factory over the desired domain of architectures, technologies, and manufacturing strategies. Each abstraction—including its purpose, the services provided, and the collaborators required to provide the services—must be defined by the CIM Framework specification. As shown in Figure 11.3, in a real CIM Framework-conformant implementation,

Minimal model to provide applications interoperability

the distributed applications inherit common structure and behavior from these abstractions.

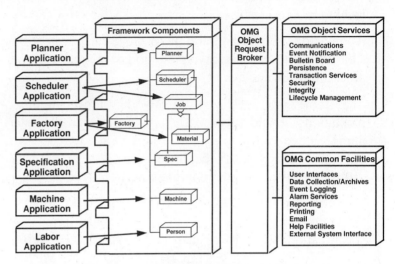

Figure 11.3 – SEMATECH CIM Framework in the
OMG Architecture Context

Maximum flexibility to support compliant implementations

The CIM Framework supports both abstract and concrete implementation approaches. In the first case, there would be no instance of the CIM Framework, as it is completely embodied in distributed applications with specialized concrete subclasses. Alternatively, a concrete CIM Framework might consist of actual CIM Framework classes that have been created on some centralized server node that can act as stubs or message forwarders to dynamically bound distributed applications. Some combination of these approaches might also be appropriate.

Objects in Industry: Case Studies in CIM

Similar to the approach taken for application implementation, the CIM Framework specification does not specify any particular implementation technology for the ORB or Object Services. CIM Framework implementations can be built on a variety of communication networks/protocols or database technologies. However, given that the CIM Framework is fundamentally object oriented, these technologies are typically accessed via an object-oriented application programming interface (API).

CIM Framework allows for diverse underlying implementation technologies

Framework Classes, Components, and Application Partitioning

For organizational purposes, the CIM Framework specification has been divided into several groups of functionally related components. The current functional groups are shown in Figure 11.4.

Functional groups of components

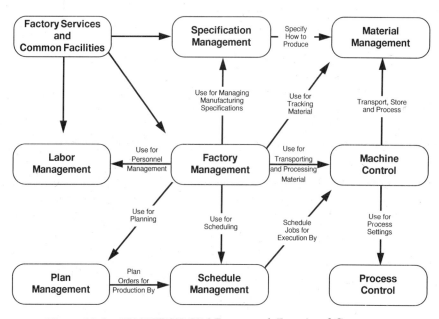

Figure 11.4 – SEMATECH CIM Framework Functional Groups

Three levels of partitioning accommodate a variety of application boundaries

In practice, these functional partitions typically correspond to one or more actual applications in a CIM Framework-conformant installation. However, to accommodate a variety of application boundaries from multiple suppliers, some of which may intersect, a more flexible approach to physical partitioning is required. The approach taken for the CIM Framework specification is to allow for a physical partitioning at three levels of granularity: *classes, components,* and *applications.*

Finest granularity: classes for consistency

Classes represent the finest level of physical granularity of the CIM Framework specification. In this case, a class is analogous to an abstract type with a well-defined interface or set of services. Objects are instances of classes, and all objects of the same class share a common interface and semantics. Classes encapsulate object state and behavior, and thus allow for multiple implementations and transparent evolution.

Intermediate granularity: components made up of classes

Components provide an intermediate unit of packaging between classes at the micro level and applications at the macro level. A component is a physical grouping of related classes into an encapsulated entity with a well-defined interface for component services. At least one class of the component is externally visible, and there may be additional encapsulated or hidden classes. The component provides the minimum unit or increment of functionality that can be added/deleted or enabled/disabled in a CIM Framework-conformant application. Thus it provides the mechanism for incrementally adding functionality and for handling collisions between overlapping applications. Components can be active or passive. From an implementation perspective, components are most closely aligned with *class categories* in Smalltalk and *modules* in C++.

Coarsest granularity: applications made up of components

Applications represent the coarsest level of physical granularity. An application consists of one or more components that may or may not be related or coupled. In general, applications correspond physically to executable program units (e.g., *images* in Smalltalk and *executables* in C++).

The CIM Framework specification also provides definitions in the following object-oriented functional areas:

- Component manager classes,
- Coordination of objects,
- Component behavior,
- Component configuration,
- Component level interface.

Interface and behavior definition to provide interoperability

Industry Acceptance

The acceptance of a unified CIM applications interoperability approach was a significant accomplishment because of the broad nature of SEMATECH's corporate membership (AMD, AT&T/NCR, Digital, H-P, IBM, Intel, Motorola, National Semiconductor, Rockwell, and TI). More than a few of these companies know a lot about software architecture, and their ideas about the best approaches often conflict. The benefit, however, is that once consensus is achieved, these companies exert tremendous influence on compliance both in the manufacturing and the software worlds. Since its first release in September 1993, the CIM Framework specification has attracted significant national and international attention, and it continues to gain acceptance as subsequent extended versions are released. A sampling of the usage and acceptance from industry includes the following observations:

The CIM Framework specification has received widespread critical acclaim

- Several consortia and accredited standards organizations are using the CIM Framework specification as the basis for their specifications and standards.

Basis for other standards

SEMI standards	• The CIM Framework is under active consideration by Semiconductor Equipment and Materials International (SEMI), the organization that generates and fosters the adoption of semiconductor industry standards. Its standards have significant influence with suppliers and users in the worldwide semiconductor manufacturing market.
SEMATECH member companies are adopting the CIM Framework	• The SEMATECH member companies, representing more than ninety percent of U.S. microelectronics manufacturing, sponsored the creation of the CIM Framework and continue to support its evolution. Depending on their current individual needs, these organizations span the spectrum from simple awareness to serious commitment in the CIM Framework adoption process. Some have incorporated it into their information technology strategies; others are at various stages of evaluation and consideration.
Suppliers are using it in their commercial applications	• Some manufacturing industry applications suppliers are developing new CIM Framework-conformant commercial applications. Others are planning the conversion of existing products so that they will be CIM Framework conformant. Yet others are evaluating current and planned products against the specification, or envisioning proposals for compatible extensions of the framework to related application areas.
NIST is working with SEMATECH to achieve standardization	• The National Institute of Standards and Technology (NIST) evaluated the CIM Framework and judged it to be "world class." NIST is working jointly with SEMATECH in multiple areas that include determining how best to extend the CIM Framework features to serve other industries, and how to promote the CIM Framework as an international manufacturing system architecture standard.
The CIM Framework is designed to be a generic manufacturing standard	NIST's point is important: the CIM Framework is not inherently limited to the semiconductor manufacturing domain. One of SEMATECH's long-standing business objectives was to provide a design that would enable semiconductor industry

suppliers to sell their products in other industries, thereby spreading their investment and strengthening the overall market for conformant products.

The Challenges Ahead

SEMATECH made a commitment to object-oriented technology in its earliest stage of CIM applications interoperability research in the late 1980s. Although the technology was very immature, the benefits seemed obvious at the time, and there has never been any wavering from that decision. On the other hand, developing CIM applications interoperability architecture in addition to a fairly new object-oriented technology was neither easy nor immediately successful. Two prior approaches were attempted over a period of three years before the process defined in late 1991 yielded the CIM Framework.

CIM Framework success is built on six years of commitment to object technology

Widespread user and supplier adoption of the CIM Framework is key to the success of this program. Although it represents multiple passes of manufacturing system design efforts by Texas Instruments and SEMATECH, the specification is still very basic in its breadth of coverage. As experience is gained from real implementations, the specification will continue to evolve in both breadth and depth, covering more industries and defining more details to strengthen interoperability.

CIM Framework specification will expand in breadth and depth

11.4 SEMATECH's SkillView

When SEMATECH began to endorse object-oriented technology in 1989, there was little hard evidence of success in an industrial context. Researchers at SEMATECH wanted to conduct a "proof-of-concept" project in object-oriented technology to test its reputation for rapid prototyping and promise of object reuse. The project's purpose was to assess the likelihood that object-oriented technology would meet SEMATECH's major software technology requirements.

SEMATECH's first significant object technology project

Selection of the object environment infrastructure

Because of this unique situation, the selection of a technology preceded the identification of the application, something that does not normally occur in the "real world." Initial steps in the project included selecting an object-oriented programming language and training a team. ParcPlace Systems' Smalltalk was chosen for its rich and extensive development environment, maturity, and its Virtual Machine (VM) feature. VM provided portability across the several platforms used at SEMATECH and enhanced the opportunity to use Smalltalk's rapid prototyping capability. In addition to the language, the team chose ParcPlace Systems' Object Behavior Analysis (OBA), an object-oriented analysis method that identifies and classifies objects in a problem domain based on behavior rather than data (see Section 4.5). The availability of training courses in Smalltalk and OBA also factored into the decision. Once the preparatory work was completed, selecting the right application for the proof-of-concept project was the next crucial step. The search for a project was guided by four criteria:

Criteria for selecting a "proof-of-concept" project

- It had to be a project that tested object-oriented concepts through the entire development cycle—from requirements analysis through the programming and deployment phases.

- It had to be a nontrivial application in order to have value in assessing the suitability of object-oriented technology for SEMATECH's requirements.

- It had to be a relatively short-term project (three months) to show value within an appropriate time frame because technology decisions were soon to be made for other larger, longer-term projects.

- The application's customer had to be deeply committed to participating in the development process. A "customer" was defined as a user with a real manufacturing problem to which object technology could be applied.

Objects in Industry: Case Studies in CIM

Early morning daily fab operations meetings were identified as an excellent venue for conducting the search. A "fab" is a semiconductor industry slang word for a wafer fabrication factory where integrated circuits are manufactured. Such meetings give technicians and management the opportunity to discuss process issues and problems and to develop corrective action plans.

Identifying a real problem to solve

The project team began attending these daily meetings, a candidate problem was identified, and a motivated supervisor volunteered to work with the development team, hoping to see his frustrations resolved in the process.

Identifying a real user to help

Briefly, the problem was that the SEMATECH consortium uses "direct hire" employees, contractors, and member company personnel that rotate through assignments in its Wafer Fab facilities. This results in a work force with frequently changing skill sets. The particular problem entailed selecting qualified technicians to repair malfunctioning equipment. The traditional process of locating a technician trained on a specific machine that was currently "down" was too time-consuming. Meanwhile, the equipment sat idle, which meant thousands of dollars of lost productivity.

Problem description

Thus, this scenario met the project search criteria perfectly, and the project named "SkillView" was born. SkillView is a query application to access the available technicians' skills to perform a specific task. It was meaningful, had a committed user, and was small enough to digest as a first object-oriented project—one of the crucial success factors in moving to object technology.

Project Results

Within three months, an application solving the problem, complete with documentation, was delivered to fab supervisors. SkillView took the form of a robust "point-and-click" query

Application delivered in three months

system, with a user interface based on the Smalltalk browser. It allowed supervisors to identify technicians and their skills on a per-shift basis. Some of the specific lessons learned from the SkillView project were as follows:

Five iterative prototypes built and reviewed with user

- **Prototyping.** Within two days of the initial requirements review meeting with the customer, a first user interface was developed by reusing browser classes. Thereafter, the interface was updated in consultation with the customer as the project progressed. During the development cycle, five prototypes were built and reviewed with the customer before the final implementation was installed on supervisors' desks.

Good reuse achieved

- **Reuse.** The project definitely demonstrated object-oriented technology's reuse. Although earlier prototypes were discarded, many of their classes were reused in subsequent versions. SkillView also took advantage of Smalltalk scanner classes, which were modified to parse input stream data.

Portability demonstrated

- **Portability.** SkillView was developed on an IBM RISC System/6000 and then installed on a PS/2. Little effort was expended in the porting process. The only significant problem encountered involved the user interface, because the team used device-dependent fonts instead of available device-independent fonts.

Application maintenance documentation took advantage of Smalltalk features

- **Documentation.** Documentation consisted of a requirements and design document, a simple user's guide, and an application maintenance document. Booch diagrams were very useful in compiling the software design document. The application maintenance document relied on Smalltalk's self-documenting feature, and served more as a navigational aid through the Smalltalk code. This approach allowed the application maintenance document to be generated as the application was developed. Moreover, it greatly

Objects in Industry: Case Studies in CIM

reduced the potential for version mismatch between code and documentation.

- **Front-loading the lifecycle.** Two-thirds of the team's time was devoted to analysis and moving up the learning curve. Much of that time was spent building OBA scripts. OBA (the analysis method) was especially useful in the most complicated, least understood functional areas. Most of the software errors occurred when only cursory use of the method was made in some of those areas. Nonetheless, the team was quite effective in employing OBA to identify "holes" in the requirements definitions, which allowed effective rework through successive prototypes. Thus, prototypes served as part of the analysis process. In fact, analysis and design were performed during the entire length of the project. Though this may appear to smack of "creeping requirements," such was not the case. Discrepancies uncovered by OBA triggered continuous dialog with the customer until a complete understanding and refinement of customer needs were achieved.

Analysis and design was two-thirds of the project

In summary, based on the project's success, SEMATECH reaffirmed its commitment to object technology. It committed itself to influencing the use of object technology for commercial as well as internal development of manufacturing applications. SkillView demonstrated that object technology provides a means for implementing new applications and for allowing technologists to solve problems quickly.

Project confirmed object technology decision

11.5 The Semiconductor Workbench for Integrated Modeling (SWIM)

The SWIM project was started by SEMATECH's Modeling and Statistical Methods division in early 1992. Its objective was to "develop a software/hardware framework in which users will be able to access a number of different modeling

SWIM is a framework for multiple modeling programs

environments sharing common data bases. Typical models that will be available through this workbench will be operational equipment and factory models, as well as technology and process models."

Mistakes in capacity planning and equipment sizes can be devastating

The cost of building a semiconductor fabrication plant has continued to escalate, breaking the $1 billion threshold in 1990. This is largely due to the high cost of individual pieces of equipment used in a fab (from a few hundred thousands of dollars for some types of equipment to several million dollars for a top-of-the-line production VLSI tester). Because of these costs, a less-than-optimal utilization of the equipment and plant can be economically devastating to chip makers. As an alternative to "hit or miss" factory or operations design, "operational" modeling packages supply the following types of information:

The capacity needs to be predicted

- What is the theoretical capacity of the factory, given the number and throughput of machines of each type, the "routings" of wafers in the factory, the reliability statistics of the various pieces of equipment (mean time between failures and mean time to repair), and the scheduling policy in use?

A factory manager needs the ability to consider the outcome of different changes

- Where are the bottlenecks? In other words, which decision would be the most likely to improve overall factory capacity or to reduce the turnaround time for a new wafer? Is it to buy one more "stepper" (a machine used to "print" the microscopic patterns that constitute the circuit paths in the silicon wafer)? Is it to change the assignment of operators to various shifts? Or is it to change the scheduling policy so that those wafers that have been in processing for the longest time get a higher priority than others? The most obvious decisions are not always the right ones. For instance, a factory manager who notices that large numbers of wafers accumulate at a certain station may conclude that one more machine in this area would greatly improve the throughput. But he cannot necessarily conclude, without performing a simulation, that the bottleneck would immediately move downstream to the next station, whose capacity must therefore be addressed, too.

- What is the total cost per produced wafer of operating the factory? What would that cost become if the "yield" (i.e., the proportion of good chips produced) increased, or if a faster turnaround time were obtained by buying another half-million dollar machine?

The cost per wafer must be known under different scenarios

Unfortunately, there is no magical software program that performs *all* the above calculations. Instead, there is a multiplicity of approaches and tools. Some programs use mathematical formulas to compute answers based on the laws of probability and queuing theory. Other programs perform extensive step-by-step simulations, reproducing in their execution the uncertainties of actual operations by randomly (but realistically) generating equipment faults and new work order arrivals. Still another approach consists of programming a spreadsheet program to perform rough calculations based on industry averages or on equipment specifications supplied by the manufacturer.

Multiple applications are required to provide these answers

Some of this software is available commercially, but many companies have developed their own ad hoc systems. Sometimes they did this because the package embodied their secret recipes to schedule the work as efficiently as possible; sometimes it is because they needed a certain capability before it became commercially available; and sometimes, of course, it was just because of the famous "not invented here" (NIH) syndrome.

Many companies have their own software

Most of the programs, "homegrown" or bought in the marketplace, require a significant amount of information to run. Furthermore, if multiple programs are required (for instance, a capacity model to determine how many machines of each type are needed, and a cost model to determine the production cost per

Large amounts of data must be translated between different formats

chip), they invariably need to exchange information; in other words, some output from program A may be required as an input to program B. Alternately, both A and B may require some of the same inputs. However, there is no standard for the representation of this type of information; therefore, different programs cannot directly read each other's data. The only data exchange or data sharing capability consists of reentering the data by hand. Not only is this reentry an error-prone process, but there is often some translation required. For example, program A may require the number of wafer starts per week and the chip size in square inches, while program B needs the wafer starts per month and the chip size in square centimeters. Several outputs from program A may also need to be combined through some formula before obtaining one figure required by B. Tedious manual intervention is also required when multiple runs need to be executed in order to converge toward a desired result, a frequent situation in a "real-world" planning problem.

The SWIM architecture provides data and presentation integration

In order to integrate a multiplicity of models in a seamless manner, SEMATECH has designed an architecture that employs some of the same concepts as the Computer-Aided Software Engineering (CASE) framework technology of Chapter 6; that is, all the information is presented consistently on the same user interface (presentation integration), and all the data is accessible through a common repository interface (data integration). Control integration, or the possibility for each program to directly invoke another or to pass messages to it, is not a strong requirement in SWIM. The conceptual architecture of SWIM is shown in Figure 11.5.

The SWIM concept is similar to the "toaster model" of CASE tool integration

It is interesting to compare this model to the "toaster model" of Figure 6.2. The Task Manager represents the same concept in both models, with the toaster model making a more explicit distinction between user interface management (the very front layer) and task management, which lies just behind it. The "tool slots" are the same, but the SWIM model shows wrappers explicitly. The idea is largely to convince suppliers of modeling software that they do not have to rewrite their product to conform to a SWIM interface. For the same reason, the "Data Loader" is shown separately to convince potential adopters that existing data sources can be reused. The SWIM Server plays the same role as the object management service layer of the toaster model, and the SWIM database is equivalent to the CASE repository.

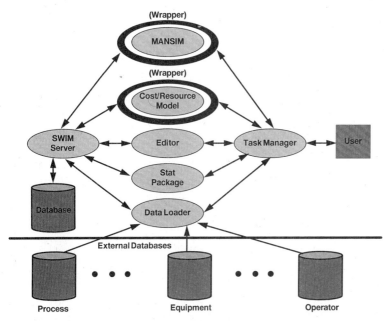

Figure 11.5 – The SWIM Architecture

The first models integrated into the SWIM framework consisted of a commercial, off-the-shelf factory simulator (ManSIM, from Tyecin Corporation) and a "cost/resource model" package developed at SEMATECH. A data loader for Consilium's Workstream package (a popular shop floor control system) was added, as well as wrappers for two popular statistical analysis tools, SAS from the SAS Institute and RS/1 from BBN, in order to analyze results.

Several commercial packages have been integrated using SWIM

The user interface presents a graphical flow chart metaphor

A very interesting aspect of SWIM is its user interface, which graphically depicts the flow of information and control between the various packages that contribute to a certain modeling operation (see Figure 11.6). The user interface is divided into a drawing area, in which the user composes the flowchart of his intended modeling strategy, and a resource area (the top part of the screen), from which he can drag simulation programs, analysis programs, data sources, and control structures (such as iterations) into the flow area.

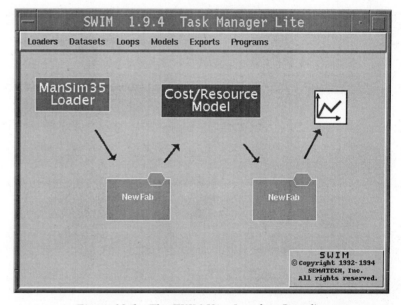

Figure 11.6 – The SWIM User Interface Paradigm

SWIM has a common database

The common database at the core of SWIM is routinely accessed in several ways:

Wrappers can be used for existing programs

- A *wrapper* for a modeling program can read data from the database, prepare the input for a modeling program in the format expected by that program, and perform the opposite transformation on the output of the model after it completes its execution. Initially, this is the most frequent form of access likely to occur, since it requires no modification to the modeling software.

- A *database editor* is provided so that the user can view and revise the data directly. Note that each individual datum (e.g., number of wafer starts per month) is held in only one place, even if multiple programs need it.

    The user can edit the data

- *Translators* can be written to take data from other programs (such as the scheduling output from a shop floor control program) and enter it into the SWIM database.

    Data translation

- Ultimately, some *modeling programs* may be modified to directly read/write the SWIM database.

    Native SWIM applications

The technical approach to the design of SWIM is resolutely object-oriented. The project started by constructing an object-oriented analysis model of its domain. Soon after, it considered technology choices such as what type of database (relational or object oriented) and what implementation language (C++ or Smalltalk) should be used. The initial prototype was written in Smalltalk/80, and the decision to potentially rewrite SWIM using C++ for production was deferred to a later stage. A commercial object-oriented database was used to implement the SWIM database. Alternatively, given the commonality of the SWIM architecture with CASE frameworks, an environment such as PCTE could have been considered.

A complete OO approach

As of mid-1995, the SWIM prototype is in use or installed at seven different SEMATECH member companies (AMD, Digital, H-P, IBM, Motorola, National Semiconductor and TI). A commercial supplier may be selected, through a request for proposals, to bring SWIM to the market.

Prototype status

Chapter 12

Transition To Objects

Any organization developing or using software periodically faces the need to make a technology transition. This may involve switching to a new operating system, acquiring and using a Computer-Aided Software Engineering (CASE) tool, committing to a software process improvement program, adding software inspections to the development process, or adopting object-oriented technology. This last option is a transition of much greater magnitude than any of the others because to obtain the promised productivity and quality improvements, the organization must adopt a complete set of concepts that affect the entire software lifecycle. Moreover, a transition of such scope requires commitment from management and a conscious cultural change by the organization.

Transition is basic to software technology

Peculiarly, the term "migration" is often substituted for "transition" in the context of moving from one technology to another, but "transition" is the preferred terminology. In non-software contexts, migration implies an eventual or periodic return to where one started—which is usually not desirable in the technology transition context.

Transition – not migration

12.1 Risk Management

Object technology transition still has risks

For those diligent enough to have read this far, it should be clear that there has not been enough experience in the implementation of an object-oriented lifecycle for anyone to develop a generic step-by-step recipe that would apply to all organizations. It follows then that there must be risks associated with adopting object-oriented technology. In reality, the risks are not related to "if" but "how" to implement the technology. To offer a good analogy to object technology in the mid-to-late 1990s and where it is going, consider the state of heart surgery in the mid-to-late 1960s. Open-heart surgery technology had only recently progressed beyond animal testing. The relatively few human procedures performed were newsworthy (but rarely successful) curiosities. Today, even heart transplants have a reasonable success rate, and bypass surgeries seem more commonplace than appendectomies.

An object technology lifecycle will become easier with experience

Object technology pioneers can expect a similar experience. Every new technology introduction must go through this learning curve before it becomes relatively risk-free. As in heart surgery, not every early project will succeed. Some will die and some that survive will not enjoy a long and pain-free life. Nevertheless, it would be imprudent for an organization to sit on the sidelines and wait until a technology fully matures before adopting it. Just like doctors, software developers must gain hands-on experience to become proficient in a technology, whether it is new or mature. Moreover, delaying productivity increases in order to minimize the pain of adopting a new technology can be just as deadly to both organizations and individual careers as deferring treatment of a disease.

No Pain, No Gain

Everyone is at least considering objects now

It seems reasonable to assume that every information technology organization is by now at least aware of object technology, and aware that achieving most of its potential benefits requires

Transition To Objects 409

adopting an appropriate systems lifecycle. In this context, each of those organizations falls into one of the following categories of transitioning to an object-oriented lifecycle:

- *Will Not Try* – With or without real knowledge and experimentation, these people think that object technology is not a judicious or viable choice, or that their organization's requirements are a special case to which it does not apply.

- *Actively Trying* – An organization is currently in some stage between the planning of a serious pilot project in object technology, and the initial implementation of an object-oriented lifecycle.

- *Currently Failing* – The *Trying* phase is in progress or complete, but the progress and/or results are disappointing (and there has probably been no thorough attempt to understand why).

- *Currently Succeeding* – The adoption of an object-oriented lifecycle is progressing satisfactorily (overall) or has been completed.

Four very different transition situations

With regard to the object-oriented lifecycle, each of these conditions has an associated set of risks and rewards (see Table 12.1). It should be clear why the material in this chapter is useful for the *Actively Trying* or the *Currently Failing* category, assuming these plan to correct the situation, but even organizations in the other two categories may find benefits. Those who are *Currently Succeeding* at implementing an object-oriented lifecycle are probably practicing most of the appropriate recommendations already, but they, too, may find some "nuggets" that will make it easier to progress or maintain that status. Those who *Will Not Try* object technology will find that in a few years—at most—every modern software tool they use (analysis and design method, language, debugger, configuration manager, etc.) will be inherently object oriented. It will then take more effort to avoid object technology than to use it. Everyone in the technical aspects of information technology will eventually have to learn object technology for defensive purposes, if for no other reason.

Transition risks can be minimized, but not eliminated

There is no way to eliminate the risks of transitioning to an object-oriented lifecycle. In most ways, good object-oriented technology usage is better than traditional procedural technology, but it can be costlier if used poorly. However, by understanding, anticipating, and managing the risks, one can mitigate their effects. It is also important to place risk in a proper perspective. At the early stages of transition, risks are not very significant because the commitment to the technology transition is minimal. Large expenditures for tools and training have not yet been made. The risks become potentially more serious when long-term tasks are committed.

Good planning is key to risk reduction

After the preliminaries have been successfully performed and the "go object-oriented" button is pushed, how can an organization minimize the risks associated with this commitment? Proper strategic and logistic planning are the keys to risk avoidance over the long term, but the highest susceptibility is to tactical errors in the selection of the components that make up the initial object-oriented lifecycle environment. This is especially true in the areas of software tools licensing, computer platforms, training, and integration. Other potential pitfalls exist in areas of performance, resource conflicts, customer acceptance, and personnel management.

Software Tool Maturity

The choice of software tools is a risk ...

The license fees for software tools range greatly, from just under $1,000 to $50,000 or more. General statements about price comparisons would be meaningless because licensing fees vary according to functionality, platform, and pricing strategy (i.e., network-based or standalone, volume discounts, etc.).

One potential short-term risk management technique consists of using multiple tools in parallel until a selection is made based on features, quality, or performance. However, this can be a costly (both in dollar and human terms) and complex solution.

... and using several tools is costly

Some of the better and more innovative tools are produced by suppliers that are very small and often undercapitalized. For example, a small supplier may be limited in the support it can provide. The need to provide support, training, and after-sales service to customers might interfere with further product development or vice versa. This is a situation that places risk not only on the project but on the supplier as well. The ultimate risk is that the supplier's business will fail and leave the customers holding "orphan" tools. Even though small companies pose a specific set of risks, it is often desirable to purchase their technically superior tools. In such a situation, it is best to collaborate with the supplier on mutually beneficial solutions that minimize the purchaser's risk. These might include:

Some of the best tools come from small companies, which implies certain risks

- Supporting the supplier's partnering with value-added resellers (VARs) who can provide marketing support, post-sales service and support, as well as (potentially) financial and human resources. VARs are often also interested in supporting integration, quality control, and customization of products.

Strategies for reducing such risks

- Applying standard business risk management techniques such as requiring the escrow of the source code and internal documentation. If the supplier were unable to continue the support or enhancement of the product, this responsibility could then be borne internally, or by some third party.

- Contracting a part-time, on-site support person from the supplier. This helps ensure that short-term support is available on a timely basis.

Analysis and Design Methods Selection

Selecting the right analysis and design methods avoids the costs of changing later

Selection of the correct analysis and design method (or methods) is very important, and this subject is covered in considerable detail in Chapter 4. Decisions about which methods to select are critical because they can influence several other choices an organization will make in populating its object-oriented lifecycle environment. For instance, if an organization changes its analysis and design methods, the body of domain knowledge previously created and documented may need to be updated, or it may become harder to exploit. Furthermore, a change in methods may require the purchase of different software tools. Finally, one of the large training expenses associated with transitioning an organization to object-oriented technology is educating its staff in the use of specific analysis and design methods. This investment is partially wasted if the methods change (although when an organization does change methods, most of the concepts and some of the features learned for one method will carry over to another). The best defense against making an improper method selection is to thoroughly study and compare methods with the organization's requirements.

Integration

It is important to face integration issues early

As an organization begins to develop complex object-oriented systems, issues involving the integration of the new systems with legacy systems (covered in Chapter 10) and other object-oriented systems will quickly become a major concern. That is the negative side. The positive side is that integrating object-oriented applications with older technology should not be any more difficult (and possibly easier) than integrating older technology systems to each other. Moreover, integrating "all object-oriented" applications can be much easier. The risk-minimization approach to integration is to purchase or build software components that conform to the prevalent interoperability standards, especially those adopted by the Object Management Group (OMG). Adherence to a set of commonly accepted standards is the only solution for minimizing integration risk.

Execution Performance

Object-oriented systems have carried the stigma of slower performance since the first appearance of the technology. In Section 10.2, we discussed how the reality has changed over time. However, there is still a real risk that any performance shortcomings, even if they are temporary, can affect the customer's or management's willingness to support an object-oriented project. Several approaches can be jointly used to minimize this risk.

Performance issues specific to object technology are diminishing

- Understand early, through prototypes, whether or not the proposed technology creates a significant problem. If not, publicize the results of this early work in order to counter uninformed rumors.

Prototype early and prototype often

- Develop clear performance metrics. List separately any aspects of the system whose performance is directly visible to the customer. This might include, for example, the time it takes to start the application, the time it takes to perform a "batch" operation (e.g., to run a simulation), or the time it takes to load a large data set. Establish early on the goals for each such facet of the system, and prepare comparable figures for a legacy system or a competitor's product (if one exists). Enforce a discipline of measuring the performance periodically. Report the evolution of the measurements as the product progresses from an early prototype toward a robust, released product.

Measure performance with metrics —not subjectivity

- If problems do appear, whether at the prototyping stage or during actual development, understand whether they come from the *architecture,* the *design,* or the *implementation.* Be aware that an architectural limitation can be overcome by a design change, and vice versa, and exploit this freedom shamelessly. If, for example, it takes much longer to

Most performance issues can be resolved with small design changes

bring up an application because it needs to launch an Object Request Broker (ORB), then maybe the ORB can be made to run all the time, and the application can bind to an already running "daemon." In other situations, steps that used to be performed in a certain sequence can be performed in parallel or in a different order so that the user's *perception* of performance is better. For instance, in an interactive application, perhaps a menu can be displayed before all the other initialization actions are complete. The necessary initializations can then be completed during the ensuing time period, when the user is no longer affected by the delay (because now he is thinking about his input or typing it). Implementation issues may be handled through the usual techniques of profiling and optimization, or they may require changes in the design. On the other hand, it is very dangerous to create a situation in which people may attack the architecture of a system—perhaps the very concept of object orientation—as a result of a design or implementation mistake.

Performance of the Development Environment

Don't under-equip developers

Another risk consists of understating the investment necessary to embark on the application of object technology. A fully productive environment may require an object database server, more expensive compilers, and additional products (such as library browsers, which are rarely used with structured methods but become critical with object-oriented systems to obtain reuse). In order to offer satisfactory performance, the development workstations may require more powerful CPUs, more memory, and local disk caches. If a shared configuration management tool is used, which is likely to be the case as soon as more than one person needs to work on the same project, the capacity of the network may be a limiting factor at every step of the development process—it can even make compilations appear to take more time. If the components of the environment are not properly sized, the developers will find their productivity to be far short of what the technology was supposed to bring them. All this comes at a cost, but hardware prices have come down to where it makes more sense to enhance the developers' environment than to lose their time and goodwill.

Competing with the Legacy

Most object-oriented projects are undertaken in the presence of a legacy system that needs to be maintained and enhanced until the new system is ready to replace it. The new and the old thus compete for development resources and human skills. It is unfortunately easy to get caught in a vicious circle, captured in the following scenario:

Object-oriented projects often compete with supporting the legacy

- Some new, urgent requirement suddenly emerges. The deadline to meet it is incompatible with the timetable of the new system. At the same time, the pressure to meet the request is too high to ignore. Therefore, a decision is made to implement the change as a "small revision" of the old system.

Resource allocation becomes a vicious circle

- This requires resources that were not budgeted. The object-oriented project, which is not yet on the critical path to a product delivery, or is not perceived as being as critical to the organization's survival, is an obvious prey. A few of its assigned resources are "borrowed" to meet the urgent need of the legacy system.

- Because of this resource shortfall, the object-oriented project slips. Its schedule is recalculated and some resources are moved around to minimize the impact, but obviously the new completion date is now later than the initial one.

- Furthermore, the new project now has one more requirement to meet: it must catch up with the functionality change just added to the legacy system. Not only are there fewer resources to do the job, but now there is more to do!

- Because of the schedule slippage, the risk increases that another new requirement will emerge before the project is complete. We are back to the first step.

Shield the project team from the impact of resource conflict, but prove its value through add-on features

In theory, the answer is simple: isolate the team in charge of the new project so that it is harder to steal its resources, and commit to the stability of the project's resources at a high enough level of management. In practice, this hard line is difficult to enforce. This is why "big bang" solutions that suddenly replace a legacy system with a new one are inherently dangerous. It is better to architect the system in such a way that the new developments yield incremental functionality that can be introduced progressively in the form of add-on or replacement modules, compatible with the legacy environment. In this manner, even if the project is downsized because it loses a temporary battle with the legacy forces, it can still win the war.

Acceptance by Management and Customers

Converting to object technology frightens many managers

The fact remains that object technology still frightens many people. Most corporate managers do not have books like David Taylor's [Taylor 91] on their nightstand, but many "know all there is to know" about object technology (which is usually summarized as follows: it is expensive and slow, and does not provide robust applications). Computer scientists and software engineers contribute to the problem by "selling technology" to their management or to customer representatives. Customers and management have a simple problem, namely, they want to know what the proposed approach will give them. They could not care less if the system used baling wire and masking tape, rather than ORBs, Object-Oriented Database Management Systems (OODBMS), and a whole lot more alphabet soup, provided it does what it needs to do in a timely, cost-effective, and high-quality manner.

Sell business benefits — not technology

The proponents of object technology must therefore sell what their audience needs—not what the technologists find exciting.

To the customer, they must emphasize the features that can be implemented more consistently or more rapidly, and the extensibility that open, standard object definitions and interfaces offer. To management, they must emphasize benefits that hit the "bottom line" of the company: quality, maintainability, productivity, and longevity of the product, as well as the ability to attract and retain valuable employees.

Personnel Impact

Since we just touched on the positive aspects of the transition to object technology in terms of people, it is also important to realize that the human aspect of this transition presents its own set of risks. Object-oriented approaches may be natural to the college graduate fresh from Stanford's or MIT's computer science department, but they are more frightening to many a seasoned professional. They threaten the senior staff developer's confidence that he masters the techniques of his job. They challenge his very status, if he finds he must walk over to someone half his age to ask for a fine point of Eiffel or C++ syntax.

Converting to object technology is a challenge to many senior developers' self-esteem

Section 12.3 will deal more extensively with the cultural factors that influence the adoption of object technology. But the first and simplest approach to mitigate this risk is, of course, education and training. For most people, training should not start with heavy technical contents, but with a gradual introduction of the *approach* to objects. It is very important to convey honestly the fact that object-oriented development entails a different way of thinking about software, and is not merely a matter of a different object syntax. If done well, this will not scare most people. On the contrary, it will get them excited about the journey and eager to sign up. After that, the type of curriculum shown on Figure 7.6 can be put in place and taught to the development staff according to their needs.

Begin education early and teach gradually

Have a plan for those who cannot make the transition

In all likelihood, not everyone will make the transition. Some consultants have said that at least ten percent of the personnel of traditional software development "shops" (e.g., COBOL-based Management Information Systems, or MIS, departments) cannot adapt to the new technology. What should one do in that case? Instead of literally destroying these people's careers in the process of changing technology, managers should identify in advance a set of tasks that remain a valid use of those employees resistant to the "paradigm shift." It may be that they will find themselves confined to the residual maintenance of a non-object-oriented application. But they may also know the application and the domain so well that they can provide more visible and less menial contributions such as teaching junior members of the team about the domain; interpreting customer requirements; designing interfaces with legacy systems; or documenting hidden functionality in the system being replaced—functionality that was introduced for a good reason years ago, but that was never captured in any requirement or design document.

12.2 Models for Transition

The transition effort must not be underestimated

Technology transition is usually much more difficult than even an organization's most conservative estimate of the process. Transition to a complex new technology such as an object-oriented lifecycle involves people at all levels of an organization. A lack of understanding of the amount of effort and commitment required to accomplish the transition can result in disillusionment, demoralization, and finger pointing within an organization. This is particularly true when expectations are not met. A lack of buy-in from any significant group within the organization can result in poor execution by staff that is masked as poor performance by the technology, a potentially dramatic setback.

All categories of personnel involved in the transition can benefit greatly from understanding the strategies, mechanics, and risks involved. When we say "all categories," we include:

Everyone will benefit from understanding and preparing for transition issues

- Managers who plan and track the work;
- Developers who need to achieve early success with demonstration prototypes;
- Training groups that must prepare new material to take users through the product changes.

Two complementary technology transition models directly applicable to moving to an object-oriented lifecycle were developed by the Software Engineering Institute (SEI) and the Software Technology for Adaptable, Reliable Systems (STARS) program. The SEI model defines the stages of technology transition and provides mechanisms for moving from one stage to the next. The STARS model is a four-step interactive process for approaching technology transition. Both models provide useful frameworks for understanding and planning technology transition.

SEI and STARS have useful transition models

All successful technology transitions start with some form of information transfer and then progress to technology implementation. Along this continuum, the SEI model identifies six phases of commitment to a new technology (see Figure 12.1) [Fowler 91]. The first three phases of the SEI model are primarily concerned with information transfer, while the last three focus on technology implementation.

SEI describes six transition phases

- **Contact** – The organization has initial contact with the technology through some means (such as briefings, marketing information, documents, etc.).
- **Awareness** – The organization becomes more broadly aware of the existence of the technology.

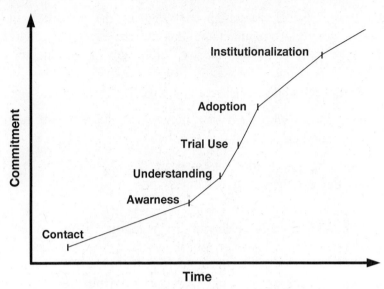

Figure 12.1 - Commitment is a Phased Process

- **Understanding** – The organization understands the technology well enough to be conversant in the relevant details.

- **Trial use** – The organization agrees to use the technology for some purpose on a trial basis, often to facilitate the adoption decision (e.g., a pilot project, prototype development, etc.).

- **Adoption** – The organization agrees to use the technology more widely for an application related to the company's business purpose.

- **Institutionalization** – The organization makes use of the technology part of the standard practices of the company (routine, everyday use of a technology).

A primary goal of transition models is to keep an organization moving from one phase of technology transition to the next. There are many mechanisms for doing this. The STARS approach (see Figure 12.2) is a pragmatic way to emphasize that function of a technology transition model.

The STARS approach contains pragmatic mechanisms for transition

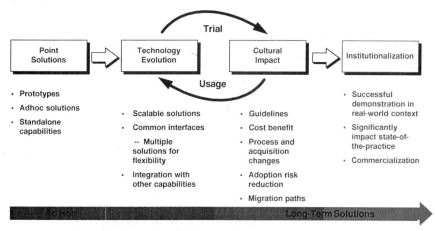

Figure 12.2 - STARS Transition Approach

The STARS model [Morin 91] begins with "point solutions" and has the same purpose as SEI's contact, awareness, and understanding phases. Point solutions include multiple, unrelated technology applications, but also include information dissemination mechanisms such as newsletters, conferences, and technology catalogs. These point solutions often have specific target audiences, and while they are defined as "disjoint," they represent a concerted effort at technology transition.

STARS begins on purpose with unrelated "point solution" projects

This model minimizes risks by adding more intermediate checkpoints

The STARS model takes an iterative approach to technology implementation. Successive evolutions of the technology are tried, and impact assessments close the feedback loop. This implement/evaluate cycle primarily benefits the organization that is testing an unproven technology, or a conservative organization that is unwilling to make a transition commitment without substantial firsthand experience. This approach moves the level of commitment toward adoption and institutionalization while minimizing the transition risks. It lowers the probability of adopting inappropriate technology for an organization's needs. On the other hand, the new technology may still be rejected because the organization cannot handle difficult (but solvable) problems encountered in the implement/evaluate cycle.

Both models are useful for transition planning

These two examples show that descriptive models can be useful as planning tools for technology transition. In SEI's case, the goal is to provide a formally defined process of technology transition that the average software engineering manager, senior engineer, or anyone planning the transition of software technology may follow without advanced training or education. The SEI process includes a practical, conceptual framework and vocabulary, and a set of models and related materials for use in planning, implementing, documenting lessons learned, and analyzing results from transition efforts. Materials that accompany the models include guidelines, checklists, templates, briefing materials, and tutorial materials.

12.3 Cultural Factors

Employees may feel their professional stature is threatened

Cultural factors should be given a high priority when transitioning to object-oriented technology. Many organizations that have adopted this technology have noted that not everyone is capable of the required paradigm shift. Even employees who are capable may feel their professional stature is based on their expertise in older technologies and fear that their worth will plummet when object-oriented technology is introduced. Other cultural factors critical to a successful transition include:

- Obtaining management commitment without overselling the benefits and/or setting unrealistic expectations;
- Possible changes in the reward system to recognize reuse with object orientation rather than emphasizing the production of new code;
- Organizational changes (even addition of new positions to foster and support reuse, as seen in Chapter 7).

Overcoming cultural barriers requires early training to increase awareness and understanding of object-oriented technology. Management should be made aware of the long-term benefits of code reuse, but at the same time, they should beware of overcommitment. The programmers need to understand that an object-oriented lifecycle will lead to improved job satisfaction through the elimination of those mundane functions of software development that currently must be repeated. Moreover, the potential for peer recognition is enhanced through the creation of components that other developers will reuse.

Training should emphasize improved job satisfaction

As with the insertion of any new technology, most individuals have more fear of change than pleasant expectations of the unknown. People are sensitive to any possible loss in position or power that change may engender. Usually, such people have established "tried and true" patterns of success that they are reluctant to abandon. And of course, change requires relearning, and almost everyone feels he has enough to do without adding the extra work of learning a new technology. It is true that for most people, the insertion of new technology means a short-term loss, rather than a gain, in productivity. It is only when they have understanding of the long-term benefits that they will begin to accept the technology change. Even then, most will enthusiastically embrace an object-oriented lifecycle

Fear of the unknown is stressful to many people

only after it has demonstrated its worth to the enterprise or to themselves personally.

Management can ease transition stress

Management of an organization can improve the probability that object-oriented technology will succeed by:

- Communicating the rationale for why the current technologies and practices are inadequate;
- Explaining why the technology change is essential to the enterprise;
- Showing commitment to the transition through the assignment of adequate resources;
- Putting procedures and controls in place to ensure progress in the technology transition.

Assign a methodologist to lead the transition

Beyond the necessity for management commitment, the assignment of a methodologist thoroughly trained in object-oriented technology is necessary. A methodologist is essential in leading the task of specifying how object-oriented techniques and tools will integrate with the current environment while preserving the technical integrity of the software lifecycle process during transition. The label of "methodologist" in this context does not carry the usual connotation of a person who develops analysis and design methods. Such an individual might be known in the organization as a Project Manager, a Senior Analyst, or a Senior Software Engineer, but for the sake of this discussion, he will be known as a methodologist.

The methodologist should be an "insider" if possible

The methodologist should be trained in analytical problem solving and be completely familiar with both the current environment and object-oriented techniques. Preferably, the methodologist should be a respected member of the current organization's team rather than an outsider (such as a consultant). We also recommend that the methodologist be responsible for

Transition To Objects

organizing or developing reference and training material to familiarize personnel, for encouraging participation in the training, and for providing ongoing guidance and advice. Further, based on assessments of success and feedback from involved participants, he should be able to assess how rapidly to move the implementation of the object-oriented technology. In summary, the methodologist must be a key participant in creating and communicating the technical roadmap so that the end goal remains clear.

Most developers exposed to object-oriented techniques intuitively feel that the applications developed with it are better for several reasons: 1) such applications are easier to understand and maintain; 2) they exhibit more tolerance to change; and 3) they have greater potential for portability and reusability. Hard empirical evidence of this is still scarce, but it is now beginning to accumulate.

Objective evidence of benefits is slowly emerging

The message to management is to prepare *now* to take advantage of these techniques, even if significant time will elapse before the organization begins its transition to an object-oriented lifecycle. With this in mind, recommendations for managing the transition to object-oriented technology in an existing development and maintenance environment follow:

Management should prepare well ahead of transition

- Start the process of educating personnel in object-oriented technology.

- Identify potential methodologists in your organization and send them through an exhaustive range of object-oriented training.

- With the cooperation of the methodologists, select object-oriented analysis and design (OOAD) methods that best fit your organization's needs.

- Select and purchase OOAD and object-oriented programming (OOP) tools that support the selected methods.

- Provide time for the methodologists to thoroughly familiarize themselves with the techniques and tools.

- Show visible management support to the organization in order to set the stage for object-oriented technology transition.

- Let support build internally without heavy pressure for immediate compliance.

- Provide forums for internal information dissemination and feedback (seminar, electronic news groups, etc.).

- Start small, with a few of the organization's respected technologists, to promote internal pull for object-oriented technology.

- Pick a pilot project. Identify a limited, but nontrivial application that is fairly modular and has simple external interfaces.

- Have the methodologists and a few key people (especially enthusiastic volunteers) design and implement the application.

- Apply the object-oriented techniques to this application only, stressing the capture and communication of lessons learned, before beginning aggressive technology transition.

- Assess and correct problems and reinforce positive aspects before moving to larger, more complex applications.

- Publicly reward the participants to encourage them to repeat the process and to encourage others to join in.

Object technology adoption requires effort to succeed

The now-famous quotation, "There is no silver bullet" [Brooks 87], applies to the object-oriented lifecycle as well as to any other technology in software engineering. These techniques must be backed up by experienced designers and implementers who can recognize poor design. Object-oriented techniques are not just buzzwords, nor are they a panacea for all problems. A sustained, incremental effort to prepare for the incorporation of these techniques can position an organization for far greater productivity in the future.

12.4 Planning the Transition

The most important ingredient in successful transition to object technology is the planning. Generally, an adequate plan includes periodic reviews of the project execution and provides for appropriate corrective action for resolving problems. The prior section, 12.3, provides an extensive list of recommended steps that should be included in the initial "get-acquainted project," that is, in the beginning of the transition to an object-oriented lifecycle. Many of these same steps are appropriate for follow-on projects, but once the commitment to object technology is made, the emphasis shifts to institutionalizing the object-oriented lifecycle process.

Plan, review, and correct where appropriate

The projects that follow an organization's initial success in object technology require just as much or more attention than the first. There is a natural tendency to relax after the flush of a first success, but the challenges of institutionalizing a process can sometimes be more difficult to overcome than the original process development itself. An organization's object-oriented lifecycle process is customarily originated and prototyped by a small, close-knit team in a carefully controlled environment. This team normally represents the most technically talented and motivated people in the organization. From this kernel, the process is institutionalized by extending outward in an ever-increasing spiral (see Figure 12.3) until it encompasses the entire organization. The implementation of the process requires that everyone in the organization be trained in the technology and committed to execute the process. This can be very stressful for many of the individuals trying to assimilate a new process and new procedures, and there are often problems in scaling the process up from a small team to a larger organization. The transition plan must therefore include contingencies for

Institutionalize success; don't relax after one successful pilot

organized assessment of problems and orderly changes as required.

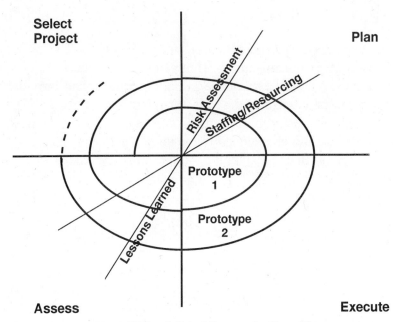

Figure 12.3 – A Spiral Process for Transition

Creating a Transition Plan

Create an individualized transition plan

The recommended steps for object technology transition through the pilot project are roughly the same, regardless of the size or type of organization. After the completion of the initial pilot, however, this uniformity ends, and each organization must create an individualized transition plan. There are many facets of a transition plan that demand customization to an individual organization. Some general categories of items to be considered are as follows:

Organize a team if this is the organizations's style

- *The Transition Team.* A successful transition to object technology does not happen without a great deal of planning and follow-through. A transition team should reflect the character and style of the organization. If the organization is normally team oriented, the team may be made up of individuals who work as a unit to make collective decisions. If the organization is more vertically structured, the team may take on the organization's normal structure of hierarchical management. If the organization is small, there may be one person who is primarily responsible for the team's functional tasks; a large organization may have multiple full- or part-time people responsible for functional tasks. Regardless of the size or structure of the team, every organization function affected by the transition should be involved in the team activities.

The STARS organization, described previously in Section 12.2, defines transition team members beyond the pilot project [Fowler 91] with the following specific responsibilities:

STARS suggests several specific team member responsibilities

- *Champion* – informally advocates the technology, calling it to the attention of others. This individual may have initially introduced the idea of the particular technology.

- *Authorizing Sponsor* – sponsors the technology at the level of upper management that provides resources, strategic and policy direction, and final approval to proceed with the implementation of a technology.

- *Reinforcing Sponsor* – sponsors the technology at the line management level that directs efforts toward planning for implementation and trial use.

- *Change Agent* – an individual or team, drawn from line management or software personnel, who does the detailed planning and implementation of the technology.

Don't proceed with an under-trained staff

- *Training.* Training is just as important in the institutionalization phase of the transition as in previous stages, but now it takes on a different complexion. In the most effective scenario, training of the non-pilot personnel would occur concurrently with the pilot project. On the other hand, many organizations will want to complete the pilot project before making a commitment to expend the monetary and people resources. At this point, the critical matter is to avoid giving in to the temptation to proceed with a full-scale object technology lifecycle implementation with an under-trained staff. Another common mistake is to train the programming staff in an object-oriented language without providing the necessary introductory training or object-oriented analysis and design training that gives them the required conceptual understanding of object technology.

Regular reviews reveal problems early on

- *Monitoring Progress, Evaluating Results, and Resolving Problems.* Regular and systematic status reviews of the progress of the object-oriented lifecycle implementation is an absolute requirement for success. These reviews should involve substantive problem-solving sessions. Reporting of progress and results should use quantitative measurement in all possible instances. One should measure such parameters as lines of code reused by each programmer, number of classes implemented in a reuse library, design components reused by analysts, etc. Individual contributors and managers alike should be required to report status and progress in their monthly or weekly organizational reports.

Resolve issues before success is threatened

Team meetings should emphasize addressing and resolving problems and fine-tuning those procedures that make up the lifecycle before they become threatening issues. Action items and specific responsibility for each item should be assigned and reviewed at every meeting. The lifecycle process should not be "poured in concrete." It should be practical and it should help—never hinder—the organization. The reason for implementing an object-oriented lifecycle is to improve productivity of the organization. Therefore, every rule and procedure must address that requirement.

12.5 Coexistence with Legacy Systems

In most cases during transition, software systems built with object-oriented technology will need to interface with older technology that may not have the characteristics of a well-behaved object-oriented system. One particular issue concerns the mixing of object-oriented techniques with old technology to enhance and maintain existing software. Most new software functionality is not implemented through the development of completely new systems, but rather as an upgrade or a replacement of portions of an existing system. Since object-oriented development is the exception rather than the rule in legacy systems, the issues of management of object-oriented technology insertion and incremental transition become key drivers in the successful implementation of mixed-technology systems.

Managing mixed technology

Object-oriented technology transition has two facets. The first is to make a commitment to object-oriented technology, build an experience base, and develop new systems using that technology. The second, and certainly the most demanding, is the integration of old and new technology, with the eventual replacement of existing systems with object-oriented systems. Coexistence with existing, or legacy, systems is an issue that must be addressed from the beginning to achieve successful technology transition. For example, we saw in Sections 11.3 and 11.5 how object-oriented approaches to Computer Integrated Manufacturing (CIM) require the design of some interactions between new, object-oriented software and legacy applications.

Integrating object and legacy technology is demanding

Many legacy systems have a ten-to-twenty-year life expectancy

It is typical, especially in large organizations, for complex systems to have a life expectancy of ten to twenty years. Most medium-to-large size companies have millions to hundreds of millions of dollars invested in their collections of existing systems. Even where there is sound financial justification for the most aggressive transitioning of legacy systems to object-oriented technology, resource requirements will typically dictate a transition period measured in years.

Develop an incremental transition architecture roadmap

One approach to integrating object-oriented technology with older technology is to transition existing systems in incremental modules. This can be facilitated when an organization develops a software architectural reference model that serves as a transition roadmap for their specific system domain. Developing such a reference model serves several purposes: 1) it creates a greater understanding of an organization's software requirements to support its business goals, and 2) it leads to a common understanding of how new software applications, whether built or purchased, fit into their overall technology vision. Organizations with an architectural reference model have fewer problems integrating new and old technologies because the model can explicitly define interfaces and interactions between systems. This enables improved quality, timeliness, and reduced cost of both the new systems and upgrades to existing systems. Software architectural models provide a greater opportunity for clear specification of interoperability and portability between the different technologies while the transition to object-oriented technology occurs.

12.6 Object-Oriented Enterprise Integration

Linking all the functions of the organization

When considering the coexistence with legacy applications, one is naturally led to examine the broader issue of what it means to address the complete range of software applications and systems required to make an entire organization function, and specifically, how to scale object technology up to this broader scope. **Enterprise integration** is defined as tying together all the functions of an organization through a smooth flow of information and a consistent set of computer-aided processes, as opposed to individual applications that create *islands of automation.*

"Enterprise integration," like "object-oriented," is a buzzword of the 1990s. It takes little cynicism to question whether combining two buzzwords represents a real concept with practical applications, or merely an exercise in esoteric thinking. We will attempt to show that combining these two concepts does indeed generate real benefits. We will also try to show that enterprise integration can be performed little by little; that is, one does not need to have reached the end of the transition to objects to attack it. This section is partially derived from strategic work started at National Semiconductor Corporation in 1992-93, though similar efforts are in progress in many other companies.

Object technology has a practical impact on enterprise integration

What is *Real* Enterprise Integration?

David Taylor defined enterprise integration in [Taylor 91] as:

Two definitions of enterprise integration

> "... the process of building and using an organization's working model to understand the process of that organization and to implement some of its functions in software."

Savant Solutions, a consulting firm specializing in manufacturing systems integration, uses similar terms to describe enterprise integration as:

> "... the alignment of the business processes, organization structure and supporting information technology with

> *each other, and with the vision and strategy of the enterprise."*

An enterprise can be more than a single company

The word "enterprise" is usually taken to mean an organization with a common sense of purpose. Like "organization," the key emphasis is indeed that enterprise integration is related to understanding and supporting the *processes* of the enterprise, not just its static *structure*. Furthermore, an enterprise may actually consist of multiple interrelated organizations. For instance, in a practical sense, the Chrysler or Ford "enterprise" consists not only of the company bearing that name, but also of a web of parts suppliers and independent car dealers that participate in the product lifecycle. Therefore, the processes that need to be supported by integrated systems, such as manufacturing or sales, span several individual companies.

Enterprises have many different requirements, but solutions remain uncoordinated

Over a long period of time, computer systems progressively took over the tracking and processing of information in organizations. Financial and payroll systems, developed in the 1960s, became commonplace in the 1970s. Computer-aided design (CAD) systems were developed in the 1970s but because of the initial high cost of interactive graphics, took almost ten years to see widespread deployment. CIM applications and systems emerged in the 1980s but are still not as pervasive as they should be. Each type of system was introduced to solve one aspect of a company's automation needs. Because there was no commonality in the installed base of prior systems, each new application tended to solve its piece of the puzzle without providing a linkage to the other functions of the enterprise. The resulting situation is pictured in Figure 12.4.

Transition To Objects 435

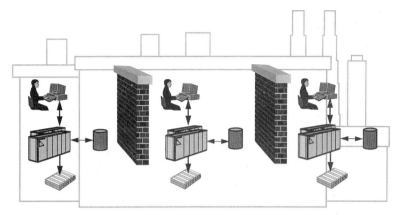

Figure 12.4 – Isolated Applications

In this situation, any information that needs to be shared across organization boundaries (the brick walls of the figure) must either be output by one system and manually reentered into another, or a special application must be written to automate the transfer of information. Over the years, this problem has become more and more complex. Many large Information Systems (IS) departments devote most of their resources to the development and maintenance of such special programs. Instead, they should be producing new functionality or improving the productivity of their organizations' design and manufacturing centers. This is the current "state of the practice" in many companies (represented in Figure 12.5) except that in "real-life" large companies, a complete diagram of this web of information systems might cover an entire wall.

The lack of integration is addressed by piecemeal solutions

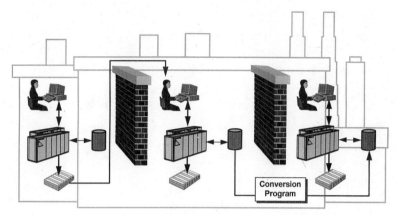

Figure 12.5 – Interconnecting Separate Applications

There is no complete or timely view of the company's data

In this case, each function in the company may work well in isolation, but it is essentially impossible to obtain accurate, near-real-time information on which to base management decisions. Also, when such a decision is made, it is very difficult to make it trickle down into other facets of the organization because automated information transfer is often hampered by a different set of computer systems and applications—hence the desire to address the concept of enterprise integration and implement an architecture that allows its realization.

The "data warehouse" solution is a one-way feed into a single repository

Traditionally, however, the challenge of enterprise integration is seen only as a need to provide decision makers with rapid access to data. With this priority in mind, a one-way transfer of information to a common "melting pot" is thought to be sufficient, as shown in Figure 12.6. This common repository of enterprise information, made accessible to decision makers through reports and on-line queries, is often called the **data warehouse.** W.H. Inmon, in his book *Building the Data Warehouse*, defines this entity as "a subject-oriented, integrated, time-variant, non-volatile collection of data in support of management's decision making."

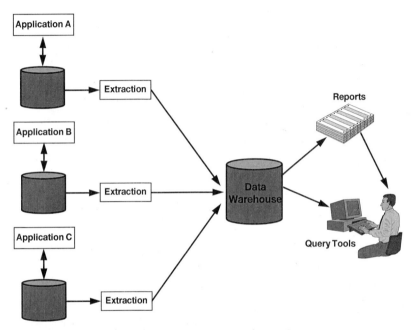

Figure 12.6 – The Data Warehouse Concept

This is all very useful to a certain extent, but products are not designed and built by management alone. The data warehouse offers a one-way information delivery mechanism, but does not help to integrate the other, directly productive processes of the organization. And even if the data warehouse improves the quality and timeliness of management decisions, these still need to be transformed into actions that affect the individual operational systems of the enterprise. Since the data warehouse concept does not include a reverse path for decisions to be transformed into commands to these systems, there is still no

There is no path to enter changes into the operational systems

effective integration. In other words, the problem is that the data in the warehouse is just that—data. It does not have any *behavior* associated with it. You cannot poke at it and make it do things. It is not made up of objects.

A Simple Example

An example based on a common management problem: processing invoices

Let us take a simple example of how a manager may control the expenses of his department by having access to billing information from suppliers (what is commonly called Accounts Payable information). In many companies, the situation is as depicted in Figure 12.4. When an invoice is received, an Accounts Payable clerk enters it into a database, and requests written approval from the appropriate manager by circulating the invoice through inter-office mail. If the invoice is approved, "A.P." issues a check to the supplier and records the payment. In reality, however, the invoice is often "accrued" against the department's account as soon as the invoice is received, in order to record the company's potential liability for the product or service received. The actual payment, which occurs later, merely balances the prior accrual in the company's books. The problem is that each department manager typically has little or no visibility of the status of his debits until the end of the month, when he gets a printed report containing all the transactions for the month—including accruals for invoices he has not yet seen. Not only does he see how much he has been spending in each category, but he may also discover that he has been billed for work that is not yet complete, or that some expenses were allocated to the wrong category or to the wrong department. This can make his situation look bad and his job more difficult.

The data warehouse does not empower the managers or streamline the update process

Let us now assume that we have created for this fictitious company a data warehouse that allows each manager to see, at any point in time, the current status of all the invoices that have been received and logged by Accounts Payable. Moreover, since the invoices refer to the corresponding purchase orders,

Transition To Objects

the manager can pull up on his screen the purchase order information to determine whether or not the invoice is legitimate, and if it was entered against the correct account number. So far, so good. But what happens if he detects an error that he wants corrected? In most companies, he still has to write a "journal voucher" or talk to somebody in the Finance department in order to get this situation corrected. This is because the path from the Accounts Payable system to his desktop computer was one-way, meaning he had no capability to stop payment or reclassify an invoice.

The truth is, however, that we do not want such a manager to have access to *all* the information contained in the operational system that controls the disposition of the invoices. We only want him to be able to do a limited set of things to an invoice: approve it, reject it, or reclassify it. It becomes clear that one cannot separate the data from the things that can be done to it—in a word, its methods.

Some changes must be permitted

Stepwise Enterprise Integration

Let us move back from this specific example to the general challenge of enterprise integration. We can now make some recommendations for a stepwise approach to this challenge:

A practical series of steps

- **Create an object-oriented model of the enterprise, or at least of an "interesting" part of it.** What we did above was to explore certain scenarios for business processes that revolve around supplier invoices. It is clear that these scenarios can be transformed into a set of *use cases* or similar analysis artifacts. Object information models and object behavior models are not far behind.

  Model!

- **Make sure that this model is very rich in behavioral information.** It is important to get away from the purely data-centric view of traditional Information System design, and to think in terms of the use of the "business object" by its consumers. This is the time to decide that an invoice not

  A business object has behavior and rules associated with it

only consists of a number, a date, an amount, a supplier name, and a purchase order reference, but that it also includes behaviors (or methods) which may be called *submit, approve, reject, defer, assign-to-account*, and *pay*. Additionally, this object class encapsulates the applicable business rules of the enterprise. In other words, not all methods can be invoked by all users, or in all states of the object. For instance, an invoice can only be approved by a manager, it cannot be paid until it is approved, and it cannot be rejected after it has been paid (but it can be rejected, even if it had been approved first, provided it has not been paid in the meantime). A key benefit of this approach is that once the new business object class is implemented, every other object or application that uses it automatically conforms to the encapsulated rules and behaviors that govern this object.

Focus on a single process and its stakeholders

- **Identify and model a business process that requires improvement.** The purpose of this process modeling is to observe one of the architectural principles seen in Chapter 2: "Understand, Simplify, and Automate." The current business process may be excessively complex, and may have become so precisely because the existing computer systems did not permit a simpler one. If the organization is considering the investment required to reengineer part of its systems, then it should certainly automate a better or simpler process, rather than an artificially convoluted one. A motivated owner of the business process should be identified. This person should have a stake in the positive outcome of the project. For instance, the financial controller may not be the best process owner for the simplification of the invoice processing because he may fear that invoices will be paid faster and the company's liquidity will suffer! The manufacturing or engineering manager, who suffers at the end of each month when he has to reconcile his actual expenses with his forecast and identify items wrongly allocated to his department, is probably a better candidate.

Business processes can be documented through "process maps" that trace the flow of information across groups. A sample process map was shown in Figure 3.7.

- **Select the development and execution environments for object-oriented applications.** Since this is the topic of several chapters of this book, we do not need to emphasize this here. There is just one additional important constraint in the case of enterprise integration. Instead of being free to choose the most productive environment for a brand new application or system, and in view of the integration steps mentioned below, one must take into account the presence of the rest of the enterprise. The installed base of operating systems, networks, databases, graphical user interfaces, query and reporting tools, and programming languages may therefore influence the selection.

    Set up the project environment

- **Design and implement one class, or perhaps a closely related set of classes.** This set of classes roughly corresponds to Meyer's concept of "cluster" (see Section 7.1). If this design includes all the classes that are necessary and sufficient to support one well-defined business process modeled in the previous step (e.g., supplier payment), then a well-bounded project with demonstrable benefits is automatically defined at the same time.

    Implement a well-bounded solution

- **Develop interfaces to the new object-oriented component.** This is the hardest part of the work, because in most cases, it brings the legacy systems back into the picture. To ensure the proper insertion of the new enterprise object into the rest of the human and system environment, three kinds of interface are typically required. These three types of interfaces (and some variants) are summarized in Figure 12.7.

    Connect the new objects to the legacy systems

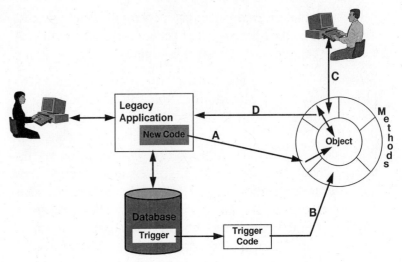

Figure 12.7 – Interfacing a new business object

Something must now invoke the new object

– Other software must be modified to invoke the methods of the new object. This change may be a simple *addition* to the content of the legacy software, or it may *replace* the current functionality (see the arrow marked "A" in Figure 12.7). If the legacy system uses a relational database with a "trigger" capability, the legacy application code may not even need to be changed. It is possible to attach a small new program to a trigger linked to the insertion, deletion, or updating of a database record. Every time the database is modified, the trigger and its associated program serve as a relay that replicates the change in the corresponding objects (arrow B).

Users must have access to the object

– Since the purpose of the new object is precisely to simplify and automate a business process, users

Transition To Objects 443

        must be given access to its contents. In most cases, this means that a client-server application is developed, allowing users to execute some of the methods of the object from their desktop computers and view the results (arrow C). If some of the methods change the object's attributes, then a higher degree of integration than that obtained with the data warehouse concept is achieved immediately!

- Finally, the changes made by the users through this new application must be fed back to the rest of the enterprise. This means that methods that change the state of some objects (or even create or delete object instances) must also invoke legacy systems in order to make consistent changes in them (arrow D). This ensures that the new capability is not a standalone system at the end of a one-way information flow, but is fully integrated with the enterprise's operational processes.   *Objects must be able to effect changes in the legacy system*

- **Deploy the new integrated feature.** This is a combination of testing, training, putting in place the necessary support mechanisms, etc. In general, this process was described in Sections 4.7 and 4.8.   *Complete the objects' lifecycle*

- **Eliminate redundant legacy functionality, if and when possible.** If the new object-oriented component fully replaces the functions of some legacy information or application, then the old subsystem may be decommissioned after some period of overlap and validation.   *Retire the old system*

- **Iterate.** Evaluate the results of the previous integration effort and identify what the next area of integration should be. It may be a process "adjacent" to the one just automated. For instance, the purchase requisition and purchase order process may be a logical segue to the invoice approval process. Or it may be an entirely different area of   *Continue with additional areas of the business*

the enterprise, as the previous integration effort may have removed the most problematic bottleneck in one aspect of the business.

A Manufacturing Integration Example

One company's experience

In one example of actual enterprise integration, National Semiconductor Corporation connected two legacy systems together:

- A commercial software package for semiconductor manufacturing *shop floor control* (Workstream from Consilium), running on multiple Digital VAX minicomputers;

- A "homegrown" *commercial invoice system* running on IBM-compatible mainframe computers.

Mimicking the transfer of wafer lots in the information system

When a wafer lot completes processing and exits the manufacturing process in one of National's factories in the United States or Europe, it must be physically shipped to another location, generally in Southeast Asia, where further processing (test, assembly, and packaging) is performed. At that point, an invoice must be generated to take care of the various consequences of this international shipment, including customs, tax liability on added value, etc. Prior to this effort, the situation was exactly the one described in Figure 12.5, in which a report was output by Workstream and the information was reentered by a clerk into the invoicing system.

Avoiding the loss of data

Another problem that National faced was the fact that as each lot disappeared from the work-in-process WIP inventory of the fab in which it had been created, it did not automatically appear in the inventory of the receiving assembly plant. As surprising as it may seem, it is actually possible to lose track of specific lots or at least of individual wafers in this transfer process.

What National did was to use a Workstream capability to add a customer-programmed "exit" to the application. This code is triggered by specific actions in Workstream and can perform anything its designer wants it to do. In this case, the code was written to invoke a service called DECmessageQ sold by Digital Equipment Corporation. DECmessageQ, or DMQ for short, is an example of a message bus as described in Section 10.3. Through National's existing worldwide DECnet network, DMQ relays the "lot move" messages generated by each implementation of Workstream to both the receiving plant's Workstream system and to an interface to the commercial invoice system. This system (somewhat simplified) is depicted in Figure 12.8. In addition to automating the creation of commercial invoices, it solves several problems in an integrated manner:

Integration between a shop floor application, a message bus, and an MIS application

- It ensures visibility of the WIP, which no longer disappears while in transit to other plants, and no longer risks being double counted in case of a system failure.

Work in process is now correctly tracked

- It prepares the record for each lot shipped to a new facility, making it much easier to track when it physically arrives there.

- The centralized factory planning system, which also runs on the mainframe computers and requires accurate information about the WIP levels in each factory, can now accomplish better planning of wafer starts.

Whether or not this example is a bona fide instance of object-oriented enterprise integration can be debated. Some consider message bus technology object-oriented, while others do not. This topic was addressed in Section 10.3.

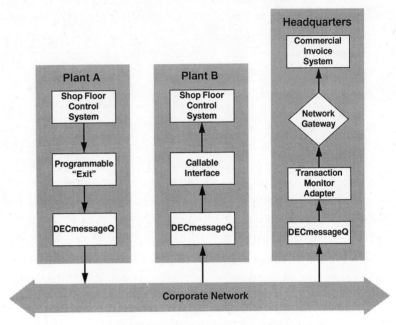

Figure 12.8 – National Semiconductor's Manufacturing Systems Integration

Reusing Enterprise Models – The Holy Grail?

Generic business object models may emerge

Intuitively, it seems that the kind of enterprise models needed to perform object-oriented enterprise integration are common to many companies and could be reused. These models often embody general business or operations principles, not trade secrets; therefore, they could be shared. Alternately, the suppliers of systems could make them available and customizable so that new integrated applications can be created more easily.

ANSI's sobering assessment

The reality is less satisfactory. The American National Standards Institute (ANSI) committee on Object Information Modeling (X3H7), in one of its meetings, concluded the following:

"Current enterprise models are very expensive

Transition To Objects

and they are done for one business at a time. They are:

- *Not easily reusable;*
- *Not complete;*
- *Not based on precisely defined concepts;*
- *Based on ad hoc approaches;*
- *Not interoperable."*

At that time, the X3H7 purpose was largely to "map the terrain" and make sure that different sources of information models are identified, compared, documented, and that the organizations that develop them, including other standards groups and consortia, talk to one another about their respective models. This may prevent unnecessary divergence, but it will not in and of itself cause a grand unified model to emerge.

X3H7 provides a mapping but not a unified model

By contrast, OMG's Business Object Management Special Interest Group (BOMSIG) hopes not only to define the problem area solved by object-oriented business models, but to cause this state of affairs to change. The mission of BOMSIG includes establishing "the use of business objects as a medium for understanding the business, managing its complexity, and delivering business solutions." In 1994-95, the BOMSIG issued two surveys, one to potential consumers of business objects (i.e., end users of computer systems) to determine their needs, and one to potential suppliers (i.e., system and software suppliers) to understand what commercial capabilities may already be available. As it processes this information, the BOMSIG will progressively submit to other OMG Task Forces draft proposals for the solicitation of common specifications in this area.

The efforts of OMG's BOMSIG

12.7 Partnering: A Potentially Low-Risk Way of Getting Started

Partnering with a compatible organization reduces risks

Ultimately, nothing will replace commitment and tenacity for an organization to make a successful transition to an object-oriented lifecycle. On the other hand, there are creative ways to potentially lower the risk relative to the "go-it-alone" approach previously described. The principal way is to find a compatible organization with mutual interests and initiate a joint object-oriented development project. Some joint venture projects are relatively easy to form, while others are so fraught with intellectual property and other business issues that they are not worth the effort.

There are benefits for partners who are both object novices

In order to determine the type of partner and joint development project that is acceptable, an organization must first determine what it expects from the relationship. Two equally novice organizations (in terms of experience with objects) may still benefit from economies of scale. They can join forces to conduct a limited application development project, or simply a "proof-of-concept" study, while sharing the cost and maximizing the learning experience. Logically, this type of project partnership works best among those organizations that have already established a working relationship.

An experienced/ novice partnering has different benefits

Another possible partnering arrangement includes an organization inexperienced in object technology with one that has already made significant progress on the object-oriented lifecycle learning curve. Since very few things in business are free, the inexperienced organization can reasonably expect to pay a proportionally higher share of the cost (people, equipment, money, etc.) to make the partnership reciprocally beneficial.

The types of organizations that can band together on a given project are almost unlimited, but there are a few that are typical. The following list is one for consideration and discussion only. It is not intended to be either all-inclusive or prescriptive.

Potential joint project partnering

- *Intra-Company Joint Projects.* If an organization is one of several within a larger organization, then it should be the natural tendency to look for other internal organizations that have a common interest in pursuing object technology. Internal joint projects are far less complex to arrange because there is usually no need for a legal contact.

  Intra-company teaming is usually the easiest

- *Supplier/Customer Joint Projects.* A supplier/customer joint project is often beneficial for both parties, especially when it involves a problem domain that is of mutual concern in their business relationship. Additionally, suppliers are usually receptive to entering into joint projects with their customers because of the potential for improving their business relationship. For the customer, improving the capability of an existing supplier is less disruptive than having to select a new supplier.

  Supplier/customer teaming often builds better relationships

- *Joint Projects Between Noncompeting Businesses.* Companies that have complementary products or services often join together in many areas such as cooperative advertising and marketing. This creates a fertile environment to explore mutual interests and needs in experimental software projects.

  Complementary businesses sometimes have preestablished relationships

- *Joint Projects Between Competitors Within a Consortium.* Consortia are a popular trend for creating a stable mechanism through which companies that are fierce competitors in certain products or services can join to cooperate in those areas where they are willing to share technology. SEMATECH and the National Center for Manufacturing Science (NCMS) are only two of a number of consortia of this type. Both consortia routinely initiate joint object technology projects that allow competitors to

  Consortia are neutral ground for the fiercest competitors

pool resources in precompetitive areas without fear of antitrust problems.

Membership in software organizations provides an excellent opportunity to look for potential partners

- *Software-Oriented Organizations.* There are several organizations that have special interests in banding together to work on software issues of some specific orientation. These various organizations have a wide range of interests and membership requirements. From an object technology viewpoint, the principal among these is OMG, but there are many others that have different primary objectives and that support object technology. A few of these are the Petrotechnical Open Software Corporation (POSC), X-Open, the Open Software Foundation (OSF), and the Workflow Management Coalition (WfMC). These groups are prime places where organizations find potential joint project partners that have common interests and needs.

Trade organizations' members usually have mutual interests

- *Trade Organizations.* There are hundreds, perhaps thousands, of national and international trade organizations. Almost every business belongs to one, and many belong to several. These groups represent organizations with common business interests and offer opportunities to investigate mutual interests in object technology projects.

12.8 Now It's Your Turn

We have now come to the end of our journey together. After describing at length the concepts of architecture, process, methods, and the specific techniques and tools that can be used to develop and execute object-oriented applications, it seems that advice on how to operate a transition to object-oriented technology is the most appropriate way to end this book.

It is said that the best teachers are the ones who tell their students what they are going to tell them, tell it to them, then tell them what they have just told them. So in closing, let us restate the use we think you can make of the material in this book.

First, we have insisted repeatedly that it is necessary to have a solid *architecture* underlying the design of an application or system. We have given several examples of architectures, including frameworks for development environments and an object management architecture for the execution environment.

Secondly, we have mentioned *planning,* not only in the natural context of project management, but also in terms of planning for the transition to a new technology. Planning is to systems development what location is to real estate: the first, second, and third most important thing.

Finally, within the selected architecture and the defined plan, there are several opportunities to *select* components of the complete approach: process definitions, analysis and design methods, CASE tools, languages, CASE environments, object databases, object request brokers, information buses, etc. As often as possible, we tried to list the important criteria for selecting a productive and consistent tool kit. But remember that the best tool kit will not help you if your work is not planned, or if you do not have a blueprint for how to construct your system.

In the end, the best case study you can possibly benefit from is not one of those mentioned in the previous chapter. It is your own successful attempt at migrating to object-oriented technology. So now, it is your turn. If this book helps you architect your systems, plan your work, and select your development and execution components better than you would have done before, then you, too, will realize the object-oriented lifecycle. And we appreciate the opportunity to have helped you do so.

Table of Acronyms

4GL	Fourth Generation Language
ACM	Association for Computing Machinery
AI	Artificial Intelligence
ANSI	American National Standards Institute
API	Application Programming Interface
ARPA	Advanced Research Projects Agency
ATIS	Atherton Tool Integration Standard
ATP	Advanced Technology Program
BOMSIG	Business Objects Management SIG
CAD	Computer-Aided Design
CASE	Computer-Aided Software Engineering
CCITT	International Consultative Committee for Telephony and Telegraphy
CDIF	CASE Data Interchange Format
CIM	Computer-Integrated Manufacturing
CM	Configuration Management
CMM	Capability Maturity Model
COCOMO	COnstructive COst Model
COM	Component Object Model
CORBA	Common Object Request Broker Architecture

CRC	Class Responsibility and Collaboration
DARPA	Defense Advanced Research Projects Agency
DBMS	Data Base Management System
DCE	Distributed Computing Environment
DFD	Data Flow Diagrams
DoD	Department of Defense
ECMA	European Computer Manufacturers Association
EDI	Electronic Data Interchange
EIA	Electronics Industry Association
ERA	Entity-Relationship-Attribute
ESPRIT	European Strategic Programme for Research in Information Technology
FBE	Framework Based Environment
FIPS	Federal Information Processing Standards
GUI	Graphical User Interface
HOOD	Hierarchical Object-Oriented Design
IDL	Interface Definition Language
IEEE	Institute for Electrical and Electronics Engineers
IPSE	Integrated Project Support Environment
IRDS	Information Resource Dictionary System
IS	Information Systems
ISD	Information Structure Diagram
ISO	International Standards Organization
IT	Information Technology
JIT	Just-In-Time

MIS	Management Information Systems
MLC	Meta-Lifecycle
MMST	Microelectronics Manufacturing Science and Technology
MRM	Manufacturing Resource Manager
NAPI	North American PCTE Initiative
NATO	North Atlantic Treaty Organization
NCMS	National Center for Manufacturing Sciences
NFS	Network File System
NII	National Information Infrastructure
NIST	National Institute of Standards and Technology
OA&D SIG	Object Analysis and Design SIG
OBA	Object Behavior Analysis
ODMG	Object Database Management Group
OID	Object Interaction Diagram
OLE	Object Linking and Embedding
OMA	Object Management Architecture
OMG	Object Management Group
OMS	Object Management System
OMT	Object Modeling Technique
OO	Object-Oriented
OOA	Object-Oriented Analysis
OOAD	Object-Oriented Analysis and Design
OOD	Object-Oriented Design
OODB	Object-Oriented Data Base
OODBMS	Object-Oriented Data Base Management System

OOP	Object-Oriented Programming
OOPL	Object-Oriented Programming Language
OOSD	Object-Oriented Structured Design
ORB	Object Request Broker
OSF	Open Software Foundation
OSI	Open Systems Interconnect
PCIS	Portable Common Interface Set
PCTE	Portable Common Tool Environment
PERT	Program Evaluation and Review Technique
POSC	Petrotechnical Open Software Corporation
QSR	Quality Systems Review
RM-ODP	Reference Model of Open Distributed Processing
RPC	Remote Procedure Call
SADT	Structured Analysis and Design Technique
SDE	Software Development Environment
SEE	Software Engineering Environment
SEI	Software Engineering Institute
SEMATECH	SEmiconductor MAnufacturing TECHnology
SEMI	Semiconductor Equipment and Materials International
SGML	Standard Generalized Markup Language
SIG	Special Interest Group
SNA	System Network Architecture
SPI	Software Process Improvement
SQA	Software Quality Assurance
SQC	Statistical Quality Control

Table of Acronyms

SQL	Structured Query Language
SSADM	Structured Systems Analysis and Design Method
STARS	Software Technology for Adaptable, Reliable Systems
STD	State Transition Diagrams
TCP/IP	Transmission Control Protocol/Internet Protocol
TIMA	Technologies for the Integration of Manufacturing Applications
UDP	Universal Datagram Protocol
UI	User Interface
V & V	Verification and Validation
VAR	Value-Added Reseller
WBS	Work Breakdown Structure
WfMC	Workflow Management Coalition
WIMP	Windows, Icons, Mouse, Pop-up Menus
WIP	Work In Process

Glossary

AD/Cycle   An application development environment offered by IBM Corp. AD/Cycle runs on the MVS operating system and uses a repository managed by the DB2 relational database management system.

Ada   An object-based, but not fully object-oriented, programming language mandated by the U.S. Department of Defense for use by its various branches and contractors in developing defense-related software. Named after Ada Augusta, countess Lovelace (the daughter of Lord Byron), and collaborator of pre-computer inventor Charles Babbage

analysis   See "requirements analysis."

ANDF   Architecture-Neutral Distribution Format. A technology developed by the Open Software Foundation (OSF) to allow executable programs to be distributed without knowing in advance on which computer architecture the program will run.

APSE   Ada Project Support Environment. An integrated project support environment that focuses on object management facilities and tool portability.

Arcadia   A confederation of research projects on software development and evolution environments, organized in 1987 and funded by the U.S. government.

architecture   A defined structure based upon a set of design principles. The definition of the structure includes its components, their functions, and their relationships and interactions.

archiving   The copying of information into long-term storage so that it remains accessible for future reference but does not occupy expensive primary storage space.

ARPA   Advanced Research Projects Agency (formerly know as DARPA.) A branch of the U.S. Department of Defense which funds research projects of military and/or industrial interest in the electronics, computer and information technology fields.

artifact   A product of a creative activity or of a manufacturing step.

ATIS   A Tool Integration Service. An object-oriented integration framework for CASE tools, initially developed by Atherton Technologies for its Software Backplane product (it was then named Atherton Tool Integration Service), and later adopted with some changes by Digital Equipment Corp. as the basis for its COHESION environment. ATIS competed with PCTE around 1990.

attribute   A real-world characteristic, property, or element of information belonging to an entity. An entity is (statically) described by its attributes. In an object database, instance variables may be considered attributes of objects. In a programming language, values associated with a type definition are attributes.

browser   A basic tool included in most repository-based development environments, allowing the user to easily and graphically "navigate" the links connecting objects in the structure. The browser may also act as a repository editor, allowing the creation, deletion, relocation, and duplication of objects or links.

Glossary

build-time
(used as an adjective) That relates to, or is used during, the creation of a software system, rather than during its operation (run-time).

C
A highly portable, procedural (non-object-oriented) language used in most commercial technical applications, and whose development and use is largely related to that of the UNIX operating system.

C++
An object-oriented language, developed by AT&T Bell Laboratories. C++ contains C as a subset.

CAIS
Common APSE Interface Set (see above the definition of APSE). A past project to define a set of common interfaces for the integration of Ada-related CASE tools.

CASE
Computer-Aided Software (or System) Engineering. The application of defined engineering processes, techniques, and tools (including computer-based tools) to the design, manufacturing, and support of software products.

CDIF
Case Data Interchange Format. An EIA standard for a representation of analysis or design models, permitting their exchange between different CASE tools.

child
In a hierarchical structure (e.g., a class hierarchy), an element situated below another one, called its parent. Also see "inheritance."

CIM
Computer-Integrated Manufacturing. The integration of multiple manufacturing functions with each other, and with business and design functions, through the use of computer hardware, networks, and software.

CIM OSA	CIM Open System Architecture. An architecture developed by a consortium of European semiconductor and computer companies under the European Community's ESPRIT program. CIM OSA is a framework for the structured description of the requirements of a manufacturing enterprise and for CIM system implementations.
class	A set of similar objects that share common methods and properties or the abstraction of that set of objects. Objects can be created as instances of a class. Depending on the authors, a class may be essentially the same as a "type" or may be a richer concept.
class hierarchy	The structure (a tree or a directed acyclic graph) that relates classes to one another through inheritance relationships.
class library	A collection of classes that implement the basic functionality of a certain application domain or service domain, and can be used as a foundation to build more complex object-oriented software (example: a graphical user interface class library). Class libraries are a major component of a software reuse policy. Class libraries can be generated internally by an organization as a means to effect internal reuse across projects; or they can be offered as commercial, off-the-shelf products.
client	In a network of computers or processes, any process that requires the presence of one or more other computers or processes, called servers, to which it delegates some common functions instead of performing them itself. See "server."
client-server	Qualifies a system architecture in which an application is divided into an application-specific part, contained in a **client** module, and generic functions which are delegated to one or more common **servers**, often located on a different computer.

coding	The phase of software development in which a programming language is used to construct new software components. Coding is sometimes used instead of "programming" when this word could be misunderstood to mean software development in general, and it is important to convey a narrower meaning without ambiguity.
COHESION	An "umbrella" program for a set of software engineering product offerings by Digital Equipment Corp. COHESION provides a framework for CASE tool integration and a repository based on ATIS.
commit	In a data base management system, cause updates to be permanently stored in the data base. To preserve the integrity of a database against possible hardware or software failures, it is often necessary to prepare a group of updates, create a "checkpoint" that can be returned to if necessary, then commit all the updates in a single transaction. If the transaction fails to complete, partial updates can be rolled back to the last consistent state.
component	A generic term for the parts of an object or system, its constituent entities.
Computer-Integrated Manufacturing	See "CIM."
conceptual design	See "external design." Also called "high-level design."

conceptual schema　　In the ANSI "three-schema architecture" for databases, the definition of the semantics of the data, regardless of how this data is going to be stored in the data base management system (the internal schema) or of how it is going to be presented to the user (the external schema). This includes the definition of data attributes and their types, any constraints about legal values for the data, the relationships between entities, and "business rules" or constraints that ensure the integrity of the database (e.g., one cannot delete a supplier record if there are outstanding purchase orders for that supplier).

construction　　A term sometimes used to refer to the assembly and integration of the components of an application.

control integration　　One of the three types of integration defined between tools in a framework (also see "data integration" and "presentation integration"). Control integration refers to the mechanisms available in the framework for tools to communicate with one another directly (e.g., invoke one another and pass data without storing it in the repository).

criterion　　A standard or test against which a judgment or decision can be made. Plural: **criteria.**

DARPA　　Defense Advanced Research Projects Agency. The name given to ARPA during the 1980s and early 1990s (see ARPA).

data base　　See "database."

Data Definition Language　　See "DDL."

Glossary

data dictionary — A collection of names of all data items used in a software system, together with their relevant properties (e.g., size of the item, data type, and representation, etc.). Also see "repository."

data integration — One of the three types of integration defined between tools in a framework (also see "control integration" and "presentation integration"). Data integration is the ability of tools to share data through the framework. The degree of integration depends on the level of granularity of the data that the framework supports (see "granularity").

data integrity — The preservation of data for its intended use, even in the presence of human or system errors.

data model — One of several alternative models used to describe the organization of data. The best-known data models are the hierarchical model, the network model, and the relational model (which is based on viewing data as a set of related two-dimensional tables).

database — A structured collection of information stored on a computer system. Also spelled "data base."

Data Manipulation Language — See "DML."

DBMS — Data Base Management System. An integrated collection of software services and utilities that provide a comprehensive method of managing data.

DCE   Distributed Computing Environment. An OSF technology providing a variety of communications services across heterogeneous networks. Components of DCE include enhanced NCS/RPC services, a distributed name service, an X.500 directory service, a security service, a distributed file system, facilities for diskless operation, PC NFS and other PC integration services, and a distributed time service.

DDL   Data Definition Language. A language used to specify the structure and organization of a database (e.g., the contents of each record and the type of each field).

deployment   The distribution and installation of an application or system. Also called "proliferation" when some limited use is an earlier part of the release process.

design   The definition and modeling of system solutions in response to requirements. Design includes formulating conceptual alternatives to solve the requirements, as well as refining the logical and physical details of the selected solution (short of the level of detail addressed by program coding).

detailed design   The second part of design as defined above (i.e., the refinement of the logical and physical details of the solution).

DFD   Data Flow Diagram. A software engineering notation used in structured analysis and design methods to depict the flow of data among processes and data stores.

dispatching   In an object-oriented system, the routing of a request for execution of a method to the appropriate implementation of that method, which depends on inheritance relationships and the possible redefinition of methods within a class hierarchy.

distributed DBMS	A data base management system with the ability to effectively manage data that is distributed across multiple computers on a network.
DME	Distributed Management Environment. An OSF technology providing the capability to remotely manage the elements of a computer network.
DML	Data Manipulation Language. A language used to specify the operations to be performed on the contents of a database (e.g., to retrieve, insert, delete, and update data records). SQL is an example of a standard DML.
domain	A particular subject or area of application (e.g., business systems, or electronic computer-aided design).
ECMA	European Computer Manufacturers Association. This organization produced the standard reference model for CASE tool integration, referred to as the "toaster model." It also created the PCTE standard.
EIA	Electronics-Industry Association. Creator of the CDIF standard.
Eiffel	An object-oriented programming language and environment supplied by Interactive Software Engineering (ISE). Eiffel has very strict type checking and a facility for the run-time verification of "assertions" about a program's objects.
enactment	The enforcement of a software process through facilities of an integrated software engineering environment.
encapsulation	An essential characteristic of object-oriented systems, whereby an object's data is only available to other objects through the object's protocol, i.e., by invoking methods on the object. Also called "information hiding."

enterprise model — A high-level representation of the functions, processes, and information of an enterprise, used to simplify the enterprise's processes and/or as input to the analysis of the requirements of a specific function or system.

entity — Anything about which information can be defined or stored. An entity can be a person, a concept, a physical construct, a place, an event, etc. This term is used when the specific technical connotation of the word "object" is not intended.

entity-relationship model — A model of information, used for structured design, in which entities and the relationships between them are both explicitly represented.

Entreprise 2 — A CASE integration framework, built using PCTE and supplied by SYSECA, a French company.

environment — In general, any combination of facilities whose integration provides significant advantages over the use of only a subset. Specifically for CASE, a complete set of facilities for software development made up of a set of tools that are integrated through a framework and a repository.

ERA — Entity-Relationship-Attribute. Another name for the entity-relationship information modeling technique (see "entity-relationship model").

ERD — Entity-Relationship Diagram. The graphical notation used to represent an entity-relationship model.

ESPRIT	European Strategic Programme for Research in Information Technology. An extensive research endeavor (10+ years, multibillion-dollar budget) of the European Community to foster joint research in computer and software technology between companies and universities located in different E.C. countries.
external design	A first stage of design that consists of identifying the externally observable behavior of a system from the perspective of its users or of other systems that interact with it, regardless of any implementation details.
flat file	A method of data storage that uses a simple succession of records in a computer file, without storing any additional structure-related information.
4GL	Fourth Generation Language. A very high-level language used to design interactive applications or reports based on the contents of a database.
framework	A set of facilities that are used to construct an environment in which several tools can interoperate; the infrastructure for tool integration. A framework provides the mechanisms for data integration, control integration, and/or presentation integration (see these terms).
granularity	The more or less fine level at which individual data elements are managed and can be accessed in a system. For instance, a CASE repository can manage data at the file level only (coarse-grained) or at the level of each individual entity within a model (fine-grained).
GUI	Graphical User Interface. A user interface where information is displayed graphically on a screen, and in which the use of icons, scroll bars, windows, menus, etc., provide a user-friendly interaction paradigm.

heterogeneous	Containing several parts that are significantly different from one another.
hierarchy	An organization of information according to a tree (or directed acyclic graph) structure.
high-level design	See "external design." Also called "conceptual design."
homogeneous	Containing parts that all share the same characteristics.
HOOD	Hierarchical Object-Oriented Design. An object-oriented design method developed by the European Space Agency, and targeted at the creation of software to be written in the Ada language.
I-CASE	A U.S. government specification for the contents of a complete software development environment.
ICSE	International Conference on Software Engineering, cosponsored once a year by the IEEE and ACM, and by the national computer societies of the host country or region (when held outside of the U.S.).
implementation	The inside view of an object or module, which specifies how it achieves its behavior, including the details of the data structures it contains.
information hiding	See "encapsulation."
information technology (IT)	The set of methods and products used to process information, including data processing services, CASE tools, and application software.

Glossary

inheritance A relationship among classes, wherein one class (the "subclass" or "child class" shares the structure and behavior defined in another class (the "superclass" or "parent class"). Additions and redefinitions can occur in the subclass structure and behavior. In **multiple inheritance,** the subclass inherits from several superclasses simultaneously. The inheritance relationship is often represented by the phrases "kind of" or "is a." Inheritance defines a hierarchy among classes.

instance An individual instance of an object.

instance variable Attribute.

instantiation The act of creating an instance. In general, the process of specifying a more specific artifact out of a generic model, (e.g., a model of a specific company's information from a general enterprise model.)

integrate To assemble a number of components into a complete, coordinated system and to verify that the system functions as a whole.

integrated environment An environment made up of a certain number of components (e.g., CASE tools) with additional facilities to help the components interoperate.

internal design A stage of the software development process in which specific methods and objects are selected to construct the solution to the requirements according to the architecture and interfaces specified during high-level or external design.

interoperability The ability of two or more systems to communicate with one another and exchange information.

IPSE Integrated Project Support Environment. An integrated environment of software engineering tools.

IRDS	Information Resource Dictionary System. A tentative standard for the architecture of an information repository.
IS	Information System. A general term for a company's computer systems (usually its business systems rather than its technical ones) and the organization that manages them.
legacy (system, application)	A previously existing system or application, often using outdated technology, which constrains the ability to adopt new technology because it cannot be retired immediately.
library	An organized collection (of subroutines, of objects, of class definitions etc.).
lifecycle	Of a software product: the succession of stages that it goes through, from requirements to retirement. Of an object: the succession of states it goes through, from instantiation to deletion. Also spelled **life cycle.**
low-level design	See "detailed design" or "internal design."
maintenance	The set of activities which ensure that an application or system continues to function in the presence of changes (e.g., a new version of the operating system) or after the discovery of a defect, or that an application or system is adapted to meet a new need.
MAP	Manufacturing Automation Protocol. A precursor of the information bus technology described in Section 10.5.
megaprogramming	A model of software engineering proposed by Barry Boehm (see Boehm '92) and espoused by the STARS (see STARS) program in particular. Megaprogramming addresses large projects through process-driven, reuse-based methods, and integrated project support environments, and allows collaborative development by geographically dispersed teams.

message	The implementation of a request made by an object to another.
metabase	The part of a database that contains the metadata (i.e., the description of the structure of the database itself).
metadata	Data that describes other data.
metamodel	A model that describes a class of models.
method	In analysis and design: a particular approach to an activity, including the techniques used to perform it, the notations used to capture its deliverables, and the rules used to verify its correctness. In object-oriented design and programming: an operation that an object can perform upon request from another object (a message causes a method to be executed). The set of methods defines the object's interface to the rest of the system.
MMST	Microelectronics Manufacturing Science and Technology. A joint program of Texas Instruments and the U.S. Air Force to develop an object-oriented semiconductor manufacturing system.
model	An abstract representation of an object, a class of objects, or a system, created for the purpose of understanding or simulating some of its aspects while hiding details that are irrelevant for a particular purpose. Many models are graphical documents, but they can also be text or an executable program.
modeling	The process of creating models.
modularity	The characteristic of a system that is made up of multiple components, each of which is easier to understand or produce than the whole system, and all of which combine to provide the functionality required of the whole system.

monolithic	Made up of a single piece.
Motif	A technology for graphical user interfaces, layered on top of the X Window system, adopted by the Open Software Foundation.
multiple inheritance	See "inheritance."
NIST	National Institute of Standards and Technology (former National Bureau of Standards). This institute, which is a branch of the U.S. Department of Commerce, has taken an active role in promoting the integration of multiple applications and tools into consistent environments. For example, it adopted the ECMA reference model for CASE tool integration, and it funds Advanced Technology Programs (ATPs) to accelerate industry research in high-risk technical areas with potentially significant economic impact.
node	Of a tree-like or graph-like model: each vertex within the model. Of a network: each individual computer within the network.
OBA	Object Behavior Analysis. An object-oriented analysis method promoted by ParcPlace Systems, based on the use of "scripts" to document and analyze the user requirements.
object	In general: anything that does something interesting, or which you can cause to do something interesting. In object-oriented systems: a bundle of information and behavior in which the information is hidden and the behavior can be triggered by sending a message requesting the performance of a certain service. An instance of a class.
object ...	*See acronyms starting with the letter "O."*
object database	See "OODB."

Glossary 475

object manager	A facility that provides operations on objects (creation, navigation, deletion, etc.), without offering the full complement of capabilities of an OODBMS. For instance, an object manager may only handle transient objects in memory, not their permanent storage.
object-oriented	That which relates to the definition and use of objects.
object-oriented...	*See acronyms starting with the letters "OO."*
Objective-C	An object-oriented language, derived from C, but not as close to the C syntax as C++. Used in the NeXT operating system.
Objectworld	A conference organized by the Object Management Group. Focuses on commercial offerings of object-oriented technology.
OCD	Object Communication Diagram. One of the notations of the Shlaer/Mellor object-oriented analysis methodology, which abstracts the set of interactions between objects are defined in the various state transition diagrams for these objects.
OID	Object Interaction Diagram. The fundamental representation of the model of an application in some object-oriented methods (see [Colbert 90]), describing the interactions among a set of objects. The OID is conceptually similar to the OCD, but the OCD is constructed after the fact from other information in the model, while the OID is constructed first and is used to derive other parts of the model.
OMA	Object Management Architecture. The architecture defined by the Object Management Group to organize the relationship between the components of an object-oriented system (object request broker, object services, common facilities, and applications).

OMG　　　　　　　Object Management Group. A consortium of object technology suppliers and users that promotes the use of object technology by creating *de facto* standards for the specification of object-oriented components and systems.

OMT　　　　　　　Object Modeling Technique: the name of the object-oriented analysis and design method invented by James Rumbaugh and his colleagues at GE (see [Rumbaugh '91]).

OOD　　　　　　　Object-Oriented Design (can be used as a generic term, or to refer specifically to Grady Booch's method—see [Booch '91]).

OODB　　　　　　Object-Oriented Data Base (also **object data base** or **ODB**). A database whose architecture follows the principles of object orientation, such as encapsulation and inheritance, by opposition to the other database architectures (hierarchical, network, or relational).

OODBMS　　　　　Object-Oriented Data Base Management System. A software product that provides the capability to create and manage OODBs.

OOPL　　　　　　Object-Oriented Programming Language (e.g., Smalltalk, Eiffel, C++, Objective-C, CLOS).

OOPS　　　　　　Object-Oriented Programming System: a complete object-oriented programming environment, usually made up of a set of tools related to a common OOPL (editor, compiler, debugger, browser, etc.).

Glossary

OOPSLA   Object-Oriented Programming Systems, Languages and Applications. The name of the most popular annual conference on object-oriented concepts and their realization. OOPSLA is sponsored by the ACM's Special Interest Group on Programming Languages (SIGPLAN) and is always held in North America, typically alternating between the U.S. and Canada.

OOSA   Object-Oriented Systems Analysis: the name given by Shlaer and Mellor to their object-based analysis method.

OOSD   Object-Oriented Structured Design: the name given by Tony Wasserman (see [Wasserman '90]) to a design method which is a blend of an object-oriented approach and of notations that originally introduced with structured analysis and design methods.

open system   A system with well-defined, public interfaces that gives it characteristics of interoperability, portability, and/or extensibility.

operating system   Control software that manages the resources of a computer.

operation   An action that an object performs upon another in order to obtain a service (often used interchangeably with "method").

ORB   Object Request Broker. The central communication mechanism between objects in the OMA.

OSF   Open Software Foundation. A consortium of computer systems and software suppliers and users involved in the selection, integration, and deployment of "open systems" technologies which its members can then sell on the market or incorporate into their products. OSF technologies include OSF/1, Motif, DCE, DME and ANDF.

OSF/1   An operating system developed by OSF circa 1991, based in part on the Mach kernel from Carnegie-Mellon University.

OSI	Open Systems Interconnect: a set of ISO standards for communication network protocols.
packaging	The process of putting together all the components needed by the customer or user of a system (executable application, documentation, release notes, etc.) so that the software can be "manufactured" and distributed.
parent	An entity situated above others (its children) in a hierarchy, and from which the children inherit structure and behavior.
PCTE	Portable Common Tool Environment. A set of services (including, but not limited to, coarse-grained repository management services) standardized by ECMA and allowing the integration of CASE tools into an integrated environment.
persistence	A property of data of objects whereby they continue to exist beyond the lifetime of the process that created them.
platform	The base technology on which something else is layered; a computer platform typically refers to the combination of a hardware architecture and its operating system.
polymorphism	The ability for a common definition to refer to different possible implementations. A polymorphic operation allows a particular method to be chosen at run-time, depending on the type of the object to which the message is sent, or on the type of the arguments provided to the operation.
portability	The ability to move an application or system from one platform to another without rewriting it.

presentation integration	One of the three types of integration defined between tools in a framework (also see "control integration" and "data integration"). Presentation integration refers to the mechanisms available in the framework for tools to display information using the same window system, and to allow the user to "cut-and-paste" or "drag-and-drop" items from one tool's window to another.
primitive	An elementary capability present in a system, as opposed to one that must be constructed by the system's user (e.g., "integer" may be a primitive type of a computer language, while "complex" may have to be defined as a composite entity by the programmer).
problem space	The set of issues, objects, attributes, etc., inherent in the real world in which a problem exists, as opposed to being artifacts of a computer system used to solve the problem.
process	A finite course of events aimed at accomplishing a given purpose or effect.
product	A commercially available, supported application or system.
property	Something associated with an object (this may be an attribute, a relationship, or an operation).
protocol	In general: a set of conventions or rules that govern the interactions between several components of a system. In object-oriented systems: the set of all valid messages that an object can receive or emit (sometimes used to designate the **public interface** of the object).
prototype	An early model that serves to validate requirements or tentative design decisions. May or may not be iteratively transformed into a product (sometimes called a "throw-away prototype" in the latter case).

rapid prototyping	The practice of developing mutiple prototypes quickly and inexpensively, so that successive iterations achieve better approximations of a desired result.
RDBMS	Relational Data Base Management System. A class of data base management systems that represent data as two-dimensional tables, with specific data values providing the link, or "join," between tables containing related information (according to the model developed by E.F. Codd).
reference model	A general model that can be used to derive a more specific one, and remains flexible enough that it does not impose a final structure.
referential attribute	In an entity-relationship model, or in a relational database, an attribute of an entity that reproduces a key attribute of another entity, and serves to link the two entities together (e.g., the department number is a key attribute of a "department" entity, and it is a referential attribute of the "employee" entity).
relationship	An association between entities or objects. In PCTE, a relationship is a pair of mutually inverse links.
reliability	The probability that a component or system will perform its intended function, within stated conditions, for a specified period of time.
repository	A data management facility (e.g., a database management system) for storing several types of information such as data definitions, project plans, CASE diagrams, process definitions, etc., and the relationships between these objects. The repository allows multiple users to locate these objects, and allows various tools to access and manipulate them, while maintaining consistency between them according to defined rules.

Glossary

requirement — A statement of a need of a system's (or an application's) customer need. A requirement may also be an external constraint imposed by some characteristic of the environment.

requirements analysis — The activity of defining and studying the problem space in order to create and document a set of agreed-upon requirements

requirements specification — The document, or set of documents (including models) that result from the requirements analysis activity.

reusability — A desirable feature of system and application components that allows them to be catalogued and used again in other systems and applications. Reusability is not limited to code but also applies to models, test plans, documents, etc.

reuse — An underlying activity of software development which consists of identifying components that can serve the purposes of multiple applications or systems, making them generic enough to be reusable, and effectively incorporating them into more applications or systems than those that triggered their initial creation.

reverse engineering — The practice of re-creating a missing artifact from another artifact that would have preceded the first one in the normal development process. For instance, re-creating a requirements specification or a design model from source code.

RFI — Request for Information. A step in the investigation process used by some organizations like SEMATECH or OMG, in which a broad request is issued to the market for the description of certain needs or capabilities, without committing to further interaction with the respondents.

RFP — Request for Proposal. A step in the selection of some desired capability from the marketplace, in which an organization publishes a focused request for a certain capability, supported by explicit requirements. Typically, the organization commits to performing a fair evaluation and selection of the best proposal in view of some realization or acquisition.

roll back — To completely undo a transaction that was started but could not be completed in its entirety (e.g., due to the unavailability of some required resources). See "commit."

run time — Refers to all the components of a computer environment which are used during the execution as opposed to its construction of the actual software or system product.

run-time binding — In an object-oriented programming system, the ability to dynamically change the method which is executed in response to a message, according to the current type of the object to which the message is addressed (opposite: compile-time binding).

schema — A set of data definitions that describe the structure of information in a data base or repository (plural: schemata).

script — A textual description of a possibly imaginary but typical sequence of events, which can serve as an input to the analysis of requirements. Scripts are the fundamental artifact used in the Object Behavior Analysis (OBA) method.

SDS — Schema Definition Set. A PCTE term for a self-contained fragment of schema.

SECS — SEMI Equipment Communication Standard. A set of protocols (SECS-I, SECS-II) for communication between manufacturing systems and manufacturing equipment, used to send commands to the equipment and to collect data and status information from the equipment.

Glossary

SEDDI
Software Engineering Data Descriptions and Interchange. A multipart standard being produced by the International Standard Organization (ISO). Compared to CDIF, it has a broader scope that covers data descriptions, not just data interchange.

SEE
Software Engineering Environment. Consists of a framework, a set of CASE tools that interoperate through this framework, and a set of policies. Often used interchangeably with IPSE (see "environment," "framework" and "IPSE").

SEI
Software Engineering Institute, a federally funded R&D center which is part of Carnegie-Mellon University. Located in Pittsburgh, Pennsylvania. SEI developed the five-level "Capability Maturity Model" for software organizations.

SEMATECH
A not-for-profit consortium of ten U.S. semiconductor manufacturers and the U.S. Advanced Research Projects Agency. Located in Austin, Texas.

SEMI
Semiconductor Equipment and Materials International. A consortium of suppliers to the semiconductor industry, whose role includes creating standards for that industry.

server
In a network of computer systems, a machine or process which is available to fulfill requests sent by other machines and processes for a specific function. Examples: file server, database server, user interface server, print server, etc. Opposite of "client." Also see "client-server." In some object-oriented design methods, an object which never initiates a message to another object, but receives requests from other objects.

SETEC
Semiconductor Equipment and Technology Center. A part of Sandia National Laboratory in Albuquerque, New Mexico.

signature   The list of types of the arguments of a subroutine or method, as well as the type of the result.

Smalltalk   An object-oriented programming language, originally developed at the Xerox' Palo Alto Research Center, and considered to be very productive for rapid prototyping because of the interpretive development environments available for it.

SoftBench   A CASE environment sold by Hewlett-Packard.

software engineering A systematic, disciplined approach to the development, operation, maintenance, and retirement of software

software process   The order and manner in which the activities of the software lifecycle are executed.

solution space   The set of issues, objects, attributes, etc., that are part of a computer system created to solve a problem, not inherent parts of the real world.

specialization   The action of defining a subclass by reference to its superclass(es), or the relationship between a subclass and its superclass(es).

specification   A detailed, precise description of a requirement, a system, a program, or a procedure.

spiral model   A software process, developed by Barry Boehm, which includes successive iterations with an increasing level of detail. During each cycle, risks and alternatives are analyzed prior to committing an increasing amount of resources to the project.

Glossary

SQL
Structured Query Language. A language (initially created by IBM in 1981 and standardized by ANSI in 1986) used to express the description and manipulation of relational data. All major RDBMS products on the market implement the standard SQL, but usually with proprietary extensions.

STARS
Software Technology for Adaptable, Reliable Systems. A program of the Department of Defense Advanced Research Projects Agency (ARPA) to develop software engineering environments with an emphasis on software process and reuse. The prime contractors of STARS are Boeing, IBM, and Paramax (a division of Unisys).

state
One of the possible distinct conditions in which an object may exist. An object's state may be characterized by the values of some (or all) of its attributes.

statechart
A particular representation of a state model, invented by David Harel to capture the dynamic behavior of complex systems. Contrary to traditional state transition diagrams, statecharts allow hierarchy relationships between states. They also include a notion of "history," whereby an object "remembers" which substate of a given state it was last in, and can return to that substate automatically when returning to the parent state. Statecharts have been adopted as part of some object-oriented analysis methods, especially OMT.

state model
A representation of all the states in which an object or system can exist, as well as the possible transitions between states, used to capture the behavior of a system. A state model often also describes the events that cause the state changes, or the actions that state changes trigger.

state transition diagram	A representation of a state model, usually "flat," (i.e., non-hierarchical). State transition diagrams have been used in structured analysis methods and have been adopted as part of some object-oriented methods.
subclass	A class that inherits the attributes and methods of another class (or of several, in the case of multiple inheritance).
superclass	The class from which a subclass derives (part of) its attributes and methods.
support	The set of activities aimed at ensuring that a customer can use a product, including processing customer requests and performing maintenance.
technique	A precise sequence of activities, or a set of rules one follows to perform an activity.
TCP/IP	Transmission Control Protocol/Internet Protocol. A set of communication protocol layers for data networks. Initially supported mostly on UNIX systems, TCP/IP is now prevalent under many different operating systems and computer hardware architectures.
technology	A generic term for the technical components that support a domain of work. For instance, "software technology" may include operating systems, user interfaces, databases, CASE methods and tools, etc.
toaster model	The nickname given, because of the shape of its graphical representation, to a reference model for CASE environments developed by Hewlett-Packard and adopted as a standard by the European Computer Manufacturers Association (ECMA). An enhanced version of the model has been adopted by the National Institute of Standards and Technology (NIST).

Glossary

tool
: A software application program that generates and manipulates other software artifacts (models, prototypes, programs, etc.). Some examples of tools are graphic modeling tools, editors, compilers, linkers, debuggers, librarians, project management software, configuration management software.

traceability
: The ability to retain and use information about the source of a particular artifact. For instance, the ability to "trace" a design feature back to the requirement(s) which it helps fulfill, or to "trace" a defect to its cause.

transaction
: In general, an exchange of messages. In database management systems, a set of individual data updates that must either be completely performed as a group, or all "rolled back" (cancelled) in order to leave the database in a consistent state.

transformation
: The creation of one software artifact from another. "Transformation" is generally used when the translation is automated, (e.g., for code generation).

type
: In procedural programming languages, a predefined category of variables that are represented in the same way and are subject to the same operators, (e.g., "integer," "float," "complex," boolean"). In object-oriented languages, a type may be the same thing as a class, or can be any set of classes, including user-defined ones, that share the same interface (i.e., the same set of operations with identical signatures).

type hierarchy
: The relationships that define an inheritance scheme in a collection of types.

UNIX
: A multi-user operating system developed at AT&T Bell Laboratories in 1969. UNIX became predominant for minicomputers and personal workstations in the 1980s, largely because it is written in a high-level language (C), which, at the time, made it more portable than its competitors.

user interface (UI)   The device management and control software that provides information to the human operator, usually by means of a bitmap display, and accepts user input, usually by means of a keyboard and a pointing device.

VAR   Value-added reseller. A company which buys a product, adds its own components to provide additional functionality and value, and resells it.

verification and validation (V&V)   An independent review of a product to ensure accuracy, quality, and conformance to specifications. Verification ensures that a product does what was specified. Validation ensures that it does what the user needs.

waterfall model   A model of software process that consists of a linear sequence of steps (such as analysis, design, coding, and maintenance) that occur strictly in the order indicated.

Bibliography

[ANSI 90] American National Standards Institute X3H4: "A Tool Integration Standard (ATIS)." Working Draft, Feb. 1990.

[AT&T 90] AT&T Software Technology Center: "CASE Jumpstarts at AT&T." Proceedings, CASE'90 Conference, Irvine (Calif.), Dec. 1990.

[Baldassari 89] Baldassari, Marco and Giorgio Bruno: "An Environment for Operational Software Engineering in Ada." Proceedings, Tri-Ada '89 Conference, pp. 126-146. ACM, New York.

[Barstow 87] Barstow, David R.: "Artificial Intelligence and Software Engineering." Sixth National Conference on Artificial Intelligence, Seattle, July 1987. Reprinted in Exploring Artificial Intelligence (R.E. Shrobe, ed.), Morgan Kaufmann, 1988.

[Bernhardt 90] Bernhardt, R.: "Manufacturing Systems Planning and Programming in a CIM Environment." Proceedings, ESPRIT '90 Annual Conference (Dordrecht), pp. 15-30. Kluwer Academic Publishers, 1990.

[Boehm 81] Boehm, Barry W.: *Software Engineering Economics*. Prentice-Hall, 1981, 767 pages.

[Boehm 88] Boehm, Barry W.: "A Spiral Model of Software Development and Enhancement." IEEE Computer, Vol. 21, No. 5, May 1988, pp. 61-72.

[Boehm 92] Boehm, Barry W. and W. Scherkis: "Megaprogramming." Proceedings, DARPA Computer Science Conference, April 1992.

[Booch 87]   Booch, Grady: *Software Engineering with Ada – Second Edition.* Benjamin Cummings, 1987.

[Booch 91]   Booch, Grady: *Object-Oriented Design with Applications.* Benjamin Cummings, 1991, 580 pages.

[Brooks 75]   Brooks, Frederick P. Jr.: The Mythical Man-Month and Other Essays on Software Engineering. Wiley, 1975.

[Brooks 87]   Brooks, Frederick P. Jr.: "No Silver Bullet: Essence and Accidents of Software Engineering." IEEE Computer, Vol. 20, No. 4, Apr. 1987, pp. 10-19.

[Brown 94]   Brown, Alan, Dave Carney, Ed Morris, Dennis Smith and Paul Zarrella: *Principles of CASE Tool Integration.* Oxford University Press, 1994.

[Buhr 84]   Buhr, R.J.A.: *System Design with Ada.* Prentice-Hall, 1984.

[Buhr 90]   Buhr, R.J.A.: *Practical Visual Techniques in System Design.* Prentice-Hall, 1990.

[Caldiera 91] Caldiera, Gianluigi and Victor R. Basili: "Identifying and Qualifying Reusable Software Components." IEEE Computer, Vol. 24, No. 2, Feb. 1991, pp. 61-70.

[CCG 90]   CASE Consulting Group: "CASE'90 Report." Proceedings, CASE'90 Conference, Irvine (Calif.), Dec. 1990.

[Chen 77]   Chen, Peter: *The Entity-Relationship Approach to Logical Data Base Design.* QED, 1977.

[Chidamber 91] Chidamber, R. ans S. Kemerer: "Towards a Metrics Suite for Object Design." In [OOPSLA 91], pp. 197-211.

[Coad 90]   Coad, Peter and Ed Yourdon: *Object Oriented Analysis.* Yourdon Press / Prentice Hall, 1990.

[Colbert 89]   Colbert, Edward: "The Object-Oriented Software Development Method: A Practical Approach to Object-Oriented Development." Proceedings, TRI-Ada 1989 Conference (Pittsburgh, Penn.), Oct. 1989.

[Cox 86]	Cox, Brad J.: *Object-Oriented Programming: An Evolutionary Approach.* Addison-Wesley, 1986.
[Curtis 89]	Curtis, Bill: "Three Problems Overcome with Behavioral Models of the Software Development Process." Proceedings, 11th International Conference on Software Engineering, Pittsburgh (Penn.), pp. 398-399. IEEE Computer Society Press, 1989.
[Dabrowski 90]	Dabrowski, Christopher E., Elizabeth N. Fong and Deyuan Yang: "Object Database Management Systems: Concepts and Features." Special Publication No. 500-179, National Institute of Standards and Technology, Gaithersburg (Md.), Apr. 1990.
[Deming 86]	Deming, W. Edwards: *Out of the Crisis.* Cambridge University Press, 1986.
[Deneubourg 91]	Deneubourg, J.-L., S. Gross, N. Franks, A. Sendova-Franks, C. Detrain and L. Chrétien: "The Dynamics of Collective Sorting: Robot-like Ants and Ant-like Robots." In [Meyer 91], pp. 356-365.
[ECMA 90]	Eurpean Computer Manufacturers Association: "A Reference Model for Comoputer Assisted Software Engineering Environment Frameworks." Technical Report TR/55, ECMA, Dec. 1990.
[EIA 91]	Electronic Industries Association: "CDIF – Framework for Modeling and Extensibility, draft version 1.5." EIA-PN2387, EIA, Washington (D.C.), 1991.
[ESPRIT 89]	European Strategic Programme for Research in Information Technology: "Open System Architecture for CIM." Springer-Verlag, 1989.
[Firth 87]	Firth, Robert, Vicky Mosley, Richard Pethia, Lauren Roberts and William Wood: "A Guide to the Classification and Assessment of Software Engineering Tools." Carnegie-Mellon University Software Engineering Institute, Technical Report CMU/SEI 87-TR-10, Aug. 1987.

[Fowler ...] Fowler, Priscilla: "Software Technology Transition: A Mini Tutorial for STARS '91" STARS '91 Proceeding, 1991.

[Freedman 82] Freedman, Daniel P. and Gerald M. Weinberg: *Handbook of Walkthroughs, Inspections and Technical Reviews.* Dorset House, 1982.

[Freeman 87] Freeman, Peter: *Software Perspectives – The System is the Message.* Addison-Wesley, 1987.

[Freitas 90] Freitas, Maria Manuel, Ana Moreira and Pedro Guerreiro: "Object Oriented Requirements Analysis in an Ada Project." Ada Letters, Vol. 10, No. 6, pp. 97-109, Jul.-Aug. 1990.

[Gibson 89] Gibson, Michael Lucas. "The CASE Philosophy." BYTE Magazine, Vol. 14, No. 4, Apr. 1989, pp. 209-218.

[Gibson 90] Gibson, Elizabeth: "Objects - Born and Bred." BYTE Magazine, Vol. 15, No. 10, Oct. 1990, pp. 245-254.

[Gilb 88] Gilb, Tom: *Principles of Software Engineering Management.* Addison-Wesley, 1988, 442 pages.

[Glassey 89] Glassey, C. Roger and S. Adiga: "Conceptual Design of a Software Object Library for the Simulation of Semiconductor Manufacturing Systems." Journal of Object Oriented Programming, Vol. 2, No. 4, Nov.-Dec. 1989, pp. 39-43.

[Goldberg 83] Goldberg, Adele and David Robson: *Smalltalk-80: the Language and Its Implementation.* Addison-Wesley, 1983.

[Harel 87] Harel, David: "Statecharts: A Visual Formalism for Complex Systems." Science of Computer Programming, Vol. 8, No. 3, June 1987, pp. 231-274.

[Harel 88] Harel, David: "On Visual Formalisms." Communications of the ACM, Vol. 31, No. 5, May 1988, pp. 514-530.

[Harmon 92]	Harmon, Paul: "Texas Instruments Choses Object-Oriented Technology for a CIM Project." Object-Oriented Strategies (Cutter Information Corporation), Vol. II, No. 10, Oct. 1992, pp. 1-12.
[Heitz 88]	Heitz, M. and B. Labreuille: "Design and Development of Distributed Software Using Hierarchical Object Oriented Design and Ada." Proceedings, Ada-Europe International Conference, New York, pp. 143-156. Cambridge University Press, 1988.
[Henderson 90]	Henderson-Sellers, Brian and Julian M. Edwards: "The Object-Oriented Systems Life Cycle." Communications of the ACM, Vol. 33, No. 9, Sep. 1990, pp. 142-159.
[Hollowell 90]	Hollowell, Glenn and R. Edwards: "Enterprise Automation in the 1990s." Texas Instruments Technical Journal, Vol. 2, No. 7, Sep.-Oct. 1990.
[Hopcroft 79]	Hopcroft, John E. and Jeffrey D. Ullman: *Introduction to Automata Theory, Languages and Computation*. Addison-Wesley, 1979.
[Humphrey 90]	Humphrey, Watts S.: *Managing the Software Process*. Addison-Wesley, 1990.
[IEEE 86]	Institute of Electrical and Electronic Engineering: "Guide for Software Quality Assurance Planning." Standard 983-1986, IEEE, New York, Jan. 1986.
[IEEE 89]	Institute of Electrical and Electronic Engineering: "IEEE Standard for Software Quality Assurance Plans." Standard P730.1-1989, IEEE, New York, Aug. 1989.
[IEEE 90]	Institute of Electrical and Electronic Engineering: "IEEE Standard Reference Model for Computing System Tool Interconnections." Standard P1175/D7, IEEE Computer Society's Task Force on Professional Computing Tools, New York, Oct. 1990.

[Jacobson 92]   Jacobson, Ivar, Magnus Christerson, Patrik Jonsson and Gunnar Övergaard: "Object-Oriented Software Engineering." ACM Press, Addison-Wesley, 1992.

[Jones 87]   Jones, C.B.: *Systematic Software Development Using VDM*. Prentice-Hall, New York, 1989.

[Krasner 92]   Krasner, Herb, Jim Terrell, Adam Linehan, Paunl Arnold and William H. Ett: "Lessons Learned from a Software Process Modeling System." Communications of the ACM, Vol. 35, No. 9, Sep. 1992, pp. 91-100.

[Kugler 90]   Kugler, P. N., R. E. Shaw, K. J. Vincente and J. Kinsella-Shaw: "Inquiry Into Intentional Systems I: Issues in Ecological Physics." Psychological Research, 1990.

[Lee 88]   Lee, K.J., M.S. Rissman, R. D'Ippolito, C. Planta and R. Van Scoy: "An OOD Paradigm for Flight Simulators – Second Edition." Carnegie-Mellon University Software Engineering Institute, Technical Report CMU/SEI-88-TR-30, 1988.

[Lippman 89]   Lippman, S.B.: *C++ Primer*. Addison-Wesley, 1989.

[Manola 95]   Manola, Frank (Ed.): "X3H7 Technical Committee (Object Information Modeling) Interim Technical Report." March 1995. Available from the editor at GTE Laboratories Inc., Waltham, Mass.

[McCabe 76]   McCabe, T.J.: "A Complexity Measure." IEEE Transactions on Software Engineering, Vol. SE-2, No. 4, Dec. 1976, pp. 308-320.

[McClatchy 91]   McClatchy, Will: "The Object of TI's Desire." Informationweek, June 10, 1991.

[McClure 89]   McClure, Carma: *CASE Is Software Automation*. Prentice-Hall, 1989.

[McClure 89a]   McClure, Carma: "The CASE Experience." BYTE Magazine, Vol. 14, No. 4, Apr. 1989, pp. 235-236.

[McGregor 90]	McGregor, John D. and Tim Korson: "Understanding Object-Oriented: A Unifying Paradigm." Communications of the ACM, Vol. 33, No. 9, Sep. 1990, pp. 40-60.
[Meyer 88]	Meyer, Bertrand: *Object-Oriented Software Construction*. Prentice-Hall (U.K.), 1988.
[Meyer 91]	Meyer, J. A. and S. W. Wilson: "From Animals to Animats." Proceedings, First International Conference on Simulation and Adaptive Behavior. MIT Press, 1991.
[Meyer 95]	Meyer, Bertrand: Object Success: Manager's Guide to Object Orientation, its Impact on the Corporatiod its Use for Reengineering the Software Process. Prentice-Hall, 1995, 192 pages.
[Micallef 88]	Micallef, Josephine: "Encapsulation, Reusability and Extensibility in Object-Oriented Programming Languages." Journal of Object Oriented Programming, Vol. 1, No. 1, Apr.-May 1988, pp. 12-38.
[Moreau 89]	Moreau, D.R. and W.D. Dominick: "Object Oriented Graphical Information Systems: Research Plan and Evaluation Metrics." Journal of Systems and Software, Vol. 10, No. 1, July 1989, pp. 23-28.
[Morin 91]	Morin, J. "Technology Transition Strategies" [2-10] STARS '91 Proceedings, 1991.
[Murata 89]	Murata, Tadao: "Petri Nets: Properties, Analysis and Applications." Proceedings of the IEEE, Vol. 77, No. 4, April 1989, pp. 541-580.
[Naecker 91]	Naecker, Philip A.: "Digital's COHESION Strategy." DEC Professional, Vol. 10, No. 3, March 1991, pg. 38.
[OMG 92]	Object Management Group (Richard Soley, ed.): *Object Management Architecture Guide*. Wiley, 1992.
[OMG 94]	Object Management Group (Andrew Hutt, ed.): *Object Analysis and Design – Survey of Methods*. Wiley, 1994.

[OMG 94a]   Object Management Group (Andrew Hutt, ed.): *Object Analysis and Design – Comparison of Methods.* Wiley, 1994.

[OMG 95]   Object Management Group: ... <about interoperability in CORBA 2>

[Orr 89]   Orr, Ken, Chris Gane, Edward Yourdon, Peter Chen and Larry Constantine: "Methodology: The Experts Speak." BYTE Magazine, Vol. 14, No. 4, April 1989, pp. 231-234.

[OOPSLA 86]   Proceedings, 1986 Conference on Object-Oriented Programming Systems, Languages and Applications. ACM SIGPLAN Notices, Vol. 21, No. 11.

[OOPSLA 87]   Proceedings, 1987 Conference on Object-Oriented Programming Systems, Languages and Applications. ACM SIGPLAN Notices, Vol. 22, No. 12.

[OOPSLA 88]   Proceedings, 1988 Conference on Object-Oriented Programming Systems, Languages and Applications. ACM SIGPLAN Notices, Vol. 23, No. 11.

[OOPSLA 89]   Proceedings, 1989 Conference on Object-Oriented Programming Systems, Languages and Applications. ACM SIGPLAN Notices, Vol. 24, No. 10.

[OOPSLA 91]   Proceedings, 1991 Conference on Object-Oriented Programming Systems, Languages and Applications (Phoenix). ACM SIGPLAN Notices, Vol. 26, No. 11.

[Parunak 94]   Parunak, H. Van Dyke: "What is an Agent?" Proceedings (Briefing Paper No. 2), Workshop on Principles of Responsible (Agent) Systems, National Center for Manufacturing Sciences, Ann Arbor, Mich., May 26-27, 1994. Published by the Industrial Technology Institute, Ann Arbor.

[Paulk 91]   Paulk, M.C., Bill Curtis and M.B. Chrissis: "Capability Maturity Model for Software." Carnegie-Mellon University Software Engineering Institute, Technical Report CMU/SEI-91-TR-24, Aug. 1991.

[Peters 91] Peters, Tom: *World-Class Quality: The Customer Will Decide.* Video Publishing House Inc., Schaumburg, Ill. 72-minute video tape with notes, 1991.

[Pressman 92] Pressman, R.S. : Software Engineering: A Practitioner's Approach. McGraw-Hill, 1992.

[Prieto-Diaz 88] Prieto-Diaz, R.: " Implementing Faceted Classification for Software Reuse." Communications of the ACM, Vol. 34, No. 5, May 1991, pp. 88-97.

[Royce 70] Royce, W.W.: "Managing the Development of Large Software Systems: Concepts and Techniques." Proceedings, WESCON, Aug. 1970. Reprinted in Proceedings, 9th International Conference on Software Engineering, Computer Society Press, 1987.

[Rubin 92] Rubin, Kenneth S. and Adele Goldberg: "Object Behavior Analysis." Communications of the ACM, Vol. 35, No. 9, Sep. 1992, pp. 48-62.

[Rumbaugh 91] Rumbaugh, James et al.: *Object-Oriented Modeling and Design.* Prentice-Hall, 1991.

[Saunders 89] Saunders, John H.: "A Survey of Object-Oriented Programming Languages." Journal of Object Oriented Programming, Vol. 1, No. 6, Mar.-Apr. 1989, pg. 5.

[Shewhart 31] Shewhart, Walter A.: *The Economic Control of Manufactured Product.* Van Nostrand Reinhold, 1931.

[Shlaer 88] Shlaer, Sally and Stephen J. Mellor: *Object-Oriented Systems Analysis – Modeling the World in Data.* Yourdon Press, 1988, 144 pages.

[Shlaer 92] Shlaer, Sally and Stephen J. Mellor: *Object Lifecycles – Modeling the World in States.* Yourdon Press, 1992, 251 pages.

[Shoham 93] Shoham, Y.: "Agent-Oriented Programming." Artificial Intelligence, Vol. 60 No. 1, March 1993, pp. 51-92.

[Spivey 89] Spivey, J.M.: *The Z Notation.* Prentice-Hall, New York, 1989.

[Stadel 91]   Stadel, Manfred: "Object-Oriented Programming Techniques to Replace Software Components on the Fly in a Running Program." ACM SIGPLAN Notices, Vol. 26, No. 1, January 1991, pp. 99-108.

[Steels 91]   Steels, Luc: "Toward a Theory of Emergent Functionality." In [Meyer 91], pp. 451-461.

[Stevens 89]   Stevens, Scott M.: "Intelligent Interactive Video Simulation of a Code Inspection." Communications of the ACM, Vol. 32, No. 7, July 1989, pp. 832-843.

[Stroustrup 90] Stroustrup, B.J.: *The Annotated C++ Reference Manual.* Addison-Wesley, 1990.

[Taylor 91]   Taylor, David A.: Object-Oriented Technology: A Manager's Guide. Addison-Wesley, 1991.

[Taylor 92]   Taylor, David A.: Object-Oriented Information Systems – Planning and Implementation. Wiley, 1992.

[TI 92]   Texas Instruments Technical Journal (multiple articles), Vol. 9, No. 5, Sep.-Oct. 1992.

[Tracz 90]   Tracz, Will: "The Three Cons of Software Reuse." Proceedings, Fourth Workshop on Software Reuse Tools, 1990.

[Wakeman 93]   Wakeman, L. and J. Jowett: *PCTE – The Standard for Open Repositories.* Prentice-Hall, 1993.

[Ward 85]   Ward, Paul T. and Stephen J. Mellor: *Structured Development for Real-Time Systems.* Yourdon Press, 1985.

[Ward 89]   Ward, Paul T.: "How to Integrate Object Orientation with Structured Analysis and Design." IEEE Software, Vol. 6, No. 2, Mar. 1989, pp. 74-82.

[Waskiewicz 92] Waskiewicz, Fred: "SEMATECH's SkillView – An Object-Oriented Development Case Study." FirstClass, Vol. II, No. 1, Apr.-May 1992, pp. 13-14.

[Wasserman 90]	Wasserman, Anthony I., Peter A. Pircher and Robert J. Muller: "The Object-Oriented Structured Design Notation for Software Design Representation." IEEE Computer, Vol. 23, No. 3, Mar. 1990, pp. 50-63.
[Wasserman 91]	Wasserman, Anthony I. and Peter A. Pircher: "Object-Oriented Structured Design and C++." Computer Language Analysis and Design, Vol. 8, No. 1, Jan. 1991, pp. 41-52.
[Williamson 90]	Williamson, M.: "Toward the Holy Grail of True CASE Integration." Digital News, May 14, 1990, pp. 59-62.
[Wirfs-Brock 90]	Wirfs-Brock, Rebecca J. and Ralph E. Johnson: "Surveying Current Research in Object-Oriented Design." Communications of the ACM, Vol. 33, No. 9, Sep. 1990, pp. 104-124.
[Wirfs-Brock 90a]	Wirfs-Brock, Rebecca J., B. Wilkerson and L. Wiener: *Designing Object-Oriented Software*. Prentice-Hall, 1990.

Index

A
abstract data types, 13
access control, 222
Accredited Standards Committee, 337
actigrams, 84-85
AD/cycle, 210
Ada, 26, 34, 134, 137-140, 290
Ada Programing Support Environment (APSE), 213-214
Advanced Research Projects Agency, see ARPA
agents, software, 297-300
AIX CASE, 218-219
American National Standards Institute (ANSI), 211, 301
analysis, see requirements analysis
application objects, 343-345
Arcadia, 213
architecture
   definition, 3-6
   review board, 28
ARPA, 11, 213, 336
artifacts, 106, 107
associations, see relationships
Atherton Tool Integration Standard (ATIS), 213-214, 216
attributes, 120, 126-127, 131
audit trail, 232

B
Booch, 119, 137-138
Broadcast Message Server (BMS), 212
browsers, 148, 173, 231
build-time components, 33

Buhr, 138-139
Business Process Reengineering, 30

C
C++, 113, 137-138, 147-150, 174, 176-177, 180, 320
Capability Maturity Model, see CMM
capsule, 208
cards, object modeling, 133
CASE
   analysis and design, 171
   costs, 166-169
   integration, 203-212
   tools, 34, 39, 45, 49, 65, 119, 140, 143-144, 201-203, 208, 218-220, 228, 232-233, 294, 319
      cost of, 166-169, 191, 410-411
      benefits, 161-166
      interoperability, 313-315
      selection, 180-200
CASE Data Interchange Format (CDIF), 187, 211, 313-314
charts
   machine, 139
   structure, 295
CIM, iv-vi, 109
   defined, 371-374
      Microelectronics Manufacturing Science and Technology (MMST) project, 374-383
      SEMATECH application framework, 383-395
classes, 120, 392
CLOS (Common LISP Object System), 150
clusters, 241-243

CMM, 11-12, 91-95, 101, 169
Coad-Yourdon, 129
COCOMO, 250, 252, 288
code generation, 183, 286-287, 291-294
Colbert, 134-135
common facilities, see CORBAfacilities
common interface specifications, 214
Common Object Request Broker Architecture, see CORBA
Common Open Software Environment (COSE), 212
competitiveness, i, iv, 2, 8
compliance, 26
components, iii, 146, 258-259, 284, 288
Computer-Integrated Manufacturing, see CIM
Computer-Aided Software Engineering, see CASE
component model, 73
configuration management (CM), 64-65, 228-229, 266
Context Object Interaction Diagram (OID), 135
control, access, 222
control integration, 202, 212
competitiveness, 2, 8
components, 146, 258-259, 284, 288
configuration management (CM), 45, 69, 77, 92, 184
construction, 69, 107
control integration, 205-206
CORBA, 175, 212, 216, 233-234, 333, 348-350, 367, 377
CORBAfacilities, 352-354
CORBAservices, 350-352
  object services roadmap, 351-352
CRC cards, 133
critical success factors, 56-60

D
DARPA, see ARPA
Data Definition Language (DDL), 33
data flow diagrams (DFD), 124, 128
data integration, 203-204, 210-211, 218
data warehouse, 437-439

Database Management System (DBMS), 57-58, 67
databases, object oriented, 346-348
datagrams, 84
debuggers, 174
defect tracking, 71
Defense Advanced Research Projects Agency, see ARPA
deliverables management, 266-269
deployment, 156-157
design, 66, 69, 137
development baseline, 267-268
drag and drop, 210
dynamic binding, 112

E
Eiffel, 113, 150
Electronic Data Interchange (EDI), 3
encapsulation, 111, 114, 141, 257
enterprise integration, 15-17, 29, 432-448
  definition, 433-434
  stepwise approach, 439-444
enterprise process, 46-48
Entity-Relationship-Attribute (ERA) model, 120, 215
environment
  development, 341
  execution, 341-370
  integrated, 176-179, 232
  standards, 212-214
ETVX (Entry, Tasks, Verification, Exit) method, 82-83
European Computer Manufacturers Association (ECMA), 206, 214, 216
European Strategic Programme for Research in Information Technology (ESPRIT), 88, 214
events, 122, 128
extended enterprise systems, 4

F
flow charts, 84-85
formal methods, 300-305
  acceptance of, 302
  limitations, 302-304

Index

VDM, 301
Z (zed), 301-304
fourth generation languages (4GL) 292-293, 319-320
framework, 203 260
   classes, 243-244
   lifecycle, 256
function point analysis, 251-252
functional decomposition, 108, 109

G
Graphical User Interface (GUI)
   builders, 172-173, 260

H
Harel statecharts, see statecharts
Hierarchial Object-Oriented Design (HOOD), 140
hyperlinks, 229

I
IDL, 175, 214, 257-258, 350, 352
information bus, 354-370
   features, 356-357
   implementation, 357-360, 445
   model, see model, information
   selection criteria, 368-370
   vs object request broker, 367
Information Resource Dictionary System (IRDS), 211
Information Structure Diagram, see ISD
inheritance, 112, 114, 141, 266, 278, 381
Integrated Project Support Environment, see IPSE
integration
   control, 206, 212
   data, 203-205, 210-211
   environment, 203-208
   facilities, 202
   presentation, 205, 209-210
   tool, 206
Interface Definition Language, see IDL
International Software Process Workshop (ISPW), 84, 88

interface specification document, 258
International Standards Organization, 217
interoperability, 13-14, 35, 233, 368, 385
IPSE, 88, 202
IRDS, 211
ISD, 126, 132
islands of automation, iv, 29
ISPW, 84, 88

K
Key Process Areas (KPAs), 91-92

L
legacy systems, 4, 415
   competing with, 415-416
   definition, 9-12
   integration, 14, 35, 39
   and object technology, 431-432, 441-443
lifecycle, 44-45, 47, 51
   architecture, 24-26
      benefits of, 39-41
      definition, 26, 47
      principles of, 27-38
   components, 54-55
   defined. 5-6, 43-45, 47
   models, 4-12, 49-77
      inadequate, 4-8
      metamodel, 55
      obsolete, 10-11
      spiral, 52-54, 81
      waterfall, 49-52, 54
   process
      designing, 79-81
   synchronization, 256

M
machine charts, 139
maintenance, 71, 157-159
megaprogramming, 288
message buses, see information buses
Meta-Lifecycle (MLC) Model, 54-55, 282
   components, 55
   conceptual design, 66-67
   construction, 69

Meta-Lifecycle (MLC) Model *(continued)*
  maintenance and support, 71
  packaging and deployment, 70-71
  project management, 72-73
  requirements analysis, 61-66
  success factors, 56-60, 66
  using, 73
metamodel, see Meta-Lifecycle (MLC) Model
    method
  assemblers, 118
  selection, 140-146
  software development, 117-119
metrics, 64, 72, 146
  code, 70
  collection, 68
  design, 68
  object-oriented, 275-279, 307-308
  process, 277
  product, 277
  project, 277-278
  software, 276
  use of, 60
Microelectronics Manufacturing Science and
    Technology project (MMST), 335, 374-383
  prototype, 378
migration, see transition
MLC, see Meta-Lifecycle Model
model-driven systems, 37-38, 40, 308-310
models,
  analysis, 135
  behavior, 121-122
  capability maturity, see CMM
  enterprise, 439-441
    reusing, 446-447
  entity-relationship-attribute (ERA), 120, 215
  function, see models, process
  information, 120, 126, 129, 132
    layers, 129
  metamodel, 55
  object, 120, 215-216
  object communication, 124, 135, 146
  object interaction, see models, object
    communication
  process, 123, 126, 128, 222

  requirements, 135
  spiral, 52-54, 81
    use of, 245
  state, 121-122, 126, 128, 132
    Mealy, 122
    Moore, 122, 128
  toaster, see toaster model
  transition, 418-422, 439-441
  waterfall, 49-52, 54
  see also lifecycle, models
mythical man-month, 263

N

National Institute of Standards and Technology,
    see NIST
NIST, 206, 214, 216, 324, 336
  and CIM Framework, 394-395
North Atlantic Treaty Organization (NATO), 214

O

Object Behavior Analysis (OBA), 130-131, 172,
    307, 396, 399
  activity script, 131
Object COBOL, 150
Object Database Management Group, see
    ODMG
object, definition of, 111
Object Information Management Technical
    Committee, see X3H7
object management, 215
Object Management Architecture, see OMA
Object Management Group, see OMG
Object Management System, see OMS
object modeling cards, 131, 133
Object Modeling Technique, see OMT
object-oriented analysis, 126, 254
  methods, see Coad-Yourdon, Colbert, Object
    Behavior Analysis (OBA), Object Modeling
    Technique (OMT), Objectory, Shlaer-
    Mellor, Wirfs-Brock
object-oriented CASE tools, 161-200
  selecting, 181-200
Object-Oriented Database Management System,
    see OODBMS

Index

object-oriented design, 136
   methods, 257
   see also Booch, Buhr, Hierarchial Object-Oriented Design (HOOD), Object-Oriented Structured Design (OOSD)
object-oriented execution environment, 341-370
   architectural components, 343-354
   development requirements, 341-343
   see also application objects, OODBMS, object-oriented programming languages, object request brokers, CORBAservices, CORBAfacilities
object-oriented methods, 284-285, 305-306
   advantages, 113-114
   overview, 108-109
   selection, 140-146, 412, 425
object-oriented programming languages (OOPLs), 147
   see also C++, CLOS (Common LISP Object System), Eiffel, Object COBOL, Object Pascal, Objective C, Simula, and Smalltalk
Object-Oriented Structured Design (OOSD), 140
object-oriented technology, 74-77
   acceptance, 317-321
   transition to, 407-451
      analysis and design methods selection, 412
      integration, 412, 432-448
      and legacy systems, 415, 431-432
      models, 418-422
      performance execution, 413-414
      phases of, 419-420
      planning, 427
      risk management, 408-418
      tool selection, 410-412
Object Pascal, 150
object repository, 202, 208
object request broker, see ORB
object services (see CORBAservices)
Object World, 332
Objective C, 113, 150, 174
Objectory, 135-136
ODMG, 321, 332-333
OMA, 330, 349, 352-353, 367, 389

OMG, 126, 175, 212, 214, 257, 284, 313-314, 328-338, 344, 447, 450
   and CIM framework, 389
   object management architecture, see OMA
   object model, 330
   object services roadmap, 350-352
   reference model, 314
OMS, 215
OMT, 119, 304
on-line backup, 231
OODBMS, 67, 346-348
Open Distributed Processing, 336-337
Open Software Foundation, see OSF
operational modeling packages, 400-401
ORB, 175, 203, 348-350, 414
   see also CORBA
OSF, 205, 332-334, 450

P

packaging, 70, 154-156
partnering, 448-450
PCTE, 209, 212, 214-220, 223-224, 233-234, 315
   environments, 218
   implementation, 218
   process enactment, 224
   purpose, 215
   repository model, 216-217
   Special Interest Group (SIG), 216
Petrotechnical Open Software Corporation (POSC), 334, 450
Polymorphism, 141
Portable Common Interface Set (PCIS), 214, 224
Portable Common Tool Environment, see PCTE
precompilers, 174
presentation integration, 205, 209-210
problem domain, 108
problem tracking, 158-159
process, iv, ix, 43, 45-46
   definition, 77-88, 902
   definition guidelines, 88
   enactment, 27, 88-91, 103, 223-224, 230, 312-313
   history, 223

process *(continued)*
  improvement (SPI), 60, 91-103
  integration, 59
  management, 202, 215
  models, 73, 128, 221-222
    criteria, 79-81
    defining, 77, 221
    documenting, 78
    see also models, process
process-driven environment, 221-224
process support, 169, 184, 230
product development process, 46
productivity, iii, 163, 165
Program Description Language (PDL), 128
Program Evaluation and Review Technique (PERT), 240
project
  management, 45, 72, 235-278
    Work Breakdown Structure (WBS), 237-244
  partnering, 448-450
  planning, 222
  tracking, 265
proof-of-concept project, see SEMATECH SkillView
prototyping, 54, 62-63, 67, 163, 145, 173, 259-262, 282-283, 398, 413

Q
quality, iii, 11, 164-166
Quality Systems Review (QSR), 101-102
quality templates, 125

R
real-time
  equipment, 64
  systems, 123, 138-139, 306-307
reengineering, i, 30
registration, role and user, 221
relationships, 120, 126-127
release notes, 155
repository, 202, 208, 210, 215, 224, 227-228, 295-296, 403
  design rationale, 310-312
  facilities, 230
  recovery service, 231
requirements analysis, 61-65, 107, 282, 399
reuse, 57, 61, 74, 114, 144, 187, 248, 255, 269-279, 283-284, 380-381, 398
  benefits, 269-270
  interactions, 276
  libraries, 69, 272
  object, 269-276
  plan, 273-275
  process, 275-276
  roles, 270-273
reverse engineering, 66,
  repository based, 295-297
  tools, 178-180
reviews, 430
risks, 264, 408-418
RM-ODP, 336-337
Rumbaug, see OMT
run-time components, 33-34

S
Sandia Laboratories, 88
scalability, 143-144
schema management, 215, 230-231
scripts
  activity, 130-131, 307
  installation, 155
SEI, 11-12, 27, 91, 94, 98, 101-102
  capability maturity model, 91-94, 169
  transition model, 419-420, 422
Semiconductor Workbench for Integrated Modeling, see SWIM
SEMATECH, iv-vi, 97, 334, 449
  CIM framework, 383-395
    architectural principles, 387-389
    goals, 385-387
    interoperability, 385, 393
    and NIST, 394
    specification components, 391-393
  six-step process, 97-100
services, 120
Simula, 67, 112, 149, 317, 325

Index

SkillView, 395-399
  criteria, 396-397
  project results, 397-399
Smalltalk, 112, 113, 137, 147, 149-150, 174, 177, 347, 376, 392, 396, 398, 405
SoftBench, 212
software
  crisis, 13
  distribution
    packaging for, 154-157
  maintenance, 157-159
  maturity, levels of, 91-93
  process, see process
  standards
    interoperability, 14
Software Development Environment (SDE), 202
  selecting, 224-234
Software Engineering Environment (SEE), 202-203
  components of, 203
  examples of, 218-220
  integrated, 221
  selecting, 224-234
Software Engineering Data Description and Interchange (SEDDI), 314-315
Software Engineering Institute, see SEI
software lifecycle, see lifecycle
Software Process Improvement (SPI), 94-103
Software Process Management System (SPMS), 90
Software Technology for Adaptable, Reliable Systems, see STARS
solution domain, 108
SPI, see Software Process Improvement
standards, 14, 23, 26, 29
  adoption, 34-35, 217, 321-328
  consortia, 325-328
    ODMG, 332-333
    OMG, 328-338
    OSF, 333-334
    RM-ODP, 336-337
    SEMATECH, 334
    X/Open, 333
  de facto, 324-325, 329

  de jure, 323-324, 326-327, 329
  participation and conformance, 233, 338-340, 370, 412-413
STARS, 27, 88-89, 91, 216
  transition model, 419, 421-422
  transition team, 429
statecharts, 123, 132, 134, 260
state transition diagrams (STDs), 122, 128
structure charts, 295
Structured Analysis and Design Technique (SADT), 84-86, 109
Structured Query Language (SQL), 210
stubs, 151
subcontracting, 258-259
subjects, 129
SWIM, 399-405
  architecture, 402-403
  common database, 404-405
  objective, 399-400
  prototype, 405
  user interface, 404
system architecture
  definition, 21, 24
  goals, 22
  purpose, 20
  scope, 23
system development process, 25-26

T

team process charts, 86-87
test baseline, 267-268
testing
  acceptance, 152-154
  alpha, 153-154
  beta, 153-154
  integration, 152
  regression, 159
  unit, 151
text editor
  syntax-driven, 175-176
toaster model, 206-207, 215, 402-403
tool slots, 208
tools, see CASE tools
Tooltalk, 212

traceability, 19, 57-59, 185
  lack of, 58
  of requirements, iii, 131, 144
tracking, project, 265-266
training, 417, 423, 430
  curriculum, 248-250
transformation engineering, 27, 285-297
  benefits, 288
  opportunities for, 286
  reference model, 288-289
transition, technology 407-451
  analysis and design methods selection, 412
  integration, 412, 432-448
  and legacy systems, 415, 431-432
  models, 418-422, 439-441
  performance execution, 413-414
  phases of, 419-420
  planning, 427-431
  risk management, 408-418
  spiral process, 428
  tool selection, 410-412
transitions, in a state model, 122, 128

U
use case, 135-136, 252
user interface prototype, 259

V
validation, 150
value added resellers, see VARs
VARs, 4, 411
verification and validation (V&V), 63-64, 68, 70, 107, 151

W
waterfall model, 49-52
Windows, Icons, Mouse, Pop-up Menus (WIMP), 190, 205
Wirfs-Brock, 133
Work Breakdown Structures (WBS), 237-241, 246-248, 250
wrapper, see capsule

X
X3H7, 337
  enterprise model reuse, 446-447
X/Open, 333, 450